Evolution of the Forebrain

EVOLUTION
OF THE FOREBRAIN

Phylogenesis and Ontogenesis of the Forebrain

Edited by

R. Hassler and H. Stephan

Max-Planck-Institut für Hirnforschung
Neuroanatomische Abteilung
Frankfurt/Main-Niederrad

295 Figures

1967

Springer Science+Business Media, LLC

This volume contains lectures delivered during a symposium on

Phylogenesis and Ontogenesis of the Forebrain

held under the Honorary Presidency of

ELIZABETH C. CROSBY, Ann Arbor, and HUGO SPATZ, Frankfurt/M.

in Frankfurt and Sprendlingen from August 15 – 19, 1965.

The meeting was sponsored by

WORLD FEDERATION OF NEUROLOGY

MAX-PLANCK-GESELLSCHAFT ZUR FÖRDERUNG DER WISSENSCHAFTEN

GESUNDHEITSMINISTERIUM DER BUNDESREPUBLIK DEUTSCHLAND

ISBN 978-1-4899-6245-4   ISBN 978-1-4899-6527-1 (eBook)
DOI 10.1007/978-1-4899-6527-1

Library of Congress Catalog
Card Number 67-17768
© Springer Science+Business Media New York 1966
Originally published by Georg Thieme Verlag, Stuttgart in 1966.
Softcover reprint of the hardcover 1st edition 1966

Preface

DARWINS theories of origin and evolution of species also applies to Homo sapiens. There is hardly a doubt, that man evolved from "animal-primates" and taxonomically belongs to the primate order. The especially strong progression of development in this order is manifest by the markedly increased size and differentiation of the forebrain and related structures. Through the enlargement of specific cortical regions, foremost in the parietal, temporal and prefrontal lobes, the human brain clearly is distinguished from the brains of all other primates, including apes.

This enlargement of the forebrain, representing the acquisition of a complicated apparatus, functionally makes possible the accumulation and correlation of the incoming information and, in the human, even communication through speech and writing. Systems such as these, which are necessary for experience acquisition and thought abstraction, have made the human brain an unique organ of mentation. The exchange of information and ideas – such as with our scientific endeavours – is only possible on the basis of the additional strong increase in growth and differentiation of these progressive parts in the human forebrain.

Furthermore, the evolutionary step from animals to human would not have been possible without a strong development of highly organized structures to inhibit and sublimate the powerful animalistic impulses and instincts, which the human has in common with all animals. Through these highly organized inhibitory structures the human is enabled to exist at a high level of social behaviour.

The human brain may be considered as representing the culmination point of a long line of evolution, starting in the lowest vertebrates with small and undifferentiated structures. A clear understanding of the many attributes of the primate brain can only be obtained through knowledge of the evolutionary processes and trends, which take place along this phylogenetic line. Such knowledge is acquired especially through the endeavours of *paleoneurology* and *comparative neuroanatomy*. The former approaches the problem through the use of endocranial casts from *fossils,* the latter through a comparative study of brains from *recent* animals. Utilization of carefully selected recent species may give information concerning the major trends in brain phylogeny, because many different stages are still represented by extant species.

Further valuable insight to trends in phylogeny may be obtained from studying the *ontogenetic development*. A relationship between phylogenesis and ontogenesis may be considered according to the "biogenetic rule" of Ernst HAECKEL, especially with respect to the brain, for which it seems to have a broad validity. This rule states that the development during ontogeny, in a relatively short period of time, frequently reflects phylogeny. It thus appears that the results of phylogenetic and ontogenetic studies can fertilize and supplement one another. Consequently both Phylogeny and Ontogeny have been combined in a symposium, the contributions of which are compiled in this publication. It is most regrettable, that the discussions could not be included in this volume on account of space limitations.

The support given by the

WORLD FEDERATION OF NEUROLOGY

which helped to make possible this publication is greatfully acknowledged.

Contents

VIII

Phylogenetische Entwicklungsregeln von Organen

Von A. Remane, Kiel

Die Verwandtschaftsbeziehungen der Lebewesen stellen wir im allbekannten Stammbaum = Dendrogramm dar, das mit Recht als Abbild des stammesgeschichtlichen Ablaufs gilt. Von geringer Bedeutung ist dabei seine Beschränkung auf den Bereich oberhalb der Art (Species) oder genauer gesagt auf den Bereich, in dem evolutionistisch erfolgreiche Verschmelzung zweier Gruppen durch Bastardierung nicht mehr vorkommt.

Das Stammbaumschema gibt zwei Probleme auf. 1. Nach welchen Methoden wird es aufgebaut? 2. Ist die ständige Divergenz nach den Gabelstellen die einzige Evolutionsart, oder gibt es in größerem Umfang Parallelitäten und Konvergenzen?

Die erste Frage ist eigentlich seit über 150 Jahren entschieden, seit erkannt wurde, daß nur eine spezielle Übereinstimmung in der Organisation, die homologe Ähnlichkeit*), echte Verwandtschaftsbeziehungen wiedergibt.

Merkwürdigerweise hat sich heute eine Methode ausgebreitet, die diese logische Struktur der Verwandtschaftsforschung völlig übersieht. Ich meine die „numerische Taxonomie", die allein durch Abzählen von Merkmalen ein „objektives" System aufbauen will. Diese Richtung, die die geistige Arbeit zweier Jahrhunderte völlig übersieht, ist als Forschungsmethode für den Aufbau des natürlichen Systems und der Phylogenie unbrauchbar, da sie die ganz verschiedene Wertigkeit der Merkmale und den subjektiven Charakter des „Merkmals" übersieht. Schon Jussieu hob 1791 hervor, daß die Merkmale nicht zu zählen, sondern abzuwägen seien. Die Methode kann nur dann brauchbare Ergebnisse bringen, wenn in einer Formengruppe zufällig eine große Zahl geringfügiger Merkmalsunterschiede von etwa gleichem Wert vorhanden ist. Sicher sind auch dann die Ergebnisse nur, wenn zwei Formenbereiche – etwa Larven und Imagines – übereinstimmende Resultate bringen.

Der Stammbaum (Dendrogramm) stellt die *divergierende* Entwicklung der Lebewesen dar. Dabei sind zu jeder Zeit viele Möglichkeiten der Weiterentwicklung gegeben. Die Mutationen sind, wie die Genetik gezeigt hat, bei jeder Species enorm vielgestaltig, so daß je nach der Umweltsituation jederzeit einige begünstigt, andere eliminiert werden können. Nach diesem Grundschema müßte die Evolution in einem komplizierten Zickzackweg verlaufen, da Erbänderungen und Umweltsituationen stark variabel sind. Seit dem vorigen Jahrhundert ist aber bekannt, daß viele Entwicklungslinien gerichtet verlaufen und die Umwandlung Schritt für Schritt in einer Reihe abläuft. Dieser Vorgang wurde als *Orthogenese* bezeichnet (Haacke, Eimer). Da in diesen Begriff die Vorstellung eines phylogenetischen Wachstums aus inneren Gründen einging, wurde die Orthogene-

---

*) Der Terminus „homolog" wurde von Richard Owen 1848 geprägt. Goethe nannte diesen Ähnlichkeitstyp „identisch", er spricht z. B. von „identischen Organen", Geoffroy bezeichnete sie als „analog", eine Bezeichnung, die seit Owen ganz anders und zum Teil entgegengesetzt zu homolog verwendet wird.

sis-Lehre oft als metaphysisch oder doch auf unbewiesenen Voraussetzungen beruhend kritisiert. Der Name Orthogenesis ist daher heute verschwunden, sein berechtigter Inhalt aber unter der Bezeichnung „*Trend*" wiederauferstanden. Die Feststellung von Trends ist besonders in der Palaeontologie verbreitet.

Trends in der Entwicklung. Eine Übersicht über die Umwandlungsreihen, die uns die Trends zeigen, gestattet folgende Feststellungen:

1. Die Trends lassen sich nicht für die Gesamtorganisation feststellen, sondern nur für einzelne Organe und Organkomplexe. Das Umwandlungstempo kann innerhalb einer Evolutionsreihe an den einzelnen Organen ganz verschieden sein, so daß altertümliche und fortgeschrittene Bildungen bei ein- und derselben Art nebeneinander existieren können.

2. Die Trends enthalten zweierlei Umbildungsreihen: a) *Anpassungsreihen,* b) *Vervollkommnungsreihen.*

3. Die *Anpassungsreihen* führen zu Spezialisierungen auf bestimmte Lebensweisen, etwa der Bewegung (Springen, Klettern usw.), des Nahrungserwerbs (Strudler, Filtrierer), der Fortpflanzung usw. Durch sie entstehen Ähnlichkeiten in verschiedenen Stammesreihen, die als Spezialisationsanalogien zu bezeichnen sind, und die die Musterbeispiele für Analogien liefern. Es sei aber betont, daß Anpassungen an eine gleiche Lebensweise zu analogen Strukturen führen können, aber nicht müssen, da die Anforderungen, die eine Lebensweise stellt, auf verschiedenen Wegen erfüllt werden können (Wühlen im Substrat, Strudeln). Nur als Regel gilt dabei der Satz, daß bei ähnlicher Grundorganisation die Anpassungen parallel und an denselben Organen ablaufen werden, bei verschiedener Grundorganisation aber andersartig und an verschiedenen Organen. Phylogenetisch können Anpassungsreihen schnell ablaufen.

Im Bereich des Nervensystems spielen derartige Anpassungsstrukturen nur eine geringe Rolle, da das Nervensystem ein Regulations- und Koordinationszentrum ist, in dem die Spezialisierungen der Lebensweise nur schwach sichtbar werden. Natürlich werden bei Tieren, bei denen das Auge stark entwickelt ist, auch die optischen Zentren im Gehirn stark ausgeprägt sein und bei Tieren mit dominierender Riechleistung die olfaktorischen Zentren. Ebenso werden bei lokomotorisch aktiven, und zwar vielseitig aktiven Tieren die motorischen Felder stark entwickelt sein, aber die Umformungen des Gehirns bei solchen Spezialisationen sind schwächer als die der entsprechenden Sinnes- und Lokomotionsorgane.

Merkwürdig und nur zum Teil deutbar sind die *Vervollkommnungsreihen* (= anagenetische Reihen). Ihnen widmete man bereits zu Beginn des vorigen Jahrhunderts besondere Aufmerksamkeit (MECKEL, GOETHE, TREVIRANUS, MIVART u. a.), eine große Zusammenfassung gab BRONN 1858 unter dem Titel: „Morphologische Studien über die Gestaltungsgesetze der Naturkörper". Diese empirisch gewonnenen Regeln wurden später als Gesetzmäßigkeiten der Phylogenese betrachtet und für den Aufbau von Stammeslinien verwendet. Wir wollen hier kurz den Grad ihrer Gültigkeit und ihrer Verwendbarkeit für die Phylogenie des Gehirns prüfen:

Gesetz der Zahlenreduktion. Ich beginne mit dem merkwürdigen Gesetz der Zahlenreduktion von Organen. Es besagt: Niedere Tiere und Pflanzen haben Organe in großer und variabler Zahl, höhere in geringer und konstanter Zahl. In ihm sind zwei Evolutionsprozesse vereinigt.

1. *Normierung der Zahl* der gleichartigen Strukturen. Die Zahl der Knochen primitiver devonischer Fische variiert von Individuum zu Individuum, das gleiche gilt für Blüten-

blätter, Staubblätter und Fruchtblätter der primitiven Blütenpflanzen wie der Magnoliaceen und Ranales. Bei höheren Wirbeltieren finden wir eine Fixierung der Schädelknochen nach Zahl und Ort, desgleichen eine solche der Blütenteile etwa bei Compositen und Orchideen. Oft können wir bei Tieren mit normierten Organen ihren Bestand in einer Formel ausdrücken, z. B. die Zahnformel der Säugetiere, eine Segmentformel für die höheren Krebse (Malacostraca), für die Insekten, die Hirudineen, Blütenformeln für Liliaceen, Campanulaceen, Compositae usw. Die Normierung erstreckt sich auf die verschiedensten Organe, besonders häufig findet sie sich bei vielzelligen, sogenannten meristischen Organen; sie kann sich aber sogar auf die Zellen des Körpers erstrecken. So bestehen der Körper der Rotatorien, Nematoden u. a. aus einer bestimmten Anzahl von Zellen, die wieder in bestimmter Zahl auf die einzelnen Organsysteme verteilt sind. Diese Zellkonstanz (Entelie) wird möglich durch eine Beschränkung aller Zellteilungen auf die frühe Embryonalzeit, sie schließt die Regeneration verloren gegangener Zellen aus. Die Zahlenkonstanz der Zellen gilt nicht nur für die Individuen einer Art, oft sind viele Arten und Gattungen nach der gleichen Zahlennorm gebaut. Diese Normierung des Baues nach Zahl und Lage der Organe zeigt ein merkwürdiges Schwanken. Sie gilt oft für einzelne Organe, für andere nicht. So ist z. B. die Zahl der Blütenteile bei Liliaceen und höheren Dicotylen normiert, nicht aber die Zahl der Blüten und Blätter, bei Säugern ist die Zahl der Zähne normiert, nicht aber die Zahl der Haare, bei Insekten kann sehr wohl die Zahl der Chitinhaare normiert sein (Chaetotaxie). Alle höheren Tiere sind in der Zahl der Organe weitgehend normiert, aber die höchste Normierung, die Zellkonstanz, finden wir bei niederen Würmern (Rotatorien, Nematoden); die Normierung der Segmentzahl finden wir bei den höheren, pterygoten Insekten, aber auch bei Hirudineen und manchen Polychaeten. Die Normierung kann einzelne Arten betreffen, sie kann für Gruppen von Tausenden von Arten in Einzelorganen konstant sein (Fünfstrahligkeit der Echinodermen, Segmentzahl pterygoter Insekten). Im Gehirn, das ja ganz spezifisch wirksame Teile erfordert, werden natürlich alle höheren Strukturen (Nuclei, Tractus) einmalige symmetrische Gebilde, aber nicht durch Zahlenminderung meristischer Organe. Die Zahl der Zellen ist, obwohl die Teilungsfähigkeit der Neuroblasten früh erlischt, bei höheren Tieren nicht fixiert, wie alle Organe bei den Rotatorien; sonst wäre die phylogenetische Entfaltung des Gehirns stark eingeengt worden.

2. Der zweite Teil des Zahlenreduktionsgesetzes, die *Zahlverminderung*, wurde früh erkannt. Schon MECKEL hob 1821 hervor, daß im Knochensystem der Wirbeltiere die Zahl der Knochen mit der Höherentwicklung abnehme. In der Tat können wir im Bereich der Deckknochen des Schädels und der Schultergürtel, ebenso im Bereich der Hand- und Fußknochen der Tetrapoden eine fast ausnahmslose Reduktion der Zahl der Knochen feststellen. Die Zahl der Blütenteile höherer Pflanzen ist nicht nur normiert, sondern auch viel geringer als die der primitiven Blütenpflanzen. Ganz deutlich zeigt das Gebiß der plazentalen Säugetiere die Zahlreduktion. Von der ursprünglichen Formel $\frac{3.1.4.3}{3.1.4.3}$, also einem Gebiß mit insgesamt 44 Zähnen, lassen sich im Laufe der Phylogenie Tausende Fälle von Zahnverlusten, aber nur sehr wenige von Zahnvermehrungen (Otocyon, Odontoceti) nachweisen; das Gesetz gilt also hier zu über 99%. Und dies geschieht an einem Organ, das seine Funktion nicht vereinfacht, sondern kompliziert (die Komplikation ist auf den Einzelzahn verlagert). Aber trotz klarer Bereiche, in denen das Gesetz hohe Gültigkeit besitzt, gibt es andere Bereiche, in denen es nicht zutrifft, ohne daß wir eine Erklärung geben können. Die Deckknochen des Schädels und des Schultergürtels lassen

es deutlich erkennen, nicht aber die Ersatzknochen des Neurokranium, hier tritt umgekehrt eine Vermehrung der Knochen im Laufe der Phylogenie auf. Die Zähne folgen, wie erwähnt, sehr stark der Reduktionsregel, die Wirbel der Wirbelsäule aber nicht, hier wechselt Verminderung und Vermehrung (Schlangen, Aale). Im Nervensystem verhalten sich die Tiergruppen verschieden. Während die Zahl der Bauchganglien innerhalb der Arthropoden bei vielen Stämmen in paralleler Entwicklung abnimmt, und zwar vorwiegend durch Verschmelzung erst getrennter Ganglien, zeigen die Spinalganglien der Wirbeltiere – ebenso wie die Wirbel – bald Vermehrung, bald Abnahme der Zahl. Da das Zahlenreduktionsgesetz vorwiegend für meristische Organe gilt, kommt es für das Gehirn selbst nicht in Betracht. Die ontogenetische Gliederung in Neuromera ist kaum die Wiederholung eines phylogenetischen Prozesses (vgl. den Beitrag von Bergquist, S. 175).

Das Gesetz der Differenzierung formuliert einen beherrschenden phylogenetischen Prozeß. Primär gleichartige Gebilde, wie etwa Zellen, meristische Organe, Individuen einer Art, wie etwa die Kasten der sozialen Insekten, werden durch Arbeitsteilung ungleich. So werden in zahlreichen Linien aus strukturell gleichartigen Gameten (Isogameten) die ungleichen Mikrogameten (Anisogamie) und schließlich Eier und Spermien, aus gleichartigen Zähnen (homodontes Gebiß) ungleichartige Schneide-, Eck-, Mahlzähne (heterodontes Gebiß), aus ähnlichen Wirbeln unähnliche usw. usw. Die Differenzierung führt dazu, daß niedere Organismen eine geringe Zahl von Zelltypen besitzen, höhere aber sehr viele verschiedenartige. Bei *Hydra* lassen sich höchstens 50 verschiedene Zellarten unterscheiden, bei den Wirbeltieren sind es Tausende. Die Differenzierung läuft in verschiedenen Stämmen parallel, z. B. bei der Ausbildung von Makrogameten und Mikrogameten, auch die Zelltypen können in entfernten Tierstämmen ganz ähnliche Formen annehmen, das gilt besonders von den Zellformen des Gehirns (Zawarzin), die bei Insekten und Wirbeltieren ganz ähnliche Formen annehmen können. Homologisierung von Zelltypen des Gehirns allein auf Grund ähnlicher Formen ist daher oft unsicher. Es ist klar, daß die Evolution durch Differenzierung um so weiter vorschreiten kann, je größer die Anzahl der Strukturen ist, die durch Arbeitsteilung differenziert werden können. Arten mit weniger Zellen oder gleichartigen Organen können also diesen Weg der Evolution weniger weit beschreiten als Arten mit zahlreichen Zellen oder verschiedenartigen Organen. In diesem Sinne ist das vorhin genannte Gesetz der Zahlenreduktion ein Antagonist zum Differenzierungsgesetz, dessen Wirkung es einengt. Daß das Gehirn der Wirbeltiere das differenzierteste Organ geworden ist, verdankt es der Tatsache, daß eine Zahlenreduktion oder Zahlenbegrenzung der Zellen nicht eingetreten ist, so daß die Differenzierung sich in ihm voll auswirken konnte. Die Differenzierung ist, wie erwähnt, einer der wichtigsten Prozesse in der phylogenetischen Weiterentwicklung. Das schließt nicht aus, daß eine bereits stattgefundene Differenzierung wieder rückgängig gemacht werden kann. Das beste Beispiel ist die Sakralregion an der Ansatzstelle des Beckens. Die Beckenwirbel nehmen bei Tetrapoden früh eine spezielle Form an (= Sakralwirbel); mit Verlust des Beckens oder Aufgabe seiner Fixierung an der Wirbelsäule können die Wirbel in diesem Gebiet wieder gleichartig werden (Schlangen, Wale). Sogar die Differenzierung der beiden ersten Halswirbel in Atlas und Epistropheus (= Axis) kann bei Schildkröten weitgehend rückgängig gemacht werden.

Synorganisation. Während die Differenzierung – ebenso wie im Bereich des Menschen die Herausbildung von Berufen – zu einer zunehmenden Mannigfaltigkeit führt, können durch einen anderen Prozeß, die „*Synorganisation*", verschiedene Teile zu „Apparaten" zusammengebaut werden. Solche Apparate können aus gleichartigen Strukturen gebildet sein, wie etwa die Blüten der Blütenpflanzen aus stark differenzierten Blättern oder die Fruchtständen vergleichbaren Corbulae von Hydroidpolypen. Oft werden aber ganz heterogene Strukturen in einen Apparat einbezogen, etwa bei der Bildung des Gehörapparates der Wirbeltiere oder bei der Bildung von Saugnäpfen vieler Fische, Insektenlarven usw. Im Gehirn werden zunehmend durch Zusammenschaltung von Gangliengruppen durch Tractus Teilapparaturen gebildet.

Während aber bei der Differenzierung oft parallele Umbildungen erfolgen, besonders wenn es sich um die Entstehung besonderer Zellformen handelt, sind die Apparatstrukturen, die durch Synorganisation entstehen, meist einmalige Gebilde und daher für die Verwandtschaftsforschung besonders wertvoll.

Internation. Weit verbreitet in Tier- und Pflanzenreich ist die Erscheinung der Internation von Organen. Auch dieses ,,Gesetz" formulierte bereits MECKEL 1821. Er stellte fest, daß bei niederen Tieren selbst wichtige Organe freiliegen, bei höheren aber ins Innere verlagert sind. Beispiele für diesen Vorgang lassen sich in Fülle geben. Das Nervensystem liegt in vielen Tierstämmen zuerst in der Körperwand (Epidermis), wird aber im Laufe der Phylogenie ins Innere verlagert. Wenn wir diesen Vorgang genauer verfolgen wollen, müssen wir die beiden Entstehungsarten von Organen berücksichtigen. Organe bilden sich bei Tieren von Geweben erstens durch Abfaltung von den Epithelien der Oberfläche (Epidermis) und des Darmhohlraumes, gelegentlich auch des Zöloms. Die Entstehung von Drüsen, Atmungsorganen wie Lungen und Tracheen, der meisten Sinnesorgane, des Nervensystems usw. sind Beispiele hierfür; zweitens durch die gewebliche Sonderung im Innern, meist im Bindegewebe oder in der Muskulatur. So entstehen Gebilde des Innenskeletts, wie z.B. Patella, Penisknochen, Herzknochen, direkt im Innern, ohne vorher an einer Grenzschicht gelegen zu haben. Dieser zweite Weg führt aber meist nur zur Bildung von Stützelementen, wie Knochen, Sehnen, Skelettachsen bei manchen Korallen. Der normale Weg der Organbildung verläuft also im Tierreich von einem differenzierten Epithelbezirk zum inneren Organ. Für diesen verbreiteten Vorgang können wir einige Erklärungen geben. Sind bestimmte Funktionen an eine Oberflächenschicht gebunden, so würde bei einer Größenzunahme des Lebewesens ihre Leistung vermindert werden, da die Oberfläche dann nur in der 2. Potenz, das Volumen in der 3. Potenz wächst. Will die Funktion also bei einer Größenzunahme Schritt halten, so muß sie die Oberfläche der Schichten vergrößern, die diese Funktion ausüben. Das kann entweder durch Einfaltung nach innen oder Auswachsung nach außen erfolgen.

Der erste Vorgang ist im Tierreich häufiger, wir können ihn bei der Entstehung großer Hautdrüsen, von Mitteldarmsäcken usw. beobachten, Auswachsung zeigen besonders die Kiemen von Wassertieren. Als Konsequenz ergibt sich, daß innerhalb einer Verwandtschaftsgruppe bei kleinen Tieren die Epithelien ungefaltet, der Darm einfach ist, bei größeren Arten vielfach durch Ein- und Ausstülpung kompliziert sind. Auf diese Beziehungen hat besonders HESSE hingewiesen. Im Gehirnbau ist bekannt, daß kleinere Tiere durchschnittlich eine glattere Oberfläche besitzen. Dieselbe Situation kann ohne Größensteigerung des Körpers eintreten, wenn eine Funktion, die an ein Epithel oder an eine Fläche gebunden ist, gesteigert wird. Dann wird die notwendige Vergrößerung des Areals leicht zu Einstülpungen (Invaginationen) oder Vorstülpungen (Evaginationen) führen. Sicher sind viele Kiemenbildungen und vergrößerte Hautdrüsen (Milchdrüsen) von Darmdrüsen auf diesem Wege in der Phylogenie ins Innere versenkt worden. Von geringer Bedeutung ist, wenn die Raumanforderungen in einer Kapsel oder einem Behälter die umschlossenen Organe zu einer Faltenbildung zwingen. Das läßt sich an den Keimblättern

(Kotyledonen) innerhalb der Samenschale beobachten, doch dürfte dieser Vorgang auch bei der Gestaltung des Oberflächenreliefs von Hirnteilen nicht ohne Bedeutung sein.

Viele Fälle einer Internation lassen sich aber nur biologisch verstehen. Wichtige Organe erhalten durch Versenkung oder Überdeckung einen Schutz. So entstehen die Kiemendeckel über Kiemenregionen. Sie sind z.B. bei Fischen dreimal in ganz ähnlicher Weise entstanden (Holocephali, Acanthodi, Osteichthyes), Kiemendeckel entstehen auch bei den Larven der Ephemeriden, vielen Malacostraken usw. Eine Versenkung des Trommelfells findet sich bei Wirbeltieren, Heuschrecken usw. Diese Ausbildung von Schutz kann in paralleler Entwicklung erfolgen wie die Kiemendeckel der Fische; es kann aber auch innerhalb einer natürlichen Gruppe die Überdeckung ganz verschieden ablaufen. Die als Sori bezeichneten Sporangienhaufen der Farne können durch zentrale oder seitliche Dachbildungen (Indusien) geschützt werden, sie können durch abgewandelte Ränder des Blattes bedeckt oder sogar versenkt werden. Hier wird die gleiche Aufgabe auf sehr verschiedenen Wegen gelöst. Auch die Versenkung des Nervensystems durch Einfaltung wie bei den Deuterostomiern oder durch direkte Versenkung wie bei den meisten Protostomiern ist wohl auf das Schutzprinzip zurückzuführen. Die verschiedenen Prozesse, die im Gehirn auf dieses Prinzip zurückführbar sind (Retraktion, Operkularisierung) hat SPATZ eingehend analysiert (s. S. 136).

Es wurden noch einige weitere Gesetze paralleler Entwicklung aufgestellt, z.B. solche der *Konzentration* und *Lokalisation*. Nach ihnen werden ursprünglich am Körper weit verteilte Organe oder Strukturen im Laufe der Phylogenese auf bestimmte Körperregionen beschränkt, oder erst getrennte Bezirke eines Funktionssystems nähern sich und verschmelzen. Daß auch diese Gesetze für das Nervensystem gelten, daß im Strickleiternervensystem in vielen Linien segmentale Ganglien sich nähern und verschmelzen, wurde oben erwähnt. Interessant ist, daß sowohl bei Arthropoden als auch bei Mollusken und Wirbeltieren Rumpfgebiete des Nervensystems dem primären Gehirn angegliedert werden und so ein Komplexgehirn bilden.

Diskussion

Die eben erwähnten Regeln oder Evolutionsprinzipien sind als bevorzugte Entwicklungswege gesichert. Sie erleichtern dem Phylogenetiker die Beantwortung der Frage, ob bestimmte Ähnlichkeiten auf Verwandtschaft oder auf paralleler Entwicklung beruhen. Überall, wo die Ähnlichkeiten dem Typ der „Entwicklungsgesetze" angehören, besteht der Verdacht einer parallelen Umbildung. Sehr wertvoll sind diese Gesetze auch für die Entwicklungsrichtung. Arten, die stärker differenziert, stärker das Zahlenreduktionsgesetz oder die Internation zeigen, sind weiter entwickelt als Arten mit geringerer Ausprägung dieser Evolutionsrichtung. Doch darf diese Feststellung nicht absolut genommen werden – das ist leider oft geschehen –, es können in allen Fällen rückläufige Entwicklungen auftreten, wenn sie auch zahlenmäßig viel seltener sind als progressive Entwicklungen im Sinne dieser Gesetze. Irreversibilität der Phylogenie gilt zwar für Arten und Gattungen, die, einmal ausgestorben, nicht wieder in identischer Form auftreten können, Irreversibilität gilt aber nicht für einzelne Merkmale, um so weniger, je einfacher diese sind.

Summary

The course of evolution can be demonstrated by a *dendrogram*, a tree with numerous branches. This kind of diagram is justified since no interbreeding occurs between the different branches. On the other hand not only divergent branches are found in phylogeny, but there are also trends of parallel or convergent lines: evolution shows special trends for long periods. These *trends* (= orthogenesis) are

1. successive steps of adaptation to a special mode of life (locomotion, nutrition, or special environment) or
2. successive parallel steps in the direction towards special complications (often named "anagenesis").

The morphologists of the last century have intensely studied the rules of improvement (e. g. BRONN 1858). Such rules are

1. Reduction in the number of multiple structures (e. g. teeth, dermal bones of the skull, leaves of flowers etc.) in the course of phylogeny.
   This rule is valid for several structures (e. g. teeth), but not for others (e. g. vertebrate spinal ganglions).
2. Differentiation of primarily identical structures (e. g. cells, meristic structures and individuals of insect states) in the course of evolution. These structures become different both in construction and function. This is one of the most effective processes in phylogeny. The mode of differentiation often proceeds parallel in different phyla. Isogamy leads to oogamy; homospory to heterospory; identical individuals of a colony become trophozoids and gonozoids; from isodont jaws evolve heterodont jaws.

In the course of evolution the organs of an organism have become completely different. Still they can be connected to new complex organs by another process, the so-called syn-organisation. By this a "functional system" or "apparatus" is formed.

There are quite a few parallels in the differentiation of organs, but very few (if any) in the functional systems, which have evolved by syn-organisation. Only if these organs have a common function, a parallel may be visible in relation to this common function.

Very common is the process of internation. Structures, which are primarily found in the surface, in an epithelium (epidermis, gastrodermis, mesodermis) begin to sink into the body or they are protected by folds of the skin.

Internation occurs parallel in many phylogenetic groups (e. g. operculum of the gills of Osteichthyes, Holocephali and Acanthodi) and is – in a different way – a factor determining the structure of the brain (cf. SPATZ).

There are further rules of minor importance, for instance the law of concentration and localisation. All these "laws" are rules and sometimes reversible. They often help us to decide whether similarities are analogous or homologous.

Literatur

BERGQUIST, H.: Neue Befunde über die frühe Ontogenese des ZNS. In: HASSLER, R., H. STEPHAN (ed.): Evolution of the Forebrain. Thieme, Stuttgart 1966, 175–179.

BRONN, H.: Morphologische Studien über die Gestaltungsgesetze. Leipzig und Heidelberg 1858.

EIMER, TH.: Die Entstehung der Arten, Bd. I, 1888; Bd. III, 1901.

GEOFFROY, E.: Philosophie anatomique. Paris 1818; Principes de Philosophie Zoologique. Paris 1830.

HAACKE, W.: Gestaltung und Vererbung. Leipzig 1893.

JUSSIEU, A.L.: Genera plantarum secundum ordines naturales disposita. 1791.

MECKEL, J.: System der vergleichenden Anatomie. Bd. 1. Halle 1821.

MIVART, ST.G.: On the use of the Homology. Ann. Mag. Nat. Hist. 6 (1870), Ser. 4.

OWEN, R.: On the archetype and homologies of the vertebrate skeleton. London 1848.

RENSCH, B.: Neuere Probleme der Abstammungslehre. Stuttgart 1947.

SPATZ, H.: Die Vergleichende Morphologie des Gehirns vor und nach Ludwig Edinger. Schr. wiss. Ges. Goethe Univ. Frankfurt, Naturwiss. Reihe, Nr. 1, 1955.

TREVIRANUS, G.R.: Biologie oder Philosophie der lebenden Natur. Göttingen 1802–1822.

ZAWARZIN, A.: Der Parallelismus der Strukturen als ein Grundprinzip der Morphologie. Z. wiss. Zool. 124 (1925).

Prof. Dr. A. REMANE
Zoologisches Institut der Universität
23 Kiel, Hegewischstraße 3

Further Observations on the Morphology of the Forebrain in Gymnophiona, with Reference to the Topologic Vertebrate Forebrain Pattern*

By H. Kuhlenbeck, T. D. Malewitz and A. B. Beasley,
Philadelphia, Pa.

Paraphrasing a remark of Ludwig Edinger, it could be said that the amphibian brain, if perhaps not in all aspects the simplest vertebrate brain, does indeed display the most distinct manifestation of the fundamental configurational pattern characteristic for the neuraxis of all vertebrates. Thus, in particular, it provides a most suitable frame of reference for an understanding of both comparative anatomy and presumptive phylogenetic evolution of the vertebrate forebrain.

Again, the three extant or recent orders of amphibia, namely urodeles, anurans, and gymnophiones, whose respective central nervous systems can be easily compared with each other, show three conspicuously different transformations of a topologically invariant set of configurations. Because the significant features of the gymnophione brain, although repeatedly pointed out by one of us (Kuhlenbeck, 1922, 1927, 1929a, 1929b, 1953), have not been adequately considered in the available literature, our present investigation intends to emphasize, first, the relationship of the gymnophione forebrain components to those of other amphibians, and second, the relationship of these generalized amphibian components to the still more generalized ones of both anamniote and amniote vertebrates.

Material and literature. Our data were obtained from formanalytic studies of several brains of the species *Siphonops annulatus* and *Schistomepum thomensis,* undertaken in order to prepare a morphologic survey of the gymnophione brain supplementing the earlier report of 1922. Among the few investigations concerning the brain of Gymnophiona (Coecilia, Caecilia, Apoda) undertaken since that time, Horne Craigie (1940, 1941) examined the structure of the capillary bed, recording a number of new and relevant data. Laubmann (1927) depicted several models showing overall features of developing and adult brains of *Hypogeophis*. Källén has likewise illustrated a few of these previously well known morphogenetic features of the gymnophione brain in a posthumous monograph by Krabbe (1962) on the amphibian brain. However, Laubmann a. Källén did not take into consideration or mention any of the here relevant morphologic problems already discussed in the preceding literature.

Again, a not inconsiderable error committed by Jacob a. Onelli (1911), namely the description of a reptilian (amphisbaenid) brain as an amphibian, gymnophione brain, and the elaboration of phylogenetic theories based on the peculiarities of this allegedly "amphibian" brain, was carried over by v. Economo in the text volume of his important work on cortical cytoarchitectonics (1925), despite the fact that one of us (Kuhlenbeck, 1922) had called attention to this mistake.

General characteristic of Gymnophiona. The Gymnophiona are wormlike, burrowing animals of the tropics with a very short tail usually resembling the head in external configuration, and without any indication of limbs. Like the other two orders of recent Amphibia (Urodela, Anura) gymnophiones differ radically from the extinct orders of Amphibia in having lost many skeletal elements. Coecilia possess large-yolked eggs; some

*) Partly supported by United States Public Health N. I. H. Grant NB 4999–01, 02.

of these are laid on land as in *Ichthyophis,* and the female guards them until the larvae hatch and take up life in the water. Other coecilians skip over the aquatic larvae stage, and a few have specialized external gills. At least one genus *(Typhlonectes)* is ovoviviparous. Generally speaking, the gymnophiones are considered, by competent naturalists, such as NOBLE (1931, 1954), to be "primitive" in many features of their anatomy. Thus, e. g., their first branchial cleft forms a spiracle directly comparable (i. e. orthohomologous) with that of sharks, of *Polypterus,* and a few other fishes, in contradistinction to the (kathomologous) non-communicating first cleft of all other Amphibia (NOBLE, 1931, 1954). Several skeletal features can be interpreted as possibly linking Gymnophiona to the extinct Lepospondyli, from which, according to NOBLE they presumably derived. Yet, the cited author justly points out that Gymnophiona, as far as is known, have no fossil representatives. The small scales found in many coecilian genera within the transverse, earthworm-like grooves are, nevertheless, most likely an "inheritance" from the Carboniferous Amphibia. Among the distinctive features of gymnophiones the protrusible tentacle should be mentioned, which is found on the side of the face between nostril and eye of all known species. In *Siphonops,* the tentacle appears as a small pit with a center that can be everted for approximately one half millimeter or more.

Gymnophiona, also known as "Blindwühler" in German, display various but considerable degrees of degeneration of the eyes, which may even be hidden under the bones of the skull. The eye musculature is modified by degeneration of some muscles and nerves and by the transfer of others to adjacent regions where the muscles assume different functions. Thus, the retractor bulbi is transformed into a retractor tentaculi, the rectus internus into a retractor of the tentacular sheath, and the levator bulbi into compressor and dilator muscles of the orbital glands.

The olfactory organ is particularly well developed in gymnophiones, and its fila olfactoria are gathered, on each side, in two distinct olfactory nerves, namely into a dorsal and a ventral one. The nasal cavity includes a fairly large sensory area representing JACOBSON's organ. According to NOBLE (1931) and other authors, odorous substances are apparently brought in contact with the motile tentacle which is believed to be a "tactile nose". It has been observed that some coecilians, such e. g. as *Siphonops,* will follow objects that have been put in contact with earthworms. Predominance of olfactory apparatus and rudimentary eyes are correlated with the burrowing habits of gymnophiones. At present, perhaps somewhat over hundred species of these amphibians have been described. In 1913, only 19 genera, with 55 species, all belonging to a single family, were known.

The gymnophione telencephalon, as e. g. exemplified by that of *Siphonops* (Fig. 1), displays, at the rostral end of each hemisphere, a well developed *olfactory bulb* connected with the two separate (dorsal and ventral) olfactory nerves. In the caudo-lateral region of the olfactory bulb, a conspicuous bulbus olfactorius accessorius is present. Roughly speaking, the olfactory bulb, including the accessory one, occupies somewhat less than one third (namely about 28 to 30%) of the total rostrocaudal hemispheric length. A *nucleus postolfactorius lateralis,* as delimited by one of us in 1922, and representing a cytoarchitecturally transitional formation between formatio bulbaris and the lateral components of the telencephalic zonal system, is clearly differentiated; it may be considered an intermediate subdivision between olfactory bulb and extrabulbar telencephalon. A *nucleus postolfactorius medialis,* recorded in other amphibians as well as in some of the gymnophiones examined in 1922, was rather indistinctly differentiated in the material now available to us, and not sufficiently well delimitable for inclusion as a separate region in our map. However, the gradient between olfactory bulb and the adjacent telencephalic zone B_4, and to a lesser extent also D_3, might be interpreted as a poorly circumscribed neighborhood either ortho- or kathomologous to nucleus postolfactorius medialis of those amphibian brains (particularly of some Anura) in which that structure is more conspicuous.

Caudally to olfactory bulb and to nucleus postolfactorius lateralis, the *periventricular cell populations* display an arrangement characterized by the longitudinal zones investigated in the early papers of our research program (KUHLENBECK, 1922, 1924, 1927, 1929a). These zones may now be conceived as open topologic subsets of the telencephalic wall, namely as topologic neighborhoods identical with those described by one of us in the two other amphibian orders. The dorsal or so-called "pallial" zones are D_1, D_2, D_3, and the ventral or "basal" zones are B_1, B_2, B_3, B_4 (cf. figs. 1, 3, 4).

The lumen of the *lateral ventricle* shows, in cross-sections, a peculiar medial concavity and lateral convexity typical for all gymnophiones which we have so far examined. The medial concavity corresponds to a considerable bulge jointly formed by the adjacent

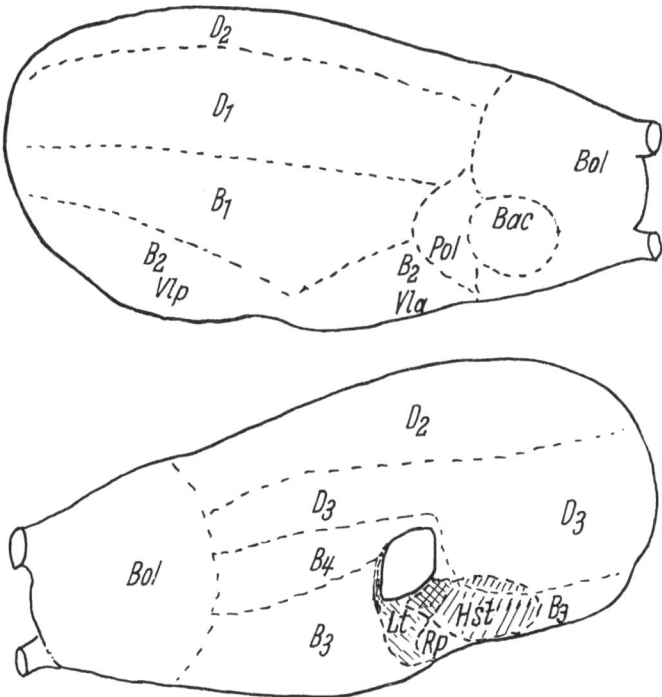

Fig. 1. Brain charts showing the regions of the gymnophione telencephalon, based on a graphic reconstruction in *Siphonops*. Combined dorsal and lateral aspect above, medial aspect below. Total length of hemisphere from tip of bulbus olfactorius to tip of polus posterior about 4.5 mm. Greatest dorso-ventral extension of hemisphere about 2.4 mm. These measurements will provide an approximate scale indicating the size of the structures shown in the photomicrographs.
Bac: bulbus olfactorius accessorius; Bol: bulbus olfactorius; Hst: hemispheric stalk; Lt: lamina terminalis; Pol: nucleus postolfactorius lateralis; Rp: preoptic recess; Vla: area ventrolateralis anterior; Vlp: area ventrolateralis posterior. Further explanations see text.

longitudinal zones D_3 and B_4, which display an internal protrusion or inversion, in some respects comparable to introversion according to Spatz, or internation according to Remane. Within the ventricular wall, at least five distinct longitudinal sulci and a few less pronounced or indistinct, accessory ones, can be recognized. Some of these sulci are approximately co-extensive with transitions or boundaries between longitudinal zones, as previously reported (Kuhlenbeck, 1929a), but the sulci need not be considered, in this abbreviated context, as essential boundary features.

The lateral ventricle contains folds of a *telencephalic choroid plexus* reaching as far rostrad as into the ventricle of bulbus olfactorius. Caudad, the choroid plexus extends toward the polus posterior of the hemisphere into the recessus posterior ventriculi lateralis, located caudally to lamina terminalis and ventriculus impar telencephali. The telencephalic choroid plexus originates from the roof plate lateral and caudal to paraphysis, and rostral to velum transversum, that is, in the roof of telencephalon medium (or impar), and protrudes into the lateral ventricles through the (secondary) interventricular foramina.

The epithelial rostral wall of the large *paraphysis* appears as a direct dorsal continuation of the telencephalic lamina terminalis; the paraphysis, representing a roof structure of the telencephalon medium, extends with its upper extremity as far as the bony wall of the

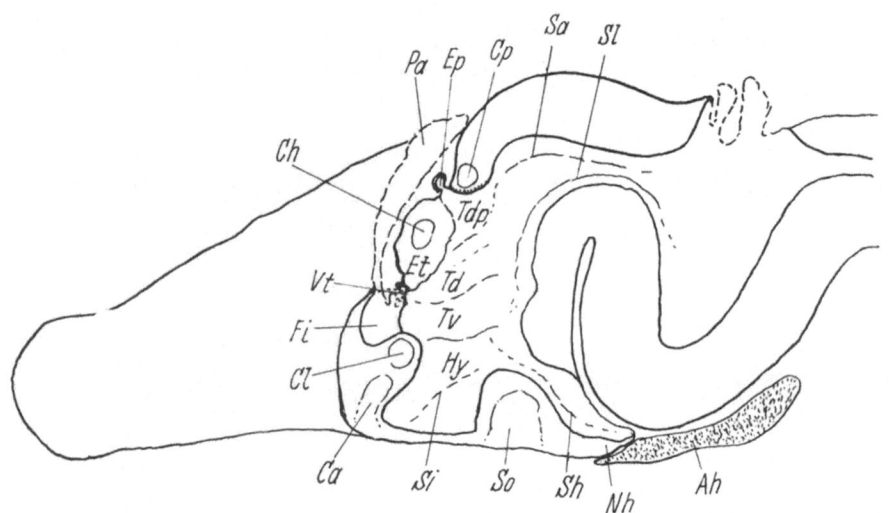

Fig. 2. Diagram of midsagittal section through brain of *Schistomepum,* showing features of third ventricle, and based on a graphic reconstruction. The three "zonal" sulci and the synencephalic sulci of third ventricle have remained unlabelled in order to avoid confusion by additional leads, but can be easily identified: sulcus diencephalicus dorsalis is the caudal border of Et; sulcus diencephalicus medius between Td and Tv; sulcus diencephalicus ventralis between Tv and Hy; sulcus synencephalicus, merging with sulcus lateralis mesencephali, ventral to Tdp; sulcus synencephalicus accessorius ventral to main synencephalic sulcus.

Ah: adenohypophysis; Ca: commissura anterior; Ch: commissura habenulae; Cl: commissura pallii ("commissura hippocampi"); Cp: commissura posterior; Ep: epiphysis; Et: epithalamus (nucleus habenulae); Fi: foramen interventriculare; Hy: hypothalamus; Nh: neurohypophysis; Pa: paraphysis; Sa: sulcus lateralis mesencephali; Sh: sulcus lateralis hypothalami; Si: sulcus intraencephalicus anterior; Sl: sulcus limitans; So: supraoptic commissure(s); Td: thalamus dorsalis; Tdp: synencephalic and pretectal subdivision of thalamus dorsalis; Tv: thalamus ventralis; Vt: region of vestigial velum transversum.

cranial cavity (Figs. 5, 6), and covers the median rostral portion of tectum mesencephali. A fold of this dorsal extension of the paraphysis was apparently mistaken by KRABBE a. KÄLLÉN (1962) for the epiphysis, which, in the gymnophiones investigated during our studies, does not extend that far dorsally (cf. also further below as well as KUHLENBECK, 1922).

The lamina terminalis, pars telencephalica, protrudes caudad with a prominent transverse ridge containing the commissura pallii and forming the floor of the ventriculus impar telencephali (Figs. 2, 5). The caudal edge of this ridge represents the basomedian portion of the telencephalo-diencephalic boundary. The preoptic recess, which, in our morphologic classification, is considered a part of the diencephalon, namely the rostral end of the hypothalamus, has its location basal to that ridge. The commissura anterior, in which fibers of medial and lateral forebrain bundle are crossing the midline, is found in the thickened part of lamina terminalis basal to commissura pallii, and within the telencephalo-diencephalic boundary zone derived from the embryonic torus transversus (or commissural plate).

As regards *some particular features* of telencephalic cytoarchitecture, the loosely cortical arrangement in zone B_2 and partly B_1 manifested within a region located in

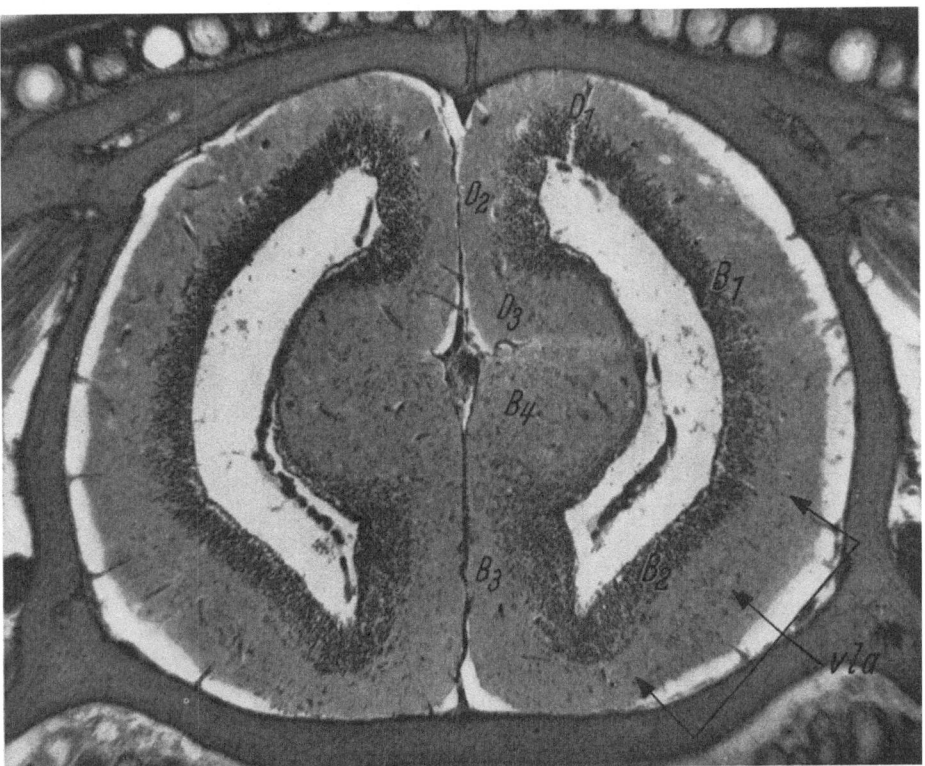

Fig. 3. Cross section through the telencephalon of *Siphonops* slightly rostral to lamina terminalis and interventricular foramen. In comparing with the lateral aspect shown in the diagrammatic brain chart of Fig. 1 it will be seen that because of the combined dorsal and lateral projection used for that chart, compensating for the lateral convexity, the Zone D_1 appears further dorsal in the cross section than in the chart. Abbreviations as in Fig. 1.

telencephalon impar and adjacent parts of the paired hemispheres rostral to lamina terminalis could here be mentioned. This rudimentary basal cortex has been designated as area ventrolateralis anterior (KUHLENBECK, 1927). Caudal to lamina terminalis, a slightly similar arrangement of scattered cells external to the densely crowded periventricular layer, and pertaining predominantly to B_2, constitutes the area ventrolateralis posterior, homologous to the true (i.e. sensu stricto) periamygdalar cortex of mammals.

Concerning the zone D_3 or *"primordium hippocampi"*, it should be noted that a rostral part, anterior and also dorsal to interventricular foramen, and a posterior, expanded part, caudal to that foramen can be distinguished. Both parts differ slightly with respect to minor details of cell arrangement, the posterior one being more "corticoid" than the anterior one.

The basolateral part of the cerebral hemisphere extending from lamina terminalis to polus posterior may appear, if individually well developed, as a slightly protruding or bulging portion of the telencephalon, and as separated from the anterior hemispheric part by an indistinctly outlined depression or shallow groove. Early investigators of the gymnophione brain had described this vaguely indicated posterior hemispheric bulge as a "temporal lobe". Since said region does indeed contain a "temporal lobe structure" of mammals, namely a kathomologon of periamygdalar cortex and of an additional amygdalar component (caudal parts of B_2 and B_3), this interpretation as a (rather defective) kathomologon of "temporal lobe" can be upheld as not entirely unjustified.

The gymnophione diencephalon. As regards diencephalic configuration (Figs. 2, 4, 5, 6), the *fundamental longitudinal zones,* represented by epithalamus, thalamus dorsalis, thalamus ventralis, and hypothalamus, are rather clearly displayed in conformity with

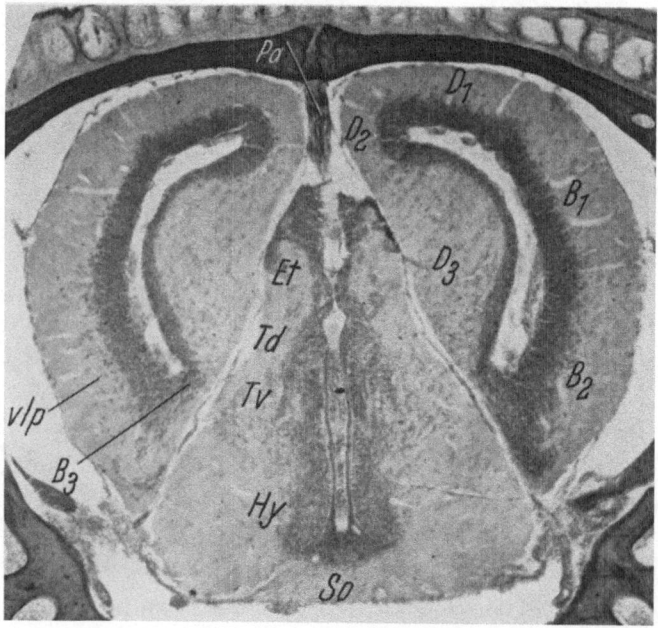

Fig. 4. Cross section through the forebrain of *Schistomepum,* showing caudal part of telencephalon (recessus posterior ventriculi lateralis), and diencephalon at rostral level of supraoptic commissure(s). Abbreviations as in Figs. 1 and 2.

the general anamniote pattern analyzed and systematized in two preceding papers (KUHLENBECK, 1929b, 1956). These zones, although representing open topologic sets, i.e. neighborhoods, without sharply outlined boundaries, can be distinguished from each other, particularly in their dorso-ventral sequence, by differences in cell-arrangement and cell population density, as well as by the course of some of the rather constant diencephalic ventricular sulci, which approximately correspond, in the adult anamniote stages, to the boundary gradients. These sulci are sulcus diencephalicus dorsalis, medius, and ventralis, of which sulcus diencephalicus medius may be rather shallow and poorly outlined.

The epithalamus consists of the fairly large nucleus (or ganglion) habenulae, extending from level of velum transversum to that of epiphysis and synencephalic recess. The commissura habenulae, in the forms studied by us, does not cross in the roof plate, but through the lumen of third ventricle (Fig. 2, KUHLENBECK, 1922). The fasciculus retroflexus, which, in the cited 1922 paper, was not sufficiently clearly delimited, in the then available cross sections, from the habenular connections with the forebrain bundles, takes an almost horizontal course toward the mesencephalic tegmentum (Fig. 6).

The epiphysis, which we consider a roof plate derivative and not a part of the epithalamus, is a short epithelial vesicle located between habenula and commissura posterior, within the roof of the synencephalic recess (cf. the comments made further above).

The thalamus dorsalis includes a narrow rostral part reaching the foramen interventriculare near the velum transversum, and a wider caudal part including a pretectal primordium. This caudal part extends into the synencephalic recess and merges, without sharp transition, into the alar mesencephalic cell masses.

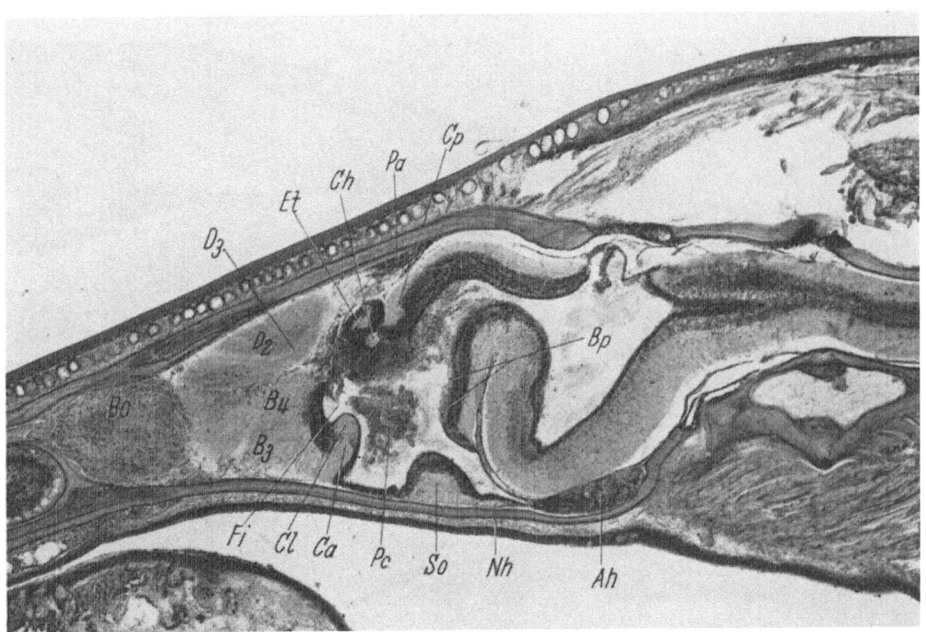

Fig. 5. Approximately midsagittal section through the brain of *Schistomepum* in situ. Figs. 2, 5, and 6 should be compared. Bp: approximate end of deuterencephalic basal plate; Pc: choroid plexus of third ventricle. Other abbreviations as in Figs. 1 and 2.

The thalamus ventralis provides rostrally the eminentia thalami forming the border of the interventricular foramen and blends caudally with the cell masses of the mesencephalic basal plate.

The hypothalamus consists of a rostral preoptic part and a caudal postoptic or parainfundibular part. The boundary between these regions is approximately indicated by the bulky ridge of the supraoptic commissure. Because of the "degenerated" optic system, an optic chiasma cannot be clearly delimited, but a few fibers, corresponding to a "chiasma rudiment" might, or might not still be included in the basal part of supraoptic commissure. Adjacent to the parainfundibular part, the neurohypophysis with an anterior infundibular portion and a more posterior part be recognized. The adenohypophysis extends caudalward to the base of the rhombencephalon.

The continuous *connectivity of diencephalon and telencephalon* is provided by the commissural plate of lamina terminalis in the midline region, and, bilaterally, by the hemispheric stalk (cf. Fig. 1). Caudally to commissura anterior, which indicates the telencephalo-diencephalic boundary zone, the lamina terminalis diencephalica extends as far as the modified chiasmatic ridge containing the supraoptic commissures. In the hemispheric stalk, the massae cellulares reunientes (KUHLENBECK, 1927) interconnect diencephalic and telencephalic structures.

In addition to the three "zonal" *sulci* enumerated above, several "accessory" sulci are present (Fig. 2). In the preoptic hypothalamus, the sulcus intraencephalicus anterior can be recognized, and in the posterior or "postoptic" hypothalamus the sulcus lateralis

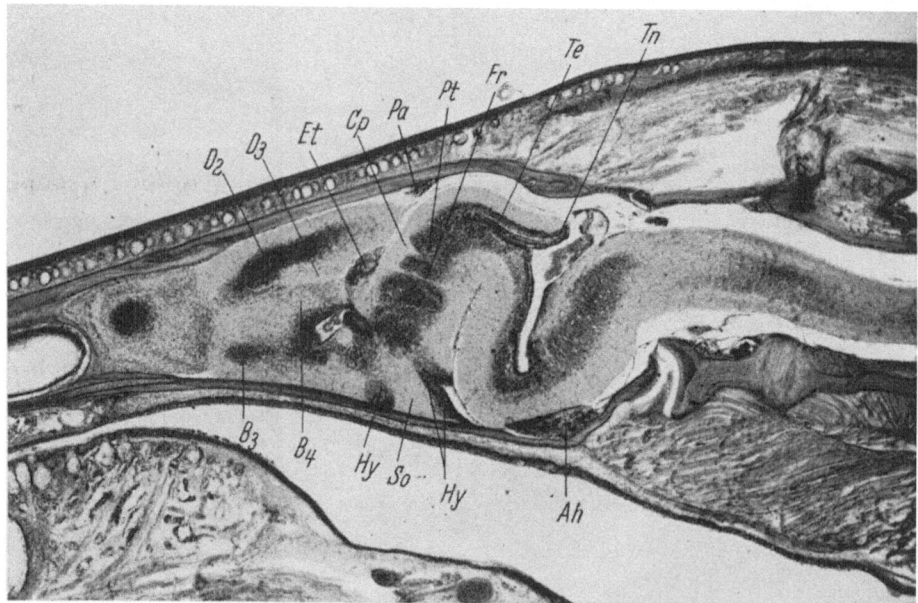

Fig. 6. Sagittal section through the brain of *Schistomepum* in situ, slightly lateral to the midsagittal plane.

Tn: cerebellar rudiment; Fr: fasciculus retroflexus (habenulo-peduncularis); Te: tectum mesencephali. Other abbreviations as in Fig. 1 and 2.

hypothalami (previously less exactly designated as sulcus lateralis infundibuli). In the ventricular wall of the dorsal thalamus, there is a sulcus synencephalicus continuous with sulcus lateralis mesencephali. A parallel ventral, less distinct sulcus synencephalicus accessorius may be present.

At the diencephalo-mesencephalic boundary, the sulcus limitans disappears, forming a pattern that can be faintly traced as a pronounced dorso-ventral curve or bend, against which the caudal neighborhoods of thalamus ventralis and dorsal hypothalamus are abutting. A noticeable depression (Figs. 2 and 5) seems to indicate the region in which the deuterencephalic basal plate may be considered to find its rostral end.

The cephalic or better *mesencephalic flexure* of the neuraxis remains, on the basal aspect, particularly salient in adult gymnophiones and represents a strikingly characteristic feature in this amphibian order. The diencephalo -mesencephalic boundary zone may be traced from the commissura posterior to the ill-defined rostral end of sulcus limitans. Particularly since, because of the "exaggerated" mesencephalic flexure, a definable tuberculum posterius cannot be seen, the inclusion of the rostral tegmental cell plate into either diencephalon or mesencephalon must remain an arbitrary procedure in attempting to define linear boundaries within a vaguely defined transitional zone characterized by low gradients.

Conclusions

Seen from the viewpoint of a *"topologization"* of morphology, this brief report on our findings could be concluded as follows: The telencephalic and diencephalic components constituting the prosencephalic neuraxis wall in Gymnophiona represent a pattern of topologically connected neighborhoods, that can be mapped, by isomorphic one-one transformations, into those of the two other extant (recent) amphibian orders. This amphibian pattern ("Bauplan"), in turn, can be mapped, either by isomorphic one-one, or by homomorphic one-many or many-one transformations of its connected neighborhoods, into the topologically equivalent, that is, morphologically homologous neighborhoods of all other vertebrates. In other words, and generalizing the results of studies begun by one of us more than forty years ago, it can be shown that the amphibian forebrain with its morphologically delimitable regions represents a simple and convenient "topologic space" whose neighborhoods, by suitable transformations, provide the required standard for the rigorous establishment of homologies in accordance with principles derived from concepts of form-analysis established by the GEGENBAUR-FÜRBRINGER school of comparative anatomy. Although such homologies are based on observations entirely independent of phylogenetic considerations, they may, however, provide, if cautiously interpreted, the required "circumstantial evidence" for phylogenetic theories.

Since it is not possible, within the scope of this short communication, to elaborate in sufficient detail and with an appropriate number of illustrations, on the problems under consideration, these questions will be dealt with in a more extensive report, now in preparation, on morphologic features of the entire brain in some Gymnophiona.

Summary

Paraphrasing a remark of LUDWIG EDINGER, it could be said that the amphibian brain, if perhaps not in all aspects the simplest vertebrate brain, does indeed display the most distinct manifestation of the fundamental configurational pattern characteristic for the neuraxis of all vertebrates. Thus, in particular, it provides a most suitable frame of reference for an understanding of both comparative anatomy and presumptive phylogenetic evolution of the vertebrate forebrain.

Again, the three extant recent orders of amphibia, namely urodeles, anurans, and gymnophiones, although easily comparable, show three conspicuously different transformations of the topologically invariant set of brain configurations. Because the significant features of the gymnophione brain, although repeatedly pointed out by one of us, have not been adequately considered in the available literature, our present communication intends to discuss and to demonstrate, first, the relationship of the gymnophione forebrain components to those of other amphibians, and second, the relationship of these generalized amphibian components to the still more generalized ones of both anamniote and amniote vertebrates.

These components, representing topological neighborhoods, are, caudally to olfactory bulb, D_1, D_2, D_3 and B_1, B_2, B_3, B_4 of telencephalic configuration, and as regards diencephalic configuration, the zones known as epithalamus, thalamus dorsalis, thalamus ventralis, and hypothalamus. Our slides illustrate the conditions obtaining in the gymnophiones Siphonops annulatus and Schistomepum thom., followed by a comparison of the relevant gymnophione components with those of other amphibian orders, and a mapping of these elements into the topologic sequences characteristic for all vertebrates.

References

CRAIGIE, E. HORNE: The capillary bed of the central nervous system of Dermophis (Amphibia, Gymnophiona). J.Morphol. 67 (1940) 477–487.

CRAIGIE, E. HORNE: The capillary bed of the central nervous system in a member of a second genus of Gymnophiona – Siphonops. J.Anat. 76: (1941) 56–64.

ECONOMO, C.v., G.N.KOSKINAS: Die Cytoarchitektonik der Hirnrinde des erwachsenen Menschen. Textband und Atlas. Springer, Berlin und Wien 1925.

JACOB, CHR., CL.ONELLI: Vom Tierhirn zum Menschenhirn. Vergleichende morphologische, histologische und biologische Studien zur Entwicklung der Großhirnhemisphären und ihrer Rinde. Lehmann, München 1911.

KRABBE, K.H.: Studies on the morphogenesis of the brain in some urodeles and gymnophions ("Amphibians"). In: Morphogenesis of

the Vertebrate Brain. IX. (completed by B. KÄLLÉN). Munksgaard, Copenhagen 1962.

KUHLENBECK, H.: Zur Morphologie des Gymnophionengehirns. Jen.Z.Naturwiss. 58: (1922) 453–484.

KUHLENBECK, H.: Über die Homologien der Zellmassen im Hemisphärenhirn der Wirbeltiere. Fol.Anat.Japon. 2: (1924) 325–364.

KUHLENBECK, H.: Vorlesungen über das Zentralnervensystem der Wirbeltiere. Fischer, Jena 1927.

KUHLENBECK, H.: Die Grundbestandteile des Endhirns im Lichte der Bauplanlehre. Anat. Anz. 67: (1929a) 1–50.

KUHLENBECK, H.: Über die Grundbestandteile des Zwischenhirnbauplans der Anamnier. Morphol.Jb. 63: (1929b) 50–95.

KUHLENBECK, H.: The telencephalic zonal system of the gymnophione Siphonops annulatus. Anat.Rec. 115: (1953) 399–400.

KUHLENBECK, H.: Die Formbestandteile der Regio praetectalis des Anamniergehirns und ihre Beziehungen zum Hirnbauplan. Fol. Anat.Japon. 28 (Nishi Festschr.): (1956) 23–44.

LAUBMANN, W.: Über die Morphogenese von Hirn und Geruchsorgan der Gymnophionen. Z. Anat. Entwickl.-Gesch. 84: (1927) 597–637.

NOBLE, G.K.: The Biology of the Amphibia. McGraw-Hill, New York 1931; Dover, New York 1954.

HARTWIG KUHLENBECK M.D.
THOMAS D. MALEWITZ Ph.D.
ANDREW B. BEASLEY Sc.D.

Department of Anatomy
Woman's Medical College of Pennsylvania
Philadelphia, Pa., USA

Vergleichende Betrachtungen am Telencephalon niederer Wirbeltiere

Von W. SCHOBER, Leipzig

Bei unseren Untersuchungen des Gehirns der Petromyzonidae, welche zu den niedersten rezenten Vertebraten gehören, ergab sich die Frage, wie weit der Bauplan des Gehirns und besonders des Endhirns als Grundlage für den Hirnbau anderer Wirbeltiere und als Ausgangspunkt für die phylogenetische Hirnentwicklung dienen kann.

Zunächst soll ein Blick auf den vermutlichen Stammbaum der niederen Wirbeltiere die phylogenetische Verwandtschaft der rezenten Klassen vor Augen führen (n. ROMER, 1959).

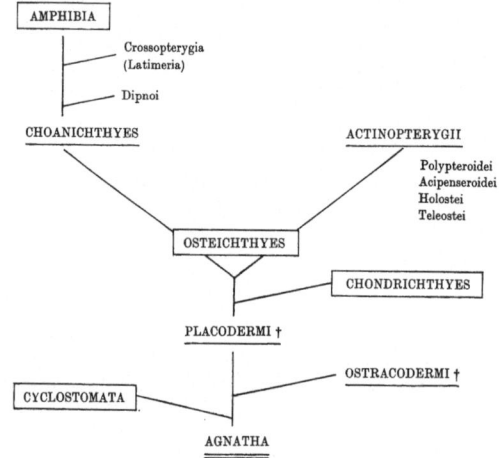

Von den ältesten Wirbeltieren, die kieferlos waren (Agnatha), leben heute noch die Cyclostomata. Sie zeigen neben primitiven Merkmalen (Fehlen der Kiefer und paarigen Flossen) auch abgewandelte Baueigentümlichkeiten (fehlendes Knochenskelett). Die Cyclostomen repräsentieren zwar eine primitive Stufe der Wirbeltierentwicklung, können aber nicht als unmittelbare Vorfahren angesehen werden. Neben ihnen gehörten noch zu den Agnatha die Ostracodermi, die über einen Knochenpanzer und ein knöchernes Endoskelett verfügten. Später traten die weiter entwickelten Formen der Placodermi auf, die echte Kiefer besaßen. Aus ihnen gingen die auch heute noch lebenden Klassen der Chondrichthyes und der Osteichthyes hervor. Letztere teilten sich schon sehr früh in zwei Gruppen: die Actinopterygii, zu denen die meisten heute lebenden Fische (Teleostei) gehören und die Choanichthyes. Erstere beschritten in ihrer Entwicklung eigene Wege und kommen für die Abstammung der Landwirbeltiere nicht in Frage. Die Choanichthyes haben nur noch wenige lebende Vertreter in der Ordnung der Dipnoi und Crossopterygii.

Unter den Vertretern dieser Ordnungen sind wohl auch die Vorfahren der Amphibien zu suchen, wobei, genau wie bei den Neunaugen, die rezenten Formen nicht als unmittelbare Vorgänger, sondern als spezialisierte Seitenzweige angesehen werden müssen.

Von den rezenten Formen der niederen Wirbeltiere ist somit keine direkter Vorgänger einer anderen, denn alle weisen irgendwelche Spezialbildungen oder Rückbildungen auf.

Trotzdem ist es nicht nur möglich, sondern auch nötig, diese Klassen miteinander zu vergleichen, um so gewisse Grundprinzipien im Bau und in der Entwicklung – in unserem Falle des Endhirns – herauszuarbeiten. Daher soll im folgenden die Morphogenese, die topographische Gliederung und der cytoarchitektonische Aufbau des Telencephalon der Petromyzonidae, der Osteichthyes (getrennt in Actinopterygii und Choanichthyes) und der Amphibia etwas näher betrachtet werden. Die Chondrichthyes finden hier keine weitere Berücksichtigung.

Morphogenese

Die frühen Entwicklungsstadien aller Endhirne sind gleich gestaltet. Es handelt sich dabei um einen *embryonalen Hirnschlauch,* an dem die dickeren Seitenabschnitte von einer dünneren Dachplatte und dem Bodenanteil unterschieden werden können. Die Seitenwände, die sich bald in eine dorsale und ventrale Area unterteilen lassen, werden im Laufe der weiteren Entwicklung recht unterschiedlich umgestaltet. Durch eine *Evagination* der Wände – verbunden mit einer Inversion – *oder* durch eine *Eversion* – verbunden mit einer Invagination – kommt es zur Herausbildung verschiedener Telencephalonformen. Dabei kann das ganze Wandgebiet verändert werden oder auch nur die Area dorsalis. In letzterem Falle finden wir neben der Bildung von Hemisphären noch ein Telencephalon medium. Der prinzipielle morphogenetische Prozeß am Wirbeltierendhirn ist wohl die Evagination. Grundsätzlich auseinanderzuhalten ist dabei die Bildung der Bulbi olfactorii und der eigentlichen Hemisphären, denn sie sind das Ergebnis getrennter Evaginationsprozesse.

In Anlehnung an die Untersuchungen von KÄLLÉN (1951) sollen einige Möglichkeiten aufgezeigt werden, wie die Wände des Endhirnschlauches umgestaltet werden können (Abb. 1).

a) Ausgehend vom frühembryonalen Hirnschlauch mit seinen Wand-, Dach- und Bodenabschnitten ist es möglich,

b) daß nur der membranöse Dachabschnitt, die Tela, vorgewölbt wird, was noch nicht zu einer Hemisphärenbildung führt (Saccus dorsalis der Neunaugen).

c) Es besteht weiterhin die Möglichkeit, daß die Wandabschnitte nach dorsal und rostral auswachsen, wobei die Tela mitgezogen wird. Man spricht dann von einer *Pseudo-Evagination,* bei der ein Septum ependymale entsteht, das *zwei Pseudohemisphären* voneinander trennt *(Neoceratodus).*

d) Eine *echte Evagination* liegt vor, wenn die Wandabschnitte nach rostral, lateral, dorsal und ventral vorgewölbt werden. Es kommt dabei zur Bildung von *Hemisphären mit Seitenventrikeln.* Diese Evagination ist oft mit einer Inversion der dorsalen, z. T. auch der ventralen Wandabschnitte verbunden.

e) Genau das Gegenteil – nämlich eine Auswärtswendung oder *Eversion* mit teilweiser Invagination der Wandabschnitte finden wir ebenfalls unter den Endhirnbauplänen. Hierbei verdicken sich die Wandpartien, wölben sich etwas in den Ventrikel vor und

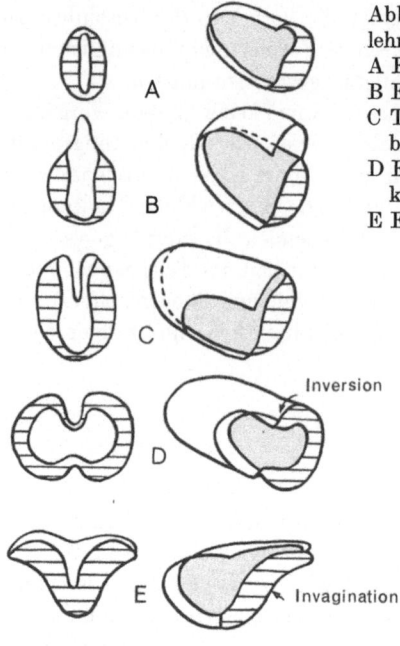

Abb. 1. Formen der Telencephalon-Morphogenese (in An-
lehnung an KÄLLÉN)
A Embryonalstadium
B Evagination der Tela; noch keine Hemisphärenbildung
C Teil- oder Pseudoevagination mit Pseudohemisphären-
    bildung
D Echte Evagination mit Hemisphären- und Seitenventri-
    kelbildung
E Eversion.

wenden sich außerdem nach auswärts und abwärts. Dorsal und oft auch lateral schließt nur die Tela chorioidea das Ventrikellumen ab. Zwischen diesen aufgezeigten Möglichkeiten gibt es verschiedene Kombinationen und Übergangsstadien, so daß die Form des Telencephalon bei den niederen Wirbeltieren recht mannigfaltig ist. Um den Grad der Veränderungen der Wandabschnitte generell und in den verschiedenen Richtungen vergleichen zu können, ist es günstig, als Bezugssystem die Commissura anterior zu wählen.

Es soll nun im einzelnen auf die Gestalt der Endhirne der niederen Wirbeltiere eingegangen werden.

Petromyzonidae (Abb. 2 A): Die Bulbi olfactorii sind evaginiert, und zwar mehr nach lateral als nach rostral. Die ventralen Wandabschnitte des übrigen Endhirnes zeigen kaum Veränderungen und sind als Telencephalon medium zu bezeichnen. Die dorsalen Wandabschnitte dagegen haben eine Umgestaltung sowohl im Sinne einer leichten Eversion als auch einer Evagination erfahren. Während der dorsale Teil des pallialen Wandabschnittes zunächst unverändert erscheint, ist er nach kaudal zu gering evertiert. Deutlich evaginiert sind die restlichen Anteile der dorsalen Area. Der membranöse Dachabschnitt ist weit nach dorsal vorgewölbt und bildet den Saccus dorsalis.

Osteichthyes – Actinopterygii (Abb. 2 B–D): Über das Endhirn der Knochenfische liegen zahlreiche neuere Arbeiten von NIEUWENHUYS (1960, 1962a, 1962b, 1963, 1964) vor. Sie unterstreichen die Befunde, die schon GAGE (1893) mitteilte, daß nämlich die endgültige Gestalt des Fischendhirnes durch eine *Eversion der dorsalen Wandabschnitte* erreicht wird. Nur die kranialen Teile des embryonalen Hirnschlauches sind zu den Bulbi olfactorii evaginiert. Dies ist bei den primitiven Formen wie *Polypterus* sehr klar zu erkennen, da hier die Bulbi noch einen deutlichen Ventrikel enthalten, während bei den höheren Fischgruppen (Teleostier) die Bulbi größtenteils massiv sind. Die Eversion der Wände wird in der aufsteigenden Reihe der Actinopterygii ständig komplizierter. Unverfälscht erscheint sie noch bei *Polypterus,* wo die Wände nicht verdickt sind, sondern in rostrokaudaler Richtung nur immer stärker nach außen und abwärts gewendet werden. Bei den höheren Formen wird der Wandabschnitt verdickt, etwas in das Ventrikellumen invaginiert, und außerdem wendet er sich in verschieden starkem Maße nach auswärts und abwärts. Dabei wird der membranöse Dachabschnitt, die Tela ependymale, mit nach außen gezogen, so daß ein T-förmiger Ventrikel entsteht.

Die ventralen Wandabschnitte sind als Telencephalon medium ähnlich wie bei den Petromyzonidae ausgebildet.

Osteichthyes - Choanichthyes (Abb. 3 A–C): Bei den zu dieser phylogenetisch wichtigen Gruppe gehörenden Lungenfischen finden wir ebenfalls nach rostral evaginierte Bulbi olfactorii mit deutlichen Ventrikeln. Vom übrigen Telencephalon sind besonders die *ventralen Wandabschnitte* stark *evaginiert*, so daß es zur *Hemisphärenbildung* kommt. Wir müssen hier allerdings, wie erst NIEUWENHUYS (1965) jüngst berichtete, streng zwischen dem Telencephalon von *Protopterus* und *Lepidosiren* einerseits und *Neoceratodus* auf der anderen Seite unterscheiden. Während bei ersteren die Hemisphären völlig aus Nervengewebe bestehen, ist bei letzterem der dorsomediale Teil der Hemisphäre nur ependymal ausgebildet (Pseudoevagination). Der nichtevaginierte kaudale Abschnitt zeigt als Telencephalon medium noch die embryonalen Verhältnisse.

Von *Latimeria* soll nur erwähnt werden, daß das Endhirn, besonders in seinen ventralen, dünnwandigen Abschnitten, ebenfalls evaginiert ist, während die dorsalen Wandanteile eher eine leichte Eversion erkennen lassen (MILLOT u. ANTHONY, 1962; NIEUWENHUYS, 1962).

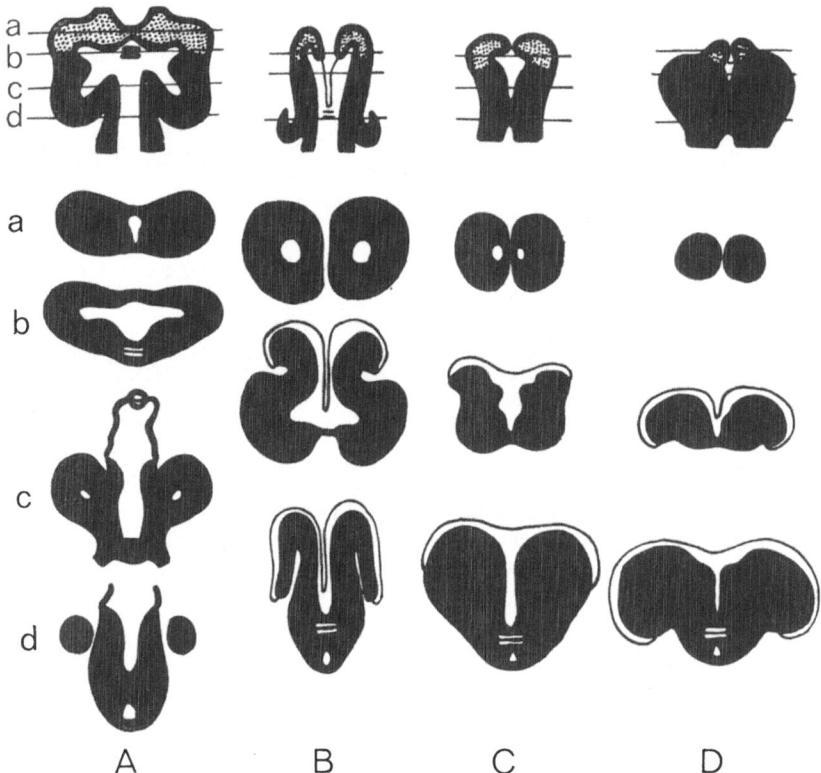

Abb. 2. Telencephalonform niederer Wirbeltiere:
A *Lampetra* (Cyclostomata, Petromyzonidae); B *Polypterus;* C *Salmo;* D *Gasterosteus*
(B–D Actinopterygii; in Anlehnung an NIEUWENHUYS).
Oben Horizontalschnitte, die darin enthaltenen Linien geben die Lage der darunter abgebildeten Frontalschnitte an.

Amphibien (Abb. 3 D): Das Endhirn der Amphibien ist vollständig evaginiert. Dabei ist nicht nur der Bulbus olfactorius weit nach rostral vorgewölbt, sondern auch die übrigen Wandanteile sind, bezogen auf die Commissura anterior, besonders nach rostral, daneben aber auch nach dorsal, ventral und lateral evaginiert. Dies hat zur Folge, daß die dorso-medialen und ventromedialen Gebiete eine Inversion zeigen, wobei es ventral zur Bildung eines massiven Septumabschnittes kommt (HERRICK, 1933a, 1948).

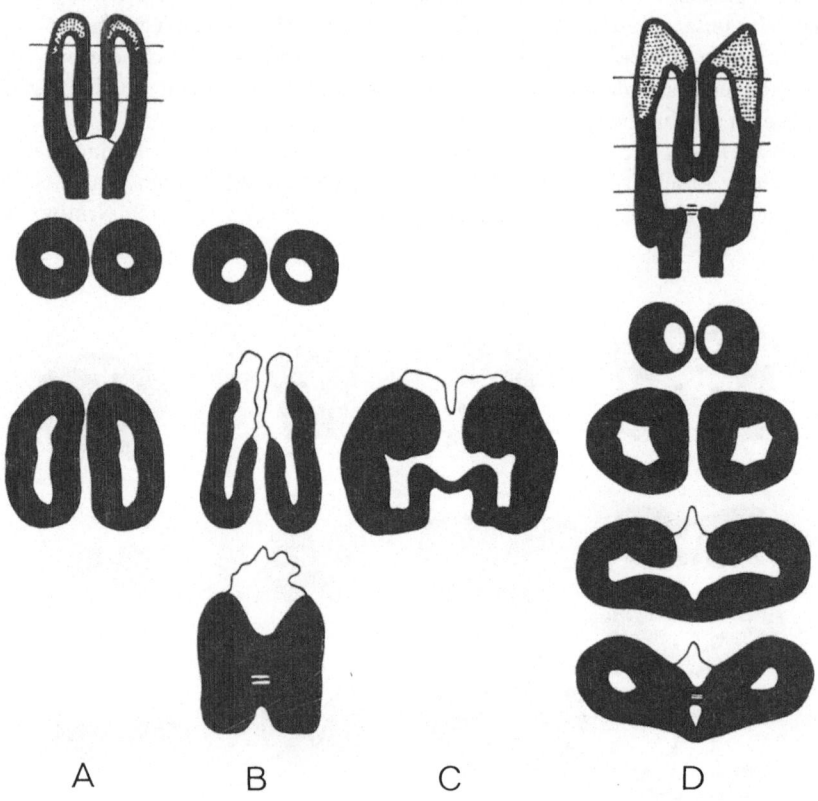

A   B   C   D

Abb. 3. Telencephalonform niederer Wirbeltiere.
A Protopterus, Lepidosiren; B *Neoceratodus;* C *Latimeria;* (A–C Choanichthyes; in Anlehnung an NIEUWENHUYS); D *Salamandra* (Amphibia, Urodela).

Topographie

Was die topographische Gliederung des Telencephalon anbelangt, so muß man fest-stellen, daß trotz der gestaltlichen Unterschiede eine gewisse Grundgliederung bei allen Formen wiederzufinden ist. Wir können am Endhirn immer die primären olfaktorischen Zentren in Gestalt der Bulbi olfactorii abgrenzen, die bisweilen stark reduziert sein kön-nen. Diesen schließen sich die sekundären und tertiären olfaktorischen Zentren an, die entsprechend der Gliederung der Endhirnwand einen dorsalen oder pallialen und ven-tralen oder subpallialen Anteil umfassen. Auf die primären Zentren folgt besonders im ventralen Gebiet noch eine Verbindungszone in Form des Nucleus olfactorius anterior (HEIER, 1948; HOFFMANN, 1960; HOLMGREN, 1922; SCHOBER, 1964).

Der palliale wie auch der subpalliale Abschnitt ist auf Grund cytoarchitektonischer Besonderheiten oft von den Autoren weiter untergliedert worden, wobei die Benennung leider recht uneinheitlich erfolgte (ARIENS-KAPPERS u. HAMMER, 1918; ARIENS-KAPPERS, HUBER u. CROSBY, 1936; DROOGLEVER-FORTUYN, 1961; HERRICK, 1933a, b; 1948; HOLMGREN, 1920; KUHLENBECK, 1924, 1929; MEADER, 1939; NIEUWENHUYS, 1963; SCHOBER, 1964). Es soll hier noch hervorgehoben werden, daß die Palliumabschnitte nicht nur an Hemisphärenbildungen gebunden sind, sondern auch einen Teil des ursprünglichen Telencephalon medium einnehmen können oder wie bei den evertierten Fischgehirnen durch die massiven, auswärts gewendeten Wandgebiete verkörpert werden. Die ventralen Wandanteile sind oft noch als Telencephalon medium erhalten und erinnern an die embryonalen Verhältnisse. Bei anderen Formen dagegen sind sie von der Evagination mit erfaßt worden und in die Hemisphären eingegangen. Wir unterscheiden hier ein Septumgebiet und eine striäre Region sowie als Übergangszone zum Diencephalon das Gebiet des Nucleus praeopticus.

Eine nähere Betrachtung der Verhältnisse bei den einzelnen Klassen zeigt uns folgendes:

Petromyzonidae (Abb. 4 A): Die primären olfaktorischen Zentren sind gut ausgebildet und erstrecken sich als Bulbus olfactorius besonders nach lateral. Ventral läßt sich als Übergangsgebiet der Nucleus olfactorius anterior ausgliedern. Eine Septumbildung kann man im ventralen Wandabschnitt nicht feststellen, so daß es fraglich ist, ob eine schmale subependymale Zellanhäufung schon als Nucleus septi bezeichnet werden kann (HEIER, 1948; SCHOBER, 1964). Das Telencephalon medium umfaßt das Primordium corporis striati und den Nucleus praeopticus. Daneben gehört aber auch noch der dorsomediale, nicht evaginierte Teil des Pallium – das Primordium hippocampi mit dem Lobus subhippocampalis – dazu. Der evaginierte Wandanteil umfaßt das restliche Pallium. Eine Unterteilung dieses Gebietes in Primordium pallii dorsale und Primordium piriforme läßt sich nicht rechtfertigen, da histologisch beide Anteile kaum voneinander zu unterscheiden sind.

Actinopterygii. Die topographische Gliederung des Actinopterygii-Endhirnes (Abb. 4 B, C) bereitet wohl die größten Schwierigkeiten, da schon eine unterschiedliche Interpretation der Morphogenese zu den verschiedensten Benennungen der einzelnen Abschnitte geführt hat (DROOGLEVER-FORTUYN, 1961; GOLDSTEIN, 1905; HOLMGREN, 1920; JOHNSTON, 1911; KUHLENBECK, 1929; NIEUWENHUYS, 1963). Neben dem Bulbus olfactorius können wir auch hier eine dorsale und ventrale Area unterscheiden. Letztere behält wie bei den Neunaugen ihre embryonale Lage bei. Sie läßt sich nochmals in eine ventrale und dorsale Zone unterteilen, wobei es schwierig ist, zu sagen, ob diese Gebiete einem Septum bzw. einem Striatum entsprechen.

Die Area dorsalis telencephali zeigt infolge der unterschiedlich starken Eversion der Wandabschnitte auch in ihrer Form und Struktur größere Unterschiede. Nach HOLMGREN (1922) und KÄLLEN (1951) ist sie homolog dem Pallium und gestattet auch – fortschreitend mit der systematischen Höherentwicklung der Fische – eine gute cytoarchitektonische Gliederung. Während bei *Polypterus* die ganze dorsale Zone in ihrer Struktur durch und durch gleich gestaltet ist und insgesamt als Pallium angesprochen werden muß, kann man bei den Teleostei einen medialen Teil (Primordium pallii pars medialis o. Lobus piriformis, HOLMGREN, 1922), einen dorsalen Teil (Primordium pallii pars dorsolateralis) und einen lateralen Teil (Primordium pallii pars lateralis o. Primordium hippocampi, HOLMGREN,

Lampetra Polypterus Gasterosteus Neoceratodus Salamandra

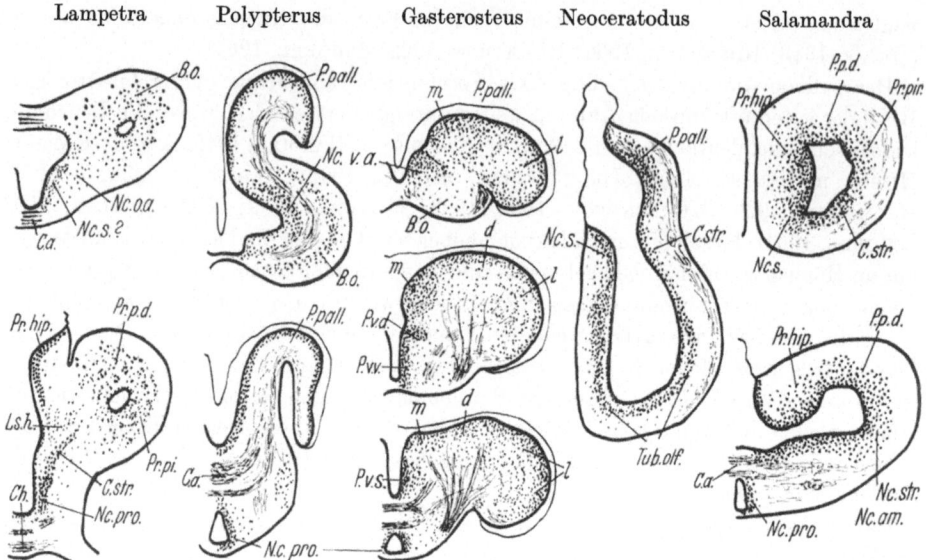

Abb. 4. Topographische Gliederung des Telencephalon verschiedener niederer Wirbeltiere.

B.o.	Bulbus olfactorius	l	lateraler Teil der P.pall.
C.a.	Commissura anterior	m	medialer Teil der P.pall.
Ch.	Chiasma	P.v.	Pars ventralis
C.str.	Corpus striatum	P.v.d.	dorsaler Teil der P.v.
L.sh.	Lobus subhippocampalis	P.v.s.	supracommissuraler Teil der P.v.
Nc.am.	Nucleus amygdalae	P.v.v.	ventraler Teil der P.v.
Nc.o.a.	Nucleus olfactorius anterior	Pr.hip.	Primordium hippocampi
Nc.pro.	Nucleus praeopticus	Pr.pi.	Primordium piriforme
Nc.s.	Nucleus septalis	P.p.d.	Primordium pallii dorsale
P.pall.	Pars pallialis (dorsalis)	Tub.olf.	Tuberculum olfactorium
d	dorsaler Teil der P.pall.		

1922) unterscheiden. Ob jedoch die einzelnen Palliumabschnitte der evaginierten Endhirne mit denen der Fische gleichgesetzt werden können ist zweifelhaft.

Dipnoi. Am evaginierten Endhirn der Dipnoi (Abb. 4 D) sind die ventralen Abschnitte
stark vorgewölbt. Den größten Teil nimmt hier das Tuberculum olfactorium ein während
medial das Septumgebiet und lateral die Striatumzone recht schmal sind. Am pallialen
Abschnitt unterscheiden HOLMGREN u. VAN DER HORST (1925) ebenfalls ein dorsales oder
hippocampales, intermediäres o. generales und ein ventrales o. piriformes Pallium.
NIEUWENHUYS (1965) meint allerdings, daß die histologischen Unterschiede in den
Palliumanteilen nicht so groß seien, daß man eine solche Gliederung durchführen könnte.

Amphibien. Bei den Amphibien (Abb. 4 E) sind äußerlich die Bulbi schwer abzugrenzen, sie zeigen aber histologisch den typischen Aufbau. An dem vollständig evaginierten
Endhirn sind an der Hemisphärenbildung die Area dorsalis und ventralis beteiligt. Die
periventrikuläre Zellschicht der dorsalen Area wird von den Autoren in drei Abschnitte
unterteilt. Dorsomedial liegt das Primordium hippocampi, dorsolateral das Primordium

piriforme und dazwischen schiebt sich das Primordium pallii dorsale oder Pallium i.e.S., das zuweilen schwierig von den angrenzenden Gebieten abzutrennen ist. Der basale Wandabschnitt läßt sich in das ventromediale Septumgebiet und das ventrolaterale Primordium corporis striati unterteilen. Nicht von der Evagination erfaßt ist der Nucleus praeopticus, der als Rest des Telencephalon medium angesehen werden kann.

Cytoarchitektonik

Ein Blick auf die cytoarchitektonische Strukturierung der Endhirnabschnitte der Anamnia zeigt uns zunächst, daß die primär olfaktorischen Zentren bei allen Formen weitgehend gleich gebaut sind (ARIENS-KAPPERS, HUBER, CROSBY, 1936; DROOGLEVER-FORTUYN, 1961; HEIER, 1948; HERRICK, 1948; HOFFMANN, 1960; KUHLENBECK, 1927; NIEUWENHUYS, 1963; SCHOBER, 1964). Sie lassen im allgemeinen eine 6-Schichtung erkennen, die von außen nach innen folgende Gebiete umfaßt: 1. Schicht der primären olfaktorischen Fasern, 2. Glomerularschicht, 3. äußere Zellschicht, die vorwiegend die großen Mitralzellen enthält, 4. Schicht der sekundären olfaktorischen Fasern, 5. innere granuläre Zellschicht, 6. Ependymschicht.

In den übrigen Hirnabschnitten liegen die Neurone meistens unmittelbar unter den Ependymzellen in einer periventrikulären Zellschicht und lassen vielfach noch keine stärkere Differenzierung erkennen. Bekanntlich spielt in der aufsteigenden Wirbeltier-reihe die Palliumevolution eine bedeutende Rolle. Sie drückt sich besonders bei den niederen Klassen darin aus, wie stark eine Zellmigration aus dem periventrikulären Grau nach peripher erfolgt, wobei es im Endeffekt zur Bildung einer Rindenschicht – des Cortex – kommt. Diese Entwicklungsrichtung finden wir bei den evaginierten und evertierten Endhirnen gleichermaßen. Es soll daher noch aufgezeigt werden, wieweit die periventri-kuläre Zellschicht, besonders im Palliumgebiet, schon bei den niederen Wirbeltieren evolutive Veränderungen erkennen läßt.

Petromyzonidae (Abb. 5 A): Bereits bei diesen niedersten Formen müssen wir fest-stellen, daß nur das Primordium hippocampi den ursprünglichen Zustand der Zellan-ordnung aufweist. Hier liegen die Zellen noch eng gepackt in 4–5 Schichten unmittelbar unter dem Ventrikelependym. Im evaginierten Palliumanteil dagegen sind schon viele Zellen nach peripher verlagert und diffus oder zum Teil als Zellnester über die Hemisphä-renwand verteilt. Zu einer Cortexbildung kommt es aber noch nicht (SCHOBER, 1964).

Actinopterygii. Bei den Actinopterygii (Abb. 5 B, C) müssen wir zunächst zwischen Polypterus und den höheren Formen trennen. Bei Polypterus zeigt der palliale Abschnitt in seinem ganzen Verlauf eine sehr ursprüngliche Zellanordnung. Alle Neurone liegen subependymal in einer eng gepackten uniformen Schicht ähnlich wie im Primordium hippocampi der Neunaugen.

Bei den höheren Fischformen haben wir im dorsalen Abschnitt in der Regel auch noch periventrikulär die größte Neuronendichte, aber daneben ist eine Verteilung der Zellen über die ganze Breite der evertierten Wand zu erkennen. Bisweilen liegen die Zellen in parallelen Schichten zum Ependym, sie zeigen aber auch bestimmte Ansammlungen, so daß bei den Teleostiern eine Gliederung in verschiedene Felder möglich ist (vgl. Abb. 4 B, C). Infolge der Eversion kommt es zu einer zentralen Abgliederung von Zellschichten, die topographisch vergleichbar mit den corticalen Zonen evaginierter Endhirne sind.

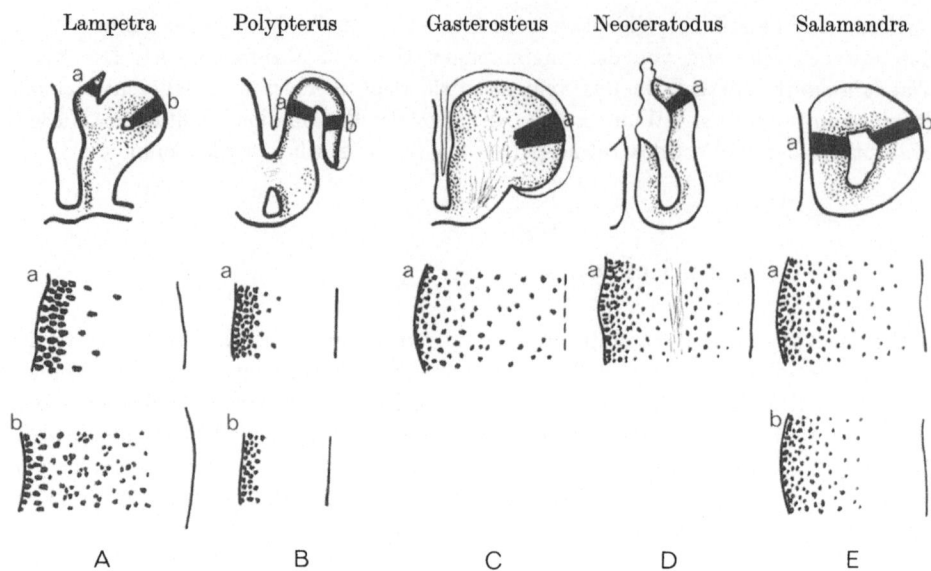

Abb. 5. Cytoarchitektonischer Aufbau des Palliumabschnittes verschiedener niederer Wirbeltiere (die innere Oberfläche zeigt bei den Zellschemata stets nach links).

NIEUWENHUYS (1962, 1963) prägte für diese Erscheinung den Begriff der „paradoxen Cortexformation".

Dipnoi. Das Pallium der Dipnoi (Abb. 5 D) ist von besonderem Interesse, da es den *ersten Schritt einer Cortexbildung* zeigt. Wir haben hier zwar noch eine gut entwickelte periventrikuläre Zellschicht, aber von dieser abgespalten und durch eine dünne Faserschicht getrennt finden wir eine Schicht nach lateral ausgewanderter, großer Zellen, die deutlich eine periphere Cortexzone bilden. Die Lungenfische sind somit die niedersten Vertebraten, die einen entwickelten pallialen Cortex aufweisen. Es soll noch erwähnt werden, daß auch im Bereich des Subpallium (Tuberculum olfactorium) die Bildung von zwei Zellschichten zu erkennen ist (HOLMGREN u. VAN DER HORST, 1925; NIEUWENHUYS, 1965; RUDEBECK, 1945).

Amphibien. Das Pallium der Amphibien (Abb. 5 E) läßt keine Cortexbildung erkennen. Der größte Teil der Zellen liegt hier als periventrikuläres Grau vor; nur einzelne Zellen sind nach peripher ausgewandert. Eine Ausnahme macht das Gebiet des Primordium hippocampi, wo bereits eine stärkere laterale Migration von Zellen zu beobachten ist, so daß fast die ganze Breite dieses Palliumabschnittes Zellen enthält.

Diskussion

Bei einer vergleichenden Betrachtung der soeben dargestellten Einzelfakten kann man feststellen, daß das Neunaugen-Endhirn, ähnlich wie das der anderen niederen Wirbeltiere, neben ursprünglichen Merkmalen auch manche Besonderheiten zeigt. So wie die rezenten Formen der niederen Wirbeltiere weitgehend Endglieder ihrer Klassen darstellen, so sind auch ihre Endhirnbaupläne das Ergebnis einer Entwicklung über Jahrmillionen.

Trotzdem lassen sich bestimmte Grundzüge in der Entwicklung erkennen. Wir finden bei den meisten Wirbeltieren in der Morphogenese das Prinzip der Evagination verwirklicht, während bei den Actinopterygii, abgesehen vom Bulbus olfactorius, das Endhirn durch die Eversion der Wände ganz anders gestaltet ist und dadurch auch eine andere histologische Differenzierung aufweist. Dabei hat sich dieses Telencephalon genau so weit vom embryonalen Zustand entfernt und zeigt die gleiche progressive Weiterentwicklung innerhalb der Knochenfische, wie die evaginierte Form der übrigen Wirbeltiere. Es ist daher wohl nicht gerechtfertigt, das Knochenfischendhirn als primitiv oder insgesamt als Telencephalon medium hinzustellen (JOHNSTON, 1911; HERRICK, 1922) oder wie DROOG-LEVER-FORTUYN (1961) davon zu sprechen, daß das Endhirn zunächst zu einer primitiven Form zurückkehrte, bevor es sich in Richtung Lungenfische und Amphibien weiterentwickelte.

Man muß hier wohl beide morphogenetischen Prinzipien gleichwertig nebeneinander stellen, da auch andere Vertreter der niederen Wirbeltierklassen (Petromyzonidae, *Latimeria*) Endhirne aufweisen, die beide Prinzipien (Evagination und Eversion) erkennen lassen. Was die histologische Differenzierung, besonders des Palliums betrifft, so weist der primitive Actinopterygier *Polypterus* das einfachste Pallium aller rezenten Wirbeltiere auf, während wir bei den Dipnoi die erste Cortexbildung feststellen können. Das Pallium der Petromyzonidae zeigt im Primordium hippocampi noch einen sehr ursprünglichen Bauplan, während die anderen Abschnitte bereits eine Zellmigration nach peripher erkennen lassen.

Alle Ergebnisse des morphogenetischen und histologischen Vergleiches deuten darauf hin, daß das Petromyzonten-Endhirn dem der Wirbeltiervorfahren recht nahekommt, jedoch nicht den ursprünglichen Bauplan verkörpert. Das gleiche gilt wohl auch für die Choanichthyes. Da das evertierte Actinopterygii-Endhirn seine eigene Entwicklungsrichtung eingeschlagen hat, erscheint als Ausgangspunkt für das Studium des evaginierten Wirbeltierendhirns und seine phylogenetische Entwicklung das Amphibienendhirn am geeignetsten. Die Hirnorganisation der Amphibien weist – unter Beachtung der Entwicklungshöhe dieser Klasse – die morphogenetisch und histologisch-topographisch klarsten Verhältnisse auf (HERRICK, 1948). Das Endhirn hat keine divergenten Entwicklungsrichtungen eingeschlagen und verfügt über keine besonderen Spezialisierungen. Damit ist es wohl am besten geeignet, als Grundlage bei einem Vergleich mit anderen Endhirnen, besonders dem der Amnioten, zu dienen.

Summary

Based on former studies on the telencephalon of Petromyzonts, an attempt was made to determine how far the brain of this group can serve as a basis for comparison with the cerebral pattern of other vertebrates and as a starting point for cerebral evolution.

In the present study, which is based on morphogenetic data, topographic organization and finer histology, the telencephalon of Petromyzonts is compared with that of Osteichthyes and Amphibians. It can be shown that the Petromyzont telencephalon is closely related to that in vertebrate ancestors; it does, however, not present the original cerebral pattern. A comparison of the architectural peculiarities of the Petromyzont telencephalon with those of the everted Actinopterygia is difficult. It is easier to establish relations to the cerebral pattern of Choanichthyes and Amphibians. However, the endbrain of Petromyzonts cannot be considered as an immediate forerunner of those endbrain patterns, since there are definite deviations according to specialization and reduction.

It would therefore seem more promising, also in future, to relate comparative investigations into the phylogeny of the telencephalon to the cerebral pattern of the tailed Amphibians. Considering their developmental stage, the brains of this group are less specialized and show clearer structural characteristics than those of Petromyzonts.

Literatur

ARIENS-KAPPERS, C.U., E.HAMMER: Das Zentralnervensystem des Ochsenfrosches (Rana catesbyana). Psych.Neurol.Blad *22* (1918), 368–415.

ARIENS-KAPPERS, C.U., G.C.HUBER, E.C. CROSBY: The Comparative Anatomy of the Nervous System of Vertebrates including Man. Hafner Publ.Comp., New York 1936.

DROOGLEVER-FORTUYN, J.: Topographical relations in the telencephalon of the sunfish Eupomotis gibbosus. J.comp.Neurol., *116* (1961) 249–264.

GAGE, S.P.: The Brain of Diemyctilus viridescens from larval to adult life and comparison with the brain of Amia and of Petromyzon. The Wilder Quarter Century Book, Ithaca 1893, p. 259–314.

GOLDSTEIN, K.: Untersuchungen über das Vorder- und Zwischenhirn einiger Knochenfische. Arch.mikr.Anat. *66* (1905), 135–219.

HERRICK, C.J.: Functional factors in the morphology of the forebrain of Fishes. Vol. I, p. 143–202. Libro en Honor de S.Ramon y Cajal, Madrid 1922.

HERRICK, C.J.: The amphibian forebrain. VI. Necturus. J.comp.Neurol. *58* (1933*a*), 1–288.

HERRICK, C.J.: The amphibian forebrain. VIII. Cerebral hemispheres and pallial primordia. J.comp.Neurol. *58*, (1933*b*) 737–759.

HERRICK, C.J.: The brain of the Tiger Salamander. The Univ. of Chicago Press, Chicago Ill. 1948.

HEIER, P.: Fundamental principles in the structure of the brain. A study of the brain of Petromyzon fluviatilis. Acta Anat. (Basel), Suppl. *8*, (1948) 1–213.

HOFFMAN, H.H.: The olfactory bulb, accessory olfactory bulb and hemisphere of some anurans. J.comp.Neurol. *120* (1960), 317–368.

HOLMGREN, N.: Zur Anatomie und Histologie des Vorder- und Zwischenhirns der Knochenfische. Acta zool. *1* (1920), 137–315.

HOLMGREN, N.: Points of view concerning forebrain morphology in lower vertebrates. J. comp.Neurol. *34*, (1922), 391–459.

HOLMGREN, N., C.J.VAN DER HORST: Contribution to the morphology of the brain of Ceratodus. Acta zool. *6* (1925), 59–165.

JOHNSTON, J.B.: The telencephalon of ganoids and teleosts. J.comp.Neurol. *21* (1911), 489–591.

KÄLLÉN, B.: Some remarks on the ontogeny of the telencephalon in some lower vertebrates. Acta anat. *11* (1951), 537–548.

KUHLENBECK, H.: Über die Homologien der Zellmassen im Hemisphärenhirn der Wirbeltiere. Fol.anat.jap. *2* (1924), 315–364.

KUHLENBECK, H.: Vorlesungen über das Zentralnervensystem der Wirbeltiere. G.Fischer, Jena 1927.

KUHLENBECK, H.: Die Grundbestandteile des Endhirnes im Lichte der Bauplanlehre. Anat. Anz. *67* (1929), 1–51.

MEADER, R.E.: The forebrain of bony fishes. Kon.Akad.v.Wetensch. (Amsterdam) Proc. sect.sc. *42* (1939), 657–670.

MILLOT, J., J.ANTHONY: Premieres précisions sur l'organisation du télencéphale chez Latimeria chalumnae. C.R.Acad. Sci. Paris *254*, (1962), 2067–2068.

NIEUWENHUYS, R.: Some observations on the structure of the forebrain of bony fishes. In: TOWER, D.B., J.B.SCHADE, Structure and Function of the Cerebral Cortex. Elsevier, Amsterdam 1960.

NIEUWENHUYS, R.: Trends in the evolution of the Actinopterygian forebrain. J.Morphol. *111* (1962), 69–88.

NIEUWENHUYS, R.: The morphogenesis and the general structure of the Actinopterygian forebrain. Acta morph. neerl.-scand. *5* (1962b) 65–78.

NIEUWENHUYS, R.: The comparative anatomy of the Actinopterygian forebrain. J. Hirnforsch. *6* (1963), 172–192.

NIEUWENHUYS, R.: Further studies on the general structure of the Actinopterygian forebrain. Acta morph. neerl.-scand., *6* (1964), 65–79.

NIEUWENHUYS, R., M. HICKEY, A survey of the forebrain of the Australian lungfish Neoceratodus forsteri. J. Hirnforsch. *7* (1965), 433–452.

ROMER, A.S.: Vergleichende Anatomie der Wirbeltiere. P. Parey, Hamburg-Berlin 1959.

RUDEBECK, B.: Contributions to forebrain morphology in Dipnoi. Acta zool. *26* (1945), 9–156.

SCHOBER, W.: Vergleichend-anatomische Untersuchungen am Gehirn der Larven und adulten Tiere von Lampetra fluviatilis (LINNÉ 1758) und Lampetra planeri (BLOCH, 1784). J. Hirnforsch. *7* (1964), 107–209.

Dr. W. SCHOBER
Hirnforschungsinstitut
701 Leipzig, Emilienstraße 14
Deutschland

The Interpretation of the Cell Masses in the Teleostean Forebrain

By R. Nieuwenhuys, Amsterdam

It has been known for a long time (Tiedemann, 1816; Stieda, 1868) that the forebrain of teleosts differs considerably from that of the other vertebrates. It consists of two solid lobes which are separated from each other by a median slit-like ventricle. Lateral ventricles surrounded by nervous tissue, which are found in almost all other vertebrates are lacking, and this is not only the case in the adult stage, but also during development. Thus, the solid condition of the teleostean telencephalic lobes is not due to an obliteration of initially present ventricular cavities, but is rather a primary feature.

The cerebral hemispheres of teleosts contain a number of well-defined nuclei, and numerous attempts have been made to homologize these cell masses with those found in other groups of vertebrates. Rabl-Rückhard (1882) regarded the forebrain lobes of teleosts *in toto* as equivalent to the basal ganglia of higher vertebrates. According to him, the pallium in teleosts, is merely represented by a thin ependymal membrane covering the forebrain lobes dorsally and being attached to their lateral sides. The interpretation of Rabl-Rückhard has been adopted by several of the early workers (Edinger, 1888, 1896; Burckhardt, 1894; Goldstein, 1905), but most later authors were of the opinion that the so-called membranous pallium of teleosts is in reality a widened roof-plate, and the teleostean forebrain lobes comprise both a pallium and a subpallium. However, the boundary between these two principal regions has been located very differently by various authors, and this has led to considerable discrepancies in regard to the interpretation of the individual cell-masses of the teleostean hemispheres. In the present account I intend to point out briefly how I arrived at my interpretation which, it should be stated at once, is by no means novel but conforms essentially with the views expressed by Holmgren (1922) and Källén (1951 a, b).

Before discussing the homologization of the nuclei present in the forebrain of teleosts, it will be necessary to touch two more general topics, (1) the taxonomy of the bony fishes, and (2) the overall morphology of the actinopterygian forebrain.

Systematical position of teleosts. Concerning the systematical relations the teleosts belong to the subclass Actinopterygii which, together with the Sarcopterygii, form the class Osteichthyes or higher bony fishes. The Sarcopterygii comprise the Crossopterygii from which the terrestrial vertebrates evolved, but the Actinopterygii merely form a side-line of vertebrate evolution. Although the actinopterygians have not given rise to anything "higher" than just fish, it may be stated that within this subclass a remarkable expansion and a continuous progressive evolution has taken place. The four superorders in which the Actinopterygii are usually subdivided, i.e. the Palaeoniscoidei, the Chondrostei, the Holostei, and the Teleostei, may be roughly regarded as stages in this evolutionary development. The teleosts form the remote end-group of this side-line of vertebrate evolution, and it will be clear that members of this end-group cannot be directly compared with species of other classes. In such a comparison the more primitive actinopterygian groups must be taken into consideration, and this is possible because the Holostei, the Chondrostei, and even the ancient Palaeoniscoidei are all represented in the recent fauna by a few species. It is important to note that the taxonomical data just outlined have been neglected by many students of the teleostean forebrain. The interpretations of, e.g., GOLDSTEIN (1905), SHELDON (1912), KUHLENBECK (1929), WESTON (1937), MILLER (1940) and DROOGLEEVER-FORTUYN (1961) are all based on direct comparisons of teleosts with non-actinopterygians. In the ensuing discussion of the general structure of the actinopterygian forebrain the remarkable telencephalon of *Polypterus*, a survivor of the Palaeoniscoidei will be taken as a starting point, and then the relations found in the Chondrostei, Holostei, and Teleostei will be considered. However, before discussing the forebrains of these groups a few remarks should be made on the form and the morphogenesis of the forebrain in general.

Morphogenesis of forebrain. It may be recalled that early in development the forebrain of all vertebrates has the shape of a simple vesicle. The side-walls of this vesicle are thickened, and from them the nervous parts of the telencephalon develop. Dorsally, rostrally and ventrally the two lateral walls are connected by a thin epithelial membrane. The principal morphogenetic event in the forebrain of almost all vertebrates is now a bulging-out or evagination of the side-walls resulting in lateral ventricles, surrounded by nervous tissue. In most vertebrate groups the evagination occurs in a rostral, lateral as well as a caudal direction (for details see: KÄLLÉN, 1951a; NIEUWENHUYS, 1966). Through the lateral evagination the dorsal (pallial) part of the telencephalic side-walls is brought in a position as to cap the ventral (basal) portion (Figs. 1a, 2a). This condition has been described by HOLMGREN (1922) and others as the inversion of the pallium. Turning now to *Polypterus*, we see in Fig. 3a that the telencephalon of this species has not evaginated. The ventral parts of the forebrain walls have maintained essentially their early embryonic position, but the dorsal parts show a bending outward or eversion; a condition which is clearly the opposite of the inversion seen in evaginated forebrains. Very remarkable relations are found in the caudal part of the forebrain of *Polypterus* (Fig. 4a). The hemispheric walls have recurved here in such a fashion that they surround a part of the extracerebral cavity (NIEUWENHUYS, 1964). It will be noted that the dorsal and lateral surfaces of the everted forebrain walls are lined with ependyma and that these surfaces, together with a wide membranous structure, surround lateral extensions of the median ventricle. The taeniae, i.e., the sites of attachment of the membranous roof, are located laterally.

In the Chondrostei, Holostei and Teleostei the hemispheres are more massive than in *Polypterus*, but embryological and comparative histological observations (NIEUWENHUYS, 1962a, 1964) have led to the conclusion that also in these groups an eversion of the telencephalic side-walls occurs. As in *Polypterus*, the dorsal surface of the chondrostean, holostean, and teleostean hemispheres is covered by a wide, ependymal membrane (Figs. 3, 4). It was demonstrated that this membranous roof develops from the originally narrow telencephalic roof plate, whose sites of attachment gradually shift from a dorsomedial toward a lateral or ventrolateral position.

The eversion was also evidenced by a comparison of the orientation of the blood vessels and the ependymal gliocytes in the amphibian and actinopterygian forebrain. In the evaginated and inverted cerebral hemispheres of amphibians the blood vessels pass

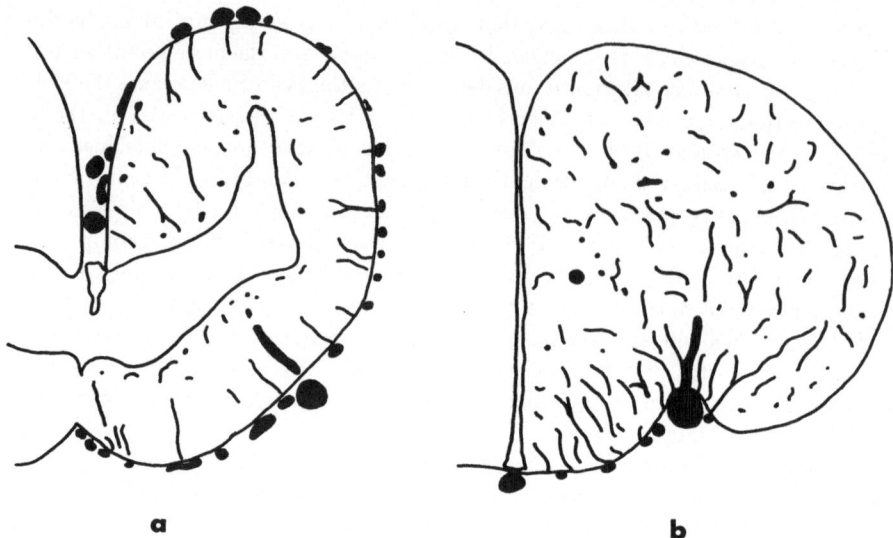

Fig. 1. The pattern of the blood vessels, as seen in a transverse hemisection through the forebrain of the amphibian *Ambystoma* (a), and the teleost *Gasterosteus* (b).

through the whole external surface and converge toward the central ventricular cavities (Fig. 1a). In the actinopterygian forebrain, however, an entirely different pattern is seen; the blood vessels enter and leave the forebrain only at the latero-basal side and fan out radially in the direction of the large ependymal surface (Fig. 1b). The processes of the ependymal cells show a pattern which conforms exactly to that of the blood vessels. In the evaginated amphibian forebrain the long processes of the ependymal cells diverge from the smaller ventricular toward the larger meningeal surface (Fig. 2a). In the forebrain of bony fishes, on the contrary, the processes of the ependymal gliocytes converge from the greatly elaborated ventricular surface toward the small, basal, meningeal surface (Fig. 2b).

A recurvature of the telencephalic side-walls with consequent widening of the roof plate is found in the Chondrostei, Holostei as well as in the Teleostei, but the extent of the eversion differs among these groups. In the Chondrostei (Figs. 3b, 4b) the recurvature is only moderate, and the same occurs in the holostean *Lepisosteus*. In *Amia*, another holostean, the eversion is very pronounced, as appears from the ventrolateral position of the taeniae and from the occurrence of deep longitudinal sulci in the meningeal surface (Figs. 3c, 4c). With regard to the extent of the eversion the teleosts conform generally to *Amia*, but in some groups, e.g. the Salmonidae, the eversion is slight as in the Chondrostei and *Lepisosteus*. It should be noted that the sections through the forebrain of the teleost *Carassius auratus*, depicted in Figs. 3d and 4d, are from a very young specimen with incomplete eversion. In the adult goldfish the taeniae are situated more ventrally and the external sulci are deep and conspicuous.

In several recent publications (DROOGLEEVER FORTUYN, 1961; SCHNITZLEIN, 1963; STORY, 1964) the lack of lateral ventricles and the presence of a thin, laterally attached roof in the actinopterygian forebrain is affirmed and, yet, the occurrence of an eversion is

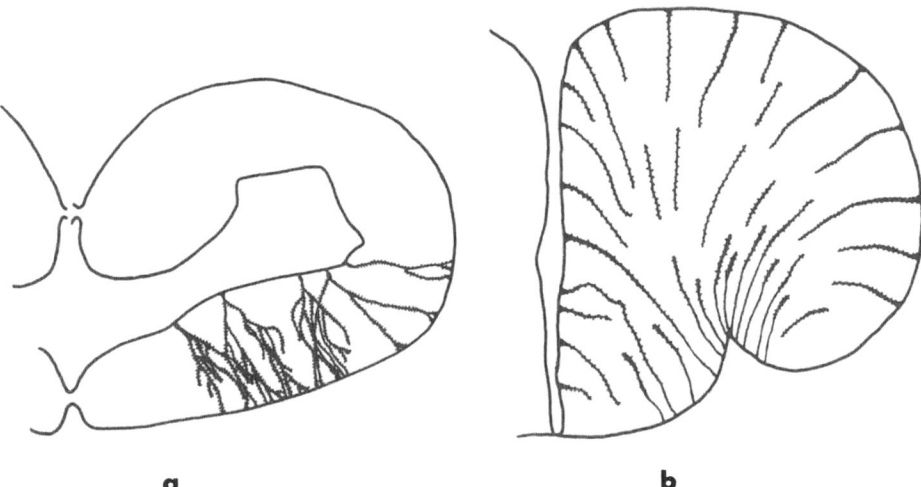

Fig. 2. Transverse hemisections through the forebrain of the amphibian *Necturus* (a) and the teleost *Gasterosteus* (b), showing the ependymal gliocytes. GOLGI method. (Fig. 2a is redrawn from HERRICK [1933]).

questioned or denied. An alternative explanation for the characteristics mentioned is not presented in these papers. Our observations justify the following conclusions: (1) The widening of the roof plate is a direct consequence of the separation and recurvature of the dorsal parts of the telencephalic side-walls. (2) The absence of lateral ventricles is due to the fact that no hemispheric evagination occurs; during development the side-walls of the actinopterygian forebrain are not going to surround a part of the telencephalic ventricular cavity, but curve in the opposite direction.

Recognition and appreciation of the eversion is of paramount importance for a correct interpretation of the parts of the actinopterygian forebrain. In general it may be stated that comparisons between everted and evaginated forebrains cannot be made on a topographical, but only on a topological basis.

Subdivision of hemispheres. Before focussing our attention on the homologization of the centres present in the actinopterygian forebrain, I should like to point out briefly to those cell masses which are usually distinguished in the (evaginated) forebrain of the other ichthyopsidan groups, i.e. the chondrichthyans, dipnoans, and the amphibians. In all these groups the telencephalic hemispheres can be divided into a dorsal, pallial, and a basal, subpallial region. These two principal regions show generally clear-cut cytoarchitectonic differences and their boundary is, moreover, often indicated by a ventricular groove and a cell-free zona limitans (JOHNSTON, 1911a, b; HOLMGREN, 1922). The pallium of amphibians and amniotes can be subdivided into three fields, namely: a dorsolateral primordium piriforme, a primordium pallii dorsalis or general pallium, and a dorsomedial primordium hippocampi. However, in fish with evaginated hemispheres I have not been able to delimit separate pallial fields. In the subpallium of ichthyopsidans with evaginated hemispheres three longitudinal zones can be distinguished: a ventrolateral striatum which, at least in amphibians and dipnoans, caudally continues into the amygdala, a

ventral olfactory tubercle, and a medial septum. The septal area contains a periven-
tricularly situated lateral septal nucleus, and a medial nucleus consisting of migrated
cells. A zone of migrated cells also occurs frequently in the ventral subpallial region; this
formation is usually designated to the cortex olfactorius or the cortex tuberculi olfactorii.

Turning now to the Actinopterygii, in Fig. 3 I have assembled cross-sections through
the middle of the telencephalon of a representative of the Palaeoniscoidei, the Chondrostei, the Holostei and the Teleostei, and Fig. 4 shows similar sections through the
caudal part of the forebrain of the same species. It appears that in the everted hemispheres of all these forms two principal regions (D and V in Figs. 3 and 4) with pronounced
structural differences can be distinguished. The boundary plane between these two regions
is often marked by a cell-free zone which, at least in the caudal part of the forebrain,
corresponds with a ventricular groove. With JOHNSTON (1911b), HOLMGREN (1922), and
KÄLLÉN (1947, 1951b) I consider the dorsal region (D) of the actinopterygian forebrain as
the pallium and the ventral region (V) as the subpallium. Due to eversion the actinopterygian pallium does not constitute the dorsal wall of a ventricular cavity, but occupies a
dorsolateral position with respect to the subpallial region.

The pallium of *Polypterus* displays an exceptionally simple histological organization
(Figs. 3a, 4a). It shows a uniform structure throughout and the great majority of its cell-
bodies is situated in a narrow periventricular layer. The pallium of the Chondrostei, Holostei and Teleostei shows an advance in differentiation over that of the Polypteriformes.

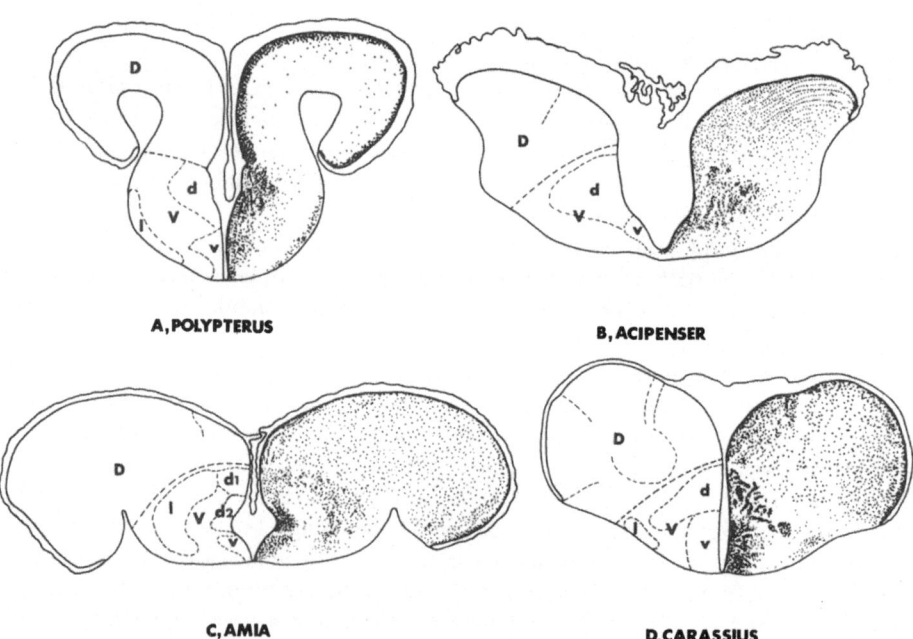

Fig. 3. Cross-sections through the middle of the forebrain of a palaeoniscid (A), a chondrostean (B),
a holostean (C), and a teleost (D).
Abbreviations: D: area dorsalis telencephali or pallium; d: dorsal subpallial nucleus; d 1, d 2: subdivisions of d; l: lateral subpallial nucleus; V: area ventralis telencephali or subpallium; v: ventral
subpallial nucleus.

The cell-bodies are dispersed through the entire width of the thickened wall and there is a tendency towards the differentiation of separate cell masses. In the Chondrostei and Holostei (Figs. 3b, c, 4b, c) two of such areas can be distinguished, but in most teleosts (Fig. 3d, 4d) four areas can be delimited: a central area, consisting of scattered, large neurons, surrounded by three periventricular fields which occupy a medial, a dorsal and a lateral position. In the literature these periventricular pallial fields are often indicated as being homologous to the hippocampal pallium, the general pallium and the piriform pallium, but in my opinion these areas are the result of a differentiation which has taken place entirely within the actinopterygian line. Hence these nuclei cannot be homologized with areas found in the pallium of tetrapods. The same holds true for a nucleus or complex of nuclei which in teleosts occupies the caudoventral part of the lateral pallial field (Fig. 4d). In the holosteans (Fig. 4c) this differentiation is present, but in the Chondrostei (Fig. 4b) and in the Polypteryformes (Fig. 4a) it is entirely lacking.

The subpallium of the actinopterygians is less variable in its structure than the pallium. Generally three nuclear masses can be discerned in this region. These nuclei have been

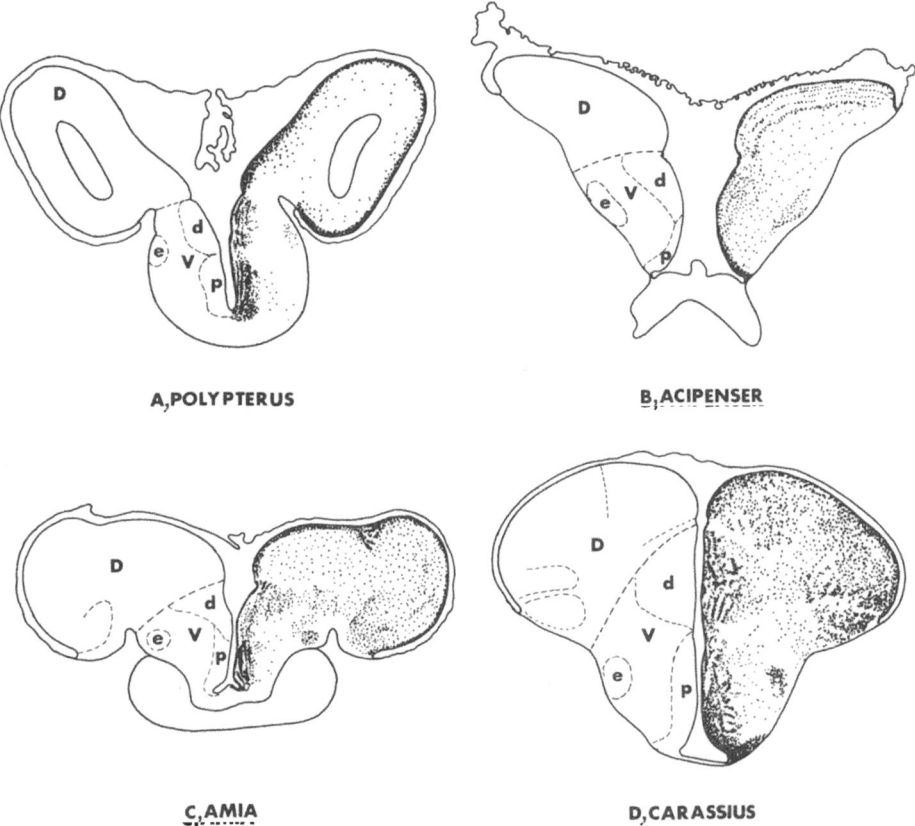

A,POLYPTERUS   B,ACIPENSER

C,AMIA   D,CARASSIUS

Fig. 4. Cross-sections through the caudal part of the forebrain of a palaeoniscid (A), a chondrostean (B), a holostean (C), and a teleost (D).
Abbreviations: D: area dorsalis telencephali or pallium; d: dorsal subpallial nucleus; e: nucleus entopeduncularis; p: nucleus praeopticus; V: area ventralis telencephali.

described in the literature under a variety of names (cf. NIEUWENHUYS, 1963). I will call them here, provisionally, the ventral, the dorsal, and the lateral subpallial nuclei (v, d, and l in Figs. 3 and 4). The two first mentioned nuclei border the median ventricle, whereas the third occupies a superficial (= submeningeal) position. In evaginated forebrains the three subpallial regions, i. e. the septal area, the olfactory tubercle, and the striatal area, surround the lateral ventricle, occupying a ventromedial, ventral and ventrolateral position respectively. However, it will be clear that in the early embryonic tube-like forebrain an-lage, the future septal area forms the ventralmost part of the side-walls, whereas the prim-ordia of the olfactory tubercle and the striatum occupy, respectively, an intermediate and dorsal position in the subpallium. For a comparison of evaginated forebrains with everted ones we have to consider these primary, early embryonic positions of the subpallial areas. Following this line of thought we come to the following interpretation of the actinop-terygian subpallial cell-masses. The ventral subpallial nucleus (v) is probably homologous to the lateral septal nucleus, whereas the dorsal cell-mass (d) may be considered the equivalent of both the olfactory tubercle and the corpus striatum. In the Holostei this dorsal subpallial column is differentiated into two separate cell masses (Fig. 3c).

The submeningeally situated lateral subpallial nucleus (l) corresponds topologically with the cortex tuberculi olfactorii of the Chondrichthyes and the dipnoans, and may also comprise the homologue of the nucleus medialis septi of evaginated forebrains, as has been suggested by KÄLLÉN. Behind the anterior commissure (Fig. 4) the ventral subpallial column (v) is replaced by the preoptic area, but the dorsal column (d) is still present. The caudolateral extension of this cell zone (i. e. the corpus precomissurale, pars intermedia of SHELDON, 1912) occupies a topological position similar to that of the dipnoan and amphi-bian amygdala.

Fibre tracts. Finally, I should like to comment on the fibre tracts of the teleostean forebrain, but the space permitted here does not allow more than a few remarks. The secondary olfactory fibres assemble in two well-defined bundles which, on account of their topographical position, are described as the medial and the lateral olfactory tracts. The lateral tract arches over a ventrally situated longitudinal furrow and terminates in the basolateral area of the forebrain. Topographically these relations are strongly remi-niscent of those found in the basolateral region of the mammalian cerebral hemispheres, and apparently this similarity has led several authors (e. g., SHELDON, 1912) to a homolo-gization of the end-stations of the teleostean lateral olfactory tract with terminal areas of the mammalian tract which bears the same name. It should be re-emphasized, however, that teleosts and mammals are not directly comparable, and it should also be kept in mind that the mammalian or tetrapod telencephalon is evaginated, whereas that of the teleosts is everted. "Lateral" and "medial" are topographical concepts, and it is by no means true that a laterally situated tract in species A of class X, is necessarily equivalent to a laterally situated tract in species B of class Y. Moreover, it is known that even within the various classes of vertebrates the secondary olfactory fibres may assemble in very different ways. To give an example: in the Holostei and in many Teleostei the medial olfactory tract has a strong component which, decussating through the anterior com-missure, terminates in the contralateral basocaudal part of the forebrain. However, in the Polypteriformes and the Chondrostei such a decussating pathway is lacking.

What has been said about the olfactory tracts also holds true for the connections between the telencephalon and the diencephalon. In most lower vertebrates the fibres which interconnect these two brainparts tend to divide into a medial and a lateral bundle,

but the composition of these bundles differs among the various groups of lower vertebrates (HERRICK, 1922). Hence, the fact that a given area or nucleus of the teleostean
forebrain sends its fibres either in the medial or in the lateral forebrain bundle does not
give much information about the homology of that nucleus.

Summary

The forebrain of teleosts cannot be directly compared with that in terrestrial vertebrates.

The walls of the telencephalon of teleosts and those of all other actinopterygian groups
have everted.

The lack of lateral ventricles and the presence of a very thin laterally attached roof
plate in the actinopterygian forebrain find their explanation in this eversion.

Comparisons between the everted actinopterygian forebrain and the evaginated cerebral
hemispheres of other groups of vertebrates cannot be made on a topographical, but only
on a topological basis.

The solid cerebral hemispheres of teleosts can be divided into a dorsolaterally situated
pallium, and a basomedially located subpallium.

The teleostean pallium and subpallium contain both several welldefined nuclear
masses. An interpretation of the subpallial cell masses is given. It is pointed out that the
teleostean pallial nuclei cannot be homologized with specific areas found in evaginated
forebrains.

References

BURCKHARDT, R.: Anat. Anz. *9* (1894), 468.
DROOGLEEVER, FORTUYN J.: J. comp. Neurol.
  116 (1961), 249.
EDINGER, L.: Abh. Senckenb. naturf. Ges.
  Frankf. a. M. *15* (1888), 91.
EDINGER, L.: Vorlesungen über den Bau der
  nervösen Zentralorgane. Leipzig 1896.
GOLDSTEIN, K.: Arch. mikr. Anat. *66* (1905),
  135.
HERRICK, C. J.: In: Libro en Honor de S. Ra-
  món y Cajal, Madrid 1922, Vol. *1*, 143.
HOLMGREN, N.: J. comp. Neurol. *34,* (1922), 391.
JOHNSTON, J. B.: J. comp. Neurol. *21* (1911a), 1.
JOHNSTON, J. B.: J. comp. Neurol. *21* (1911b),
  489.
KÄLLÉN, B.: Kungl. Fysiogr. Sällsk. Förhandl.
  17 (1947), 203.
KÄLLÉN, B.: Acta anat. *11* (1951a), 537.
KÄLLÉN, B.: Kungl. Fysiogr. Sällsk. Handl. *62*
  (1951b), Nr. 5.

KUHLENBECK, H.: Anat. Anz. *67* (1929), 1.
MILLER, R. N.: J. comp. Neurol. *72* (1940), 149.
NIEUWENHUYS, R.: Acta morphol. neerl. scand.
  5 (1962), 65.
NIEUWENHUYS, R.: J. Hirnforsch. *6* (1963), 171.
NIEUWENHUYS, R.: Acta morphol. neerl. scand.
  6 (1964), 65.
NIEUWENHUYS, R.: Progress in Brain Research
  15 (1966),
RABL-RÜCKHARD, H.: Arch. Anat. Physiol. 111
  (1882).
SCHNITZLEIN, H. N.: Proc. XVI Int. Congress of
  Zool. Vol. *3* (1963), 229.
SHELDON, R. E.: J. comp. Neurol. *22* (1912), 177
STIEDA, L.: Z. wiss. Zool. *18* (1868), 1.
STORY, R. H.: J. comp. Neurol. *123* (1964), 285.
TIEDEMANN, F.: Anatomie und Bildungsge-
  schichte des Gehirns (1816).
WESTON, J. K.: Proc. Kon. Ned. Akad. Weten-
  schaft *40* (1937), 894.

Dr. R. NIEUWENHUYS
Netherlands Central Institute for Brain Research
Ijdijk 28, Havens Oost, Amsterdam, The Netherlands

The Primordial Amygdaloid Complex of the African Lungfish, Protopterus*

By H. N. Schnitzlein, Birmingham, Ala.

A few weeks ago a manuscript was accepted from our laboratory (Schnitzlein a. Crosby, in press) on the nuclei and the fiber systems of the telencephalon of the African lungfish, *Protopterus*. The present paper expands the material previously presented on the primordial amygdaloid complex in this lungfish.

Although the living *Dipnoi*, or lungfish, are no longer generally regarded as the direct ancestors of the amphibians (Fig. 1), the telencephalon of *Protopterus* has not undergone the extensive changes characteristic of those of most fish. Prominent lateral telencephalic ventricles, lacking in most of the Actinopterygii, but otherwise typical for vertebrates, are present in the African lungfish.

The chordate phylogenetic tree illustrated in Fig. 1, (prepared by Dr. H. R. Steeves III., of the Department of Anatomy, University of Alabama Medical Center) was based largely on information other than the characteristics of the brain. The known anatomy of the brain of the various vertebrates, however, lends support to such a scheme. The nuclei and tracts of the more generalized telencephalon of the lungfish (Elliot Smith, 1908; Herrick, 1921; Rudebeck, 1944, 1945) are more easily homologized with comparable regions of the amphibians and the reptiles than with the telencephalic areas of the teleosts and the ganoids.

Specimens and Methods. The material which had been used previously for a publication on the morphology of the telencephalon of the lungfish (Schnitzlein a. Crosby, in press) included sections cut in transverse, horizontal, and sagittal planes and stained by thionin, Weil, Protargol, pyridine silver, or Golgi methods. For the present paper, further Golgi-impregnated preparations have become available.

The specimes of *Protopterus* were kindly collected by Dr. D. W. Ewer of the University of Ghana and Dr. Wilfried Yeargan of the University of Alabama Medical Center. The fish ranged from small specimens of approximately six inch length with external gills to some nearly three feet long.

List of Abbreviations

area preop.	preoptic area
amyg.	primordial amygdaloid complex
amyg. ant.	primordial anterior amygdaloid nucleus
amyg. basolat.	primordial basolateral amygdaloid nucleus

*) This investigation was supported in part by a research grant (NB-04295) from the National Institute of Neurological Diseases and Blindness, the National Institutes of Health, U.S. Public Health Service.

amyg. corticomed.	primordial corticomedial amygdaloid nucleus
b. olf.	olfactory bulb
b. olf. acc.	accessory olfactory bulb
comm. ant.	anterior commissure
comm. hab.	habenular commissure
corn. amm.	primordial cornu ammonis
gen. pall.	primordial general pallium
gy. dent.	primordial dentate gyrus
hab.	habenula
hip. ant.	anterior continuation of the hippocampus
hyperstr.	hyperstriatum
l. f. b.	lateral forebrain bundle
neostr.	neostriatum
nuc. olf. ant.	anterior olfactory nucleus
paleostr.	paleostriatum
pir.	primordial piriform cortex
st. med.	stria medullaris
st. term.	stria terminalis
striat.	neostriatum
thal. dors.	dorsal thalamus
tr. corticohab. lat.	lateral corticohabenular tract
tub. olf. z. intermed. s. caud.	olfactory tubercle, intermediate zone, caudal subdivision
tub. olf. z. intermed. s. mid.	olfactory tubercle, intermediate zone, middle subdivision
tub. olf. z. intermed. s. rost.	olfactory tubercle, intermediate zone, rostral subdivision

Primordial Amygdaloid Complex

The amygdaloid complex (or archistriatum) of lower vertebrates has received considerable attention in the past half century. It has elements which receive the lateral olfactory tract, parts related to the piriform and the general cortex, and portions which are more intimately associated with the deeper striatal nuclei within the telencephalon. The amygdaloid complex of mammals (Fox, 1940) is commonly divided into three general regions: (1) the anterior amygdaloid nucleus, which constitutes an area usually extending anterior to the other amygdaloid areas along the lateral olfactory tract, (2) the corticomedial nuclear group, which characteristically receives the lateral olfactory tract fibers, and (3) the basolateral part, which is more intimately related to the primordial general cortex and which has commissural connections with the contralateral amygdaloid area.

The mammalian amygdala is homologous to the epistriatum of some investigators (Ariens Kappers, 1928a & b). The term epistriatum was first used by Edinger (see Johnston, 1915). However, the various interpretations attached to it (see Elliot Smith, 1919) suggest that in the use of this term an homology with the amygdala is not always indicated. Crosby (1917) used the term dorsolateral area (caudal part) for part of the primordial amygdaloid region of the alligator.

Johnston (1923) divided the dorsal ventricular ridge of reptiles into a cephalic and a caudal portion, and emphasized the fact that the two parts were of different derivation. The caudal part, which is the amygdaloid ridge, he interpreted as the result of the ingrowth of the piriform lobe cortex. Crosby (1917) had earlier shown in the alligator that the lateral olfactory tract fibers project to this more caudal portion of the dorsolateral area. The caudal dorsal ventricular (amygdaloid) ridge has been called the archistriatum and discussion has been carried on over the years as to whether this region is "cortical" (ELLIOT SMITH, 1910), or whether it is "striatal" (ARIENS KAPPERS, 1928a). As ELLIOT SMITH pointed out, both may be true, depending on the meaning intended.

The primordial amygdaloid complex of *Protopterus* is more dorsal in its position than is the comparable area in many other vertebrates, probably due to the increased size of the olfactory tubercle (Fig. 8) and to the very limited primordial general pallial area (Figs. 2, 4, 6, 7). The primordial amygdaloid complex in this fish extends from the caudal part of the accessory olfactory bulb to the caudal pole of the hemisphere (Fig. 8) and is divisible into a primordial anterior amygdaloid nucleus, a primordial corticomedial area and a primordial basolateral area. These nuclear subdivisions are not entirely homologous with comparable regions in all mammals since alterations and particularly additions in the fiber connections of nuclei obviously occur in phylogeny.

Primordial anterior amygdaloid nucleus. The primordial anterior amygdaloid nucleus (Figs. 2, 7 C) of *Protopterus* lies along the lateral olfactory tract at the base of the pri-

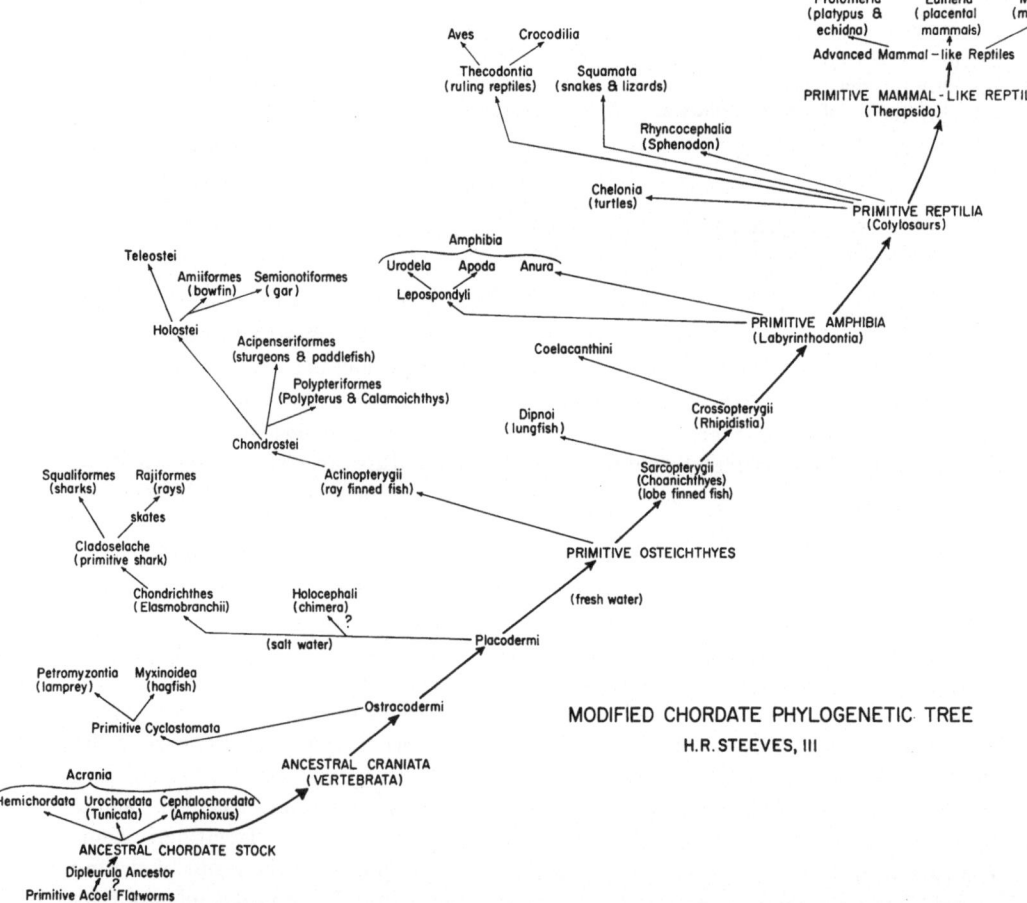

Fig. 1. A modified Chordate phylogenetic tree illustrating a probable relationship of the African lungfish to other vertebrates.

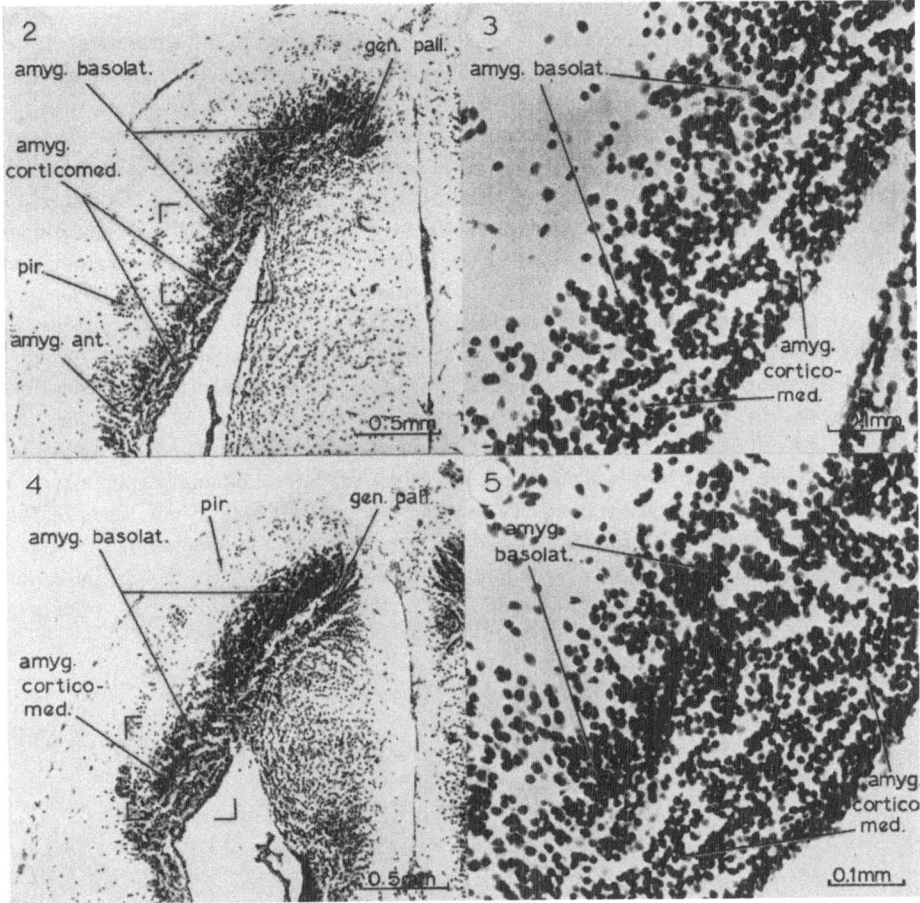

Fig. 2. Photomicrograph of a transverse section stained with thionin through the dorsal part of the hemisphere of *Protopterus*. The section is at the caudal pole of the accessory olfactory bulb. The location of Fig. 3 is indicated by the outlined area.

Fig. 3. Photomicrograph of a higher magnification of the portion of Fig. 2 indicated by the outline. The differences in cell types between the primordial basolateral and primordial corticomedial regions of the amygdala may be noted.

Fig. 4. Photomicrograph of a transverse section stained with thionin through the dorsal part of the hemisphere ot *Protopterus*. This section is caudal to that illustrated in Fig. 2 and through a plane just rostral to the anterior commissure. The location of Fig. 5 is indicated by the outlined area.

Fig. 5. Photomicrograph at a higher magnification of the portion of Fig. 4 indicated by the outline. This section is through the caudal part of the primordial basolateral nucleus of the amygdala. The magnification is the same as that of Fig. 3.
For abbreviations s. p. 40.

mordial piriform cortex (Fig. 2) and is rather long rostrocaudally, although it is not a large nuclear mass. The cell bodies of these neurons are small (Figs. 2, 6). RUDEBECK (1945) suggested that the caudal part of his nucleus olfactorius lateralis in *Protopterus* corresponded to the ventrolateral nucleus of urodeles (which HERRICK homologized with the amygdala). The caudal part of the nucleus of the lateral olfactory tract of the alligator (CROSBY, 1917) may be homologous with the anterior amygdaloid nucleus of *Protopterus*.

Primordial corticomedial area. The primordial corticomedial area of the amygdala appears at levels caudal to the accessory olfactory bulb (Figs. 2–8) being continuous rostrally with the periventricular layer of the accessory olfactory bulb (Fig. 7) and with the cells of the dorsolateral part of the anterior olfactory nucleus (Fig. 7 A). The corticomedial area gradually shifts ventrally (Fig. 4) as the caudal end of the hemisphere is attained and the nucleus does not extend to the caudal pole of the hemisphere.

The primordial corticomedial area consists of smaller cells than those of the basolateral part (Figs. 2–6). These densely arranged neurons are situated adjacent to the ventricle and are not sharply delimited from those of the basolateral region (Figs. 2–5).

The corticomedial portion of the amygdala receives the lateral olfactory tract after the fibers have passed through or along the anterior amygdaloid area. Some fibers of this tract penetrate the basolateral part of the amygdala. These components of the lateral olfactory tract are largely unmyelinated and include axons from the olfactory bulb, and particularly from the accessory olfactory bulb. HERRICK (1921) reported numerous

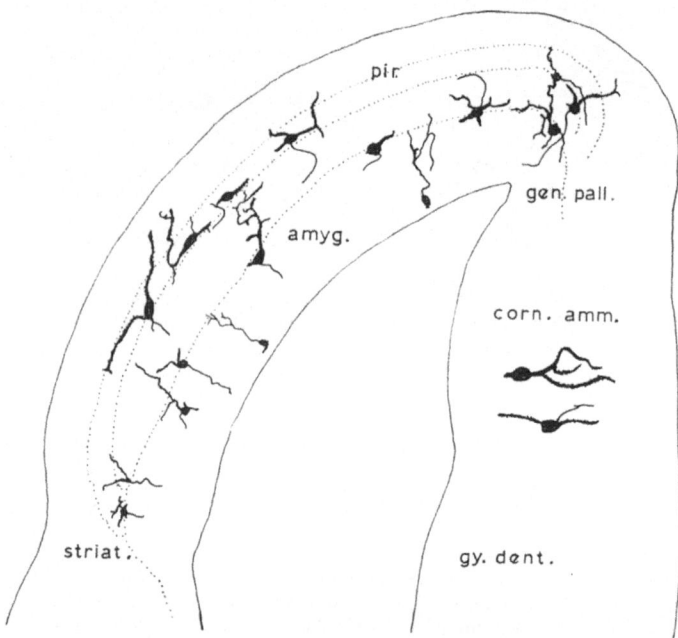

Fig. 6. Drawings of GOLGI impregnated neurons in the dorsal portion of the hemisphere of *Protopterus*. The larger cells of the primordial basolateral area project their axons medially into the corticomedial area, ventrally toward the basal regions of the hemisphere, and dorsomedially into the primordial general pallial area. Cells of the primordial piriform lobe and the primordial cornu ammonis are shown for comparison with those of the primordial amygdaloid complex.
For abbreviations s. p. 40.

non-myelinated nerve fibers in the dorsolateral olfactory tract of amphibians. Fibers interrelate the corticomedial nuclear group with the primordial basolateral area, the piriform lobe and the lateral zone of the olfactory tubercle. From the corticomedial portion of the amygdala, fibers of the stria terminalis (Fig. 9 A), which pass dorsal and medial to the lateral forebrain bundle, connect with the bed nuclei of the anterior commissure and with the preoptic area (Fig. 9 A) on the same and on the opposite side, and accompany the medial forebrain bundle into the hypothalamus and perhaps into more

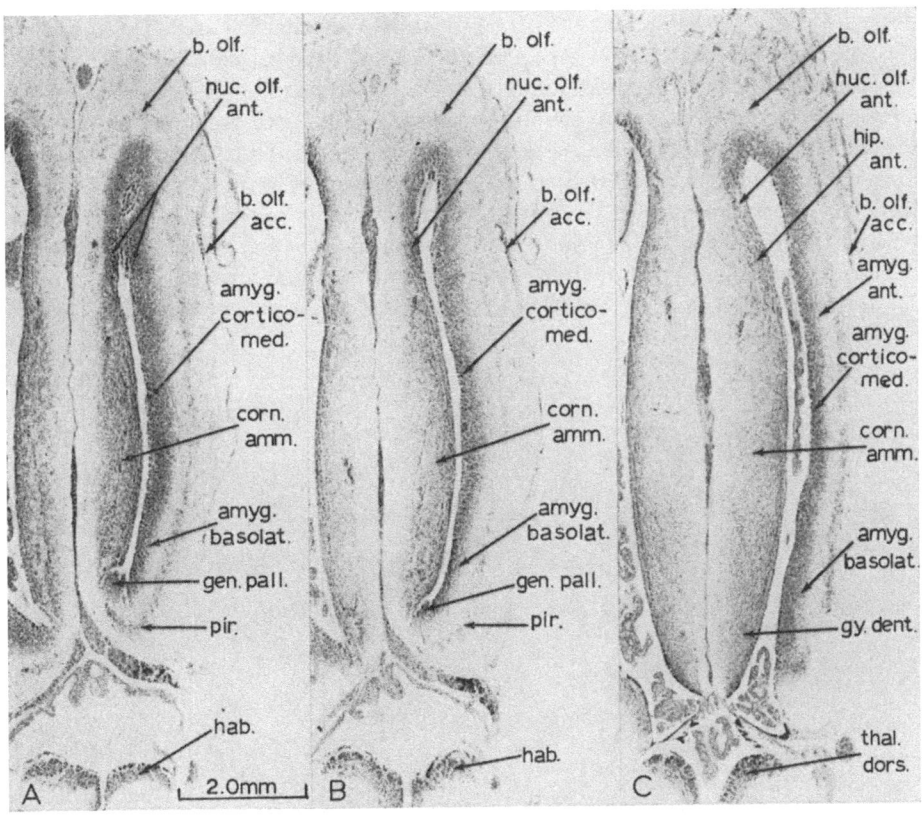

Fig. 7A. Photomicrograph of a horizontal section stained with thionin through the brain of *Protopterus*. The plane of the cut is just dorsal to the junction of the lateral ventricle and the olfactory ventricle. The relations of the anterior olfactory nucleus and other telencephalic structures to the primordial amygdaloid complex may be noted. The magnification for Figs. 7A, B, and C is indicated by the scale on this figure.

Fig. 7B. Photomicrograph of a horizontal section stained with thionin through the brain of *Protopterus*. The plane of the cut is just ventral to that of Fig. 7A. The caudal relationships of the primordial piriform cortex, primordial general pallium, primordial cornu ammonis, and primordial basolateral part of the amygdala may be noted.

Fig. 7C. Photomicrograph of a horizontal section stained with thionin through the brain of *Protopterus*. The plane of the cut is through the ventral part of the primordial amygdaloid complex to illustrate a part of the primordial anterior amygdaloid nucleus.
For abbreviations s. p. 40.

caudal areas. Some fibers, which have been called the ventral olfactory projection tract of CAJAL by JOHNSTON, pass lateral and ventral to the lateral forebrain bundle. These have connections similar to the stria terminalis and may be considered as functionally a part of it. Similar fascicles course transversely through the lateral forebrain bundle. The cortico-medial portion of the amygdala also contributes to the lateral corticohabenular tract (Fig. 9 A), projecting axons to the habenula via the stria medullaris (Fig. 9 B). Many of these are interhemispheric connections which cross in the habenular commissure (Fig. 9 C) and project to the amygdala and piriform areas of the opposite hemisphere.

Primordial basolateral amygdaloid area. The basolateral portion of the amygdala (Figs. 2–8) lies, in general, lateral to the primordial corticomedial area. The basolateral area in *Protopterus* has poorly separated rostral and caudal subdivisions. The caudal enlargement in other vertebrates was pointed out by ARIENS KAPPERS (1928). The present knowledge of the cell types and connections of the two parts of the primordial basolateral region of *Protopterus* is not extensive enough to permit making homologies with the subdivisions recognized in higher vertebrates. The basolateral part of the amygdala is continuous with the primordial piriform lobe cortex, with the primordial generall pallium, and with the primordial hippocampal formation near the caudal pole of the hemisphere (Fig. 7). All of these areas come into continuity with each other at the dorsal part of the hemisphere caudal to the anterior commissure – a relationship which is rather general for vertebrates.

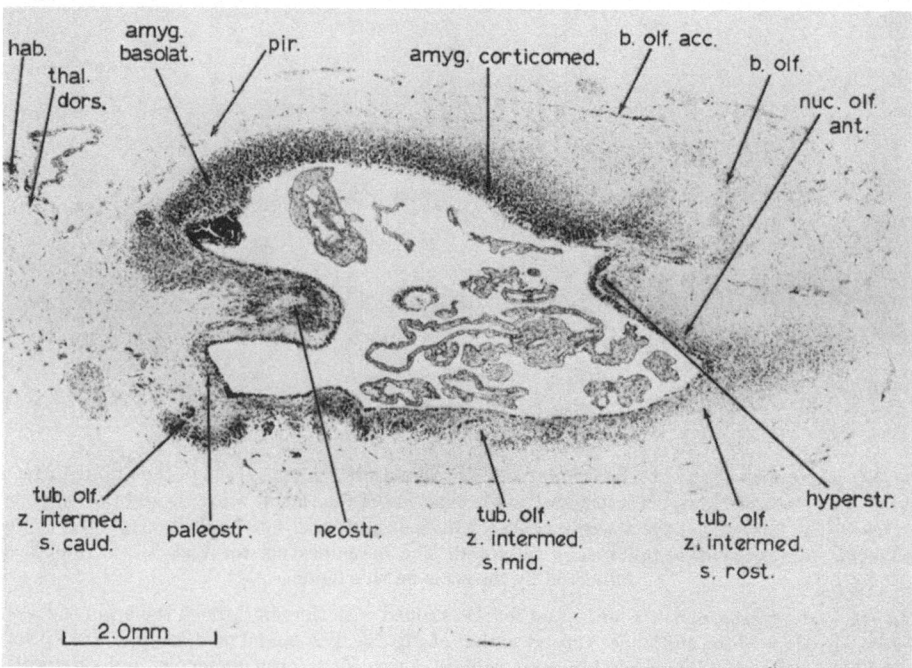

Fig. 8. Photomicrograph of a parasagittal section through the hemisphere of *Protopterus*. The plane of the section is through the lateral part of the lateral ventricle. This illustration shows particularly some of the relationships of the olfactory tubercle and the primordial amygdaloid complex.

For abbreviations s. p. 40.

The cells of the basolateral area (Figs. 2–6) are larger than those of the primordial corticomedial region. They are similar to those of the primordial general pallium.

The basolateral part of the amygdala receives olfactory tract fibers, some of which have projected through the piriform lobe and interconnect with it. These probably come largely from the accessory olfactory bulb, rather than directly from the olfactory bulb. ARIENS KAPPERS (1928) confirmed the earlier findings of HERRICK in amphibians that the epistriatum receives fibers primarily from the accessory olfactory bulb. This nucleus is interrelated with the lateral tuberculum and is associated with the primordial piriform cortex. Other association fibers of the primordial basolateral part of the amygdala include connections with the primordial general pallium which lies along the dorsomedial border of the caudal part of the basolateral nucleus.

The caudal basolateral nucleus also contributes axons to the lateral corticohabenular tract (Fig. 9 A). This bundle, with fascicles from the primordial piriform area, may be termed a caudal lateral corticohabenular tract which, with its rostral component, joins the stria medullaris (Fig. 9 B). Some of the fibers may end in the habenula of the same side, others terminate in the contralateral habenula. The majority of the fibers of the lateral

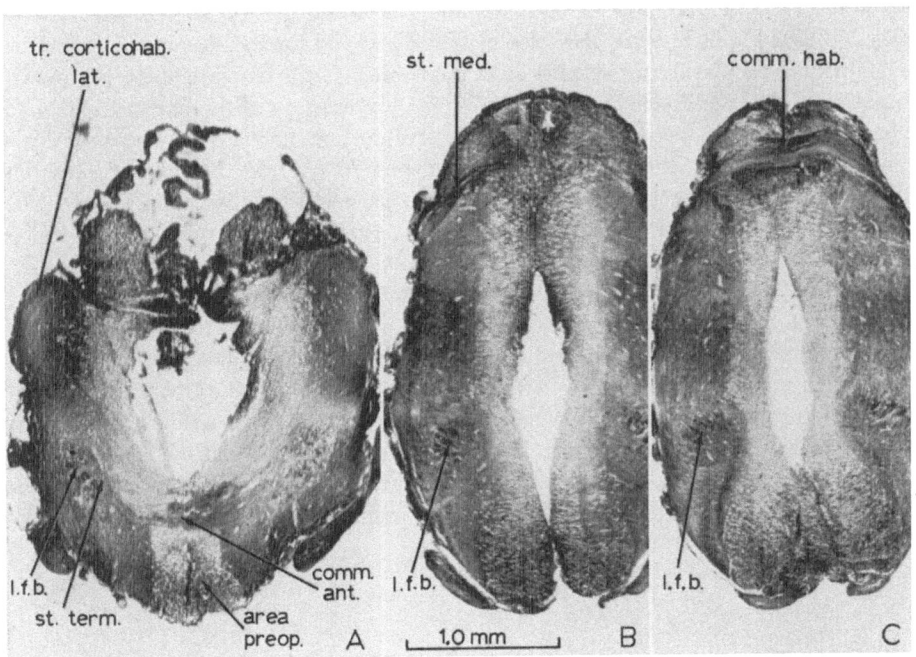

Fig. 9A. Photomicrograph of a transverse section of the brain of *Protopterus* at the level of the anterior commissure. Pyridine silver impregnation. The magnification for Figs. 9A, 9B, and 9C is the same and is indicated on Fig. 9B.

Fig. 9B. Photomicrograph of a transverse section caudal to Fig. 9A. The plane of this section is through the rostral diencephalon. Pyridine silver impregnation.

Fig. 9C. Photomicrograph of a transverse section through the habenular commissure. Pyridine silver impregnation.
For abbreviations s. p. 40.

corticohabenular tract cross in the habenular commissure (Fig. 9 C) and course to the opposite telencephalic hemisphere constituting a true commissural connection through the habenular commissure. The basolateral nuclei are also connected with each other through the anterior commissure. This is demonstrable in pyridine silver preparations (Fig. 9 A). These interhemispheric connections (commissura archistriatica) of the caudal amygdala have been reported in other vertebrates by ARIENS KAPPERS (1928), and others.

Summary and Conclusions

The anterior amygdaloid nucleus is in part a relay nucleus along the lateral olfactory tract. The corticomedial part of the amygdala has many olfactory-visceral interrelationships in the lateral olfactory tract, the preoptic area, the hypothalamic nuclei, and the habenular nuclei. A part of the primordial basolateral region of the amygdala is related to the lateral olfactory tract but it has more intimate association with adjacent primordial pallial areas, which include the primordial piriform lobe, the primordial hippocampal formation, and (perhaps less extensively) the primordial general cortex. It might be suggested that the increase in these association fibers and the increase in the primordial general pallial area result in the splitting of the laminae of the basolateral regions and the formation of the caudal dorsal ventricular ridge in reptiles. In all probability the amygdala, particularly the basal part of the basolateral portion, increases phylogenetically in conjunction with neocortical and neothalamic regions (ARIENS KAPPERS, 1928). The differences in the phylogenetic development of the parts of the amygdaloid complex may well account for some of the diverse functional patterns elicited from stimulation, or ablation, of all or parts of the amygdala (ZBROZYNA, 1963).

JOHNSTON (1923) suggested that the amygdala might be a complex in which olfactory, gustatory, and general somatic sense impressions are brought into correlation. ARIENS KAPPERS (1928) also pointed out that the archistriatum, or amygdala, was very large at birth, which was a strong argument in favor of the conception that the archistriatum is not concerned only with olfaction. He suggested that this lateral area was also interrelated with secondary trigeminal systems. Recently, CREUTZFELDT, BELL and ADEY (1963) concluded that, in the cat, the main olfactory input into the amygdala is into the anterior area and the basal and the cortical nucleus. The multisensory convergence came largely into units in the central nucleus.

Acknowledgements

The preparation of this presentation was made possible through the generous contributions of Dr. ELIZABETH CROSBY. The author is also especially appreciative of the efforts of Dr. D. W. EWER, University of Ghana, and Dr. WILFRED YEARGAN, University of Alabama, in obtaining the specimens used in this study. The technical assistance of Mrs. EMILY HOLT, Mrs. LUCILLE HILL, Mrs. JAY GREEN, and Miss PATRICIA BROWN is gratefully acknowledged.

Note: This paper represents a part of the material presented at the Symposium on the Phylogenesis and Ontogenesis of the Forebrain. The portion deleted because of necessary publication limitations has been published in the Alabama Journal of Medical Sciences, 1966, vol. 3, p. 39.

References

ARIENS KAPPERS, C. U.: The corpus striatum, its phylogenetic and ontogenetic development and functions. Acta Psychiat. Neurol. *3* (1928a), 92–113.

ARIENS KAPPERS, C. U.: The development of the cortex and the functions of its different layers. Acta Psychiat. Neurol. *3* (1928b), 115–132.

CREUTZFELDT, O. D., F. R. BELL, W. R. ADEY: The activity of neurons in the amygdala of the cat following afferent stimulation. Progress in Brain Research, Elsevier Publishing Company *3* (1963), 31–40.

CROSBY, E. C.: The forebrain of alligator mississippiensis. J. comp. Neurol. *27* (1917), 325–402.

ELLIOT SMITH, G.: The cerebral cortex in Lepidosiren, with comparative notes on the interpretation of certain features of the forebrain in other vertebrates. Anat. Anz. *33* (1908), 513–540.

ELLIOT SMITH, G.: Some problems relating to the evolution of the brain. Lancet *1* (1920), 147–153.

ELLIOT SMITH, G.: Preliminary note on the morphology of the corpus striatum and the origin of the neopallium. J. Anat. (London) *53* (1919), 271–291.

FOX, C. A.: Certain basal telencephalic centers in the cat. J. comp. Neurol. *72* (1940), 1–62.

HERRICK, C. J.: A sketch of the origin of the cerebral hemispheres. J. comp. Neurol. *32* (1921), 429–454.

JOHNSTON, J. B.: The cell masses in the forebrain of the turtle, Cistudo Carolina. J. comp. Neurol. *25* (1915), 393–468.

JOHNSTON, J. B.: Further contributions to the study of the evolution of the forebrain. J. comp. Neurol. *35* (1923), 337–481.

RUDEBECK, B.: Does an accessory olfactory bulb exist in Dipnoi? Acta zool. *25* (1944), 89–96.

RUDEBECK, B.: Contributions to forebrain morphology in Dipnoi. Acta zool. *26* (1945), 1–156.

SCHNITZLEIN, H. N., E. C. CROSBY: The telencephalon of the lungfish, Protopterus. J. Hirnforsch. (in press).

ZBROZYNA, A. W.: The anatomical basis of the patterns of autonomic and behavioral response effected *via* the amygdala. Progress in Brain Research, *3* (1963), 50–68.

H. N. SCHNITZLEIN, Ph. D.
Department of Anatomy
University of Alabama Medical Center
1919 Seventh Avenue, South
Birmingham 3, Ala., USA

L'Organisation générale du prosencéphale de Latimeria chalumnae Smith (Poisson crossoptérygien coelacanthidé)

Par J. MILLOT et J. ANTHONY, Paris

Avec NIEUWENHUYS nous avons récemment présenté à l'Académie des Sciences de Paris (1964) une note dans laquelle nous avons montré quelques aspects transversaux du diencéphale de *Latimeria* et reconstitué, sur un schéma sagittal, les principaux faisceaux du prosencéphale. De nombreuses préparations transversales et sagittales figurent également dans notre tome II de l'*Anatomie de Latimeria chalumnae*, (1965). Nous avons pensé qu'il serait intéressant, dans la présente communication, de compléter la documentation déjà publiée par un choix de préparations horizontales de nos exemplaires C 11 et C 17. Il nous arrivera cependant, à l'occasion, et pour des raisons de commodité, d'avoir recours à quelques coupes d'orientation différente.

L'organisation du prosencéphale ne diffère pas fondamentalement de celle que l'on connaît chez les autres Poissons. Les principaux territoires constitutifs sont les mêmes et se repèrent aisément (Figs. 1 à 4).

Le télencéphale montre une partie dorsale massive, le pallium, de structure uniforme, où aucun indice ne différencie un archipallium d'un paleopallium. L'extrémité antérieure du pallium, effilée, représente le lobe olfactif et reçoit le «nerf olfactif». Celui-ci réunit les fibres du second neurone de l'olfaction; il s'étire sur une longueur d'une quinzaine à une vingtaine de centimètres car le bulbe olfactif, qui lui donne origine, se loge naturellement dans la capsule nasale, tandis que l'encéphale, où il aboutit, n'occupe que la partie la plus reculée de la cavité crânienne. Dans sa région moyenne, le pallium offre son diamètre transversal maximum. Fortement bombé vers la ligne médiane, il paraît exécuter une inversion, impression renforcée par l'emplacement, ventro-latéral, de son attache à la base de l'hémisphère (Fig. 5). Pourtant nous préférons parler d'une simple protrusion palléale en direction du plan de symmétrie. Les coupes transversales montrent en effet que la cavité du ventricule latéral n'y envoie pas de diverticule notable, et que le point de fixation du toit épithélial au pallium est très largement décalé vers l'extérieur, ce qui évoque une croissance transversale du pallium plutôt qu'un enroulement. Chaque ventricule latéral rejoint par une large fente la cavité, relativement spacieuse, du télencéphale primaire, ou ventricule télencéphalique médian; il n'y a donc pas à proprement parler de *trous* de MONRO. Le toit épithélial recouvre les hémisphères et s'infléchit en arrière en formant le versant rostral du *velum transversum;* il ne comporte pas de paraphyse.

Dans l'étage ventral du télencéphale, on note une inversion indiscutable du septum. Le ventricule latéral s'enfonce profondément entre le striatum et lui. Au niveau le plus dorsal du striatum, là où il se raccorde intérieurement au pallium (sillon strio-palléal), nous n'avons pas relevé de saillie appréciable et constante pouvant mériter le nom d'epistriatum.

Le diencéphale se compose des quatre étages habituels de HERRICK (Fig. 6), séparés par des sillons longitudinaux. Dans l'epithalamus, les plans horizontaux font apparaître

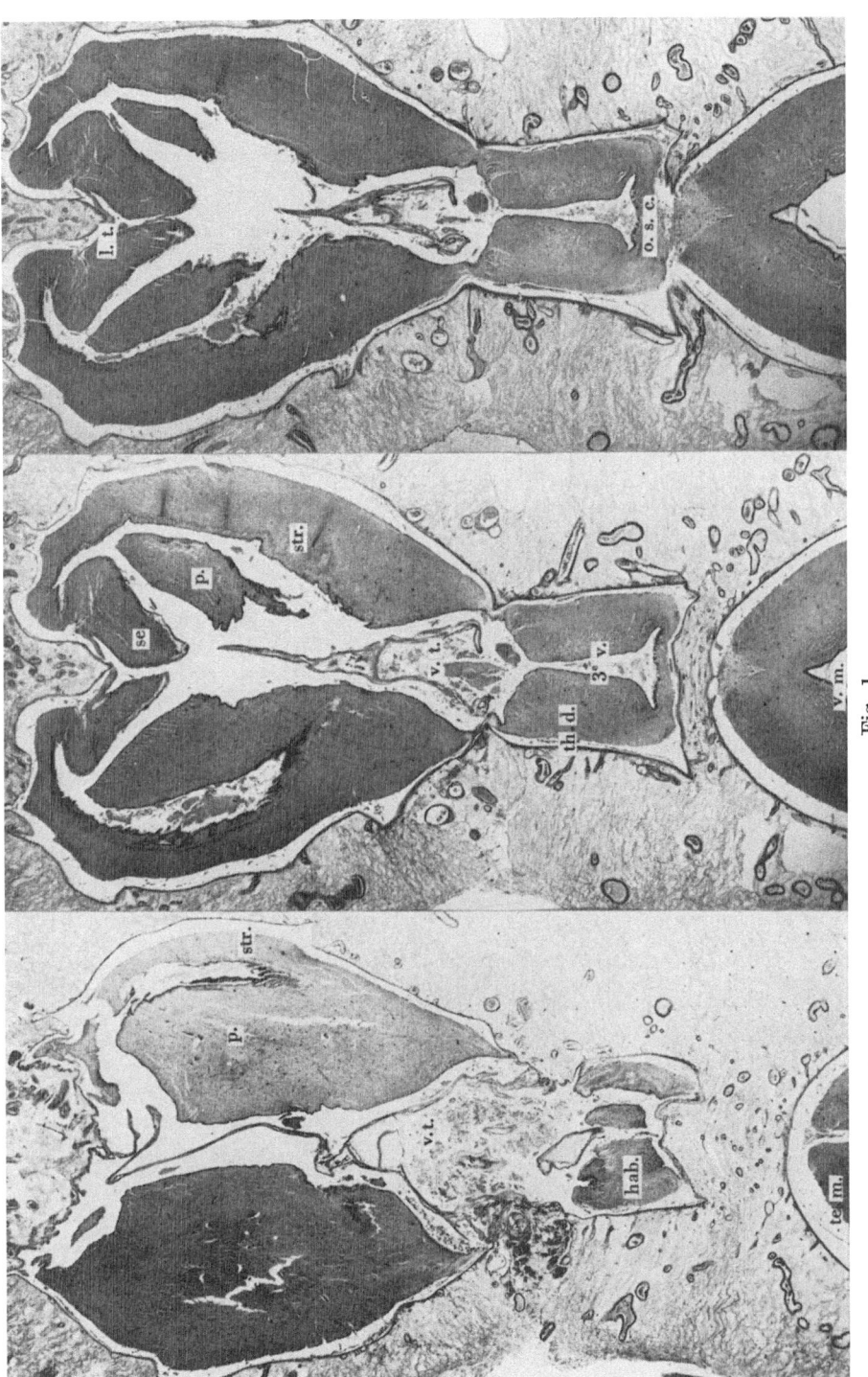

Fig. 1.

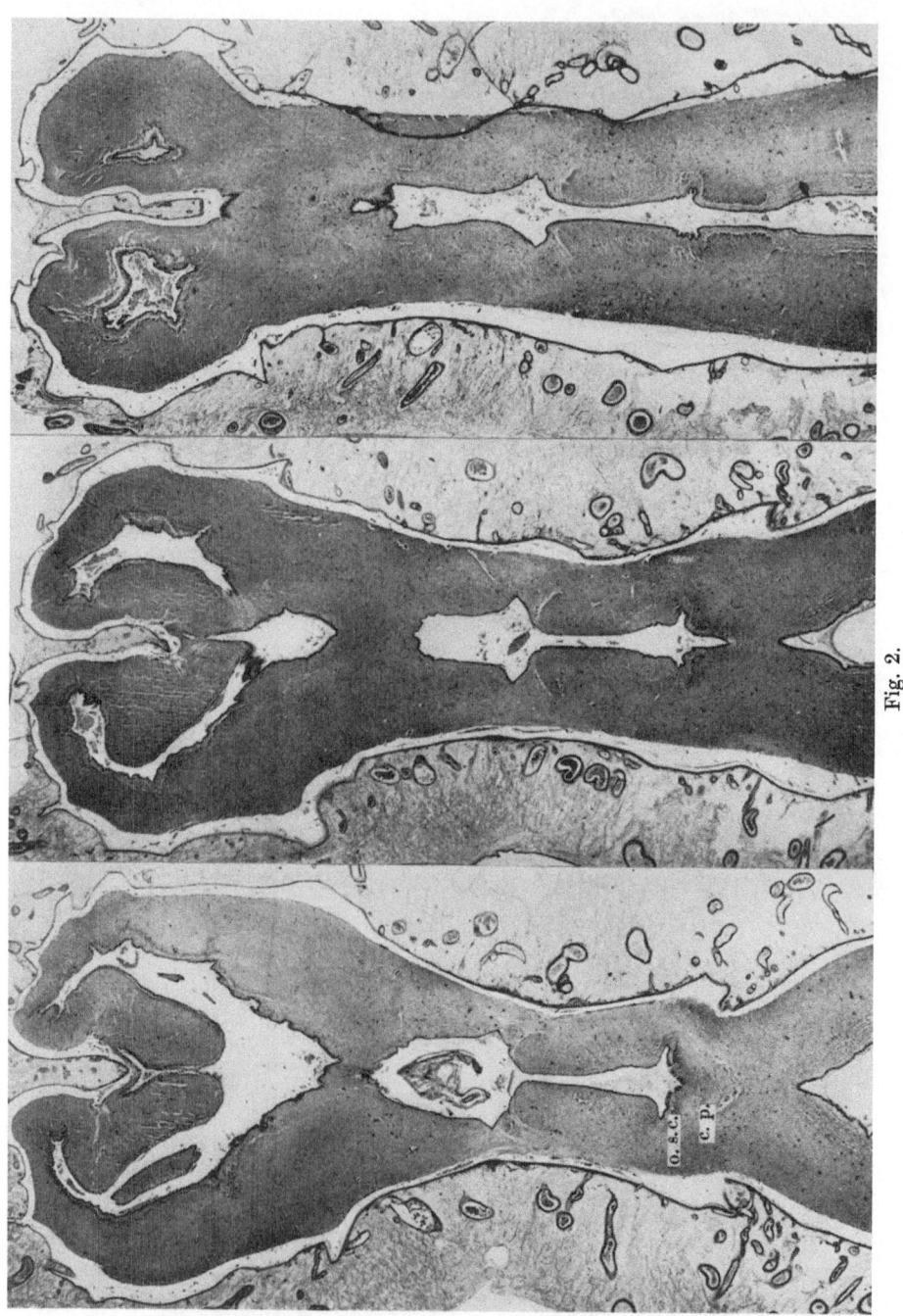

Fig. 2.

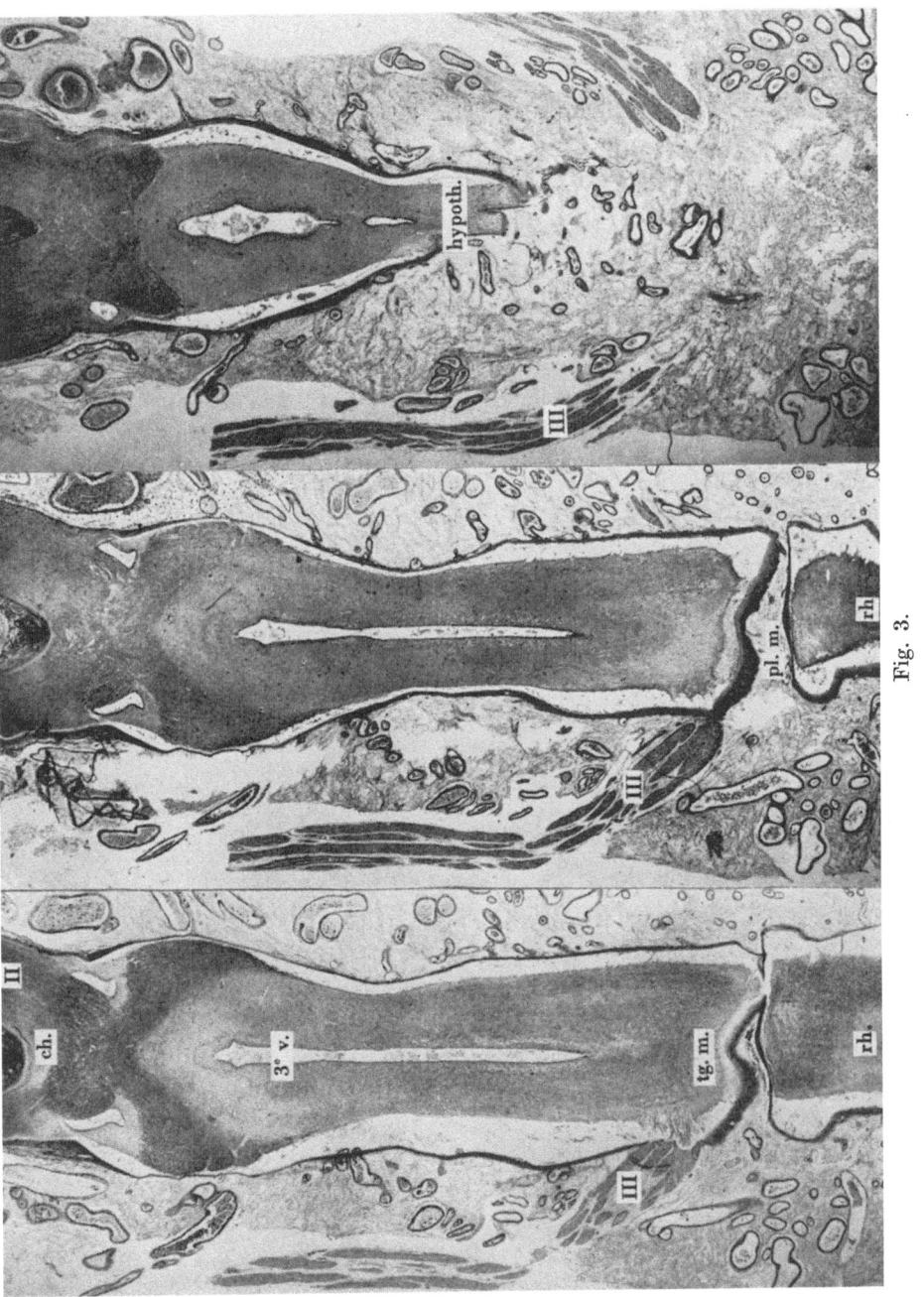

Fig. 3.

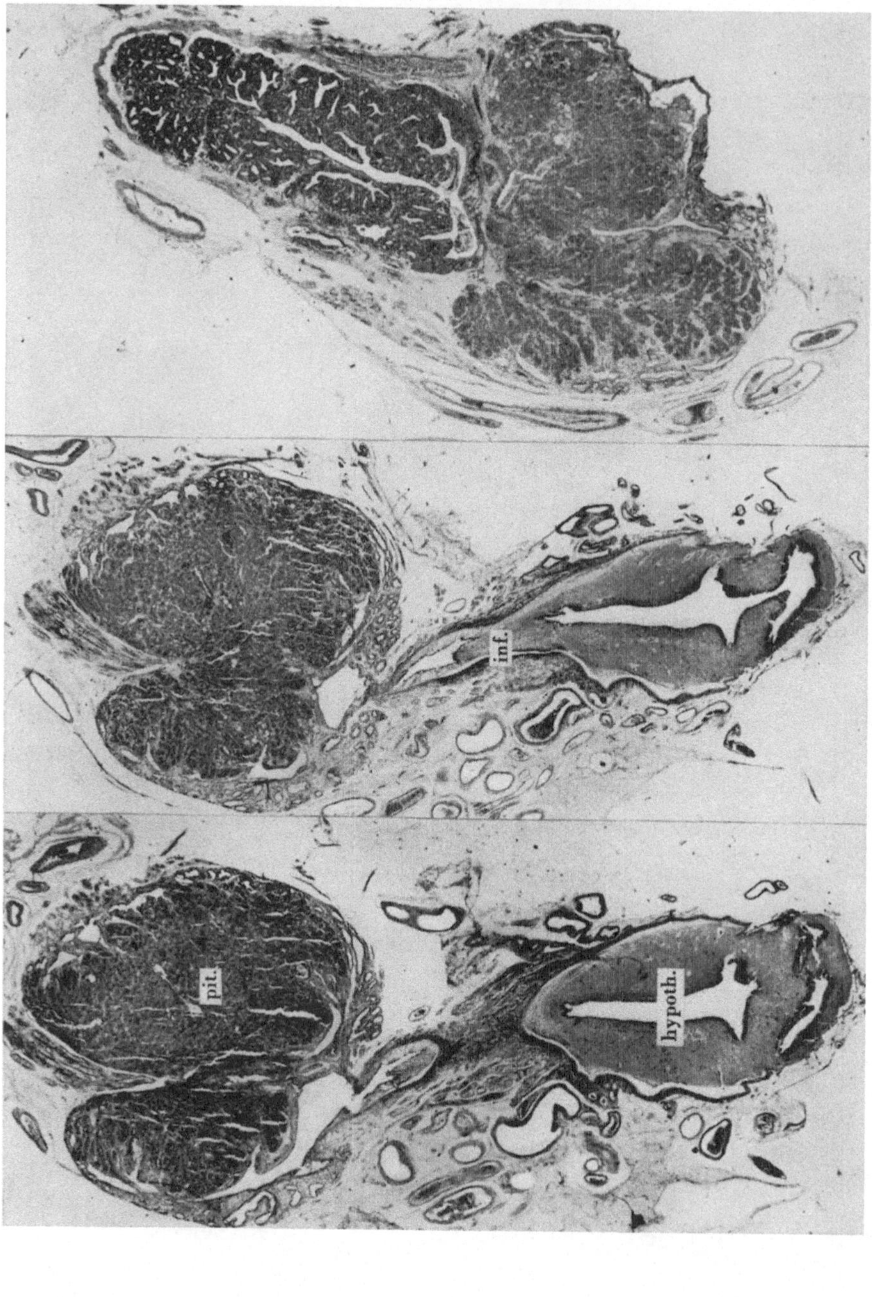

Fig. 1 à 4. Divers aspects horizontaux du prosencéphale de *Latimeria*. De la Fig. 1 à la Fig. 4, et de gauche à droite sur chaque figure, les préparations s'échelonnent dans le sens dorso-ventral. G.N. x 7 environ.

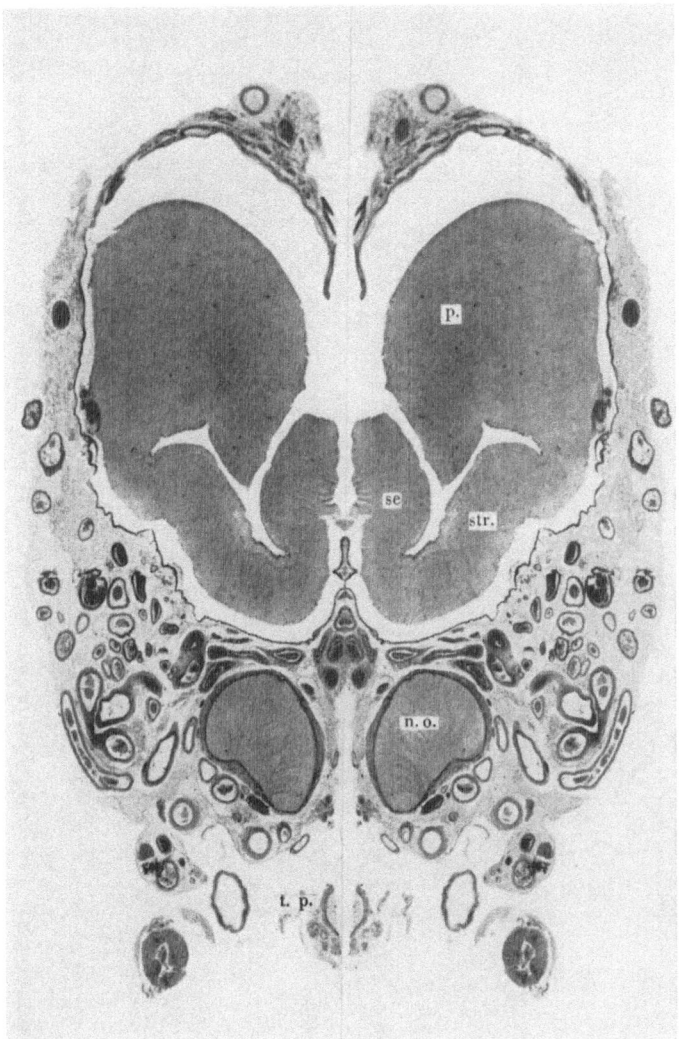

Fig. 5. Coupe transversale du télencéphale, au niveau du tiers antérieur des hémisphères. C 3.
G.N. x 7 environ.

d'abord la commissure habénulaire, puis le ganglion de l'habenula. Celui-ci décrit une saillie, par son bord interne, dans le troisième ventricule, et par son contour antérieur, vers le *velum transversum;* du pôle postérieur de l'epithalamus s'élève une courte tige épiphysaire, bientôt dilatée en une modeste épiphyse, simple vésicule épithéliale reposant sur le côté gauche du toit. La masse arrondie du thalamus dorsal, bien délimitée par les sillons diencéphaliques dorsal et moyen, marque la région la plus étroite du troisième ventricule; elle vient presque au contact de son homologue dans sa région antérieure. Le thalamus ventral est à la fois plus long, puisqu'il atteint en avant la région sus-optique, plus large au niveau des fibres optiques, et aussi un peu plus haut. De sa région postérieure se détache l'hypothalamus, sous le sillon diencéphalique ventral.

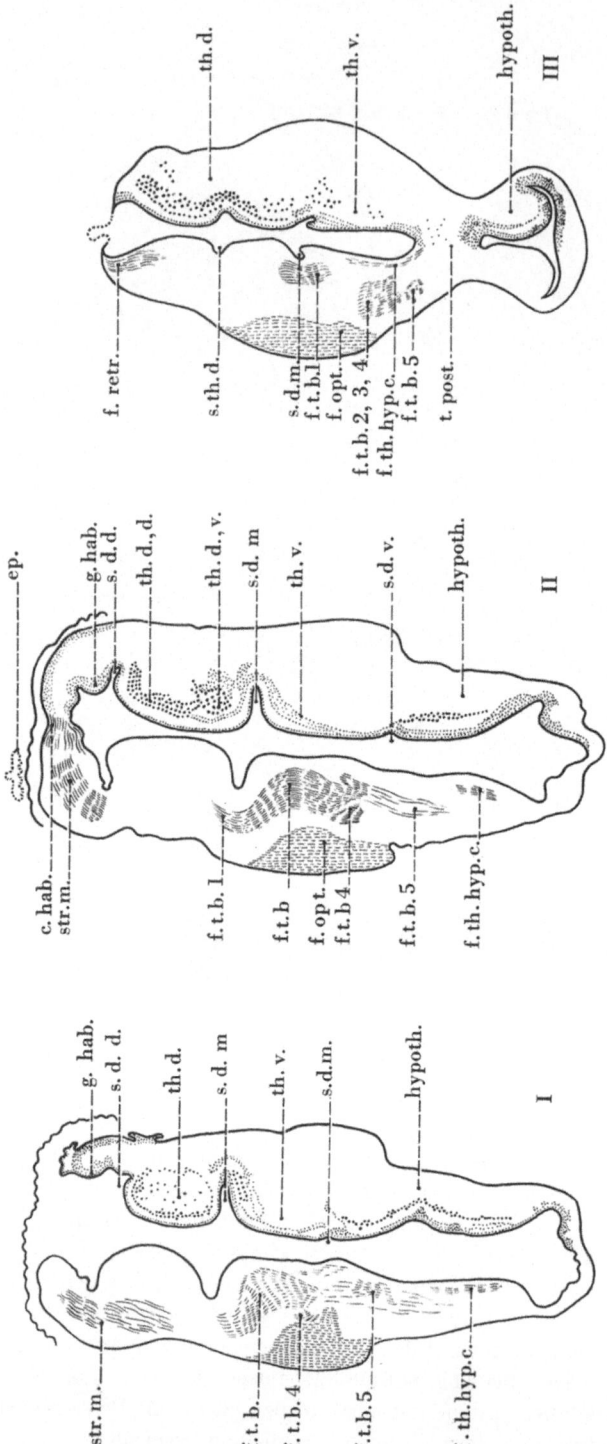

Fig. 6. Coupes transversales schématiques du diencéphale. Les corps cellulaires sont figurés dans la partie droite, les fibres dans la partie gauche. (MILLOT, J., R. NIEUWENHUYS, J. ANTHONY, 1964).

L'hypothalamus forme l'étage le plus vaste du diencéphale de *Latimeria*. Très long à sa base, il s'étend du *tuberculum posterius*, sous lequel il pousse un diverticule, au chiasma optique. Il se dirige presque directement vers l'avant, s'étire sous les nerfs optiques et sous le télencéphale en une tige conique et s'évase finalement en donnant attache au complexe hypophysaire. L'hypophyse (lobe nerveux et glande pituitaire) a une disposition aberrante. D'une longueur insolite, elle joint le plancher du diencéphale à la voûte buccale. Plus précisément, du lobe nerveux de l'hypophyse part un cordon conjonctivo-vasculaire, partiellement creux, d'une dizaine de centimètres de longueur, qui décrit par dessus le *dorsum sellae* une trajectoire incurvée vers le bas pour se rendre au fond de la fosse hypophysaire du basisphénoïde contre le parasphénoïde; il représente sans aucun doute la poche embryonnaire de Rathke. Le cordon contient un chapelet d'îlots pituitaires, dont deux principaux: l'un postérieur, partiellement pénétré par les digitations du lobe nerveux issu de l'infundibulum, correspond à la glande pituitaire classique; l'autre, antérieur, demeure fixé dans la fosse hypophysaire. Ainsi étirée vers l'avant, l'hypophyse entraîne un décalage des éléments du plancher diencéphalique; la région optique vient surmonter le chiasma; le *tuberculum posterius* est franchement déjeté en avant. On s'attendrait à voir, annexé à l'infundibulum, un sac vasculaire bien développé, comme il est de règle chez les Poissons de profondeur. En réalité, le sac vasculaire de *Latimeria* n'existe pas en tant qu'organe indépendant: l'infundibulum ne montre rien, ni dans sa partie caudale, ni dans sa partie ventrale, qui puisse l'évoquer. Par contre, on observe accolée à, ou plus ou moins incluses dans l'extrémité postérieure de l'hypophyse, des vésicules irrégulièrement divisées et plissées, tapissées par un épithémium épendymaire, qui semblent ne pas faire directement partie de la *pars nervosa* et pourraient être des vestiges du sac vasculaire. Nous ne saurions l'affirmer avec certitude, n'ayant pu mettre en évidence dans ces vésicules les cellules à couronne caractéristique.

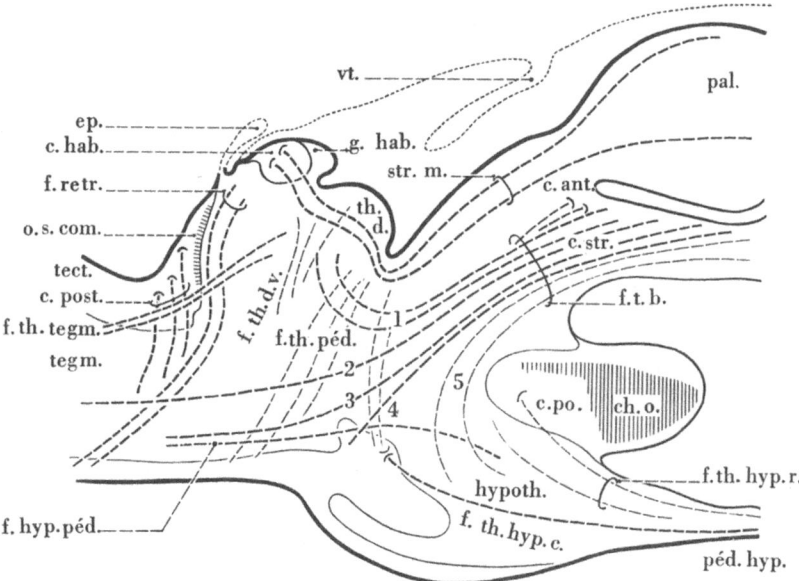

Fig. 7. Diagramme des faisceaux diencéphaliques, projetés sur le plan sagittal médian. (Millot, J., R. Nieuwenhuys, J. Anthony, 1964).

Pour la description des principaux faisceaux du prosencéphale il est indispensable d'ajouter aux préparations horizontales, des coupes transversales (Fig. 6) et sagittales (Fig. 7).

Les faisceaux du télencéphale. Les afférences télencéphaliques sont essentiellement olfactives. Elles prennent naissance dans le bulbe olfactif, dont la structure, banale dans son ensemble, se signale tout de même par la répartition désordonnée des cellules mitrales (Fig. 8), contrastant avec l'alignement régulier que l'on observe d'ordinaire. Les fibres olfactives ont une longueur exceptionnelle chez *Latimeria,* puisqu'elles relient la cavité nasale à l'arrière crâne, où l'encéphale se trouve relégué. Elles aboutissent au lobe olfactif, où elles se scindent en deux contingents: l'un, antéropostérieur, s'enfonçant dans le pallium, l'autre descendant, destiné à la région basale, septum et striatum. Du pallium, on voit se dégager en direction du diencéphale un volumineux faisceau, la *stria medullaris,* sans que l'on puisse situer le niveau du relais probable des fibres olfactives secondaires. La *stria medullaris* se rend à l'habenula. Elle apparaît composée de deux ensembles de fibres, bientôt convergents. L'un dorsal suit le bord dorsal, ou taenia, du pallium, tandis que l'autre, plus abondant, court au centre de la zone la plus épaisse du pallium. Tous deux s'élèvent, accolés, dans la partie rostrale du thalamus dorsal et atteignent l'epithelium, où ils s'engagent dans une vaste *commissura habenularum,* commune à l'ensemble des Vertébrés. Du striatum émane vers l'arrière un *faisceau télencéphalique* basal fortement développé. Il forme d'abord un système compact, puis déploie dans le diencéphale ses divers composants. Un nombre important de ses fibres s'infléchissent dorsalement et viennent s'épanouir dans le thalamus dorsal (f. t. b. l.). La plupart des autres traversent le thalamus ventral en se séparant en trois groupes: le *tractus strio-tegmentalis* (f. t. b. 2), le *tractus strio-peduncularis* (f. t. b. 3) et le *tractus strio-tubercularis* (f. t. b. 4); ce dernier remarquable par sa compacité, se rend probablement à l'hypothalamus postérieur. Enfin, un composant tout à fait ventral se détache très précocement du tronc commun et, par un trajet arqué à concavité antérieure, atteint l'hypothalamus; certaines de ces fibres traversent le tractus optique (f. t. b. 5). Les connexions intra-hémisphériques, sans doute assez pauvres, n'ont pu être définies.

Les faisceaux du diencéphale. Le diencéphale émet plusieurs faisceaux vers le tronc cérébral. L'un des plus apparents est le *faisceau retroflexe* de Meynert, issu du ganglion de l'habénula. Il descend dans la partie postérieure du thalamus dorsal, et se rend au tegmentum mésencéphalique. Son point de terminaison demeure imprécis. C'est aussi au tegmentum qu'aboutit le *tractus thalamo-tegmentalis,* né du thalamus dorsal. Nous avons individualisé deux autres faisceaux destinés au mésencéphale basal: le *tractus thalamo-peduncularis* et le *tractus hypothalamo-peduncularis;* celui-ci part de l'hypothalamus dorsal et passe audessus du *tuberculum posterius* pour atteindre le pédoncule. En outre de nombreuses fibres à orientation générale verticale unissent les étages du diencéphale entre eux. En partie diffuses, elles s'ordonnent parfois en fascicules. C'est ainsi que pour certaines nous proposons le terme de *tractus dorsoventralis thalami* (Fig. 7). D'autres vont au *tuberculum posterius* et, nous semble-t-il, s'entrecroisent dans la *commissura tuberculi posterius,* puis se portent en avant pour former de petits filets bien distincts dans l'hypothalamus dorsal et pénétrer ensuite dans l'infundibulum: le système entier peut être nommé *tractus thalamo-hypothalamicus caudalis* (Fig. 7). Plus en avant et plus dorsalement, des axones traversent la paroi hypothalamique. Ils s'entrecroisent, pour une part, dans la commissure

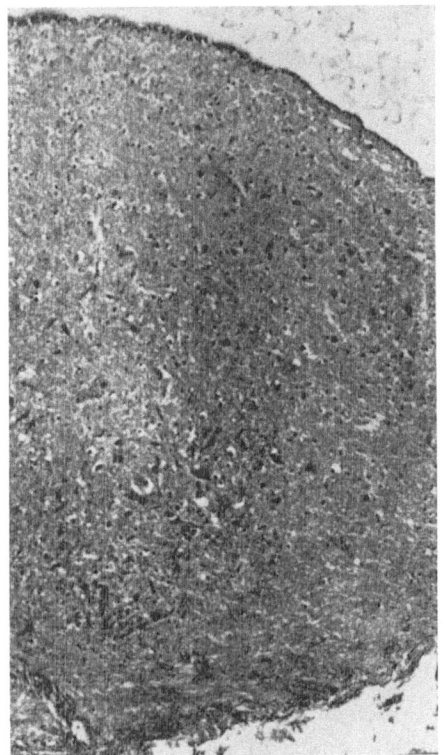 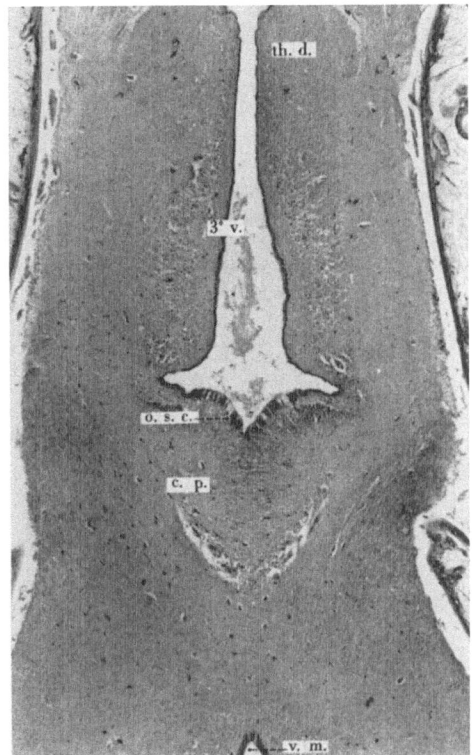

Fig. 8.   Fig. 9.

Fig. 8. Coupe horizontale du bulbe olfactif, montrant la dispersion des cellules mitrales (grandes
cellules sombres).

Fig. 9. Coupe horizontale de l'organe sous-commissural. G.N. x 20 environ.

post-optique. Nous n'avons pu reconstituer leur trajet complet, mais leur origine pourrait
être thalamique, ce qui les désignerait comme un *tractus thalamo-hypothalamicus rostralis*,
semblable à celui de divers Ichthyopsidés. Il est à noter que les éléments cellulaires du
diencéphale se répartissent sur le pourtour du ventricule, ou à faible distance de lui, mais
ne forment pas de noyaux superficiels sous-méningés.

Le diencéphale reçoit les fibres optiques au moment où elles se croisent. La paroi
diencéphalique englobe en grande partie le chiasma, qui, d'autre part, se présente comme
une véritable décussation semblable à celle des Vertébrés plus élevés. Les nerfs optiques
des autres Poissons s'entrecroisent au contraire en bloc ou par fascicules.

Une épaisse commissure postérieure limite le prosencéphale vers le tectum mésencéphalique. Elle embrasse dans sa concavité l'angle postérieur du 3ème ventricule, bordé
par un très bel organe sous-commissural (Fig. 9).

Liste des Abréviations

c.ant.	commissure antérieure	c.po.	commissure postoptique
c.hab.	commissure habénulaire	c.str.	striatum
c.p., c.post.	commissure postérieure	ch., ch.o.	chiasma optique

ep.	épiphyse	p., pal.	pallium
f.hyp.ped.	faisceau hypothalamo-pédonculaire	pit.	glande pituitaire
		pl.m.	plica rhombomesencephalica
f.opt.	fibres optiques	ped.hyp.	pédoncule hypothalamique
f.t.b.	faisceau télencéphalique basal	rh.	rhombencéphale
f.t.b. 1	faisceau strio-thalamique	s.d.d.	sillon diencéphalique dorsal
f.t.b. 2	faisceau strio-tegmental	s.d.m.	sillon diencéphalique moyen
f.t.b. 3	faisceau strio-pédonculaire	s.d.v.	sillon diencéphalique ventral
f.t.b. 4	faisceau strio-tuberculaire	s.th.d.	sillon du thalamus dorsal
f.t.b. 5	faisceau strio-hypothalamique	str.	striatum
f.th.d.v.	faisceau thalamique dorsoventral	str.m.	strie médullaire
f.th.hyp.c.	faisceau thalamo-hypothalamique caudal	t.post.	tuberculum posterius
		te.m., tect.	tectum mésencéphalique
f.th.hyp.r.	faisceau thalamo-hypothalamique rostral	tegm., tg.m.	tegmentum mésencéphalique
		th.d.	thalamus dorsal
f.th.ped.	faisceau thalamo-pédonculaire	th.d., d.	thalamus dorsal, partie dorsale
f.th.tegm.	faisceau thalamo-tegmental	th.d., v.	thalamus dorsal, partie ventrale
g.hab.	ganglion habénulaire	th.v.	thalamus ventral
hab.	niveau approximatif de la commissure habénulaire	v.m.	ventricule mésencéphalique
		vt	velum transversum
hypoth.	hypothalamus	II	nerf optique
inf.	infundibulum	III	nerf moteur oculaire commun
mes.	mésencéphale	3è v.	troisième ventricule
o.s.c., o.s.com.	organe sous-commissural		

Summary

The prosencephalon of *Latimeria* is not fundamentally different in structure from that at present known in fishes. It contains the same main fibre bundles. Among the characteristics however, which give *Latimeria* a certain peculiarity, are a frontal shift in the region of the hypothalamus; the extreme diminution and want of differentiation of the vascular sac, which in deep sea fishes is well developed; the presence of a real chiasma partially lodged in the diencephalic wall, and the irregular distribution of the mitral cells inside the olfactory bulb. At the level of the hemisphere, there is a strong protrusion of the pallium towards the symmetry axis, which suggests an inversion; but we believe that this is only an apparent one.

Références

MILLOT, J., J. ANTHONY: Considérations préliminaires sur le squelette axial et le système nerveux central de *Latimeria chalumnae* Smith. Mém. Inst. Sci. Madag., sér. A, XI (1956), 167–87.

MILLOT, J., J. ANTHONY: Anatomie de *Latimeria chalumnae*, t. II, Système nerveux et Organes des sens. 130 p., 57 fig., 77 pl. h.t. Edit. C.N.R.S., Paris 1965.

MILLOT, J., R. NIEUWENHUYS, J. ANTHONY: Le diencéphale de *Latimeria chalumnae* Smith (Poisson coelacanthidé). C. R. Acad. Sci. Paris, 258 (1964), 5051–55.

Prof. Dr. J. MILLOT
Musée de l'Homme
Place du Trocadéro
Paris XVIᵉ, France

Prof. Dr. J. ANTHONY
Laboratoire d'Anatomie Comparée
55, rue de Buffon
Paris Vᵉ, France

The Hippocampal and Septal Formations in Anurans*

By H. H. Hoffman, Birmingham, Ala.

The present paper represents the results of a morphologic study of the medial hemisphere wall of the forebrain of the tailless amphibians. In an earlier publication concerned with the telencephalon of anurans (Hoffman, 1963), the hippocampal and the septal formations were identified and some of their connections reported. This current study has provided the opportunity for a further elaboration of these cellular configurations and will afford a background for subsequent experimental investigations of such areas.

Materials and Methods. This study is based on serial sections of the brains of the common frogs, *Rana pipiens* and *Rana catesbiana*. Other anuran brains (of *Pipa pipa, Xenopus laevis, Hyla cinerea* and *Bufo marinus*) have been prepared in like manner. Buffered formalin-fixed specimens were embedded, sectioned transversely, sagittally or horizontally at twenty microns and stained with thionin. The pyridine silver method of Ranson (1911), as modified for anurans (Hoffman, 1963), was utilized in the study of the fiber systems. Other buffered formalin preparations were stained with the protargol technique in a further study of fiber connections. The neuron types in the various areas were demonstrated through the use of a rapid Golgi technique.

List of Abbreviations

amyg.	amygdala
b. olf. acc.	accessory olfactory bulb
bed nuc. comm. hip.	bed nucleus of the hippocampal commissure
comm. ant. + bed nuc.	bed nucleus of anterior commissure
comm. hip. + bed nuc.	bed nucleus of hippocampal commissure
d. b. B.	diagonal band of Broca
f. interv.	interventricular foramen
l. f. b.	lateral forebrain bundle
l. mit.	mitral cell layer
N. O.	optic nerve
N. olf.	olfactory nerve (fila)
N. vm.	vomeronasal nerve
nuc. acc.	nucleus accumbens
nuc. olf. ant.	anterior olfactory nucleus
nuc. olf. ant. p. dorsolat.	anterior olfactory nucleus, dorsolateral part
nuc. olf. ant. p. post.	anterior olfactory nucleus, posterior part
nuc. olf. ant. p. ventrolat.	anterior olfactory nucleus, ventrolateral part
nuc. olf. ant. p. ventromed.	anterior olfactory nucleus, ventromedial part
nuc. preop. perivent. magnocell.	large celled periventricular preoptic nucleus

*) This investigation was supported in part by research grant (NB-01804) from the National Institute of Neurological Diseases and Blindness, the National Institute of Health, U.S. Public Health Service.

nuc. preop. perivent. parvocell.	small celled periventricular preoptic nucleus
nuc. sept. hip.	septohippocampal nucleus
nuc. sept. lat.	lateral septal nucleus
nuc. sept. med.	medial septal nucleus
nuc. sept. med. p. dors.	medial septal nucleus, dorsal part
nuc. sept. med. p. vent.	medial septal nucleus, ventral part
parahip. area	parahippocampal area, primordial subiculum (prim. subic.)
prim. hip.	hippocampal primordium (rostral extent)
prim. hip. a	primordial hippocampus a, primordial cornu ammonis (prim. corn. amm.)
prim. hip. ant.	primordial hippocampus, anterior continuation
prim. hip. b	primordial hippocampus b, primordial gyrus dentatus (prim. gy. dent.)
prim. pall. dors.	primordial dorsal pallium
prim. pir.	primordial piriform cortex
prim. pir. ant.	primordial piriform cortex, anterior continuation
prim. pir. s	superpositioned primordial piriform cortex
rec. preop.	preoptic recess
s. lim. hip. sept.	hippocamposeptal limiting sulcus
s. lim. lat.	lateral limiting sulcus
s. lim. med.	medial limiting sulcus
s. lim. sept. tub.	septotubercular limiting sulcus
striat.	striatum
striat. p. dors.	striatum, dorsal part
striat. p. vent.	striatum, ventral part
tub. olf.	tuberculum olfactorium
vent. lat.	lateral ventricle
X	continuation of amygdala with primordial cornu ammonis
Y	posterior part of nucleus olfactorius anterior
z. lim. med.	medial limiting zone

General Considerations

The gross brain of *Rana pipiens* is typical of those brains of all the tailless amphibians thus far examined. Features of the rostral portions of the brain common to all forms are: fused olfactory bulbs, a prominent vomeronasal nerve coursing to a ventrolaterally situated accessory olfactory bulb, a very slight ridge indicative of the region related to the tuberculum olfactorium, a smooth cylindrical hemisphere wall, and a circular fissure separating the olfactory bulb and the hemisphere regions.

Transverse sections through the rostral portions of the olfactory bulb demonstrate a ventrally situated layer of mitral cells, a ventrolateral vomeronasal nerve and the anterior olfactory nucleus forming a ring of cells which completely surrounds the rostral portion of the olfactory ventricle (HOFFMAN, 1963). Dorsomedial, ventromedial, dorsolateral, and ventrolateral quadrants are demonstrable in anurans comparable to those described by OBENCHAIN (1925) in the anterior olfactory nucleus of South American opossums. This division is made on the basis of the position of the nuclear gray with respect to the ventricle and not on differentiating cellular characteristics. At this level the quadrants are quite symmetrical and give no indication of the predominance of the medial hemisphere wall in anurans.

A study of the sulci of the medial wall of the ventricle gives a fairly clear outline of the shifting of the nuclear patterns in anurans. Such a pattern has previously been pointed out by Röthig (1912). The hippocamposeptal limiting sulcus (Figs. 3, 4, 5, 6, 8) appears in the dorsal hemisphere at the rostralmost limit of the differentiating primordial hippocampal formation. As this formation increases in size, the sulcus moves ventrally (Figs. 4, 5), and, at midhemisphere levels, is situated at almost the ventral limit of the ventricle wall (Fig. 6). More caudally, with the increasing size of the precommissural septum, the sulcus again shifts dorsally (Fig. 8) and, at its caudalmost extent, terminates by merging into the ventromedial segment of the interventricular foramen. The septotubercular limiting sulcus (Figs. 3, 4, 5), by contrast, is first seen, and remains at the ventromedialmost extent of the ventricle. The slight prominence just rostral to it (Fig. 5) is formed by the cluster of cells comprising the posterior part of the nucleus olfactorius anterior. At the caudal end of the septotubercular sulcus, the posterior part of the nucleus olfactorius anterior differentiates to form the septohippocampal nucleus and, at this level, the sulcus blends into the ventricle wall.

Nuclear Patterns of the Medial Hemisphere Wall

The Primordial Hippocampal Formation. Rostrally, the anterior continuation of the primordial hippocampal formation appears in the dorsomedial segment of the hemisphere wall, just caudal to a clearly recognizable anterior olfactory nucleus (Fig. 1). Cells of this dorsomedial segment begin to thin out and are gradually replaced by slightly larger, more lightly stained cells (Figs. 1, 2). The area occupied by these cells increases in size until, at midhemisphere levels, such cells have replaced the entire medial portion of the anterior olfactory nucleus (Figs. 5, 6). Caudally from this midpoint, the formation diminishes in size but extends in the same relative position for the entire length of the hemisphere (Figs. 7, 8, 9, 10). At all levels, the formation is bounded ventrally by a hippocamposeptal limiting sulcus and, at most mid and caudal levels by a medial cell-free zone (Figs. 7, 9). Dorsally, the formation merges indistinguishably with the primordial general pallium. Medially, dorsomedial olfactory fibers separate the formation from the hemisphere wall and laterally its boundary is formed by the ependyma of the medial wall of the ventricle.

The primordial hippocampal formation may be characterized by three cellular configurations. Dorsally, the area of cells merging in part with the primordial general pallium will be referred to as a parahippocampal area. This group of cells is best demonstrated at the more rostral limits of the hemisphere (Figs. 4, 5, 6, 12) though identifiable caudally as far as the rostral limits of the anterior commissure (Figs. 9, 10). The constituent cells are small, with a minimal number of visible dendrites. They are probably primarily associational elements as pointed out by Herrick (1933a) in *Necturus*. Centrally, an expanded area of large, lightly staining cells is demonstrable from the rostralmost limits of the formation (Figs. 2, 4, 6, 7). This cell group may be readily identified at all levels of the hemisphere and, at the most caudal limits of the hemisphere, merges with the primordial piriform area (Fig. 10). Ventrally, at this level, the group likewise becomes continuous with the amygdaloid complex (Figs. 10, 12). The most characteristic cells of this group are large elements with widely branched dendrites directed toward the surface of the brain. Both dendrites and axons are directed toward the ventricle. The axons may pass dorsalward to the parahippocampal area, the primordial dorsal pallium and the lateral hemisphere wall (alveus). Other fiber bundles may be traced ventrally along the wall of the

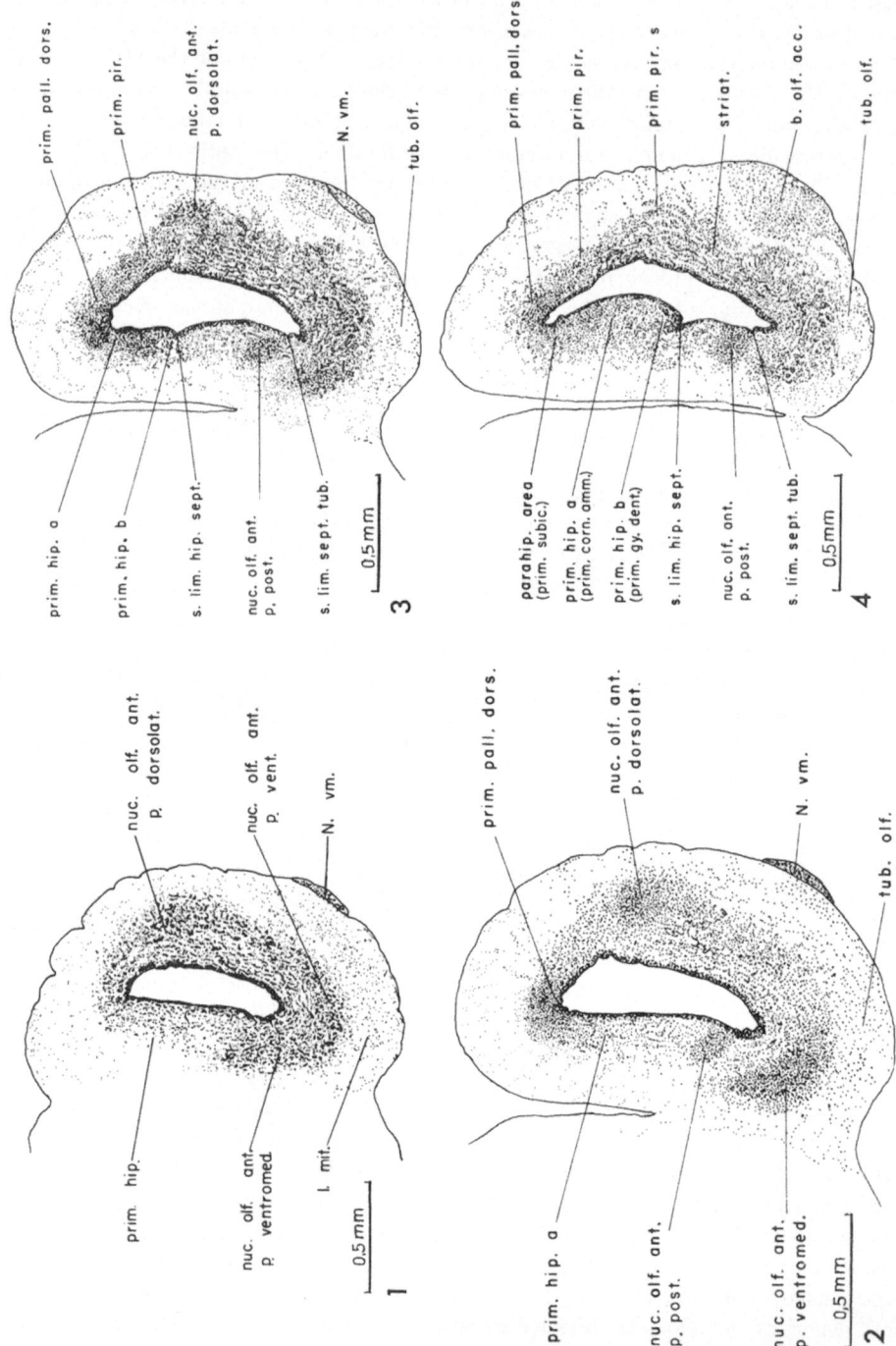

ventricle as fibers of the fornix and as commissural fibers. Still others pass ventralward toward the precommissural septal area (hippocamposeptal fibers). Ventrally, a more densely packed cluster of smaller cells, interspersed with a minimal number of the larger elements, is demonstrable in the anuran brain (Figs. 3, 4, 5, 6, 7, 8). Again the dendritic aborizations are primarily directed toward the surface although in the anuran, the cell bodies lie in the deep cellular layer. Axons arising from these cells may course ventrally as hippocamposeptal fibers, or may pass toward the ventricle wall to join the fornix, and also undoubtedly contribute to the medial corticohabenular tract and to the medial forebrain bundle.

RöTHIG (1912, 1926) contrasted the primordial amphibian with the primordial reptilian hemisphere structures and suggested that the area of large, lightly staining cells in the dorsomedial wall (prim. hip. a, Figs. 2–10) might be homologized with the cornu ammonis of higher forms and that the area of small, more densely packed cells central to it (prim. hip. b, Figs. 3–8) might be homologous to the gyrus dentatus. Since a third and dorsal area (parahip. area, Figs. 4–10) is also recognizable in some anurans and since it is properly located, as regards to such an area in other vertebrate forms, it may be representative of a primordial subiculum.

In anurans, a precise delimitation of nuclear masses is not always possible due to the minimal degree of differentiation exhibited. Some striking cellular differences are apparent as in the cornu ammonis (prim. hip. a, Figs. 2–10), however, and the location of cell groups with respect to each other throughout the amphibian series is remarkably consistent. The best and probably most reliable criteria, however, are afforded by an identification of such fiber connections as can be demonstrated. Since fiber connections, comparable to those in many higher forms, can be related to the primordial subiculum, the primordial cornu ammonis and the primordial gyrus dentatus of the anuran brain, the homologies here made appear to be justified.

The Precommissural Septum. The prominence of the precommissural septum is a unique feature of the anuran brain. At all levels (Figs. 5–9) the area is separated from the

Fig. 1. Drawing of a transverse section through the junction of the olfactory bulb and the hemisphere. Note the replacement of the dorsomedial quadrant of the anterior olfactory nucleus by the primordium hippocampi. The remaining quadrants of the anterior olfactory nucleus are undifferentiated. Magnification indicated.

Fig. 2. Drawing of a transverse section through the rostral portion of the cerebral hemisphere. Note the general differentiation of the anterior olfactory nucleus and particularly the presence of a distinct posterior part of this nucleus (nuc. olf. ant. p. post.). The olfactory tubercle has appeared. Magnification indicated.

Fig. 3. Drawing of a transverse section at the level of the terminating vomeronasal nerve (N. vm.). The rostral portion of the hippocamposeptal limiting sulcus is illustrated and two distinct cell groups appear in the primordial hippocampal formation at this level. Magnification indicated.

Fig. 4. Drawing of a transverse section through the rostral portion of the accessory olfactory bulb. The hippocamposeptal limiting sulcus is seen ventral to its position in the previous section and three cellular groups, (parahip. area, prim. hip. a, and prim. hip. b), are apparent in the primordial hippocampal formation. Magnification indicated.

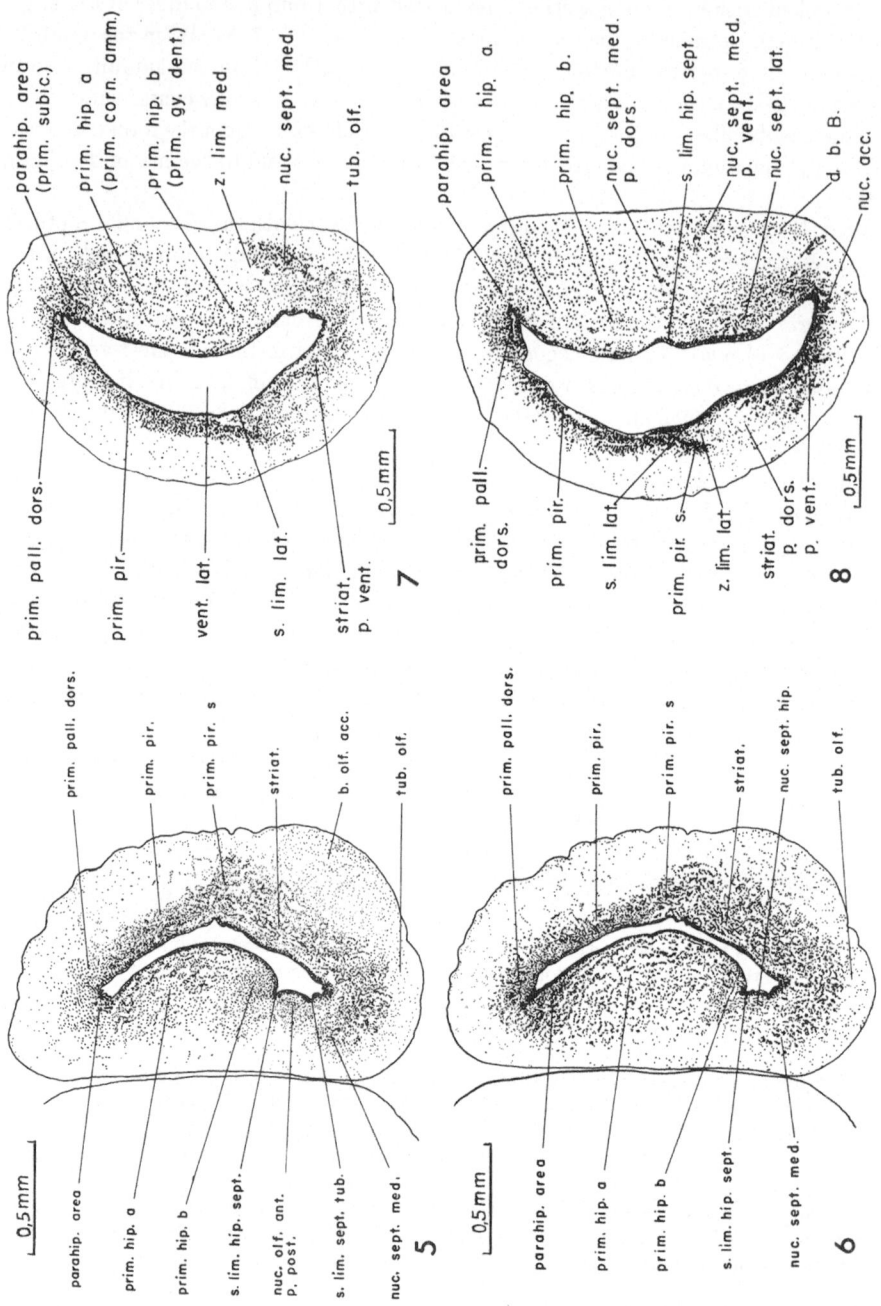

primordial hippocampal formation by a hippocamposeptal limiting sulcus and by a relatively cell-free zone. At rostral and midhemisphere levels, the posterior part of the anterior olfactory nucleus (nuc. olf. ant. p. post., Fig. 6) is situated between the hippocampal and the septal formations. This portion of the anterior olfactory nucleus is bounded dorsally by the hippocamposeptal limiting sulcus and ventrally by the septotubercular limiting sulcus. It terminates caudally as the posterior part of the anterior olfactory nucleus which is present in many forms (CROSBY a. HUMPHREY 1939). The septohippocampal nucleus thus lies at the rostral level of the precommissural septum. The precommissural septal area, following the terminology of ANDY a. STEPHAN (1964) for mammals, has the following components: (1) a septohippocampal nucleus, (2) a medial septal nucleus with dorsal and ventral division and (3) a lateral septal nucleus merging with (4) the nucleus accumbens septi, (5) the bed nuclei of the commissures, (6) a nucleus triangularis and (7) possibly the bed nucleus of the stria terminalis. The anterior continuation of the hippocampus terminates in a nuclear mass which is the septohippocampal nucleus (Fig. 6) of the present terminology but which probably falls within the primordium hippocampi of JOHNSTON (1913).

Two large nuclear masses (a medial septal nucleus and a lateral septal nucleus) can be clearly identified (Fig. 8). This is in decided contrast to the precommissural septal area of the tailed amphibians, particularly *Necturus* where this area is much more limited (HERRICK, 1933a). ANDY a. STEPHAN (1964), in describing the septum of the cat, divided the nuclei of the septal area into dorsal, ventral, medial and caudal groups. Such a general pattern of division is evident in the anuran brain though the area is not so highly organized.

The medial septal nucleus has a distinct dorsomedial segment with its cells seemingly laminated due to the passage of septohippocampal and hippocamposeptal fibers (Fig. 8). This dorsomedial part must be emphasized as it is the most prominent portion of the entire precommissural septal complex. The cells forming this nuclear segment can be

Fig. 5. Drawing of a transverse section through the mid-portion of the accessory olfactory bulb illustrating recognizable primordial areas in all quadrants of the hemisphere. Note the ventral position of the hippocampal limiting sulcus (s. lim. hip. sept.) and the first appearance of the medial septal nucleus (nuc. sept. med.). Magnification indicated.

Fig. 6. Drawing of a transverse section through the level at which the primordial hippocampal formation encompasses nearly the entire medial hemisphere wall. The medial septal nucleus is clearly illustrated as is the almost acellular tuberculum olfactorium. There are some hints at this level of a cortical polymorph layer on the medial side of the tuberculum. The septohippocampal nucleus is pointed out (nuc. sept. hip). Magnification indicated.

Fig. 7. Drawing of a transverse section at the level of the rostral extent of the medial septal nucleus. The primordial hippocampal formation is somewhat reduced in extent and the septal area now undergoes marked differentiation. The tuberculum olfactorium fails to show the differentiation illustrated in Fig. 6. Magnification indicated.

Fig. 8. Drawing of a transverse section at the level of maximal differentiation of the precommissural septal area. Dorsal and ventral parts of the medial septal nucleus and its continuity with the diagonal band of BROCA are illustrated. The lateral septal nucleus as it overlies the nucleus accumbens septi is demonstrated. Magnification indicated.

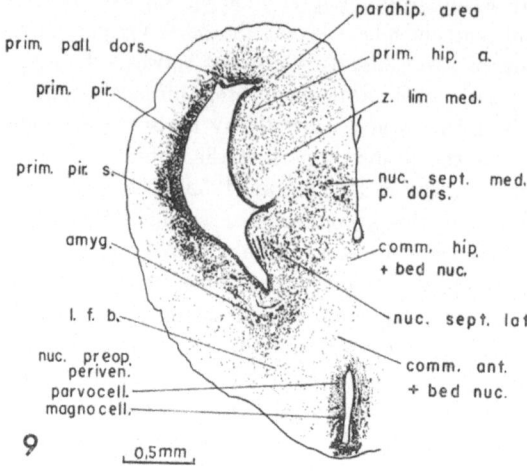

prim. pall. dors.

prim. pir.

prim. pir. s.

amyg.

l. f. b.

nuc. preop.
periven.
parvocell.
magnocell.

9   0,5mm

parahip. area

prim. hip. a.

z. lim med.

nuc. sept. med.
p. dors.

comm. hip.
+ bed nuc.

nuc. sept. lat.

comm. ant.
+ bed nuc.

Fig. 9. Drawing of a transverse section through the rostral part of the anterior and hippocampal commissures. Note that the dorsal part of the medial septal nucleus extends dorsal to the hippocampal commissure and that the lateral septal nucleus is continuous with the bed nuclei of both commissures. The amygdala is illustrated. Magnification indicated.

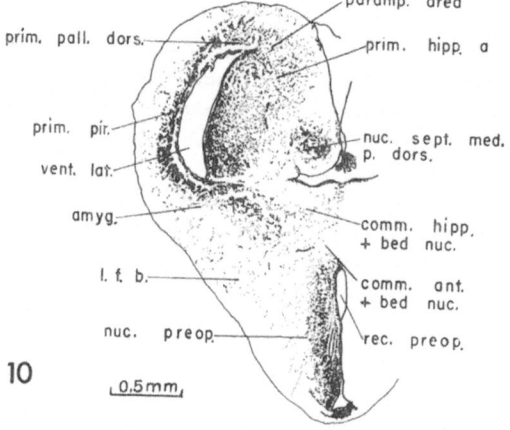

prim. pall. dors.

prim. pir.

vent. lat.

amyg.

l. f. b.

nuc. preop.

10   0,5mm

parahip. area

prim. hipp. a

nuc. sept. med.
p. dors.

comm. hipp.
+ bed nuc.

comm. ant.
+ bed nuc.

rec. preop.

Fig. 10. Drawing of a transverse section through the caudalmost extent of the dorsal part of the medial septal nucleus. At this level the amygdalar complex is clearly illustrated and the amygdala merges dorsally with the primordial cornu ammonis. The primordial piriform cortex is in continuity with the primordial cornu ammonis at the ventromedial angle of the lateral ventricle. Magnification indicated.

traced from the most rostral limits of the precommissural septum (nuc. sept. med., Fig. 5) to its most caudal extent where they lie dorsal to the hippocampal commissure (nuc. sept. med. p. dors., Fig. 10). At this caudalmost level, the medial septal nucleus merges indistinguishably with the bed nucleus of the hippocampal commissure (Fig. 11). HERRICK (1910) referred to this caudalmost portion of the nucleus as the corpus precommissurale or as the nucleus medialis septi. This dorsomedial portion of the medial septal nucleus apparently differentiates to become the dorsal septal nucleus of higher forms. Ventrally, near the midline, the medial septal nucleus becomes continuous with the diagonal band (of BROCA) and its nucleus, (Fig. 8). The anterior and the posterior parts of the medial septal nucleus, as described by ANDY a. STEPHAN (1964) in the cat, are not discernible in anurans and, although the relationship with the nucleus of the diagonal band is clearly evident, its septal and its tubercular portions are not distinct cytologically as they apparently were in the material studied by these observers.

The lateral septal nucleus lies along the wall of the ventricle. It is separated from the ventral segment of the medial septal nucleus and the nucleus of the diagonal band by the medial forebrain bundle (Fig. 8). Its cells are, in general, rather densely packed and vary considerably in size. The major cells of the lateral septal nucleus appear somewhat larger than the similar neurons of the medial septal nucleus. Numerous small bipolar cells are found in GOLGI preparations throughout this septal area and are probably intrinsic in nature, conveying impulses from one septal area to the other. It is probable that in anurans, the lateral septal nuclear complex incorporates both the ventral and the caudal groups described by ANDY a. STEPHAN (1964) in the cat.

In our preparations, the ventral portion of the lateral septal nucleus merges quite indistinguishably with the gray of the nucleus accumbens septi and thus falls into relationship with the striatal complex of the ventrolateral hemisphere wall (Fig. 8). Caudally, a merging occurs between the lateral septal nucleus and the bed nuclei of the anterior and hippocampal commissures (Figs. 9, 11). It is not possible, however, to define a nucleus septalis triangularis as a distinct entity. It is probably incorporated in the band of cells directed toward the bed nucleus of the anterior commissure. A thin column of cells sweeping dorsal to the caudalmost portion of the anterior commissure may be homologous to the anterior part of the bed nucleus of the stria terminalis as described by ANDY a. STEPHAN (1964).

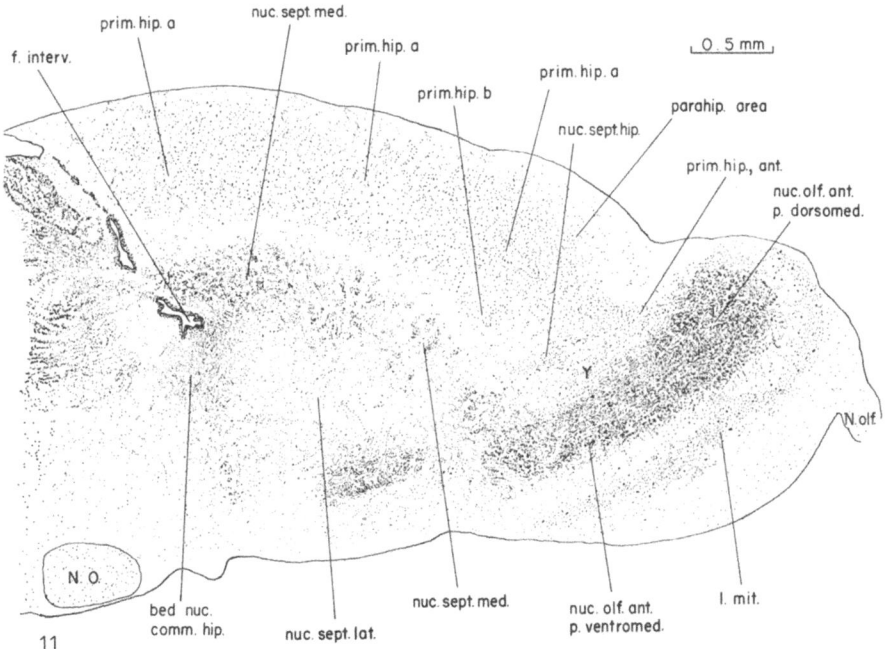

Fig. 11. Drawing of a sagittal section, thionin stained, taken medial to the ventricular formation of the hemisphere to illustrate the primordial cellular groups with respect to their rostrocaudal relationships. Note particularly the anterior continuation of the primordial hippocampal formations and (Y) the posterior extent of the anterior olfactory nucleus. Note also the dorsal position of the dorsal part of the medial septal nucleus with respect to the interventricular foramen. Magnification indicated.

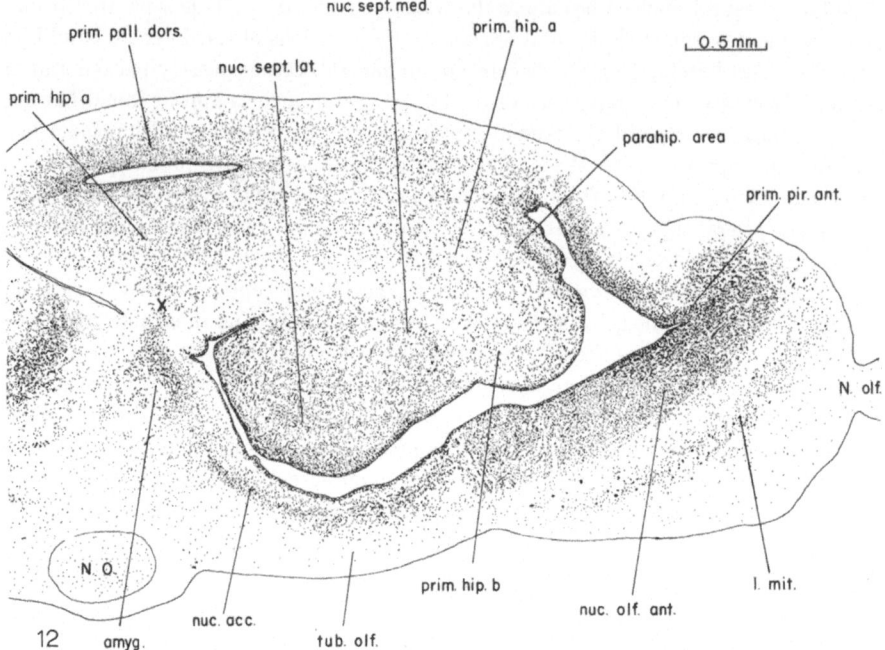

Fig. 12. Drawing of a sagittal section through the lateral ventricle. To be emphasized particularly are the mitral cell layer, the primordial hippocampus a, its continuity at X with amygdaloid complex and the ventrally situated nucleus accumbens septi and the poorly developed tuberculum olfactorium. Magnification indicated.

Discussion

It must be pointed out that the medial hemisphere wall in anurans, when contrasted with the lateral hemisphere wall, is quite disproportionately developed. Both the primordial hippocampal and precommissural septal areas are unusually large with respect to the overall size of the brain and exhibit a remarkable degree of differentiation. HERRICK (1933b) has called attention to the amphibian primordium hippocampi as being further advanced toward cortical structure than any other pallial region. Such an advance is particularly well demonstrated in the levels of the cornu ammonis (prim. hip. a) shown in Figs. 5, 6, and 7. Here the migration of some cells toward the wall of the hemisphere is suggestive of a developing cortex. Such a formation is in decided contrast to the pattern of differentiation in other parts of the anuran brain and particularly in the striatal complex of the lateral hemisphere wall. In this lateral striatal area, it is barely possible to distinguish between the dorsal and the ventral parts. No marked degree of differentiation then is evident in the lateral wall. This fact emphasizes that in anuran line development there is a disproportionate differentiation of the medial wall of the hemisphere.

The medial and the lateral septal nuclei occupy a prominent position in the ventromedial wall of the anuran hemisphere. Their degree of differentiation is such that, as has been previously emphasized, the area is strikingly similar to the precommissural septal area of many higher vertebrates. The septal nuclei are intimately related through correlative fiber tracts with all their neighboring structures. They probably receive many ascending fibers through the medial forebrain bundle, which is correspondingly more highly developed and organized than is the lateral forebrain bundle. Ventral and lateral connections of the medial hemisphere wall with the amygdaloid, the striatal, and possibly the piriform areas are made possible by fascicles of the diagonal band (of Broca). Many of these same connections have been outlined for *Ambystoma* by Herrick (1927) but in this latter amphibian, the development of the medial hemisphere wall does not differ so strikingly from that of the lateral hemisphere wall. In *Necturus*, Herrick (1933a) pointed out that there is no clear differentiation of individual medial and lateral septal nuclei either from each other or from surrounding structures.

In the anuran brains thus far examined, a paucity of nerve cells would appear to be characteristic of the tuberculum olfactorium. A plexiform layer can be identified throughout the extent of the area so designated, but no cortical migration has been noted. In a very limited area (Fig. 6), some suggestion of a polymorph layer can be identified. In keeping with the pattern of the anuran hemisphere, the cells of this layer are oriented medially toward the precommissural septal area rather than toward the lateral part of the tubercle and the striatal complex of the lateral hemisphere wall, where no real differentiation is apparent.

Summary

The medial hemisphere wall of the anuran brain is characterized by two cellular masses, the primordial hippocampal formation and the precommissural septal area. These masses are separated by a medial cell free zone and by a medial limiting sulcus in the medial ventricular wall.

The primordial hippocampal formation is composed of three fairly distinct cell groups. These groups may be homologous to the dentate gyrus, Ammon's horn and the subiculum of higher forms. The hippocampal formation extends the full length of the hemisphere and is bounded ventrally, at all levels, by a medial limiting sulcus. Just rostral to the level of the anterior commissure it merges with the amygdaloid complex.

The precommissural septum is unusually prominent in the anuran brain and two large nuclear masses, a medial septal nucleus and a lateral septal nucleus, can be clearly defined. The medial septal nucleus has a medial and a dorsal position with reference to the lateral septal nucleus and is continuous ventrally with the septal part of the diagonal band of Broca. The lateral septal nucleus lies adjacent to the ventromedial wall of the ventricle. It is bordered medially by the fornix and the medial forebrain bundle, ventrally by the accumbens nucleus and caudally is continuous with the bed nuclei of the hippocampal and the anterior commissure.

Acknowledgments

The author expresses appreciation to Dr. Elizabeth C. Crosby for guidance and counsel during the course of this investigation. Gratitude for technical assistance and photography is expressed to Dr. H. R. Steeves, III., L. Hill, J. Green, E. Holt, R. Henning, and S. Hooton.

References

ANDY, O.J., H.STEPHAN: The septum of the cat. p. 1–84, Thomas, Springfield/Ill. 1964.

CROSBY, E.C., T.HUMPHREY: Studies of the vertebrate telencephalon. I. The nuclear configurations of the olfactory and accessory olfactory formations and the nucleus olfactorius anterior of certain reptiles, birds, and mammals. J. comp. Neurol., *71* (1939), 121–213.

HERRICK, C.J.: The morphology of the forebrain in amphibia and reptilia. J. comp. Neurol., *20* (1910), 413–457.

HERRICK, C.J.: The amphibian forebrain. IV. The cerebral hemispheres of Amblystoma. J. comp. Neurol. 43 (1927), 231–326.

HERRICK, C.J.: The amphibian forebrain. VI. Necturus. J. comp. Neurol., *58* (1933a), 1–288.

HERRICK, C.J.: The amphibian forebrain VIII. Cerebral hemispheres and pallial primordia. J. comp. Neurol. *58* (1933b), 737–759.

HOFFMAN, H.H.: The olfactory bulb, accessory olfactory bulb and hemisphere of some anurans. J. comp. Neurol., *120* (1963), 317–368.

JOHNSTON, J.B.: The morphology of the septum, hippocampus, and pallial commissures in reptiles and mammals. J. comp. Neurol., *23* (1913), 371–478.

OBENCHAIN, J.B.: The brains of the South American Marsupials, Caenolestes and Orolestes. Field Mus. of Nat. Hist., pub. 224, Zool. Series 14 (1925), 175–232.

RANSON, S.W.: Nonmedullated nerve fibers in the spinal nerves. Am. J. Anat. *12* (1911), 67–88.

RÖTHIG, P.: Beiträge. 5. Die Zellanordnungen im Vorderhirn der Amphibien. Verh. Kon. Akad. Wetensch. Amsterdam, Sec. 2, Deel. *17* (1912), 1–23.

RÖTHIG, P.: Beiträge zum Studium des Zentralnervensystems der Wirbeltiere 10. Über die Faserzüge in Vorder- und Zwischenhirn der Anuren. Jb. Morphol. Abt. 2, Z. mikr. anat. Forsch. *5* (1926), 23–58.

HENRY H. HOFFMAN Ph. D.
Department of Anatomy
University of Alabama Medical Center
1919 Seventh Avenue, South
Birmingham 3, Alabama, USA

The Nuclear Pattern of the Telencephalon of the Blacksnake, Coluber constrictor constrictor*

By J. H. CAREY, Hamburg, N. Y.

The brain of the blacksnake, *Coluber c. constrictor,* a species distributed throughout the Eastern United States, was used for the study. The material was fixed in 10% formalin and stained with thionin. This material is the property of the Department of Anatomy, University of Alabama and loaned to the author for study.

List of Abbreviations

Ac. b.	bulbus olfactorius accessorius	M. C.	mitral cell
Amg.	amygdala	N. Acc.	nucleus accumbens
CA	cornu ammonis	N. Bed.	bed nucleus
C. H.	commissura hippocampi	Ne.	neostriatum
D. V. R.	dorsal ventricular ridge	N. h. Sept.	nucleus hippocamposeptalis
D. V. R. C.	dorsal ventricular ridge, caudal part	N. olf. a.	nucleus olfactorius anterior, pars dorsalis
Glom.	glomerulus	N. Sept. Dor.	nucleus septalis dorsalis
G. P.	general pallium	N. Sept. Lat.	nucleus septalis lateralis
Gy. D.	gyrus dentatus	N. Sept. Med.	nucleus septalis medialis
Hip. a.	hippocampus, anterior continuation	N. Vom.	n. vomeronasalis
		O. V.	olfactory ventricle
Hip. d.	hippocampus, dorsal part	Pal. aug.	paleostriatum augmentatum
Hip. dm.	hippocampus, dorsomedial part	Pal. prim.	paleostriatum primitivum
Hyp.	hyperstriatum	Sub.	subiculum
L. F. B.	lateral forebrain bundle	T. olf.	tuberculum olfactorium
L. pir.	piriform cortex		

Results

Entering nerves. In the snake the telencephalon has three entering nerves, n. terminalis, fila olfactoria, and n. vomeronasalis. The latter two approach each olfactory bulb from its rostal area in the form of a ring and enclose the anterior tip in a cone. The vomeronasal fibers are dorsal and pass over the upper surface of the bulb and then turn downward entering the rostral tip of the accessory bulb medially. The anterior part of each accessory bulb (Fig. 1a) appears on the medial surface of each olfactory bulb immediately behind the tip of the olfactory formation. The accessory bulbs are large and lie close together

*) This study was financed in part by the Sister Elizabeth Kenny Neurological Center at the Children's Hospital, Buffalo, New York.

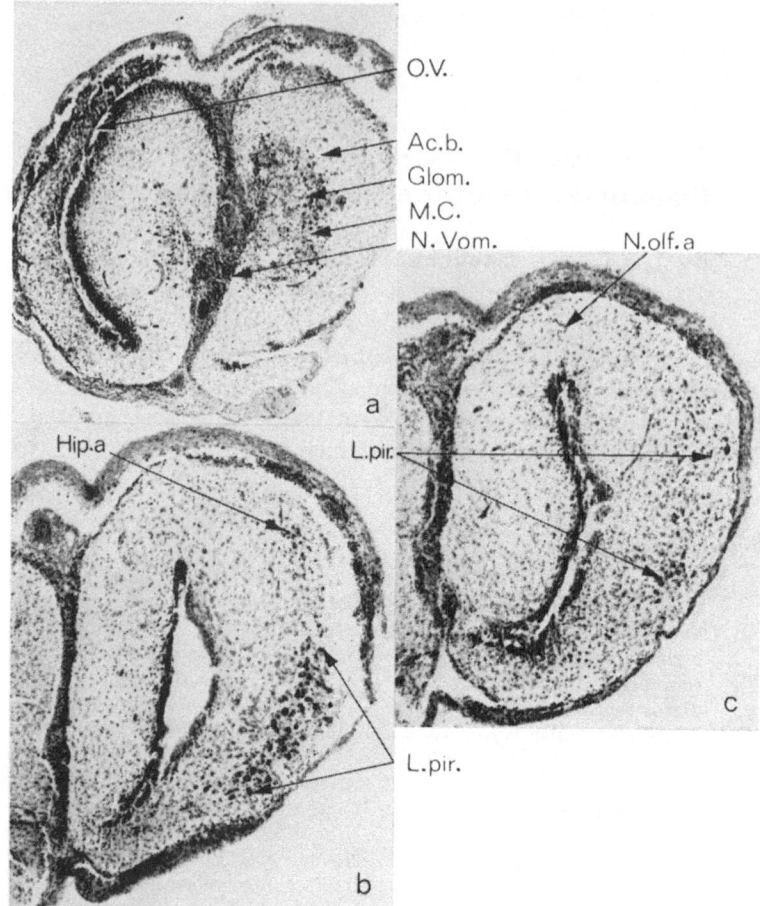

Fig. 1. Transverse sections through the telencephalon of the blacksnake, *Coluber c. constrictor*. Thionin stain. For abbreviations see page 73.

a) Medial location of the accessory bulb with its entering nervus vomeronasalis. 37,5 x.

b) Level of the olfactory crus. 60 x.

c) Level of the anterior part of nucleus olfactorius anterior. The thinning of the large cells in the prepiriform area is shown. 60 x.

just posterior to their rostral tips. There is a short olfactory crus attaching the bulb to the tip of the hemispheres.

Nucleus olfactorius anterior. The nuclear pattern of the telencephalon of snakes conforms with those of vertebrates above and below them in the evolutionary scale. The obliquity of the ring formed by the nucleus olfactorius anterior around the ventricle is due to the extension caudomedially of the olfactory formation. The ring is substituted for in the dorsomedial and dorsal areas of the hemisphere by the hippocampal formation, in the dorsolateral region by the general pallium, on the lateral side by piriform cortex, ventromedially by the septal areas, and ventrally by the tuberculum olfactorium.

The nucleus olfactorius anterior has been described for the snake (MEYER, 1892), the turtle (JOHNSTON, 1915), the alligator (CROSBY, 1917), and the chameleon (SHANKLIN, 1930). The undifferentiated gray of the nucleus olfactorius anterior appears rostrolaterally at the level of the posterior tip of the accessory bulb (Fig. 1 a). Its cells are oval in outline with large nuclei and there is a single large nucleolus in each cell.

Hippocampal formation. The available descriptions of the reptilian hippocampal formation are not in entire agreement with each other. The hippocampal formation has not been divided in *Sphenodon* (ELLIOT SMITH, 1919; HINES, 1923; CAIRNEY, 1926; DURWARD, 1930). SHANKLIN (1930) recognized three parts to the hippocampal area in the chameleon and designated a large dentate gyrus. In the rostrodorsal part of the nucleus olfactorius anterior of the blacksnake a sheet of dark staining cells gradually becomes visible (Fig. 1 b). This column of cells represents the anterior continuation of the hippocampus. Farther caudally this anterior continuation lies more medially (Fig. 3 a) and is continuous with the main hippocampal area. This latter area extends caudally through the dorsomedial and then the medial hemisphere wall to the caudal pole. There it lies adjacent to but is not continuous with the amygdala. It is difficult to subdivide it into the dorsomedial (gyrus dentatus) and the dorsal (cornu ammonis) regions which are so easily recognized in the alligator (CROSBY, 1917) and lizards (SHANKLIN, 1930). It seems probable that with the great reduction in the olfactory formation in this snake, there is a comparable decrease in the gyrus dentatus, which is represented only in the ventral tip of the dorsomedial portion of the hippocampal formation. The rest of the hippocampal area is representative of the cornu ammonis of other forms except for its lateral tip which may be compared with the subiculum of such forms.

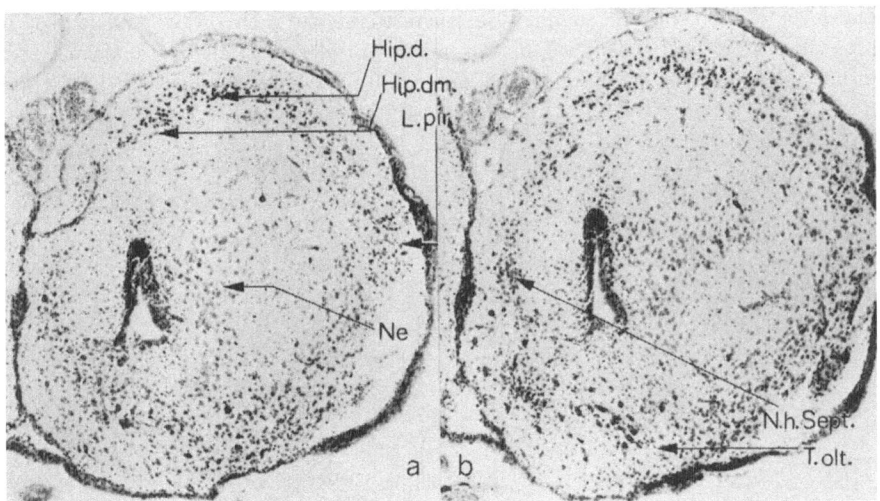

Fig. 2. Transverse sections. Thionin. 60x. For abbreviations see page 73.

a) Anterior tip of the hemisphere showing the rostral portion of the neostriatum and the dorsal and dorsomedial portions of the hippocampal formation.

b) Middle of the olfactory tubercle illustrating the anterior tip of the nucleus hippocamposeptalis. The clumping of the cells to form simple islands is visible in the tuberculum olfactorium.

Septum. In various reptiles – for example, in the turtle (JOHNSTON, 1915, 1923), the alligator (CROSBY, 1917), and the chameleon (SHANKLIN, 1930) – the septal areas have been divided secondarily. These observers recognized nuclei lateralis and medialis septi, a nucleus accumbens, a nucleus of the diagonal band, and nuclei of the anterior and hippocampal commissures. A so-called primordium hippocampi (now nucleus hippocamposeptalis) was recognized in the alligator (CROSBY, 1917), following the identification in the turtle (JOHNSTON, 1915). MEYER (1892) illustrated a medial parolfactory area, a nucleus of the diagonal band of BROCA and nuclei of the commissures in the snake.

In the septal area of the blacksnake, seven nuclei can be distinguished; (1) a lateral septal nucleus, (2) a medial septal nucleus, (3) a dorsal septal nucleus, (4) an accumbens nucleus, (5) a nucleus of the diagonal band of BROCA, (6 & 7) bed nuclei of the hippocampal and the anterior commissures, and (8) nucleus hippocamposeptalis, actually intermediate between the septum and the hippocampus (see Figs. 2–4).

At the posterior end of the tuberculum olfactorium, cells of the nucleus hippocamposeptalis appear below the tip of the dorsomedial portion of the hippocampal formation (Fig. 2b). They are continuous with the anterior continuation of the hippocampal area and are separated from the septal area by a cell free zone. This nucleus hippocamposeptalis has a position similar to the area termed primodium hippocampi in the turtle by JOHNSTON (1915) and in the alligator by CROSBY (1917) but is better developed somewhat farther caudalward in the snake. It is comparable to the nucleus of similar name described in higher forms. Nuclei lateralis and medialis septi (Figs. 3b, 4b) extend through the septum, the nucleus medialis septi being larger than the lateral one. The nucleus lateralis septi is not sharply demarcated from nucleus accumbens at all levels. A nucleus dorsalis septi is well developed in the caudal half of the septal area. Nucleus accumbens septi (Fig. 3b) extends as far forward as the nucleus lateralis septi but becomes indistinct at the level of the anterior commissure where it merges with the bed nucleus of the commissure. It is well developed in the snake. A nucleus triangularis has not been identified. The nucleus of the diagonal band of BROCA is continuous with the ventral

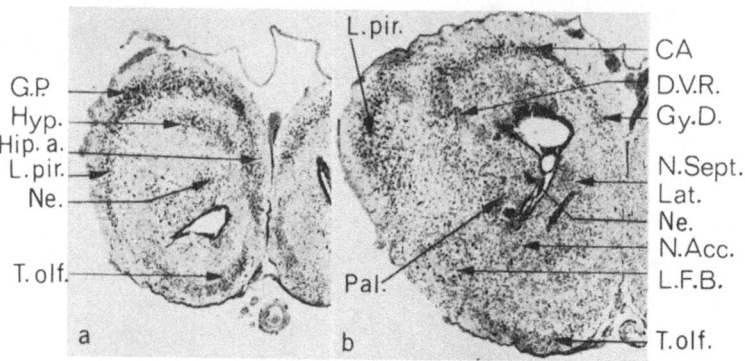

Fig. 3. Transverse sections. Thionin. 20x. For abbreviations see page 73.

a) The section passes through the tuberculum olfactorium, the anterior tip of the general pallium and the hyperstriatum.

b) The telencephalon at the level of the septum. The anterior part of the dorsal ventricular ridge and the nucleus septalis lateralis are present. Two divisions of the hippocampal formation, cornu ammonis and gyrus dentatus, can be recognized.

tip of the medial septal nucleus and extends over the caudal border of the tuberculum olfactorium and laterally toward the ventral tip of the piriform area and the neostriatum (Figs. 2b, 3b). The bed nuclei of the hippocampal and anterior commissures can be distinguished in sections through the caudal part of the nucleus lateralis septi (Fig. 4c). They are similar in position to the comparable nuclei in the alligator and the turtle.

Tuberculum olfactorium. The tuberculum olfactorium (Figs. 2a, 2b) begins behind the crus replacing the ventral part of nucleus olfactorius anterior and extends to the preoptic region from which it is separated by the diagonal band of BROCA and its nucleus. At its widest extent it forms a crescent which can be followed from the nucleus medialis septi to the ventral tip of the piriform lobe. The tuberculum has (1) a zonal layer, (2) a cortical layer, and (3) a polymorph layer as does the corresponding area in the turtle (JOHNSTON, 1915) and the alligator (CROSBY, 1917). The cells of the cortical layer (Fig. 2b) are collected into small groups. Such a group does not have a hilum but might suggest a very simple island. The polymorph layer of the tuberculum is indistinctly separated from the nucleus olfactorius anterior and, at some levels, from the paleostriatum.

Piriform cortex. The piriform cortex has been described in *Sphenodon* (ELLIOT SMITH, 1919; HINES, 1923; CAIRNEY, 1926; DURWARD, 1930), in the turtle (JOHNSTON, 1915), in the alligator (CROSBY, 1917), and in the lizard (ARIENS KAPPERS, HUBER, and CROSBY, 1936). Both JOHNSTON and CROSBY discussed the relationship of the piriform cortex to its forward continuation, the nucleus of the lateral olfactory tract or prepiriform area. MEYER (1892) called the area in the snake the lateral mantle zone.

In the blacksnake such a prepiriform area (Figs. 1b, 1c) is present. Rostrally the piriform lobe is adjacent to the hippocampal formation. However, it is separated from the hippocampus for much of the extent of the hemisphere by the general pallium but caudoventrally the piriform cortex and hippocampus again approach each other.

Striatum. The striatal areas in reptiles have been variously designated. KUHLENBECK (1924, 1925, 1926) divided the area into a nucleus basalis (paleostriatum), a nucleus basalis accumbens, a nucleus epibasalis and a nucleus centralis. ELLIOT SMITH (1919) described paleostriatal and hypopallial areas in *Sphenodon* and, in the same form, DURWARD (1930) recognized a somatostriatum and an olfactostriatum. In the turtle (JOHNSTON, 1915) separated the dorsal ventricular ridge into an anterior portion, related to overlying general cortex, and a posterior segment, related to the amygdaloid area. CROSBY (1917) divided the area in the alligator into ventrolateral, intermediolateral and dorsolateral areas. WARNER (1931) designated a neostriatal and a paleostriatal area for the rattlesnake, *Crotalus atrox.*

In the blacksnake, the striatal areas occupy the basal portion of the lateral wall of the hemisphere. Three major regions are recognizable, (1) a paleostriatum, (2) a neostriatum, and (3) a poorly developed hyperstriatum. In the anterior part of the hemisphere, a group of cells appears dorsolateral to the ventricle which represents the anterior end of the neostriatum (Fig. 2a). As the general pallium thickens (Fig. 3a), the neostriatal area lies medial. Beneath the thickening a column of cells forms a ridge, the dorsal ventricular ridge of JOHNSTON (1915) or the anterior hypopallial area of ELLIOT SMITH (1919). Medial to the ridge and dorsal to the ventricle is a group of cells that is related to the hyperstriatum (Fig. 2a). In the lateral wall of the hemisphere a few large cells adjacent to the lateral forebrain bundle suggest a paleostriatum primitivum (Fig. 4a). Dorsal to the lateral forebrain bundle is a group of smaller cells, the paleostriatum augmentatum.

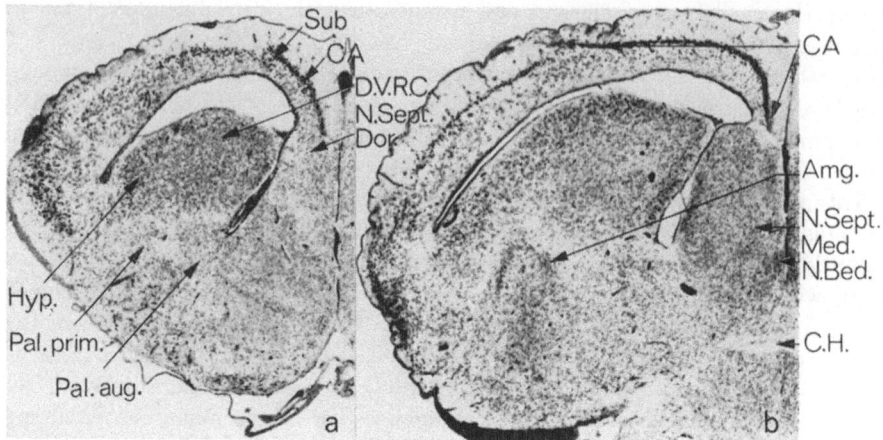

Fig. 4. Transverse sections. Thionin. 20x. For abbreviations see page 73.

a) The level of the precommissural septum. The subiculum, caudal part of the dorsal ventricular ridge, and nucleus septalis dorsalis are illustrated in the medial wall of the hemisphere. The three parts of the striatum are visible in the lateral wall.

b) The telencephalon at the level of the hippocampal commissure. In the septal area the nucleus septalis medialis and the bed nuclei of the hippocampal and anterior commissures are illustrated in the ventromedial wall of the hemisphere. The anterior tip of the amygdaloid complex is visible in the caudoventral part of the lateral hemisphere wall.

General pallium. MEYER (1892) divided the cortex into medial, lateral and dorsal mantle zones. His figures show the area designated as general pallium by later observers although he has not labelled it as such. CAIRNEY (1926) and DURWARD (1930) recognized a general cortex in *Sphenodon*. In the turtle, JOHNSTON (1915) designated the area between the piriform cortex and the hippocampal cortex as a general pallium and showed its relation to the anterior part of the dorsal ventricular ridge (hypopallium anterius). CROSBY (1917) described the area in the alligator where the hypopallium anterius is present but much smaller than in the turtle or snake. In the blacksnake the general pallium is recognizable in the anterior part of the hemisphere as a group of cells interposed between the hippocampal and piriform areas (Fig. 3a). This general pallium borders either side of the anterior hypopallial ridge and overlies it (Fig. 3b).

Amygdala. In general the amygdaloid complex in the blacksnake is similar to that discussed in the literature for other reptiles. ELLIOT SMITH (1919) suggested that the hypopallium posterius is related to the amygdaloid area. JOHNSTON (1913, 1923) described the hypopallium posterius (caudal dorsal ventricular ridge) and related it to the amygdaloid area. CAIRNEY (1926) divided the amygdaloid complex into (1) a posterior hypopallial ridge, (2) a nucleus anterior amygdalae (nucleus of the lateral olfactory tract of JOHNSTON, 1923), and (3) a nucleus medialis amygdalae (large celled nucleus of the turtle, JOHNSTON, 1923). CROSBY (1917) related a portion of the posterior part of the dorsolateral area of the alligator to the amygdaloid area. CURWEN (1935) divided the amygdaloid area into seven nuclear areas in *Tupinambis nigropunctatus*. She suggested that the medial group was older phylogenetically and had olfactovisceral connections.

The posterior hypopallial ridge of the snake, regarded as part of the amygdaloid complex, is a curved band of cortex-like cells forming a ventricular eminence in the caudal part of the snake hemisphere. It is connected rostrally with the ventral tip of the piriform cortex by a strand of less compactly arranged cells. This ridge continues far caudally into the caudal pole of the hemisphere where it forms an oval ring of cortex-like gray (Figs. 5a, 5b). Ventral to it but appearing at somewhat more caudal levels than the rostral end of the posterior hypopallial ridge is a mass of gray (Fig. 5a). This gray occupies the caudoventral part of the lateral hemisphere wall. It is secondarily divisible into several nuclear groups, which are similar to those described for the amygdaloid area in other reptiles but not entirely comparable to them. They will not be analyzed further for the blacksnake until their fiber connections can be studied.

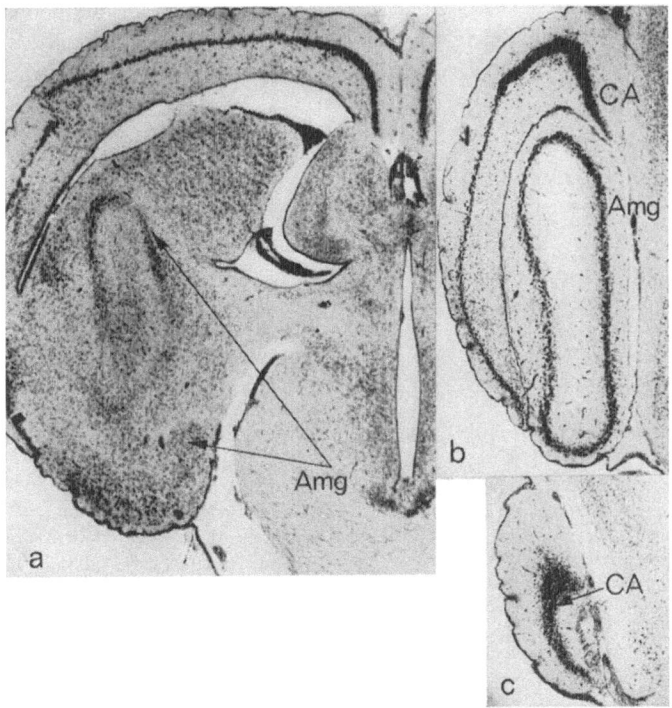

Fig. 5. Transverse sections through the caudal part of the telencephalon. Thionin. 20x. For abbreviations see page 73.

a) The anterior end of the amygdaloid ridge is shown.

b) The level of the caudal part of the amygdaloid ridge showing the relationship to the cornu ammonis.

c) Caudal tip of the cornu ammonis and the amygdala.

Summary and Conclusions

The nuclear pattern in the telencephalon of the blacksnake, *Coluber c. constrictor*, is similar to those in the vertebrates above and below reptiles in the evolutionary scale. Certain relations described for the blacksnake may be recapitulated.

The bulbus olfactorius accessorius is larger than the bulbus olfactorius.

The nucleus olfactorius anterior is short and has an oblique tilt.

The tuberculum olfactorium has three layers and there are small islands of Calleja.

The septum is well developed.

The cornu ammonis is relatively large in comparison with the gyrus dentatus. There is a small lateral area in the hippocampal formation which has been compared with the subiculum.

The piriform cortex has a large celled anterior extension into the posterior crus.

There is a well developed striatal complex differentiable into a paleostriatum, a hyperstriatum, and a neostriatum.

The general pallium is a thin sheet over the dorsolateral portion of the hemisphere and is associated anteriorly with the dorsal ventricular ridge.

The amygdala is well developed and has a dorsal caudal part and a more ventral part. The dorsal part is represented in the posterior part of the dorsal ventricular ridge. No definite infolding of the piriform cortex was observed in relation to the posterior hypopallial ridge.

References

ARIENS KAPPERS, C.U., G.C. HUBER, E.C. CROSBY: Comparative Anatomy of the Nervous System of Vertebrates, including Man. Macmillan New York 1936.
CAIRNEY, J.: J. comp. Neurol. 42 (1926), 111.
CROSBY, E.C.: J. comp. Neurol. 27 (1917), 325.
CURWEN, A.O.: Anat. Rec. 61 (1925), 13.
DURWARD, A.: J. Anat. 65 (1930), 8.
ELLIOT SMITH, G.: J. Anat. 53 (1919), 271.
HINES, M.: J. comp. Neurol. 35 (1923), 483.
JOHNSTON, J.B.: J. comp. Neurol. 23 (1913), 371.

JOHNSTON, J.B.: J. comp. Neurol. 25 (1915), 393.
JOHNSTON, J.B.: J. comp. Neurol.35 (1923), 337.
KUHLENBECK, H.: Folia Anat. Japon. 2 (1924), 325.
KUHLENBECK, H.: Anat. Anz. 60 (1925), 33.
KUHLENBECK, H.: Folia Anat. Japon. 4 (1926), 111.
MEYER, A.: Z. wissensch. Zool., 55 (1892), 63.
SHANKLIN, W.M., Acta zool. 11 (1930), 425.
WARNER, F.J.: Kon. Akad. Wetensch. Amsterdam, Proc. sect. 34 (1931), 1156.

JOSHUA H.CAREY, M.D.
University of Buffalo
229 Main Street
Hamburg, N.Y. 14075, USA

A Study of the Hippocampal Formation in the Opossum, Didelphis virginiana

By E. G. HAMEL, JR., Birmingham, Ala.

The medial telencephalic wall is composed of cortical and nuclear areas forming part of the rhinencephalic ring or limbic lobe of the cerebrum. This lobe includes the septum, the hippocampal formation, and the amygdala, the organization of which in the marsupial is quite simple but sets a pattern for higher mammalian forms. In the marsupial, this pattern remains relatively uncomplicated because the rostrocaudal extent of the dorsal portion of the ring is not interrupted by the corpus callosum. The present paper deals with the hippocampal formation of the opossum, *Didelphis virginiana*. This area, along with the amygdala and the septal area, is also under investigation in certain kangaroos and South American marsupials, with the findings to be presented at a later date.

Literature. The hippocampal formation of the marsupials has been considered by ZIEHEN (1898), ELLIOT SMITH (1894), RETZIUS (1898), JOHNSTON (1913), HERRICK (1924), and LOO (1931). The hippocampus of *Caenolestes* and that of *Orolestes* were described by OBENCHAIN (1925). In his study of the cortex of the opossum, GRAY (1924) described the areas adjacent to the hippocampal formation. Reference to special details from the literature will be made in the text.

Materials and Methods. The descriptive information in this presentation is drawn from the collection of opossum brains at the University of Alabama. These include transverse, horizontal, and sagittal cut serial sections of brains which have been stained by thionin, WEIGERT and pyridine silver methods. Several special preparations have been made to demonstrate the hippocampus in cross section throughout its entire length, and some GOLGI material is available. Certain fiber connections have been demonstrated by MARCHI preparations of brains after various cerebral lesions.

Description of Normal Hippocampal Formation

Classically, the hippocampal formation of marsupials is divided into an anterior continuation of the hippocampus, a gyrus dentatus, a cornu ammonis, and a subiculum. The last three areas make up the major part of the formation for most of its extent. The hippocampal formation in the opossum can also be divided into three general rostrocaudal regions: an undifferentiated anterior continuation which is related to the anterior olfactory nucleus and the septum, an intermediate portion which occupies most of the medial hemisphere wall and shows the classical pattern, and an inferior portion which is related to the amygdaloid complex and is, in a certain sense, continuous with it. It is this rostrocaudal arrangement which will be considered in the present account.

Rostral Region. The anterior continuation of the hippocampal formation appears rostrally in serial sections as a mass of cells at the dorsomedial aspect of the olfactory

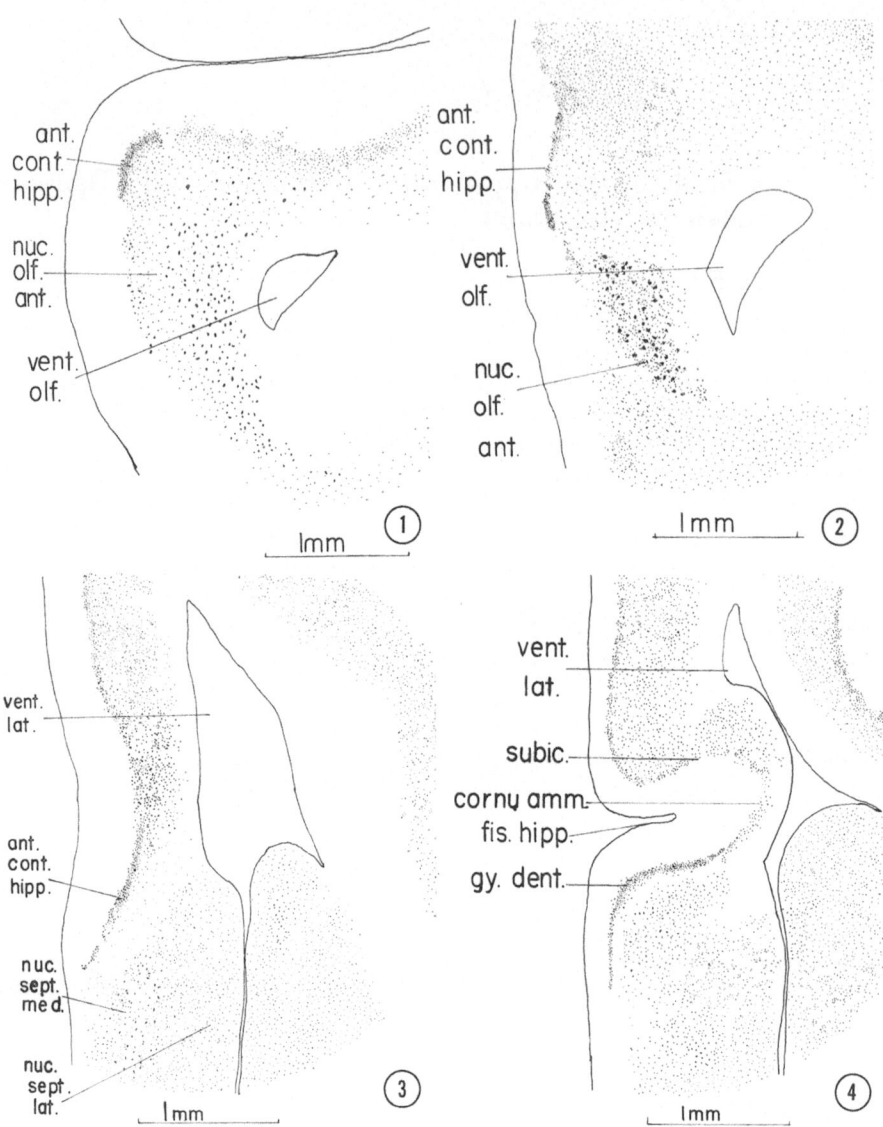

Figs. 1–4. All figures are from cross sections of an opossum brain stained with thionin. Fig. 1 at base of olfactory bulb, Fig. 2 where olfactory peduncle meets the cerebrum, Fig. 3 several sections caudal to Fig. 2, Fig. 4 at level of rostral part of the hippocampal fissure.

peduncle (Figs. 1–4). The anterior olfactory nucleus (Figs. 1, 2) forms a ring of gray (Herrick, 1924) about the ventricle at this level and the anterior continuation differentiates as a portion of this nucleus which has been displaced toward the surface of the peduncle. Gray (1924) suggested a resemblance between the anterior continuation of the hippocampus and the gyrus dentatus. Our material indicates that the anterior continuation is morphologically continuous posteriorly with the gyrus dentatus. The cells of this

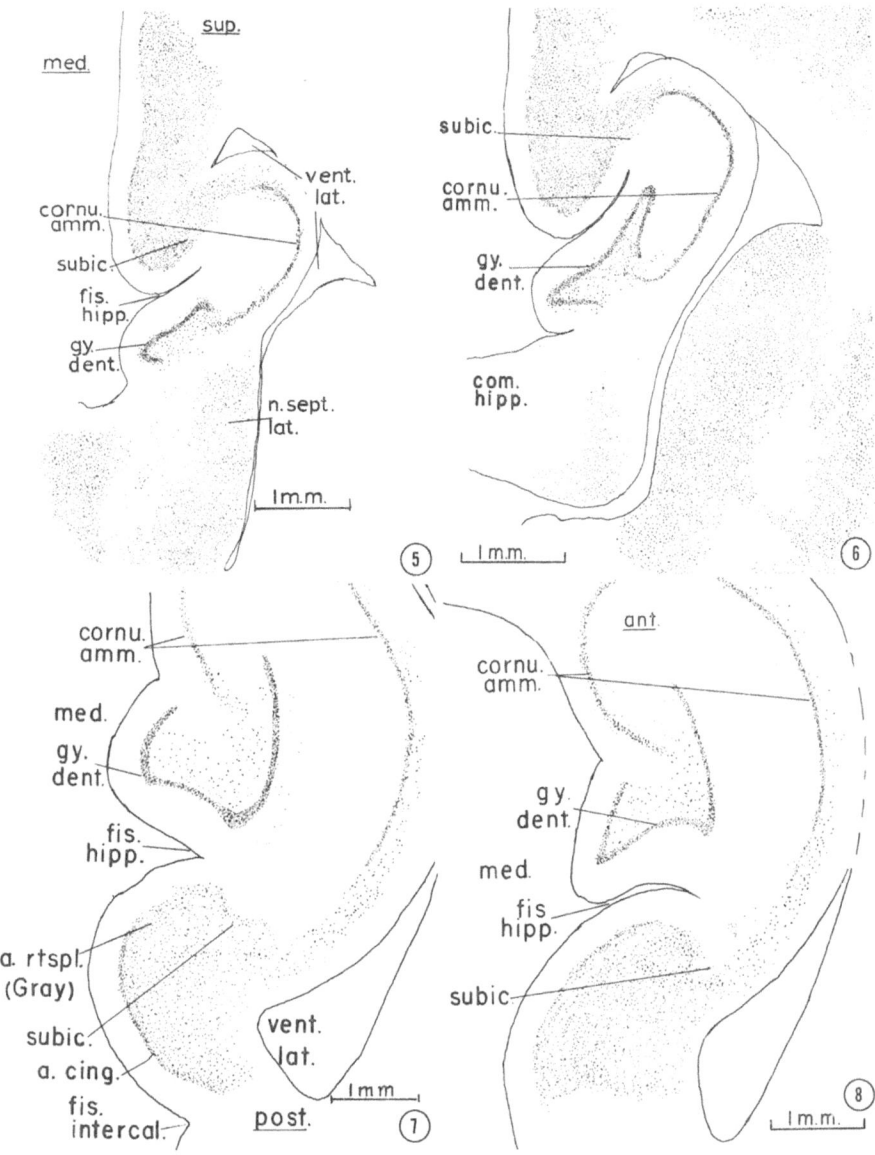

Figs. 5–8. All figures are drawn from cross sections of the opossum hippocampal formation stained with thionin. Fig. 5 is at the level of rostral end of the hippocampal commissure. Fig. 6 is through the caudal part of the hippocampal commissure. Fig. 7 is at the dorsocaudal part of the hemisphere, diagonal to the longitudinal axis of the brain. Fig. 8 is inferior to the lowest point of the hippocampal fissure.

region were described by HERRICK (1892) as mostly pyramidal, with which the present observations agree.

Intermediate Region. The intermediate region of the hippocampal formation (Figs. 5–8) extends from the caudal end of the anterior continuation to a plane through the inferior

extent of the hippocampal gray. In this region the hippocampal fissure and the hippocampal formation begin to flatten (Figs. 1–5).

A true cellular differentiation into a gyrus dentatus, a cornu ammonis and a subiculum
becomes obvious at the rostral end of the intermediate region with the definition of a
hippocampal fissure about which the various parts of the formation are arranged (Fig. 4).
Indeed, only with the formation of the medial interposition at the level of the hippocampal commissure is it possible to distinguish completely the area of merging between
the cornu ammonis and the gyrus dentatus (Fig. 5). The intermediate region is the area of
greatest development of the various portions of the hippocampal formation in the opossum
(Figs. 4–8).

The laminar pattern of the hippocampal formation in the intermediate region in the
opossum is similar to that described for other mammalian forms (Figs. 9–12). Certain
variations in the degree of development of some layers, especially in the cornu ammonis,
have been observed.

The dentate gyrus is a very prominent structure in this intermediate region. It demonstrates molecular, granular, and polymorph layers, the last showing some regional
variations (Fig. 12). A transverse section of the hippocampal formation at the point where
it begins to curve downward toward the amygdala shows changes in the dentate gyrus.
The midportion of the gyrus becomes concave in a direction opposite to the trough formed
by the entire structure (Figs. 8, 9). The cells of the medial or free end of the pyramidal
layer have a looser arrangement than those of the lateral part (Fig. 12c).

The appearance of the cornu ammonis in the intermediate region of the hippocampal
formation, as seen under low powers of the microscope, is illustrated in Fig. 9 (and 11).
The layers described by CAJAL (1909) can be defined although there is a paucity of cells
in both the stratum oriens and the stratum lacunosum (Fig. 11a, b). Variations in the cell
type and in the density are also evident along the extent of the pyramidal layer of the
cornu ammonis. The cells nearest the dentate gyrus (Fig. 10c) are arranged in a broad
loose band which gradually becomes more compact, finally narrows to a single row of
cells, and then ends abruptly at the subiculum (Fig. 12c).

The subiculum (Fig. 12s) extends from the definitive cornu ammonis to the retrosplenial area of GRAY from which it is poorly delimited. The subiculum has a zonal or
plexiform layer and an underlying fairly wide, somewhat irregular mass of cells, interrupted and dispersed by fibers between the subcortical white matter and the concave aspect of
the cornu ammonis (Fig. 12c). GRAY (1924) recognized a comparable arrangement. The
cells of the subiculum are pyramidal and spindle shaped in type.

Nearest the cornu ammonis, the cells of the subiculum are the large pyramidal cells
typical of that area (Fig. 12c), but they are rapidly displaced by smaller pyramids and
spindle shaped cells toward the piriform cortex (Fig. 12p). These two parts correspond to
the area subicularis and area presubicularis of GRAY (1924).

Inferior Region. The inferior region of the hippocampal formation extends downward
in the hemisphere to lie in relation with the medial aspect of the amygdala (Fig. 15). In
this inferior region the three major subdivisions – the gyrus dentatus, the cornu ammonis
and the subiculum – have representation, but gradually diminish in size (Figs. 14, 15).
The cornu ammonis extends ventrally into relation with the amygdala. This extension
begins at the most ventral tip of the cornu ammonis (Fig. 16b). From this point, the
pyramidal layer of the cornu can be traced as a thin strand of cells into the medial portion
of the amygdala. In more caudal sections (Fig. 16 V), it merges with the basal accessory II

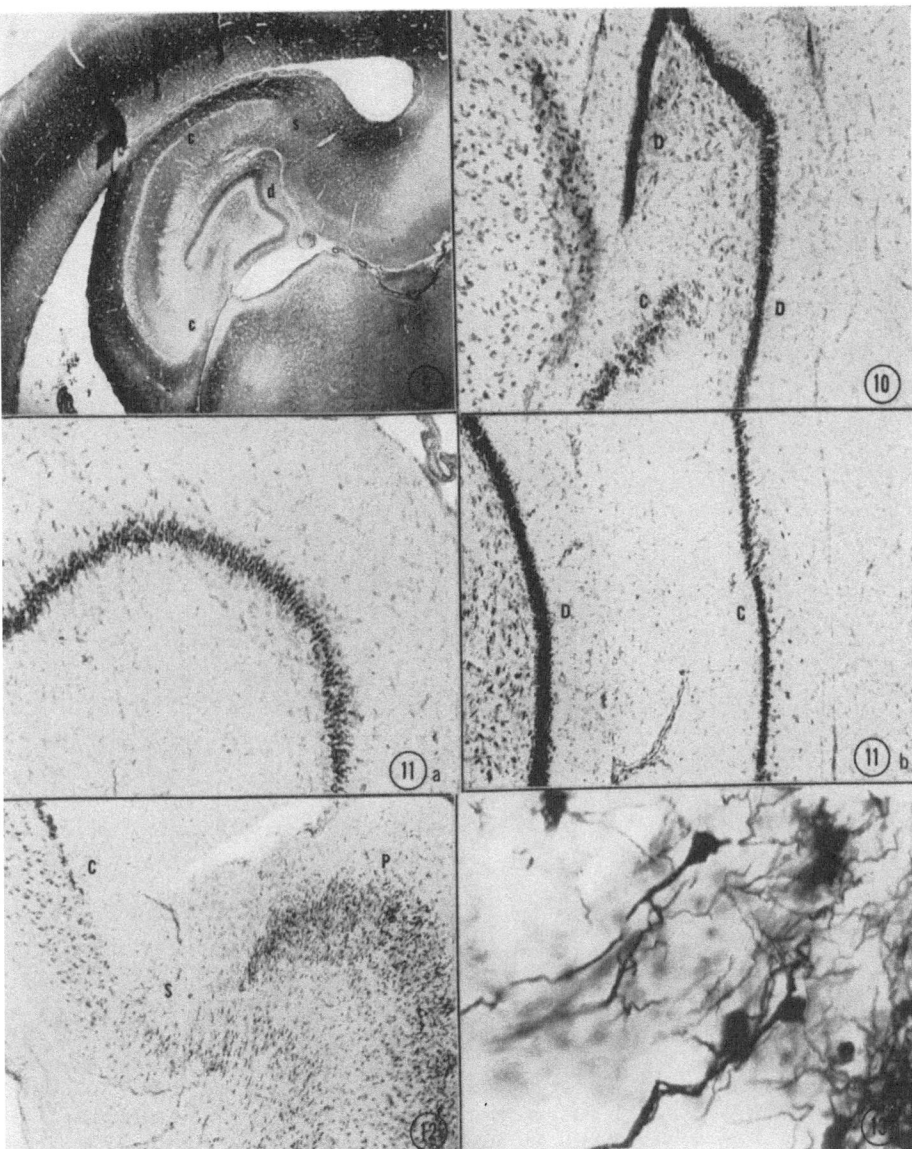

Figs. 9–13. Fig. 9. Photograph of a cross section of opossum brain through the mid-portion of the hippocampal formation. Pyridine silver preparation (35 x). Fig. 10. Cross section of dentate gyrus, mid-portion of hippocampal formation, thionin preparation. Area indicated by C is the medial end of the cornu ammonis (100 x). Fig. 11a cells of the cornu ammonis at point of greatest curvature, 11b is also cornu ammonis (C) showing all layers in a thionin preparation. Cells of dentate gyrus (d) at left (100 x). Fig. 12, section through subiculum in a thionin preparation; C is cornu ammonis, S is subiculum, P is piriform (retro-splenial) cortex (100 x). Fig. 13 is a Golgi preparation of cells of the pyramidal layer of cornu ammonis (130 x).

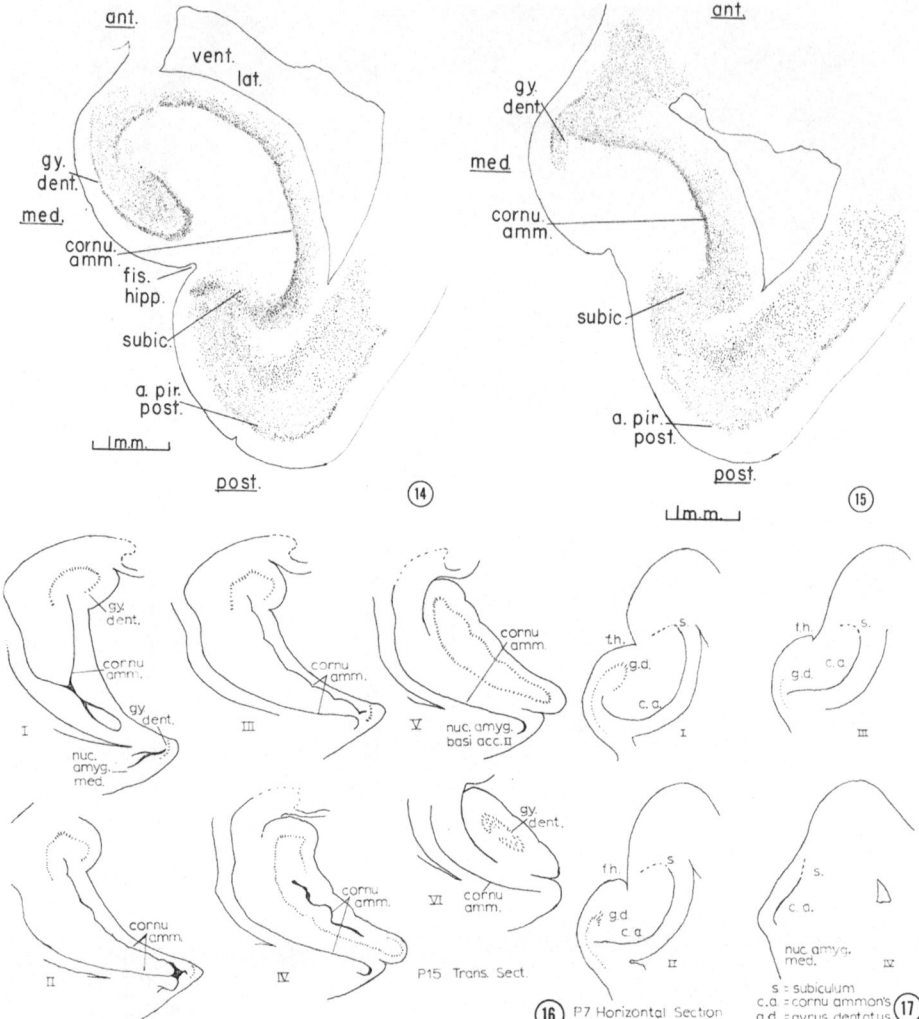

Figs. 14–17. Figs. 14 and 15. Cross sections through the inferior part of the hippocampal formation. Drawings are made from horizontal sections of the cerebrum. Fig. 15 is approximately 150 microns inferior to figure 14. Thionin preparation. Fig. 16. Outline drawings from cross sections of the opossum brain showing hippocampal amygdala relation. Drawing I is rostral, VI is caudal. Fig. 17. Hippocampal–amygdala relations as seen in horizontal section. Fig. 17 I corresponds approximately to Fig. 14.

nucleus (van der Sprenkel, 1927). The most ventrocaudal portion of the gyrus dentatus ends blindly and definitively except for a tiny band which parallels the portion of the cornu ammonis just described as this latter projects to the amygdala and merges with it (Figs. 16, 17 II). These relations of the cornu ammonis and the dentate gyrus with the amygdala are illustrated in Fig. 21 of the van der Sprenkel paper (1927), but are neither described nor considered in that paper. Only by following the cells of this mass toward the recurved rostral tip of the temporal pole is it possible to see the slight connection with the medial amygdaloid nucleus. The subiculum in the inferior region of the hippocampal formation decreases in size and gradually merges with adjacent cortex of the piriform lobe (Fig. 17 IV).

Certain Experimental Results

The connections of the cerebral hemisphere of the opossum are being studied with the aid of MARCHI and NAUTA degeneration techniques. Certain selected findings relating especially to the septal area and the hippocampal formation are presented here to complete the consideration of the morphology of these structures.

Cortico-hippocampal Connections. Small areas of the cerebral cortex were removed by suction in three opossums. The brains of these animals were removed and stained by the MARCHI technique two weeks after the surgery. Examination of these specimens shows degenerated fibers extending from the lesion into the subcortical white matter where they pass superior to the uppermost edge of the ventricle and arch ventrally into the cingulate area and the subiculum (Fig. 18). Most of the affected fibers end in these areas, but a few extend into the cornu ammonis, where scattered MARCHI granules can be seen amongst the pyramidal cells. Presumably, there is a secondary discharge to the cornu ammonis from the cingulate and the subiculum areas.

Hippocampal Connections. A large lesion was made in the brain of one opossum by removing the cerebral cortex from the lateral aspect of the hemisphere several millimeters

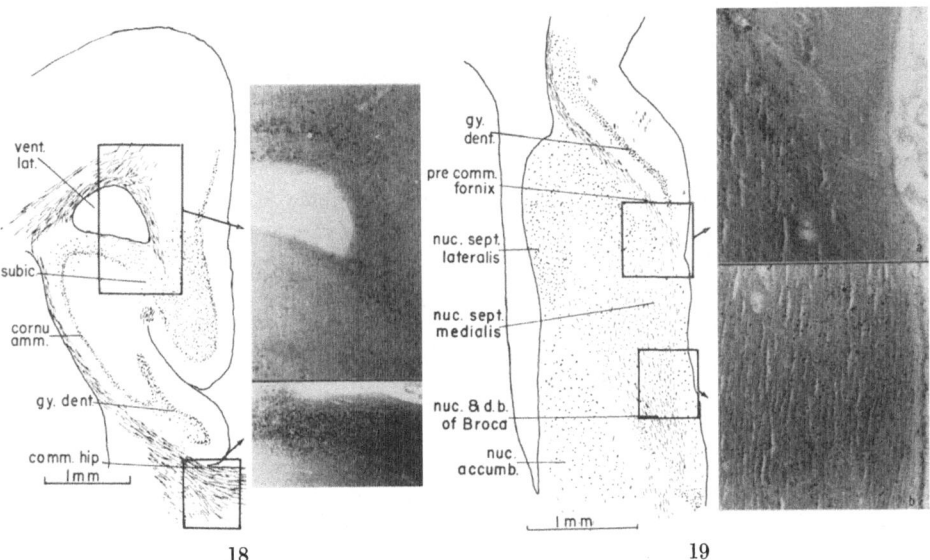

18   19

Figs. 18–19. Fig. 18. The drawing at left establishes the regions from which the photographs of MARCHI preparations are made. Degenerated fibers arch over the ventricle into the cingulate area (a) and also pass into the hippocampal commissure. (b), after lesion of cortex. Fig. 19 MARCHI degeneration granules extending into septal areas indicated by drawing, after lesion of cornu ammonis.

below its superior margin. The lesion extended medially into the lateral ventricle, involving the lateral aspect of the cornu ammonis and, more caudally, damaging the fimbria and the fornix. In the more rostral sections through the hemisphere, in this MARCHI series, degenerated fibers can be traced through the fimbria and then ventrally by the precommissural fornix into the septum, chiefly into the medial septal nucleus (Fig. 19). Similar granules can also be seen at more rostral levels of the anterior continuation of the hippocampus. These represent connections from more caudal parts of the hippocampal formation or from the cingulate gyrus into this most anterior part of the hippocampus.

In the series described in the last paragraph, damage to the fornix just above the anterior commissure produced degeneration of some fibers in the postcommissural fornix. Such degenerated fascicles can be followed ventralward into the hypothalamus. There, a few fibers reach the mamillary body of the same side, and a very few decussate in the supramamillary commissure to enter the midbrain. In the transverse series, MARCHI granules can also be seen posterior to the lesion in the fimbria and in the cornu ammonis. In sections which are posterior and inferior to the lesion, MARCHI granules are present in the cingulate area and in the posterior piriform cortex. These degenerated fibers reach this area through the cingulum, for its fibers were likewise interrupted by the lesion.

Bulbar Connections. In one opossum, the olfactory bulb was exposed from above and approximately the upper third of the bulb was removed by suction. The accessory bulb and the anterior olfactory nucleus were not involved in this lesion since they were found to be uninjured upon histologic examination. In MARCHI material made from this brain, degenerated fibers can be traced caudalward in the lateral olfactory tract (Figs. 20, 22). These are reduced gradually as the fibers end in prepiriform and piriform cortices, where the degeneration granules are present in the pyramidal cellular layers. In the more caudal sections of the piriform cortex, MARCHI granules occur in its cellular layer even though they are not in the superficial fiber layer. As the olfactory bulb joins the cerebral peduncle, degenerated fibers leave the ventral part of the lateral olfactory tract to pass dorsomedially to the anterior continuation of the hippocampus. These fibers are components of the medial olfactory tract. MARCHI granules can be seen amongst the cells of this part of the hippocampal formation (Figs. 23, 24), as well as among the cells of the dorsal part of the anterior olfactory nucleus. There are also some MARCHI granules present among the cortical cells of the extreme rostal end of the cerebral hemisphere. Their origin is not known.

Relatively few degenerated fibers can be seen joining the intermediate olfactory tract to pass into the anterior commissure (Fig. 21). Such fibers as are present suggest a very small, but direct, interbulbar connection in the opossum.

Discussion and Summary

The septal area and the hippocampal formation of the opossum compose a morphologically and functionally related system in the medial hemisphere wall. As such, they are a part of the limbic lobe, a ring of gray in the wall of the telencephalon. In the present account, the hippocampal formation has been considered in the light of three rostrocaudal divisions: the anterior continuation of the hippocampus, the intermediate region which presents a classical pattern, and a caudal or inferior portion related to the amygdala.

From its position at the dorsomedial aspect of the anterior olfactory nucleus, the anterior continuation of the hippocampus, itself an ontogenetic derivative of the anterior olfactory nucleus, extends into relation with the definitive dentate gyrus at more caudal levels. In MARCHI preparations, fibers of bulbar origin have been traced into this anterior continuation of the hippocampal formation from both the medial and the dorsomedial olfactory tracts. This corroborates the studies of ELLIOT SMITH (1894) and HERRICK (1924) which had been made on normal material. These findings indicate a common olfactory function for this line of gray matter. The MARCHI preparations show a moderate inter-bulbar connection through the intermediate olfactory tract and the anterior commissure. Our preparations do not permit such a detailed analysis of bulbar projections as were described by LOHMAN a. LAMMERS (1963) for the guinea pig. However, the absence of any obvious involvement of the anterior olfactory nucleus in our lesions indicates that the degenerated fibers in the anterior commissure arise largely from the bulbar formation in this form.

The intermediate portion of the hippocampal formation is the region of its greatest development and differentiation. The division into subiculum, cornu ammonis, and gyrus dentatus is typical of mammalian forms. Since the opossum has no corpus callosum, the hippocampal formation lies dorsal to the diencephalon rather than in the temporal pole of the hemisphere. The minimal development of neocortex also has left the hippocampal

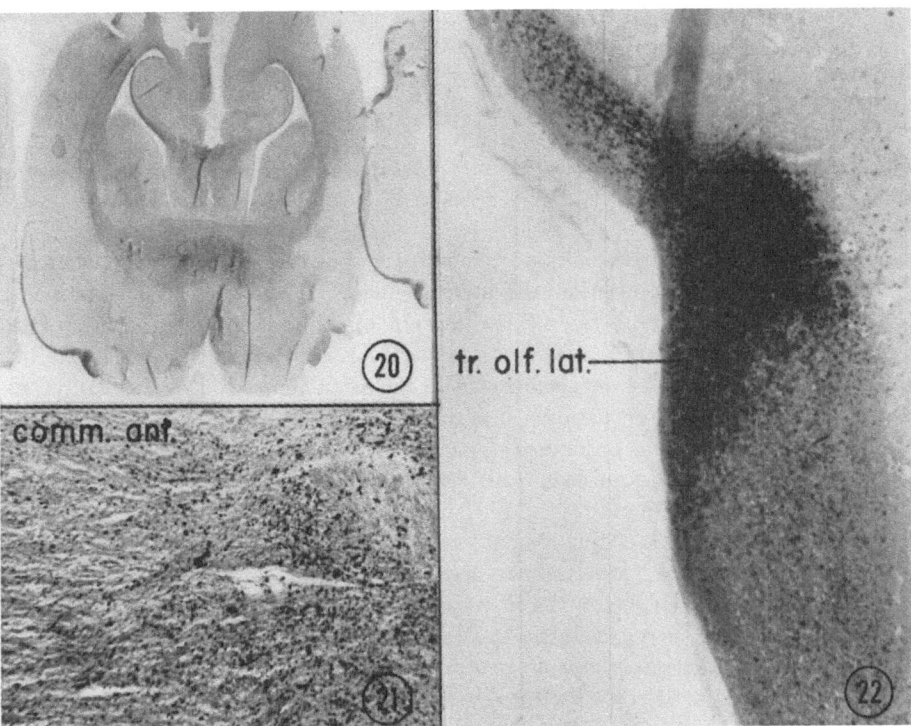

Fig. 20. Cross section of opossum brain in MARCHI preparation after bulbar lesion. High power view of anterior commissure of brain in Fig. 21. Fig. 22. High power view of lateral olfactory tract of brain in Fig. 20.

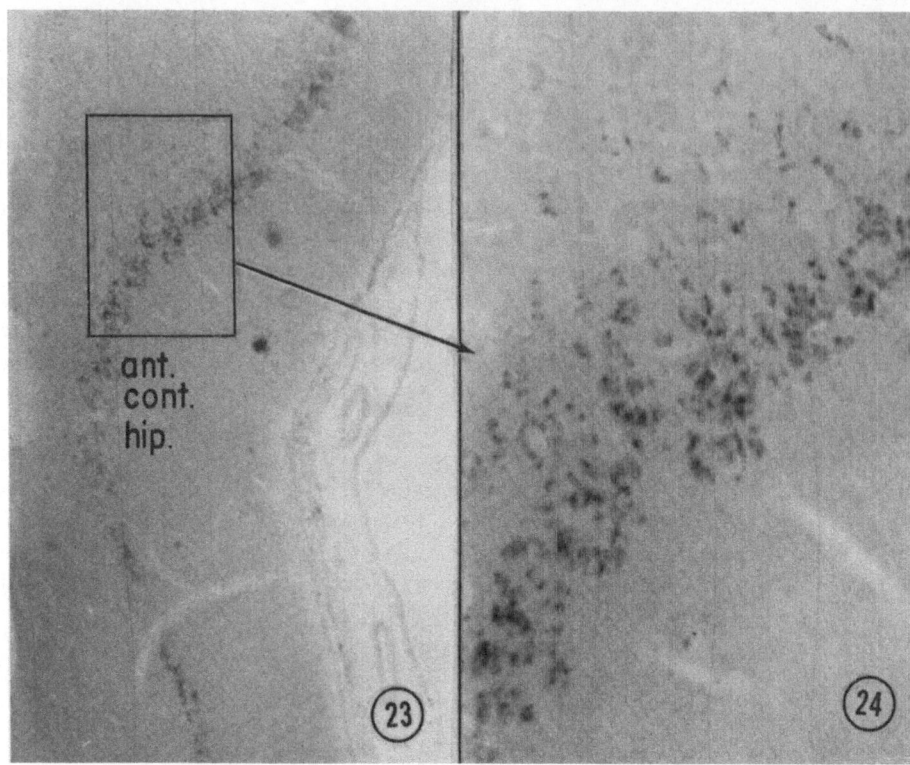

Fig. 23 and 24 are low and high power pictures of MARCHI granules in the anterior continuation
of the hippocampus.

formation in a superficial position, as opposed to its position in other mammals where
this area has been rolled deeper into the lateral ventricle. The MARCHI material presented
indicates that a corticohippocampal relationship of the intermediate portion of the
hippocampal formation to the neocortex is probably established chiefly in the subicular
and cingulate areas, but a few of the fibers enter the cornu ammonis directly. SCHNEIDER,
CROSBY a. KAHN (1963) demonstrated such fibers in MARCHI preparations of the parietal
and preoccipital regions in primate material. Long ago, OBENCHAIN (1925) illustrated
such fibers in South American marsupials describing them as "perforating fibers of the
great temporo-ammonic tract."

The inferior portion of the hippocampal formation is not as outstanding as are its other
parts. The most interesting aspect of this division is its relationship with the amygdaloid
complex. A gradual flattening of the three parts of the hippocampal formation against
the medial wall of the amygdala gives it a certain continuity with this area. The cornu
ammonis forms a bridge as it extends ventrally and medially to blend with the medial
amygdaloid nucleus and the basal accessory II nucleus. A relationship is thus established
between the cornu ammonis and both corticomedial and basolateral divisions of the
amygdala. The dentate gyrus wanes to a thin band of cells with little evidence of continuity with the amygdala, although the subiculum ends by blending with the posterior
piriform cortex.

The present degeneration studies indicate a projection from the hippocampal formation, especially the cornu ammonis, to the medial septal nucleus and, to a lesser extent, to the lateral septal nucleus. These fibers are a part of the precommissural fornix pictured by RÖTHIG (1909) and defined by LOO (1931). Postcommissural fornix fibers were also observed in the experimental studies. They follow the well established pathways into the diencephalon.

Acknowledgements

The author wishes to acknowledge the invaluable criticism of Dr. ELIZABETH C. CROSBY and the suggestions of Dr. TRYPHENA HUMPHREY in the preparation of this paper. The technical and secretarial assistance of Miss ANNE BALL, Miss ROSEMARY HENNING and Mr. CHARLES YATES is gratefully acknowledged.

This work was supported by Grant NB-05250, US PHS.

References

ELLIOT SMITH, G.: Anat. Anz. *10* (1894). 470–474.

GRAY, P.A.: J. comp. Neurol., *37* (1924), 221.

HERRICK, C.J.: J. comp. Neurol. *37* (1924), 317–359.

HERRICK, C.J.: J. comp. Neurol. *2* (1892), 1–20.

JOHNSTON, J.B.: J. comp. Neurol., *23* (1913), 371–478.

JOHNSTON, J.B.: J. comp. Neurol., *35* (1923), 337–481.

LOHMAN, A.H.M., H.J. LAMMERS: Progress in Brain Research. *3* (1963), 149–161.

LOO, Y.T.: J. comp. Neurol., *51* (1930), 13–64.

LOO, Y.T.: J. comp. Neurol., *52* (1931), 1–148.

OBENCHAIN, J.B.: Zool. Ser. *14* (1925). 175–232.

RAMÓN Y CAJAL, S.: Histologie du Système Nerveux de l'homme et des vertébrés. Maloine, Paris 1909–1911.

RETZIUS, G.: Biol. Untersuch., (Stockholm) N.F., *8* (1898), 23.

RÖTHIG, P.: Im Selbstverlage der Senckenbergischen Naturforschenden Gesellschaft. 1–19, Frankfurt 1909.

SCHNEIDER, R.C., E.C. CROSBY, E.A. KAHN: Progress in Brain Research *3* (1963), 192–214.

VAN DER SPRENKEL, H. BERKELBACH: J. comp. Neurol., *42* (1926), 211–254.

ZIEHEN, T.: Denkschr. d. Med.-naturwissensch. Gesellsch. zu Jena, *6* (1908), 789. Semon's Forschungsreisen, *3* (1908).

Earl G. HAMEL, Ph. D.
Department of Anatomy
University of Alabama Medical Center
1919 Seventh Avenue South
Birmingham, Ala., USA

Some Aspects of the Early Development of the Hippocampal Formation in Certain Insectivorous Bats

By J. W. BROWN, Birmingham, Ala.

The fascinating configuration of the adult mammalian hippocampal formation has attracted the attention of many investigators (CAJAL, 1911; ELLIOT SMITH, 1896 a, 1897 a and others). Among the studies on *Chiroptera* are accounts of various portions of the hippocampal formation (ELLIOT SMITH, 1897 b, 1898; JOHNSTON, 1913; ROSE, 1926; HUMPHREY, 1936; MANN, 1963; BROWN, 1965).

Little information is available concerning the embryonic development of the hippocampal formation in mammals, including *Chiroptera*. Among the embryologic studies are the accounts of HINES (1922), ELLIOT SMITH (1910, 1923), MACCHI (1951), and HUMPHREY (1964 and at this symposium, 1966) for man, ELLIOT SMITH (1896 b) for *Ornithorhynchus* and TILNEY (1938) for the opposum and the pig. The paucity of information concerning the embryology of the mammalian hippocampal formation prompted the present investigation of its development in insectivorous bats. Because of space limitations only certain aspects of the early development of the hippocampal formation through stages in which many but not all adult characteristics are present will be considered.

Materials. The series of 23 embryos of *Tadarida brasiliensis mexicana* to be considered range from 6 mm to somewhat over 12.5 mm CR length. These were sectioned at 10 μ, the majority in the frontal plane, the others transversely and sagittally and stained with MALLORY's quadruple stain.

List of Abbreviations

A	alveus	M	molecular layer
CA 1, CA 2, CA 3	subdivisions of cornu ammonis	MC	marginal cells
CF	concavity for fimbria	P	pyramidal cell layer
D	area where gyrus dentatus will develop	PD	primordial gyrus dentatus
		PL	polymorphic layer
DZ	diffuse zone	PP	primordial pyramidal cell layer
EM	external limiting membrane	SL	stratum lacunosum
F	fimbria	SO	stratum oriens
FAS	forerunner of cornu ammonis-subicular region	SR	stratum radiatum
		UD	undifferentiated portion of gyrus dentatus
FD	forerunner of gyrus dentatus		
FDS	fimbriodentate sulcus	X	greatest number of migrating cells into PD
HS	hippocampal sulcus		
LG	lamina granularis		

Observations

In this brief paper it will be necessary to confine the description of the hippocampal formation to that portion which is best developed for each embryo considered even though differences in the degree of development of the other parts appear embryologically. In general the best developed portion of the hippocampal formation shifts somewhat posteriorly at older ages. In embryos of 6 to 10.5 mm CR length this portion of the hippocampal formation begins just anterior to the paraphysis and continues posteriorly through planes of the anterior portion of the interventricular foramen. In the 10.5 mm embryo this portion of the hippocampal formation has a greater extent, nearly reaching the posterior border of the interventricular foramen and the level of the anterior commissure, which begins to form in the 10 mm embryo.

In 11 and 12 mm embryos, the best differentiated part of the hippocampal formation begins at posterior levels of the paraphysis and extends somewhat beyond the anterior commissure and the interventricular foramen. In the 12.5 mm embryo, this portion begins at posterior levels of the hippocampal commissure as it does in all older embryos and the adult. It extends farther posteriorly at 12.5 mm than at 12 mm and continues to be better differentiated farther posteriorly in each successively older embryo. However, the postero-ventral extremity of the hippocampal formation remains relatively less well differentiated even in the adult.

The terminology used in this paper is that of PEELE (1961) and CROSBY, HUMPHREY a. LAUER (1962) already used for the structure of the hippocampal formation of the adult bat (BROWN, 1965). The term hippocampal formation used by these investigators includes the gyrus dentatus (fascia dentata), cornu ammonis (Ammon's horn) and subicular region. The gyrus dentatus consists of a molecular layer, a granular layer (lamina granularis) and a region of polymorph cells lying within the hilus of lamina granularis. The cornu ammonis is characterized by the layer of Ammon's pyramids or double pyramids and can be subdivided into three parts (CA 1, CA 2 and CA 3) using LORENTE DE No's (1934) terminology. Space limitation prevents consideration of the subicular area. Although the terms hippocampus and primordium hippocampi have been used for the cornu ammonis of the human embryo (e.g. by HINES, 1922), they have been applied in other ways as well. To prevent misinterpretation primordial gyrus dentatus and primordial cornu ammonis will be used in this paper.

Initial Development, 6 to 8 mm. The area along the dorsomedial surface of the hemisphere giving rise to the hippocampal formation can be recognized in the 6 mm embryo (Fig. 1 A) and is better represented in the 8 mm embryo (Fig. 1 B). More dorsally in this area of the 6 mm embryo is a narrow marginal layer into which a few cells (FAS) are migrating from the ependymal layer (medullary epithelium). These cells increase in the older embryos so that at 8 mm (Fig. 1 B) clusters of small cells (FAS) are present in the marginal layer. These primary cells foreshadow the development of the combined cornu ammonis-subicular area. Ventromedial to this area, an even more narrow cell-free marginal layer (D), lying immediately adjacent to the ependymal layer, indicates the area where gyrus dentatus develops (6 mm, Fig. 1 A). This narrow cell-free marginal layer (D) is a little wider in the 8 mm embryo (Fig. 1 B).

Stage of Layer Formation in Cornu Ammonis, 9 mm. The hemispheric wall in the 9 mm embryo is broader and the thickening or ingrowth of the primordial hippocampal

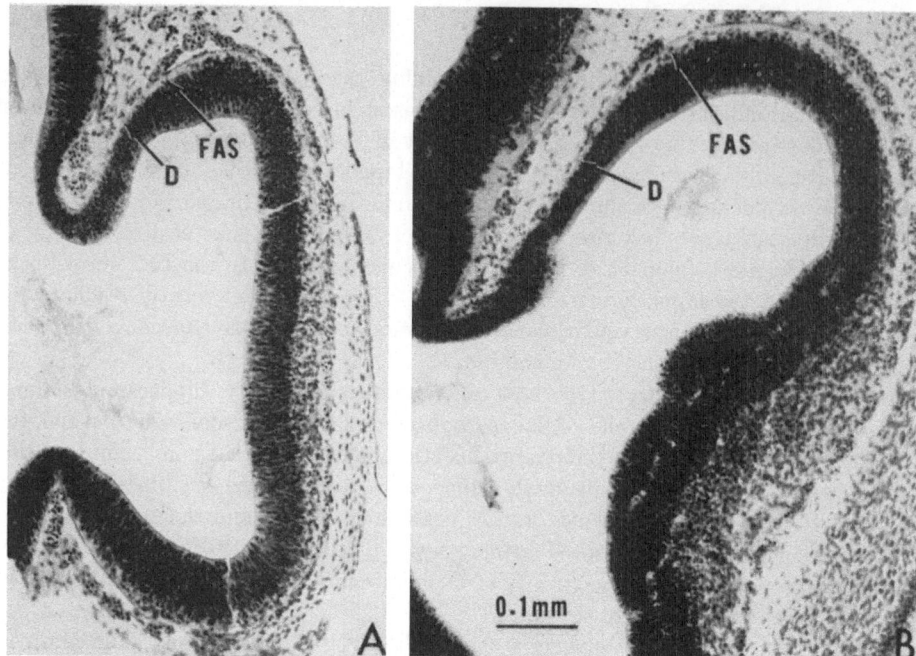

Fig. 1. Photomicrographs, all at the same magnification (see Fig. 1B), through the telencephalon of frontally sectioned *Tadarida mexicana* embryos at levels of the anterior portion of the interventricular foramen to show the initial stages of development of the hippocampal formation. *A*. 6 mm embryo (section No. BE 20-3-3-5). Only a few scattered cells foreshadow the combined cornu ammonis-subicular area and a narrow cell-free marginal area indicates the site where gyrus dentatus will develop. *B*. 8 mm embryo (BE 1-3-1-1). Small cell clusters foreshadow the combined cornu ammonis subicular area and the cell-free marginal layer indicating the site of formation of gyrus dentatus is a little wider than at 6 mm. For abbreviations see page 92.

formation has produced a slight bulge on the ventricular wall (Fig. 2 A). The ependymal layer is wider, but the greatest increase is in the marginal layer of the primordial hippocampal formation, which now contains scattered cells. It is wider than that of all other cortical areas, a characteristic noted by HINES (1922) for the human embryo and by TILNEY (1938) for the opposum embryo. In the dorsal part of the developing hippocampal formation, continued cell migration produces a distinct primordial pyramidal cell layer (PP) in cornu ammonis. This cell layer, separated slightly from the ependymal layer, broadens dorsally as it becomes continuous with the developing subicular area and tapers ventrally into a region where there is no cell layer but only scattered cells migrating from the medullary epithelium (ependymal layer) to form the forerunner of the gyrus dentatus (FD).

Adjacent to the forerunner of gyrus dentatus is a cell-poor concavity in the ependymal layer (CF) from which many cells appear to migrate into the marginal layer. In older embryos fibers of the fimbria collect in this concavity. Most of the cells (MC) along the outer surface of the broad marginal layer of the hippocampal formation appear to have migrated from the cell-poor concavity. These superficially placed cells, identified by HINES (1922) for human embryos as fascia dentata or gyrus dentatus, will be refered to as mar-

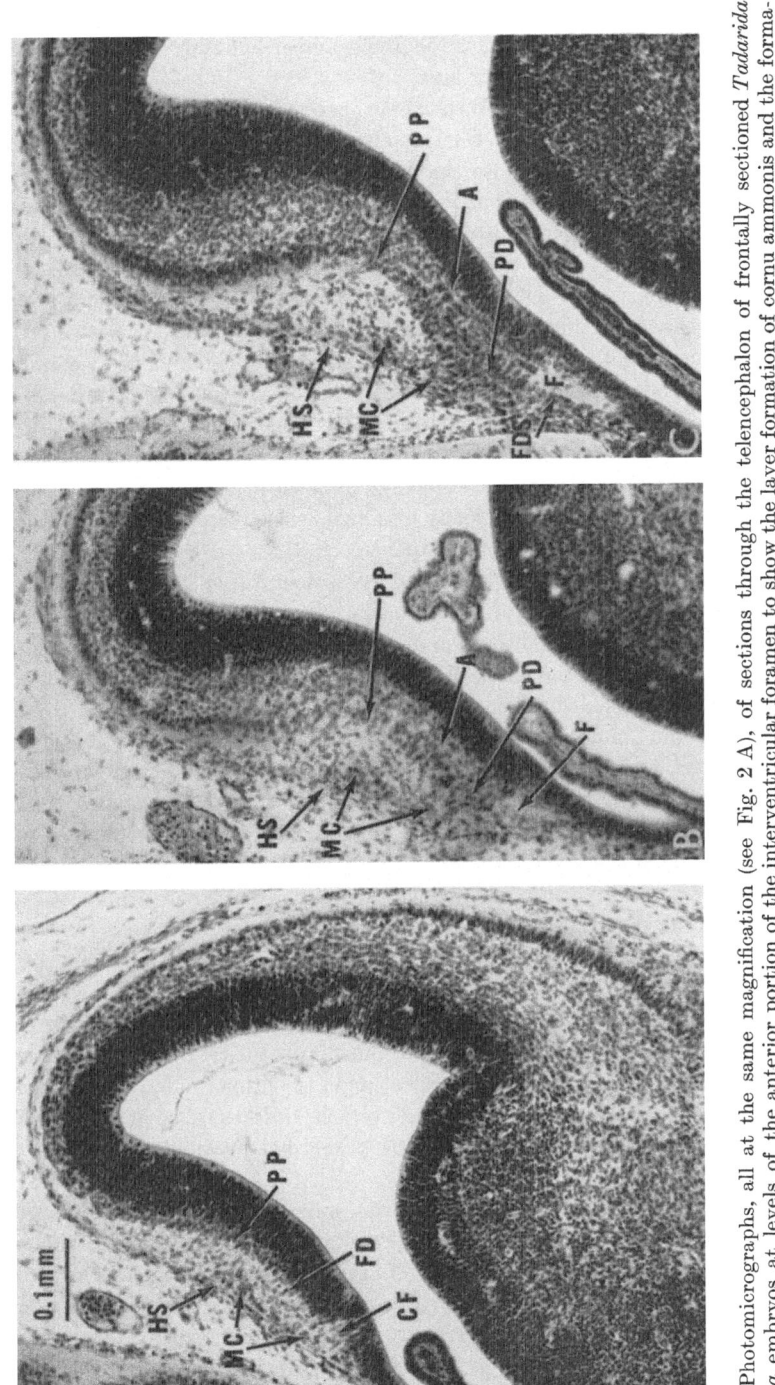

Fig. 2. Photomicrographs, all at the same magnification (see Fig. 2 A), of sections through the telencephalon of frontally sectioned *Tadarida mexicana* embryos at levels of the anterior portion of the interventricular foramen to show the layer formation of cornu ammonis and the formation of the primordial gyrus dentatus. *A.* 9 mm embryo (BE 5·3·3·5). The initial stage in the formation of the primordial pyramidal cell layer of cornu ammonis. Note the marginal cells under the external limiting membrane. *B.* 10 mm embryo (BE 16·4·2·1). The primordial pyramidal cell layer is farther from the ependymal layer and the primordial gyrus dentatus is present. *C.* 10.5 mm embryo (BE 4·3·2·5). A well developed layer of primordial Ammon's pyramids is present. Note the linear arrangement of cells along their path of migration into the enlarged primordial gyrus dentatus. For abbreviations see page 92.

ginal cells (MC) in this paper. The shallow indentation on the surface of the primordial cornu ammonis (see arrow at HS, Fig. 2 A) is the hippocampal sulcus or fissure. A surface blood vessel is usually associated with this area.

10 mm. In the 10 mm embryo (Fig. 2 B) the hemispheric wall of the primordial hippocampal formation has become markedly thicker, although the ependymal layer is narrower, and forms a pronounced bulge into the ventricle. The primordial pyramidal cell layer (PP) of the developing cornu ammonis, which is much wider than at 9 mm, contains more cells, some of which are larger. This cell layer is farther from the ependymal layer but in continuity with it by a loosely arranged zone of migrating cells. In this cell zone deep to the portion of the primordial pyramidal cell layer that is nearest the fimbria, fibers of the alveus (A) are collecting and can be followed into the fimbria (F). Between the primordial cornu ammonis and the fimbria is a rather large mass of small, darkly staining cells (PD) which are migrating from the ependymal layer. This cell mass is the primordial gyrus dentatus. On its surface, where it is in continuity with the primordial pyramidal cell layer, it is more dense, but not sufficiently so to form a cell layer as in cornu ammonis. Just under the external limiting membrane, extending from the primordial gyrus dentatus dorsally and to some extent ventrally are many marginal cells (MC). An accumulation of these cells lies opposite to the shallow hippocampal sulcus (HS).

In the 10.5 mm embryo (Fig. 2 C) the primordial gyrus dentatus (PD) is larger than at 10 mm, due to the continued migration of cells from the ependymal layer. Its cells exhibit a marked linear arrangement along their path of migration. The primordial gyrus dentatus now produces a definite bulge on the external surface of the hemisphere with the hippocampal sulcus (HS) more marked superior to it and the fimbriodentate sulcus (FDS) well defined inferiorly. In cornu ammonis the increased number of larger primordial pyramidal cells forms a well defined layer (PP). External to this layer is a wide, pale staining primordial molecular layer containing the marginal cells (MC) along its external border. Small fiber bundles of the alveus (A) are present along the entire extent of the cellular zone deep to the primordial pyramidal cell layer.

11 mm. In the 11 mm embryo (Fig. 3 A), the larger primordial gyrus dentatus (PD) forms a greater bulge on the external surface than before. Migration of cells from the ependymal layer is less evident because of the greater accumulation of alveus fibers (A), but the greatest migration is into its ventral and external border (arrow at X). The majority of the cells of the primordial gyrus dentatus stain darkly and are round, whereas most of the marginal cells (MC) adjacent to its dorsomedial border stain lightly and are oval. So far, there is no superficial cellular concentration in the primordial gyrus dentatus to indicate the beginning of lamina granularis. Moreover, the dorsal surface of the primordial gyrus dentatus is directly continuous with the primordial pyramidal cell layer of cornu ammonis.

In the primordial pyramidal cell layer (PP) of cornu ammonis the neurons are larger than at 10.5 mm. Just superficial to the pyramidal cell layer there is an accumulation of fibers to form a primordial stratum radiatum (SR). A large clear diffuse zone (DZ) remains in the primordial molecular layer between the stratum radiatum and the surface of the primordial gyrus dentatus. It contains the majority of the marginal cells of this region. In the cell layer deep to the primordial pyramidal cell layer the increased number of alveus fibers are concentrated along the ependymal layer.

Stage of Layer Formation in Gyrus Dentatus, 12 mm. A distinct advance in the differentiation of the hippocampal formation has appeared in the 12 mm embryo (Fig.

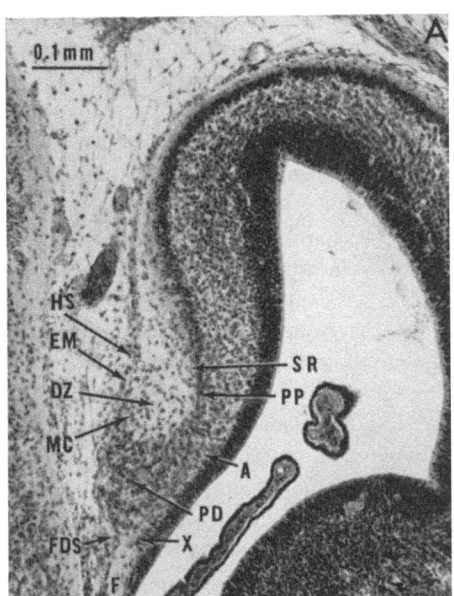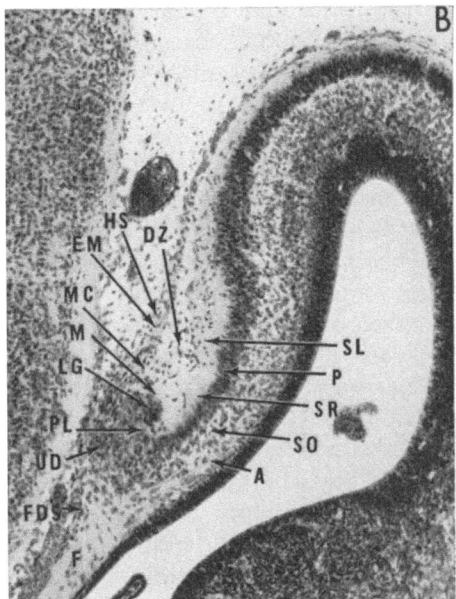

Fig. 3. Photomicrographs of frontal sections through the telencephalon of *Tadarida mexicana* embryos. *A*. 11 mm embryo (BE 8-4-1-3). A section through the level of the anterior part of the interventricular foramen. The primordial gyrus dentatus is larger and the primordial pyramidal cell layer is better defined. *B*. 12 mm embryo (BE 2-5-2-7). A section at the level of the hippocampal commissure. The differentiation of the granular and polymorphic layers is beginning in the primordial gyrus dentatus adjacent to cornu ammonis. Note the separation of lamina granularis from the pyramidal cell layer of cornu ammonis. The stratum lacunosum, the pyramidal cell layer and the stratum oriens are identifiable in cornu ammonis. For abbreviations see page 92.

3 B). Additional inward growth has increased the curvature of the cornu ammonis and its expansion into the ventricle. Growth of the primordial gyrus dentatus has produced a greater bulge on the surface so that the hippocampal sulcus (HS) and the fimbriodentate sulcus (FDS) are better defined. Furthermore, the large size of dentate gyrus has resulted in its separation from the adjacent portion of the pyramidal cell layer of cornu ammonis.

Changes in the finer structure are also evident. The neurons of the pyramidal cell layer of cornu ammonis (P) are larger and begin to resemble the double pyramids of the adult (Fig. 3 B). In some sections the pyramidal cell layer is narrower adjacent to the gyrus dentatus, foreshadowing the development of the subdivisions of the cornu ammonis. Between the pyramidal cell layer and the ependymal layer the stratum oriens (SO) and the alveus (A) are now clearly defined. The latter is still traversed by small bands of cells migrating through it into the medioventral part of the gyrus dentatus in particular, but to some extent into cornu ammonis. External to the pyramidal cell layer is a zone corresponding to stratum radiatum (SR) and outside of it the stratum lacunosum (SL). However, no superficial or definitive molecular layer is identifiable at this age or in older embryos and even in the adult it is not well defined.

It is noteworthy that in the dorsolateral region of the primordial gyrus dentatus, adjacent to cornu ammonis but separated from it, a small segment has taken on the characteristics of the definitive lamina granularis (LG) and of the underlying polymorph layer (PL) (Fig. 3 B). Lamina granularis consists of a small cluster of closely packed, darkly staining small cells having a cortex-like appearance. The polymorph layer contains cells which are larger and more loosely arranged than elsewhere in the gyrus dentatus. Along the medial surface of the undifferentiated part of gyrus dentatus (UD) the cells are more dense but do not yet have the characteristic arrangement of lamina granularis. Superficial to lamina granularis is a narrow molecular layer (M).

At the dorsomedial tip of the developing dentate gyrus, and in lesser numbers along its external surface, are numerous marginal cells (MC). The greater number of these cells occupy a pale-stained, triangular, diffuse zone (DZ) just deep to the hippocampal sulcus and external to stratum lacunosum of Ammon's horn. On superficial examination this zone looks like a fissure, but the external limiting membrane (EM) is continuous over the surface except where blood vessels penetrate it.

12.5 mm. The hippocampal formation bulges into the ventricle to a greater degree in the 12.5 mm embryo (Fig. 4) than previously. In cornu ammonis the stratum lacunosum (SL) is wider and the pyramidal cell layer (P) is consistently narrower in the region adjacent to the gyrus dentatus. The differentiation of the definitive lamina granularis (LG), the polymorphic layer (PL) and the molecular layer has progressed farther medialward (externally) leaving a proportionately smaller part of gyrus dentatus undifferentiated (UD). With the increased differentiation of gyrus dentatus and increased width of stratum lacunosum of cornu ammonis the clear diffuse zone (DZ) opposite the hippocampal sulcus (HS) is narrowed and the number of marginal cells relatively reduced. Bands of migrating cells still extend from the ependymal layer through the enlarging alveus into cornu ammonis and, more especially, into the undifferentiated portion of the gyrus dentatus.

In a *Tadarida* embryo somewhat over 12.5 mm in CR length the cornu ammonis has undergone considerable ingrowth toward the ventricle so that it lies farther from the surface (Fig. 5). The total length of cornu ammonis has increased and it is assuming a marked crescent shape thus approaching the classical U-shaped configuration of the adult. The neurons of the pyramidal cell layer are larger and better differentiated in all its subdivisions. In the mid-region of the crescent (CA 2) they are larger than in the adjacent portions. In addition, the pyramidal cell layer narrows toward the gyrus dentatus (CA 3) and broadens (CA 1) as it approaches the subicular area, so that its three definitive parts (CA 1, CA 2, CA 3) are identifiable. Also stratum lacunosum (SL) is wider and contains more cells than in the younger embryos.

There is a marked increase in the size of the gyrus dentatus at this age (Fig. 5). The lamina granularis (LG), the polymorph layer (PL) deep to it and the molecular layer (M) superficial to it have all increased in extent medially so that only the most external portion of gyrus dentatus remains undifferentiated (UD). Moreover, partly as a result of the addition of large numbers of neuroblasts to the undifferentiated portion, there has been a shift or rotation of the whole structure internally (laterally) so that the molecular layer of gyrus dentatus is brought into apposition to the stratum lacunosum of cornu ammonis. This brings the lamina granularis closer to and somewhat parallel with the pyramidal cell layer of the cornu ammonis. As a result of these changes the clear diffuse zone (DZ) is reduced to a narrow line and carried deeper into the hemispheric wall along with a few marginal cells and associated blood vessels. Accompanying these changes there is a reduction in the number of marginal cells. In this embryo as in the younger specimens the external limiting membrane (EM) forms an unbroken line across the surface of the

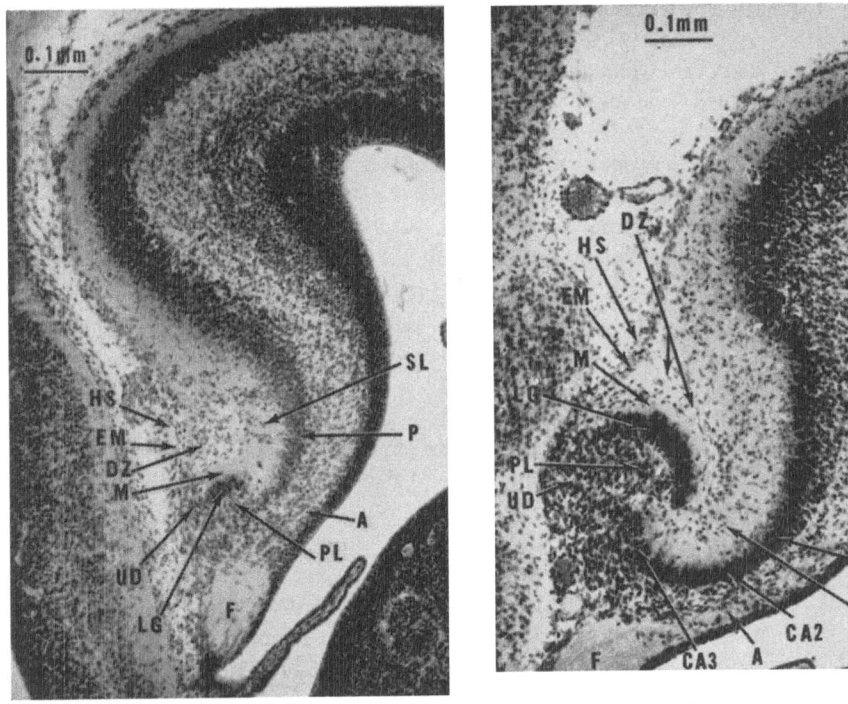

Fig. 4.   Fig. 5.

Fig. 4. Photomicrograph of the dorsomedial portion of the telencephalon of a frontally sectioned 12.5 mm *Tadarida mexicana* embryo (BE 18-13-1-4) at a level just caudal to the hippocampal commissure. Further differentiation has occurred in the gyrus dentatus. For abbreviations see page 92.

Fig. 5. Photomicrograph of the dorsomedial part of the telencephalon at a level just caudal to the hippocampal commissure from a frontally sectioned *Tadarida mexicana* embryo which is somewhat longer than 12.5 mm (BE 19-1-4). Most adult characteristics of cornu ammonis are identifiable and the differentiation of gyrus dentatus has spread to include a greater portion of the primordial cell mass. The pale staining diffuse area extends from the hippocampal sulcus deeply between the molecular layer of the gyrus dentatus and stratum lacunosum. Note the large number of marginal cells in this zone. For abbreviations see page 92.

pale diffuse zone, and there is no evidence of a fissure extending into the brain substance from the shallow superficial hippocampal sulcus (HS, compare EM in Figs. 3 and 4).

Only a few cells extend from the ependymal layer through the more prominent alveus (A) into the hippocampal formation (Fig. 5). The ependymal layer adjacent to the hippocampal formation is much narrower than in younger embryos and contains only an occasional mitotic figure.

Discussion and Conclusions

In the initial stage of development of the hippocampal formation in the insectivorous bats studied, scattered cells migrate into the combined cornu ammonis-subicular region before they do into the gyrus dentatus area (Fig. 1). A continued cellular migration forms a distinct cell layer in the primordial cornu ammonis before it forms the primordial gyrus dentatus (Fig. 2 A). At the time of formation of the primordial pyramidal cell layer in cornu ammonis, marginal cells derived from the ependymal layer spread along the surface of the marginal zone of the primordial hippocampal formation. They are followed by a massive migration of the cells of the primordial gyrus dentatus (Fig. 2 B) with which they are in contact, especially in the 10.5 and 11 mm embryos (Figs. 2 C and 3 A). Only a few of the marginal cells may be incorporated into the superficial part of the primordial gyrus dentatus (see below).

After the formation of the primordial pyramidal cell layer in cornu ammonis, its cells gradually assume the characteristics of definitive pyramidal cells (11 to 12 mm). Further differentiation (12 to more than 12.5 mm, Figs. 3 B to 5) forms the subdivisions of the pyramidal cell layer (CA 1, CA 2 and CA 3). Coincidently the other layers of cornu ammonis appear.

Accompanying the establishment of a clearly defined pyramidal cell layer in cornu ammonis (12 mm, Fig. 3 B), differentiation of laminae begins in the primordial gyrus dentatus which was present for only a brief period (10 to 11 mm). Simultaneously the "break" between cornu ammonis and gyrus dentatus referred to by ELLIOT SMITH (1923) develops. When the definitive laminae of gyrus dentatus first appear a superpositio medialis is present, comparable in some respects to that in certain reptiles (ARIENS KAPPERS, HUBER and CROSBY, 1936). The initial differentiation of the granular and polymorphic layers begins in the region adjacent to cornu ammonis, the first part of the primordial gyrus dentatus to develop. Laminar formation spreads progressively around the undifferentiated cell mass so that it will occur at a later stage in the medial or ventromedial part of the gyrus dentatus, which develops last from the ependymal layer. The cells on the surface of the primordial cell mass form lamina granularis and those in the central part of the mass form the polymorphic layer. The common embryonic origin of the lamina granularis and the polymorphic layer occupying its hilus indicates clearly that the polymorphic cells are a part of gyrus dentatus as believed by CAJAL (1911) and others and do not constitute a part of the cornu ammonis as interpreted by LORENTE DE NO (1934) and ROSE (1926) in their studies of the adult structure.

Moreover, this study clearly indicates that, for the bat, the marginal cells occupying the position of the cells HINES (1922) identified as fascia dentata (gyrus dentatus) for the human embryo do not contribute in any significant degree to the formation of either lamina granularis or the polymorphic layer of gyrus dentatus, although some of those superficial to the primordial gyrus dentatus when it first appears (Fig. 2 B) are probably included with the gyrus dentatus as it reaches the external surface (Figs. 2 C–3 B). Most of the marginal cells occupy a position just deep to the hippocampal sulcus in the area which retains a clear diffuse appearance as the stratum lacunosum of cornu ammonis and the molecular layer of gyrus dentatus develop. As these two laminae increase in size the clear diffuse zone is reduced and carried more deeply into the brain substance as the whole hippocampal formation enlarges. As this occurs the marginal cells decrease in number and few are identifiable in the oldest embryos studied. Some of these marginal cells may be incorporated into the stratum lacunosum and at least some of them probably degenerate and disappear.

The formation of gyrus dentatus and cornu ammonis takes place entirely within the wall of the telencephalon, which itself increases in width as the hippocampal formation becomes more extensive. As cornu ammonis increases in size it forms a bulge on the ventricular surface and the increase in size of gyrus dentatus produces a bulge on the external surface. Although the shallow hippocampal sulcus dorsal to the latter becomes more clearly defined throughout the embryonic period, it remains only a superficial indentation. In no case does it extend deeply into the telencephalic wall. In the older bat embryos a pale-stained diffuse zone, containing a few marginal cells and in most instances blood vessels, extends deeply from the hippocampal sulcus. On superficial examination it could easily be misinterpreted as a fissure (Fig. 5). No doubt, separation along this zone would occur readily on dissection as ELLIOT SMITH (1923) believed in his interpretation of the human hippocampal fissure as an artifact.

It is difficult to compare the observations made in this study with those of other investigators for other vertebrates. This is due, in part, to differences in terminology and emphasis and lack of a closely spaced embryonic series (ELLIOT SMITH, 1923, HINES, 1922, and MACCHI, 1951 for man; TILNEY, 1938 for pig and opposum). However, the recent study by HUMPHREY (1964, 1966) of the development of the human hippocampal formation is sufficiently comprehensive to permit a satisfactory comparison. In general, the sequence of development of the hippocampal formation in insectivorous bats and in man is similar; e.g., in both forms the cornu ammonis, after appearance, develops more rapidly in the early stages and gyrus dentatus appears later. However, there are definite differences, some of which appear to be the result of the omission in the bat of stages seen in human development, e.g., in the bat there is no initial cell layer formation in the primordial gyrus dentatus, as there is in the human embryos, but a cell mass develops in continuity with the pyramidal cell layer of cornu ammonis with only a surface concentration of cells indicating this intermediate stage. In addition, in *Tadarida* there is no primordial granular cell layer in direct continuity with the pyramidal cell layer of cornu ammonis as in man but rather a definitive granular layer with a cortex-like appearance develops without a preliminary condensation of cells and, as soon as it appears, is separated from the layer of Ammon's pyramids. On the contrary, the migration of cells into the external part of gyrus dentatus is marked in the bat and less evident in the human embryo.

The gyrus dentatus and cornu ammonis differentiate in different ways in the insectivorous bats studied. The primordial gyrus dentatus develops as a mass of cells which migrates from the ependymal layer without forming a definite cell layer (Fig. 2 B), whereas the primordial cornu ammonis appears initially as a cell layer which is slightly but distinctly separated from the ependymal layer (Fig. 2 A).

Summary

The description of the development of the hippocampal formation is confined to its best differentiated part for each stage considered in a series of *Tadarida brasiliensis mexicana* embryos ranging from 6 mm to somewhat over 12.5 mm CR length.

In the 6 to 8 mm embryos, the cornu ammonis is foreshadowed by the migration of a few cells from the ependymal layer into the narrow marginal layer. Ventral to these cells, the future gyrus dentatus is indicated by an even narrower cell-free marginal layer. In the 9 mm embryos a distinct cell layer constitutes cornu ammonis whereas only a few scattered cells migrating into the marginal layer foreshadow gyrus dentatus. Adjacent to gyrus dentatus, many cells (designated marginal cells) migrate from the ependymal layer to the outer part of the marginal layer leaving a concavity in which the fimbria will later develop.

In the 10 mm embryo, between the better defined primordial pyramidal cell layer of cornu ammonis and the fimbria, a large mass of migrating cells constitutes the primordial gyrus dentatus. In this cell mass the granular layer and the polymorphic layer begin to develop adjacent to cornu ammonis at 12 mm. Differentiation of the definitive granular layer, occurring in the periphery of the cell mass, spreads medially and is completed at a later stage (19 mm). The central cells become the polymorphic layer. The marginal cells do not significantly contribute to gyrus dentatus.

All layers of cornu ammonis are recognizable in the 12 mm embryo. The subdivisions of cornu ammonis are clearly identifiable in the embryo somewhat over 12.5 mm in length.

Acknowledgements

The author wishes to express his appreciation to Dr. E. CARL SENSENIG for the use of the *Tadarida mexicana* embryos and for the aid offered by his grant No. NB-00951-08 from the National Institute of Neurological Diseases and Blindness. Also the author gratefully acknowledges the aid offered by grant No. HD-00230 from the National Institute of Child Health and Human Development to Dr. TRYPHENA HUMPHREY whose advice on the problem is greatly appreciated. The author also wishes to express his appreciation to Dr. ELIZABETH C. CROSBY for her suggestions and encouragement in the preparation of this paper.

References

ARIENS KAPPERS, C.U., G.C.HUBER, E.C. CROSBY: The Comparative Anatomy of the Nervous System of Vertebrates, including Man. The Macmillan Company, New York 1936.

BROWN, J.W.: Anat. Rec., *151* (1965), 444 (Abstract),

CAJAL, S. RAMÓN Y: Histologie du Système Nerveux. Vol. II. Maloine. Paris 1911.

CROSBY, E.C., T.HUMPHREY, E.W.LAUER: Correlative Anatomy of the Nervous System. The Macmillan Company. New York 1962.

ELLIOT SMITH, G.: J. Anat. (Lond.), *30* (1896a), 157–167, 185–205.

ELLIOT SMITH, G.: Quart. J. Micr. Sci., *39* (1896b), 181–206.

ELLIOT SMITH, G.: J. Anat. (Lond.), *32* (1897a), 23–58.

ELLIOT SMITH, G.: Anat. Anz. *13* (1897b), 23bis 27.

ELLIOT SMITH, G.: J. Anat. (Lond.), *32* (1898), 231–246.

ELLIOT SMITH, G.: Lancet *I* (1910), 147–153.

ELLIOT SMITH, G., A. ROBINSON (Ed.): In Cunningham's Text-book of Anatomy, 5th Ed. WilliamWood and Company. New York 1923

HINES, M.: J. comp. Neurol., *34* (1922), 73–171.

HUMPHREY, T.: J. comp. Neurol., *65* (1936), 603–710.

HUMPHREY, T.: Trans. Amer. neurol. Ass., *89* (1964), 207–209.

HUMPHREY, T.: In: HASSLER, R., H.STEPHAN (ed.): Evolution of the Forebrain. Thieme, Stuttgart 1966, 104–116.

JOHNSTON, J.B.: J. comp. Neurol., *23* (1913), 371–478.

LORENTE DE NÓ, R.: J. Psychol. Neurol. (Lpz.)
  46 (1934), 113–117.
MACCHI, G.: J. comp. Neurol., 95 (1951), 245
  bis 305.
MANN, G.: The Rhinencephalon of Chiroptera.
  Univ. de Chile. Santiago 1963.

PEELE, T.L.: The Neuroanatomic Basis for
  Clinical Neurology, 2nd. Ed., McGraw-Hill
  Book Company, Inc. New York 1961.
Rose, M.: J. Psychol. Neurol. (Lpz.), 34 (1926),
  1–111.
TILNEY, F.: Bull. neurol. Inst. N.Y., 7 (1938),
  1–77.

JERRY W. BROWN, Ph. D.
Department of Anatomy
University of Alabama Medical Center
1919 Seventh Avenue South
Birmingham, Ala. 35233, USA

The Development of the Human Hippocampal Formation Correlated with some Aspects of its Phylogenetic History* ** ***

By T. HUMPHREY, Birmingham, Ala.

The few extensive developmental studies on the human hippocampal formation either have dealt primarily with early differentiation (HINES, 1922) or emphasized later development (MACCHI, 1951). Consequently they have not completely demonstrated the manner in which the gyrus dentatus develops or its layers differentiate. Correlations between developmental stages of the parts of the human hippocampal formation and its subdivisions in submammalian vertebrates are few (ROSE, 1926; TILNEY, 1938), although more general correlations are often made (e.g., KUHLENBECK a. DOMARUS, 1920). Before attempting correlations, however, it is necessary to consider the terminology to be used and the most significant literature.

The terminology of KAPPERS et al. (1936), PEELE (1961) and CROSBY et al. (1962) will be employed. As these references state, the hippocampal formation includes the gyrus dentatus (fascia dentata of ELLIOT SMITH, HINES, ABBIE, MACCHI and others), the cornu ammonis (or Ammon's horn), the subiculum with its various subdivisions (LORENTE DE NÓ, 1934; ROSE, 1926), the anterior continuation of the hippocampus, the fasciolar gyrus and the supracallosal gyrus (or induseum griseum). Hippocampus, when applied to microscopic material, usually refers to the cornu ammonis (e.g., HOCHSTETTER, 1919) but is used grossly for the entire hippocampal formation. To prevent confusion, gyrus dentatus, cornu ammonis (or Ammon's horn) and hippocampal formation will be used. Primordium hippocampi also has been applied in different ways by HOCHSTETTER (1919), HERRICK (1921), HINES (1922), JOHNSTON (1913, 1923), HOFFMAN (1963) and others and will be avoided. The more specific designations – such as primordial hippocampal formation, primordial gyrus dentatus (or anlage of gyrus dentatus) and primordial cornu ammonis – will be substituted.

For correlations with the hippocampal formation of other vertebrates, the stages in human development to be investigated should include the appearance of the corpus

* This publication covers only part of the paper presented at the symposium on Phylogenesis and Ontogenesis of the Forebrain. Due to limitations of space, the material on the correlations between the development of the olfactory bulbs and the gyrus dentatus will be published elsewhere (HUMPHREY, 1966).

** This investigation was supported by a U.S. Public Health Service research career program award, NB-K6-16716, from the National Institute of Neurological Diseases and Blindness.

*** Aided by grant HD-00230, National Institute of Child Health and Human Development, National Institutes of Health. The present paper is publication No. 43 in a series of physiological and morphological studies on human prenatal development. Most of the embryologic material on which this paper is based was prepared under the support of previous grants from The Penrose Fund of the American Philosophical Society, The Carnegie Corporation of New York, The University of Pittsburgh, The Sarah Mellon Scaife Foundation of Pittsburgh, and Grant B-394 from the National Institute of Neurological Diseases and Blindness, National Institutes of Health.

callosum, not identified below mammals, and the differentiation of the granular layer of the gyrus dentatus, not recognized below monotremes (KAPPERS et al., 1936). The corpus callosum first develops between 60 mm CR (HAMILTON et al., 1962) and 84 mm CR (HEWITT, 1962). The granular layer of the gyrus dentatus appears by 79 mm CR (13.5 weeks, HUMPHREY, 1964) and is well represented by 110 mm CR (MACCHI, 1951) or 15.5 weeks (HUMPHREY, 1964). To include sufficient age levels, then, embryos and fetuses 9 mm long to 84.5 mm in CR length were selected (6 to 13.5 weeks of menstrual age, according to the tables of STREETER, 1920). Mostly NISSL and protargol silver preparations of transversely, horizontally and sagittally sectioned series were used. Limitations of space prevent the inclusion of additional data on the material.

Brief references to the development of the human hippocampal formation have been made by HIS (1904), HOCHSTETTER (1919), ELLIOT SMITH (1923), ESCOLAR (1959) and others. The accounts of HINES (1922) and MACCHI (1951) are more extensive. HOCHSTETTER (1919) identified the anlage of the hippocampus (i.e., cornu ammonis) in a 14.8 mm embryo in the area designated later as primordium hippocampi (cornu ammonis) by HINES (1922) in 11.8 and 14 mm embryos and by BARTELMEZ a. DEKABAN (1962) in a 10 mm embryo. MACCHI (1951, Fig. 13) recognized the scattered cells in the "hippocampus (gyrus arcuatus)" of a 23 mm fetus as an "intermediate layer" but did not identify either the gyrus dentatus or the cornu ammonis until much later (63–84 mm fetuses). However, both a primordial gyrus dentatus and a primordial cornu ammonis were identified for a 32 mm fetus by HUMPHREY (1964).

List of Abbreviations

CA 1, CA 2, CA 3	regions of the cornu ammonis according to the terminology of Lorente de Nó, 1934.
ch. pl.	choroid plexus
corn. am.	cornu ammonis
corn. am. prim.	primordium of cornu ammonis
ep. cell cluster	cell cluster resembling ependymal cells
epiph.	epiphysis
ep. l.	ependymal layer
f.-d. fis.	fimbriodentate fissure
fimb.	fimbria
fis. hip.	fissura hippocampi
gy. dent.	gyrus dentatus
gy. dent. anl.	anlage of gyrus dentatus
l. gran., gy. dent.	lamina granularis, gyrus dentatus
l. migr. cells	layer of migrating cells
l. migr. cells in gy. dent. anl.	layer of migrating cells in anlage of gyrus dentatus
l. pyr., corn. am.	lamina pyramidalis, cornu ammonis
l. term.	lamina terminalis
marg. cells	marginal cells
migr. cells	migrating cells
M, L, R, P, O, A, E	strata of the cornu ammonis: marginale (M), lacunosum (L), radiatum (R), pyramidale (P), and oriens (O) respectively, and the alveus (A) and ependymal layer (E) of the cornu ammonis
par.	paraphysis
prim. cell l., gy. dent. anl.	primordial cell layer in anlage of gyrus dentatus
prim. gran. l.	primordial granular layer

prim. l. pyr.	primordial lamina pyramidalis
prim. polym. l.	primordial polymorphic layer
str.	striatum
SZ	zone of cells separating from the migrating layer of cells

Observations

Only a description of the gyrus dentatus and the cornu ammonis in the best developed region of the hippocampal formation for each age level is possible in this brief account. Until the septal region begins to differentiate at 18 to 21 mm, the best developed portion is located over the roof of the telencephalon lateral to the lamina terminalis (Figs. 1 A and 2 A). As soon as the medial septal nucleus can be recognized (19.1 mm, HINES, 1922), the best developed area is at paraphyseal arch levels (Fig. 2 C). The best developed area then shifts farther caudalward and by 10.5 to 11.5 weeks lies along the dorsolateral border of the diencephalon (Fig. 4 A). For the 12- to 13.5-week fetuses the best developed region lies lateral to the diencephalon (Fig. 4 B).

9 to 11 mm (6 weeks). In the dorsomedial wall of the hemisphere of a 9 mm embryo (Fig. 1), a definite marginal layer has appeared rostrally, adjacent to the lamina terminalis in the region which HINES (1922) and BARTELMEZ a. DEKABAN (1962) identified as the primordium hippocampi (or primordial cornu ammonis) of 11.8 mm and 10 mm embryos respectively. The present investigation demonstrates that the gyrus dentatus is formed by cells that migrate directly from the ependymal layer into this cell-free marginal zone (Figs. 1, 2) or marginal velum of HINES. Thus this area constitutes the anlage of the gyrus dentatus and will be so designated hereafter. For this 9 mm embryo, no comparable marginal layer has appeared elsewhere in the hemisphere (Fig. 1 A).

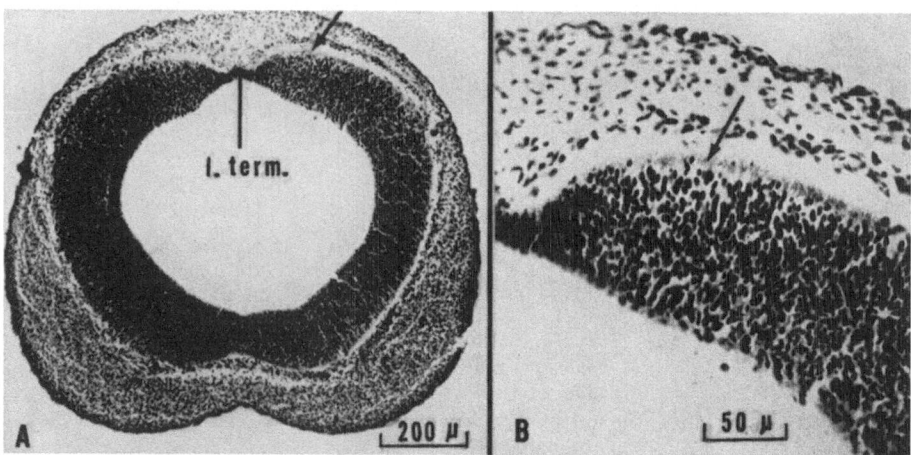

Fig. 1. Photomicrographs of a section through the telencephalon of a 9 mm human embryo prepared by the HOLMES' silver technique. (The author is grateful for the loan of this series, Embryo *ES*, by Dr. DONALD L. KIMMEL of the Chicago Medical School.) The area indicated by the *arrow* in both *A* and *B* is the cell-free marginal zone into which the neuroblasts of gyrus dentatus migrate later. For abbreviations see page 105.

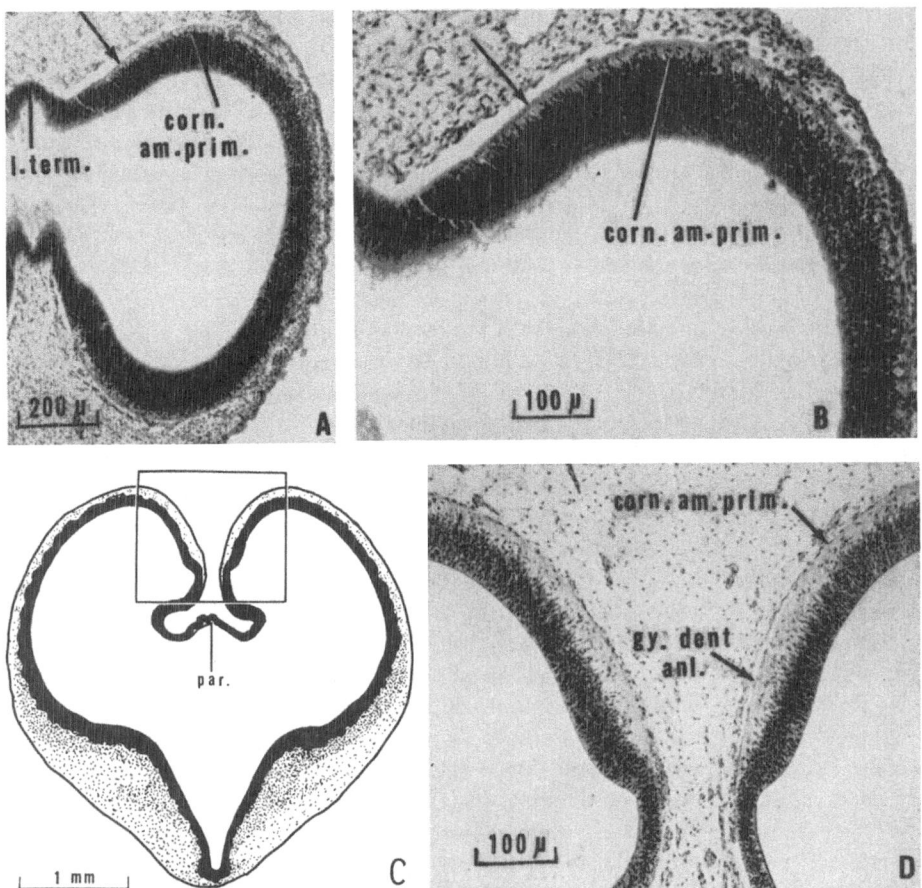

Fig. 2. *A* and *B* are photomicrographs of a section through the telencephalon of an 11 mm human embryo (I 2), hematoxylin and eosin stain. The *arrows* in *(A)* and *(B)* point to the area constituting the anlage of the gyrus dentatus in the primordial hippocampal formation. *(C)* and *(D)* are from an 18 mm human embryo (♯ 142) prepared by the protargol technique of BODIAN. Part *(D)* is a photomicrograph of the area enclosed in the lines on the orientation drawing in *(C)*. For abbreviations see page 105.

For the 9.5 and 10 mm embryos, the gyrus dentatus anlage is slightly more extensive and at 11 mm has again increased in size (Fig. 2 A, B). Laterodorsal to it a few cells begin to migrate from the ependymal layer to form a primordial cornu ammonis in which the migrating neuroblasts are grouped in clusters of 3 to 5 cells to constitute a combined marginal-mantle layer (Fig. 2 B).

14 to 18 mm (6.5 to 7 weeks). The most marked change in the primordial hippocampal formation in the 14 mm embryo is its great increase in extent with little added differentiation. In other words, the major change is an increase in size mainly due to cell proliferation. By 18 mm, however, the characteristic cell clusters of the primordial cornu ammonis increase in size and number and separate slightly from the ependymal layer (Fig. 2 D).

At 18 mm also, the marginal zone of the gyrus dentatus anlage is relatively wide and distinct, although neuroblasts in it are still rare (Fig. 2 D).

20.5 to 27.1 mm (7,5 to 8.5 weeks). In the 20.5 and 20.7 mm embryos, the primordial hippocampal formation begins to resemble that of older fetuses (compare Fig. 3 A with Fig. 3 B–D). In the primordial cornu ammonis, the cell clusters are more numerous than at 18 mm, extend a greater distance along the wall, and begin to separate from the underlying ependymal layer to form a primordial pyramidal layer (Fig. 3 A). In the gyrus dentatus anlage, there is a slight increase in the width of the marginal zone (Fig. 3 A). This zone also contains a few more scattered neuroblasts than at 18 mm, with some near the external limiting membrane, but there is no indication of a cell layer.

The primordial pyramidal cell layer of the cornu ammonis region at 27.1 mm appears compressed and is indistinct (Fig. 3 B), in part because migrating cells are scattered deep to it. Near the ependymal layer, however, other neuroblasts are forming a layer of migrating cells that is just separating from the ependymal layer. In the gyrus dentatus anlage neuroblasts are scattered through the marginal zone and some have collected beneath the external limiting membrane (marginal cells, BROWN, 1965). More neuroblasts are migrating from the ependymal layer than earlier (compare Figs. 3 A and 3 B), the greatest number being near the cornu ammonis.

32 mm (9 weeks). By 32 mm (Fig. 3 C), the hippocampal formation has increased both in thickness and in dorsoventral extent. In the cornu ammonis, the pyramidal layer is distinct and essentially continuous, but the cell clusters that form it remain quite evident. Although it does not yet have the subdivisions characteristic of the mammalian pyramidal layer, because it is a continuous lamina it will be designated as the definitive pyramidal layer henceforth. The layer of neuroblasts migrating from the ependymal layer is more distinct than at 27.1 mm (compare Figs. 3 B and 3 C) and is continuous ventrally with a wider layer of migrating neuroblasts in the gyrus dentatus anlage (Fig. 3 C). There the layer of migrating cells is larger, with more loosely arranged cell clusters than in the cornu ammonis. Toward the fimbria, many neuroblasts are migrating into it from the ependymal layer (Fig. 3 C). In addition, a more superficial zone of migrating cells (*SZ*, Fig. 3 C), aligned with the pyramidal layer of the cornu ammonis, is appearing in the wider and more cellular marginal zone of the gyrus dentatus anlage.

37 mm (10 weeks). Although still narrow, the pyramidal layer of the cornu ammonis is more distinctly separated from both the layer of migrating cells deep to it and the external limiting membrane superficially than at 32 mm (compare Figs. 3 C and 3 D). The migrating cell layer is less clear, partly because the neuroblasts scatter as they begin to join the pyramidal layer.

In the primordial gyrus dentatus, the layer of migrating cells is even more diffuse than in the cornu ammonis, because the more superficial cells have migrated into alignment

Fig. 3. Photomicrographs of sections through the developing hippocampal formation of fetuses of: *(A)*, 20.7 mm CR, # 93 A; *(B)*, 27.1 mm CR, # 116; *(C)*, 32 mm CR, # 33; *(D)*, 37 mm CR, # 34; *(E)* and *(F)*, 44 mm CR, # 85; *(G)*, 48.6 mm CR, # 119. All of these photographs were taken at the same magnification, that shown on Fig. 3 G. *(A)* to *(D)* inclusive are from protargol preparations by the method of BODIAN. The general location of the area photographed in each case is shown by the labels *corn. am.* and *gy. dent.* on the *right* side of Fig. *4 A*. *(E)* and *(F)* are from a toluidin blue and erythrosin preparation and *(G)* is from a thionin and erythrosin series. The general location of the area photographed in *(E)*, *(F)* and *(G)* is indicated by the two *large arrows* on the right side of Fig. *4 A*. For abbreviations see page 105.

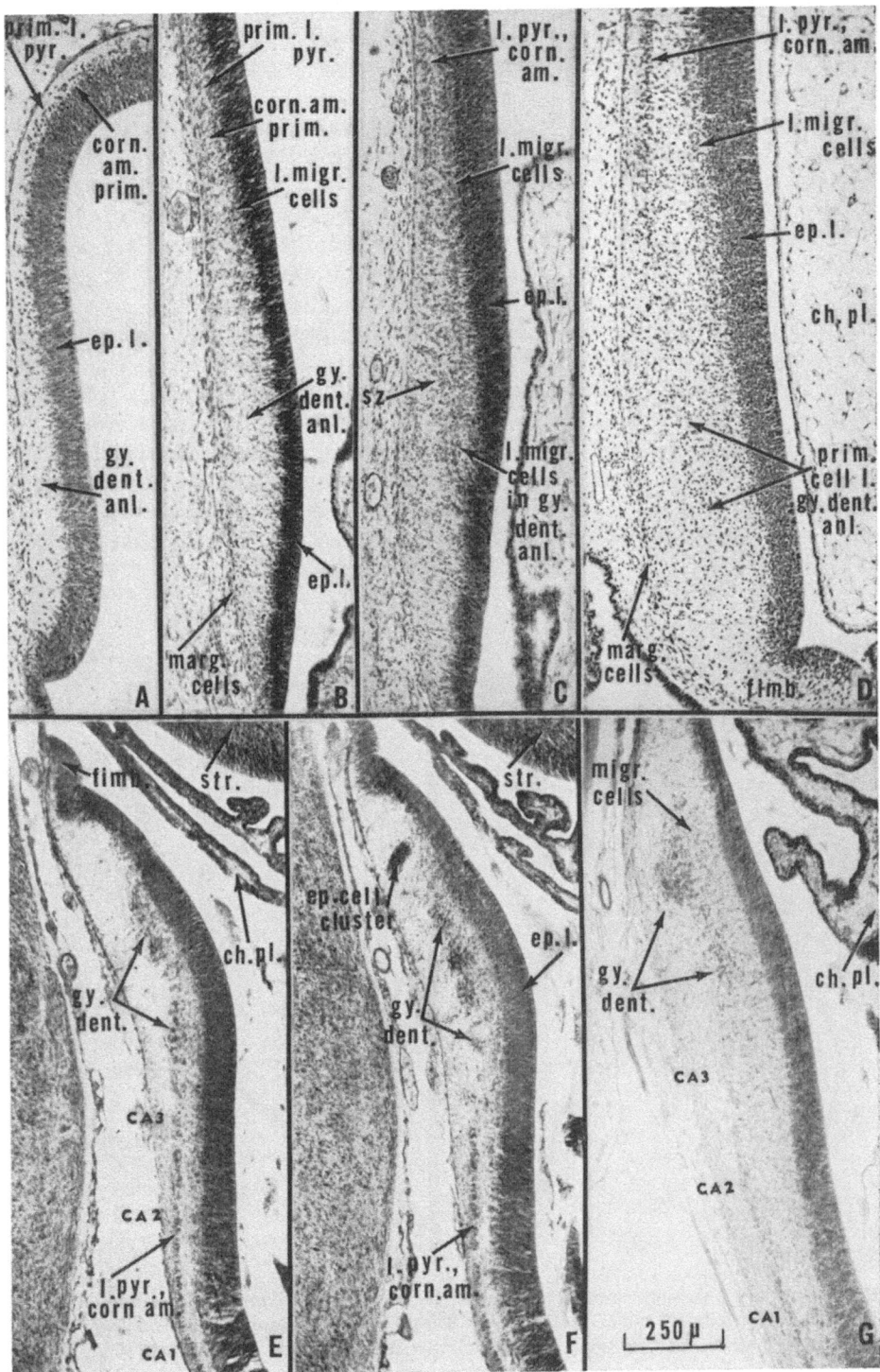

with the pyramidal layer of the cornu ammonis to constitute the beginning of a cell layer in the gyrus dentatus anlage (Fig. 3 D). Neuroblasts are scattered between the indistinct cell layers and many marginal cells are concentrated beneath the external limiting membrane.

44 mm (10.5 weeks). In the cornu ammonis at 44 mm, except near the gyrus dentatus, the layer of migrating cells seen earlier is disappearing as these cells join the pyramidal layer (compare Figs. 3 D and 3 E). The pyramidal layer, although narrow and compact, is distinct. It may be more or less in continuity with the cell layer in the gyrus dentatus (Fig. 3 E) or separated from it (Fig. 3 F). Variations in the arrangement and degree of development of the cells in the cornu ammonis indicate the location of CA 3, CA 2 and CA 1 of LORENTE DE NÓ (1934).

The changes in the gyrus dentatus are more striking than those in the cornu ammonis. Its cell layer may consist of nearly continuous large patches of cells (Fig. 3 E) or more or less separated smaller cell clumps (Fig. 3 F). Near the fimbria, where the cell layer is frequently in contact with the ependymal layer (Fig. 3 E), many neuroblasts are migrating into it. Often dense cell clusters, with the characteristics of ependymal epithelium, have separated from the ependymal layer *en masse* to form semi-isolated cell clusters (Fig 3 F).

48.6 mm (11 weeks). The pyramidal layer of the cornu ammonis, although better represented than at 10.5 weeks, is still a narrow cell band with few neuroblasts migrating

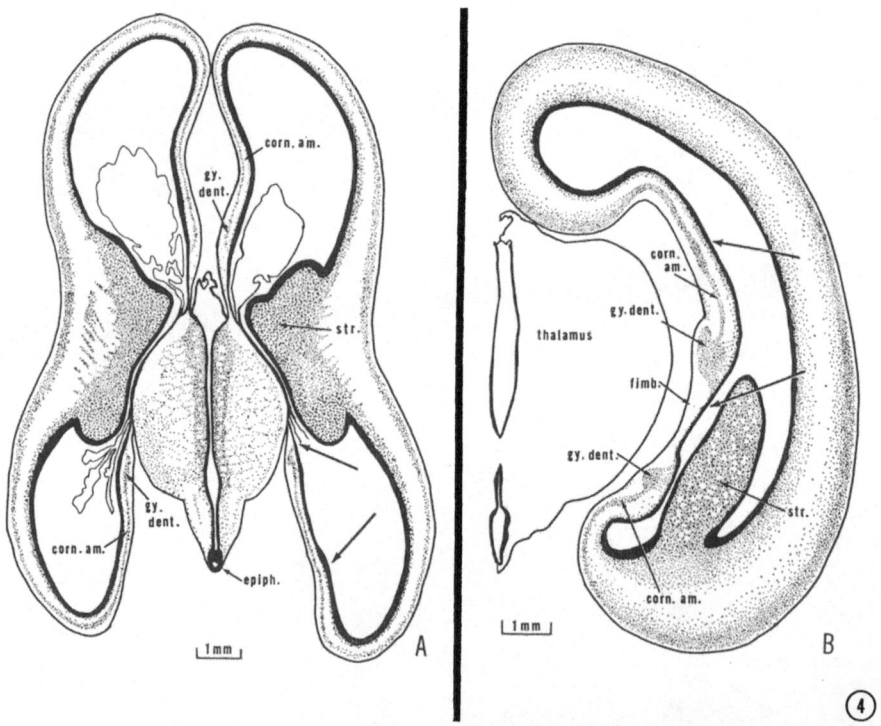

Fig. 4. Drawings of sections through the telencephalon of *(A)* a 48.6 mm CR human fetus (# 119, 11 weeks menstrual age) and *(B)* an 84.5 mm CR fetus (# 95, 13.5 weeks menstrual age) to show the general position of the hippocampal formation photographed in Figs. 5 and 6. For abbreviations see page 105.

into it from the ependymal layer (Fig. 3 G). All traces of a layer of migrating cells are lost, the migrating cells being scattered between the ependymal layer and the pyramidal layer. The diffuse, irregular, poorly developed part of the pyramidal layer near the gyrus dentatus is area CA 3 of LORENTE DE NÓ (1934). The adjacent part of the pyramidal layer, CA 2, is a more distinct but also an interrupted cell band with better developed cells. The pyramidal layer of CA 1 is continuous and compact (Fig. 3 G). Stratum moleculare, stratum lacunosum and stratum radiatum are identifiable, but not demonstrable at the magnification of Fig. 3 G, as they are slightly later in development (Fig. 5 A).

The gyrus dentatus also has undergone significant changes. The large cell patches resembling medullary epithelium, like those in Fig. 3 F, are less frequent. Consequently

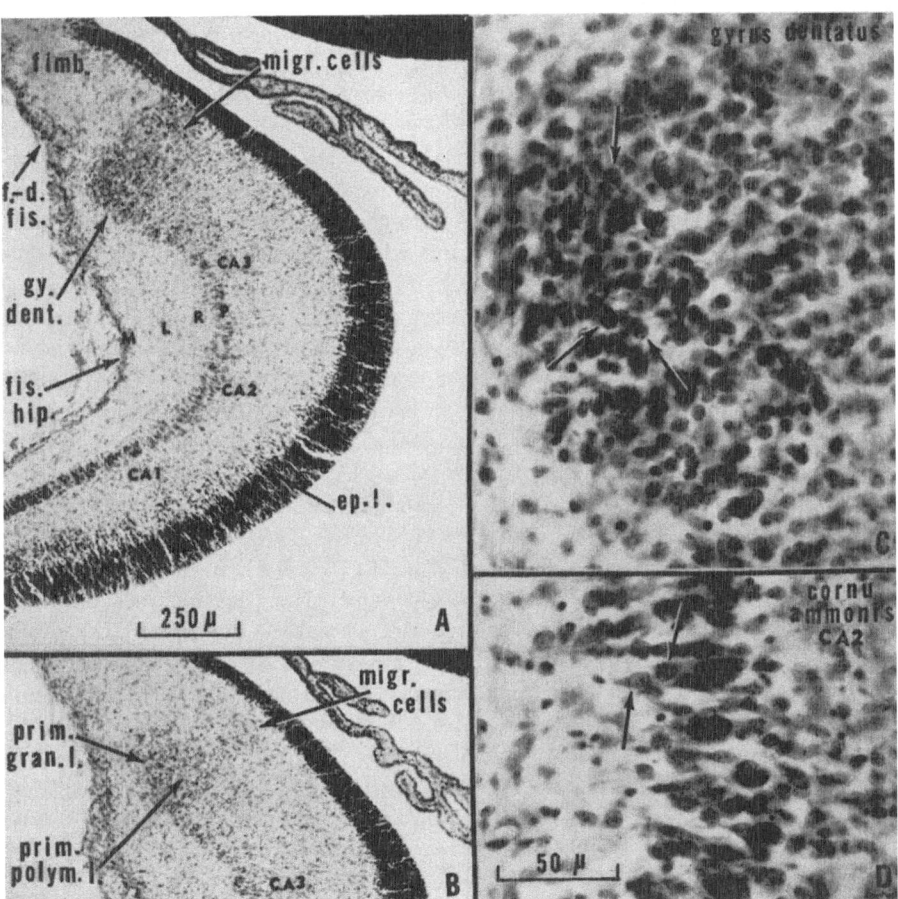

Fig. 5. Photomicrographs of sections through the hippocampal formation of a human fetus of 56 mm CR length (# 148, 11.5 weeks of menstrual age, thionin and eosin stain). *(A)* and *(B)* are taken at the magnification shown on *(A)* from sections 135 μ apart to show the variation in cell migration into the gyrus dentatus. The location of the region photographed is approximately that between the large arrows on Fig. *4 A*. *(C)* and *(D)* are higher power photomicrographs (see scale on 5 D) of the gyrus dentatus and cornu ammonis from the section illustrated in *(A)*. The arrows on *(C)* and *(D)* indicate some of the best developed cells. For abbreviations see page 105.

the cells are more uniformly distributed (Fig. 3 G). An increased migration of neuroblasts thickens the fimbrial end of the gyrus dentatus and may give it a hook-like appearance (Fig. 3 G) as well as increase its cell density. Some cells scattered through it stain intensely and so serve to delimit the gyrus dentatus. Throughout its extent, the migration of neuroblasts into the gyrus dentatus is greater than into the adjacent cornu ammonis. Near the fimbria the cell migration is the most marked.

56 mm (11.5 weeks). The pyramidal layer in the cornu ammonis is much wider than at 11 weeks (Fig. 5 A) and most of the neurons are now oriented perpendicular to the surface, with a process at each end (Fig. 5 D), a characteristic of the double pyramids of Ammon's horn in the adult. Those in CA 3 are most widely spaced (Fig. 5 A) and those in CA 2 are the best developed. In CA 1 the cell lamina is narrower and more compact than elsewhere and the neurons least well developed. Into all three of these areas, neuroblasts are migrating from the ependymal layer, with the greatest intensity of migration into CA 1 and the least into CA 2 (Fig. 5 A).

In the gyrus dentatus (Fig. 5 A–B) the neuroblasts migrating from the ependymal layer are most numerous near the fimbria where an ovoid, ball-like cell mass, continuous particularly on its external surface with area CA 3 of the cornu ammonis, has appeared (HUMPHREY, 1964). Where this cell mass joins Ammon's horn there is an abrupt change from the double pyramid cell type of CA 3, with an oval nucleus and some chromophilic granules (like those in CA 2, Fig. 5 D) to densely arranged, smaller cells with rounded nuclei, no evidence of chromophilic material and no special orientation (Fig. 5 C). These cells constitute a primordial lamina granularis (Fig. 5 B). The deeper, less dense area represents a primordial polymorphic layer. Near the fimbria, the migration of neuroblasts from the ependymal layer may be very conspicuous (Fig. 5 A), as BROWN (1965) found for insectivorous bat embryos, but usually the migrating cells are less numerous (Fig. 5 B).

In the well developed parts of the hippocampal formation, a fimbriodentate fissure and a wide, shallow hippocampal fissure (or sulcus) have appeared (Fig. 5 A). Later the hippocampal fissure is better defined, but at 13.5 weeks it is still a shallow groove (Fig. 6 B).

60.5 and 61.5 mm (12 weeks). There is relatively little change in the cornu ammonis of fetuses between 11.5 weeks (Fig. 5 A) and 12 weeks (Fig. 6 A). By 12 weeks, however, fibers of the alveus have collected along the ependymal layer near the gyrus dentatus so that all layers of the cornu ammonis are represented and the pyramidal layer is wider (compare Figs. 5 A and 6 A). In the gyrus dentatus (Fig. 6 A), the primordial granular layer is more extensive and more distinct than at 11.5 weeks (Fig. 5 A) and the primordial polymorphic layer is less dense (compare Figs. 5 A and 6 A).

79 and 84.5 mm (13.5 weeks). All layers and subdivisions of both the gyrus dentatus and the cornu ammonis are represented at 13.5 weeks in fetuses with the hippocampal formation well developed. The layers are distinct in the cornu ammonis and the alveus is present throughout (Fig. 6 B). The double pyramids are identifiable in all three zones, but best differentiated in CA 2 and least developed in CA 1.

In the gyrus dentatus, the most obvious change is the beginning differentiation of a definitive granular layer as the cells become aligned parallel with each other (Fig. 6 B). As this occurs, the granular layer separates from the pyramidal layer of Ammon's horn and superpositioning occurs. Also accompanying the separation is a marked migration of neuroblasts from the ependymal layer into the still undifferentiated portion of the primordial granular layer near the ependymal layer. As this migration continues, the ependymal layer becomes thin in a small area near the fimbria (Fig. 6 B), as noted by

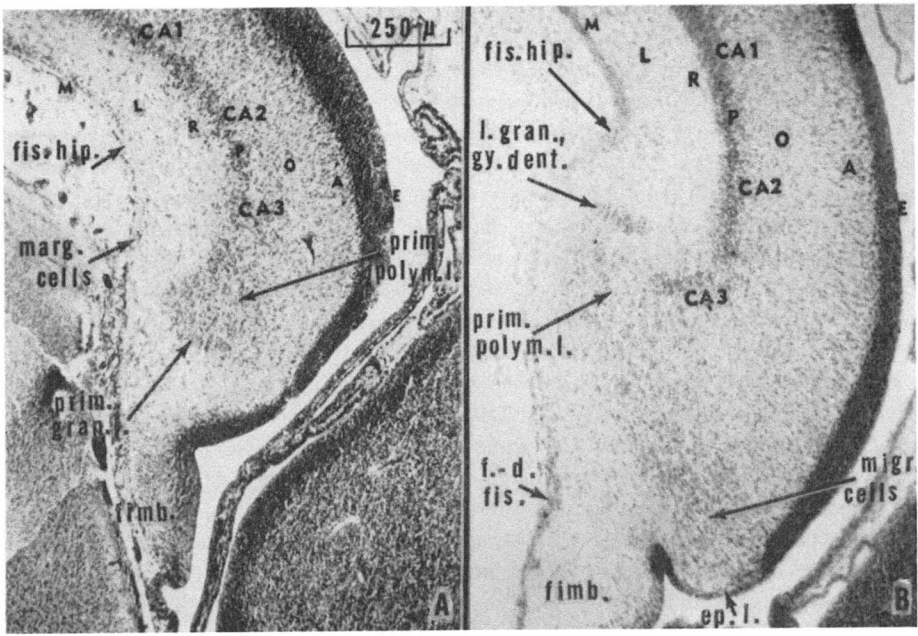

Fig. 6. Photomicrographs of sections from *(A)* a 12-week human fetus (# 39, 61.5 mm CR length, protargol preparation) and *(B)* a 13.5-week human fetus, (# 118, 79 mm CR length, thionin and eosin stain). Both parts of this figure are at the same magnification, the scale shown on *(A)*. The region photographed is approximately that between the two large, unlabeled arrows on Fig. *4 B*. For abbreviations see page 105.

BROWN (1965). The cells of the primordial polymorphic layer have not yet begun to differentiate and it remains in continuity with the pyramidal layer of Ammon's horn after the definitive granular layer begins to overlie it (Fig. 6 B).

Discussion

The human hippocampal formation is first represented by a cell-free marginal zone in the dorsomedial wall of the hemisphere (9 mm). This area is the anlage of the gyrus dentatus and not the primordial cornu ammonis as stated by HOCHSTETTER (1919, anlage of hippocampus), HINES (1922, primordium hippocampi) and BARTELMEZ and DEKABAN (1962). However, the earliest migration of neuroblasts from the ependymal layer is into the adjacent primordial cornu ammonis (about 11 mm). Scattered neuroblasts appear in the cell-free marginal zone of the gyrus dentatus anlage at 18 mm, but only after small cell clusters have migrated into the primordial cornu ammonis. Thus the gyrus dentatus anlage appears first but development does not progress beyond an enlargement of its cell-free marginal zone until after clusters of neuroblasts develop in the primordial cornu ammonis area. The marginal cells (BROWN, 1965) that collect under the external limiting membrane in the region ventral to the primordial cornu ammonis do not "slip along the marginal velum" to develop into the gyrus dentatus as stated by HINES (1922). These cells are only rarely in contact with the cell lamina of gyrus dentatus and contribute few cells to it, if any. By 13.5 weeks, part of the marginal cells become incorporated into the stratum lacunosum, as noted also by BROWN (1965).

Although the gyrus dentatus is represented by a cell-free marginal zone before any changes appear in the region that develops into the cornu ammonis, the earliest migration of neuroblasts, the first cell layer formation and the earliest differentiation within the cell lamina of the hippocampal formation all occur in the cornu ammonis. At 11 weeks, before the cell layer of gyrus dentatus has thickened into an undifferentiated ball-like cell mass, the three subdivisions of the cornu ammonis (CA 1, CA 2, and CA 3) and the superficial strata (stratum moleculare, stratum lacunosum and stratum radiatum) are all identifiable. However, the subdivision of the cornu ammonis adjacent to the gyrus dentatus (CA 3) is not so well represented as the other two zones (CA 2 and CA 1) until a ball-like cell mass begins to form in the gyrus dentatus (see Figs. 3 G, 5 A–B and 6 A). Also, by 13.5 weeks, when a definitive lamina granularis is only beginning to appear in the gyrus dentatus, all layers of the cornu ammonis are present.

Before neuroblasts migrate far enough from the ependymal layer to form a definite cell lamina in the human hippocampal formation of 18 to 22 mm embryos, it resembles the hippocampal formation of the lungfish (SCHNITZLEIN and CROSBY, 1966) and of tailless amphibians like the frog (HERRICK, 1921; HOFFMAN, 1963, 1965) in which a few neurons are separating from the ependymal layer in the cornu ammonis region. After a continuous cell band is formed in the cornu ammonis and gyrus dentatus at 44 mm (Fig. 3 E–F) the embryonic human hippocampal formation is comparable to that of reptiles, such as *Lacerta* (LEVI, 1904a), the alligator (CROSBY, 1917), the chameleon (SHANKLIN, 1930), and the snake (LEVI, 1904a; CAREY, 1965) and to that of those birds nearer the main evolutionary line, like the water birds (CRAIGIE, 1930; KAPPERS et al., 1936; CROSBY et al., 1965) that have a pars dorsalis (cornu ammonis) and a pars dorsomedialis (gyrus dentatus) in continuity with each other.

For some reptiles, the gyrus dentatus overlies the cornu ammonis to form the super-positio medialis of KAPPERS (1929), KAPPERS et al. (1936). In human development, no comparable stage appears. Instead the earlier developing pyramidal layer of the cornu ammonis overlies the later appearing layer of migrating cells in the gyrus dentatus anlage (Fig. 3 C) and neuroblasts move outward from the layer of migrating cells to form a cell layer in the gyrus dentatus in alignment and in continuity with the pyramidal layer of cornu ammonis. This continuity is maintained until the definitive granular layer of the gyrus dentatus appears (Fig. 6 B). The superpositio medialis then formed, with the definitive granular layer overlying the pyramidal layer of the cornu ammonis, is characteristic of the monotreme, marsupial and higher mammalian hippocampal formations.

In the development of the human gyrus dentatus, an intermediate stage appears that has not been described for any adult vertebrate. In this stage the cell layer of the gyrus dentatus thickens into an undifferentiated ball-like cell mass. Along the outer surface of this mass (illustrated also by MACCHI, 1951), the cells concentrate to form a primordial granular layer. As the definitive granular layer develops, it breaks away from the pyramidal layer of the cornu ammonis. For the brain of a fetal monotreme *(Echidna)*, a drawing by ELLIOT SMITH (1910) illustrates the ball-like cell mass stage in development. However, in an adult monotreme (*Ornithorhynchus*, HILL, 1893; HINES, 1929; ABBIE, 1939) and in marsupials (LOO, 1931; HAMEL, 1965; ABBIE, 1939), the characteristic mammalian type of gyrus dentatus is present. Some adult vertebrate not yet studied (or perhaps now extinct like those mentioned by EDINGER at this symposium) may have a ball-like cell mass for the gyrus dentatus, like that found in 11.5- and 12-week human fetuses.

The polymorphic layer of gyrus dentatus develops from the deep part of the undifferentiated cell mass forming gyrus dentatus at 11.5 to 12 weeks, but remains in continuity with the pyramidal layer. On the basis of the developmental history of the polymorphic layer as part of the gyrus dentatus, as well as its embryonic and adult position, this layer is the polymorphic layer of the gyrus dentatus (CAJAL, 1911; KAPPERS et al., 1936 and others; nucleus of the fascia dentata of ELLIOT SMITH, 1896, 1910) and not part of the cornu ammonis as might be indicated by its continuity with the pyramidal layer in the adult (CA 4 of LORENTE DE NÓ, 1934 and h 5 of ROSE, 1926).

After the gyrus dentatus enlarges and begins to rotate toward the cornu ammonis, a shallow hippocampal fissure is consistently identifiable opposite the cornu ammonis. There is no comparable fissure in reptiles and birds and this fissure is not homologous with that labeled hippocampal fissure for lower vertebrates by JOHNSTON (1923), HINES (1923) and others, but not by CROSBY (1917). Likewise, it is not the hippocampal fissure identified by HINES (1922) in early stages of human development for her hippocampal fissure lies opposite the gyrus dentatus anlage. It could not be identified in the 20.7 to 37 mm embryonic material studied (see Fig. 3 A–D).

The cell layer in the gyrus dentatus first appears adjacent to the pyramidal layer of the cornu ammonis and develops progressively toward the fimbrial region (Fig. 3 A–D). However, it is neither formed by the growth of cells downward from the cornu ammonis, as implied by ELLIOT SMITH (1910), or by the migration of cells dorsalward along the cornu ammonis, as stated by HINES (1922). Instead, the cells of the gyrus dentatus migrate directly outward from the ependymal layer, as LEVI (1904 b) found for the mouse and the dog and BROWN (1965) for insectivorous bats.

Summary

The human hippocampal formation first appears as a cell-free marginal zone, identified by HINES in 1922 as the primordium hippocampi (i.e. primordial cornu ammonis). However, an indefinite cell layer is present in the primordial cornu ammonis before neuroblasts migrate from the ependymal layer into this marginal zone to form the gyrus dentatus. The cell layer of the gyrus dentatus first develops next to the cornu ammonis and then progresses toward the fimbria. Marginal cells under the external limiting membrane (gyrus dentatus of HINES) contribute to the gyrus dentatus only slightly, if at all.

The hippocampal fissure is not identifiable until the cells of gyrus dentatus form a ball-like mass (11.5 weeks). This gyrus then rotates toward the cornu ammonis and the hippocampal fissure appears. This fissure has no submammalian homologue.

The pyramidal layer of the cornu ammonis is still in continuity with the surface of the cell mass constituting the gyrus dentatus when an increase in its density superficially forms a primordial granular layer. Continuity is lost when the primordial granular layer develops a cortex-like appearance (13.5 weeks). The granular layer then overlaps the cornu ammonis. The cells deep to the granular layer form a primordial polymorph layer. By 10.5–11 weeks the parts of the cornu ammonis (CA 1, CA 2, CA 3) and its superficial laminae are identifiable; by 13.5 weeks all laminae are present. The characteristic interlocking of the gyrus dentatus with the cornu ammonis develops later.

References

ABBIE, A.A.: J. comp. Neurol., *70* (1939), 9–44.

BARTELMEZ, G.W., A.S. DEKABAN: Contr. Embryol. Carneg. Inst., *37* (1962), 13–32.

BROWN, J.W.: In: HASSLER, R., H.STEPHAN (ed.), Evolution of the Forebrain. Thieme, Stuttgart 1966, 92–103.

CAJAL S. RAMÓN Y: Histologie du Système Nerveux. Vól. II A. Maloine. Paris 1911.

CAREY, J.H.: In: HASSLER, R., H.STEPHAN (ed.), Evolution of the Forebrain. Thieme, Stuttgart 1966, 73–80.

CRAIGIE, E.H.: J. comp. Neurol., *49* (1930), 223–357.

CROSBY, E.C.: J. comp. Neurol., *27* (1917), 325–402.

CROSBY, E.C., B.R. DE JONGE, R.C.SCHNEI-DER: In: HASSLER, R., H.STEPHAN (ed.), Evolution of the Forebrain. Thieme, Stuttgart 1966, 117–135.

CROSBY, E.C., T.HUMPHREY, E.W. LAUER: Correlative Anatomy of the Nervous System. Macmillan. New York 1962.

EDINGER, T.: In: HASSLER, R., H.STEPHAN (ed.), Evolution of the Forebrain. Thieme, Stuttgart 1966, 153–161.

ELLIOT SMITH, G.: Anat. Anz., *12* (1896), 119–126.

ELLIOT SMITH, G.: Lancet, *1* (1910), 147–153.

ELLIOT SMITH, G.: Cunningham's Textbook of Anatomy ed. by Arthur Robinson. 5th ed., pp. 622–627. William Wood & Co., New York 1923.

ESCOLAR, J.: An. Anat., *8* (1959), 215–231.

HAMEL, E.: In: HASSLER, R., H.STEPHAN (ed.), Evolution of the Forebrain. Thieme, Stuttgart 1966, 81–91.

HAMILTON, W.J., J.D. BOYD, H.W. MOSSMAN: Human Embryology, 3rd ed. Williams and Wilkins Co. Baltimore, Maryland 1962.

HERRICK, C.J.: J. comp. Neurol., *33* (1921), 213–280.

HEWITT, W.: J. Anat. (Lond.), *96* (1962), 355–358.

HILL, A.: Phil. Trans. Roy. Soc. London, Ser. B, *184* (1893), 367–387.

HINES, M.: J. comp. Neurol., *34* (1922), 73–171.

HINES, M.: J. comp. Neurol., *35* (1923), 483–537.

HINES, M.: Phil. Trans. Roy. Soc. London, Ser. B, *217* (1929), 155–287.

HIS, W.: Die Entwicklung des menschlichen Gehirns während der ersten Monate. Hirzel. Leipzig 1904.

HOCHSTETTER, F.: Beiträge zur Entwicklungsgeschichte des menschlichen Gehirns. Vol. I. Deuticke. Wien-Leipzig 1919.

HOFFMAN, H.H.: J. comp. Neurol., *120* (1963), 1–52.

HOFFMAN, H.H.: In: HASSLER, R., H.STEPHAN (ed.), Evolution of the Forebrain. Thieme, Stuttgart 1966, 61–72.

HUMPHREY, T.: Trans. Amer. neurol. Ass., *89* (1964), 207–209.

HUMPHREY, T.: Ala. J. med. Sci., *3* (1966).

JOHNSTON, J.B.: J. comp. Neurol., *23* (1913), 371–478.

JOHNSTON, J.B.: J. comp. Neurol., *35* (1923), 337–481.

KAPPERS, C.U. Ariens: The Evolution of the Nervous System. Bohn. Haarlem 1929.

KAPPERS, C. U. ARIENS, G. C. HUBER, E. C. CROSBY: The Comparative Anatomy of the Nervous System of Vertebrates, including Man. Macmillan Co, New York 1936; Hafner, New York 1960.

KUHLENBECK, H , E. V. DOMARUS: Anat. Anz., *53* (1920), 316–320.

LEVI, G.: Arch. ital. anat. embriol., *3* (1904a), 235–247.

LEVI, G.: Arch. mikr. Anat. Entwickl. *64*(1904b), 389–404.

LOO, Y.T.: J. comp. Neurol., *52* (1931), 1 bis 148.

LORENTE DE NÓ, R.: J. Psychol. Neurol (Lpz.), *46* (1934), 113–177.

MACCHI, G.: J. comp. Neurol., *95* (1951), 245–305.

PEELE, T.L.: The Neuroanatomic Basis for Clinical Neurology. 2nd ed. McGraw-Hill Book Co., Inc. New York and London 1961.

ROSE, M.: J. Psychol. Neurol., *34* (1926), 1–111.

SCHNITZLEIN, H.N., E.C. CROSBY: J. Hirnforsch., (1966), in press.

SHANKLIN, W.M.: Acta zool., *11* (1930), 425 to 490.

STREETER, G.L.: Contr. Embryol. Carneg. Inst., *11* (1920), 143–170.

TILNEY, F.: Bull. neurol. Inst. (N.Y.) *7* (1938), 1–77.

TRYPHENA HUMPHREY, M. D.
Department of Anatomy, University of Alabama Medical Center
1919 – 7th Avenue South, Birmingham, Alabama, USA

Evidence for some of the Trends in the Phylogenetic Development of the Vertebrate Telencephalon*

By E. C. Crosby, B. R. De Jonge and R. C. Schneider,
Ann Arbor, Mich.

In this brief account of the shifts in position and changes in size and in differentiation of certain structures in the medial and lateral walls of the vertebrate telencephalon, it is understood that one is not dealing with a direct line of evolution of these structures. The brains considered are those of representative forms which occupy positions on side branches and in some cases on almost terminal branchlets of the phylogenetic tree. Of these forms, the lungfish is probably the nearest and the higher teleosts and higher birds the farthest away from the main trunk. Nuclear patterns and relations and fiber tracts appearing in many divergent forms are supposed to be common characteristics of the brains of animals no longer known to exist, which held positions along the main trunk of this phylogenetic tree.

Hippocampal Formation

The cyclostome hippocampal formation (42, 49, 60–61, 82) lies in the dorsomedial unevaginated portion of the telencephalon (Fig. 1 A). In many fishes (78, 80, 84) (Fig. 1 B–C), it occupies the medial and then the dorsomedial telencephalic wall, bordering on the unpaired median ventricle. In lungfishes (81), amphibians (7, 13, 46–48, 50, 53–54), reptiles (11, 18, 29, 39, 63, 73, 83), birds (14–16, 31–32, 56), and mammals lacking a corpus callosum (33, 41), the hippocampal formation extends along the medial and dorsomedial wall of the evaginated cerebral hemisphere (Figs. 2, 3 B, 4 A–C, 5 A, 6 A–B).

In cyclostomes (42, 72, 82) (Fig. 1 A) and tailed amphibians (13, 46–48) the hippocampal formation is substantially alike throughout. In other submammals (53–55, 78, 80) (Figs. 1 B–C, 2, 4 A–B, 5 A), it may usually be subdivided into areas broadly comparable to the mammalian (Fig. 3) a) anterior continuation, b) cornu ammonis, c) dentate gyrus, and d) subiculum but these are not so well differentiated as, and differ in certain functional respects from, the mammalian subdivisions. In cyclostomes, fishes, and amphibians, the hippocampal gray (Figs. 1, 2) is periventricular in position but has modified pyramidal cells in fishes (80), frogs (54) and reptiles (18). A tendency toward cortex formation (48, 54) is suggested by some more peripherally located cells (Fig. 2 A) in the anuran hippocampal formation. A cortical arrangement (18) appears in the dentate gyrus of various reptiles (Figs. 4 A–B, 6 A–B) and some birds (16) (Fig. 5 B) and is present in all major subdivisions of the hippocampus in most mammals (Fig. 3 D). In some forms such as higher birds

*) This work was supported by a grant from Mr. Alvin M. Bentley, to whom the authors wish to express their sincere appreciation.

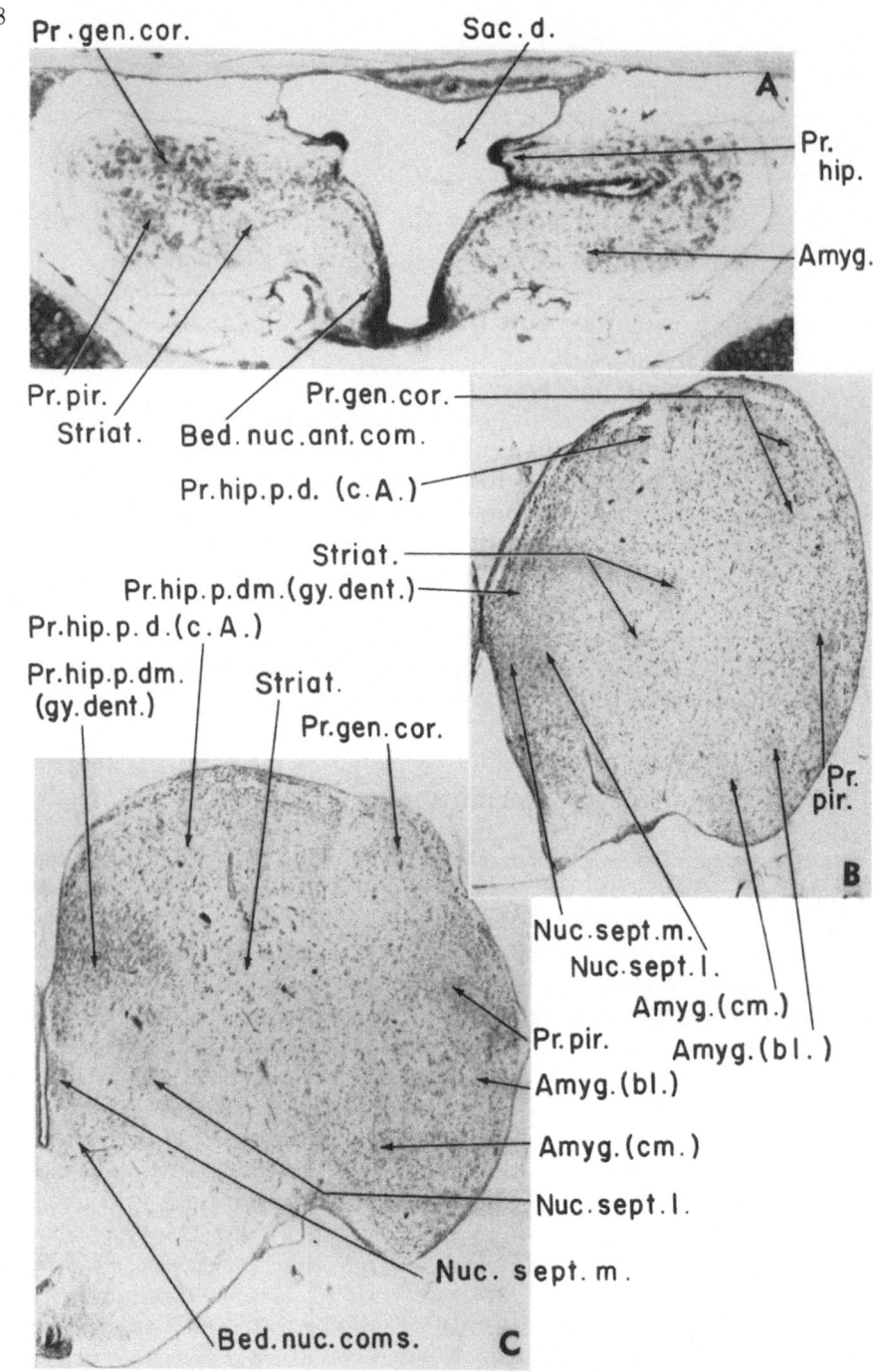

Fig. 1. A Photomicrograph through caudal level of the telencephalon of a cyclostome *(Petromyzon fluviatilis)*. Thionin preparation. x 83.
B and C Photomicrographs through the more rostral (B) and more caudal (C) levels of the telencephalon of the yellow perch *(Perca flavescens)*. Thionin preparation. x 16.5. For abbreviations see page 132.

[sparrow (56), bluejay] where the olfactory bulbs are very small, the hippocampal formation, and especially the gyrus dentatus, is small or lacking rostrally. Since the gyrus dentatus is a correlative, rather than a primarily receptive, area for olfactory, a lessened amount of gustatory or other visceral impulses relayed to it over the medial forebrain bundle must also be considered in assessing causes for its reduction or lack in such vertebrates as higher birds and some anosmatic mammals as certain whales (1, 10).

In mammals with a corpus callosum, the hippocampal formation shifts posteriorly, its most posterior inferior portions being drawn downward and forward with the development of the temporal pole in higher mammals. Remnants of the formation above the corpus callosum constitute the mammalian induseum griseum (Fig. 3 C). A septohippocampal nucleus (Fig. 3 C) extends through the precommissural septal area and septum pellucidum from the anterior continuation toward the retrosplenial portion of the hippocampus.

The anterior continuation of the hippocampus (Figs. 3 B, 4 B) receives terminal fibers of the medial olfactory tract in vertebrates in general (for example, opossum) (41). Olfactory impulses may discharge to the primordial gyrus dentatus directly over the medial olfactory tract in some lower forms but certainly, in higher vertebrates, reach this gyrus only after relay in the anterior continuation of the hippocampus or in septal or tubercular areas (21). Rostrally, the hippocampal formation (4, 18, 22, 37, 55, 71, 74, 76, 78) is interrelated with the septal areas, the olfactory tubercle, and the preoptic and the anterior hypothalamic areas by fascicles of the precommissural fornix (which pass through the septal areas rostral to the anterior commissure) and by fascicles accompanying the medial forebrain bundle (Fig. 6 A). The medial forebrain bundle (Fig. 6) interrelates particularly gustatory and visceral receptive areas of the caudal hypothalamus with septal, tubercular, and hippocampal regions (18, 22, 25, 53, 55, 59, 65, 69, 81). The postcommissural fornix is represented by scattered bundles in cyclostomes. It is easily identifiable in most vertebrates at commissural levels (8, 21) (Figs. 6 B–C) but is difficult to trace to caudal hypothalamic levels in many submammals [although it has been described to the mammillary body in *Varanus* (68)]. The part of the fornix arising from cortical regions adjoining the hippocampal area can be followed in reptiles to its crossing in the supramammillary decussation but is lost in the midbrain tegmentum. The well known avian (cortico)- septomesencephalic tract (15, 32, 56) (Fig. 5 A), which combines various precommissural fornix and diagonal band components of other forms, arises from parahippocampal and hippocampal areas. The postcommissural fornix has been traced to hypothalamus (particularly to the mammillary body) in mammals (21, 23, 28, 74–76, 89). It carries afferent and efferent fibers with respect to the hippocampus, often with a septal relay in course (88, 89). Space permits no further description here. Fornix fibers from the mammalian cingulate gyrus, and also from parietal cortex, decussate in the supramammillary commissure and enter midbrain (23, 74–75, 88). In the rat, NAUTA (74) traced them to the ventral tegmental nucleus. A hippocampal commissure (Figs. 6 B–C) has been described in many vertebrates. Medial corticohabenular fibers (18, 21, 40, 53, 56, 78, 81), components of the stria medullaris, are present in essentially similar relations in most vertebrates (Figs. 6 B–C). They connect hippocampal and septal areas with the habenula. Association fibers interconnect the hippocampal formation and other primordial or developed cortical areas in all vertebrates (18, 53, 81). They increase in number and specificity in mammals, and especially in primates.

Rostral portions of the hippocampal formation are concerned especially with olfactogustatory correlations. Caudal (or, in mammals, temporal) portions are dominated by the

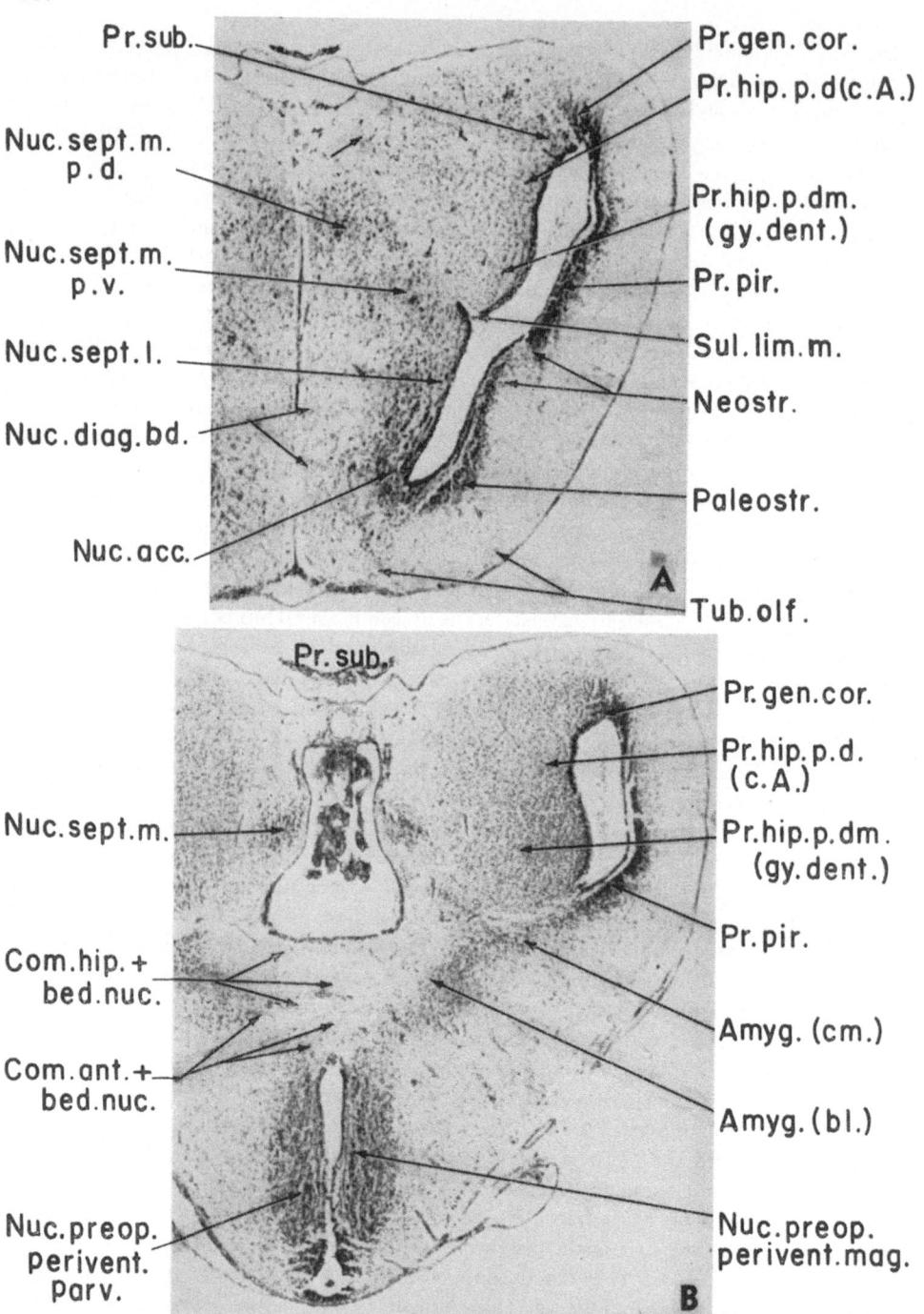

Fig. 2. A and B Photomicrographs of sections through the telencephalon of the frog *(Rana pipiens)*. A illustrates a plane rostral to the anterior commissure and B a plane through the commissure. In A and B *Pr.sub.* is dorsal to *Pr.hip.p.dm.* The arrow in A indicates the outwandered cells of the hippocampus. Thionin preparations. x 18. For abbreviations see page 132.

amygdaloid-piriform and, in primates, amygdaloid-temporal interrelations (77). The hippocampal formation increases between these rostral and caudal regions as other neopallial (temporal, occipital, parietal) and hippocampal interconnections become increasingly important through mammals to man. By means of such interconnections impulses from somesthetic, visual and auditory association areas are placed under the effects of discharges from the hippocampus and also relay to the hippocampus, often by paths synapsing in the hippocampal gyrus.

Septal Area

The septal areas are variously developed in vertebrates. Bed nuclei of the anterior commissure are present in representatives of all vertebrate orders and a bed nucleus of the hippocampal commissure in all such representatives except possibly cyclostomes (Figs. 1, A, C, 2 B, 4 D, 6 B, C). The rest of the septal area of petromyzonts (82) is a small mass of undifferentiated gray. In many lower submammals (Figs. 1 B–C, 2) medial and lateral septal nuclei, a nucleus accumbens, a nucleus of the diagonal band, and scattered cells (potentially bed nuclei) along the stria terminalis and the stria medullaris are demonstrable but variously developed.

In the lungfish (81), the medial septal nucleus and the nucleus of the diagonal band are obscured at some levels by their infiltration by islands of the olfactory tubercle. In urodeles KUHLENBECK (67) identified medial and lateral septal nuclei, periventricular cells and a pars lateralis septi. The medial and lateral septal nuclei and nucleus accumbens and the relations of these masses to the bed nuclei of the commissure were discussed for *Amblystoma* by HERRICK (46). In anurans, the septal nuclei are proportionately large (Fig. 2) and the medial septal nucleus is divisible into dorsal and ventral parts (53–54). A septohippocampal nucleus and a possible representative of part of the mammalian nucleus triangularis have been tentatively identified in anurans.

Medial and lateral septal or parolfactory nuclei (Figs. 4A, C, 6 A) have been recognized in *Sphenodon* (11, 52), turtle (62, 63), alligator (18), lizard (39) and snake. Dorsal septal (12) and septohippocampal nuclei and a possible representative of part of the mammalian nucleus triangularis have been seen in snakes (Figs. 4, C–D) and caimen. A nucleus accumbens septi has been widely recognized (18, 26, 35, 63). In many birds [chicken, dove, humming bird (15)] medial and lateral septal nuclei, a nucleus of the diagonal band, a nucleus accumbens, and bed nuclei of the commissures and the striae are present although not highly developed. In birds such as the bluejay and the sparrow (56) with very small (or fused) olfactory bulbs, the rostral ends of the medial and the lateral septal nuclei and the nucleus accumbens are very small or even lacking at some levels. Often avian septal areas are better developed at or near commissural levels (Fig. 5 A) but neither a triangular nor a dorsal septal nucleus has been found in birds.

All the submammalian septal nuclei thus far mentioned have been identified and several of them subdivided in the shrew (3), the cat (4) and *Galago* (2) by ANDY a. STEPHAN. Their papers provide a convenient pattern for interpretation of the mammalian precommissural septal area. Accounts of the major septal nuclei of many mammals, including man (21), are available, although the dorsal septal, the septohippocampal and often the triangular nuclei, and the secondary subdivisions of the medial and the lateral septal nuclei have not often been identified. Because of limitations of space, no further description of them or of the septum pellucidum will be given.

In all vertebrates, the septal areas are interconnected with the hippocampus and the adjacent primordial cortical areas by fascicles of the precommissural fornix (8, 21–22, 71, 88–89) (Fig. 6 A). The well known medial forebrain bundle (Fig. 6) interrelates the septal areas and the hippocampus with preoptic and hypothalamic regions in all forms except possibly cyclostomes. This bundle is large in anurans (53), moderately developed in reptiles (18) (Fig. 6), small in lower, and very small in higher, birds (56), variable in development in subprimate mammals (8, 37, 40), and reduced in size in primates (21–22). Except in cyclostomes, fibers of the anterior commissure interconnect the septal areas of the two sides. The bed nuclei of the commissure and the nearlying portions of the medial septal nucleus (37–40, 53, 78, 81, 87) are interconnected bilaterally with the amygdaloid areas by fascicles of the striae terminales (Fig. 6 C) and project to the habenulae over the stria medullaris (18, 21–22, 40, 43, 53, 78, 81). The diagonal band (Fig. 6 A) interconnects the medial septal nucleus with the striatal region and probably with the anterior amygdaloid area of the lateral hemisphere wall (18, 53, 81).

Olfactory Tubercle

The olfactory tubercle has not been identified in petromyzonts or in tailed amphibians. In many fishes (78, 80), it has few cells and no layering. However, in the lungfish (81) the very large tubercle projects rostrally, laterally, and caudally beyond the remainder of the telencephalon and its islands infiltrate the striatal areas laterally and the septal areas medially. It shows major subdivisions comparable to those in mammals (see next paragraph). It has molecular, pyramidal, and polymorphic layers and large islands of CALLEJA. In anurans (Fig. 2 A) the olfactory tubercle (54) lacks typical layers and has no islands. The turtle has deeply staining cell clusters without a hilus, which may be considered potential islands. The available material of the olfactory tubercle of the turtle shows no differentiation of cortical and polymorphic layers. Molecular or plexiform, pyramidal, and polymorphic layers (evident but not labelled in Fig. 4 B) and small potential islands without a hilus are demonstrable in the alligator (18) and, at midtubercular levels at least, in the snake (12). The olfactory tubercle is somewhat better represented in lower than in higher birds. No islands are present in the available material. In the sparrow and the bluejay brains, the tubercle is undifferentiated.

The mammalian olfactory tubercle [and that of lungfish (81)] has been divided rostro-caudally into rostral, middle, and caudal subdivision (9). These are evident early in its development in man (58). The rostral subdivision is well developed in the lungfish and in mammals with large olfactory systems; the middle subdivision has correlative functions; the caudal subdivision depends for its development on ascending connections from lower centers. Mediolaterally (9), the mammalian olfactory tubercle (Figs. 3 B–C) shows medial, intermediate, and lateral zones (m, i, l, Fig. 3 C). Such subdivisions are also present in the lungfish (81) and a medial and a combined intermediate and lateral zone are fore-shadowed in anuran and snake brains. Typical layers and islands of CALLEJA (consisting of pyramidal cells, or granule cells, or both cell types) have been widely recognized in mammals but the differences from one form to another are considerable.

In the opossum (41), the tubercle has a greatly folded cortical layer and three types of islands; in rodents, it is well developed and in the rat (40), twelve islands of CALLEJA as well as a large medial island have been identified. Carnivores (36) show a like pattern. Edentates have a highly

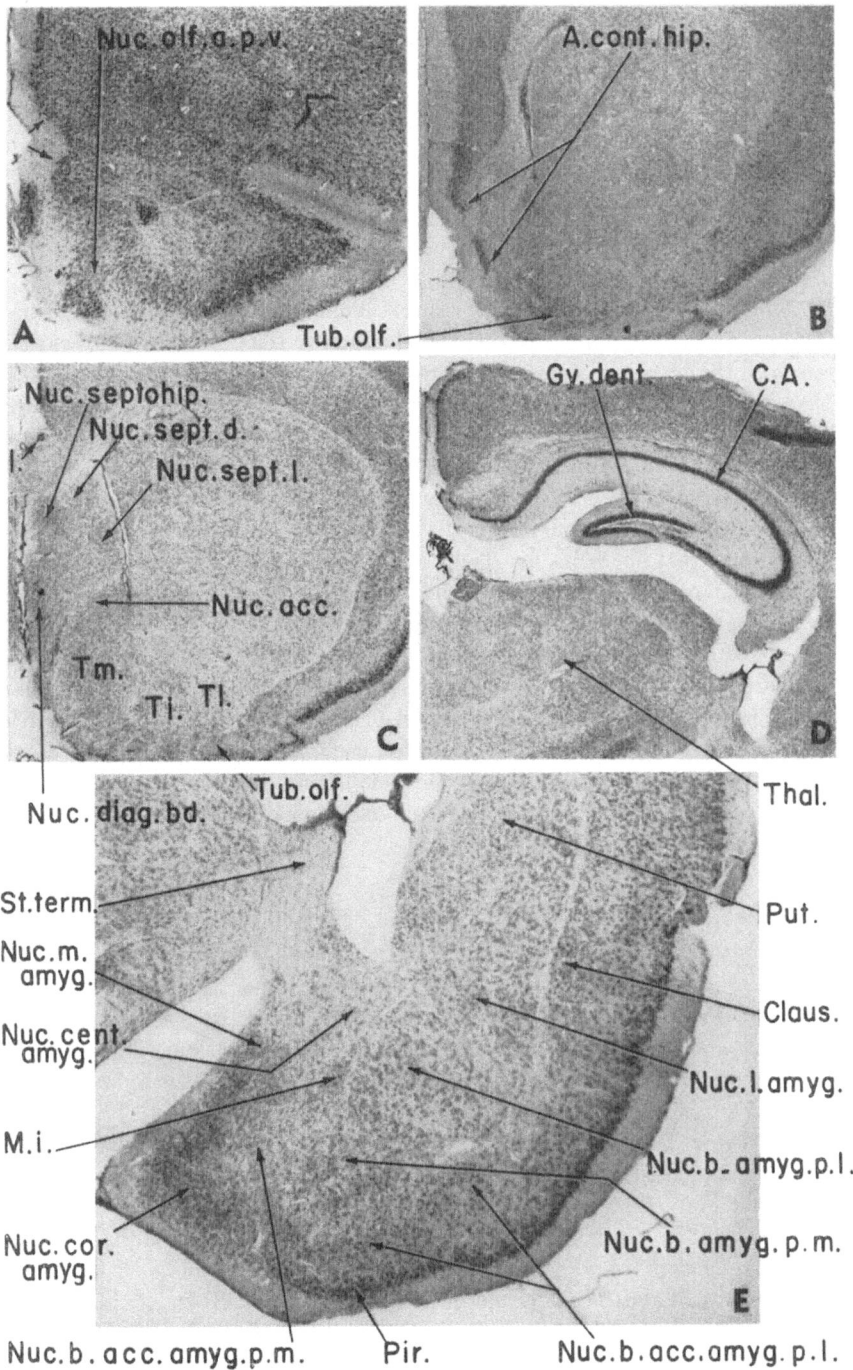

Fig. 3. A–D Photomicrographs of sections from various levels of the white mouse telencephalon to illustrate portions of the hippocampal formation or its vestiges. A and B anterior continuation of hippocampus. x 19. C induseum griseum (I) and septohippocampal nucleus. x 14.5. D well developed hippocampus overlying dorsal thalamus. x 12.4. Thionin preparation. E Photomicrograph illustrating some of the major amygdaloid nuclei in the white mouse. Thionin preparation. x 25. For abbreviations see page 132.

developed olfactory tubercle with secondary subdivisions. In the bat (57) *(Tadarida mexicana)* the rostral part of the tubercle is poorly developed with only the lateral part having layers but caudally the usual lamination occurs with the cortical layer convoluted and the medial island large. In the macaque (69), rostrally, layering occurs only in the lateral region. Farther caudally, layering is present throughout but folding occurs only laterally. All subdivisions of the mammalian olfactory tubercle have been identified in the human embryo (58). Through the midportions and the caudal part of the adult human tuberculum (21) a typical layering and islands can be demonstrated but the layering is not constant and not present all across any one field. It is better developed laterally (where folding sometimes occurs).

In vertebrates in general, where there is a differentiable medial part of the olfactory tubercle, this part receives fibers of the medial olfactory tract. As stated earlier, it interconnects with the precommissural septal area and the hippocampus (Fig. 6 A) and, also, with the contralateral olfactory tubercle (81) by way of the anterior commissure (Fig. 6 B) and receives and distributes fibers to preoptic and hypothalamic areas through the medial forebrain bundle. Functionally it belongs with the other structural areas of the medial hemisphere wall, being a part of this medial olfactovisceral correlative region.

The lateral part (lateral and intermediate zones) of the olfactory tubercle receives lateral olfactory tract fibers (53, 69–70, 80–81) rostrally except in anosmatic forms. It is interconnected with the piriform lobe and with striatal areas in many submammals [fishes (80–81), anurans (53), reptiles] and probably in mammals. At least in the lungfish (81), fascicles from the chief sensory nucleus of V comparable to the avian quintofrontal tract (90) project to its intermediate part. Its lateral part is related to the cerebellum in these same fishes (81), although the direction of conduction is unknown. The relative development of the caudal third of the tubercle depends on the richness of these ascending fascicles. The lateral part of the olfactory tubercle belongs functionally with other structures in the lateral hemisphere wall related to olfactosomatic correlations.

Amygala

The cyclostome amygdala (Fig. 1 A) is a small, undifferentiated cell mass near the caudal end of the telencephalon. It merges with the primordial piriform gray through scattered cells and, caudally, is approached by the primordial hippocampus. In tailed amphibians, a periventricular gray mass, under a ventricular eminence, which is related laterally to the piriform lobe and medially to the hippocampus, constitutes the nucleus amygdalae (47–48). An amygdala has been recognized for some time in fishes (91) and has been subdivided in these forms into an anterior amygdaloid nucleus and basolateral and corticomedial groups (79, 81) (Figs. 1, B–C).

In the frog, GAUPP (38) identified a distinct posterior hemisphere mass and this was later termed the epistriatum and then the amygdala. In frogs and toads, HOFFMAN (54) identified basolateral and corticomedial nuclear groups (Fig. 2 B) and an anterior amygdaloid area similar to those described for fishes and with like connections.

In the brains of various reptiles [*Sphenodon* (11, 29, 34), turtle (64), lizard (39), caiman (Fig. 4 B), snake (Fig. 4 D)] a cortex-like band of cells, often under a ventricular eminence, has been termed the posterior hypopallial ridge or the posterior part of the dorsal ventricular ridge (Figs. 4 A, D). This ridge belongs to the amygdaloid complex, as does the gray ventral and lateral to it (Fig. 4 D). Various subdivisions have been recognized.

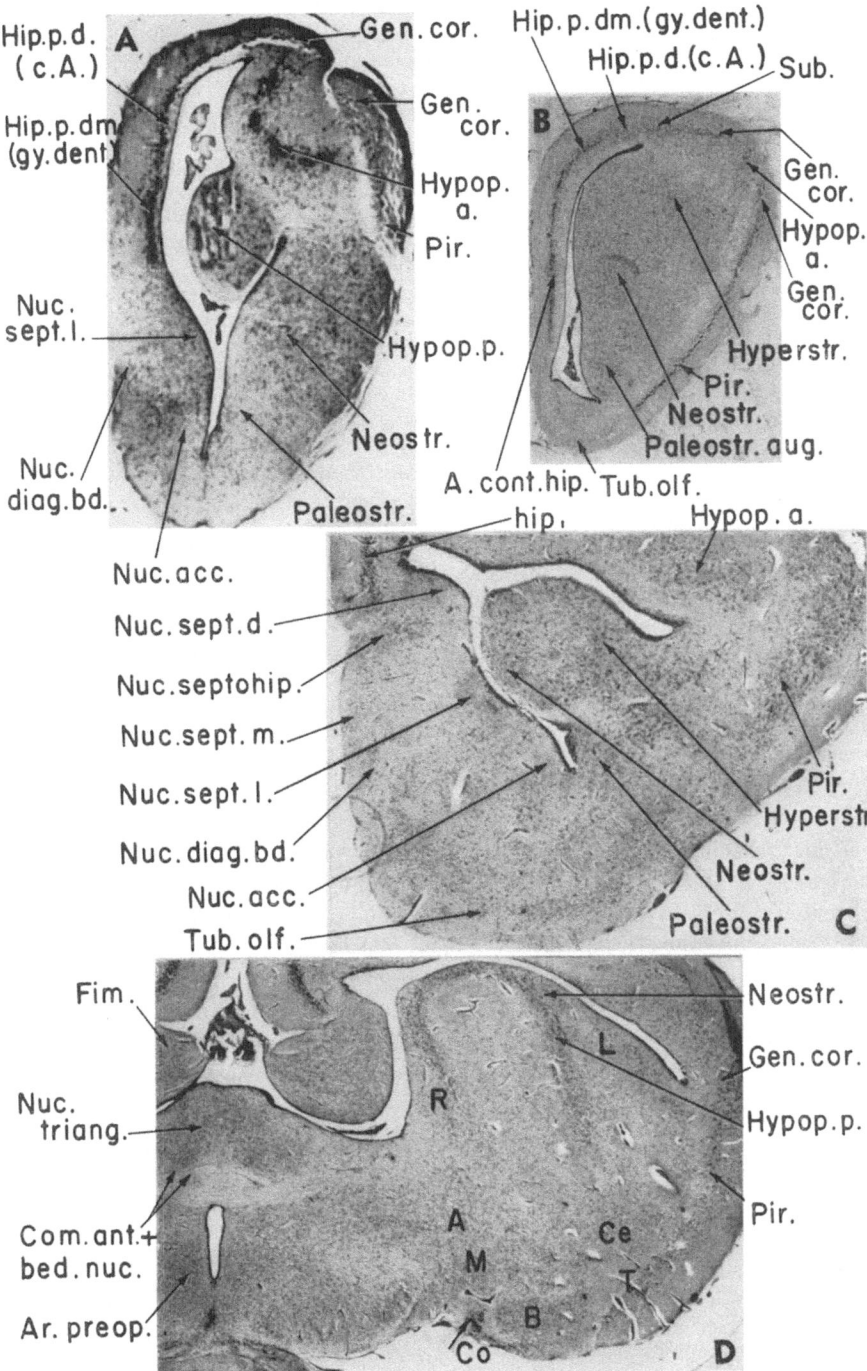

Fig. 4. A–C Photomicrographs illustrating the arrangement of the telencephalic gray at selected levels of brains of: A turtle; B caiman; C snake *(Boa constrictor)*. Magnifications 20.7, 7.4, 16.5 respectively. D Photomicrograph through the amygdaloid complex of the snake *(Boa constrictor)*. x 16.5. Thionin preparations. For abbreviations see page 132.

From the material available it is not possible to compare successfully all the secondary sub-divisions of the amygdaloid complex in reptiles such as the snake with those identified in *Tupi-nambis* by CURWEN (27) or with those described in mammals by various observers. The nucleus of the lateral olfactory tract as identified in the turtle by JOHNSTON (64) is present in the *Boa* (Fig. 4 D, *T*). It receives fibers of the lateral olfactory tract and commissural fibers interconnect the area on the two sides of the brain, characteristic connections which identify the nucleus sufficiently. The cortical nucleus (Fig. 4 D, *Co*) becomes more prominent at more caudal levels and forms the ventral loop of the posterior hypopallial ridge near the caudal end of the hemi-sphere. It contributes fascicles to the stria terminalis. A prominent cell mass in the ventral part of the amygdaloid complex (*B* in Fig. 4 D), medial to the nucleus of the lateral olfactory tract, has been termed the medial nucleus by JOHNSTON (63) (turtle) and GOLDBY (39) and the basal nucleus by JOHNSTON (64). It is probably the nucleus anterior amygdalae of CAIRNEY (11). It contributes fibers to the stria terminalis. This mass is easy to identify but the present authors do not have sufficient evidence to establish its homologies. Dorsomedial to the nucleus basalis of JOHNSTON (64) is a nuclear area (Fig. 4 D, *M*) which contributes fibers to both the stria terminalis and to the amygdalohabenular component of the stria medullaris. It is representative of the ventromedial [alligator (18) and *Sphenodon* (52)] or the medial amygdaloid nucleus [*Sphenodon* (24)] and probably comparable, in part at least, to the bed nucleus of the stria terminalis of JOHNSTON for the turtle and of the avian nucleus taeniae (56). Probably it may be compared with portions of the medial amygdaloid nucleus of mammals (19).

An anterior amygdaloid area (*A* in Fig. 4 D), which receives lateral olfactory tract fibers, and a central nucleus (Fig. 4 D, *Ce*), comparable to that recognized by JOHNSTON (64) and GOLDBY (39) in the turtle and lizard respectively, have been identified in the *Boa*. The prominent commissu-ral connection of the posterior hypopallial ridge (Figs. 4 A, D) through the anterior commissure makes this ridge functionally comparable, in part at least, to the basolateral complex of the mammalian amygdala. Probably to this same subdivision should be allocated the small nuclear masses (*L* and *R* in Fig. 4 D) on either side of the ridge. Still other nuclear groups are determinable in the amygdaloid area but their connections and so their significance are not known at present. However studies are under way to determine more clearly the differentiating fiber connections of the various reptilian amygdaloid nuclei.

The amygdala has been studied in many mammals (Fig. 3 E, mouse). The amygdaloid nuclei are frequently allocated to two major groups (19–21, 57, 92) – a) a corticomedial group, consisting of the medial and the cortical nuclei and the nucleus of the lateral olfactory tract, and b) a basolateral group, containing lateral, basal and basal accessory nuclei, any or all of which may be subdivided. The central amygdaloid and the anterior amygdaloid nuclei occupy rather equivocal positions and are sometimes regarded as independent subdivisions. Intercalate cell groups appear in the zones between the nuclei. The corticomedial group is usually better developed in mammals with relatively large olfactory systems. The basolateral group is larger in higher mammals and overshadows the corticomedial group in man.

The anterior amygdaloid nucleus, present in many mammals, is lacking in some bats (57). The nucleus of the lateral olfactory tract may be a single gray mass [cat (36)] or may be subdivided into rostroventral and caudodorsal parts [shrew (20)], or dorsal and ventral parts [rabbit (92)] or rostral and caudal parts [bat (67), man (19)]. The cortical amygdaloid nucleus has secondary divisions caudally in such mammals as shrew (20) and man (19). The medial amygdaloid nucleus, large in carnivores and well represented in the free-tailed bat (57), fuses with the hippocampus in the fruit-eating bat (86), the rabbit (92), the mole (20), and the shrew (20). It is small in higher forms. It has been identified in man (19) by some but not all observers (51). The central amygdaloid nucleus, often difficult to delimit, is present in many mammals including such forms as the shrew (20) and man (19). The lateral amygdaloid nucleus, identifiable in all mammals, may merge with the basal nucleus [bat (57), man (19)].

In all vertebrate orders from fish to man, lateral olfactory tract fibers distribute to the corticomedial amygdaloid group (21, 57, 69, 79, 81, 87). This distribution is to the anterior amygdaloid area (where present), the medial amygdaloid nucleus, the nucleus of the lateral olfactory tract and the cortical amygdaloid nucleus. The lateral olfactory tract has also been said to discharge to the basal nucleus (17) of the basolateral amygdaloid group. Where present, the accessory bulb contributes fibers to the lateral olfactory tract and these have been traced to the amygdala (but not to specific nuclei) in lungfish (81) and anurans (44, 53) and probably distribute to this complex in all forms in which they occur. The stria terminalis, having a partial crossing in the anterior commissure, interrelates the amygdala with septal, preoptic and anterior hypothalamic areas from fishes to man (8, 18, 21–22, 37, 40, 53, 66, 81). Where differentiable, the dorsal and the ventral olfactory projection tracts (18, 64) (Figs. 6, B–C) have like functions. The nuclei of the lateral olfactory tract in the two hemispheres are interconnected by the anterior commissure (57). A large interamygdaloid connection through this commissure interrelates portions of the basolateral amygdaloid group of the one side with corresponding areas on the other side (21, 40, 81). The interrelations established through such commissural connections and through the association fibers and cell strands which interrelate it with other cortical areas give to the basolateral group the functional character, although not the anatomic structure, of cortex. It is regarded, then, as a vicarious cortex. Amygdalohabenular (or lateral corticohabenular) fibers (8, 14, 18, 21, 53, 55, 81), present in many vertebrates, often arise from the medial amygdaloid nucleus. They pass to the habenula as components of the stria medullaris. Some of them end in the habenula before or after crossing in the habenular commissure. Others are true commissural fibers interconnecting amygdaloid areas (81).

Striatum

The petromyzont striatum (72, 82) (Fig. 1 A) or paleostriatum is a small gray mass extending from the medial part of the evaginated telencephalon into the unevaginated portion where it lies near the ventricle. Caudally, a few larger cells are intermingled with the smaller cells of this unremarkable gray. The area contributes fibers to the ventral peduncle of the lateral forebrain bundle. The existence of fibers to this area from the dorsal thalamus is uncertain. The so-called somatic area (45) or striatum of higher fishes is probably a paleostriatal area, since it is functionally related with both peduncles of the lateral forebrain bundle. The olfactosomatic area (45), also a part of the striatum (Fig. 1, B–C) but overlying the somatic or paleostriatal portion and connected with it by fiber bundles, is interrelated with the dorsal thalamus through the dorsal peduncle of the lateral forebrain bundle but also contributes fibers to its ventral peduncle. It may represent a primitive neostriatal area.

In the brains of lungfishes (81) and anurans (54), differentiating paleostriatal and neostriatal areas can be recognized (Fig. 2 A). The paleostriatal area may be divided into a smaller-celled paleostriatum augmentatum merging with the nucleus accumbens and a paleostriatum primitivum represented by small clusters of larger cells which contribute fibers to the ventral peduncle of the lateral forebrain bundle. The neostriatum of both orders lies dorsal to the paleostriatum and, together with the paleostriatum augmentatum, is interconnected with the dorsal thalamus by fibers of the dorsal peduncle of the lateral forebrain bundle and contributes fibers to the ventral peduncle. A small hyperstriatal area has been suggested for lungfish (81) and anurans (54). HERRICK (46) regarded

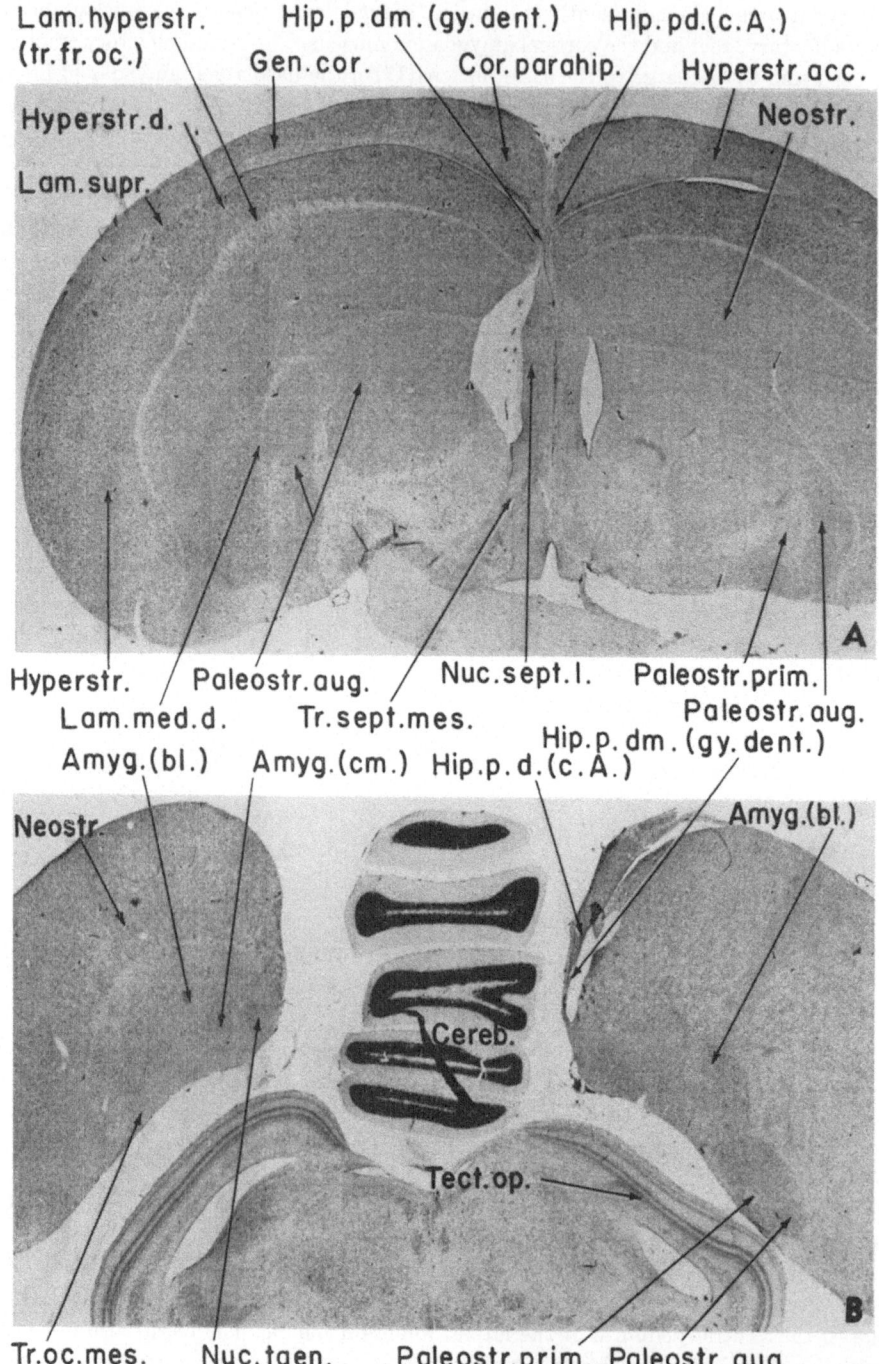

Fig. 5. A and B Photomicrographs of sections through more rostral (A) and more caudal (B) levels of the telencephalon of the bluejay *(Cyanocitta cristata)*. Thionin preparation. x 6.8. For abbreviations see page 132.

a striatal region related to the nucleus accumbens of *Amblystoma* as representative of the head of the caudate nucleus, and dorsal and ventral striatal masses, as representative of the lenticular nucleus. The striatal relations in *Necturus* (47) are comparable but somewhat more obscure. Probably the dorsal striatal region in urodeles corresponds to the anuran neostriatal area, and the area related to the nucleus accumbens, together with the ventral part of the striatal mass, to the paleostriatal areas of the frog and the toad.

The marked differentiation of the lateral telencephalic wall in such reptiles as caiman, lizard (24) and snake foreshadows particularly the development of avian rather than mammalian cerebral hemispheres, although there is a certain parallelism in development. In the interests of brevity, then, the reptilian and the avian striatal areas will be handled together. Paleostriatal, neostriatal and hyperstriatal areas are differentiable in reptilian and avian (5, 8, 56) cerebral hemispheres in relations similar to those seen in the lungfish (81), but much better developed (Figs. 4 A–C, 5). The paleostriatum (Figs. 4 A–C, 5, 6) is divisible into a paleostriatum primitivum and a paleostriatum augmentatum, with the former containing some larger neurons and the latter merging with the nucleus accumbens. The paleostriatum primitivum (Figs. 5, 6 B–C) discharges through the ventral peduncle of the lateral forebrain bundle to the ventral thalamus and midbrain tegmentum. The paleostriatum augmentatum (Figs. 5, 6) is interconnected with the dorsal thalamus by the dorsal peduncle and discharges to the ventral thalamus and the tegmentum by the ventral peduncle of the lateral forebrain bundle. The neostriatum (Figs. 4–6) of reptiles is better developed than that in anurans and less well developed than the comparable area in birds (Fig. 5). It is proportionately very large in some of the highest birds.

In the rostral part of the reptilian lateral hemisphere wall, there is a cortex-like band of cells, related to the primordial general cortex but arching toward the ventricular wall where it often underlies a ventricular elevation (turtle, snake). This band (Fig. 4 A–C) is the anterior part of the dorsal ventricular ridge [turtle (63)] or anterior hypopallial ridge [*Sphenodon*, (11, 34) lizard (39), caimen (Fig. 4 B), snake (12)]. It may very well represent a primordial cortical layer in forms like the alligator (18), but in the chameleon (83) and in birds (55), where the area is separated from the surface by a thin corticoid layer, it is often termed the accessory hyperstriatum (Fig. 5 A). It has been regarded as cortex by some observers (15) and a layering proposed. The layering fell somewhat into disfavor when the layers were described as parallel to the surface by one investigator (15) and subdivisions of the area were found to be from medial to lateral (so perpendicular to the surface) by another (30). Several observers (31, 56), including the senior author, have considered the accessory hyperstriatum to be vicarious cortex, that is an area with cortex-like associative functions but not the position or layering of true cortex.

The accessory hyperstriatum (7, 30–31, 56) and the hyperstriatum [(6–7, 56); hyperstriatum superius (5)] are interrelated with the nearlying cortical fields by association fibers (56). They are interconnected with each other by similar fascicles and intrinsic fibers interrelate the various parts of a hyperstriatal area (56). There are also interconnections with underlying neostriatal and probably paleostriatal areas. Such evidence as is available indicates that specific parts of the dorsal thalamus are connected through the lateral forebrain bundle (Fig. 6) with special parts of the hyperstriatum and that the hyperstriatum is composed of various regions, each with its special functional significance (56). This is supported by recent experimental work (85).

The general pattern of the mammalian striatal areas is well known (21). Space does not permit discussion of it. It is suggested that the submammalian paleostriatum primitivum

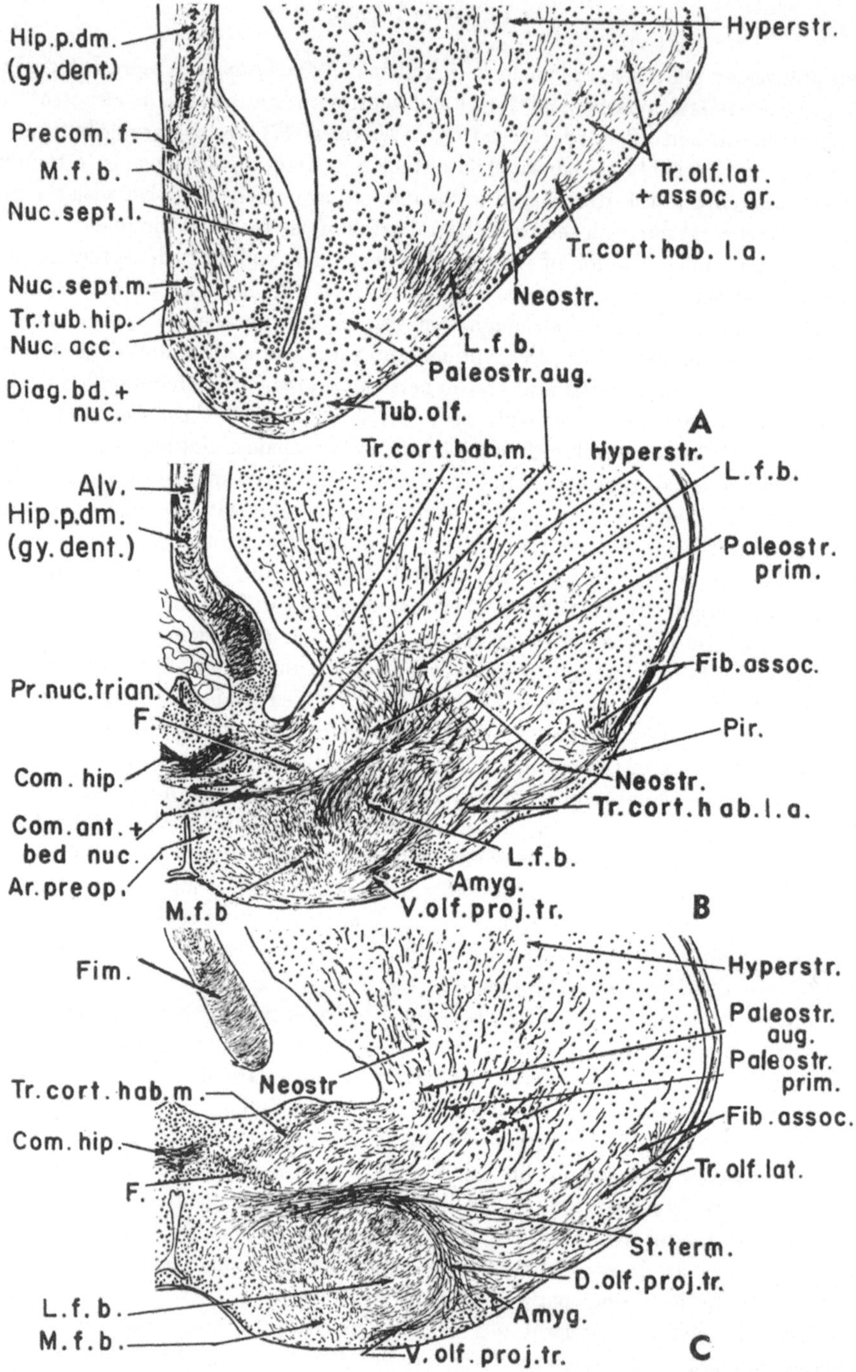

130

Fig. 6. Photographs of three drawings showing fiber paths at precommissural telencephalic (A) and commissural (B, C) telencephalic levels, C being from a plane slightly caudal to that of B. Alligator. These drawings by the senior author are comparable in part to figures 583 and 584 in the book by ARIENS KAPPERS et al. but have been cut down and slightly modified for the present paper. Preparation stained in pyridine silver. x 13. For abbreviations see page 132.

is represented in the inner segment of the globus pallidus. The globus pallidus increases in general from lower through higher submammals to mammals. The paleostriatum augmentatum of lower forms is comparable to those portions of the mammalian caudate nucleus lying in relation to the anterior commissure and merging with nucleus accumbens. The neostriatum of submammals may be comparable to the neostriatum (possibly part of putamen and a large part of caudate) of mammals. The hyperstriatum of reptiles and birds may have some representation in the mammalian putamen. In considerable part the hyperstriatal areas are an attempt on the part of the reptilian-avian line to form highly correlative centers by increasing the gray along the ventricular wall. Such an increased gray together with the accessory hyperstriatal gray would further elaborate correlations and perhaps complex cyclic behavior. It would not provide so excellent a basis for the specific localization of modalities as does mammalian cortex, which represents a fundamental departure at perhaps some primitive reptilian level from the reptilian-avian line of development. Possibly the accessory hyperstriatum may be compared with the claustrum (83).

Some General Conclusions

Because of the limitations of space only some general conclusions will be mentioned. The first is the constancy of appearance in vertebrates of many of the main fiber bundles – such as various components of the stria medullaris and the stria terminalis, the ventral and the dorsal peduncles of the lateral forebrain bundle, the medial forebrain bundle, the precommissural fornix, the anterior commissure, and various other systems. These bundles vary in size, in number of fibers, in relative development of their components, and even in the precise significance of the impulses which they carry. Nevertheless their appearance in almost all vertebrates suggests an underlying stability and constancy of the developmental pattern. In contrast is the variability in size, nuclear and cellular differentiation and sometimes location of the nuclear masses demonstrable in passing from one order and sometimes one species to another.

Of considerable interest are the differences in connections and functions of the primarily olfactovisceral rostromedial hemisphere wall as compared with the more clearly developed somatic and olfactosomatic character of the lateral hemisphere wall. Concomitant with this is the functional division of the olfactory tubercle into medial and lateral (including lateral and intermediate) zones.

Finally the suggestion might be repeated that the rostral end of the submammalian and lower mammalian hippocampal formation, which is also its subcallosal part, in highest mammals is an olfactovisceral correlative area interrelated with the hypothalamus, the septal areas, and the medial part of the olfactory tubercle. The posterior part of the hippocampal formation of lower forms, which is its temporal region in higher forms, is related to amygdaloid areas and, particularly in primates, to rostral temporal cortex. Intermediate parts of the hippocampal formation develop with the differentiation of various neopallial areas with which they have pattern relations (77). The hippocampal formation, then, is not a single entity.

List of Abbreviations

A	reptilian anterior amygdaloid area
A. cont. hip.	anterior continuation of hippocampus
Ar. preop.	preoptic area
Alv.	alveus
Amyg.	amygdala
Amyg. (bl.)	amygdala (basolateral group)
Amyg. (cm.)	amygdala (corticomedial group)
Ar. preop.	preoptic area
B	basal amygdaloid nucleus in reptiles
Bed nuc. ant. com.	bed nucleus of anterior commissure
Bed nuc. coms.	bed nucleus of commissures
C. A.	cornu ammonis
Ce	central amygdaloid nucleus in reptiles
Cereb.	cerebellum
Claus.	claustrum
Co	reptilian cortical amygdaloid nucleus
Com. ant. + bed nuc.	anterior commissure + bed nucleus
Com. hip.	hippocampal commissure
Com. hip. + bed nuc.	hippocampal commissure + bed nucleus
Cor. parahip.	parahippocampal cortex
D. olf. proj. tr.	dorsal olfactory projection tract (Cajal)
Diag. bd. + nuc.	diagonal band and nucleus
F.	fornix
Fib. assoc.	association fibers
Fim.	fimbria
Gen. cor.	general cortex
Gy. dent.	dentate gyrus
Hip.	hippocampus
Hip. p. d. (c. A.)	hippocampus, dorsal part (cornu ammonis)
Hip. p. dm. (gy. dent.)	hippocampus, dorsomedial part (gyrus dentatus)
Hyperstr.	hyperstriatum
Hyperstr. acc.	accessory hyperstriatum
Hyperstr. d.	dorsal hyperstriatum
Hypop. a.	anterior hypopallium
Hypop. p.	posterior hypopallium
I.	induseum griseum
L	amygdaloid group (laterally disposed) of reptiles
L. f. b.	lateral forebrain bundle
Lam. hyperstr. (tr. fr. oc.)	hyperstriate lamina (fronto-occipital tract)
Lam. med. d.	dorsal medullary lamina
Lam. supr.	supreme lamina
M	medial amygdaloid nucleus of reptiles
M. f. b.	medial forebrain bundle
M. i.	intercalate mass
Neostr.	neostriatum
Nuc. acc.	accumbens nucleus
Nuc. b. acc. amyg. p. m.	basal accessory amygdaloid nucleus, medial part
Nuc. b. acc. amyg. p. l.	basal accessory amygdaloid nucleus, lateral part
Nuc. b. amyg. p. l.	basal amygdaloid nucleus, lateral part
Nuc. b. amyg. p. m.	basal amygdaloid nucleus, medial part
Nuc. cent. amyg.	central amygdaloid nucleus
Nuc. cor. amyg.	cortical amygdaloid nucleus
Nuc. diag. bd.	nucleus of the diagonal band
Nuc. l. amyg.	lateral amygdaloid nucleus
Nuc. m. amyg.	medial amygdaloid nucleus

Nuc. olf. a. p. v.	anterior olfactory nucleus, ventral part
Nuc. preop. perivent. mag.	magnocellular periventricular preoptic nucleus
Nuc. preop. perivent. parv.	parvocellular periventricular preoptic nucleus
Nuc. sept. d.	dorsal septal nucleus
Nuc. sept. l.	lateral septal nucleus
Nuc. sept. m.	medial septal nucleus
Nuc. sept. m. p. d.	medial septal nucleus, dorsal part
Nuc. sept. m. p. v.	medial septal nucleus, ventral part
Nuc. septohip.	septohippocampal nucleus
Nuc. taen.	taenial nucleus (probably comparable to medial amygdaloid nucleus of other forms)
Nuc. triang.	triangular nucleus
Paleostr.	paleostriatum
Paleostr. aug.	paleostriatum augmentatum
Paleostr. prim.	paleostriatum primitivum
Pir.	piriform
Pr. gen. cor.	primordial general cortex
Pr. hip.	primordial hippocampus
Pr. hip. p. d. (c. A.)	primordial hippocampus, dorsal part
Pr. hip. p. dm. (gy. dent.)	primordial hippocampus, dorsomedial part
Pr. nuc. trian.	primordial triangular nucleus
Pr. pir.	primordial piriform
Pr. sub.	primordial subiculum
Precom. f.	precommissural fornix
Put.	putamen
R	amygdaloid group of unknown homology in reptiles
Sac. d.	dorsal sac
St. term.	stria terminalis
Striat.	striatum
Sub.	subiculum
Sul. lim. m.	medial limiting sulcus
T	nucleus of lateral olfactory tract in reptiles
Tect. op.	optic tectum
Thal.	thalamus
Tr. cort. hab. l. a.	lateral corticohabenular tract, anterior portion
Tr. cort. hab. m.	medial corticohabenular tract
Tr. oc. mes.	occipitomesencephalic tract
Tr. olf. lat.	lateral olfactory tract
Tr. sept. mes.	(cortico) septomesencephalic tract
Tr. tub. hip.	tuberohippocampal tract
Tub. olf.	olfactory tubercle
Ti.	intermediate zone of olfactory tubercle
Tl.	lateral zone of olfactory tubercle
Tm.	medial zone of olfactory tubercle
V. olf. proj. tr.	ventral olfactory projection tract

References

1. ADDISON, W. H. F.: J. comp. Neurol., 25 (1915), 497–522.
2. ANDY, O. J., H. STEPHAN: J. comp. Neurol., 111 (1959), 503–545.
3. ANDY, O. J., H. STEPHAN: J. comp. Neurol., 117 (1961), 251–274.
4. ANDY, O. J., H. STEPHAN: The Septum of the Cat. Charles C. Thomas, Springfield, Ill. 1964.
5. ARIENS KAPPERS, C. U.: Kon. Akad. Wet. (Amst.) Proc. sect. sci., 26 (1923), 135–158.
6. ARIENS KAPPERS, C. U.: The Evolution of the Nervous System. Bohn, Haarlem 1929.
7. ARIENS KAPPERS, C. U., E. HAMMER: Psychiat. neurol. Bl. (Amst.), 22 (1918), 368–415.

8. ARIENS KAPPERS, C. U., G. C. HUBER, E. C. CROSBY: The Comparative Anatomy of the Nervous System of Vertebrates, including Man. Macmillan Co., New York 1936.

9. BECCARI, N.: Arch. ital. anat. embriol., 9 (1910), 173–220.

10. BREATHNACH, A. S., F. GOLDBY: J. Anat. (Lond.), 88 (1954), 267–291.

11. CAIRNEY, J.: J. comp. Neurol., 42 (1926), 255–348.

12. CAREY, J. H.: In: HASSLER, R., H. STEPHAN (ed.): Evolution of the Forebrain. Thieme, Stuttgart 1966, 73–80.

13. CLAIRAMBAULT, P.: J. Hirnforsch., 6 (1963), 87–121.

14. CRAIGIE, E. H.: J. comp. Neurol., 49 (1930), 223–357.

15. CRAIGIE, E. H.: J. comp. Neurol., 56 (1932), 135–168.

16. CRAIGIE, E. H.: J. comp. Neurol., 61 (1935), 563–591.

17. CREUTZFELDT, O. D., F. R. BELL, W. R. ADEY: Progress in Brain Research, 3 (1963), 31–49.

18. CROSBY, E. C.: J. comp. Neurol., 27 (1917), 325–402.

19. CROSBY, E. C., T. HUMPHREY: J. comp. Neurol., 74 (1941), 309–352.

20. CROSBY, E. C., T. HUMPHREY: J. comp. Neurol., 81 (1944), 285–305.

21. CROSBY, E. C., T. HUMPHREY, E. W. LAUER: Correlative Anatomy of the Nervous System. Macmillan Co., New York 1962.

22. CROSBY, E. C., R. T. WOODBURNE: Res. Publ. Ass. nerv. ment. Dis., 20 (1940), 52–169.

23. CROSBY, E. C., R. T. WOODBURNE: J. comp. Neurol., 94 (1951), 1–32.

24. CURWEN, A. O.: Anat. Rec., 61 (Suppl.) (1935), 13.

25. CURWEN, A. O.: J. comp. Neurol., 66 (1937), 375–404.

26. CURWEN, A. O.: J. comp. Neurol., 69 (1938), 229–247.

27. CURWEN, A. O.: J. comp. Neurol., 71 (1939), 613–636.

28. DAITZ, H. M.: Brain, 76 (1953), 509–512.

29. DURWARD, A.: J. Anat. (Lond.) 65 (1930), 8–44.

30. DURWARD, A.: J. Anat. (Lond.) 66 (1932), 437–477.

31. DURWARD, A.: J. Anat. (Lond.) 68 (1934), 492–499.

32. EDINGER, L., A. WALLENBERG, G. HOLMES: Abhandl. d. Senckenb. nat. Gesellsch., Frankfurt am Main, 20 (1903), 343–426.

33. ELLIOT SMITH, G.: Quart. J. Micr. Sc., 39 (1896), 181–206.

34. ELLIOT SMITH, G.: J. Anat. (Lond.), 53 (1919), 271–291.

35. FILIMONOFF, I. N.: J. Hirnforsch., 7 (1964), 229–251.

36. FOX, C. A.: J. comp. Neurol., 72 (1940), 1–62.

37. FOX, C. A.: J. comp. Neurol., 79 (1943), 277–295.

38. GAUPP, E.: Anatomie des Frosches. F. Vieweg u. Sohn, Braunschweig 1899.

39. GOLDBY, F.: J. Anat. (Lond.), 68 (1934), 157–215.

40. GURDJIAN, E. S.: J. comp. Neurol., 38 (1925), 127–163.

41. HAMEL, E.: In: HASSLER, R., H. STEPHAN (ed.): Evolution of the Forebrain. Thieme, Stuttgart 1966, 81–91.

42. HEIER, P.: Acta anat. (Basel) (Suppl.) 8 (1948), 1–213.

43. HERRICK, C. J.: J. comp. Neurol., 20 (1910), 413–547.

44. HERRICK, C. J.: J. comp. Neurol., 33 (1922), 213–280.

45. HERRICK, C. J.: Libro en honor de S. Ramón y Cajal, Madrid, 1 (1922), 143–204.

46. HERRICK, C. J.: J. comp. Neurol., 43 (1927), 231–325.

47. HERRICK, C. J.: J. comp. Neurol., 58 (1933), 1–288.

48. HERRICK, C. J.: J. comp. Neurol., 58 (1933), 481–505.

49. HERRICK, C. J., J. B. OBENCHAIN: J. comp. Neurol., 23 (1913), 635–675.

50. HESS, W. N.: J. comp. Neurol., 68 (1937), 161–171.

51. HILPERT, P.: J. Psychol. Neurol., 36 (1928), 44–74.

52. HINES, M.: J. comp. Neurol., 35 (1923), 483–537.

53. HOFFMAN, H. H.: J. comp. Neurol., 120 (1963), 317–368.

54. HOFFMAN, H. H.: In: HASSLER, R., H. STEPHAN (ed.): Evolution of the Forebrain, Thieme, Stuttgart 1966, 61–72.

55. HUBER, G. C., E. C. CROSBY: J. comp. Neurol., 40 (1926), 97–227.

56. HUBER, G. C., E. C. CROSBY: J. comp. Neurol., 48 (1929), 1–225.

57. HUMPHREY, T.: J. comp. Neurol., 65 (1936), 603–711.

58. HUMPHREY, T.: Proc. VIIIth Internat. Congress of Anatomists, Wiesbaden (1965), p. 55.

59. INGRAM, W. R.: Res. Publ. Ass. nerv. ment. Dis., 20 (1940), 195–244.

60. JANSEN, J.: J. comp. Neurol., 49 (1930), 359–507.

61. JOHNSTON, J.B.: J. comp. Neurol., *22* (1912), 341–404.
62. JOHNSTON, J.B.: J. comp. Neurol., *23* (1913), 371–478.
63. JOHNSTON, J.B.: J. comp. Neurol., *25* (1915), 393–468.
64. JOHNSTON, J.B.: J. comp. Neurol. *35* (1923), 337–481.
65. KRIEG, W.J.S.: J. comp. Neurol., *55* (1932), 19–89.
66. KUHLENBECK, H.: Anat. Anz., *54* (1921), 304–316.
67. KUHLENBECK, H.: Confin. neurol. (Basel), *14* (Suppl.) (1954), 1–230.
68. DE LANGE, S.J.: Folia neuro-biol. *7* (1913), 67–138.
69. LAUER, E.W.: J. comp. Neurol., *82* (1945), 215–254.
70. LAUER, E.W.: J. comp. Neurol., *90* (1949), 213–241.
71. LOO, Y.T.: J. comp. Neurol., *52* (1931), 1–148.
72. MAGEE, K.: Personal communication.
73. MEYER, A.: Z. wiss. Zool, *55* (1892), 63–133.
74. NAUTA, W.J.H.: J. comp. Neurol., *104* (1956), 247–271.
75. NAUTA, W.J.H.: Brain, *81* (1958), 319–340.
76. RÖTHIG, P.: Arch. Anat. Physiol., Anat. Abt., (1911), 49–56.
77. SCHNEIDER, R.C., E.C. CROSBY, E.A. KAHN: Progress in Brain Research, *3* (1963), 191–217.
78. SCHNITZLEIN, H.N.: J. comp. Neurol., *118* (1962), 225–267.
79. SCHNITZLEIN, H.N.: In: HASSLER, R., H. STEPHAN (ed.): Evolution of the Forebrain. Thieme, Stuttgart 1966, 40–49.
80. SCHNITZLEIN, H.N.: Ala. J. med. Sc. *3* (1966), 39–45.
81. SCHNITZLEIN, H.N., E.C. CROSBY: J. Hirnforsch. (1966). In press.
82. SCHOBER, W.: J. Hirnforsch., *7* (1964), 9–209.
83. SHANKLIN, W.M.: Acta zool., *11* (1930), 425–490.
84. SHELDON, R.E.: J. comp. Neurol., *22* (1912), 177–339.
85. SHOWERS, M.J.: Anat. Rec., *151* (1965), 416.
86. SPIEGEL, E.: Arb. neurol. Inst. Univ. Wien, *22* (1919), 418–497.
87. VAN DER SPRENKEL, H. BERKELBACH: J. comp. Neurol., *42* (1926), 211–254.
88. VALENSTEIN, E.S., W.J.H. NAUTA: J. comp. Neurol., *113* (1959), 337–363.
89. VOTAW, C.L.: Anat. Rec., *127* (1957), 382–383.
90. WALLENBERG, A.: Anat. Anz., *24* (1904), 142–155; 357–369.
91. WESTON, J.K.: Kon. Akad. Wet. (Amst.), Proc. sect. sci., *40* (1937), 894–904.
92. YOUNG, M.W.: J. comp. Neurol., *65* (1936), 295–401.

ELIZABETH C. CROSBY, Ph. D.
Neurosurgical Research Laboratory
Department of Surgery
University of Michigan Medical Center
Ann Arbor, Mich. and
Department of Anatomy
University of Alabama Medical Center
Birmingham, Ala., USA

BUD R. DEJONGE, M. D.
RICHARD C. SCHNEIDER, M. D.
Neurosurgical Research Laboratory
Department of Surgery
University of Michigan Medical
Center
Ann Arbor, Mich., USA

Gehirnentwicklung (Introversion-Promination) und Endocranialausguß

Von H. Spatz, Frankfurt

Übersicht über drei cerebrale Entwicklungsprinzipien

Es soll von drei untereinander zusammenhängenden Prinzipien der Hirnentwicklung die Rede sein, von zwei bekannten und von einem wesentlich neuen, auf das näher eingegangen werden soll.

Das 1. Prinzip. Es betrifft die bekannten *Parallelen von Phylogenie und Ontogenie.* Daß es sich dabei um eine Regel mit vielen Ausnahmen handelt und nicht um ein Gesetz, ist bekannt (REMANE 1960). Doch gerade für die Hirnentwicklung ist die Regel besonders wichtig. Der alte Anatom FRIEDRICH TIEDEMANN hat bereits 1816 (das ist genau 50 Jahre vor der Aufstellung des sogenannten „Biogenetischen Grundgesetzes" durch ERNST HAECKEL) in einer denkwürdigen Monographie solche Parallelen aufgezeigt, wenn er auch die Worte „Phylogenie" und „Ontogenie" noch nicht kannte. Diese Monographie betrifft *speziell* die Entwicklung des Gehirns, verglichen bei Tieren und bei menschlichen Embryonen.

Bezüglich der *Phylogenie* beschränken wir uns hier bewußt vorwiegend auf eine bestimmte Reihe innerhalb der plazentalen Mammalier, nämlich auf die Ordnungen der *Insectivoren* und der *Primaten* sowie auf die quasi Zwischenform zwischen beiden, nämlich auf die heute oft bereits zu den Primaten gerechneten *Tupaiiden,* welche REMANE (1956) in seiner „Grenzgruppe" der „Subprimaten" unterbringt. – Bezüglich der *Ontogenie* beschränken wir uns auf die Neuroembryologie des Menschen.

Das 2. Prinzip. Das zweite ebenfalls bekannte Prinzip der Hirnentwicklung besagt: In Phylogenie und in Ontogenie laufen Wachstum und Differenzierung nicht überall zur gleichen Zeit ab, sondern an verschiedenen Orten zu verschiedenen Zeiten. Dies nennen wir das außerordentlich verbreitete „*Prinzip der Heterochronie der Entwicklung der cerebralen Teile*"[1]). Es gibt *frühere* Teile, die in der Entwicklung vorauseilen; sie dienen elementaren, mehr automatischen Funktionen. Und es gibt *spätere* Teile, die in der Entwicklung nachfolgen; diese sind meist mit differenzierteren, teilweise mehr willkürlichen Leistungen betraut. (JACKSON: "Evolution is a passage from the most automatic to the most voluntary.")

Der *Neocortex* ist in Phylogenie wie in Ontogenie ein *später* Teil der Großhirnrinde. Bei den Primaten besitzt dieser *Spätling* in der Entwicklung außerdem jene höchst merk-

---

[1]) Im höchsten Abschnitt des Hirnstammes, im Diencephalon, braucht sich die Matrix im Bereich des zugehörigen III. Ventrikels schon bei menschlichen Keimlingen vom Ende des 3. intrauterinen Monats auf, während sich der Matrixaufbrauch im Bereich der telencephalen Seitenventrikel erst einige Zeit nach der Geburt – also rund ein halbes Jahr später – vollzieht (SPATZ, 1925, 1960; KAHLE, RICHTER). Auf die relativ retardierte Myelogenese im Telencephalon braucht nur hingewiesen zu werden.

würdige, beim Menschen einen Höhepunkt erreichende *Tendenz*[1]) zu einer *progredienten*
Entfaltung. Dabei sind innerhalb des Neocortex wieder frühere (oder primäre) und spätere
(sekundäre) Abschnitte zu unterscheiden. Wenn es zu dieser Progression des Neocortex
kommt, verhalten sich die in der Entwicklung vorausgeeilten, also früheren Hirnteile
konservativ; sie bleiben in der Entwicklung stehen, oder sie bilden sich sogar zunehmend
zurück. Letzteres trifft bei den Primaten, z.B. für den rhinencephalen Palaeocortex, die
,,Frührinde", zu, die bei den Insectivoren besonders gut ausgebildet war. –

In der Phylogenie nennt man die ,,früheren" Hirnteile bekanntlich vielfach die ,,älteren" und
die ,,späteren" die ,,jüngeren"; so werden auch wir die Bezeichnungen ,,palaeo" und ,,neo" nicht
vermeiden können. Doch der Nachteil ist, daß diese Benennungen in der Ontogenie nicht im
selben Sinn gebraucht werden können wie in der Phylogenie.

Das 3. Prinzip. Jetzt kommen wir zu dem dritten Entwicklungsprinzip, welches die
beiden zuerst genannten voraussetzt. Es ist in der Neurologie bisher kaum beachtet worden; zwar ist ein Teil der zugrunde liegenden Tatsachen bekannt, aber *die Einordnung
in ein umfassendes und sehr verbreitetes Prinzip ist neu.* Eben deshalb soll dieses hier etwas
näher erklärt werden. Kurz formuliert, lautet das neue Prinzip: *In Phylogenie und
Ontogenie werden die jeweils früheren Hirnteile mit fortschreitender Entwicklung in der
Reihenfolge ihrer Entstehung gegen das Innere zu verlagert.* Das nennen wir **Introversion**.
Jeweils später hinzugetretene Teile dagegen ragen an der Oberfläche frei hervor. Das ist
Promination[2]). Was hier mitgeteilt wird, ist als ein unvollkommener Anfang zu betrach-
ten. Weitere Forschung wäre erwünscht. Sicher ist, daß an dieses Prinzip öfters gedacht
werden sollte.

Ein der *Introversion* entsprechender Entwicklungsvorgang ist aus der allgemeinen Biologie her bekannt. REMANE (1952) spricht dabei unter Bezugnahme auf J. MECKEL (1821)
von ,,*Internation*". So wird z.B. das Trommelfell in der Phylogenie der Wirbeltiere von
seiner ursprünglich freien Situation an der Oberfläche zunehmend in das Innere verlagert.
(Man hat diesen Prozeß auch unter dem Gesichtspunkt eines *Schutzes* für die in die Tiefe
verlagerten Teile betrachtet.)

Bei der Entwicklung des *Gehirns* unterscheiden wir – das ist wichtig – zwei eng miteinander verbundene Arten der Introversion. Die eine nennen wir ,,Retraction", die andere
,,Suppression".

Bei der *Retraction* sinkt die Oberfläche früherer Hirnteile ein, *ohne dabei verdeckt zu
werden*. Die Gehirnoberfläche rückt von der Arachnoidea nach innen zu ab. Der unter der
Arachnoidea frei werdende Raum wird durch das ,,Liquorkissen" der Zisternen einge-
nommen. Retrahiert sind z.B. bei den meisten Mammaliern die Unterfläche des Hirnstammes, die an die Cisterna pontocerebellaris grenzt, und beim Menschen der Neocortex
auf Höhe der Konvexität der Hemisphären, worauf wir zurückkommen. Die Bildung der
Zisternen ist nach dieser Konzeption ein sekundärer Vorgang, der durch das lokale Einsinken der Gehirnoberfläche hervorgerufen wird. Bei der pathologischen Hirnatrophie
wird dieser Vorgang gesteigert.

---

[1]) Bei den Erinaceiden besteht diese Tendenz nicht, wie aus dem Vergleich von rezenten und
fossilen Formen hervorgeht. – Sehr deutlich ist dagegen die Progredienz des Neocortex in der
Pferdereihe (T. EDINGER 1948).
[2]) Die Vermutung liegt nahe, daß bei der Bildung von Furchen und Windungen allgemein das
Prinzip von Introversion und Promination eine Rolle spielt. Das würde besagen, daß die Stellen,
welche zum Tal geworden sind, im Wachstum zurückgeblieben sind, während diejenigen, die zu
prominenten Kuppen geworden sind, sich in Ausdehnung befunden haben.

Zur Suppression kommt es dann, wenn retrahierte Hirnteile infolge des stärkeren Wachstums umgebender, späterer Teile *verdeckt* werden. Bei der *Insula* sind Retraction und Suppression aufeinanderfolgende Phasen der Introversion. Andere Hirnteile verharren auf dem Zustand der Retraction, offenbar weil es in ihrer Umgebung keine sich stärker ausdehnenden Hirnteile gibt. Das gilt z.B. für die erwähnte Unterfläche des palaeencephalen Hirnstammes. Öfters geht die Introversion (sei es durch Retraction, sei es durch Suppression) mit einer relativen oder auch absoluten Verkleinerung der betreffenden Hirnteile einher (ohne daß lebensnotwendige Gebiete ihre funktionelle Bedeutung verlieren würden). Zu innerst kommen früheste Hirnteile zu liegen, wie der Hypothalamus und andere vegetative Zentren des Hirnstammes.

Die *Promination* ist der Gegensatz zur Introversion. Prominent sind jeweils späte Hirnteile, deren Oberfläche frei vorragt, d.h. weder an ein Liquorkissen grenzt, noch von anderen Hirnteilen bedeckt wird. Hirnteile, die bei früheren Formen prominierten, verfallen bei späteren Formen der Introversion, während dann andere Teile prominent werden[1]. So prominiert z.B. bei den Insectivoren und auch noch bei Tupaiiden und Prosimiern das Palaeocerebellum. Bei den Affen wird dieses zunehmend introvertiert, während das Neocerebellum zu prominieren beginnt (Abb. 1).

Impression und Endocranialausguß

Im adulten Zustand berühren die prominenten Hirnteile – das ist von großer Bedeutung – die Arachnoidea und gelangen dann meist in Kontakt mit der Endocranialwand. Hiermit stehen wir vor dem *Vorgang der Impression, welcher die Promination zur Voraussetzung hat.* Die Impression kommt dadurch zustande, daß prominente Hirnteile (nicht allenfalls nur Windungen) bei den höheren Wirbeltieren (Vögeln und Säugern), vermutlich infolge des Wachstumsdruckes, die Fähigkeit erlangen können, sich durch Arachnoidea und Dura hindurch an der Endocranialwand abzuformen. Die Dura ist an solchen Stellen relativ dünn. Die introvertierten Hirnteile dagegen haben den Kontakt mit der Endocranialwand – durch Retraction[2]) oder Suppression – verloren und damit die Fähigkeit zur Impression eingebüßt. Die Dura ist in diesem Fall relativ dick. Das Negativ der Impressionen wird durch den **Endocranial-Ausguß** in das Positiv des äußeren Hirnreliefs zurückverwandelt. Der Ausguß hat bekanntlich, auf fossile Tiere und fossile Hominiden angewandt, zur Wissenschaft der Palaeoneurologie geführt; infolge seiner Fähigkeit zur Impression an der Endocranialwand ist das Gehirn das einzige Weichorgan, über dessen Oberfläche die Palaeontologie Aussagen machen kann, lange nachdem das Organ selber zugrunde gegangen ist (T. EDINGER). Für uns ist der Endocranialausguß deshalb besonders wichtig, weil er das genannte Wechselspiel zwischen der Introversion von jeweils früheren und der Promination von jeweils späteren Hirnteilen auch bei rezenten Geschöpfen widerspiegelt. Das Vorhandensein der Impression ist ein Merkmal jeweils späterer Hirn-

---

[1]) Es dürfte nicht wahrscheinlich sein, daß introvertierte Hirnteile später wieder prominent werden können. Ursprünglich mögen die Anlagen aller Hirnteile oberflächlich gelegen sein.

[2]) Es gibt, wie z.B. bei Gorilla und anderen Großtieren, eine „totale Retraction" mit fast völligem Fehlen der Impressionsfähigkeit. Solche Sonderfälle erfordern eine besondere Interpretation, auf die hier nicht eingegangen werden kann. Auf die Flachheit der vorderen und mittleren Schädelgrube bei Gorilla sei mit Hinsicht auf den „Basalen Neocortex" aufmerksam gemacht.

teile. Das Fehlen der Impression, infolge Retraction oder Suppression, weist meist daraufhin, daß es sich bei den betreffenden Hirnteilen um in der Entwicklung vorausgeeilte, also frühere handelt. Es sieht so aus, als ob sich prominente und impressionsfähige Hirnteile aktiv, introvertierte passiv verhalten. Das ist eine gut gestützte Hypothese.

Kurz zusammengefaßt, ergibt sich folgende *Konzeption:* Frühe Hirnteile, die ursprünglich an der Außenseite lagen, werden nach innen verlagert, also *introvertiert,* während jeweils spätere Teile *prominieren.* Nur die letzteren können die Fähigkeit zur *Impression* besitzen, während die früheren Hirnteile diese Fähigkeit infolge Introversion (sei es durch Retraction, sei es durch Suppression) nicht haben. Dies veranschaulicht das unten stehende Diagramm.

Frühere Hirnteile		Spätere Hirnteile
I. INTROVERSION		II. PROMINATION
(= Internation)		
1. Retraction $\Big\}$	keine Impression	*Impression*
2. Suppression		

Unser besonderes Interesse gilt demjenigen Abschnitt des Neocortex, der sich in Phylogenie und Ontogenie am *spätesten* entfaltet und der beim Menschen konstant impressionsfähig ist. Dies ist der bisher wenig beachtete „*Neocortex basalis*" im Bereich der vorderen und mittleren Schädelgrube.

Die Einführung einiger neuer Termini hat sich für eine kurze Verständigung unerläßlich erwiesen. In der englischen Sprache würde Promination etwa „exposure", Suppression „overlap" entsprechen.

Einige Beispiele zur Erläuterung

(Abb. 1). Wir folgen, trotz gewisser Einwände, der auf Phylogenie und Ontogenie beruhenden Einteilung von L. EDINGER (1911). Demgemäß unterscheiden wir zwischen dem früheren *Palaeocerebellum* (Vermis + Flocculus) einerseits und dem erst bei den Mammaliern sich hinzugesellenden *Neocerebellum* (Hemisphären) andererseits. In der Ontogenie des Menschen erweist es sich, daß die palaeocerebellaren Teile früher in die heterochrone Myelogenese (FLECHSIG) eintreten als die neocerebellaren Teile, was ich bestätigen kann. Die Figuren der Abb. 1 sind der Phylogenie der Primaten entnommen, und zwar handelt es sich um Endocranialausgüsse(!). Die Lage des Vermis, als Prototyp des Palaeocerebellum, wird durch Pfeil bezeichnet.

Bei der lissencephalen *Tupaia glis* (1a) prominiert der stark ausgebildete Wurm; einzelne Lobuli werden durch Impression hervorgehoben. Der Flocculus kommt ganz außen zum Vorschein. Die neocerebellaren Hemisphären sind kleiner als der Wurm. Bei dem madagassischen *Propithecus diadema* (b) sind die neocerebellaren Hemisphären größer geworden; sie prominieren aber nicht so stark wie der Vermis. Zu beachten ist außerdem, daß sich der Neocortex der *Konvexität* sehr deutlich imprimiert (beim Menschen wird er die Fähigkeit zur Impression durch Retraction weitgehend verlieren). Nach Überspringung vieler Stufen zeigen wir (auf gleiche Größe gebracht) den Endocranialausguß des Gibbon *Symphalangus* (c). Jetzt sind die neocerebellaren Hemisphären *erheblich* größer geworden und prominieren deutlich. Zwischen ihnen befindet sich eine Einsenkung; hier liegt der kleiner gewordene palaeocerebellare Vermis. Es ist *retrahiert.* Nun wieder ein

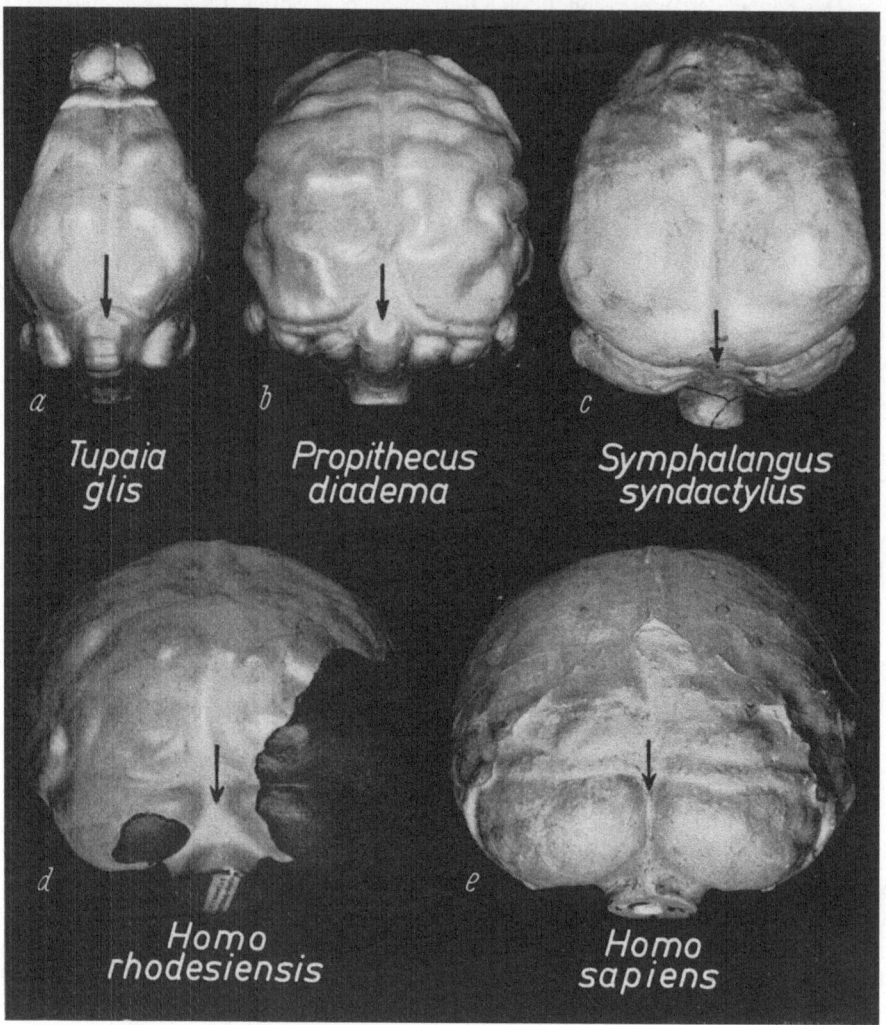

Abb. 1. Palaeocerebellum – Neocerebellum

Sprung: der Endocranialausguß vom fossilen *Homo rhodesiensis* (d). Auch hier sieht man (trotz des Schädeldefektes) einen durch Pfeil bezeichneten ziemlich *breiten Zwischenraum* zwischen den ausgedehnten Hemisphären. Er entspricht der Lage des retrahierten Vermis. Dieser Befund ist schon Boule aufgefallen beim Neandertaler von La Chapelle aux Saints. Wir haben ihn auch bei weiteren fossilen Hominiden wiedergefunden. Man sollte immer darauf achten. Beim *Homo sapiens* (e) sind die neocerebellaren Kleinhirnhemisphären nochmals größer geworden. Der Zwischenraum zwischen ihnen hat sich nun zu einem Spalt verengt[1]. Die Untersuchung des Kleinhirns selber lehrt, daß der palaeocere-

---

[1] Eine gewisse individuelle Variationsbreite ist zu berücksichtigen. Nur beim Menschen liegt in dem Spalt die Crista occipitalis interna.

bellare Wurm fast ganz *supprimiert* ist; dadurch hat er die Fähigkeit zur Impression eingebüßt. Zu beachten ist außerdem, daß die Konvexität der Großhirnrinde des Homo
sapiens fast glatt ist. Darauf kommen wir zurück.

Abb. 1 zeigt also, daß der Vergleich des Endocranialausgusses in einer aufsteigenden
Reihe der Primaten das Wechselspiel zwischen der Introversion früherer Kleinhirnteile
und der Promination späterer wiedergibt.

Vom Hirnstamm und vom optischen System

(Abb. 2). Der Hirnstamm gehört zum weitaus größten Teil zum Palaeencephalon von
L. EDINGER. Seine ventrale Oberfläche grenzt an die Cisterna ponto-cerebellaris und ist
retrahiert. Eine Sonderstellung nimmt der Pons ein. Wie aus Phylogenie und Ontogenie
hervorgeht, ist dieser ein später Anteil und gehört zum Neencephalon. Damit stimmt
überein, daß die Oberfläche der Brücke prominiert. Doch obwohl sie nahe an die Arachnoidea der Zisterne heranreicht, kommt es doch nur ausnahmsweise zur Impression des
Pons (eine solche Ausnahme findet sich beim Pferd).

An der dorsalen Oberfläche des Hirnstammes liegt im Bereich des Mesencephalon in der
Vierhügelplatte das frühe, subcorticale Zentrum des optischen Systems, nämlich das
Tectum opticum. Dieses erlangt bekanntlich bei niederen Wirbeltieren dauernd eine
erstaunliche Ausdehnung (die anfänglich sogar zur Verwechslung mit den Großhirnhemisphären geführt hatte). Bei den adulten Mammaliern tritt das Tectum opticum, von
wenigen Ausnahmen abgesehen, zurück, während jetzt die corticale Station des optischen
Systems, die Sehrinde, auftritt. Doch in einer Phase der Ontogenie bei frühen menschlichen Embryonen prominiert das Mittelhirndach (MD) und erinnert, wie schon TIEDE-
MANN wußte, an den Dauerzustand bei niederen Wirbeltieren. Dies ist also ein altes Beispiel des Prinzips der Parallelität von Phylogenie und Ontogenie. Abb. 2a demonstriert
die hochgradige Promination des durch Pfeil bezeichneten, das Tectum opticum enthal-
tenden Mittelhirndaches (MD) bei einem menschlichen Embryo von 27 mm (nach HOCH-
STETTER). Abb. 2b soll daran erinnern, wie tief das kleingewordene Mittelhirndach beim
Erwachsenen supprimiert ist.

Die höchste Station des optischen Systems, die früh myelinisierte *Sehrinde* (Area 17),
die zu den neocorticalen Primärgebieten (KUHLENBECK) gehört, ist bei den Insectivoren,

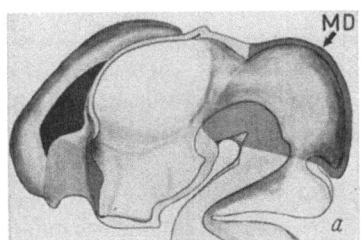
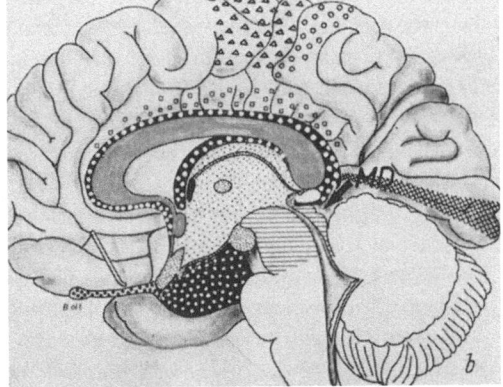

Abb. 2. Mittelhirndach (MD). a Gehirnmodell eines menschlichen Embryos von 27 mm auf dem
Medianschnitt nach F. HOCHSTETTER Ha 3. b Das gleiche beim Erwachsenen.

einschließlich der makroptischen Macroscelididen, noch sehr klein und liegt – was wir hervorheben – noch völlig frei an der äußeren Oberfläche der Hemisphären (vgl. Abb. 279 A in POLYAKS "Vertebrate Visual System"). Schon bei Tupaia ist die nun erheblich größer gewordene Sehrinde teilweise über die Mantelkante hinweg in die Fissura interhemisphaerica supprimiert worden und kommt an die Medianseite des Occipitallappens zu liegen. Dieser Vorgang schreitet in der Reihe der Primaten fort, während sich das spätere Feld 19, das mit differenzierteren Funktionen betraut ist, größtenteils an der äußeren Oberfläche ausbreitet. Beim Menschen liegt bekanntlich nur mehr ein kleiner Anteil des Feldes 17 an der Außenseite des Occipitalpols; der Hauptteil ist von außen nicht sichtbar, ist also supprimiert (Abb. 2 b).

Unter den Insectivoren fanden wir (STEPHAN u. SPATZ) nur bei den tagaktiven *makroptischen* Macroscelididen ausgesprochene Promination und neuerdings auch Impression (!) des großen Tectum opticum. Bei den primitiveren *mikroptischen* Insectivoren (Igel, Maulwurf, u. a.) ist das viel kleinere Tectum opticum supprimiert. Dies ist eine Bestätigung für die Angaben von T. EDINGER (1964) und STARCK (1963), daß „exposure" bzw. „Freiliegen" – wir sagen „Promination" – des Tectum mesencephali bei den Mammaliern keineswegs immer ein Zeichen einer niedrigeren Evolutionsstufe ist und daß „overlap", also „Suppression", nicht immer ein Zeichen einer höheren Stufe darstellt. In unserem Fall steht die Suppression des Tectum opticum der mikroptischen Insectivoren offenbar mit der geringeren Ausbildung des Sehvermögens dieser Tiere im Zusammenhang. Bei den Macroscelididen mit ihrem prominenten Tectum opticum ist es umgekehrt.

Ein Pendant zur primären Sehrinde ist die früh myelinisierte, primäre *Hörrinde*. Diese ist im Bereich des Gyrus transversus des Temporallappens in die Fissura Sylvii *supprimiert*. Das sekundäre akustische Gebiet des Gyrus temporalis superior ist beim Menschen konstant *retrahiert*, wie der Endocranialausguß lehrt. Die mehr basal liegenden, spät myelinisierten Schläfenlappenwindungen, die zum Basalen Neocortex gehören, imprimieren sich dagegen besonders gut.

Vom Archicortex und vom limbischen System

(Abb. 3 a und b). Um etwas über die Phylogenie des Archicortex aussagen zu können, müssen wir bis zu den Reptilien zurückgehen. Auf Abb. 3 (abgeändert nach L. EDINGER) sehen wir, daß bei der Riesenschlange Python (a) der Archicortex noch teilweise an der Außenseite der Hemisphäre freiliegt. Die kleine Anlage des Neocortex befindet sich hier noch völlig an der Außenseite. Bei einem Marsupialier, der Känguruhratte (b) hat sich die Anlage des Neocortex sowohl nach basal als in Richtung auf die Fissura interhemisphaerica ausgedehnt (s. Pfeile). Hierbei sind frühe Rindenanteile über die Mantelkante hinweg in der Reihenfolge ihrer Entstehung in die Fissura interhemisphaerica supprimiert worden. Zutiefst liegt hier der zusammengerollte Archicortex. Ihm ist ein neocorticales Primärgebiet (I) gefolgt, nämlich das „spleniale Segment" nach der Segmentlehre von CHR. JAKOB (vgl. Abb. 24 dieses Autors). Das darauf folgende „ektomarginale Segment" (II) ist zum Teil supprimiert, zum Teil liegt es noch an der Außenseite.

Das spleniale (oder viscerale) Segment des Neocortex und der Archicortex sind beide Abschnitte des großen „*limbischen Systems*". Dieses mit elementaren, autonomen Funktionen betraute, frühe System ist bei den Mammaliern in seiner ganzen Ausdehnung supprimiert(!).

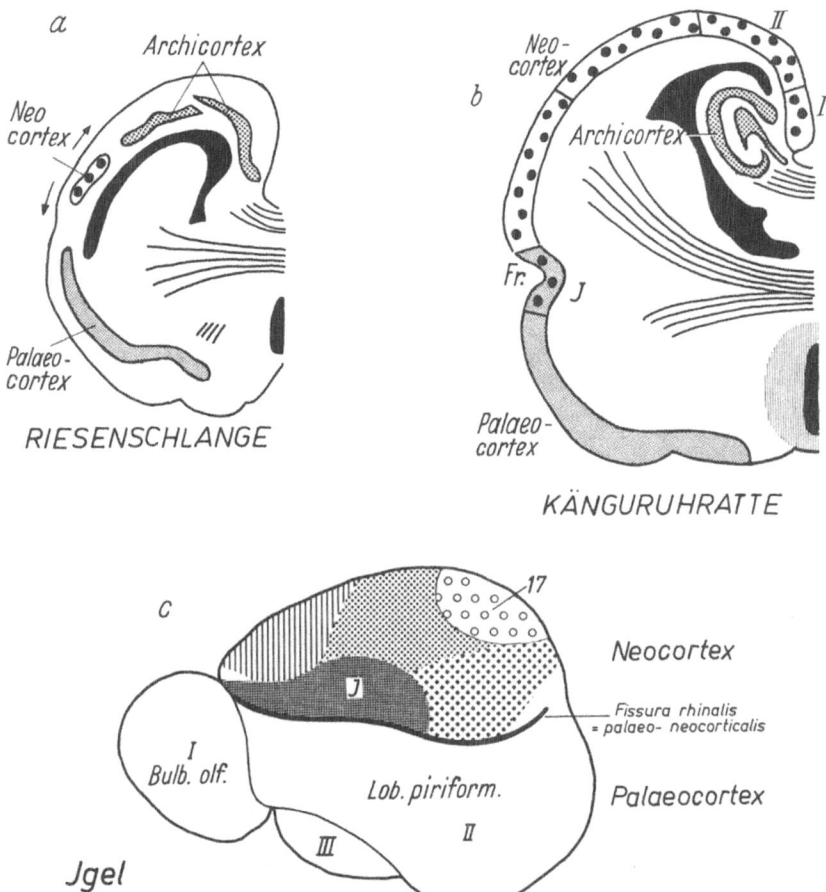

Abb. 3. a und b Suppression des Archicortex und früher Segmente des Neocortex (I und II).
c Igelgroßhirn von der linken Seite; vergrößert. J = Insula; 17 = Area striata.

Von der Insula und vom Palaeocortex

(Abb. 3 c und 4). Die Insel ist bei den Insectivoren ein bemerkenswert ausgedehntes Gebiet, das z. B. beim Igel (Abb. 3 c) bis zum oralen Pol der Hemisphären reicht und frei an der Außenseite liegt, also prominiert. Unmittelbar ventral davon breitet sich, durch die Fissura palaeo-neocorticalis (rhinalis) geschieden, der teilweise zum Rhinencephalon gehörige *Palaeocortex* aus, der mehr als die Hälfte der Hemisphäre des Igels ausmacht, also größer ist als der Neocortex. Erst von den eigentlichen Primaten an wird die Insel, gleichzeitig mit der Ausbildung der (den Tupaiiden noch fehlenden) Fissura Sylvii und im Zusammenhang mit der „Hemisphaerenrotation" um die Insel als Achse (JAKOB, SPATZ 1949), zunehmend in den Grund der Fissura Sylvii supprimiert. *Die sogenannte „Opercularisation" der Insel ist ein Sonderfall der Suppression.* Während man früher an eine Beziehung

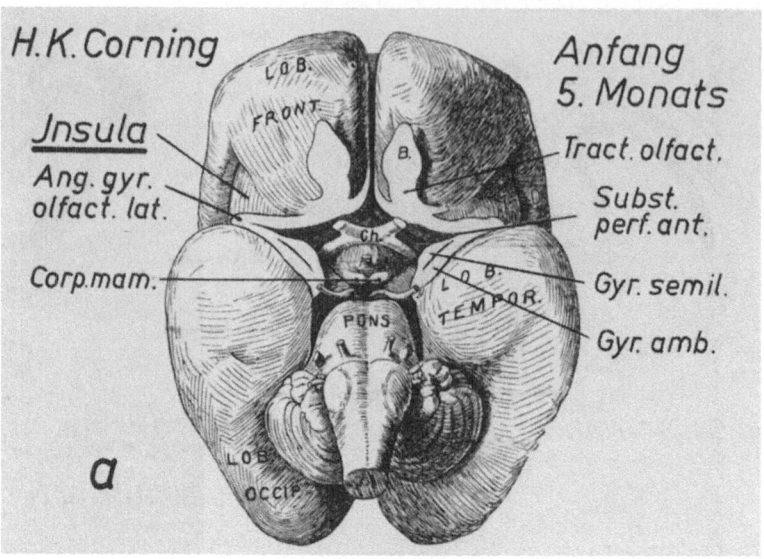

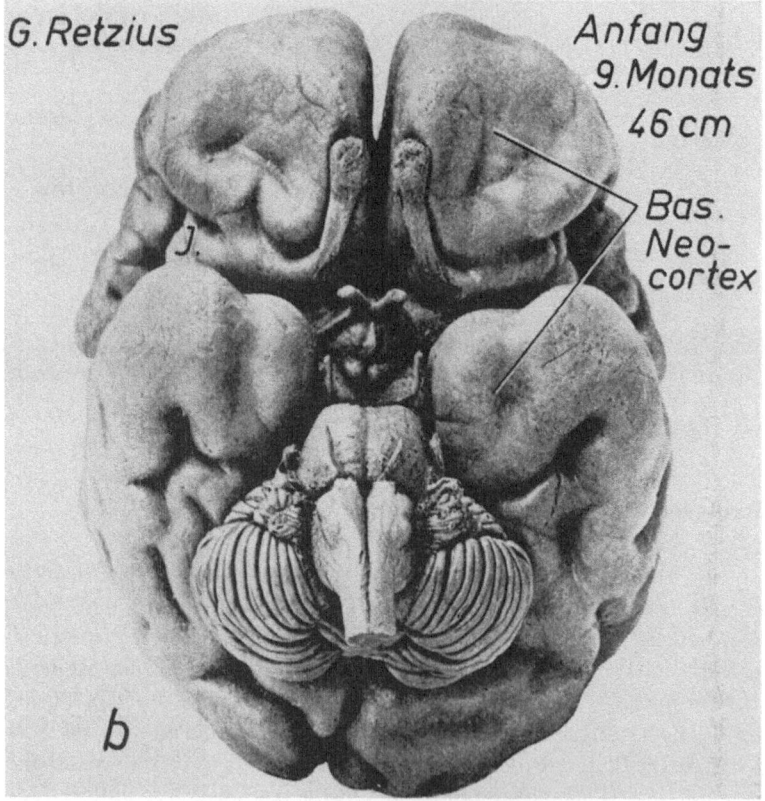

Abb. 4. Insula. Rhinencephalon.

der Insel zur Sprache[1]), also zu einer ausgesprochenen humanen Leistung dachte, erwägte schon KLEIST eine Beteiligung an den Geschmacksleistungen. Neuerdings hat die Zuordnung des Geschmacks zur Insula durch die Experimente von AKERT u. Mitarb. eine exakte Stütze erhalten. Die von Anfang an bestehenden, engen topographischen Beziehungen zwischen der Insel und dem Rhinencephalon werden dadurch begreiflich. Auf die gleichzeitige Suppression des Palaeocortex kommen wir auf S. 146 unten zurück.

Während der *Ontogenie* kommt es zuerst zu einem Einsinken (Retraction) der noch glatten Oberfläche des Inselareals. Dies ist auf der Abb. 4a (CORNING) zu sehen bei einem Fetus vom Anfang des 5. Monats. (Ähnlich ist die Abb. 394 von CLARA, die einen Fetus aus dem 6. Monat betrifft.) Offenbar zeigt die Retraction an, daß das Wachstum der Insel hinter dem der umgebenden Operkularwindungen zurückgeblieben ist. Dann folgt zunehmend die Verdeckung, also Suppression, der jetzt in Gyrifikation befindlichen Insel. Dies demonstriert Abb. 4b bei einem Fetus vom 9. Monat (nach RETZIUS, Tafel XX, 2). *Bei der Entwicklung der Insel sind also Retraction und Suppression zeitlich aufeinanderfolgende Phasen der Introversion* (S. 138).

Die Abb. 4a und 4b zeigen wieder die topographischen Beziehungen zwischen der Insel und dem Rhinencephalon über den Tractus olfactorius lateralis. Abb. 4a läßt die überraschend große Ausdehnung und Überblickbarkeit des fetalen Rhinencephalon (im Gegensatz zu dem des Erwachsenen) gut erkennen. *Beim Erwachsenen sind die Zentren der beiden elementaren Sinne, also des Geruchs und des Geschmacks, tief supprimiert.* Der breite Abstand, der im fetalen Zustand zwischen dem vorderen Rand des Temporallappens und dem hinteren Rand des Orbitalhirns klafft, ist dann völlig verschwunden. Infolge der Hemisphärenrotation bedeckt der Temporalpol dann sogar hintere Anteile des Orbitallappens.

Bei Abb. 4b fällt uns auf, daß der in der Entwicklung zurückgebliebene Basale Neocortex noch fast glatt ist, während der übrige Neocortex bereits in die Phase der Gyrifizierung eingetreten ist (s. S. 146).

Zur Phylogenie und Ontogenie des Basalen Neocortex

(Abb. 3c u. 5). Der lange Zeit kaum beachtete „Basale Neocortex" (SPATZ)[2]) ist phylogenetisch und ontogenetisch der späteste Teil der Neurinde. Beim Igel (Abb. 3c nach BRODMANN) und den meisten mikroptischen Insectivoren ist der *Neocortex auf die Konvexität der Hemisphäre beschränkt*, wo sich sein Ursprung befand (s. Abb. 3a). Der Neocortex des Igels, der sich im wesentlichen noch aus Primärgebieten (KUHLENBECK) zusammensetzt, liegt wie eine Kappe dem Palaeocortex auf, der die größere Hälfte der Hemisphäre ausmacht. Auch die Grenzfurche, die Fissura palaeo-neocorticalis (= F. rhinalis), liegt oben (Abb. 3c). Wenn man das Igelhirn von der Basis betrachtet, so hat man nur Palaeocortex vor sich. *Ein „Basaler Neocortex" existiert noch nicht.*

Abb. 5 (nach SPATZ u. STEPHAN) zeigt, wie sich mit der progredienten Ausdehnung des Neocortex auf die Basis der Hemisphäre zunehmend derjenige Hirnteil ausbildet, den wir

---

[1]) FLECHSIG hat 1896 (S. 25 u. S. 43) die Insel wegen des späten Zeitpunktes der Myeologenese zu seinen „Assoziationszentren" (Terminalgebieten) gerechnet und Beziehungen zur Sprache angenommen. Dies hat heute wohl nur mehr historisches Interesse.

[2]) Meine ursprüngliche Bezeichnung „Basale Rinde" (1937) wurde, da sie zu Verwechslungen mit dem Palaeocortex Anlaß gab, durch die Benennung „Basaler Neocortex" ersetzt.

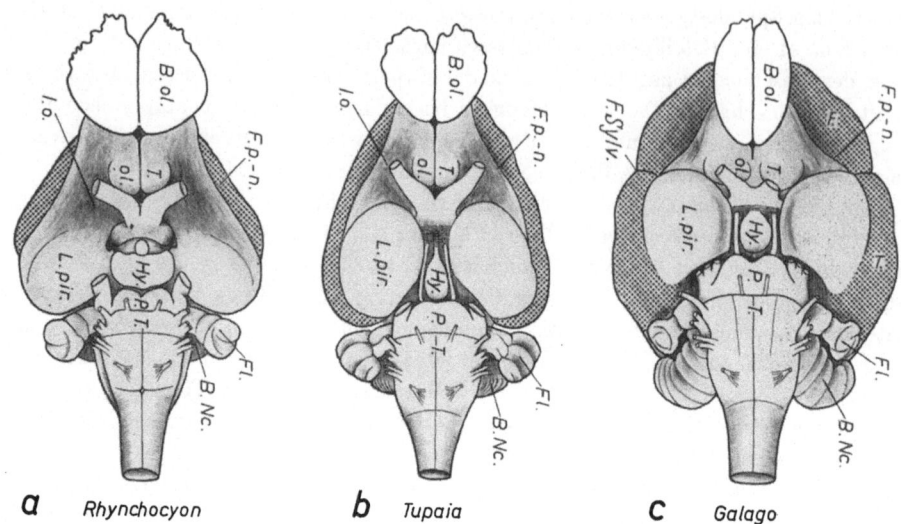

a   *Rhynchocyon*   **b**   *Tupaia*   **c**   *Galago*

Abb. 5. Basaler Neocortex: schraffiert. Palaeocortex: hell. F.p.n. = Fissura palaeo-neocorticalis (= rhinalis); T.olf. = Tuberculum olfactorium; F.Sylv. = Fissura Sylvii.

„*Basalen Neocortex*" nennen (schraffiert). Bei dem makroptischen Insectivor *Rhynchocyon stuhlmanni* (a) befindet sich der Basale Neocortex noch ganz im Beginn; er ist ein schmales lokales Segment außerhalb des ausgedehnten Palaeocortex. Bei *Tupaia glis* (b) ist er größer geworden und bildet nun eine zusammenhängende Randzone. Bei dem lissencephalen Prosimier *Galago demidovii* (c) ist ein deutlicher Fortschritt erreicht: Der Basale Neocortex ist erheblich größer geworden und die Grenzfurche, d.i. die Fissura palaeo-neocorticalis (F. p.-n.) wurde dabei nach basal und medial verschoben. Außerdem wird der Basale Neocortex jetzt durch die Ausbildung der bei den Tupaiiden noch fehlenden (!) Fissura Sylvii in einen frontalen und einen temporalen Abschnitt zerlegt. Dieses wichtige Merkmal kehrt bei allen rezenten und auch bei fossilen Primaten wieder. Die fortschreitende Zunahme des Basalen Neocortex in der Reihe der Primaten kann hier nicht dargestellt werden. *Weitaus die höchste Stufe wird beim Menschen erreicht.*

Auch in der *Ontogenie* erweist sich der Basale Neocortex als Spätling in der Entwicklung. Mikroskopische Untersuchungen bei frühen menschlichen Embryonen führten zu der Feststellung, daß sich die Anlage des Neocortex noch nicht auf die Basis der Hemisphären ausgebreitet hat; der Palaeocortex ist entsprechend weiter ausgedehnt. Die heterochrone Myelogenese tritt sowohl im frontalen als im temporalen Abschnitt des Basalen Neocortex auffällig spät in Erscheinung und auch bei der Windungsbildung bleiben die beiden Abschnitte zurück, wie Abb. 4b (S. 144) gezeigt hat.

Abb. 5 illustriert ferner, wie gleichzeitig mit der Ausdehnung des Basalen Neocortex der Palaeocortex successiv kleiner wird. Dies kann man auch bei der zunehmenden Rückbildung des wohl charakterisierten, bei den Insectivoren sehr großen Tuberculum olfactorium (Abb. 5, T. olf.) verfolgen. Bei der vergleichenden Untersuchung des Endocranialausgusses erwies sich, daß bei *Rhynchocyon* eine sehr deutliche Impression stattfindet, bei *Tupaia* ist sie schwächer; bei *Galago* und in der ganzen Reihe der eigentlichen Primaten, mit Ausnahme von *Daubentonia,* fehlt die Impression des Tuberculum völlig. Dieses ist mit zunehmender Verkleinerung introvertiert, nämlich zuerst retrahiert und dann supprimiert. Beim Menschen findet sich nur ein winziger Rest im Bereich der Substantia perforata anterior ganz in der Tiefe.

Vom Basalen Neocortex der Hominiden bei Betrachtung des Endocranialausgusses

(Abb. 6). Abb. 1 führte zu dem Ergebnis, daß der Endocranialausguß das Wechselspiel zwischen der Introversion des Palaeocerebellum und der Promination des Neocerebellum widerspiegelt. Wir kommen jetzt abschließend zum Endocranialausguß des Großhirns des *Homo sapiens recens* mit Hinweisen auf einige fossile Hominiden. Es geht wesentlich um die Frage, welche Teile des Menschenhirns imprimieren sich und welche imprimieren sich nicht ?

Die Impressiones gyrorum der Großhirnrinde haben zwar beim Menschen eine erhebliche individuelle Variationsbreite, aber es gibt zwei die Verteilung betreffende Tatsachen, die – normale Bedingungen vorausgesetzt – *konstant* sind:

1. Auf der Höhe der Innenseite des Schädeldaches (Calvaria, Kalotte) finden sich *keine* Impressionen. Dies hat SMITH AGREDA an 400 eröffneten Schädeln von erwachsenen Menschen festgestellt. Dementsprechend ist der Endocranialausguß von oben gesehen (norma verticalis) normalerweise *glatt* (abgesehen von Abformungen der Duragefäße, der PACCHIONISchen Granulationen und von einem Grenzgebiet zur Basis). Das ist deswegen so bemerkenswert, weil bei den Ungulaten und Carnivoren sowie auch bei vielen Primaten (s. z.B. Abb. 1 c) gerade der Neocortex der Konvexität durch besonders gute Impressionsfähigkeit ausgezeichnet ist. Das Fehlen der Impression beim Menschen auf der Höhe der Kalotte erklärt sich dadurch, daß die diesbezüglichen Windungen der Konvexität des Neocortex, welche die früh myelinisierten neocorticalen *Primärgebiete der Motorik und der Sensibilität* enthalten, beim Menschen durch ein zisternenartiges Liquorkissen von der

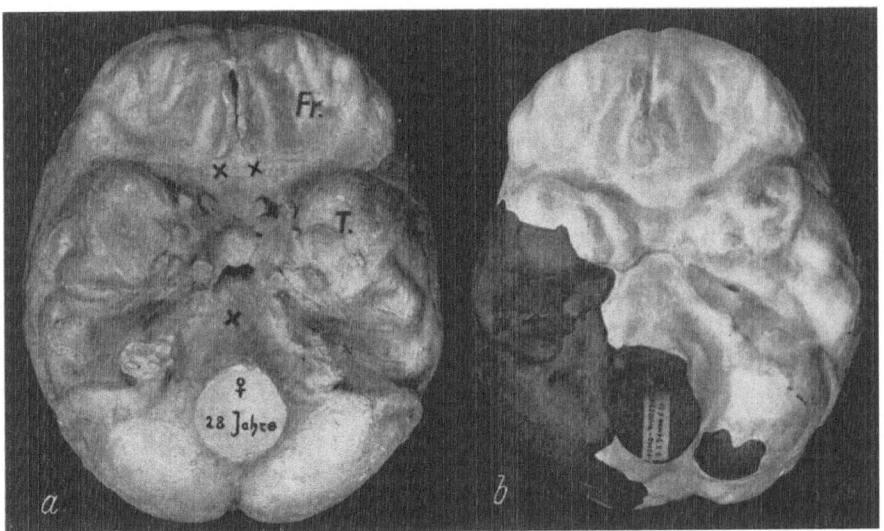

Abb. 6. a Endocranialausguß vom Homo sapiens. b Endocranialausguß vom Homo rhodesiensis.

Arachnoidea geschieden werden. *Sie sind retrahiert.* Dies entspricht eben ihrer frühen Entstehung. Die Dura ist relativ dick.

2. Im schroffen Gegensatz hierzu finden sich beim Menschen konstant die stärksten Impressionen basal, im Bereich der vorderen und mittleren Schädelgrube. Hier ist die Dura relativ dünn. Der Endocranialausguß (Abb. 6a) zeigt die deutlichen Impressionen, die den Windungen des frontalen (Fr.) und des temporalen (T) Abschnitts des *Basalen Neocortex* entsprechen. Früher hat man einmal daran gedacht, daß sich diese Hirnteile infolge ihrer Schwere am Schädelgrund abformen sollen. Doch wie wäre damit zu erklären, daß bei so vielen Tieren die Impressionen, wie oben gesagt, gerade an der Innenseite der Kalotte und nicht an der Basis so deutlich hervortreten?

Bei der Betrachtung des menschlichen Endocranialausgusses von der Seite ist festzustellen, daß im Bereich des Übergangs von der Basis zur Konvexität *seichte* Impressionen von mehr variabler Verteilung vorkommen. Bevorzugt sind die Frontal- und Occipitalpole, der caudale Gyrus frontalis inferior mit dem BROCAschen Gebiet (im weiteren Sinne) und die Windungen des Lobulus parietalis inferior. Dagegen imprimiert sich, wie gesagt, nicht der Gyrus temporalis superior im Bereich der Cisterna fissurae Sylvii.

Wenn man die Basalansicht des Endocranialausgusses in Abb. 6a näher betrachtet, stellt man fest, daß es neben den hochgradigen Impressionen, welche durch die Windungen beider Abschnitte des Basalen Neocortex hervorgerufen werden, auch Hirnregionen gibt, die keine Impressionen verursachen. Das ist einmal die erwähnte Unterseite des palaeencephalen Hirnstammes (x), welche im Bereich der Cisterna pontocerebellaris retrahiert ist. Sodann ist ein mit xx bezeichnetes Gebiet im Bereich der Cisterna chiasmatis konstant retrahiert und deshalb nicht impressionsfähig. Hier befinden sich der Tractus olfactorius und caudale Abschnitte des Orbitalhirns (xx). Aus neueren architektonischen Untersuchungen von SANIDES geht hervor, daß die letzteren in Beziehung stehen zum basalen olfactorischen Cortex. Auch KLEIST hatte auf diese Relationen hingewiesen. Die Dura ist hier wieder dick im Gegensatz zur dünnen Dura über den benachbarten, sich stark imprimierenden vorderen zwei Drittel des Orbitalhirns. Wenn man weiterhin an die supprimierten Teile, wie Palaeocortex, Insel u.a. denkt, so zeigt sich, daß alles, was introvertiert ist, durch die *frühe* Entstehung gekennzeichnet ist. Diese introvertierten Hirnteile kontrastieren mit dem sich so deutlich imprimierenden Basalen Neocortex, der *spätester* Anteil der Neurinde ist.

Der Endocranialausguß gestattet auch bei *fossilen Hominiden* gewisse Aussagen[1]) über die äußere Gestalt der längst vergangenen Gehirne zu machen. Abb. 6b zeigt den Endocranialausguß von dem relativ gut erhaltenen Schädel des *Homo rhodesiensis.* Besonders fällt die Schmalheit des temporalen Anteils des Basalen Neocortex auf (vgl. mit Abb. 6a), was bereits von K.B.SCHULTZ festgestellt worden ist. Bei einigen anderen fossilen Hominiden (aber nicht beim Neandertaler) hat sich uns dieser bemerkenswerte Befund bestätigt. Der Vergleich von Abb. 6b mit Abb. 6a ergibt ferner, daß auch der frontale Anteil des Basalen Neocortex (das Orbitalhirn) offenbar noch nicht voll ausgebildet gewesen ist. (Vom rezenten Homo sapiens standen 40 Endocranialausgüsse zum Vergleich zur Verfügung.)

Bei den Ausgüssen der Kalotte des Pithecanthropus erectus I (DUBOIS), des Düsseldorfer Neandertalers, sowie auch einiger Vertreter der Sinanthropus-Gruppe ist schon früher aufgefallen, daß sich die frontalen Impressionen ungewöhnlich weit scheitelwärts erstrecken. Doch läßt sich dieser Befund nicht verallgemeinern (CONOLLY, OVERHAGE). Weiteres über den Endocranialausguß fossiler Hominiden s. bei ARIENS KAPPERS, STARCK (1965) u.a.

---

[1]) Auf die weitgehenden Versuche einer Zuordnung (z.B. die Sprache betreffend) früherer Autoren wird hier nicht eingegangen.

Von der Bedeutung des Basalen Neocortex für den Menschen

Der ,,Neocortex basalis" ist in Phylogenie und in Ontogenie ein besonders später, prominenter Anteil der neocorticalen Secundärgebiete. Sein frontaler und sein temporaler Abschnitt sind, im Gegensatz zum Palaeocortex und zum Neocortex der Konvexität, durch ihre konstante hochgradige Impression ausgezeichnet.

Lange Zeit hat der Basale Neocortex in der Lehre von der Zuordnung von Funktionen zu einzelnen Hirnteilen (sogenannte ,,Lokalisationslehre") als ,,stumme" Region gegolten. Erst 1934 wurde durch KLEIST[1]) auf Grund von Beobachtungen bei Hirnverletzten die allgemeine Aufmerksamkeit wenigstens auf den frontalen Anteil des Basalen Neocortex, d. i. auf das ,,*Orbitalhirn*", hingelenkt. Spätere Untersuchungen, die sich teilweise auch auf Tumoren und lokale Atrophien (PICKsche Krankheit) bezogen, stammen u. a. von BOSTROEM u. SPATZ, DUUS, GRÜNTHAL, HEYGSTER, SPATZ (1937), WALCH u. WELTE.

Ein wichtiges Ergebnis dieser Untersuchungen ist, daß bei Patienten mit doppelseitiger Läsion des Orbitalhirns, wenigstens am Anfang, keine Störungen der formalen Intelligenz (einschließlich der Merkfähigkeit, der Auffassung und der Rechenfähigkeit) nachweisbar sind[2]). Das Wesentliche ist vielmehr (auch nach eigenen klinischen Beobachtungen) eine *Veränderung des Charakters*[3]). Diese macht sich initial öfters durch eine Störung des Taktgefühls und durch andere Abweichungen im Verhalten gegenüber den Mitmenschen bemerkbar. Später ist das Syndrom der ,,Persönlichkeitsstörung" (personality change) mehr oder weniger offenkundig und oft mit einem Verlust der ethischen Hemmungen verbunden. KRETSCHMER spricht von einer Störung der ,,sphärischen Integrierung" und der ,,dynamischen Steuerung". Ich wiederhole meinen Satz: ,,Hier wird der Mensch in seinem innersten Kern getroffen."

Wenn wir vom *frontalen* Abschnitt des Basalen Neocortex bzw. vom Orbitalhirn sprechen, so denken wir an die vorderen Zweidrittel desselben, d. i. an das Gebiet, welches durch die starke Impressionsfähigkeit ausgezeichnet ist, und nicht an das hintere Drittel, das retrahiert ist (Abb. 6a, xx) und zum Palaeocortex in Beziehung steht.

Als *temporalen Abschnitt* des Basalen Neocortex bezeichnen wir das Gebiet des Temporallappens, welches mit dem frontalen Abschnitt zusammen ursprünglich eine einheitliche Zone gebildet hat (Abb. 5b bei *Tupaia*) und welches ebenso wie der frontale Abschnitt phylo- und ontogenetisch sich spät entwickelt sowie sich hochgradig imprimiert. Von den klinischen Folgen doppelseitiger Läsionen dieses temporalen Abschnittes des Basalen Neocortex ist noch wenig bekannt. Auf Grund von einigen Feststellungen bei den seltenen Fällen von reiner Temporallappenatrophie (auf dem Boden der PICKschen Krankheit) neige ich zu der Annahme, daß solche Läsionen ebenfalls Persönlichkeitsstörungen hervorrufen können (LUERS u. SPATZ, JAKOB).

Befunde der Pathologie haben also neuerdings zu dem bemerkenswerten Ergebnis geführt, daß durch lokale Läsionen des Basalen Neocortex (d. i. besonders seines frontalen

---

[1]) Ältere Arbeiten von WELT (1888) und BERGER (1923) waren in Vergessenheit geraten.

[2]) Begreiflicherweise fehlen auch die motorischen und sensiblen Paresen, die bekanntlich bei Läsionen der betreffenden Zentren an der Konvexität des Neocortex auftreten.

[3]) SCHOPENHAUER, ein guter Kenner des neurologischen Schrifttums seiner Zeit, vertrat noch die Meinung, daß Störungen des Charakters, im Gegensatz zu solchen des Intellekts, nicht von Veränderungen des Gehirns abhängig seien.

Abschnittes) höchste, *spezifisch humane* Vermögen gestört werden. Der Basale Neocortex ist hiernach keine stumme Hirnregion; im Gegenteil, er ist vom anthropologischen Standpunkt aus gesehen bedeutungsvoll. Dies scheint gut damit übereinzustimmen, daß der Basale Neocortex ein besonders später Hirnteil ist, der erst beim Homo sapiens seine höchste Ausbildung erreicht hat.

Es erhebt sich die Frage: Hat die biologische Evolution des Basalen Neocortex, die bei bestimmten fossilen Hominiden noch im Gange war, heute etwa ihren Höhepunkt erreicht, oder ist hier in Zukunft ein weiterer Fortgang der Evolution im Sinne der Hypothese ECONOMOS von der ,,progressiven Neocorticalisation" denkbar? Auf diese Frage, mit der ich mich wiederholt beschäftigt habe (SPATZ, 1955, 1961, 1964, 1965), kann hier nicht eingegangen werden.

Summary

According to the principle of heterochronicity in the development of different cerebral divisions, which is valid in phylogeny as well as in ontogeny, *"earlier"* and *"later"* parts of the brain can be distinguished.

Another principle, which so far has not received due attention, implies that in phylogeny and in ontogeny earlier parts, originally lying freely on the surface of the brain, are gradually removed from the surface towards the interior of the organ, as its development advances. We have characterized this process as *"introversion"*. Here we distinguish: (1) "retraction", if the cerebral surface withdraws from the arachnoid (formation of cisternae) without being covered by other parts of the brain, and (2) "suppression", if the retracted part is subsequently overlapped by adjacent "later" and more intensively growing parts. The insula is first "retracted" and subsequently "suppressed".

"Introversion" is contrasted with *"promination"*. "Prominent" are such exposed parts of the brain which are "later" in phylogeny and ontogeny. In the adult state in mammals such parts are able to impress themselves through the arachnoid and dura mater into the endocranial wall (due to their "growth-pressure"). Earlier introverted parts have lost this faculty. The endocranial cast reflects the play of "introversion" and "promination".

Earlier (accelerated) parts of the brain	Later (retarded) parts of the brain
I. INTROVERSION (= Internation)	II. PROMINATION
1. Retraction 2. Suppression } No impression	Impression

Attention is called particulary to the "basal neocortex" of the frontal and temporal lobes, as a prominent part of the cortex with a constant faculty of impression. Its highest stage of evolution in the primate series is reached in man. It represents a particularly late cerebral region, both in phylogeny and in ontogeny. Pathology has furnished evidence that it is responsible for the performance of higher faculties of human personality.

Literatur

AKERT, K.: zit. bei BENJAMIN u. ROBERTS.

ARIENS KAPPERS, C.U.: Anatomie comparée du système nerveux, Masson. Paris 1947.

BENJAMIN, R.M., K.AKERT: Cortical and thalamic areas involved in taste discrimination in the albino rat. J. comp. Neurol. *111* (1959), 231–251.

BERGER, H.: Klinische Beiträge zur Pathologie des Großhirns. Herderkrankungen der Präfrontalregion. Arch. Psychiat. *69* (1923), 1–46.

BOSTROEM, A., H. SPATZ: Über die von der Olfactoriusrinne ausgehenden Meningeome. Nervenarzt *2* (1929), 506–521.

BOULE, M., H.VALLOIS: Fossile Menschen. S. 210. Baden-Baden 1954.

CLARA, M.: Das Nervensystem des Menschen. 2.Aufl. J.A.Barth, Leipzig 1953.

CONOLLY, C.J.: External Morphology of the Primate Brain. Ch. Thomas. Springfield, Ill. 1950.

CORNING, H.K.: Lehrbuch der Entwicklungsgeschichte d. Menschen. Bergmann, München 1921.

DUUS, P.: Über psychische Störungen bei Tumoren des Orbitalhirns. Arch. Psychiat. *109* (1939), 596–648.

v. ECONOMO, C.: Der Zellaufbau der Großhirnrinde und die progressive Cerebration. Ergeb. Physiol. *29* (1929), 83–128.

EDINGER, L.: Bau der nervösen Zentralorgane. Bd. I. Vogel. Leipzig 1911.

EDINGER, T.: Evolution of the Horse Brain. Geol. Soc. Amer. Memoir 25 (1948).

EDINGER, T.: Die Palaeoneurologie am Beginn einer neuen Phase. Experientia 6 (1950), 250–258.

EDINGER, T.: Midbrain exposure and overlap in mammals. Am. Zool. *4* (1964), 5–19.

FLECHSIG, P.: Gehirn und Seele. Veit & Co., Leipzig 1896.

FLECHSIG, P.: Anatomie des menschlichen Gehirns und Rückenmarks auf myelogenetischer Grundlage. Thieme. Leipzig 1920.

GRÜNTHAL, E.: Über die Erkennung der traumatischen Hirnverletzungen. Karger. Berlin 1936.

HEYGSTER, H.: Die psychische Symptomatologie bei Stirnhirnläsionen. Hirzel. Leipzig 1948.

JAKOB, CHR.: Vom Tierhirn zum Menschenhirn. J.E. Lehmann. München 1911.

JAKOB, H.: Zur Pathologischen Anatomie der Pickschen Krankheit. Arch. Psychiatr. *201* (1960), 260–297.

KAHLE, W.: Studien über die Matrixphasen und die örtlichen Reifungsunterschiede im embryonalen menschlichen Gehirn. Dtsch. Z. Nervenheilk. *166* (1951), 273–302.

KLEIST, K.: Gehirnpathologie. S. 1159ff. J.A. Barth. Leipzig 1934.

KRETSCHMER, E.: Die Orbitalhirn- und Zwischenhirnsyndrome nach Schädelbasisfrakturen. Arch. Psychiat. *182* (1949), 452–477.

KRETSCHMER, E.: Medizinische Psychologie. 10.Aufl., S. 39. Thieme. Stuttgart 1950.

KUHLENBECK, H.: Vorlesungen über das Zentralnervensystem der Wirbeltiere. G.Fischer. Jena 1927.

LUERS, TH., H.SPATZ: Picksche Krankheit. (Progressive umschriebene Großgehirnatrophie). Handb. d. spez. path. Anat. Bd. XIII, Teil I, Bdt. A 614–715, Springer. Berlin 1957.

OVERHAGE, P.: Zur Frage einer Evolution der Menschheit während des Eiszeitalters. Acta biotheoret. II.T. *16* (1962), 27–56.

POLYAK, ST.: The Vertebrate Visual System. The Univ. of Chicago Press. Chicago 1957.

REMANE, A.: Die Grundlagen des natürlichen Systems der Vergleichenden Anatomie und der Phylogenetik. Akad. Verlagsges. Leipzig 1952, S. 244ff.

REMANE, A.: Palaeoontologie und Evolution der Primaten. Primatologia I (1956), 267 bis 378.

REMANE, A.: Die Beziehungen zwischen Phylogenie und Ontogenie. Zool. Anz. *164* (1960), 306–337.

RETZIUS, G.: Das Menschenhirn. II.Tafeln. Stockholm 1896.

RICHTER, E.: Die Entwicklung des Globus pallidus und des Corpus subthalamicum. Monograph. ges. Geb. Neurol. Psychiat., Heft 108, Springer. Berlin-Göttingen-Heidelberg 1965.

ROBERTS, TH.G., K.AKERT: Insular and opercular cortex and its thalamic projection in Macaca mulatta. Schweiz. Arch. Neurol. Neurochir. *92* (1963), 1–43.

SANIDES, F.: The cyto-myeloarchitecture of the human frontal lobe and its relation to phylogenetic differentiation of the cerebral cortex. J. Hirnforsch. *6* (1964), 269–282.

SANIDES, F.: Structure and function of the human frontal lobe. Neuropsychologia *2* (1964), 209–219.

SCHULTZ, K.B.: Der Innenraum des Schädels in stammesgeschichtlicher Betrachtung mit besonderer Berücksichtigung des Rhodesiafundes. Verh. Ges. Phys. Anthropol. (1931), 30–39.

SMITH-AGREDA, V.: Über die Verteilung der Impressiones gyrorum an der Innenseite des Gehirnschädels des Menschen. Dtsch. Z. Nervenheilk. *173* (1955), 37–68.

SPATZ, H.: Über die Bedeutung der basalen Rinde, auf Grund von Beobachtungen bei Pickscher Krankheit und bei gedeckten Hirnverletzungen. Z. ges. Neurol. *158* (1937), 208–232.

SPATZ, H.: Gegensätzlichkeit und Verknüpfung bei der Entwicklung von Zwischenhirn und „basaler Rinde". Allg. Z. Psychiat. *125* (1949), 166–177.

SPATZ, H.: Menschwerdung und Gehirnentwicklung. Nachr. Giessener Hochschulges. *20* (1952), 32–55.

SPATZ, H.: Gehirn und Endocranium. Homo *5* (1954), 49–52.

SPATZ, H.: Die Evolution des Menschenhirns und ihre Bedeutung für die Sonderstellung des Menschen. Nachr. Giessener Hochschulges. *24* (1955), 52–74.

SPATZ, H.: Gedanken über die Zukunft des Menschenhirns. In: Benz, E., Der Übermensch, S. 319–383, Rhein-Verlag. Zürich 1961.

SPATZ, H.: Über die Anatomie, Entwicklung und Pathologie des „basalen Neocortex". Acta med. belg. (1962), 766–779.

SPATZ, H.: Der basale Neocortex und seine Bedeutung für den Menschen. Ber. Phys. Med. Ges. Würzburg, *71* (1964), 7–17.

SPATZ, H.: Vergangenheit und Zukunft des Menschenhirns. Jb. Akad. Wiss. Lit. Mainz. Steiner. Wiesbaden 1965.

SPATZ, H., G.J. STROESCU: Zur Anatomie und Pathologie der äußeren Liquorräume des Gehirns. Nervenarzt *7* (1934), 425–437.

STARCK, D.: „Freiliegen des Tectum mesencephali" ein Kennzeichen des primitiven Säugergehirns? Zool. Anz. *171* (1963), 350–359.

STARCK, D.: Die Neencephalisation (Die Evolution zum Menschenhirn). Menschl. Abstammungslehre. Fischer. Stuttgart 1965.

STEPHAN, H., H. SPATZ: Gehirne afrikanischer Insectivoren. Morphol. Jb. *103* (1962), 108–174.

TIEDEMANN, F.: Anatomie und Bildungsgeschichte im Foetus des Menschen nebst einer vergleichenden Darstellung des Hirnbaues in den Thieren. Steinische Buchhandl. Nürnberg 1816.

VERSLUYS, J.: Hirngröße und hormonales Geschehen bei der Menschwerdung. Maudrich. Wien 1939.

WALCH, R.: Orbitalhirn und Charakter. In: Rehwald, E., Das Hirntrauma, S. 203–213. Thieme, Stuttgart 1956.

WELT, L.: Über Charakterveränderungen des Menschen infolge Läsionen des Stirnhirns. Dtsch. Arch. klin. Med. *42* (1888), 339–404.

WELTE, E.: Über die Zusammenhänge zwischen anatomischem Befund und klinischem Bild nach stumpfem Schädeltrauma. Arch. Psychiat. *179* (1948), 234–315.

Prof. Dr. H. SPATZ
Max-Planck-Institut für Hirnforschung
Neuroanatomische Abteilung
Deutschordenstraße 46
6 Frankfurt-Niederrad, Deutschland

Brains from 40 Million Years of Camelid History

By T. Edinger, Cambridge, Mass.

The first word in the title of this symposium was a compulsory invitation to paleo-neurology. Phylogeny is the science which traces the descent from each other of the organisms that lived in consecutive times of the Past. The documents of those histories are fossils. These do not include brains; but Prof. Spatz has just shown you that brain research is possible by means of a fossilized skull. An endocast of its brain cavity will reproduce the brain's form. Such "fossil brains" – either artificial or natural endocasts: brain-shaped stones prepared out of skulls – have been studied since the beginning of the past century. Even before a report on horse brain evolution gave a new stimulus to research in paleoneurology, C. Judson Herrick could write that "The positive paleonto-logical evidence regarding the phylogeny of the brain is more extensive and illuminating than is generally recognized."

I can also use words of Herrick to explain why, to report conscientiously on phylogeny, my topic today can only be brains from only one of several evolutionary lines in only one family, Camelidae, of one suborder, Tylopoda, of the vast mammalian order Artiodactyla. It is because the "historic approach to the problems of brain morphogenesis" can be "accepted only as checked against other lines of evidence, particularly the known sequence of evolutionary history as revealed by fossilized skeletal evidence". From such evidence, bones and teeth, the divergent evolutionary lines have been established in the Tylopoda.

Material. A plaster endocast of a miocene *Procamelus* was first described in 1877, a natural endocast of the oligocene camelid *Poëbrotherium* in 1883, and both types of brain have since been remarked upon in many papers and textbooks. Due to the increased interest in paleoneurology, seven American institutions enabled me to study specimens from the entire main line of the Tylopoda.

Until a few years ago, there was complete agreement that the line leading to camel and llama can be traced back to the four-toed, hare-sized *Protylopus*. Its remains are found in strata of the Upper Eocene. During that period, believed to have lasted from 45 to 36 million years ago, the various early artiodactyls were in most respects still similar to each other. In our context, recent doubts whether *Protylopus* was the ancestral tylopod are irrelevant. Brains known of other late eocene, American and European ungulates are much like that of *Protylopus,* differing from later brains in the same primitive characters. The camel brain has evolved from this type (Fig. 1, B, D).

Besides the eocene brain, and part of a brain laid open in a Lower Oligocene skull from a sideline of the Tylopoda, the following specimens were studied: Two natural endocasts of a still unnamed genus from lowest Oligocene (the locality from which Hofer and Wilson will show you a primate brain), and six of the Middle Oligocene two-toed, goat-sized *Poëbrotherium;* the plaster endocasts of one late Miocene and two early Pliocene *Procamelus* (Fig. 3). I know of no Pleistocene cameline endocranial casts. Of the three surviving genera, I had none of the two llamas, but of *Camelus* a duplicate of the ancient endocast of *C. bactrianus* of the Royal College of Surgeons of England, and the excellent one of *C. dromedarius* described by SIMON (Acta anat. 60 (1965) 122–151; our Fig. 1 A, C).

For you who can work with the limitless material of extant brains, I must say that in paleontology such a series of sixteen endocranial casts, some of them not replicas of whole brains, from times separated by millions of years, is wealth. Its value lies in the fact that these specimens stem from consecutive phases of camel phylogeny, and can throw light on true phylogenetic changes. Arranged in order of time, these few samples from the evolutionary process tell a detailed story, from a primitive artiodactyl brain, via clearly structurally intermediate phases, to the camel stage known to neontology.

We who have come together for this symposium illustrate so overwhelmingly the diverse nature of the disciplines now participating in brain research that you could only be bored if I mention all those phases. A detailed description, with references, is in preparation, for the new Frankfurt-published journal for zoological systematics and evolution-research. I will here report on such observations as have, I believe, significance beyond that one evolutionary line.

Eocene and Present

The *Protylopus* brain is incomplete insofar as it is the natural endocast of only the cranium proper. Its anterior end is a transverse break that shows olfactory peduncles in cross section. As the olfactory bulbs thus must have lain anterior to the cerebrum, it is possible to estimate total length of the brain by means of the bulbus: cerebrum proportions in an oligocene specimen. While the cast is 48 mm long, total brain length can be assumed to have been 56 mm. The two *Camelus* endocasts are (167 and 185:) 176 mm long, so we can say that brain length increased x 3.1; skull length, from (111 and 124:) 117 to around 600 mm, x 5.1. If one judges, again, only from the three specimens at hand, disregarding intermediates and variations, one has the surprise to find that maximum cerebral length, breadth and height have all become times 3.3 as much as they were in *Protylopus.* (Size increase of the pituitary body is more in correlation with the increase in stature. On the eocene cast it is represented by a slight elevation, delimited only laterally, breadth being 2 mm which compares to cerebral breadth as 1:14.5. On the dromedary endocast it is a very prominent, circumscript hillock, and its breadth, 15 mm, compares to that of the cerebrum as 1:6.2 – as 1:6.0 in SIMON's photograph of a brain –).

Not those diameters, of course, but the configuration of the specimens shows that the cerebrum became more and more dominant, vaulted high above though not over the olfactory bulbs. Its neocortex expanded in two ways, and I wonder whether one of these advances can be accounted for by the difference in size of ancestor and descendents.

In *Protylopus* the cerebral vault is hardly higher than that of the cerebellum. The neocortex has prominent convolutions separated by sulci which all are longitudinal and apparently had no branches. Progressive gyrification, reflecting additions of new areas, has led to the complicated system of infolding in *Camelus.* As usual in mammals so large, sulcus details are blurred on the endocasts, but the pattern is easily recognized. In a ventral view of the 29 mm broad *Protylopus* cerebrum, ½ mm of neocortex is visible on

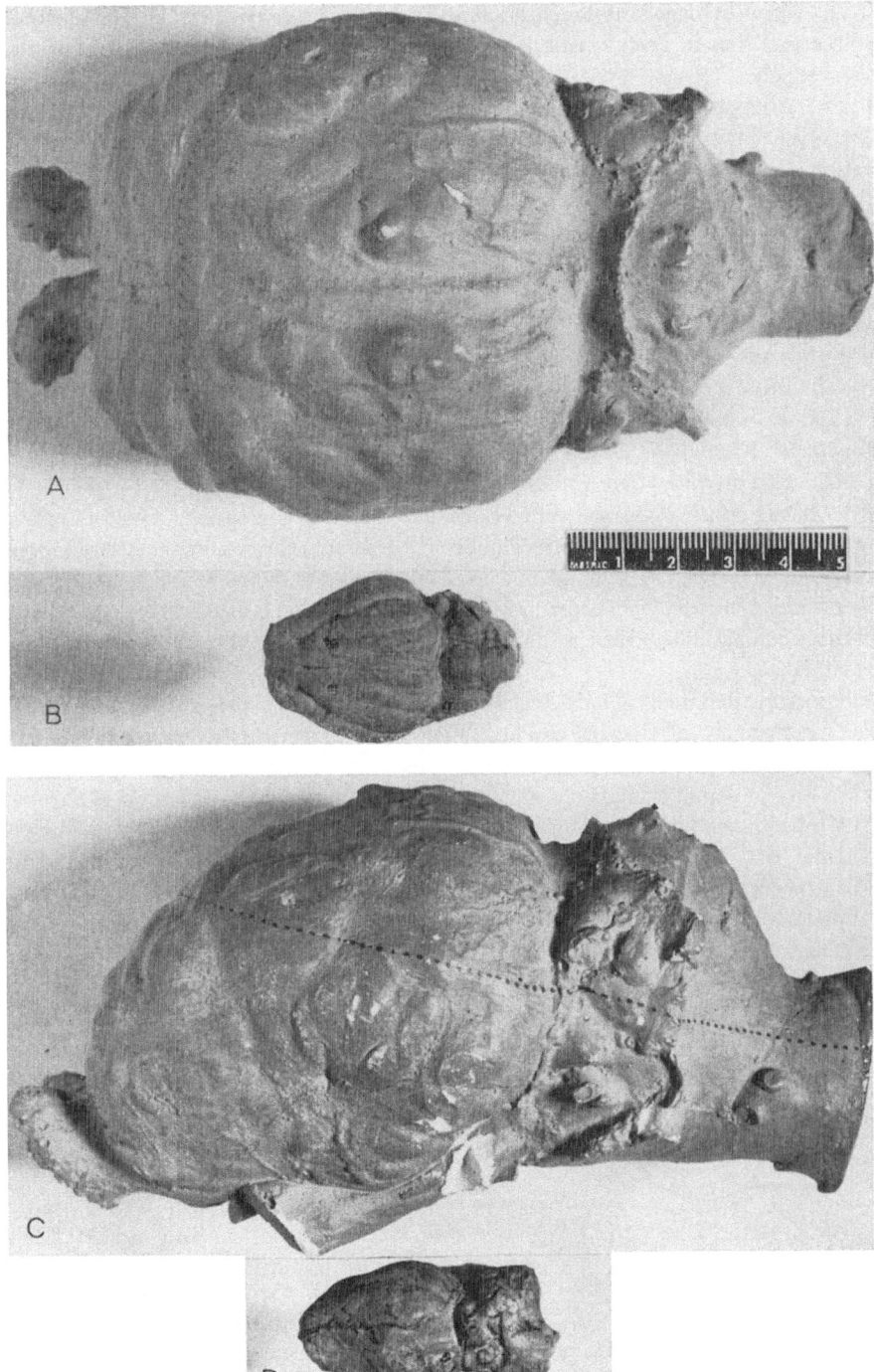

Fig. 1. *Camelus* and *Protylopus,* dorsal and lateral views, x 0.72.

each side where in camels its breadth is one-fourth of the ventral surface. In the lateral view it is seen that in *Protylopus* the fissura palaeo-neocorticalis lies at the midheight of the hemisphere.

One is inclined to relate these differences in extent of the neocortex with the difference in size of the two animals. But they are not just another case of BAILLARGER's law in an evolution. Compensation between expansion of cortical sheet and white matter was not only by increased fissuration. There was also a downward shift of the neocortical border. Unfortunately, what knowledge I have of modern brains comes mostly from the literature and a few endocranial casts. Am I wrong? To me it seems that that shift was not linked, or not only linked with size. It appears to be linked with time: a process which can be observed only in evolution through the ages.

In the hooved and herbivorous predecessors of the Ungulata s. str., paleocene and early eocene endocranial casts show a fissura rhinica above mid-height; but those were lissence-phalic brains. In many eocene and oligocene ungulate brains, smaller and larger than that of *Protylopus*, I found the fissure in just above to just below midheight while it limits a neocortex with prominent convolutions. To me, at present, this seems to be their neuro-logical "Zeitsignatur". Among extant brains, I have seen a comparable situation only on two of three figured, far larger brains of an utterly different, a large mammal, the aardvark. I have been extensively reading up on extant mammals whose bodies and brains are close to the size those of *Protylopus* must have been. My impression is that where today there is gyrification comparable to that in the Eocene, the neocortex is not restricted to the dorsal convexity[1]).

Among extant mammals of hare-size, the Hyracoidea are best suited for that comparison. These coneys are classified either as Paenungulata, or between Artiodactyla and Perisso-dactyla, they share with Tylopoda rubbery padded toes and possibly rumination, and with the first probable tylopod the small size (skull lengths: 71 to 117 mm; *Protylopus*: 111, 124). From the figures of four authors who studied seventeen hyracoid brains plus endocranial casts, and one cast at hand, I learnt that there is of course more variability than paleoneurology knows of, and that fissuration is not richer than it was in *Protylopus*; but the share of the piriform lobe in cerebral height, although ⅓ in one case, is on the average close to ⅛ (Fig. 2).

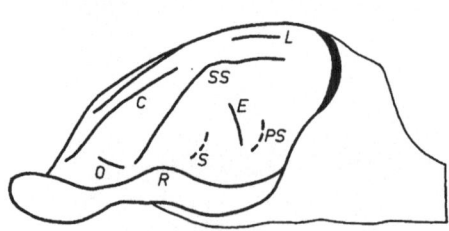

Fig. 2. *Procavia capensis*, endocranial cast, x 0.95. (From WELLS, S. Afr. J. Sci. 1939.)

You have just seen on one of SPATZ's slides L. EDINGER's demonstration of the neopallium in extant lower versus higher mammals. To show the situation in the former, my father happened to choose a mammal of hare-size. That "kangaroo-rat" is a marsupial; but that *Protylopus*-like situation, the mid-height fissure limiting an only dorsal, ± smooth neocortex, also exists in primitive, small placental mam-

[1]) After the symposium I did, however, see in the Max-Planck-Institut a 35 mm-long unpublish-ed endocranial cast with a gyrified neocortex that extends over only half of the cerebral height. Its similarities with and considerable differences from the eocene ferungulate type are especially interesting because it reproduces the brain of a *Rhynchocyon*: that genus which the extensive calculations of BAUCHOT and STEPHAN have revealed as the most encephalized of extant Insec-tivora.

mals. What the diagram shows so strikingly well, opposing marsupial and dog, is the great truth long established by comparative brain anatomy; when put in comparison, the neocortex "becomes" increasingly complicated in the "passage" from lower to higher mammals. The ancestral series of camelid brains is new evidence of two additional facts: a) that assumed process has actually taken place within families of higher mammals, with the passage of time; b) in their remote Past, the mantle was thrown into folds when still in the dorsal position which is known, smooth or almost smooth, in small lower mammals of the Present.

Cerebellum. The Eocene-to-Present changes of the forebrain are so clearly reflected in the likewise enlarging hindbrain that I am not digressing from our topic if I say a few words about the cerebellum. The details of its surface are perfectly reproduced on the eocene and oligocene endocasts. As lobulation further increased, the cauliflower-like formation failed and fails to imprint the endocranium, and brains are needed to show the whole way of transformation of the cerebellum. In the eocene specimen, straightly sagittal furrows set off from each other the divisions for which even JANSEN a. BRODAL allow that gross morphology uses the antiquated terms vermis, hemispheres, and flocculi. The latter did not keep pace with the overall enlargement of the cerebellum, nor did the vermis. Its share in cerebellar breadth was one-third in the Eocene and is only about one-fifth now. Thus surface configuration allows a paleontologist to practically see that the quantity of fibers increased in the cortico-ponto-cerebellar system. The enlargement of the cerebellum was mainly by increase of its hemispheres: influenced by the neocortex. Thus we see in the camel line the evolutionary course imagined, by comparing different extant cerebella, when the hemispheres were distinguished as neocerebellum by L. EDINGER.

A Famous Miocene Brain

The two other samples from the fossil series (Fig. 3) are plaster casts representing brains of modern types. The smaller specimen is COPE's (1877) endocast of a late miocene *Procamelus occidentalis;* the larger was recently prepared from a cranium of an early pliocene *Procamelus*. These were brains that functioned perhaps 15 and 10 million years ago, respectively. You see that great evolutionary progress had occurred, and that advance continued within the genus. The neocortex had in the early Oligocene added transverse sulci in the frontal area, and the two *Procamelus* phases show increasedly complex patterns of infolding. Cerebral height in the geologically younger individual was greater by 15 mm, whereas piriform lobe height was the same as in the older brain on one side, less on the other side.

But it is not because of their anatomy that I introduce you to the two different brains of *Procamelus*. It is because the ancient specimen has acquired a nimbus which it does not deserve. This is one of the fossils by which paleoneurology has become involved in research for which it can not offer a sound basis: determination of exact brain/body size relationships.

I have mentioned such quantitative studies as we can do; the diameters of parts of brains reveal proportional changes in phylogenies. We paleontologists measure what material we have. Bodies we do not have (and complete skeletons of fossil mammals are rare). Apparently we have been dangerously successful envisaging our lifeless material as

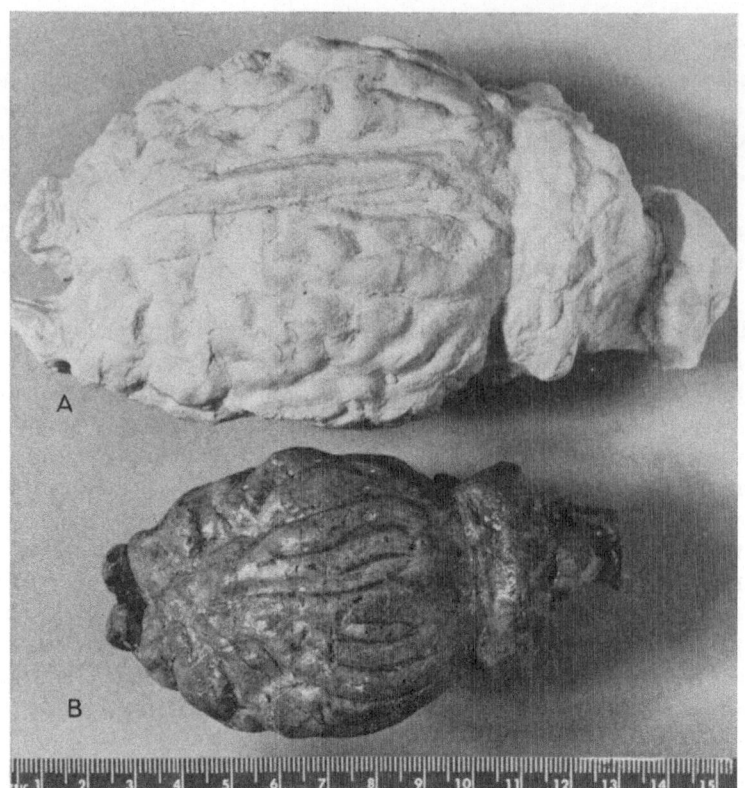

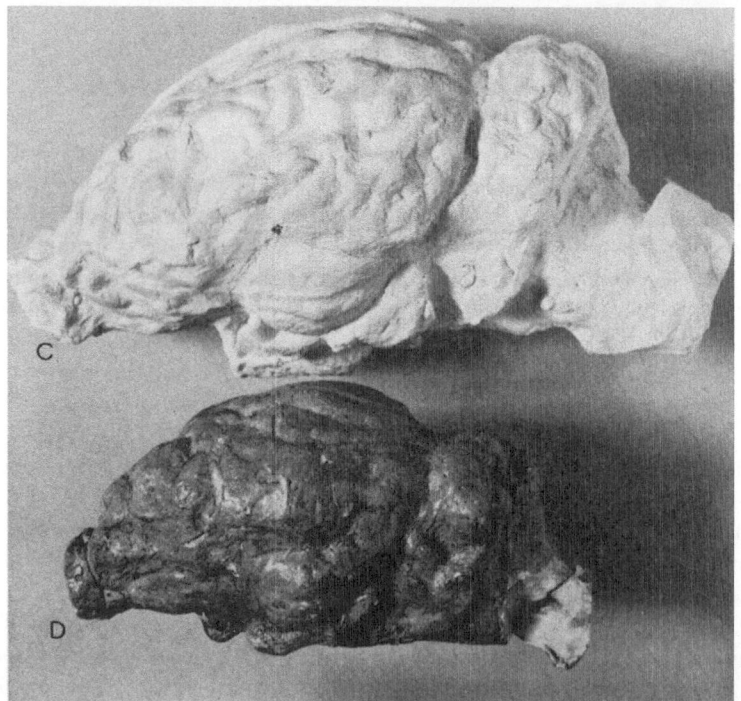

Fig. 3. Miocene and Pliocene *Procamelus*, dorsal and lateral views, x 0.61.

remains of living animals. EUGENE DUBOIS believed he had fossil proof that in phylogenies the proportion of brainweight to bodyweight increased by leaps, and that brain quantity can increase only in geometrical progression, by duplication, etc. Now that so much allometric study is done by neozoologists, it is urgent that I show you the gradual progress in brain evolution and plead against the continuing belief that paleozoology has demonstrated that natura facit saltus. Only a few years ago, this was written: "Bei vollkommen gleicher Körpergröße und fast übereinstimmender Skeletspezialisation hatte *Procamelus gracilis* ein nur halb so großes Gehirn als *Lama guanaco*". This is stated in a paper which not only critically surveys results obtained by other zoologists in metric studies and warns that applying them to evolution is speculation, but draws its own conclusions from material so rich that it includes, e.g., more than 200 skulls of 5 species of one family, and 75 skulls, adult skulls only, of another species. Evidently, the various authors who saw, in the camelid case of "DUBOIS' law", "une des plus belles et plus sûres" of the "paläontologische Bestätigung des Cephalisationsgesetzes", have not been aware of the meager basis of DUBOIS' calculation.

To begin with, what was it that has been accepted as typical brain size of ancestor and descendent? First, that of one individual of a species that flourished all over North America for some 5 million years; then, that of one individual of a group now living in South America – a group of which HERRE has reported endocranial capacities varying from 175 to 335 cc. –. There was a jump between the two, but a jump in time. Yet it was assumed that all that time there were no grades in brain size!

Next – what has been believed to mean that body size was the same in the miocene species as it is in those extant camelids must be COPE's remark that *Procamelus occidentalis* "equalled the llama in general proportions". The scientists who weigh dozens of bodies of one species and cite DUBOIS' interpretation of COPE, they obviously do not know the basis of COPE's guess. From the area in which the skull was found, the following other specimens were assigned to that species: one jaw fragment, a few vertebrae, and one almost entire forelimb with the proximal end of the scapula. There is no evidence that these were bones of one individual, or from one herd; they may well be remains of animals which did not live at the same time.

Misinterpreted as an exact statement, COPE's vague remark was actually a necessary complement to his figures and description of those few fossils. It was likewise quite proper that, in another paper, he called *P. occidentalis* of deer-size. As one never knows a whole mammal long extinct, paleontologists must use such vague terms to get a mental picture of the animal. Thus COPE was also justified describing another of the many species of *P.* as sheep-sized.

That species is *P. gracilis*. I mention it to point to another grotesqueness in the supposed camelid brain size duplication without body enlargement. The specimen used by DUBOIS in his calculations (1928) is a duplicate of COPE's endocast of the type skull of *P. occidentalis*. By some lapsus calami, he wrote of it as the brain of *P. gracilis*. I have come across seven papers which, like that I have cited, repeat that error, e.g. "The lama has twice the ... cephalization of *Procamelus gracilis* of the same size and general build"!

There are also some minor oddities. The endocast supposedly providing exactly a brain size important in evolution is not a complete one. Anteriorly, both the slope rather than rise of its dorsal surface and the lack of an imprint of the cribriform plate show that in the skull the ethmoidal chamber had not been cleaned out when the cast was made, so that this includes hardly more than the olfactory tracts, not the whole olfactory bulbs.

Further, most of DUBOIS' disciples cite him concerning the ratio of the size of the body to the size of the whole brain. DUBOIS failed to make clear which parts are the "Psycho-encephalon", whose proportions to the body he believed to have doubled in "cephali-sation leaps".

Conclusion

Not "reduced to mathematical formulation" (HERRICK), the phylogeny of the camel brain can be said to have followed the course common to the majority of the Mammalia during the Neozoic Era. From small ancestor to large survivor, the brain became a lesser part of the body – that is, cephalization decreased, while neëncephalisation increas-ed. Much is made of this trend, one of many which ran a parallel course in numerous mammals of that era. In horses, for example, as in camels, toe reduction has gone as far as is possible in the two orders; and the orbits, at first open posteriorly, have become closed sockets for the eyes. Why, then, the emphasis on progressive evolution or mere size increase of the most complexly composed element of the vertebrate body as "necessary phylogenetic perfection"? Cui bono are the phases of "brain value" believed to have been so important during evolution that they must be mathematically substantiated?

I can not get rid of the suspicion that the brain value concept rests on the 19th century mixup of all other phylogenies with that of the peculiar, hitherto short-lived peak of one line of the order Primates that is so successful in its unnatural way, Homo sapiens. Only so can I understand that the latest of the papers on hyracoid brains suggests that "the slightly inferior cerebral development" of a pleistocene brain "may have sufficed to determine the extinction of the species". Too often one forgets the "Dauertypen", the animals that persist almost or quite unchanged through long times of earth history. In North America recently, rats and mice have been acknowledged as the dominant mam-mals; they reproduce faster than man's intelligence can destroy them. Extinct are the camels on the continent of their origin and long bloom, where as late as during the Pleis-tocene they ranged from Alaska to Florida.

They had, as you have seen, from an "inferior" brain, evolved just the kind of neocort-icalized and absolutely large brain which is believed to perform best "die höchsten Hirn-leistungen" and thereby to be an advantage in nature, even in the Past under conditions we can but vaguely imagine.

An evolutionist can judge success only from wide distribution of a species, and from long-continued existence through the ages. Bones of the last North American Tylopoda have been carbondated. Some are 8000 years old, others possibly much younger. These large creatures had survived the presumed rigors of the Ice Age, and became extinct before there was a man who could shoot them.

Summary

From a series of endocranial casts demonstrating evolutionary advances in the surviving line of the Camelidae, the oldest has been shown as a contrast to the extant brains; in the forebrain, the neocortex was convoluted but only dorsal, in the cerebellum the vermis was dominant. Specimens from two intermediate phases have been used to warn that reliable contributions to brain/body studies should not be expected from paleoneurology. And I used the North American end phase to indicate that paleoneurology also can not help to solve the great enigma of extinctions. Its object is collecting brain facts from the Past. Its literature now consists of more than 1500 papers and books. In not one phylogeny has it been proven – can it be proven? – that survival or extinction was due to either relative or absolute brain size, or to unprogressive or progressive forebrain evolution.

Dr. TILLY EDINGER
Museum of Comparative Zoology
Harvard University
Cambridge, Mass., USA

Einige Bemerkungen zur Modifikabilität,Vererbung und Evolution von Merkmalen des Vorderhirns bei Säugetieren*

Von W. HERRE, Kiel

Zur Kennzeichnung von Eigenarten des Vorderhirns der Säugetiere in phylogenetischer Sicht spielen Größenbeziehungen, Gestaltbesonderheiten sowie das Furchenbild eine besondere Rolle (EDINGER 1950; STARCK 1962). Aus diesem Grunde scheint es notwendig, die Beeinflußbarkeit dieser Strukturen durch Umweltfaktoren und ihre Abhängigkeit von der erblichen Grundlage beurteilen zu können. Das Schrifttum lehrt, daß das sichere Wissen über diese Problematik gering ist. Daher haben wir unsere Aufmerksamkeit solchen Fragen zugewandt. Über einige bisherige Ergebnisse unserer Untersuchungen seien kurze Bemerkungen gemacht.

Unser Ziel ist die Klärung von allgemeinen Problemen der Hirnevolution: Wir bearbeiten Hirneigenarten verschiedener Tiergruppen (RÖHRS 1955, 1959, 1965; SAMTLEBEN 1958; NETELER 1963; SCHUMACHER 1963; THIEDE 1964, 1965; HERRE 1965). Es zeigte sich, daß die Lösung von Grundproblemen gefördert wird, wenn die Variabilität in niederen taxonomischen Einheiten bekannt ist. Innerhalb von Arten ist die Fülle der Ausformungsmöglichkeiten und ihre Ursachen zu erforschen. In dieser allgemein-biologischen Sicht bieten Haustiere ein geeignetes Untersuchungsmaterial, weil sich im Hausstand die Variabilität der Wildarten erweiterte und sich bei Haustieren sehr unterschiedliche Formen entwickelten (HERRE 1958, 1959, 1961, 1964).

Unterschiede zwischen Wild- und Haustiergehirnen

Kapazität, Hirngewicht. Über Gehirneigenarten von Haustieren hat als erster DARWIN (1868) auf Grund von Hirnschädelkapazitätsmessungen berichtet. Seine Angaben wurden übersehen, bis KLATT 1912 an ihnen anknüpfte (KLATT 1912, 1921, 1928), was die Forschung belebte. Die Angaben von DARWIN wurden vielfach bestätigt (z. B. BOLZANO 1924; BAHRAMI 1926; HERRE 1952; RÖHRS 1955; STOCKHAUS 1965; GORGAS 1965). Wir haben diese Studien ausgebaut (HERRE 1936, 1956; HERRE u. THIEDE 1965) und können aussagen, daß beim Übergang von der Wildart zur Hausform ganz allgemein eine Verringerung des Hirngewichtes bei vergleichbaren Körpergrößen eintritt; sie beträgt ungefähr 20%, kann aber bis zu 30% ansteigen. Diese Feststellung gilt für Haussäugetiere und Hausgeflügel (HERRE 1956; SENGLAUB 1959). Nur wenige Farmtiere bilden Ausnahmen; die Gründe sind noch unbekannt. Von diesen Tieren seien Rentier, Nerz und Labormaus genannt (HERRE 1955, 1964; BÄHRENS 1960; FRICK u. NORD 1963). Es ist von Interesse,

---

*) Herrn Professor Dr. ERICH REISINGER, Graz, zum 65. Geburtstag in herzlicher Verbundenheit.

daß auch wichtige Sinnesorgane beim Haustier eine geringere Entwicklung zeigen als bei der Wildart. Für das Auge von Wild- und Hausschweinen legte WIGGER (1939) Befunde vor, über Veränderungen am Ohr berichtete HERRE (1953). RÖHRS (1957) und SCHUMACHER (1963) haben über die Auswirkungen der Hirnverringerung auf das Verhalten allgemeinere Angaben gemacht.

Unterschiedliche Größenänderung der Hirnteile. Es erhebt sich die Frage, welche Teile des Gehirns beeinflußt werden. Beispielhaft sei aus den Ergebnissen von SCHUMACHER (1963) über den Iltis, *Mustela putorius* Linnaeus 1758, und seine Hausform, das Frettchen, *Mustela putorius* f. furo, (zur Nomenklatur der Haustiere vgl. BOHLKEN 1961) berichtet, daß die Medulla oblongata die geringste Abnahme mit 9% zeigt; es folgen Mittelhirn mit 11%, Kleinhirn mit 16% und Endhirn mit fast 32%. Prinzipiell gilt diese Reihenfolge auch für die anderen Fälle, welche wir analysierten. Dies ist nicht nur in phylogenetischer, sondern auch in funktioneller Sicht von Interesse. Das Endhirn ist der stammesgeschichtlich jüngste Hirnteil, aber auch das höchste Überwachungs- und Integrationszentrum. Die Hirngewichtsabnahme wird vor allem durch eine Verkleinerung des Endhirns bewirkt.

Weitere Erhebungen in unserem Arbeitskreis zeigten beim Frettchen im Vergleich zum Iltis, ebenso wie bei Alpaka und Lama, *Lama guanaco* f. glama, im Vergleich zum Guanaco, *Lama guanacoë* MÜLLER 1776 eine stärkere Minderung des Isokortex im Vergleich zum Allokortex. Nach den Studien von STEPHAN (1951) über Wildschweine, *Sus scrofa* Linnaeus 1758, und Hausschweine, *Sus scrofa* f. domestica, ist bekannt, daß auch die einzelnen Regionen des Isokortex nicht einheitlich betroffen werden. Während die insuläre, die postzentrale und parietale Hauptregion eine Abnahme beim Haustier von nur 5% zeigen, wird die okzipitale Hauptregion um 25–30% kleiner und in ihr die Area striata um 37%. Weitere Änderungen im Feinbau sind zu beobachten.

Ursachen der Größenänderung. Welches sind die Ursachen dieser Veränderungen? Handelt es sich um umweltbedingte Modifikationen oder spielen erbliche Steuerungen mit, welche dem Menschen in bewußter oder unbewußter Auslese die Möglichkeit gaben, Besonderheiten zu erzüchten? Kann das Phänomen als Modell für Evolutionsvorstellungen erörtert werden? Dazu: Schon bei Hirnschädelkapazitätsmessungen fiel KLATT (1912) auf, daß im Zoo aufgewachsene Wildcaniden beträchtlich geringere Hirnkapselkapazitäten zeigen als Tiere freier Wildbahn. Umgekehrt zeigten verwilderte Hauskatzen, *Felis silvestris* f. catus, eine Zunahme der Hirnschädelkapazität; sie erreichen aber die Werte der Wildkatze, *Felis silvestris* Linnaeus 1758, nur zum Teil. RÖHRS (1955) hat diese Angaben bestätigt und erweitert. Das gleiche Ergebnis zeitigten neue Studien an Kaninchen. Hauskaninchen haben kleinere Gehirne als die wilde Stammform, wie zuletzt CHOINOWSKI (1958) belegte. HÜCKINGHAUS (1965) ermittelte an verwilderten Hauskaninchen, *Oryctolagus cuniculus* f. domestica, von den Kerguelen, daß deren Hirnschädelkapazitäten deutlich höher liegen als bei Haustieren, aber im unteren Bereich der Wildkaninchenwerte bleiben. Das weist auf Umwelteinflüsse hin. Doch Studien an Hirnen selbst sind zu fordern.

KLATT (1932) hatte die Möglichkeit, im Zoo aufgewachsene Rotfüchse, *Vulpes vulpes* Linnaeus 1758, zu untersuchen; wir zogen mehrfach blind in Gefangenschaft gekommene Rotfüchse auf. Es zeigte sich im allgemeinen bereits in der ersten Gefangenschaftsgeneration eine bemerkenswerte Hirngrößenabnahme. Kamen die Tiere etwas älter in Gefangenschaft, ließen sich die Veränderungen nicht feststellen. Es kann aber kein Zweifel an einem Umwelteinfluß geäußert werden, nur müssen wir versuchen, Aufschluß über den Zeitpunkt der Beeinflußbarkeit, die sensible Phase, zu gewinnen (HERRE 1965). STEPHAN

(1954) hat eine cytoarchitektonische Analyse der Gefangenschaftsfüchse begonnen und ermittelt, daß im wesentlichen jene Bereiche kleiner werden, welche auch bei Haustieren die Minderungen zeigen.

RÖHRS u. HERRE (1958, 1964) konnten 1956 in Südamerika einige verwilderte Hausesel, *Equus africanus* f. asinus Linnaeus 1758, erlegen und mit Hauseseln vergleichen, welche „im Dienst" standen. Es zeigte sich, daß verwilderte Hausesel deutlich höhere Hirn-gewichte bei vergleichbaren Körpergrößen als gewöhnliche Hausesel haben. Die Zu-nahme wirkt sich vor allem in einer Ver-breiterung und größeren Höhe in kauda-len Teilen des Endhirns gestaltlich aus. Nach all diesen Befunden kann als sicher gelten, daß das Gehirn so hochentwickel-ter Säugetiere, wie es Lagomorpha, Carni-vora oder Ungulata sind, in starkem Maße umweltmodifikabel ist, so daß seine Größe allgemein um 20% abnimmt und sich ein-zelne Teile bis über 30% verringern kön-nen. Diese Feststellung ist wichtig bei funktioneller Betrachtung von Hirneigen-arten, sie ist aber auch im Auge zu behal-ten bei einer Bewertung von Hirnbeson-derheiten in phylogenetischer Sicht, wenn bei solchen Betrachtungen Größenbeziehungen zugrunde gelegt werden.

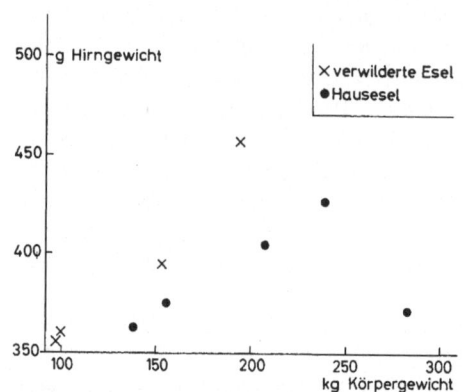

Abb. 1. Relative Hirngewichte südamerikani-scher Esel.

Unterschiede in der Hirnform. Die Gehirne von Haustieren zeigen eine Reihe von erheblichen Gestaltunterschieden gegenüber der Wildart (KLATT 1921; HERRE 1936, 1956; RAWIEL 1939; STARCK 1954; HERRE u. THIEDE 1965). Da Haustiere zur gleichen biologi-schen Art gehören wie die Stammform, handelt es sich um innerartliche Ausformungs-erscheinungen, deren Kenntnis zur Beurteilung der zwischenartlichen Unterschiede, also der evolutiven Schritte, von Interesse sind (HERRE 1964). Gestaltverschiedenheiten stehen mit Unterschieden in der absoluten Größe im Zusammenhang. Die absoluten Größen-unterschiede der Organe sind bemerkenswert; sie stehen zu Unterschieden in der Körper-größe in einem Verhältnis, welches durch allometrische Berechnungen in Zahlenwerten ausgedrückt werden kann (STEPHAN 1954, 1960; RÖHRS 1959, 1961). Als Beispiel für die absoluten Werte sei aus unseren Pudelzuchten genannt: Zwergpudel haben ein um 60 g schweres Hirn bei rund 4 kg Körpergewicht, Hirne von Königspudeln wiegen 90–100 g bei ca. 30 kg Körpergewicht. In unserem Material an ausgewachsenen Hausschweinen schwanken die Gewichte der Hirne zwischen 59 g bis 170 g. Die Gehirne der kleinen Tiere sind rundlicher, relativ breiter und höher als die der großen (KLATT 1921; RAWIEL 1939; STARCK 1954; VOLKMER 1956). Bei großen Gehirnen springt der Stirnpol zapfen-artig vor das Großhirn, bei kleinen liegt das rostrale Ende des Frontallappens wie ein Keil zwischen den Riechlappen. Ganz ähnlich sind die Unterschiede der Gehirne von Tieren verschiedener Wuchsformen (zum Begriff der Wuchsform vgl. HERRE, FRICK, RÖHRS 1961) innerhalb von Arten. Als Beispiel von Haushunden seien Boxer und Barsoi gleicher Körpergröße genannt (HERRE 1956). Die Unterschiede sollen hier nicht im ein-zelnen erörtert werden. Auch bei den südamerikanischen Hauscameliden Lama und Alpaka wirkt das Gehirn gestauchter, vor allem im hinteren Drittel, als beim Guanaco

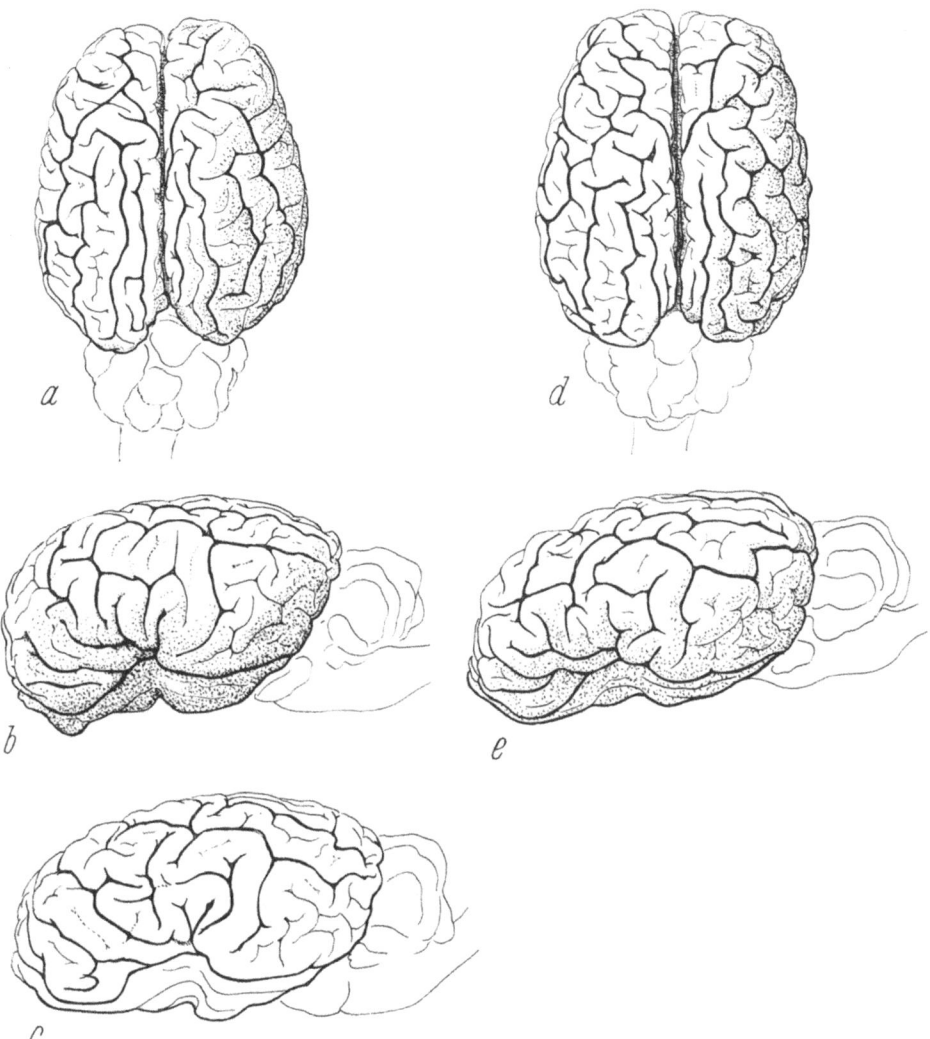

Abb. 2. Eselgehirne aus Südamerika, a, b Hausesel Nr. 3, Hirnfrischgewicht 363 g, c, d verwilderter Esel Nr. 5, Hirnfrischgewicht 300 g, e Hausesel Nr. 2, Hirnfrischgewicht 375 g.

(HERRE u. THIEDE 1965). Das gleiche gilt für Schwein, Schaf, Katze (HERRE 1956). RÖHRS (1955) hat daher pointiert gesagt: Beim Haustiergehirn „fehlen" vor allem occipitale Bereiche. Diese innerartlichen Wandlungsmöglichkeiten der allgemeinen Form verdienen der Beachtung auch bei phylogenetischen Diskussionen.

Wie wirken sich diese doch manchmal recht auffälligen Gestaltveränderungen auf die inneren Bauteile, auf die Größen der Areae usw. aus? Unser Arbeitskreis analysierte Gehirne von Zwergen und Riesen gleicher Hunderassen (VOLKMER 1956) sowie von Rassen (zum Begriff der Rasse vgl. HERRE 1961) bei Hunden (vgl. HERRE 1956) und Cameliden

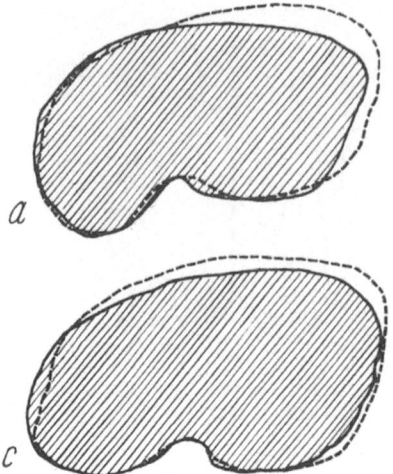

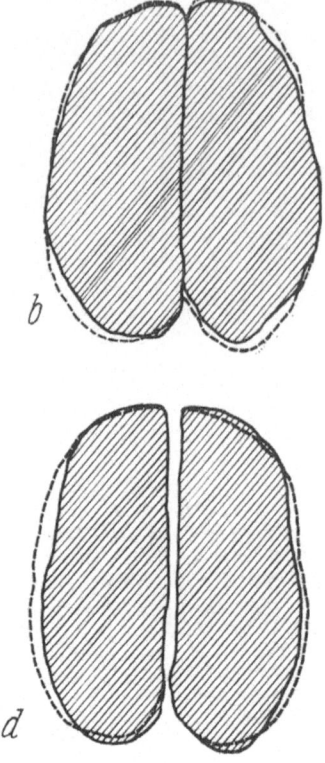

Abb. 3. Schematische Darstellung der Umrißlinien von Hirnen südamerikanischer Esel. Ausgezogene Linie = Hausesel; gestrichelte Linie = Verwilderte Esel.
a, b Hausesel Nr. 3 verglichen mit verwild. Esel Nr. 5,
c, d Hausesel Nr. 1 verglichen mit verwild. Esel Nr. 7.

(HERRE u. THIEDE 1965). Für Zwerge und Riesen, also Schläge gleicher Rasse, ergibt sich prinzipiell die gleiche relative Größe der Untereinheiten auf den verschiedenen Größenstufen trotz der Formunterschiede. Zwischen Rassen lassen sich Differenzen ermitteln, die von der Wuchsform unabhängig sind und auf erbliche Eigensteuerungen hinweisen. Einen ähnlichen Hinweis auf erbliche Sondersteuerungen geben auch Analysen der relativen Hirngröße sowie der äußeren Struktur von Hundehirnen durch STEPHAN (1954) und HERRE u. STEPHAN (1955).

Unterschiede im Furchenbild. Doch ehe das Erbproblem erörtert sei, muß die Furchung ins Auge gefaßt werden, weil sich zwischen Wildarten und deren Hausformen Unterschiede in der Furchung zeigen. Das ist wichtig, weil im Prinzip für Säugetiere gilt, daß mit steigender phylogenetischer Ranghöhe die Intensität der Furchung zunimmt, und weiter, daß im gleichen Verwandtschaftskreis von Säugern größere Hirne im allgemeinen stärker gefurcht sind als kleine. Haustiere fallen durch ein sehr viel lebhafteres Furchenbild auf als die Wildform. Das gilt für alle untersuchten Arten (KLATT 1921; HERRE 1936, 1965; RAWIEL 1939; RÖHRS 1955; STEPHAN 1951). Es gibt jedoch Ausnahmen von dieser Regel. Das Frettchen hat ein einfacheres Furchenmuster als das Wildtier Iltis (SCHUMACHER 1963). Es gibt hochdomestizierte Hausschweine mit sehr klarem Furchenbild (HERRE 1936, RAWIEL 1939), und auch hochdomestizierte Haushunde mit einfachem Furchenmuster sind bekannt (STARCK 1954). In allen Fällen liegt aber prinzipiell der gleiche Sachverhalt vor. Dazu: Als wir in unserem Arbeitskreis die Furcheneigenarten von Wildtieren und Hausformen zu untersuchen begannen, zeigte sich bald, daß bei Haustieren die eingesenkte Oberfläche geringer ist als bei Wildtieren der gleichen Art (STEPHAN 1951; SCHUMACHER 1963; HERRE u. THIEDE 1965). Die Furchenminderung trifft alle Bereiche, aber in etwas unterschiedlichem Ausmaß (STEPHAN 1951). Im Prinzip zeigen alle Haustiere gegenüber ihren Wildarten die Minderung der eingesenkten Oberfläche, es lassen sich aber einige Eigenarten bei Rassen von Hunden, Schweinen, Schafen, Cameliden nachweisen, die wieder auf eigene Erbsteuerungen hinweisen.

Die im Zusammenhang mit der Domestikation, also im innerartlichen Bereich von Tierarten auftretenden eigenartigen Furchenbilder bedürfen weiterer Analyse, weil Furchungseigenarten bei stammesgeschichtlichen Erwägungen eine Rolle spielen. Die innerartlichen Veränderungen bei Haustieren können vielleicht zum Verständnis der evolutiven Vorgänge und Ereignisse beitragen. Die Tatsache, daß bei allen Wildarten, die in den Hausstand kamen, ähnliche Besonderheiten auftreten, deutet auf Pluripotenz hin (HERRE 1962).

Kreuzungsexperimente

Wildcaniden. Erbstudien schienen uns dringend erwünscht, um zu Problemen der Gewichtsrelationen, der Formbesonderheiten und des Furchenbildes weitere Anhaltspunkte zu gewinnen. Ein Zufall half weiter. In den Zoologischen Gärten von Nürnberg und Děčin glückten Kreuzungen von Canidenarten, die sich bei ähnlichem Körpergewicht in Hirneigenarten unterschieden: Schakal, *Canis aureus* Linnaeus 1758, und Kojote, *Canis latrans* Say 1823 (HERRE 1965). Die Kojoten haben ein wesentlich schwereres Gehirn als die Schakale. In der Form des Gehirns zeigt sich bei den Schakalen eine bemerkenswerte Variabilität. Es gibt spindelförmige Hirne, die sich zum Lobus rhinalis hin verjüngen mit einem schwach ansteigenden Profil und geringer Aufwölbung im parietalen und occipitalen Bereich. Außerdem sind birnenförmige Hirne vorhanden mit starker Ausformung in der parietalen und occipitalen Region. Diese Formunterschiede sind innerartlich größenabhängig; der spindelförmige Typ findet sich nur bei leichten Gehirnen, wie HARRIES (in Vorbereitung) feststellte. Als artkennzeichnend sind nach HARRIES beim Schakalgehirn die geringe zentrale Ausladung des Lobus sigmoideus und die recht kleinen Bulbi olfactorii zu werten. Bei *Canis latrans,* dem Kojoten, ist innerartlich die Hirnform einheitlicher, wenigstens in unserem Material. HARRIES fand eine gestreckte Ovalform mit meist starkem Anstieg der Profillinie am Lobus orbitalis. Sowohl Lobus parietalis als auch Lobus sigmoideus und Lobus temporalis posterior treten stärker hervor. Artkennzeichnend ist die starke Ausladung des Lobus sigmoideus, vor allem des Gyrus centralis nasalis und des Gyrus centralis caudalis. Dadurch erscheint das Kojotengehirn etwas kastenförmig. Außerdem ist die starke Entwicklung der Bulbi olfactorii im Vergleich zum Schakal hervorzuheben. Im Furchenbild gleichen sich Schakal und Kojote im Grundtyp, doch zeigt nach den Feststellungen von HARRIES der Schakal eine geringere Furchenausprägung als der Kojote, was mit der absoluten Größe in Zusammenhang gebracht werden kann. Das gesamte Oberflächenrelief tritt beim Kojoten schärfer hervor als beim Schakal. Es entsteht der Eindruck einer kompakten Zusammenlagerung von Windungen und Lappen, weniger der einer Furchung. Besonders kennzeichnend beim Kojoten ist der weite Bogen der Fissura coronalis als Folge der starken Ausbildung des Lobus sigmoideus. Durch die Ausdehnung der Fissura coronalis hat auch die Fissura cruciata mehr Raum, sie ist bei Kojoten deutlich länger als bei Schakalen. Der Gesamtverlauf der Furchen erweckt beim Kojoten den Eindruck einer stärkeren Stauchung in der Längsachse als beim Schakal, dessen Furchenbild gestreckter und flacher wirkt. Auch bei der quantitativen Analyse zytoarchitektonischer Einheiten konnte HARRIES Unterschiede zunächst zwischen den Arten ermitteln; die Kreuzungsindividuen sind noch nicht bearbeitet; solche Analysen sind sehr zeitraubend.

Kreuzungsexperimente mit Säugetieren, bei denen Eigenarten des Hirns in ihrem Erbgang erforscht wurden, liegen kaum vor. Die Kreuzungen von Hunderassen gegensätzlicher Wuchsform sind zu nennen, welche KLATT durchführte. Aber dabei wurden domestizierte Tiere als Ausgangsmaterial herangezogen und die Untersuchung auf Form und Furchenbild begrenzt (KLATT 1941–1944; OBOUSSIER 1942). Wir haben erstmalig Wildcaniden benutzt, von den Ausgangsfamilien der Arten reichlich Vergleichsmaterial gezüchtet, reziproke Kreuzungen in erster und zweiter Generation studiert und auch Rückkreuzungen erzielt; wir stießen bis zur Erfassung zytoarchitektonischer Einheiten vor.

Bei den Kreuzungstieren der ersten Generation zwischen Schakal und Kojote liegen die Gewichtswerte, bei bemerkenswerter Streuung, etwa intermediär. Auch in den Furchungseigenarten und Gestaltbesonderheiten nehmen die Individuen erster Generation eine Zwischenstellung ein. In der zweiten Generation zeigen sich Spaltungen. Es fallen Tiere, deren Hirne gewichtsmäßig im oberen Bereich der Kojoten liegen, neben solchen, welche den oberen Bereich der Schakale erreichen. Insgesamt sind die Werte merkwürdig – und noch unerklärlich – zum Kojoten hin verschoben. Das ist nicht einfach als materneller Einfluß zu deuten, weil die Verschiebung gleichermaßen in reziproken Kreuzungen beobachtet werden kann. HARRIES kann eine weitere interessante Tatsache mitteilen: Die schwersten Hirne der Tiere zweiter Kreuzungsgeneration – im oberen Kojotenbereich – zeigen nicht die Formbesonderheiten der Kojoten, sie gleichen vielmehr kleinen, schlanken Schakalgehirnen, während kleine, leichte Gehirne eine kojotenähnliche Kastenform haben können. Dies läßt den Schluß zu, daß Gewicht und Form nicht streng korreliert sind, sondern durch voneinander unabhängige Faktoren die Ausprägung bestimmt wird. Einen

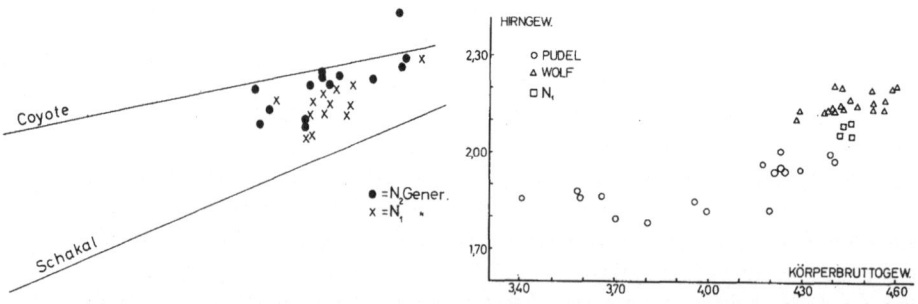

Abb. 4. Durchschnittliche relative Hirngewichte von Schakalen und Coyoten sowie von 1. und 2. Nachzuchtgeneration.

Abb. 5. Relative Hirngewichte von Wolf und Pudel sowie von 1. Nachzuchtgeneration.

ähnlichen Schluß ließen die Studien an Hundehirnen durch KLATT zu. Noch interessanter sind Furchenbesonderheiten bei den Tieren zweiter Kreuzungsgeneration. Bei einigen von ihnen zeigt sich eine vermehrte Brückenbildung und zahlreichere Nebenfurchen in einer Form, die bei keiner der Elternarten zu beobachten ist. Die Furchenenden neigen zu Verzweigungen. Man gewinnt den Eindruck, daß irgendwie Erbanlagenkonstellationen zustande kamen, die zu Gestaltungen führten, welche in den Arten ungewöhnlich sind. Die Feststellungen legen den Gedanken nahe, daß zu den Eigenarten des Furchenbildes bei Haustieren, neben dem Umwelteinfluß, besondere Rekombinationen von Erbanlagen beitragen, die sich im Laufe der Domestikation einstellten. Auf die Bedeutung von Rekombinationsvorgängen zum allgemeinen Verständnis von Domestikationsmerkmalen hat RÖHRS (1961) besonders nachdrücklich hingewiesen. In unserem Arbeitskreis sind weitere Analysen im Gang, um diese Probleme in entwicklungsphysiologischer Sicht klären zu helfen.

Kreuzung zwischen Wild- und Hauscaniden. Nach diesen Ermittlungen schien es wünschenswert, auch den erblichen Steuerungen in den Unterschieden der Gehirne von Wolf, *Canis lupus*, und Haushund, *Canis lupus* f. familiaris, nachzugehen. Dazu erzüchteten wir einen Stamm südeuropäischer Wölfe und Stämme eingetragener Pudel verschiedener Größe, weil sich Pudel auch in einer Reihe äußerer Merkmale vom Wolf deutlich unterscheiden. Wir verpaarten eine Wölfin und einen Königspudel (HERRE 1965). Sie erbrachten bislang in vier Jahren Nachzucht, und es gelang bereits auch von zwei weiblichen Kreuzungstieren weitere Nachzucht, also zweite Kreuzungsgeneration, zu bekommen. Zur Untersuchung kamen bislang vier Gehirne von Tieren erster Kreuzungsgeneration, die anderen Tiere sind noch zu jung. Die Gehirne von Wölfen und Pudeln unterscheiden sich deutlich in den absoluten und relativen Gewichten. Die Gehirne der ersten Kreuzungsgeneration sind im Gewicht intermediär. Unsere Wölfe und Pudel haben unterschiedliche Hirnformen; die Kreuzungstiere sind intermediär. Wölfe haben ein recht klares Furchenbild, Pudel zeichnen sich durch ein unklareres Furchenbild mit vielen Nebenfurchen und Brückenbildungen aus. Die „Puwos" scheinen den Pudeln ähnlich, das Musterbild der Pudel erweist sich als ziemlich dominant. Auf die Eigenarten der Gehirne der 2. Kreuzungsgeneration sind wir sehr gespannt. Deutlich wird aber bereits, daß Größe, Form und Furchung genetisch nicht streng miteinander korreliert sind, sondern voneinander unabhängigen Erbanlagen ihre Ausprägung verdanken. Das sind Feststellungen, welche bei Spekulationen über die Evolution der Gehirne von Säugetieren im Auge zu behalten sind.

Diskussion

Insgesamt ergibt sich, daß Größe und Gestalt sowie der Anteil der eingesenkten Oberfläche am Vorderhirn von Säugetieren bemerkenswert umweltmodifikabel ist. Das muß beachtet werden bei der Deutung von Strukturen in funktioneller Betrachtung, jedoch ebenso bei Erwägungen über evolutive Vorgänge und Ereignisse, die auf solchen Eigenarten aufbauen. Größe, Gestalt, Furchung werden im wesentlichen durch zahlreiche eigene Erbanlagen gestaltet. Nach Artkreuzungen, also bei ungewöhnlichen Erbanlagenkombinationen, lassen sich am Gehirn Besonderheiten beobachten, welche von den arttypischen Normen abweichen. Die Gehirne von Haustieren zeichnen sich gleichfalls gegenüber der Norm der Wildart durch oft starke Abweichungen aus. Diese können durch

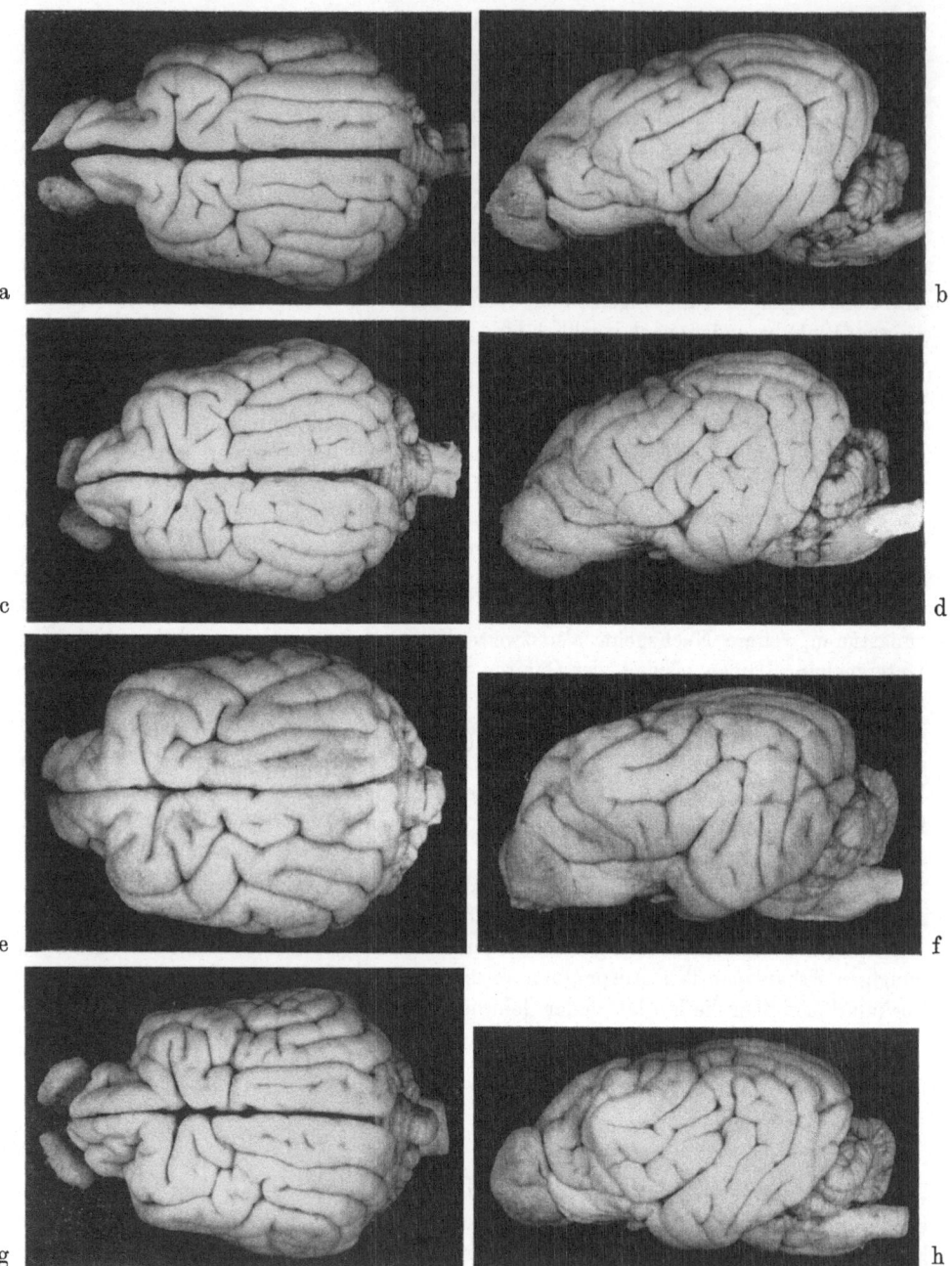

Abb. 6. Gehirne aus der Sammlung des Instituts für Haustierkunde, Kiel, a, b europäischer Wolf,
Nr. 4995, Körpergewicht brutto 31000 g, netto 23300 g, Hirngewicht 132 g; c, d Königspudel,
männlich, Nr. 6559, Körpergewicht brutto 23000 g, netto 15900 g, Hirngewicht 86,5 g; e, f Zwergpudel, männlich, Nr. 4924, Körpergewicht brutto 4400 g, netto 3000 g, Hirngewicht 54,2 g;
g, h erste Kreuzungsgeneration Pudel ♂ x Wolf ♀ (Puwo), männlich, Nr. 4965, Körpergewicht
brutto 28700 g, netto 21100 g, Hirngewicht 114 g.

besondere Rekombinationen von Erbeinheiten gedeutet werden, die zu Entwicklungs-
bedingungen führen, welche sich in unklaren Furchenbildern auswirken. Das Problem der
Ordnung von Einzelelementen zu den artkennzeichnenden Gestaltungen wird damit auch
im Bereich der Merkmale des Gehirns aktuell.

Zur Klärung wichtiger Fragen des Verhältnisses innerartlicher Ausformung zu zwischen-
artlichen Unterschieden (HERRE 1964) im Gehirn von Säugetieren tragen Betrachtungen
von Form und Furchen bei. Als besonders bedeutsam schält sich aber die Klärung von
Größenbeziehungen heraus. Diese erfordert eingehende und zeitraubende Analysen der
Einzelquanten, ihrer Verteilung im Ganzen und die Beurteilung ihrer Beziehungen zum
Ganzen mit biologisch sinnvollen Methoden. In solcher Sicht gewinnen allometrische
Betrachtungen und Berechnungen auch für Erörterungen über die Hirnevolution eine
hohe Bedeutung (RÖHRS 1959, 1965; STEPHAN 1960). Als bemerkenswertes Ergebnis
dieses Forschungsbereiches in den letzten Jahren erscheint mir die Feststellung, daß sich
die Beziehungen zwischen Hirngewicht und Körpergewicht im innerartlichen Bereich in
anderer Weise durch Zahlenwerte kennzeichnen lassen als zwischenartliche Unterschiede,
und daß sich bedeutsame evolutive Ereignisse auch in der Evolution des Gehirnes in einer
Durchbrechung von Allometrien auswirken (RÖHRS 1959, 1965; THIEDE 1965). Doch
darauf kann in dieser Übersicht nicht ausführlicher eingegangen werden, es sollte nur
herausgestellt werden, daß vielseitige Studien notwendig sind, wenn begründete Vorstel-
lungen zur Phylogenese des Vorderhirns entwickelt werden sollen und daß eine Kenntnis
der Variabilität auch von Gehirneigenarten in den niederen systematischen Einheiten
wichtig ist, wenn ein Verständnis der evolutiven Vorgänge und Ereignisse, die sich in
zwischenartlichen Unterschieden auswirken, gewonnen werden soll.

Summary

In order to judge the evolution of the brain, correlations of the size, singularities in
shape as well as the patterns of fissures were analysed. The modificability of the brain
structures, the genetics of these structures and its intraspecific variations compared with
the interspecific differences of the special groups should be known before studying the
evolutionary processes and events. Studies of our team brought forth the result that the
brain weight is reduced up to 20% already in the first generation living in captivity.
Domesticated animals yielded important material for general biological investigations,
because they set an appropriate example for a very great variability within the same
biological species. We studied the singularities of the brains of domesticated animals in
comparison with those of their wild living ancestors in details. The brain of domesticated
animals is 20–30% lighter than that of their ancestors. The size of brain and body-
weight displays a notable variation. Allometric methods are efficient aids to see the
correlations in a correct manner. In the brain of domesticated animals the forebrain
decreases most; the area striata decreases to ca. 40%. The brain of domesticated animals
generally is much more fissured than that of their wild ancestors. But this means not a
greater surface; our investigations revealed that the infold surface is comparativly
smaller. Domesticated animals show particular patterns of fissuring, which are unknown
with their wild ancestors. Cross breedings of wild Canidae, *Canis aureus* and *Canis
latrans*, give valuable hints for the genetic analysis of singularities of the brain. *Canis*

latrans with an equal body-weight has heavier brain than *Canis aureus*. Further peculiarities of the brains of both species are described. The analyses of the first and second cross-breeding generations has shown that size, shape and pattern of fissuring are influenced by genetic factors which are independent from each other. In the second generation singularities appeared which were not noticed within the species of the grand-parents. They were interpreted as the consequences of recombination. This phenomenon contributes also to the appreciation of peculiarities of domestiated animals. We have also interbred male poodle, *Canis lupus* forma familiaris, and wolf, *Canis lupus lupus*. The peculiarities of the brains of both and the crossbred were described. This experiment shows too, that size, shape and fissuring patterns are not strongly correlated with each other. The importance of these data for general problems of the evolution of the forebrain is discussed.

Literatur

BAEHRENS, D.: Über den Formenwandel des Mustelidenschädels. Allometrische Untersuchungen an Schädeln von *Mustela vison, Mustela lutreola, Mustela nivalis* und *Martes martes*. Morph. Jb. *101* (1960), 279–369.

BAHRANI, T.: Untersuchungen über die Schädelkapazität verschiedener Equiden. Diss., Halle/S. 1926.

BOHLKEN, H.: Haustiere und zoologische Systematik. Z. Tierzüchtg. u. Züchtungsbiol. *76* (1961), 107–113.

BOLZANO, O.: Untersuchungen über die Schädelkapazität verschiedener Schafrassen. Diss. Halle/S. 1924.

CHOINOWSKI, H.: Vergleichende Messungen an Gehirnen von Wild- und Hauskaninchen. – Zool. Anz. *161* (1958), 259–271.

DARWIN, CH.: Das Variieren der Tiere und Pflanzen im Zustande der Domestikation. London 1868.

EDINGER, T.: Die Palaeoneurologie am Beginn einer neuen Phase. Experientia (Basel) *6* (1950), 250–258.

FRICK, H., H.-J.NORD: Domestikation und Hirngewicht. Anat. Anz. *113* (1963), 307–316.

GORGAS, M.: Betrachtungen zur Hirnschädelkapazität zentralasiatischer Wildsäugetiere und deren Hausformen. Zool. Anz. *176* (1965) 227–235.

HERRE, W.: Untersuchungen an Hirnen von Wild- und Hausschweinen. – Verh. Dtsch. Zool. Ges. Freiburg (1936), 200–211.

HERRE, W.: Studien über die wilden und domestizierten Typlopoden Südamerikas. Zool. Garten (N.F.) *19* (1952), 70–98.

HERRE, W.: Studien am Skelett des Mittelohres wilder und domestizierter Formen der Gattung *Lama* Frisch. Acta anat. *19* (1953), 271–289.

HERRE, W.: Das Ren als Haustier. Leipzig 1955.

HERRE, W.: Fragen und Ergebnisse der Domestikationsforschung nach Studien am Hirn. Verh. Dtsch. Zool. Ges. Erlangen (1956), 143–214.

HERRE, W.: Einflüsse der Umwelt auf das Säugetiergehirn. Dtsch. med. Wschr. *83* (1958), 1568–1574.

HERRE, W.: Abstammung und Domestikation der Haustiere. In: HAMMOND, J., J. JOHANNSON, F. HARING, Handbuch der Tierzüchtung Bd. I, (1958) 1–58.

HERRE, W.: Domestikation und Stammesgeschichte. In: HEBERER, G.: Evolution der Organismen II, 801–856. Jena 1959.

HERRE, W.: Der Art- und Rassebegriff. In: HAMMOND, J., J. JOHANNSON, F. HARING: Handbuch der Tierzüchtung Bd. III/1, (1961) 1–25.

HERRE, W.: Zum phylogenetischen Pluripotenzbegriff. In: KURTH, G.: Evolution und Hominisation. S. 36–48. Jena 1962.

HERRE, W.: Neues zur Umweltbeeinflußbarkeit des Säugetiergehirns. Naturwiss. Rundschau *16* (1964), 359–364.

HERRE, W.: Zur Problematik der innerartlichen Ausformung bei Tieren. Zool. Anz. *172* (1964), 403–425.

HERRE, W.: Neue Erkenntnisse über Abstammung und Entwicklung von Haustieren. Math.-naturw. Unterricht (Frankfurt) *17* (1964), 1–7.

HERRE, W.: Demonstration im Tiergarten des Instituts für Haustierkunde der Universität Kiel, insbesondere an Wildcaniden und Canidenkreuzungen. Verh. Dtsch. Zool. Ges. Kiel (1965), 622–635.

HERRE, W., H. FRICK, M. RÖHRS: Über Ziele, Begriffe, Methoden und Aussagen der Haustierkunde. Z. Tierzüchtg. u. Züchtgsbiol. 76 (1961), 114–124.

HERRE, W., H. STEPHAN: Zur postnatalen Morphogenese des Hirnes verschiedener Haushundrassen. Morph. Jb. 96 (1955), 210 bis 264.

HERRE, W., U. THIEDE: Studien an Gehirnen südamerikanischer Tylopoden. Zool. Jb. Anat. 82 (1965), 155–176.

HÜCKINGHAUS, F.: Craniometrische Untersuchungen an verwilderten Hauskaninchen von den Kerguelen. Z. wiss. Zool. 171 (1965), 181–196.

KLATT, B.: Über die Veränderung der Schädelkapazität in der Domestikation. S. H. Ges. Naturf. Frd. Berlin (1912), 153–179.

KLATT, B.: Studien zum Domestikationsproblem. – I. Untersuchungen am Hirn. Bibliotheca genet. 2 (1921), 1–181.

KLATT, B.: Vergleichende Untersuchungen an Caniden und Procyniden. Zool. Jb. allg. Zool. Phys. 45 (1928), 217–292.

KLATT, B.: Gefangenschaftsveränderungen bei Füchsen. Jen. J. Naturwiss. 67 (1932), 452 bis 468.

KLATT, B.: Kreuzungen extremer Rassetypen des Hundes. Z. menschl. Vererb.- und Konst.-lehre Bd. 25–28 (1941–1964).

LANDAU, E.: Beitrag zur Kenntnis des Katzenhirns (Hirnfurchen). Morph. Jb. 38 (1908), 1–35.

LUNAU, H.: Vergleichend-metrische Untersuchungen am Allocortex von Wild- und Hausschweinen. Anat. Anz. 62 (1956), 673 bis 698.

NETELER, B.: Vergleichende Untersuchungen an Gehirnen südamerikanischer Caniden. Ing. Diss. Kiel 1963.

OBOUSSIER, H.: Zur Frage der Erblichkeit von Hirnfurchen. Z. menschl. Vererb. u. Konst.-lehre 29 (1942), 831–864.

RAWIEL, F.: Untersuchungen an Hirnen von Wild- und Hausschweinen. Z. Anat. 110 (1939), 344–370.

RÖHRS, M.: Vergleichende Untersuchungen an Hirnen verschiedener Urodelen. J. wiss. Zool. 158 (1955), 341–420.

RÖHRS, M.: Vergleichende Untersuchungen an Wild- und Hauskatzen. Zool. Anz. 155 (1955), 53–69.

RÖHRS, M.: Neuere Ergebnisse und Probleme der Allometrieforschung. Z. wiss. Zool. 162 (1959), 1–95.

RÖHRS, M.: Ökologische Beobachtungen an wildlebenden Tylopoden Südamerikas. Verh. Dtsch. Zool. Ges. Graz (1957), 538–554.

RÖHRS, N.: Allometrische Untersuchungen an Canidengehirnen. Verh. Dtsch. Zool. Ges. Frankfurt (1959), 295–307.

RÖHRS, M.: Biologische Anschauungen über Begriff und Wesen der Domestikation. Z. Tierzüchtg. u. Züchtgsbiol. 76 (1961), 7–23.

RÖHRS, M.: Allometrieforschung und biologische Formenanalyse. Z. Morph. Anthrop. 51 (1961), 289–321.

RÖHRS, M.: Vergleichende Untersuchungen zur Evolution des Gehirns von Edentaten. Z. Zool. System. u. Evol.-Forschg. (1966), im Druck.

SAMTLEBEN, H.: Untersuchungen an Hirnen verschiedener Anuren. Z. wiss. Zool. 161 (1958), 38–83.

SCHUMACHER, U.: Quantitative Untersuchungen an Gehirnen mitteleuropäischer Musteliden. J. Hirnforsch. 6 (1963), 137–163.

SENGLAUB, K.: Vergleichende metrische und morphologische Untersuchungen an Organen und am Kleinhirn von Wild-, Gefangenschafts- und Hausenten. Morph. Jb. 100 (1959), 11–62.

STARCK, D.: Die äußere Morphologie des Großhirnes zwergwüchsiger und kurzköpfiger Haushunde, ein Beitrag zur Entstehung des Furchentyps. – Gaz. méd. portug. VII (1954), 209–224.

STARCK, D.: Die Evolution des Säugetiergehirnes. Steiner. Wiesbaden 1962.

STEPHAN, H.: Vergleichende Untersuchungen über den Feinbau des Hirnes von Wild- und Haustieren. I. Studien am Schwein und Schaf. Zool. Jb. Anat. 71 (1951), 487–586.

STEPHAN, H.: Vergleichend-anatomische Untersuchungen an Hirnen von Wild- und Haustieren. II. Die Oberfläche des Allocortex bei Wild- und Hausform von Epimys norvegicus Erxl. Morph. Jb. 93 (1954), 425–471.

STEPHAN, H.: Vergleichend-anatomische Untersuchungen an Hirnen von Wild- und Haustieren. III. Die Oberflächen des Allocortex bei Wild- und Gefangenschaftsfüchsen. Biol. Jb. 73 (1954), 95–115.

STEPHAN, H.: Die Anwendung der Snell'schen Formel h = k . p auf die Hirn-Körpergewichtsbeziehungen bei verschiedenen Hunderassen. Zool. Anz. 153 (1954), 15–27.

STEPHAN, H.: Methodische Studien über den quantitativen Vergleich architektonischer Struktureinheiten des Gehirns. Z. wiss. Zool. *164* (1960), 143–172.

STOCKHAUS, K.: Metrische Untersuchungen an Schädeln von Wölfen und Hunden. Z. zool. System. u. Evol. Forsch. *3* (1965), 157–258.

THIEDE, U.: Formbesonderheiten an Gehirnen südamerikanischer Säuger. Verh. Dtsch. Zool. Ges. Kiel (1965), 613–618.

THIEDE, U.: Zur Hirngestalt mitteleuropäischer und südamerikanischer Musteliden; ihre innerartliche Ausformung und zwischenartlichen Unterschiede. Z. zool. System. u. Evol.-forschung. (1966), im Druck.

VOLKMER, D.: Cytoarchitektonische Studien an Hirnen verschieden großer Hunde (Königspudel und Zwergpudel). Z. mikr.-anat. Forsch. *62* (1956), 267–315.

WIGGER, H.: Vergleichende Untersuchungen am Auge von Wild- und Hausschwein unter besonderer Berücksichtigung der Retina. Z. Morph. Oek. Tiere *36* (1939), 1–20.

Prof. Dr. WOLF HERRE
Institut für Haustierkunde
Neue Universität
Olshausenstraße 40–60
23 Kiel, Deutschland

Neue Befunde über die frühe Ontogenese des ZNS

Von H. Bergquist, Göteborg

Als ein Beitrag zur Diskussion der phylogenetischen Prinzipien kann folgendes angeführt werden: Monophyletische, diphyletische oder polyphyletische Stammbäume geben zweifellos ein gewisses Gerüst für phylogenetische Spekulationen. Reale, einwandfreie Stammbäume konnten aber noch nicht aufgestellt werden.

Anstatt „fehlender Glieder" spricht man heute lieber von fließenden Übergängen zwischen verschiedenen Stammformen, von Mikromutationen oder von ganz dramatischen Makromutationen verursacht.

Evolutionistische Kräfte wirken ohne Ziele und in verschiedenen Richtungen in diesem Augenblick wie vor Millionen Jahren. Die Natur solcher Kräfte ist uns aber ebenso unbekannt wie den Entwicklungsmechanikern Anfang dieses Jahrhunderts.

Ein äußerst wichtiger Fortschritt ist die Entdeckung des DNA-Moleküls in den Chromosomen. Damit ist eine exakte, biochemische Grundlage für ein tieferes Verständnis des phylogenetischen Problems gegeben und eine Möglichkeit, gerichtete Mutationen und damit die ersten experimentell dargestellten Evolutionen zu erhalten. Anstatt der biogenetischen Regel wird dann sicherlich von Veränderungen des genetischen Codes mit begleitenden Informationen in verschiedenen Richtungen gesprochen.

In jedem Falle ist heute die Entwicklungsdiskussion viel vorsichtiger als früher, weil der Entwicklungsmechanismus ganz offenbar sehr verwickelt ist (Willmer, 1960; Løvtrup, 1965).

Für die vergleichenden Anatomen sind die embryologischen Untersuchungen meistens eine Voraussetzung. Früher war die vergleichende Anatomie eine Hilfsforschung für die Phylogenie. Heute erscheint der Zusammenhang zwischen diesen Forschungsrichtungen als sehr unsicher. Einen Beleg für diese Behauptung stellt zum mindesten die Entwicklung der frühen Ontogenese des zentralen Nervensystems dar (Weiss, 1960).

Klassische Lehre. In den meisten Handbüchern steht noch folgendes: Die Neuralplakode bildet eine Neuralrinne und die Neuralrinne schließt sich zu einem Neuralrohr. Vorne im Neuralrohr entsteht ein Zwei-Blasen-Stadium, aus den Blasen Archencephalon und Deuteroencephalon bestehend. Das Deuteroencephalon teilt sich durch die Isthmusfalte in Mesencephalon und Rhombencephalon. Der Name Archencephalon wird gleichzeitig oft durch die Bezeichnung Prosencephalon ersetzt. Das ist das Drei-Blasen-Stadium. Am Ende folgen schließlich die fünf Gehirnbläschen: Telencephalon und Diencephalon aus Prosencephalon. Das Mesencephalon ändert sich nicht. Im Rhombencephalon tritt dorsal eine Metencephalonspalte auf und der Rest wird oft als Myelencephalon bezeichnet.

Neuromerien. Schon Anfang des neunzehnten Jahrhunderts wurde eine Querbandung des Neuralrohres beobachtet. Hill (1900) sah sie in lebenden Lachsembryonen. Eine Pro- und Kontradiskussion folgte. Seit 25 Jahren sind umfassende morphogenetisch-topographische und experimentelle Untersuchungen über dieses Problem von BERGQUIST (1932, 1964 b, c), KÄLLÉN (1953 b, 1955 u. a.) durchgeführt worden. Das Ergebnis ist ziemlich überraschend.

Ich spreche hier nicht vom allerfrühesten neurogenetischen Geschehen, der neuralen Induktion. Diesbezüglich soll auf folgende Arbeiten hingewiesen werden: NIEUWKOOP (1962), SAXÉN u. TOIVONEN (1962), JACOBSON (1962, 1964), KÄLLÉN (1965) und HAMBURGER u. LEVI-MONTALCINI (1950).

Im ersten Abschnitt eines im Centrul de Embriologie Normala si Patologica, Timisoara (MENKES, 1958) gemachten Zeitrafferfilmes sieht man drei paar Zellengruppierungen schon im Neuralwall (BERGQUIST, 1963), die vermutlich LEHMANNS (1945) archencephalen, deuteroencephalen und spinalen Zonen der Neuralrinne entsprechen.

Dieses tritt also vor dem Schließen der Neuralrinne ein. Unmittelbar vor diesem Schließen entstehen aber neue Gruppierungen des Zellenmaterials dadurch, daß die ersten Querbandungen („transverse bulges"), die Proneuromeren, in rostro-caudaler Richtung sichtbar werden. Diese sind bei Beginn der Bildung des Neuralrohres wieder abgeschwächt oder beinahe ganz ausgelöscht (die *Interneuromerie I*).

Es folgt die am besten entwickelte Querbandung des ganzen Neuralrohres, die Neuromerie. Auch sie wird von einer Interneuromerie, also von einer nicht segmentierten Phase abgelöst *(Interneuromerie II)*. Schließlich entwickelt sich eine *Postneuromerie* nur im Gehirnteil des Neuralrohres, die beinahe unmittelbar nach ihrem Entstehen von den HIS-HERRICKschen Längsbandungen überquert wird.

Durch die Überquerung der Postneuromeren von HIS-HERRICKschen Zellsäulen entsteht ein Karomuster mit sogenannten „Grundgebieten" (BERGQUIST, 1932) oder „migration areas" (BERGQUIST u. KÄLLÉN, 1954), die Auswanderungsgebiete von Zellen in die Neuralrohrwand bilden. Jene radiale Migration kann später die sogenannten sukzessiven Migrationen I und evtl. II verursachen und schließlich Cortex und Gehirn-Nuclei entwickeln. Migration I und die Postneuromerie entstehen zeitlich gleichzeitig und Migration II entsteht selten (BERGQUIST, 1956).

Mitoseaktivität. Die drei verschiedenen Querbandungen, das Karomuster und die successiven Migrationen stehen in Korrelation zu einer erhöhten Mitosenaktivität im Neuralepithel (BERGQUIST, 1932, 1957; KÄLLÉN, 1956 b, c). Einspritzen von Kolchizin in gravide Tiere bewirkt das Aufkommen von noch mehr exponierten Neuromeren (KÄLLÉN, 1956 a, b, 1962). Den Zusammenhang zwischen Proliferation und Formbildung im Neuralrohr bestätigten auch HUGOSSON (1957 a) und RÜDEBERG (1961).

Durch die obengenannten Zeitrafferfilme gab BERGQUIST (1963, 1964 a) weitere Beweise für die verschiedenen Querbandungsphasen und die nicht mitosenaktiven interneuromeren Phasen.

Vergleichende Anatomie. Sowohl die drei Neuromerensequenzen als auch die Migrationsareale und die successiven Migrationen sind in allen größeren Vertebratengruppen seriell homolog (BERGQUIST, 1932, 1954; KÄLLÉN, 1953 a; SAETERSDAL, 1953; WEDIN, 1955). In Embryonen mit massivem ausgebildeten Nervenrohr, z. B. bei Cyclostomen, Teleostiern und Amphibien ist aber die Neuromerie nicht so ausgeprägt.

KUHLENBECK (1926, 1954, 1956) u. a. meinen, daß die Furchenbildungen in der Nervenrohrwand als wichtige Grenzen zwischen verschiedenen Gebieten des embryonalen Gehirns angesehen werden können. BERGQUIST, KÄLLÉN u. a. haben gefunden, daß so der Fall in späteren Embryonalstadien sein kann, nicht aber in den frühesten Stadien. Wegen der

Variabilität der Furchen, die offenbar rein mechanisch von der Mitosenaktivität abhängig sind, muß angenommen werden, daß die Furchen morphologisch eine mehr sekundäre Bedeutung besitzen.

Schlußfolgerungen. Nach dem soeben erfolgten Vergleichen der Blasenstadien mit den nun angegebenen Verhältnissen wird festgestellt:

a) Das Zwei-Blasen-Stadium besteht nicht. Vielleicht entspricht dieses von VON KUPFFER (1906) beschriebene Embryonalstadium der Interneuromeren-Phase I.

b) Im Drei-Blasen-Stadium entspricht „Mesencephalon" wahrscheinlich der Neuromere c und die übrigen „Blasen" entsprechen den anderen encephalen Neuromeren.

c) Das Fünf-Blasen-Stadium wird durch die sekundären Ausbuchtungen, beispielsweise die Hemisphären, die Augenblasen, die Epiphyse sowie den Hypothalamus charakterisiert. Man kann sich aber fragen, welchen Blasen entsprechen eigentlich die fünf Blasen in dem „Fünf-Blasen-Stadium"? Vermutlich entspricht dieses Blasenstadium dem Migrations-Area-Stadium.

Übersicht über die frühe ZNS-Entwicklung

Neuralplatte
↓
Neuralrinne
↓
Neuralrohr   Proneuromeren
↓
Interneuromeren-Phase I
↓
Neuromeren
↓
Interneuromeren-Phase II
↓
Postneuromeren
↓
Migrationsareale
↓
Successive Migrationen
↓
Cortex und Gehirnganglien

Summary

The so-called brain vesicle stages are not relevant. The course of morphogenesis of the encephalic part of the central nervous system is characterized by the appearance of "transverse bulges" in the following sequence: proneuromery, neuromery, postneuromery with two interneuromeric phases. Longitudinal cell columns (HIS-HERRICK) cross over the postneuromeres and give rise to the migration areas. From these areas cells migrate lateralwards in successive migration-waves and afterwards the cell masses can be arranged in cortical and nuclear structures. These ontogenetic phases are well established through topographic-morphogenetic, experimental and cinematographic observations. There seems to be no relation between the ontogenesis of the neural tube and the vertebrate phylogenesis.

Literatur

BERGQUIST, H.: Zur Morphologie des Zwischenhirns bei niederen Wirbeltieren. Acta zool., 13 (1932), 57–303.

BERGQUIST, H.: Ontogenesis of diencephalic nuclei in vertebrates. Lunds Univ. Arsskr. Avd. 2, 65 (1954), 1–34.

BERGQUIST, H.: On early cell migration processes in the embryonic brain. J. Embryol. exp. Morphol. 4 (1956), 152–160.

BERGQUIST, H.: Mitotic activity during successive migrations in the diencephalon of chick embryos. Experientia 13 (1957), 1–16.

BERGQUIST, H.: Neuromerele in stadiile timpurii ale dezvoltarii sistemului nervos central. Stud. cerc. stiinte med. Ac. R.P.R. Baza Timisoara 10 (1963), 7–17.

BERGQUIST, H.: Mitotic Activity in early phases of CNS-development. Proc. Symp. Prague 1962 (1964a), 1–15.

BERGQUIST, H.: Die Entwicklung des Diencephalons im Lichte neuer Forschung. In: Progress in Brain research. Vol. 5. (1962). Lectures on the Diencephalon 223–229, 1964b.

BERGQUIST, H.: The formation of the front part of the neural tube. Experientia 20 (1964c), 92–97.

BERGQUIST, H., B. KÄLLÉN: Notes on the early histogenesis and morphogenesis of the central nervous system in vertebrates. J. comp. Neurol. 100 (1954), 627–660.

HAMBURGER, V., R. LEVI-MONTALCINI: Some aspects of neuroembryology. In: (Ed. P. Weiss), Genetic Neurology Univ. Chicago Press, Chicago, pp. 128–160, 1950.

HILL, CH.: Developmental history of primary segments of the vertebrate head. Zool. Abt. Anat. Ontog. Thiere 13 (1900), 393–446.

HUGOSSON, R.: Morphologic and Experimental Studies on the Development and Significance of the Rhombencephalic Longitudinal Cell Columns. Lund 1957.

JACOBSSON, C.-O.: Cell migration in the neural plate and the process of neurulation in the axolotl larva. Zoo. Bidrag Uppsala 35 (1962), 433–449.

JACOBSSON, C.-O.: Motor nuclei cranial nerve roots and fibre pattern in the Medulla oblongata after reversal experiments on the neural plate of axolotl larvae. 1. Bilateral operations. Zool. Bidr. Uppsala 36 (1964), 73–160.

KÄLLÉN, B.: On the segmentation of the central nervous system. Lunds Univ. Arsskr., Avd. 2, 64 (1953a), 1–10.

KÄLLÉN, B.: On the significance of the neuromeres and similar structures in vertebrate embryos. J. Embryol. exp. Morph. 1 (1953b), 387–392.

KÄLLÉN, B.: Contribution to the knowledge of the regulation of the proliferation processes in the vertebrate brain during ontogenesis. Acta Anat. 27 (1956a), 351–360.

KÄLLÉN, B.: Studies on the mitotic activity in chick and rabbit brains during ontogenesis. Lunds Univ. Arsskr., Avd. 2, 26 (1956b), 1–14.

KÄLLÉN, B.: Regulation of proliferation processes during ontogenesis of the brain. Progr. Neurobiol. (1956c), 353–357.

KÄLLÉN, B.: Mitotic patterning in the central nervous system of chick embryos; studied by a colchicine method. Z. Anat. Entwickl.-Gesch. (Basel), 123 (1962), 309–319.

KÄLLÉN, B.: Early Morphogenesis and pattern formation in the central nervous system (1966), in press.

KUHLENBECK, H.: Betrachtungen über den funktionellen Bauplan des Zentralnervensystems. Folia anat. jap., 4 (1926), 111–135.

KUHLENBECK, H.: The human diencephalon. A summary of development, structure, function and pathology. Confin. neurol. Suppl. 14 (1954), 1–230.

KUHLENBECK, H.: Neuromery and longitudinal pattern in the vertebrate prosencephalon. Anat. Rec., 124 (1956), 417.

KUPFFER, C. VON: Die Morphogenie des Zentralnervensystems. Handbuch vergl. experim. Entw.-Lehre Wirbeltiere. II, 3, Hertwig Jena, 1906.

LEHMANN, F. E.: Die Morphogenese in ihrer Abhängigkeit von elementaren biologischen Konstituenten des Plasmas. Rev. Suisse Zool. 57 (1950), 141–151.

LØVTRUP, S.: Morphogenesis in the amphibian embryo. Zool. Gothob., 1 (1965), 1–139.

MENKES, B.: Cercetari de Embriologie Experimentala. Acad. Rec. Populare Romine, Timisoara 1958.

NIEUWKOOP, P. D.: The organisation centre. I. Induction and determination. Acta Biotheor. 16 (1962), 57–68.

RÜDEBERG, S.: Morphogenetic studies on the cerebellar nuclei and their homologization in different vertebrates including man. Lund 1961.

SAETERSDAL, T. A. S.: On the development of the neural chord in Trichogaster trichopterus (Pallas) and Gadus callarias. Univ. Bergen Årb. Naturvit. 5 (1953), 1–30.

SAXÉN, L., S. TOIVONEN: Primary embryonic induction. Academic Press, London 1962.

WEDIN, B.: Embryonic segmentations in the head. Diss. Malmö 1955.

WEISS, P.: Nervous system (neurogenesis). In: VILLIER, WEISS, HAMBURGER: Analysis of development. Philadelphia, London, 1955, 346–401.

WILLMER, E. N.: Cytology and Evolution. New York, London 1960.

Prof. Dr. HARRY BERGQUIST
Laboratorium of Zoology
University of Göteborg
Fjärde Langgatan 7 B
Göteborg SV, Sweden

Applications of Light and Electron Microscopic Autoradiography to the Study of Cytogenesis of the Forebrain

By S. Fujita, Lafayette, Indiana

(Stage I) Proliferation of the Matrix Cell

In the early stages of ontogeny, the wall of the prosencephalon, later developing into highly specialized structures is still made up of relatively simple medullary epithelium, as are the other portions of the neural tube. No neuroblast appears yet in this stage. Recent investigations (21) have elucidated in detail the functional and morphological characteristics of the matrix cell that constitutes the medullary epithelium.

Concept of the matrix cell

The matrix cells carry out an *elevator movement* (Fig. 1) during their generation cycle (15, 16, 29, 30, 33, 37, 52, 53, 54, 55, 56, 57, 64); at the time of DNA synthesis nuclei of the matrix cells are located in the deeper half of the matrix layer, the S-zone (zone of DNA synthesis), and when they finish DNA synthesis the nuclei ascend toward the ventricular surface. During the postsynthetic (premitotic) time, t_2, they pass through the I-zone (intermediate zone). Matrix cells divide on the ventricular surface in the M-zone (zone of mitosis) and, after mitotic time, t_M, both nuclei of the daughter cells shift back to the I-zone where they spend the postmitotic (presynthetic) time, t_1. When they reach the S-zone again a new generation cycle begins (16). On the basis of this model of the elevator movement of the matrix cell, the length of the various fractions of the generation cycle,

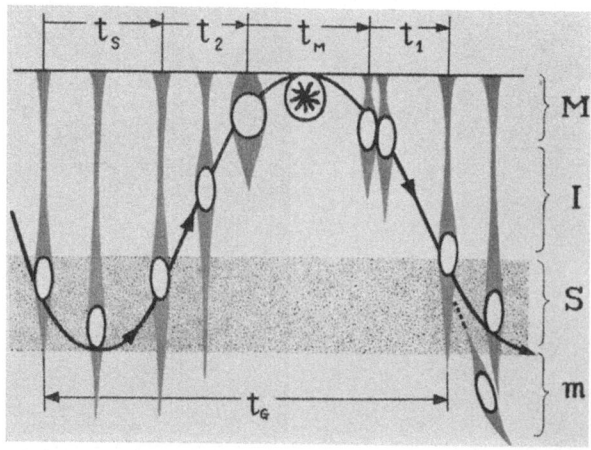

Fig. 1. Schematic representation of an *elevator movement* of the matrix cell. M, I and S indicate M-zone (zone of mitosis), I-zone (intermediate zone) and S-zone (zone of DNA synthesis) respectively; m, mantle layer; t_S, DNA synthetic time; t_2, postsynthetic time; t_M, mitotic time; t_1, presynthetic time, and t_G, generation time.

viz. the transit times of nuclei in the M-, I- and S-zones were calculated in the cerebral vesicle of mouse embryos (23) and in various portions of the chick neural tube (16) by means of the cumulative labeling method of tritiated thymidine autoradiography. The mechanism of this up-and-down-movement of the nuclei is explained by cyclic changes of the cytoplasm and cytomembrane system during the generation cycle (18, 25, 27). Recent organculture study of human embryonic brain (48) suggests that this elevator movement plays an essential role in the early morphogenesis of the central nervous system.

Although it is generally accepted that the matrix cell exists in the wall of the neural tube and performs an elevator movement (33, 43, 44), it has been believed that the neural tube consists of various kinds of cells (5, 35), such as germinal cells, (primitive) spongioblasts, neuroblasts, ependymoblasts, etc. A question now arises: Is the matrix layer composed solely of the matrix cell? Two kinds of experiments (9, 17) were carried out to answer this question, and results obtained in both cases confirmed the cellular homogeneity of the matrix layer. First, morphological observations using the electron-microscope (9, 10, 44, 45, 65) revealed the homogeneous nature of the cells in the matrix layer. The matrix cells possess discrete cell boundaries and are arranged just as a usual columnar epithelium. The perikaryon of the matrix cell is characterized by abundant free ribosomes aggregating in a polysomic pattern and by poorly developed membrane systems (9, 10, 45, 61, 65). Another evidence of cellular uniformity of the matrix layer came from cumulative labeling experiments of tritiated thymidine autoradiography (17, 23). During continuous labeling with tritiated thymidine, the percentage of labeled cells in the matrix layer increase linearly with time (17) and soon reach 100%, as seen in Fig. 2. This fact indicates that the rate of influx of cells into the DNA synthetic compartment is constant with time, and all of the cells in this population proceed in the generation cycle at a constant rate. Thus, it is right to conclude that these cells form a single homogeneous population. I designated the cells of this population matrix cells, and the layer they form, the matrix layer (16, 17, 48). Recently LANGMAN (43) and ATLAS a. BOND (4) confirmed the homogeneity of the matrix cells using, respectively, cumulative and pulse-labeling techniques of tritiated thymidine autoradiography.

Previous investigators have used various synonyms to designate this cell population; for example, germinal cell (5, 43, 66, 67, 68), primitive spongioblast (5, 35), ependymal cell (4, 6, 59, 60, 61, 62), primitive ependymal cell (33, 57, 58), and neuroepithelial cell (44). However, the author prefers the name "matrix cell" to the other synonyms because they have frequently been used to designate various ill-defined cell types. This question was discussed in detail in previous papers (14, 15, 16, 17, 18, 20, 21).

Observations by various methods indicate that the matrix cell forms a homogeneous

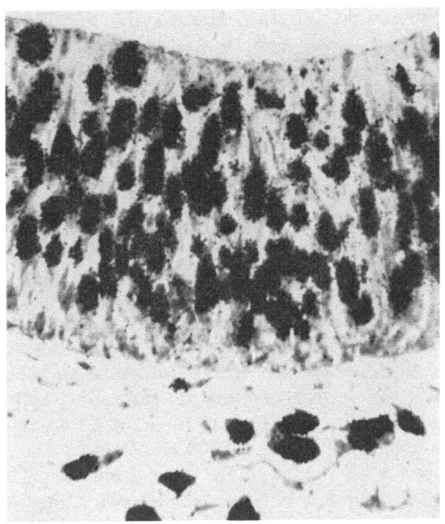

Fig. 2. Autoradiograph of neural tube wall of 2-day chick embryo, labeled for 6 hours with $100 \mu C$ of tritiated thymidine. Resin-embedded, 0.2μ section. x 500.

population from both the functional and morphological points of view. However, our recent electron microscopic studies have revealed the existence of some heterogeneity within this cell population; a few matrix cells possess a cilium on their apical process and some have a few membranes of the endoplasmic reticulum in peripheral processes while others do not. Should this slight morphological heterogeneity be interpreted as incompatible with the concept of the homogeneity of the matrix cell population? We know that the homogeneity of the matrix cell is proved by a satisfactorily quantitative method (17, 23). To avoid this paradox, we may speculate that the slight heterogeneity might come from the presence of matrix cells in different phases of the generation cycle during which the matrix cells perform different biophysical and biochemical processes; at t_M, the cells round up and divide without synthesizing RNA, and at t_S they extend their cell bodies and replicate their chromosomes synthesizing both DNA and RNA at the same time. In order to test this possibility, we should first identify matrix cells in various phases of the generation cycle separately in an electron micrograph and attempt to correlate their morphological variations to the phases of the cells.

Identification of matrix cells in different phases of generation cycle and their ultrastructural characteristics

Matrix cells in the mitotic period (t_M): They can be identified as rounded cells in the M-zone showing mitotic chromosomes.

Matrix cells in the DNA synthetic phase: They are identified as labeled cells in the S-zone in autoradiographs taken shortly after administration of tritiated thymidine.

Matrix cells in t_1 and t_2 phases: In the I-zone these two types of cells exist intermingled. We should separate these cells. Fortunately, we know that matrix cells form a homogeneous population distributed at random in the phases of the generation cycle. Therefore, if we labeled them continuously for $(t_2 + t_M)$ hours with tritiated thymidine (Fig. 1), all the ascending cells (i.e. t_2-cells) in the I-zone should have incorporated tritiated thymidine while the descending cells (i.e. t_1-cells) should remain unlabeled.

Materials. The author selected the optic lobe of 4-day chick embryo as a model of cerebral vesicle typical of the neural tube wall and labeled it with 100 μC of tritiated thymidine for three hours. At this stage, the premitotic time, t_2, of the matrix cell is approximately equal to two hours, and the mitotic time about one hour. Thus, $t_2 + t_M = 3$ hrs.

Methods. Optic lobes of 4-day chick embryos were cut out after three hours of cumulative labeling, fixed in PALADE's buffered osmic acid solution, dehydrated and embedded in either epon or Vestopal W with special reference to their orientation. For light microscopy, 0.2 μ sections, and for electron microscopic autoradiography, gold sections were cut by the PORTER-BLUM II ultramicrotome. Sections for light microscopic autoradiography were mounted on glass slides, coated with Kodak NTB 3 emulsion by the dipping method, exposed for 2–4 weeks and developed with Dektol at 18° C for 5 minutes. They were then fixed, rinsed and stained with warm 0.5% toluidine blue solution at pH 9. Sections for electron microscopic autoradiography were placed on formvarcovered grids, coated with carbon and then with de-gelatinized Kodak NTE emulsion. This method is a combination of the methods of PRZYBYLSKI (50) and SALPETER a. BACHMAN (51). The grids are stored in helium gas for 1 month, treated with gold latensification solution (SALPETER a. BACHMAN 1964), developed in Dektol at 18° C for 2 minutes, fixed, rinsed and dried. Grids were examined with a Philips EM 200 microscope.

Fig. 3. Autoradiograph of optic lobe of 4-day chick embryo, labeled for 3 hours with 100 μC of tritiated thymidine. Resin-embedded, 0.2 μ section. Nuclei of matrix cells in the M-zone and those of ascending t_2-cells in the I-zone are labeled. Unlabeled nuclei in the I-zone represent matrix cells at t_1-period. x 850.

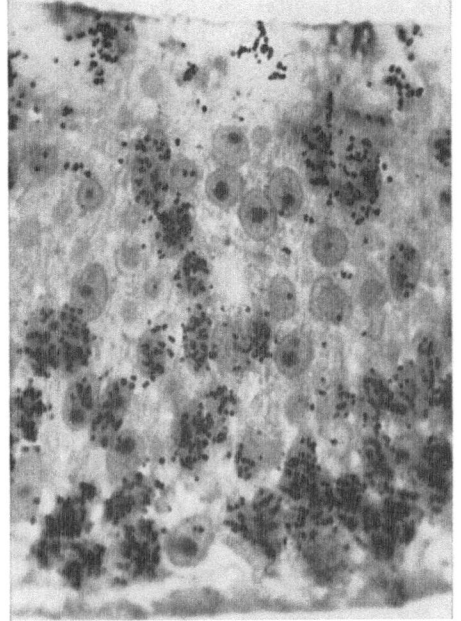

The light microscopic autoradiography with thin resin-embedded section has the following definite advantages (Fig. 3) in comparison with usual paraffin section autoradiography. Resolution is greatly improved; most grains are localized well in the limit of nuclear membrane. The nuclei of matrix cells appearing heavily overlapped in paraffin sections are now separate so that correct counts of labeled and unlabeled cells can be easily carried out.

Light microscopic examination of the autoradiographs (Fig. 3) revealed that about one-third of the nuclei in the M- and I-zones had incorporated radioactive thymidine, and that the other two-thirds remained unlabeled. Since there are two postmitotic cells corresponding for each premitotic cell, the cells must spend almost the same length of time descending as they do ascending and dividing; i.e., $t_1 = t_2 + t_M = 3$ hrs.

The descending, or t_1-cells, as seen unlabeled in the light microscope seem to differ from the ascending cells in the smaller size of their nuclei. Observing the sections from the same block with electron microscope (Fig.4), I can similarly identify descending t_1-cells as unlabeled and ascending t_2-cells as labeled cells (Fig. 5). This investigation is still in progress and more observations are necessary before we can present a complete picture of the phase-specific characteristics of the matrix cell. Preliminary observations, however, suggest that the following ultrastructural changes take place during one generation cycle of the matrix cell.

a. During the t_S and t_2 periods, the endoplasmic reticulum and GOLGI complex are confined in the apical process of the matrix cell, but in the descending t_1-cell some vesicles or small parts of the membrane complex migrate into the peripheral process. During cell division, the amount of internal membrane decreases, and what is left desintegrates into small vesicles. These changes parallel those of the nuclear membrane.

b. The distribution of mitochondria changes as the nucleus of the matrix cell moves up and down; with the ascending and descending movements of the nucleus of the matrix cell they tend to flow into the apical and peripheral processes respectively, though their migration seems to lag behind the movement of the cytoplasm. No cyclic changes of the morphology of individual mitochondrion have yet been detected.

c. The cilium is also one of the structures that suffer cyclic changes; it disappears as the matrix cell enters the t_2 period and reappears at the late t_1 period, at least in some cells. The cilium seems to persist into t_S period but I have not yet obtained direct evidence,

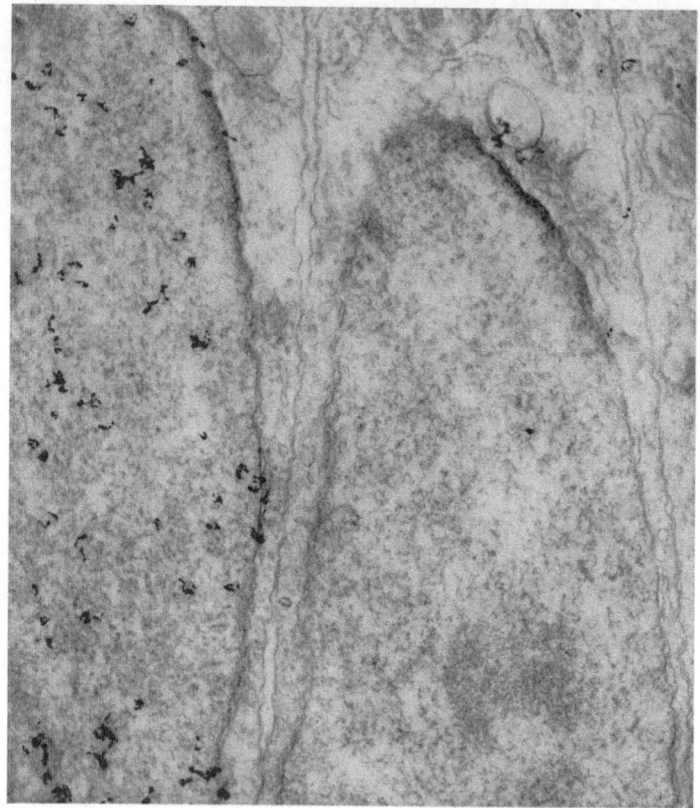

Fig. 4. Electron microscopic autoradiograph from the same block as shown in Fig. 3. Cells in the I-zone. Note labeled nucleus of ascending t_2-cell on the right and unlabeled nucleus of descending t_1-cell on the left. x 15.000.

since it is very difficult to follow in one section, the cytoplasmic process from the ventricular surface to the perikaryon of the cell that is synthesizing DNA in the S-zone. However it is probable that the cilium having grown at the late t_1 stage remains intact during the t_S period.

The disappearance of the cilium in the t_2-period seems to be an atrophy rather than depilation. The likelihood of finding a cilium in the apical process of a matrix cell appears to be much higher in cells with a long t_1 time such as floor plate cells. If we may assume that ciliary development requires a certain length of time, the fact that the cilium is very rare in rapidly proliferating cells such as alar and basal plate matrix cells can also be reasonably explained.

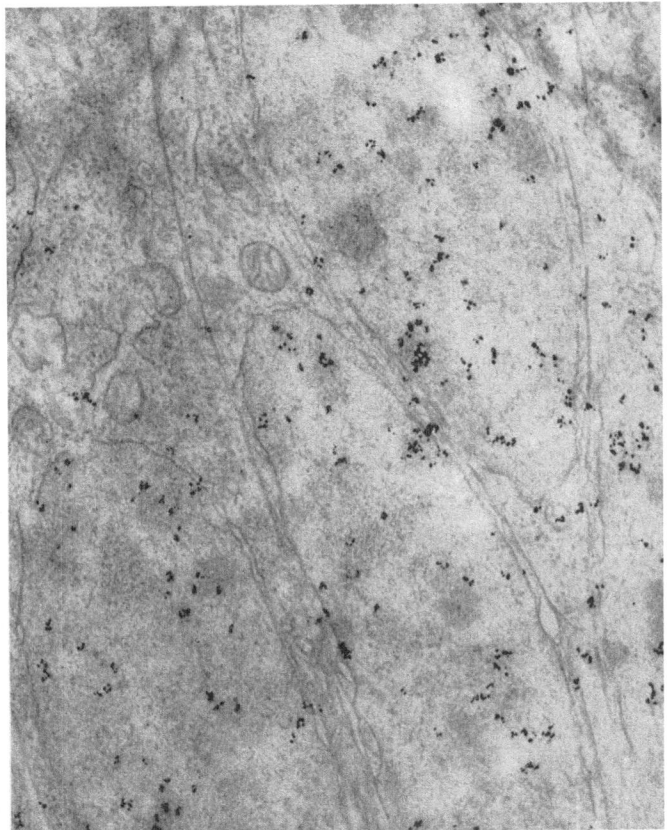

Fig. 5. Electron microscopic autoradiograph from the same material as in Figs. 3 and 4. Junction between M- and I-zones, showing 4 ascending matrix cells. x 10.000.

d. The centriole (basal body, blepharoblast) in the apical process seems to disappear and reappear with the cilium, but evidence so far obtained is too little to draw any definite conclusion. However, it may be worthwhile mentioning that the centriole of the matrix cell, when it is present, is always recognizable in the most apical part of the ventricular process, as if it is bound to the terminal bars in both sides.

The preliminary nature of the conclusions presented above should be stressed. They are mere inferences spun from incomplete fragments of observation. Work is now in progress in our laboratory aimed at substantiating these working hypotheses.

The most important contribution of autoradiography to the understanding of cellular organization of the early neural tube is establishment of the concept of the matrix cell. The neural tube at its early stage is composed solely of the matrix cell and matrix cells proliferate only to produce cells of their own type. This stage is called stage I or stage of matrix cell proliferation.

(Stage II) Neuroblast Differentiation

As we discussed in the previous chapter, the early neural tube is composed solely of the matrix cells. But soon the differentiation of neuroblasts from the matrix cells commences, and stage I or the stage of pure matrix cell proliferation terminates. Commencement of stage II is recognized in autoradiographs by the appearance of unlabeled cells in the periphery of the neural tube wall (Fig. 6). The presence of this type of cell is more conspicuous after the cumulative labeling by which all the cells in the matrix layer are uniformly labeled, but all the cells in the mantle layer (i.e. neuroblasts) remain completely free from the radioactivity as seen in Fig. 6. This finding also indicates that one can establish correct identification of the neuroblast in light and electron micrographs by means of cumulative labeling autoradiography. This principle was applied to the study of the fine structure of the neuroblast in the chick embryo (9, 10, 13).

Fine structure of the neuroblast

I have compared the ultrastructural characteristics of the neuroblast in the mantle layer of the 5-day chick embryo to those of the matrix cell (9, 10). Electron microscopic examination of these unlabeled cells (i.e. neuroblasts) revealed the following features of neuroblasts in the mantle layer; the neuroblast is characterized by developing rough- and smooth surfaced endoplasmic reticulum in the perikaryon, an accumulation of cytoplasm and cytoplasmic organelles in the peripheral pole, a deflection of the peripheral cytoplasmic process (the growing axon) so that it lies parallel to the outer surface of the neural tube, and frequent nuclear indentations in the peripheral pole. Sometimes the centriole is found in the peripheral part of the perikaryon. In the matrix cell, by contrast, this structure is usually confined to the most apical part of the inner process. The migration of the centriole to the peripheral pole reflects the drastic changes the neuroblast undergoes as it is differentiated from the matrix cell. The circumstances which enable the centriole in the neuroblast to migrate to the periphery appear to be related to the following changes of the cell. The atrophy of the inner process, the disappearance of terminal bars, the detachment from the ventricular surface, the fixation of the cell polarity toward the periphery, and the outward migration of the cell from the matrix layer into the mantle layer.

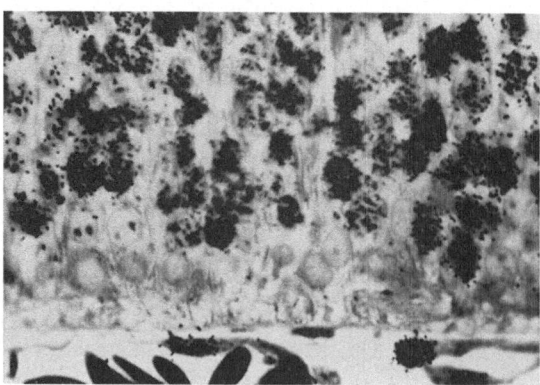

Fig. 6. Light microscopic autoradiograph from 5 day chick embryo, labeled for 9 hours with 100 μC of tritiated thymidine. Resin-embedded, 0.2 μ section. External half of transverse section of the optic lobe. Note sharp separation of labeled matrix cells and unlabeled neuroblasts in the mantle layer. x 900.

The neuroblast in the mantle layer, just described, cannot be regarded as the neuroblast at the earliest stage of differentiation, since neuroblasts are differentiated from the matrix cell in the I-zone while they are descending during the t_1 period (17, 19). In order to study earliest stages in the differentiation process, it is desirable to identify and examine the neuroblast at this very early moment. The following is an attempt to do this.

A neuroblast once differentiated from the matrix cell does not carry out further DNA synthesis and remains in the $2n$ state containing the presynthetic and postmitotic amount of DNA. Thus, DNA synthesis in a daughter cell that is produced in the M-zone and destined to become a neuroblast should be suppressed and the cell should pass through the S-zone without synthesizing DNA. Therefore, we can expect to detect in the S-zone a neuroblast just differentiated from the matrix cell by means of a cumulative labeling autoradiography. If we label a neural tube at stage II cumulatively with tritiated thymidine for $t_2 + t_M + t_1$ hours, all the cells in the I-zone would be labeled (Fig. 1) and all the matrix cells in the S-zone would take up the label while neuroblasts in the S-zone should be the only cells that do not incorporate tritiated thymidine. The earliest changes in ultrastructure occurring at the time of neuroblast differentiation were studied by this technique.

Four-day chick embryo was labeled with 100 μC of tritiated thymidine for 6 hours ($= t_2 + t_M + t_1$). Optic lobe was cut out, embedded in epon or in Vestopal W, autoradiographed as described in previous chapter, and examined by light and electron microscope.

Identification of neuroblasts in the S-zone and earliest changes of ultrastructure in neuroblast differentiation

Observations with light microscope revealed (Fig. 7) that all the cells in the M- and I-zone are labeled and that all the cells in the mantle layer, where it is present, are

unlabeled. In the S-zone a few unlabeled cells are found as expected. They are regarded as neuroblasts at the earliest stage of differentiation. Electron microscopic examination of sections cut from the same block revealed that the unlabeled cells in the S-zone are of intermediate nature between the matrix cell and the neuroblast in the mantle layer (Fig. 8). This investigation is now in progress and we should like to reserve a more detailed description of early neuroblast differentiation until a complete image of the morphological changes are obtained.

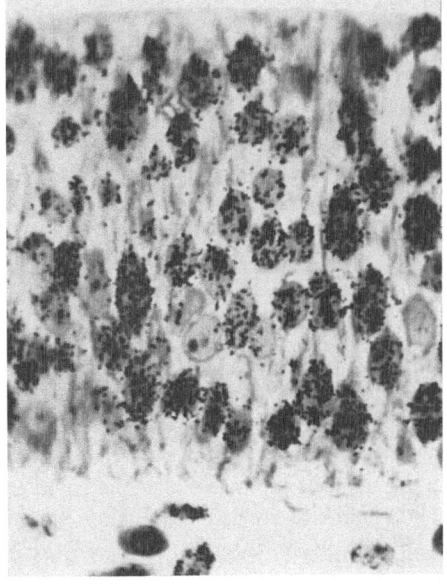

Fig. 7. Light microscopic autoradiograph of optic lobe from 4-day chick embryo, labeled for 6 hours with 100 μC of tritiated thymidine. Note presence of unlabeled cells in the S-zone. They are presumably neuroblasts. x 850.

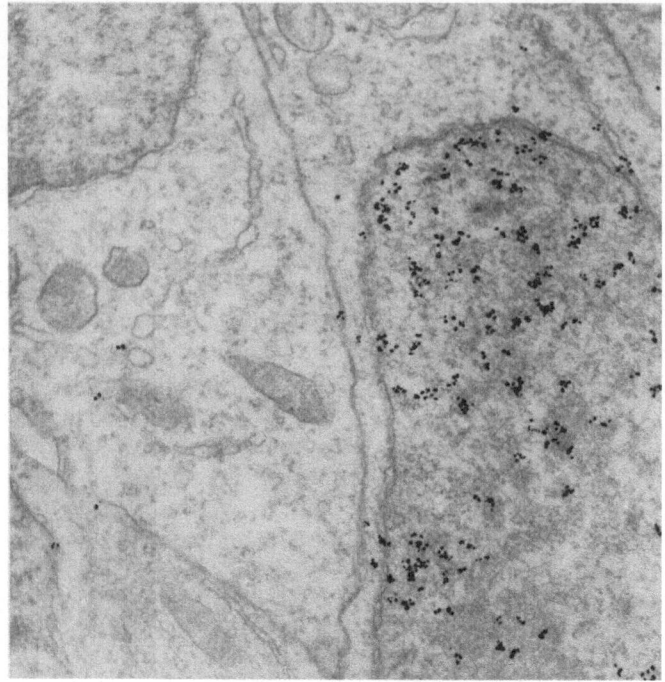

Fig. 8. Electron microscopic autoradiograph from the same block as Fig. 7. Cells in the S-zone. Presumable neuroblast is on the left. Many of the silver grains scattered outside the nucleus are probably due to alpha particles emitted by uranyl acetate with which the section is stained. x 15.000.

Chronological analysis of neuron differentiation

As described in previous sections, the labeled matrix cells increase in number as the time of cumulative labeling elapses and soon reach 100%. On the other hand, the neuroblasts already produced and accumulated in the mantle layer do not take up any label since they do not undergo further mitosis. Following cumulative labeling for more than one generation time of the matrix cell (Figs. 2, 9), the DNA of all the matrix cells becomes uniformly labeled. The neuroblasts that are produced by subsequent mitosis are labeled as homogeneously as the matrix cells, their direct progenitors. The label in the neuroblasts remains permanently undiluted, since they do not carry out further DNA synthesis. On the other hand, the label in the matrix cells is diluted by half in each successive mitosis. As a result, the neuroblasts produced from the matrix cell in subsequent generations contain gradedly decreasing amounts of the label (19, 26). By continuous infusion of labeled thymidine or by repeating injections of the label, it is always possible to produce a condition in which all the neurons produced after the commencement of cumulative labeling contain detectable amount of labeled DNA when examined. The condition produced by this method greatly facilitates interpretation; all the negative neurons were produced earlier and all the radioactive ones were produced later than the beginning of cumulative labeling. Thus,

Fig. 9. Schematic representation of
neuroblast differentiation and formation of labeled neuroblasts. ³HTDN
→ shows the time of cumulative labeling. The black dots represent silver
grains reduced by the radioactivity
of the nuclei. The label in the matrix
cell is diluted by half as the generations proceed. But the labeled neuroblast retains the label until it matures
to the neuron.

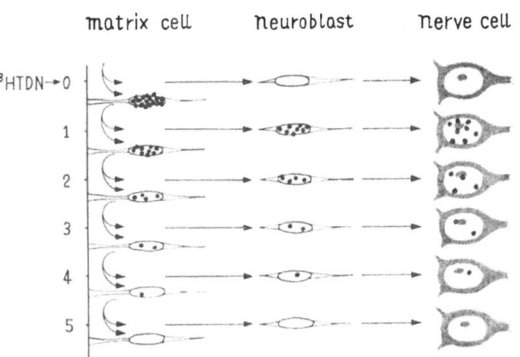

starting with the cumulative labeling at various stages of development and examining the
distribution of labeled neurons at later stages, we can map out the chronology of neuron
differentiation in the central nervous system. Several reports of this type of analysis have
appeared (2, 3, 7, 18, 19, 22, 24, 47, 58, 59, 63) and BERRY and his associates deal specifically
with this problem in this Symposium. The author will leave a detailed description of the
applications of tritiated thymidine autoradiography to the analysis of neuron differentiation to their paper (6).

Termination of the stage II

Analysis of neuron differentiation by the cumulative labeling method also enables one to
determine the time at which the matrix cell ceases to produce neuroblasts; if the label is
injected later than this time, labeled neurons do not appear. Though commencement and
termination of the stage II varies from place to place even in the same animal, it is
generally found that the stage of neuroblast differentiation or stage II is restricted to a certain period of ontogeny. Neuroblast production in the chick brain begins a few
days after incubation and ceases in the middle of the incubation period (19). In the mouse,
neuron production in the superior colliculus ends at 14 days of gestation (7) and similarly
in the rat cerebrum this process appears to end at about the time of birth (34). In the
human cerebrum, morphologic observations (18, 31) indicate that neuron production
commences in the eighth week and comes to an end by the 22nd week of pregnancy.
With cessation of neuroblast production the remaining matrix cells change into either
glioblasts or ependymal cells (19, 20). This is stage III or the stage of the differentiation
of neuroglia and ependyma.

(Stage III) Differentiation of Neuroglia and Ependyma

When neuroblast production terminates and stage III commences, the matrix layer
becomes thin and most matrix cells transform into *glioblasts* characterized by small,
round nuclei, and the others differentiate into ependymal cells.

Characteristic labeling pattern of stage III in flash labeling autoradiography

The glioblasts rapidly migrate out from the ventricular layer, maintaining their DNA synthetic activity. Thus, following this period many radioactive cells are found outside the matrix layer in flash-labeled autoradiographs. Such pulse-labeled positive cells in the mantle and marginal layers are not found prior to stage III. The appearance of flash-labeled positive cells outside the matrix layer signifies the beginning of stage III. Simultaneously some matrix cells that have lined the central canal transform into ependymal cells, in which developing cilia (plural!) and abundant vacuole systems are found. Rough-surfaced endoplasmic reticulum appears in the perikaryon. Microvilli and blepharoblasts (or centrioles) also increase in number (11, 12, 45).

Like stage II, commencement of stage III varies from place to place and from region to region. In the chick cerebrum the transition from stage II to stage III takes place from 9 to 11 days of incubation; there are regional differences, but, in general, the frontal portion undergoes this transition earliest, i.e. at 9 days and the occipital region latest, i.e. at 11 days.

Electron microscopic observation of the cellular elements in the cerebrum at stage III

Electron microscopic observations of the glioblasts in the chick cerebrum at stage III (11, 12) revealed that the glioblasts possess a morphology similar to the so-called oligo-dendroglia cells of FAHRQUHAR a. HARTMAN (8). We have observed transitional forms of the glioblast into immature astrocytes (11). SMART a. LEBLOND (60) labeled the glioblasts in the cerebrum of newborn mice with tritiated thymidine and were able to follow their maturation into oligodendroglia and astrocytes. A similar experiment was performed by the author on the glioblasts of chick spinal cord (20). Both conclusions are in full agreement.

We designate a glioblast as a cell differentiated from the matrix cell and as one that may potentially become either an oligodendrocyte or an astrocyte. Our glioblast may correspond to the spongioblast (bipolar and migrating) of previous investigators (49, 60). However, the author prefers the name of glioblast to spongioblast because the latter has also been used erroneously to designate resting forms of the matrix cell (the so-called primitive spongioblast). Furthermore, the very name of spongioblast seems inadequate since this term means an immature cell forming the syncytial neuro-spongium whose presence is completely excluded by recent investigations with the electron microscope (9, 61, 62, 65).

On the origin of subependymal and subpial glioblasts

In the mammalian cerebrum, glioblasts, when produced at the later half of the fetal life, accumulate in the subependymal layer and persist after birth. In some regions other than the cerebrum, for example in the spinal cord, cells similar to the glioblasts appear in the subpial position. These cells are homologous with the subependymal glioblasts in the cerebrum except for their site of origin; both types of cell first appear at about the end of the stage of neuroblast differentiation and both have a similar morphology.

DEL RIO HORTEGA (36) believed that both come from pial cells, though he failed to find their precursors in the leptomeninx. KERSHMAN (38) supported HORTEGA's view and, having studied a series of human fetuses, emphasized the role of the choroid plexus as a source of the subependymal round cells which he called microglioblasts. However, we have studied the cerebrum of the chick, fetal and neonatal mouse, rabbit, cat and man, and found (32) that the subependymal (glioblast) layer and choroid plexus are spatially separated from each other from the very beginning of the appearance of the subependymal layer. In human brains both structures come into a very close (but parallel) position in the parieto-temporal region, but in other forms, especially in the chick, they are separated so far from each other that it is difficult to accept the assumption that those cells have migrated from the choroid plexus to the subependymal layer. The author would

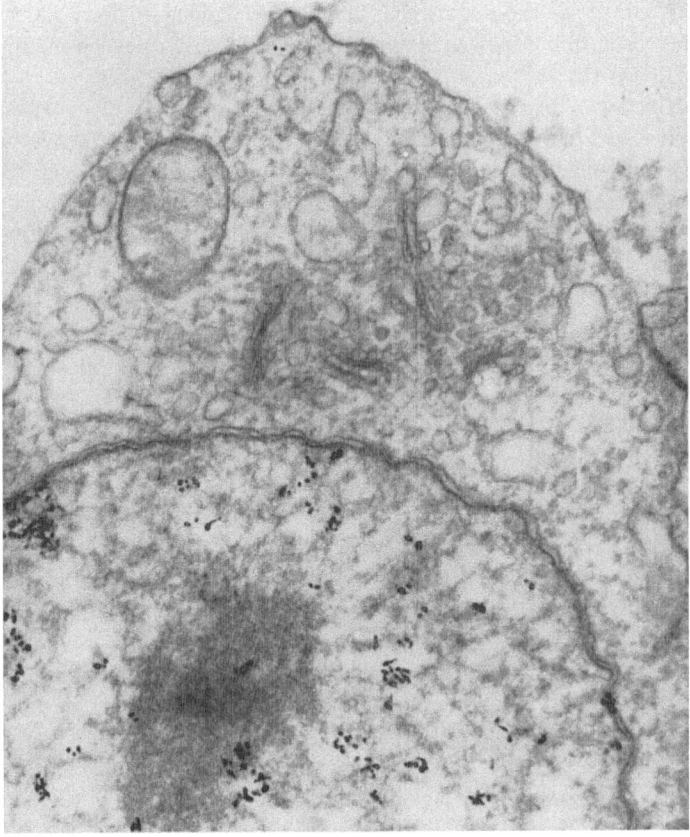

Fig. 10. Electron microscopic autoradiograph of mesenchymal cell surrounding the neural tube. Two-day chick embryo, labeled for 2 hours with 100 μC of tritiated thymidine. Note abundance of vesicles and Golgi apparatus. Silver grains are localized on filamentous structures in the nucleus. x 30.000.

like to point out that there is no direct or indirect evidence to support this hypothetical but extensive migration. Further evidence against the HORTEGA-KERSHMAN view is found in autoradiographic studies with cumulative labeling of newborn mouse brain (28). In the experiments, newborn mice were labeled continuously for up to 3 days by multiple injections of tritiated thymidine with intervals of 4 hours. By 3 days of cumulative labeling nearly all of the glioblasts in the subependymal and subpial layers had taken up the label, while cells in the leptomeninx and in the choroid plexus remained unlabeled in many places. This finding indicates that both subependymal and subpial glioblasts proliferate rapidly with a generation time of about 3 days, and, that cells in the leptomeninx and in the choroid plexus multiply much more slowly at this stage. It is, therefore, inconceivable that the great increase in the large population of the glioblasts in the brain parenchyma could be supplied by the small and slowly proliferating population of the extra-parenchymal cells. The author has discussed in detail in previous papers (18, 20) other lines of evidence that excludes their leptomeningeal origin.

Observations with the electron microscope also support their matrix cell origin (11, 12); the morphology of the subependymal cells in the cerebrum and subpial cells in the spinal cord are similar to the matrix cells and to the oligodendroglial cells, but not to the mesenchymal cells surrounding the brain (Fig. 10) nor to the so-called microglial cells of FAHRQUHAR a. HARTMAN (8).

DEL RIO HORTEGA (36) based his reasoning on the specificity of his silver carbonate staining method, he identified these glioblasts as the same type as the microglial cells of the adult brain because he could stain both cell types by the same method. However, there is no good reason to believe his silver carbonate method is specific to mesodermal cells. METZ a. SPATZ (46) already pointed out the arbitrary nature of the results of this method and stressed that the cells stained by this carbonate method are also of neurectodermal origin.

We studied brains of newborn mice and rabbits and found the same pattern of staining (32) as HORTEGA described 33 years ago, that the subependymal and subpial cells are stained fairly specifically with his silver staining method. However, our interpretation of this finding is different since we know these cells cannot be of mesodermal origin. The conclusion is that, though HORTEGA's silver carbonate method stains macrophages positively both inside and outside of the brain, it also stains some forms of neurectodermal neuroglial cells; immature neuroglial cells seem to be stained rather selectively by this method. Recently several investigators working independently have applied autoradiographic techniques to the study of the origin of macrophages in the brain and all came to the same conclusion that the macrophages in the injured brain are derived from mononuclear cells of circulating blood (1, 32, 39, 40, 41, 42).

These two lines of investigation, i.e. histogenetic studies on the origin of glioblasts in embryonic and neonatal brains, on the one hand, and on the origin of macrophages in adult injured brains, on the other, seem to shed light on the long disputed problems of the origin and metamorphotic potentiality of microglial cells. It now appears not so outrageous to assume that there might be no free cells of mesenchymal origin in the parenchyma of the *normal* brain except circulating blood cells in the vascular system, which is separated from the parenchyma by the basement membrane, and that only pathologic alterations of the neurectodermal tissue produce a space for free mesenchymal cells to infiltrate into the brain parenchyma.

Summary

Tritiated thymidine autoradiographic and electron microscopic studies on the cellular organization of the neural tube revealed that the wall of the early neural tube is composed solely of the matrix cells, which form a homogeneous population from both the morphological and functional points of view. Combination of autoradiography and electron-microscopy enables us to examine ultrastructural changes of the matrix cell as it proceeds through various phases of the generation cycle performing an elevator movement. Following the stage of pure matrix cell proliferation that is called stage I, some matrix cells differentiate into neuroblasts. This is stage II. The neuroblasts, as they differentiate from the matrix cells lose the capacity for further mitosis. They pass through the S-zone of the matrix layer and migrate out into the mantle layer. Using electron microscopic autoradiography, the author demonstrated ultrastructural characteristics of neuroblasts at their early stages of differentiation. After a period of cumulative labeling with tritiated thymidine, all the matrix cells become uniformly labeled and produce labeled neuroblasts by subsequent mitoses while the neuroblasts that have been produced earlier remain completely unlabeled. Examining the distribution of labeled neurons at later stages, one can determine the precise time of neuroblast differentiation in various parts of the brain. The same method also enables one to detect the time of cessation of neuroblast production from matrix cells. When the neuroblast differentiation comes to an end, stage III begins, i.e., the remaining matrix cells change into glioblasts and ependymal cells. Some glioblasts accumulate in the subependymal layer, a number migrate rapidly into deep parenchyma and others reach the subpial or marginal region. These glioblasts have sometimes been called "microglioblasts" and have been regarded as being of mesodermal origin. However, present investigations exclude the possibility of mesodermal origin. They mature, as development proceeds, into oligodendroglial cells and astrocytes. The macrophages in the brain seem to be derived from circulating blood cells. In conclusion, cytogenesis in the central nervous system is summarized in the following schema:

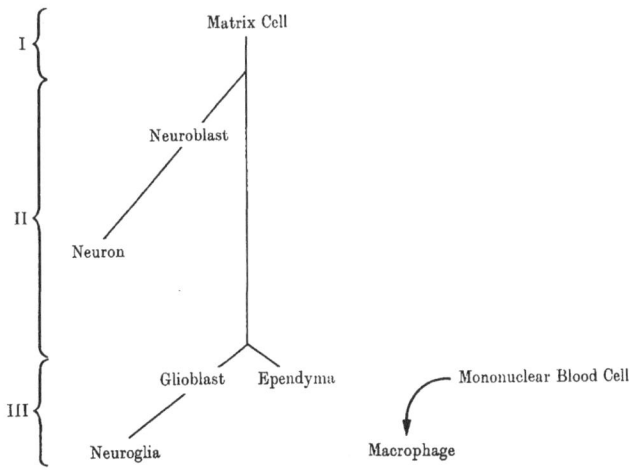

This work is supported in part by the National Science Foundation under grant 22 122.

References

1. ADRIAN, E.K., B.E.WALKER: Incorporation of thymidine-H³ by cells in normal and injured mouse spinal cord. J. Neuropath. exper. Neurol. *21* (1962), 597–609.
2. ANGEVINE, J.B.: Autoradiographic study of histogenesis in the area dentata of the cerebral cortex in the mouse. Anat. Rec. *148* (1964), 225.
3. ANGEVINE, J.B., R.L.SIDMAN: Autoradiographic study of cell migration during histogenesis of cerebral cortex in the mouse. Nature *192* (1961), 766–768.
4. ATLAS, M., V.P.BOND: The cell generation cycle of the eleven-day mouse embryo, J. Cell Biol. *26* (1965), 19–24
5. BAILEY, P., H.CUSHING: Tumors of the Glioma Group. Lippincott, Philadelphia 1926.
6. BERRY, M., A.W.ROGERS: Histogenesis of mammalian neocortex. In: HASSLER, R., H.STEPHAN (ed.): Evolution of the Forebrain. Thieme, Stuttgart, 1966, 197–205.
7. DELONG, G.R., R.L.SIDMAN: Effects of eye removal at birth on histogenesis of the mouse superior colliculus. An autoradiographic analysis with tritiated thymidine, J.comp.Neurol. *118* (1962), 205–224.
8. FAHRQUHAR, M.G., J.F.HARTMANN: Neuroglial structure and relationships as revealed by electron microscopy, J. Neuropath. exper. Neurol. *16* (1957), 18–39.
9. FUJITA, H., S.FUJITA: Electron microscopic studies on neuroblast differentiation in the central nervous system of domestic fowl, Z.Zellforsch. *60* (1963), 463–478.
10. FUJITA, H., S.FUJITA: Electron microscopic observations on the histogenesis (I) On the nerve cell differentiation. Acta anat. Nippon. *38* (1963), 85–94.
11. FUJITA, H., S.FUJITA: Electron microscopic studies on the histogenesis of the central nervous system of the domestic fowl (II), Neuroglia, Acta anat. Nippon. *38* (1963), 95–108.
12. FUJITA, H., S.FUJITA: Electron microscopic studies on the differentiation of the ependymal cells and the glioblasts in the spinal cord of domestic fowl. Z. Zellforsch. *64* (1964), 262–272.
13. FUJITA, H., J.OCHI, N.OTSUKA, S.FUJITA: Some problems on the histochemical studies of the nervous tissue, Rec. Adv. Res. Nerv. System (Tokyo) *8* (1964), 471–484.
14. FUJITA, S.: Medulloepithelioma: Its place in the histogenetic classification of neurec-

todermal tumors. Acta Pathol. Japon. *8* (1958), 789–794.
15. FUJITA, S.: Mitotic pattern and histogenesis of the central nervous system, Nature *185* (1960), 702–703.
16. FUJITA, S.: Kinetics of cellular proliferation. Exper. Cell Research *28* (1962), 52–60.
17. FUJITA, S.: The matrix cell and cytogenesis in the developing central nervous system, J.comp.Neurol. *120* (1963), 37–42.
18. FUJITA, S.: Histogenesis of the central nervous system and classification of neurectodermal tumors. Rec. Adv. Res. Nerv. System (Tokyo) *7* (1963), 117–143.
19. FUJITA, S.: Analysis of neuron differentiation in the central nervous system by tritiated thymidine autoradiography, J. comp. Neurol. *122* (1964), 311–328.
20. FUJITA, S.: An autoradiographic study on the origin and fate of the sub-pial glioblast in the embryonic chick spinal cord, J. comp. Neurol. *124* (1965), 51–60.
21. FUJITA, S.: The matrix cell and histogenesis of the central nervous system, Laval Méd. 36 (1965), 125–130.
22. FUJITA, S., M.HORII: Analysis of cytogenesis in chick retina by tritiated thymidine autoradiography, Arch. Histol. Japon. *23* (1963), 359–366.
23. FUJITA, S., M.HORII, T.TANIMURA, H.NISHIMURA: H³-thymidine autoradiographic studies on cytokinetic responses to X-ray irradiation and to thio-TEPA in the neural tube of mouse embryos, Anat. Rec. *149* (1964), 37–48.
24. FUJITA, S., A.KOJIMA, H.KAKU: Analysis of cellular differentiation and proliferation in the central nervous system by H³-thymidine autoradiography, Proc. Japan Conf. Radioisotopes *4* (1963), 178–180.
25. FUJITA, S., H.MIO, J.KIKKAWA, M.TEGOSHI, M.OMATSU: Cytoplasmic changes during mitosis, J. Kyoto Pref. Univ. Med. *68* (1960), 1001–1007.
26. FUJITA, S., S.MIYAKE: Selective labeling of cell groups and its application to cell identification, Exper. Cell Res. *28* (1962), 158–161.
27. FUJITA, S., Y.OMATSU, H.MIO, T.MORIKUBO, J.KIKKAWA: Mitotic pattern of epithelium *in situ* and *in vitro*, J. Kyoto Pref. Univ. Med. *68* (1960), 950–953.
28. FUJITA, S., M.SHIMADA: Cellular proliferation in the brain of early postnatal mouse. In preparation.

29. FUJITA, S., K. SUWA, M. TEGOSHI, S. MIYA-KE, O. TAKEOKA: Emigration of mitotic cells in single-layered cylindrical epithelium and neural tube: Its relationship to a new concept of the cytogenesis of the central nervous system, Acta pathol. Japon. *10* (1960), 202.

30. FUJITA, S., O. TAKEOKA: Identity of "Keimzellen von His" and primitive spongioblast revealed by colchicine technique and trypaflavin administration, Acta pathol. Japon. *11* (1961), 256–257.

31. FUJITA, S., O. TAKEOKA: Morphologic and quantitative studies on the cellular proliferation and differentiation in the central nervous system of human embryo, Acta pathol. Japon. *11* (1961), 256.

32. FUJITA, S., O. TAKEOKA: Studies of gliogenesis using tritiated thymidine autoradiography, electron microscopy and silver impregnation techniques, with special references to the origin and fate of Hortega's microglioblasts. Trans. Soc. Pathol. Japon. *53* (1964), 192–194.

33. HICKS, S. P., C. J. D'AMATO, M. A. COY, E. D. O'BRIEN, J. M. THURSTON, D. L. JOFTES: Migrating cells in the developing central nervous system studied by their radiosensitivity and tritiated thymidine uptake. In: Fundamental Aspects of Radiosensitivity. Brookhaven Symposium in Biology No. 14, Upton, New York 1961.

34. HICKS, S. P., M. C. CAVANAUGH, E. D. O'BRIEN: Effects of anoxia on the developing cerebral cortex in the rat. Amer. J. Path. *40* (1962), 615–635.

35. HIS, W.: Die Neuroblasten und deren Entstehung im embryonalen Mark. Z. Anat. Entwicklungsgesch. (1889), 249–300.

36. HORTEGA, DEL RIO P.: Microglia. In: PENFIELD'S Cytology and Cellular Pathology of the Nervous System. Vol. 2, 483–534. Paul P. Hoeber, New York 1932.

37. KERSHMAN, J.: The medulloblastoma and neuroblastoma: A study of human embryos, Arch. Neurol. Psychiat. *40* (1938), 937–967.

38. KERSHMAN, J.: Genesis of microglia in the human brain, Arch. Neurol. Psychiat. *41* (1939), 24–50.

39. KONIGSMARK, B. W., R. L. SIDMAN: Origin of gitter cells in the mouse brain, J. Neuropath. exper. Neurol. *22* (1963), 327–328.

40. KONIGSMARK, B. W., R. L. SIDMAN: Origin of Macrophages in the mouse. J. Neuropath. exper. Neurol. *22* (1963), 643–676.

41. KOSUNEN, T. U., B. H. WACKSMAN: Radioautographic studies of experimental allergic

encephalomyelitis (EAE) in rats, J. Neuropath. exper. Neurol. *22* (1963), 324–326.

42. KOSUNEN, T. U., B. H. WAKSMAN, I. K. SAMUELSON: Radio-autographic study of cellular mechanisms in delayed hypersensitivity. II. Experimental allergic encephalomyelitis in the rat, J. Neuropath. exper. Neurol. *22* (1963), 367–380.

43. LANGMAN, J.: The formation of the spinal cord in the chick embryo, examined by means of radioautography. 8th Int. Congr. Anatomists, Wiesbaden, p. 69 (1965). Cf. Martin, A., J. Langman, The development of the spinal cord examined by autoradiography. J. Embryol. exp. Morphol. *14* (1965), 25–35.

44. LYSER, K. M.: Early differentiation of motor neuroblast in the chick embryo as studied by electron microscopy. I. General aspects., Developmental Biol. *10* (1964), 433–466.

45. MELLER, K., W. WECHSLER: Elektronmikroskopische Befunde am Ependym des sich entwickelnden Gehirns von Hühnerembryonen, Acta Neuropath. *3* (1964), 609–626

46. METZ, A., H. SPATZ: Die Hortegaschen Zellen (= das sogenannte dritte Element) und über ihre funktionelle Bedeutung, Z. ges. Neurol. *89* (1924), 138–170.

47. MIALE, I. L., R. L. SIDMAN: An autoradiographic analysis of histogenesis in the mouse cerebellum, Exper. Neurol. *4* (1961), 277–296.

48. MIYAKE, S., K. ARAKI, M. SUGAHARA, S. FUJITA: Growth pattern of "matrix cells" of the central nervous system and formation of tubular structure simulating neural tube in cultures of human fetal brain, Arch. Histol. Japon. *22* (1961), 117–121.

49. PENFIELD, W.: Neuroglia, normal and pathological. In: PENFIELD'S Cytology and Cellular Pathology of the Nervous System. Vol. 2, 421–479. P. Hoeber, New York 1932.

50. PRZYBYLSKI, R. J.: Electron microscope autoradiography, Exper. Cell Res. *24* (1961), 181–184.

51. SALPETER, M. M., L. BACHMAN: Autoradiography with the electron microscope. A procedure for improving resolution, sensitivity, and contrast, J. Cell. Biol. *22* (1964), 469–477.

52. SAUER, F. C.: Mitosis in the neural tube, J. comp. Neurol. *62* (1935), 377–405.

53. SAUER, M. E., A. C. CHITTENDEN: Desoxyribonucleic acid content of cell nuclei in the neural tube of the chick embryo; evidence for intermitotic migration of nuclei, Exper Cell Res. *16* (1959), 1–6.

54. SAUER, M.E., B.E. WALKER: Radioauto-graphic study of interkinetic nuclear migration in the neural tube, Proc. Soc. exp. Biol. *101* (1959), 557–560.

55. SCHAPER, A.: Die frühesten Differenzierungsvorgänge im Centralnervensystem, Roux' Arch. Entwickl.-Mech. *5* (1897), 430 bis 431.

56. SCHAPER, A.: The earliest differentiation in the central nervous system of vertebrates, Science *5* (1897), 430–431.

57. SIDMAN, R.L., I.L. MIALE, N.FEDER: Cell proliferation and migration in the primitive ependymal zone; an autoradiographic study of histogenesis in the nervous system, Exper. Neurol. *1* (1959), 322–333.

58. SIDMAN, R.L.: Histogenesis of mouse retina studied with thymidine-³H. In: Smelser, C.K.: The Structure of the Eye, pp. 487 to 506. Academic Press, New York 1961.

59. SMART, I.: The location of mitosis-capable neuron precursor cells in the mouse neural epithelium, J. Anat. *99* (1965), 212.

60. SMART, I., C.P. LEBLOND: Evidence for division and transformations of neuroglia cells in the mouse brain, as derived from radioautography after injection of thymidine-H³, J. comp. Neurol. *116* (1961), 349 to 367.

61. TENNYSON, V.M.: Electron microscopic observations of the development of the neuroblast in the rabbit embryo. Proc. Intern. Congr. Electron Microscopy, 5th, Philadalphia, Vol. II, p. N-8. Academic Press, New York 1962.

62. TENNYSON, V.M., G.D. PAPPAS: An electron microscope study of ependymal cells of the fetal, early postnatal and adult rabbit, Z. Zellforsch. *56* (1962), 595–618.

63. UZMAN, L.L.: The histogenesis of the mouse cerebellum as studied by its tritiated thymidine uptake, J. comp. Neurol. *114* (1960), 137–160.

64. WATTERSON, R.L., P.VENEZIANO, A.BERTHA: Absence of a true germinal zone in neural tubes of young chick embryos as demonstrated by colchicine technique, Anat. Rec. *124* (1956), 379.

65. WECHSLER, W.: Elektronmikroskopische Untersuchung der Entwicklung von Nervenzellen und Nervenfasern bei Hühnerembryonen, Anat. Anz. *115* (1965), 287–302.

66. YAMAMOTO, Y., S. FUJITA, H. MIO, J. KIKKAWA, S. MIYAKE, O. TAKEOKA: Ontogenetic and phylogenetic study on the distribution of germinal cells and ependymal cells in the central nervous system, Acta pathol. Japon. *10* (1960), 202–203.

67. YAMAMOTO, Y., H. MIO, J. KIKKAWA, M. TEGOSHI, Y. OMATSU, Y. OKUMURA, S. FUJITA: Morphological studies on the distribution of germinal cells in the central nervous system of *Gecko japonicus*, J. Kyoto Pref. Univ. Med. *68* (1960), 1544–1547.

68. YAMAMOTO, Y., Y. OMATSU, H. MIO, J. KIKKAWA, Y. OKUMURA, M.TEGOSHI, S.FUJITA: Morphological studies on the distribution of germinal cells in the central nervous system of bony fish, J. Kyoto Pref. Univ. Med. *68* (1960), 1617–1621.

Dr. SETSUYA FUJITA
Department of Biological Sciences
Purdue University
Lafayette, Indiana, USA

Histogenesis of Mammalian Neocortex

By M. BERRY and A.W. ROGERS, Birmingham

TILNEY (1933) suggested that the definitive six layered neocortex was formed by three consecutive mass migrations of cells forming the supragranular, the granular and the infra-granular layers respectively. TILNEY's interpretation remained unchallenged until recently when the attention of the present authors centred on this subject after a study of the effects of x-irradiation on foetal neocortex (BERRY et al, 1963, BERRY a. EAYRS, 1963). The results suggested that the infra-granular layers were formed first and that the granular and supragranular layers were produced later in ontogeny. This hypothesis was not only entirely contrary to that proposed by TILNEY but also implied that the movements of cells were more complex than had previously been envisaged for, if the deep layers were established first, then cells forming the more superficial layers must migrate through the first formed laminae to attain their adult location. The technique of autoradiography was subsequently applied to the study of this problem (BERRY et al, 1964, BERRY a. ROGERS, 1965). The results obtained confirmed the earlier findings and suggested an outline of the pattern of migratory movements of neuroblasts throughout cortical histogenesis. Standard histological preparations were also studied and a tentative theory was formulated to explain the mode of genesis and migration of neuroblasts. It was suggested that the nuclei but not the cytoplasm of germinal, or ependymal cells, divide. One nucleus moves along the superficially directed process of the ependymal cell to attain a subzonal or intrazonal position where cytoplasmic division occurs and the neuroblasts become independent. The nucleus remaining at the ventrical pole of the ependymal cell divides again to repeat the cycle.

The results presented in this report are of tissue culture experiments, designed to obtain more information concerning the morphology of cells at various phases of development, and of an autoradiographic study of the genesis of glial cells.

Materials and Methods. *Tissue Culture.* Pregnant rats were killed on the 15th, 16th, 17th, 18th or 19th day of gestation, timed from the finding of spermatozoa in the vaginal smear. Their foetuses were removed and decapitated, and the heads immediately placed into a calcium and magnesium free solution buffered at pH 7.4 (PAUL, 1959). The cerebrum was dissected free from the head and the pallium separated from the corpus striatum and hippocampal complex and chopped into small pieces. These fragments of brain tissue were transferred into a similar solution as before but containing Versene, in a concentration of 1:5,000, and gently agitated for 20 mins. in a water bath, set at 37° C., to obtain a cell suspension.

Slides were prepared by building up the edges with glass rod and wax, sterilizing firstly in 70% alcohol and later in ultra violet light. The trough formed on the slide was filled with HANKES solution, containing 1% lactalbumin, and the slides placed in an oven at 37° C.

The cell suspension was added to the solution on the slides. After incubating for 6 hrs. the fluid was gently poured off, the glass rods and wax removed and the slides placed in 10% formalin for 5 mins, washed in distilled water and stained by the conventional eosin – HARRIS's haemotoxylin method.

Two litters at each age were studied.

Autoradiography. The slides studied were identical to those used by BERRY a. ROGERS (1965). The material was gathered from foetal rats injected with tritiated thymidine, in utero, on either the 16th, 17th, 18th 19th, 21st or 22nd day after conception. Autoradiographs of specimens killed at 30 days *post partum* were examined in this work. The glial cells were identified as astrocytes, oligodendroyctes or microglia according to the classification by SMART a. LEBLOND (1961).

Results

Tissue Culture

The degree of tissue fragmentation and dispersion varied in all slides. Where fragmentation was greatest the cells were widely dispersed, rounded in shape and possessed few processes. Some cultures contained only small aggregates of unbroken tissue with very

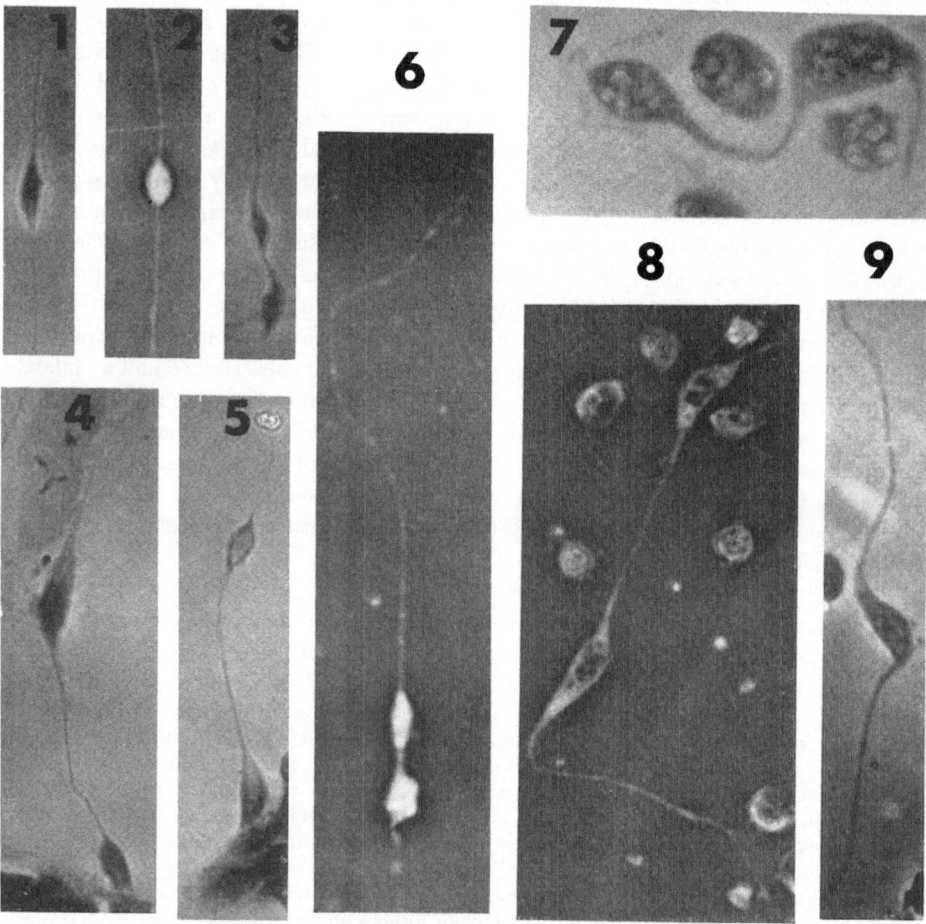

Fig. 1. Cells in tissue culture showing the bipolar configuration of nuclei in cultures prepared on the 15th (1), 16th (2) and 19th (9) days of gestation and two nuclei in one process from cultures prepared on the 15th (3), 16 (4 and 5), 17th (6) and 19th (7 and 8) days of gestation. (Phase – contrast, eosin-haematoxylin stained, x 1.000 – The differences in the sizes of cells appears to be due to the different degrees of flattening of the cells in each culture).

few free cells. Those most suitable for study, and fortunately most frequently obtained, showed partial cell dispersion. In these specimens the processes of free cells were easily seen and within the fragments of brain tissue the neural elements were often loosely arranged so that their interrelationships could be made out. Furthermore similarities in morphology existed between certain cells in these cultures and those seen in NISSL, silver and GOLGI-COX specimens. These latter preparations are described below.

15, 16 and 17 days post-conception. The slides contained four types of structural elements apart from fragments of blood vessels, nucleated red blood cells and occasional fibroblasts. The two most numerous types were rounded cells containing little cytoplasm and slender bipolar cells with small densely staining nuclei and very little cytoplasm (Fig. 1). The third was similar to the latter except that two nuclei were connected by a fine protoplasmic thread which often extended beyond one or both nuclei. The fourth type, probably a developing neurone or neuroglial cell, possessed more cytoplasm, several fine beaded processes and a vacuolated nucleus (Fig. 2). This cell was rarely seen in the earlier aged specimens.

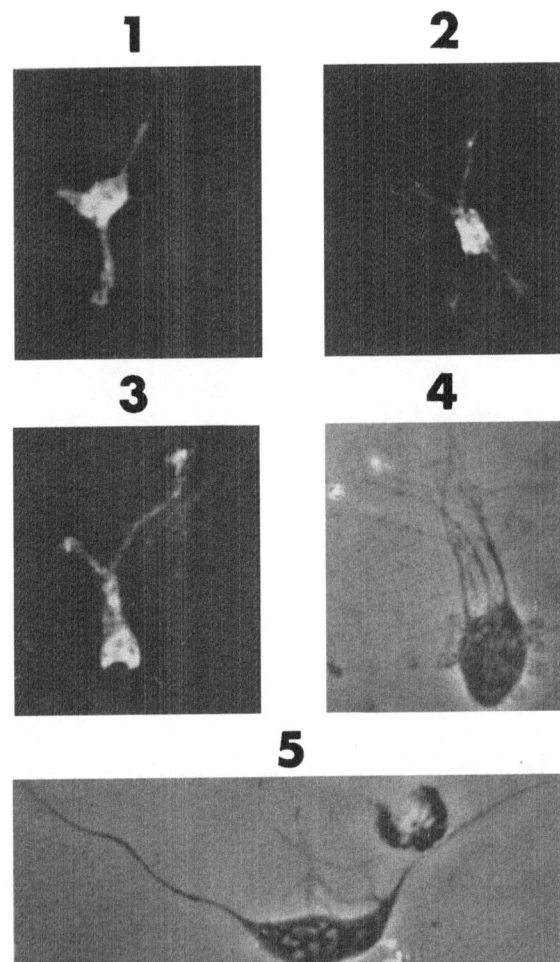

Fig. 2. Cells in tissue culture showing the degree of differentiation attained by cells on the 16th (1 and 2), 17th (3) and 19th (4) days of gestation. A bipolar cell from a 19 day culture with processes stemming from the perikaryon is shown in 5. (phase-contrast, eosin-haematoxylin stained, x 1.000).

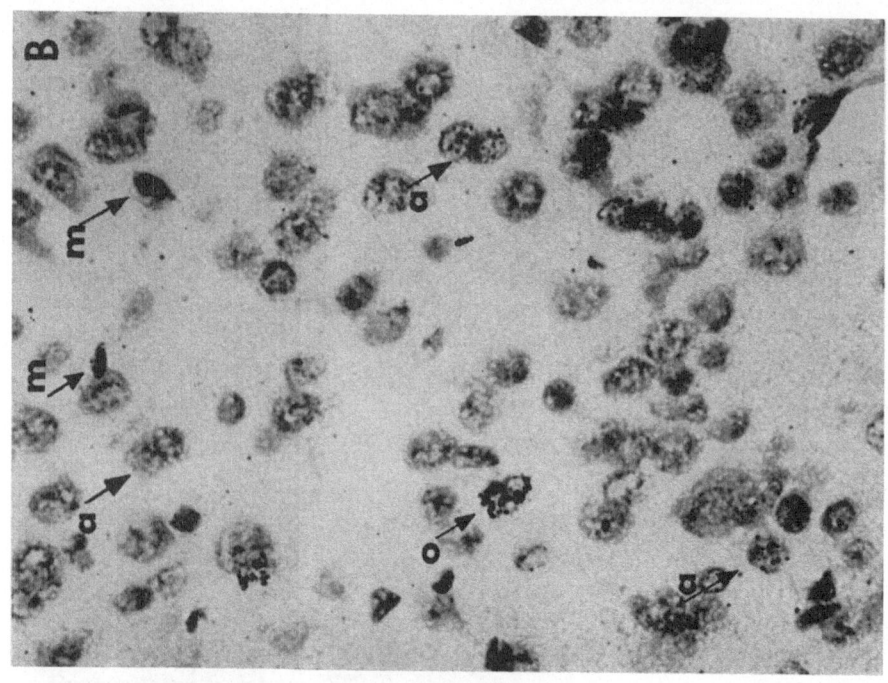
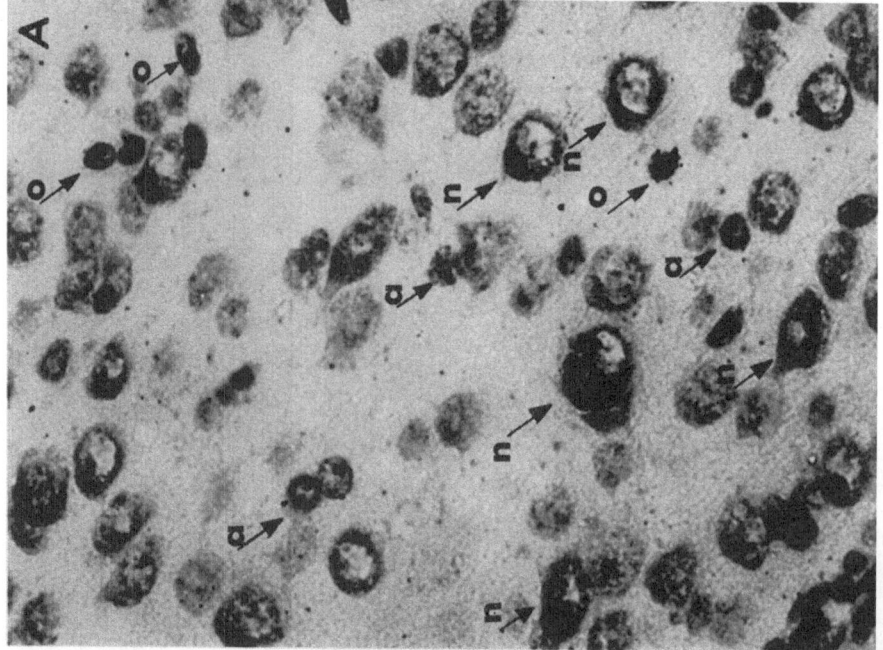

Fig. 3. Cerebral cortex of rat at 30 days *post partum* after injection of H$_3$-thymidine on the 19th day of gestation. A: superficial cortex. B: deep cortex. Labelled cells are marked n-neurone, a-astrocyte, o-oligodendroglia and m – microglia (autoradiograph haematoxylin stained x 1.000).

18–19 days post-conception. The four structures described above were also present in these later specimens but in addition many more differentiating cells were seen which showed evidence of a greater degree of maturity than in previous slides. Such cells had abundant strongly basophilic cytoplasm, large vacuolated nuclei and numerous processes which often extended for long distances. In addition a fifth type of cell had appeared which was very similar in morphology to the bipolar cell but was larger, having abundant deeply staining cytoplasm, additional processes, and a vacuolated nucleus (Fig. 2).

No mitotic figures were seen in any of the preparations.

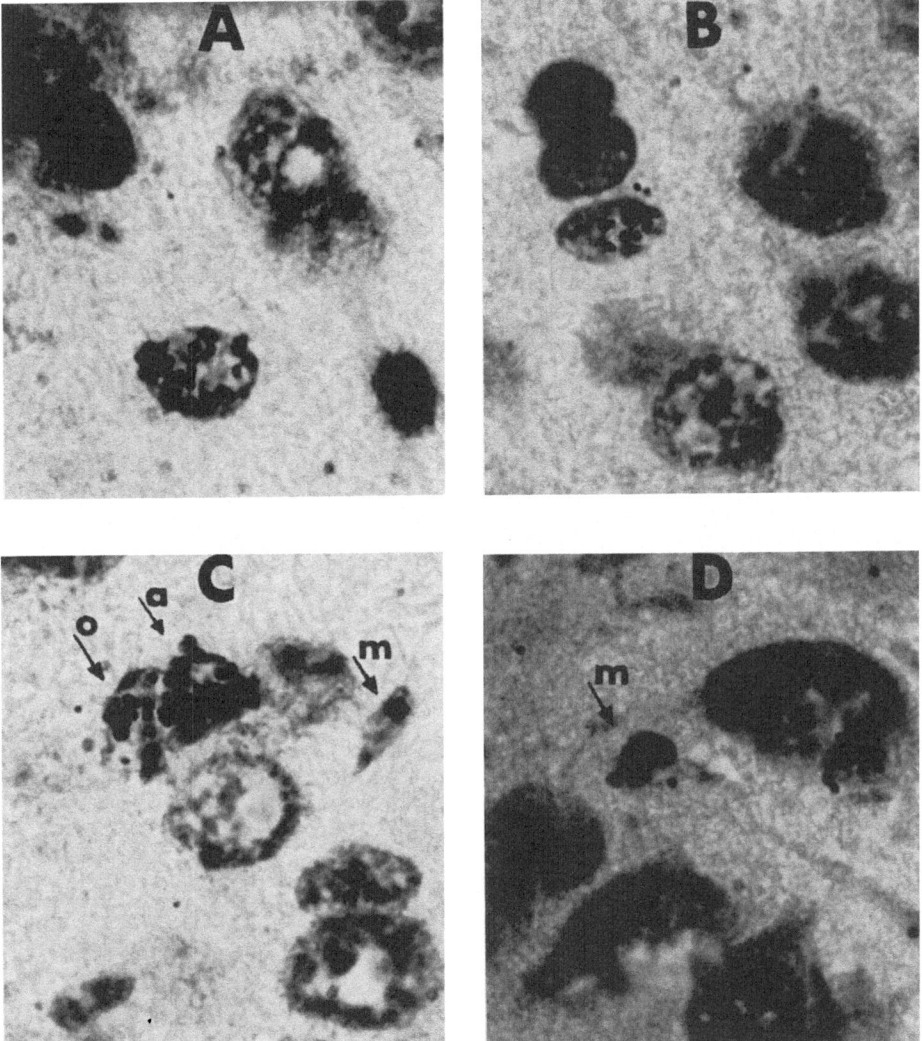

Fig. 4. Representative micrographs from the cortex of a rat aged 30 days *post partum* injected with H₃-thymidine on the 21st day *post conception*. A: labelled astrocyte in layer IV, B: labelled oligodendroglia in layer III, C: labelled astrocyte (a) oligodendroglia (o) and microglia (m) in layer VI, D: labelled perinuclear cell in layer Vb, note also heavily layer microglia (m). (Auto-radiographs haematoxylin stained, x 2.500).

Genesis of Glial Cells

After injection on the 16th and 17th days of gestation. No glial cells were labelled at 30 days *post-partum* in the cortex of rats injected with H_3 thymidine on the 16th and 17th days of gestation.

After injection on the 18th and 19th days of gestation. Microglia, astrocytes and oligo-dendroglia were labelled in specimens from both litters although the greater number of labelled cells were found in the cortices of later injected animals. The labelled cells were distributed randomly throughout the cortex and not concentrated in any particular lamina (Fig. 3).

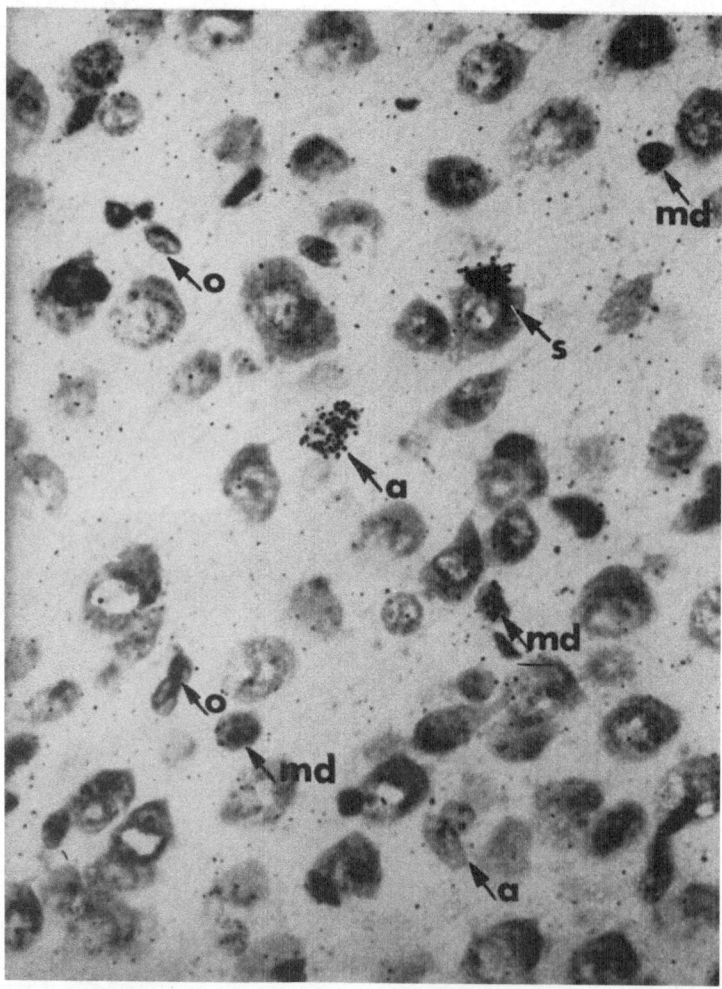

Fig. 5. Layer II of the cortex of a rat killed at 30 days *post partum* and injected with H_3 thymidine on the 22nd day *post conception*. Satellite cells (s), oligodendroglia (o), astrocytes (a) and microglia (m) are labelled. Two medium dark cells are also labelled (md.) (see SMART a. LEBLOND, 1961). Note the absence of label over the neurones. (Autoradiograph, haematoxylin stained, x 1.000).

After injection on the 20th and 21st days of gestation. The number of glial cells taking up the label was greater than that seen in earlier injected animals. The three types of glial cells were labelled and distributed evenly throughout the substance of the cortex (Fig. 4).

After injection of the 22nd day of gestation. At this age more glial cells than at any other time were labelled and, as before, were randomly scattered throughout the cortex (Fig. 5). In all specimens studied the amount of the label seen over all types of glial cells varied from a very low to a very high concentration.

Ependymal and subependymal layers. Ependymal and subependymal cells were labelled at all times after injection. In the subependymal layer the labelling was often scanty, while ependymal cells were almost invariably heavily labelled (Fig. 6). In both structures the incidence of labelling increased with age.

Discussion

Tissue Culture

The morphology of the cells of the developing cortex is difficult to make out because they are so closely packed together. Nuclear types can be distinguished after basic dye staining and some evidence of the presence of cytoplasmic processes can be obtained from

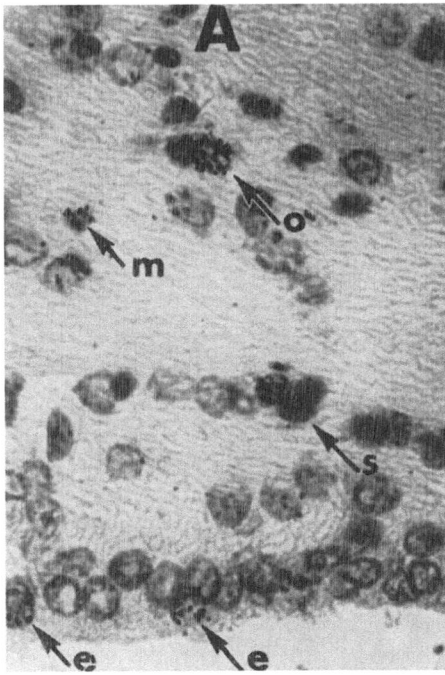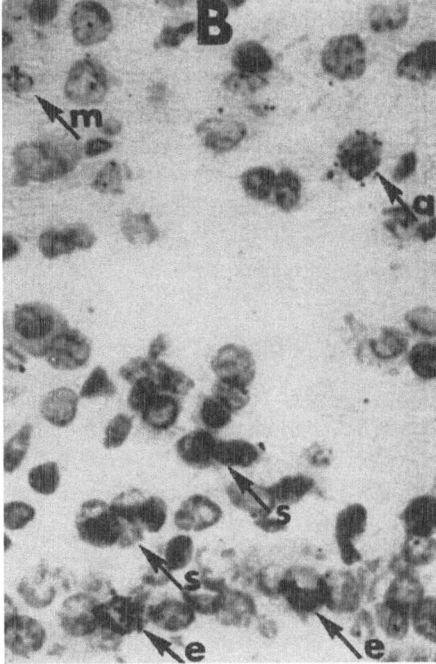

Fig. 6. Labelled cells in the ependymal (e) and subependymal (s) regions of the cortex obtained from rats injected with H₃-thymidine on the 19th (A) and 22nd (B) days *post conception* and killed at 30 days *post partum*. Note also the oligodendroglia (o), astrocytes (a) and microglia (m) labelled in the corpus callosum. (Autoradiographs haematoxylin stained, x 1.000).

a study of silver impregnation preparations. The GOLGI-COX and CAJAL silver methods have produced the most detailed pictures but these stains may be selective and therefore of limited application. The tissue culture method described in this work facilitates the study of cells separated from one another and thus may be of help in obtaining a better understanding of the histology of the developing neocortex. However, the disadvantages in growing cells in artifical media *in vitro* are twofold. Firstly the cells tend to lose their characteristic morphology and secondly their inter-relationships with other cells and their location within the tissue under study are lost. The shortness of the incubation time used in the present study may have overcome the former, while a knowledge of the position of the cell types in histological sections may help to overcome the latter objection.

The separation of cells by Versene in a calcium and magnesium free solution is routinely used in the culture of cells from many tissues and was successful here for embryonic brain cells. As judged by the shape of cells in the GOLGI-COX preparations (BERRY a. ROGERS, 1965) the morphology of the cells was not greatly disturbed by this technique. The method confirmed several points already alluded to in previous publications. In the slides prepared from foetuses on the 15th, 16th and 17th days after conception the cortex is very primitive. The cellular elements are mostly germinal or migratory in type and very few had completed migration by the 17th day of gestation (BERRY a. ROGERS, 1965) while the axon content is very scanty; thalamo-cortical fibres appear to grow into the cortex at about the 16th day of gestation (HICKS, 1958). In tissue culture the cells were indeed mostly of the migratory type i. e. a nucleus contained within a fine cytoplasmic process. The frequent occurrence of two nuclei within one protoplasmic filament supports the thesis of delayed cytoplasmic division. Identification of germinal cells was made difficult by the absence of mitotic figures. The free darkly staining nuclei were taken to be migratory cells in which the processes had broken away since in the loosely arranged tissue-clumps all cells with darkly staining nuclei were seen to be bipolar. No evidence of cytoplasmic division, similar to that hypothesisized by BERRY a. ROGERS (1965) was found.

Differentiating cells were present in great numbers in the specimens prepared from foetuses killed on the 18th and 19th day of gestation. They appeared to be of two types, a multipolar cell and a bipolar cell. It is impossible to say which elements these two forms are destined to become.

Histogenesis of Glia

The results of the autoradiographic study suggest that: (1) mitosis in the ependymal layer does not finish abruptly but falls off from about the 16th day onwards (2) the pattern of histogenesis and migration of neuroglia is dissimilar to that of neurones (3) the neuroglia found in the various layers of the cerebral cortex are not formed at the same time as are the neurones which occupy these laminae (4) the number of labelled glial cells present in the cortex at 30 days *post partum* progressively increases after injection on the 17th day to attain a maximum by the 22nd day after conception (5) glial cells only are formed on the 22nd day after conception – histogenesis of neurones has been shown to be completed by about the 21st day of gestation (BERRY a. ROGERS, 1965) (6) astrocytes, oligodendroglia and microglia are equally labelled in the cortex at 30 days *post partum* following injection over the period studied.

These conclusions suggest several alternative hypotheses. Ependymal cells which are to form glial cells are either demarcated early in cortical histogenesis and glial elements are produced *pari passu* with neuroblasts or alternatively, ependymal cells may lie dormant in the ependymal layer until all neurones have been formed and gliogenesis proceeds after birth in a manner similar to that described by SMART a. LEBLOND (1961).

In the first alternative the paucity of label in the earlier injected specimens may be correlated with a vigorous mitotic activity of the formed glial cells so that little or no label remains at 30 days *post partum*. In the second hypothesis the paucity of label may be correlated with a correspondingly small number of ependymal cells possessing the ability to take up the label i.e. the premitotic state may be attained by these cells in an exponential fashion as the 22nd day of gestation is approached.

A third hypothesis however could equally serve, namely that ependymal cells give rise to neurones during the first stage of their life history and later switch to the production of neuroglia, the transition beginning on about the 17th day of gestation and being completed by the 22nd day of gestation.

In the present state of knowledge there is little evidence to favour any one of these hypotheses.

Summary

The histogenesis of glial elements and the morphology of developing neocortical cells in tissue culture are being studied using albino rats. The preliminary findings suggest that the genesis of glial cells is quite different from that of neurones in respect of both their pattern of migration and the time of their formation. Observations of primitive cortical cells in tissue culture appear to support the nuclear migration hypothesis of histogenesis outlined in earlier publications.

Acknowledgements: The authors wish to thank Professor J.T. EAYRS for reading the manuscript and for his interest and helpful criticism of this work and Dr. M.H. JEYNES for his help with the tissue culture method.

References

BERRY, M., B.G. CLENDINNEN, J.T. EAYRS: Electrical activity in the rat x-irradiated during early development. E.E.G.clin.Neurophysiol. *15* (1963), 91–104.

BERRY, M., J.T. EAYRS: Histogenesis of the cerebral cortex. Nature (Lond.), *197* (1963), 884–885.

BERRY, M., A.W. ROGERS: The migration of neuroblasts in the developing cerebral cortex J. Anat. *99* (1965), 691–709.

HICKS, S.P.: Radiation as an experimental tool in mammalian developmental neurology, Physiol. Rev. *38* (1958), 337–356.

PAUL, J.: Cell and Tissue Culture. E. and J. Livingstone Ltd., Edinburgh-London 1959.

SMART, I., C.P. LEBLOND: Evidence for division and transformation of neuroglia cells in the mouse brain as derived from radioautography after injection of thymidine-H³. J. comp. Neurol. *116* (1961), 349–366.

TILNEY, F.: Behaviour in its relation to the development of the brain. Part II. Correlation between the development of the brain and behaviour in the albino rat from embryonic states to maturity. Bull. Neurol. Inst. N.Y., *3* (1933), 252–358.

M. BERRY, Ph. D.
Department of Anatomy
University of Birmingham
Birmingham, England

A.W. ROGERS, Ph. D.
Department of Human Anatomy
University of Oxford
South Parks Road
Oxford, England

Electron-microscopic Studies on Developing Foetal Cerebral Cortex of the Rat

By R. L. Holmes, Leeds and M. Berry, Birmingham

Studies with the optical microscope have recently thrown some doubt on the hitherto accepted theory of the mechanism of cortical development proposed by Tilney (1933) (see Berry et al. 1964). It was considered that studies using the electron-microscope might be of value in elucidating this problem; and accordingly the investigation of the ultrastructure of the cerebral cortex of foetal rats described below was begun. The preliminary results reported here are concerned mainly with the general pattern of cortical development during the last third of gestation.

Materials and Methods. Rat foetuses aged 14–21 days post-coitum were killed by decapitation. The outer coverings of the brain were immediately removed, leaving the pia-arachnoid in place. Pieces of the dorsolateral cortex were cut out and immersed in a cold solution of glutaraldehyde, buffered to a pH of 7.4 as previously described (Holmes 1964). After fixation the specimens were cut into smaller pieces, washed, treated with osmium tetroxide and later embedded in Vestopal W. Sections were cut from some blocks in a plane at right-angles to the ventricular and pial surfaces, and these, mounted on Sjöstrand type grids, enabled the whole thickness of the cortex to be examined in single sections. Sections were cut from other blocks in planes at various depths tangential to the cortical surface. In the following description the term 'vertical' is used to denote the plane extending at right-angles to the ependymal and pial surface, and 'horizontal' to denote a plane parallel to these surfaces.

Observations

Structure of the cortex of 14–15 day foetuses. The cortex of the youngest foetuses examined, aged 14–15 days, consists mainly of close-packed cells (Fig. 1). The nuclei are round or oval, and except in the most superficial zone tend to be elongated vertically. They are surrounded by relatively sparse cytoplasm, which contains numerous granular inclusions made up of small particles of ribosomal size, and mitochondria. The perikarya of some of the cells extends into processes, although these are not a striking feature at this early stage. When present however such processes contain fine fibrils, or in some instances microtubules, lying along the length of the process. Processes are usually poorer in granules than the perinuclear cytoplasm, but do contain elongated mitochondria. A few small blood vessels lie within the cortex at this stage, but there is as yet no clear differentiation between perivascular tissue and the rest of the cortical elements.

The structure of the periventricular and subpial zones of the cortex differs somewhat from that of the middle zone. The layer of tissue immediately bounding the ventricle consists of superfically-lying cells interspersed with processes of other cells whose nuclei lie at a deeper level (Fig. 2). These various elements are generally closely packed, and many of the cells are in mitosis. The ventricular surface is somewhat irregular and processes of adjacent cells are separated by a narrow cleft which penetrates between them for a short distance (Fig. 2). Deep to this cleft the processes are closely applied to each other, and terminal bars, or maculae adherens, are prominent in this zone, particularly in horizontal sections (Fig. 3). Many cells bear short cilia which project into the ventricular lumen.

For the most part the cytoplasm of the periventricular cells resembles that of more deeply lying elements, and contains small granules, fibrillae and mitochondria. The apical cytoplasm at the level of the terminal bars contains in addition electron-dense granules measuring up to ca. 1000 Å in diameter.

The outermost zone of the cortex, which at this stage constitutes about one fifth of its thickness, is looser in structure (Fig. 1). Some cells in this zone have irregular or elongated nuclei, and are orientated in a horizontal plane. Small bundles of narrow fibres pass vertically between groups of cells, and some of these fibres can be seen to be continuous with expanded 'feet' which form a continuous external layer to the cortex. The surface of this layer is bounded by a double membrane, whose inner component is closely applied to the outer one, except at junctions between adjacent processes. At these points the inner membrane follows the surface of the 'feet', and dips away from the outer, so that a small triangular space is left between the two.

Occasional cells resembling phagocytes occur in the outermost layer of the cortex (Fig. 4); and superficially lie blood vessels and connective tissue elements.

Changes during development. The structure of the innermost zone of the cortex appears to change little during the developmental period studied. A few canaliculi of endoplasmic reticulum develop in the cells in later foetal stages, but even at 21 days these are relatively inconspicuous. Numerous mitotic figures persist in this zone throughout the period.

By contrast major changes occur in the middle zone of the cortex during the later foetal stages. Its thickness increases greatly, due both to an increase in the number of cells, and also to the appearance of numerous bundles of fibres. These first appear at the 16th day; and at 18 days, while the innermost layer of the middle zone still consists predominantly of closely packed cells similar to those described earlier, the rest of the zone contains numerous bundles of fibres, each single fibre measuring c. 0.5μ in diameter (Fig. 5). Many of the bundles run parallel to the cortical surface, but others run ver-

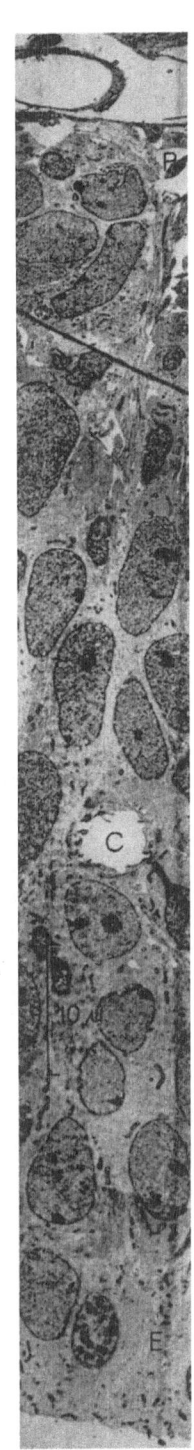

Fig. 1. Montage of sections through the whole thickness of the cortex of a 15 day rat foetus. The cortex is predominantly cellular. E – ependymal zone; P – sub-pial zone; C – capillary.

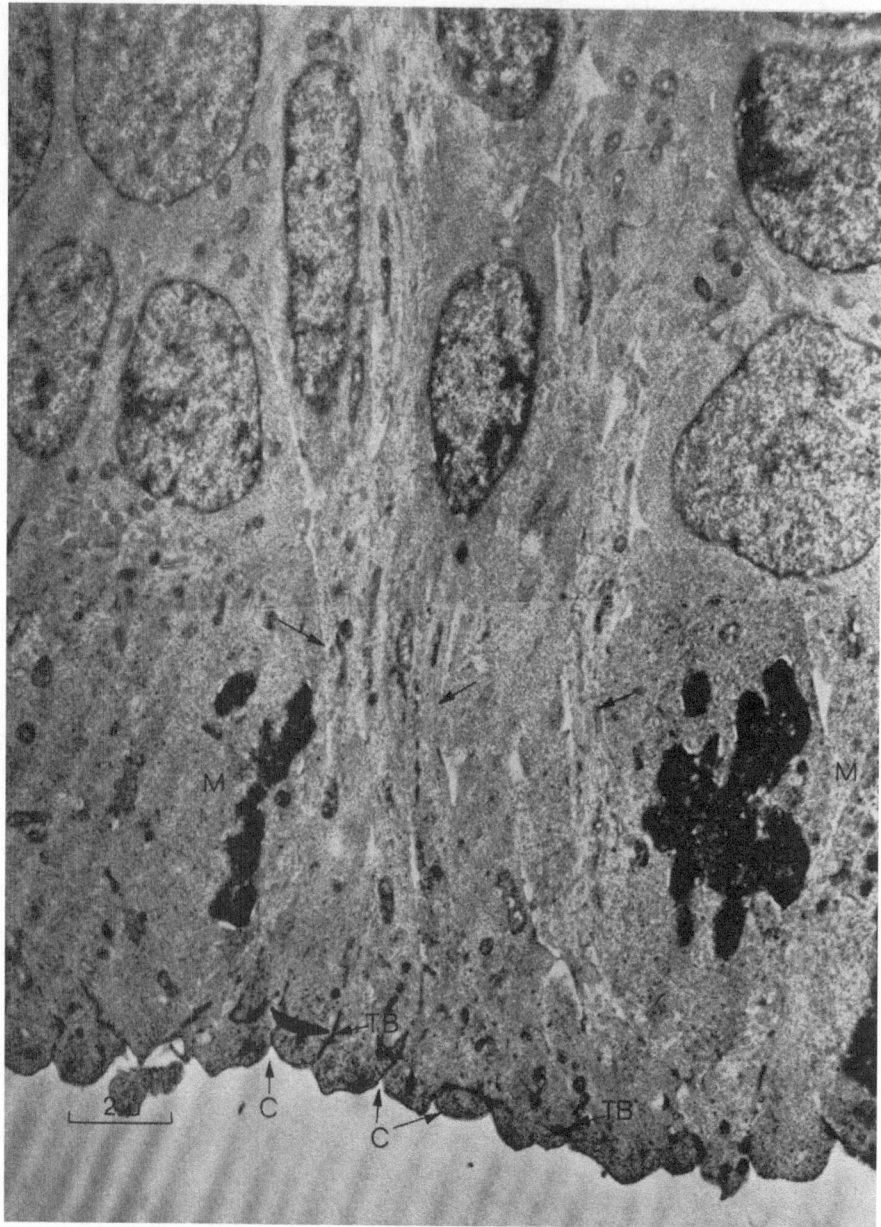

Fig. 2. Vertical section through the ependymal zone of a 14 day foetus. Cells in the innermost layer in mitosis (M) are separated by processes (↑) of cells lying deeper. At the surface, cell processes are separated by narrow clefts (C), deep to which lie terminal bars (TB).

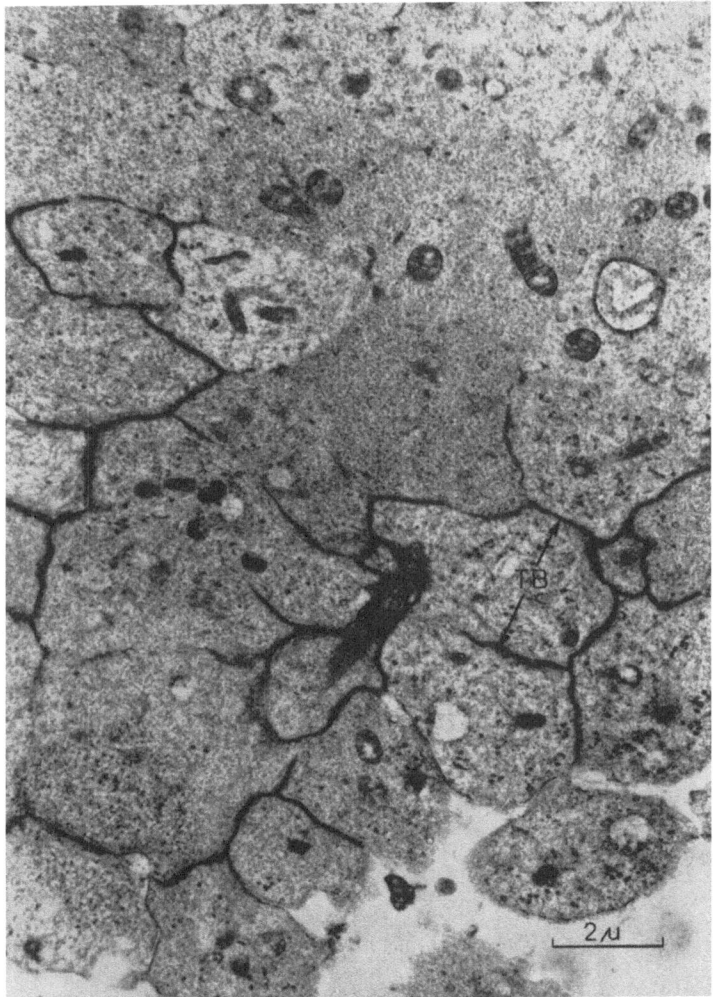

Fig. 3. Horizontal section through the ependymal zone of a 14 day foetus. Terminal bars (TB) are prominent. Note the electron-dense granules in the apical cytoplasm of the cells.

tically, and some in other directions. Cells in association with the bundles have often lost their vertical orientation, and tend to lie along the plane of the fibres.

Structurally the fibres differ from perinuclear cytoplasm. Many contain longitudinally arranged parallel fibrillae or microtubules and elongated mitochondria. Others contain only sparse fibrillae or tubules lying in a matrix of low electron density, and hence appear paler on electron micrographs.

This general pattern of cells and fibres persists in the later stages of development of the foetal cortex, but as the thickness increases, so do the number and complexity of the bundles of fibres. In some areas these form a loose irregular meshwork, with pronounced spaces between the fibres.

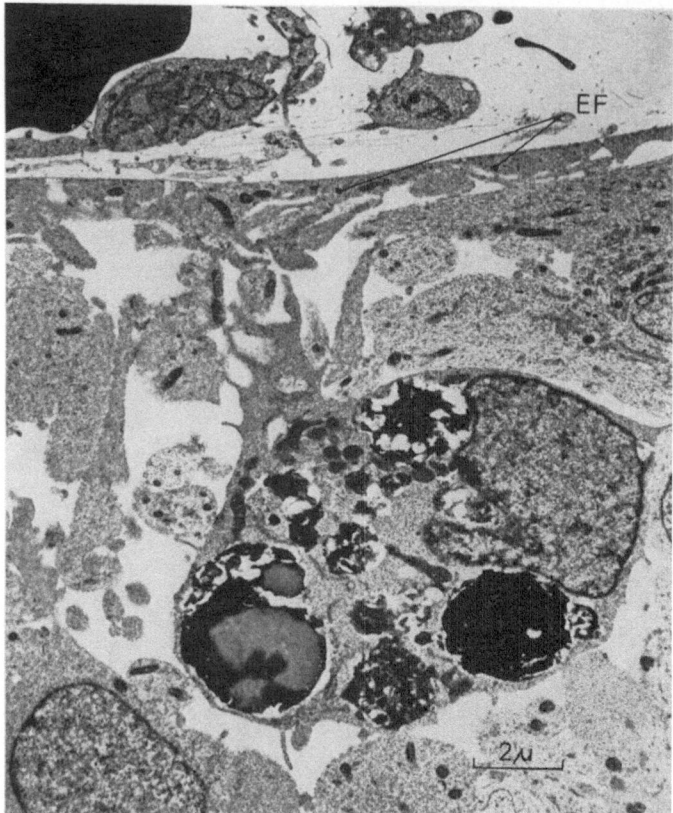

Fig. 4. Cell resembling a phagocyte in the sub-pial zone of the cortex of a 15 day foetus. Part of a pial capillary lies to the upper left; below this (EF) is the limiting layer of the cortex, formed by expanded 'feet' which are not yet well developed. (Compare 'feet' in Fig. 6).

Fig. 5. Bundles of fibres, mainly lying horizontally, in the middle zone of the cortex of an 18 day foetus.

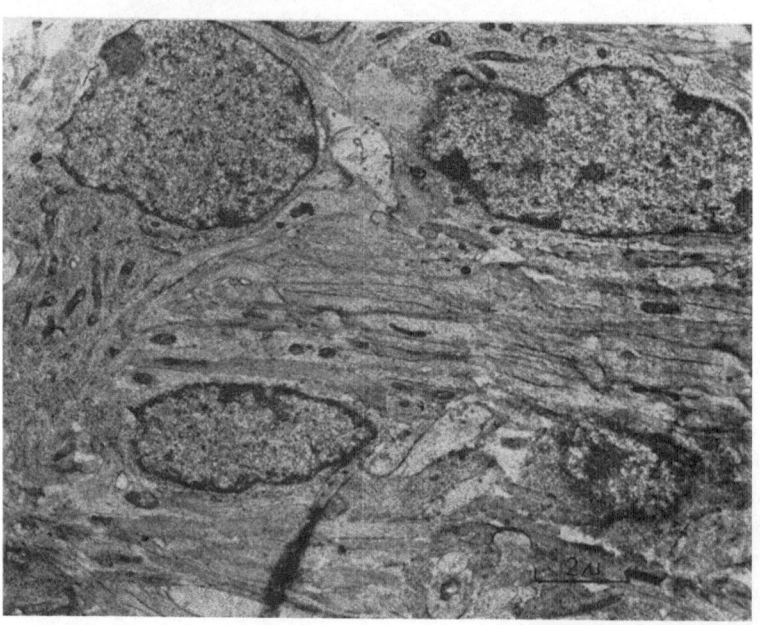

Fig. 6. Montage of sections showing the outer zone of the cortex of an 18 day foetus. Note the deeper zone of cells and vertical fibres which is replaced superficially by a layer of loosely arranged fibres and cells. One cell (A) lies horizontally. Others (B) are rounded and have electron-dense cytoplasm. Expanded feet (EF) forming the bounding layer of the cortex are well developed.

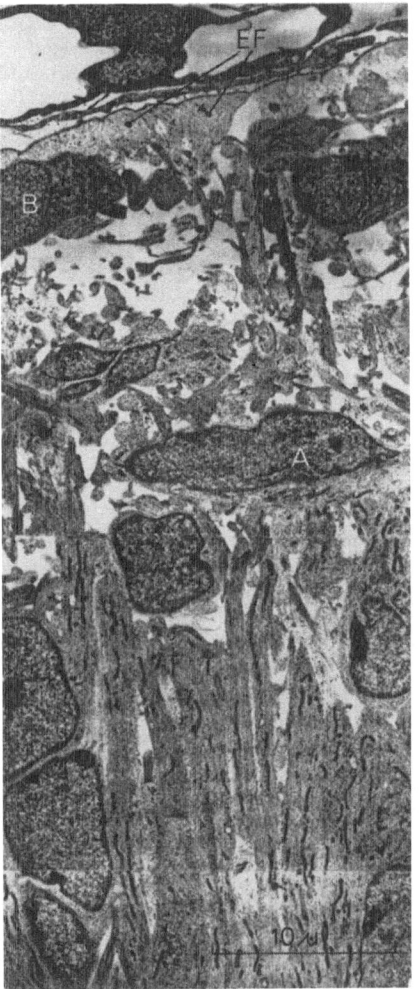

The outer zone of the cortex, with the exception of the immediately sub-pial region, remains richly cellular during the later stages of development, but there is an increasingly striking development of vertical fibres passing between groups of cells. Some of these appear thicker than fibres lying deeper in the cortex (Fig. 6). Many of the cells maintain a vertical alignment; others lying superficially have rounded nuclei.

The sub-pial zone, which consists of a mixture of cells and fibres, remains loosely organized in the later phases of development. Some cells lie horizontally (Fig. 6), and resemble in some respects distorted cortical cells. Others are smaller, more rounded, and have cytoplasm of greater electron density, containing many mitochondria. Fibres run in all planes, but a number of vertical ones are still present, and some of these are clearly in continuity with the superficial expanded feet which form the outer boundary of the cortex, and are more prominent than at earlier stages.

Discussion

Few studies of the ultrastructure of the immature cerebral cortex have so far been published, and this brief survey of the general configuration of the cortex of the foetal rat is intended to be no more than a preliminary to more detailed studies. The observations reported above have been confined to foetal specimens; but development continues after birth, and a complete picture can only be obtained by study of material taken from all developmental stages.

Differentiation of the various elements of the mature cortex has not proceeded very far in the specimens described above, and clearly from RAMSEY's brief report on the cortex of the post-natal rat (1961) considerable changes occur in the early post-natal period. VOELLER et al. (1963) observed that distinction between neuroglial cells and neuronal cell bodies was difficult in the cortex of the late foetal kitten, and also, in contrast to the findings in the rat, that very few cell processes were present even in neonatal specimens.

In material from a 12 week human foetus Pappas a. Purpura (1964) noted dense packing of cells similar to that seen by the present authors in the earlier stages of the foetal rat.

It is hoped that ultrastructural studies will eventually throw some light on the mechanism of cortical development. Clearly it will be difficult to demonstrate two widely separated nuclei lying within a single cytoplasmic mass, but it should be possible to observe earlier stages of the nuclear migratory process postulated by Berry et al.(1964). The well developed terminal bars between processes at the ependymal surface might suggest that these parts of the cells at any rate are firmly anchored at this site, and that with these points fixed movements of nuclei could occur within the cytoplasm. Combined studies using both optical and electron microscopes, as well as experimental techniques, are essential to any further progress in the elucidation of this problem.

Summary

The ultrastructure of the cerebral cortex of the foetal rat between the 14th and 21st days has been studied, using material fixed in glutaraldehyde, treated with osmium tetroxide, and embedded in Vestopal W. At 14 and 15 days the cortex consists mainly of close-packed cells showing some distinction between finely granular perinuclear cytoplasm and less granular elongated cellular processes containing longitudinal fibrillae. The ependymal zone consists of tightly packed processes showing well-developed terminal bars in their superficial contact zones. The sub-pial layer is looser, and vertically oriented fibres expand here to form a continuous layer at the surface. Mitotic figures are common in the inner zones of the cortex. From 16 days horizontally-arranged fibres are conspicuous in the middle zones, while fibres in the outer zone maintain their vertical orientation. Development during the rest of foetal life is characterized by a great increase in cortical thickness, particularly in the layers of horizontal fibres, accompanied by increasing differentiation of cellular and fibrous structures.

References

Berry, M., A. W. Rogers, J. T. Eayrs: Pattern of cell migration during cortical histogenesis. Nature (Lond.) *203* (1964), 591–593.

Holmes, R. L.: Comparative observations on inclusions in nerve fibres of the mammalian neurohypophysis. Z. Zellforsch. *64* (1964), 474–492.

Pappas, G. D., D. P. Purpura: Electron microscopy of immature human and feline neocortex. In: Progress in Brain Research, 4, 176-186. Elsevier, Amsterdam-London-New York 1964.

Ramsey, H.: Electron microscopy of the devel-

oping cerebral cortex of the rat. Anat. Rec. *139* (1961), 333.

Tilney, F.: Behaviour in its relation to the development of the brain. II. Correlation between the development of the brain and behaviour in the albino rat from embryonic states to maturity. Bull. Neurol. Inst. N.Y. (1933), 252–358.

Voeller, K., G. D. Pappas, D. P. Purpura: Electron microscope study of development of cat superficial neocortex. Exper. Neurol. *7* (1963), 107–130.

R. L. Holmes, D. Sc.
Department of Anatomy
School of Medicine
University of Leeds
Leeds, Yorkshire, England

M. Berry, Ph. D.
Department of Anatomy
University of Birmingham
Birmingham, England

Electron-Microscopy of the Cytodifferentiation in the Developing Brain of Chick Embryos

By W. WECHSLER, Köln

Electron-microscopy (EM) is a link between morphology and cytochemistry. The electron-microscope may elucidate the fine structure of cells and the reciprocal relationships of the different cellular constituents in both the embryo and the adult brain. As a consequence there are new sources of information for cytogenetic and histogenetic aspects of neuroembryology.

The different cellular elements of the CNS are derived from neuroectodermal matrix cells, which are uniform and undifferentiated. These cells will become the neural plate, neural tube and the primitive brain vesicles. The specialization of the matrix cells is called cytodifferentiation. Although this differentiation has been observed with the light-microscope, the actual alterations are biochemical.

An analysis of the fine structure of these cells is fundamental for a better understanding of many aspects of *cytodifferentiation*. This report is a summary of our observations. The fine structure of the undifferentiated matrix cells, the differentiation of neurons, and the glial differentiation are discussed. Developmental problems of histogenesis of the brain are only briefly considered.

General Remarks

Conventional terminology of the developing cells of the CNS lags behind the modern neuroembryological outlook.

His (1889, 1890) distinguished 3 types of cells: (1) *germinal cells,* mitotic figures beneath the ventricular surface, (2) the elongated bipolar *spongioblasts,* extending from the inner to the outer surface of the brain vesicles, and (3) the *neuroblasts.* CAJAL (1929, 1952) described 5 different stages in the developmental cycle of a neuron: (1) cellule germinative de HIS, (2) cellule apolaire ou polygonale, (3) cellule bipolaire, (4) cellule unipolaire ou neuroblaste de HIS, and (5) cellule multipolaire. CAJAL's cytodifferentiation theory of the ependyma was that of: (1) spongioblaste primitif de HIS, (2) cellule épithéliale primordiale ou corpuscule de GOLGI, (3) cellule épithéliale jeune ou ramifiée, (4) cellule épendymaire ou épithéliale définitive. The other neuroglial elements are ascribed to spongioblasts retracting and carrying the processes from the surfaces into the CNS. Similar concepts have been proposed by many other authors (BAILEY a. CUSHING, 1930, RANSON, 1941; ROMANOFF, 1960; etc.).

Recently, with tritiated thymidine autoradiography several scientists have demonstrated the mitotic and migration cycles of undifferentiated cells and neuroblasts in the developing brain of various species. In light of these findings it is difficult to assess the differences in early stages of neurogenesis among germinal cells, primitive ependymal cells, and the apolar and bipolar neuroblasts and spongioblasts.

Fujita (1965), whose concepts we have adopted, proposed that the undifferentiated neuroectodermal cells of the neural tube and the brain vesicles be called *"matrix cells"* (Matrixzellen). The entire layer composed of these cells should be called "matrix layer" (Matrixschicht von Kahle 1951; Fig. 1/I and II). Our concept of the cell lineage in the CNS is the following:

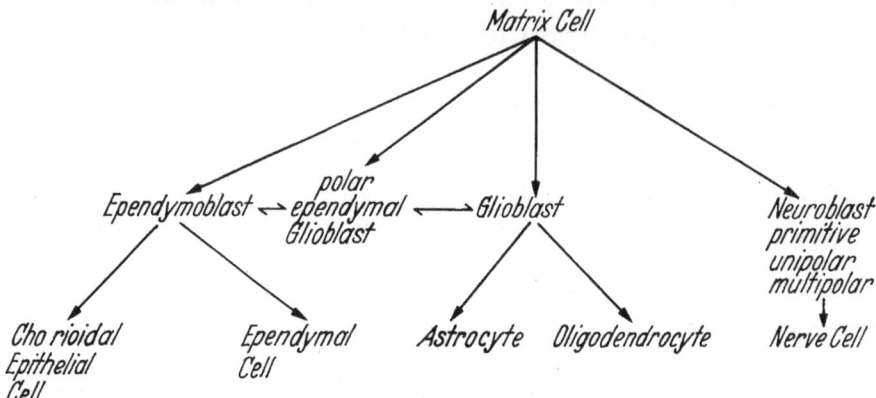

In this scheme the differentiation of nerve cells is restricted to multipolar neurons. The development of the microglia is excluded because there is disagreement whether these cells are of neuroectodermal or mesodermal origin. Fig. 1/I in the upper line represents the phylogenetic differentiation of nerve cells according to Cajal (1911), and the lower left line represents the ontogenetic development of multipolar nerve cells. In the following chapters the fine structural changes occuring during the development of the different cellular elements of the CNS are described.

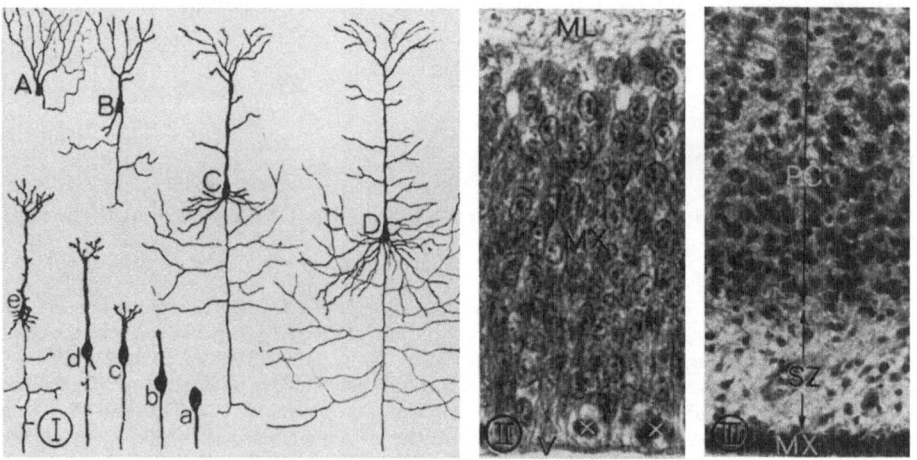

Fig. 1. The general principles of cytogenesis and histogenesis in the chick brain are represented: I. The upper line shows the phylogenetic development of cortical pyramidal neuron, according to Cajal (1952): A – frog, B – reptile, C – mouse, D – man. The line in the lower left represents ontogenetic development: a – primitive neuroblast, b – bipolar neuroblast, c – unipolar neuroblast, d and e – progressive differentiation of the cortical pyramidal neurons.
II. The matrix layer (MX) of the telencephalon, chick embryo of 3 ½ days. ML – marginal layer, V – ventricular surface, x – mitosis.
III. The hemispheral cortex with a small matrix layer (MX), the migration zone (SZ "Schwärmzone"), and a broad primary cortex (PC) in a chick embryo of 13 days.

The Differentiation of Cells

Cytodifferentiation is a long and complex biochemical and morphological process. The differentiation and the changes in the fine structure of the developing cells are: (1) phase of undifferentiated matrix cells, (2) phase of primordial cytodifferentiation, and (3) phase of definitive differentiation and maturation. There are little differences in fine structure between phase I and II, but significant changes between phase II and III. This means, there are only minor ultrastructural differences observed among matrix cells, primitive neuroblasts and glioblasts; but the primitive neuroblasts and mature nerve cells show remarkable differences in the cell morphology and fine structure.

Materials and Methods. Our light- and electron-microscopic studies were performed on chick-embryos incubated for 2–21 days and on young chickens. The material was fixed in a 1% osmium-tetroxyde-potassium-bichromate-mixture (WOHLFAHRT-BOTTERMANN, 1957) with and without glutaraldehyde-prefixation, and embedded in Vestopal W. Thin sections were examined under a ZEISS EM 9 microscope.

The Fine Structure of the Matrix Cells

The neuroectodermal matrix cells have a primitive cytoplasmic fine structure (Fig. 2). Numerous free ribosomes but few profiles of endoplasmic reticulum are observed in the perikaryon (cell-body) of the matrix cells (Fig. 2a). This is a common observation in embryonic cells of different origin and may represent a general principle of cytogenesis. – "First, it should be noted that as a rule the endoplasmic reticulum is not a prominent structure in the relatively undifferentiated cell. The cytoplasmic matrix of such cells is always dense with RNP-particles, but there are only few profiles of the ER except for the nuclear envelope. These features are so constant for rapid proliferating, undifferentiated cells as to be useful for identification" (PORTER, 1961).

Using light-microscopy techniques, no differences in fine structure of matrix cells are seen, but using the techniques of EM, striking differences are seen between perikaryon and primary processes of matrix cells: a greater number of organelles may be in one compartment of a cell, although there are some organelles in each compartment. The mitochondria of the matrix cells are well differentiated. A small GOLGI apparatus is present but lysosomes are absent (Fig. 2). At this stage the GOLGI apparatus is usually situated in the apical primary processes. When the material is prefixated with glutaraldehyde, microtubuli are found in the matrix cells (Fig. 2 b). Microtubuli are frequently present in the (elongated, bipolar) primary processes, but not in the perikaryon. On the contrary, mitochondria and ribosomes are prominent in the perikaryon, but less in the primary processes. Throughout the matrix cells the endoplasmic reticulum is immature. The steric distribution of organelles suggests a primordial polarity of the matrix cells.

During the formation of the prosencephalon and delimination of the telencephalon and diencephalon the elongated processes of the matrix cells can reach the outer and the inner surface of the primitive brain vesicles. Electron-microscopy demonstrates the different features of the so-called internal and external limiting membranes. The internal surface structures of the matrix layers have signs of primitive epithelium with some cilia, micro-

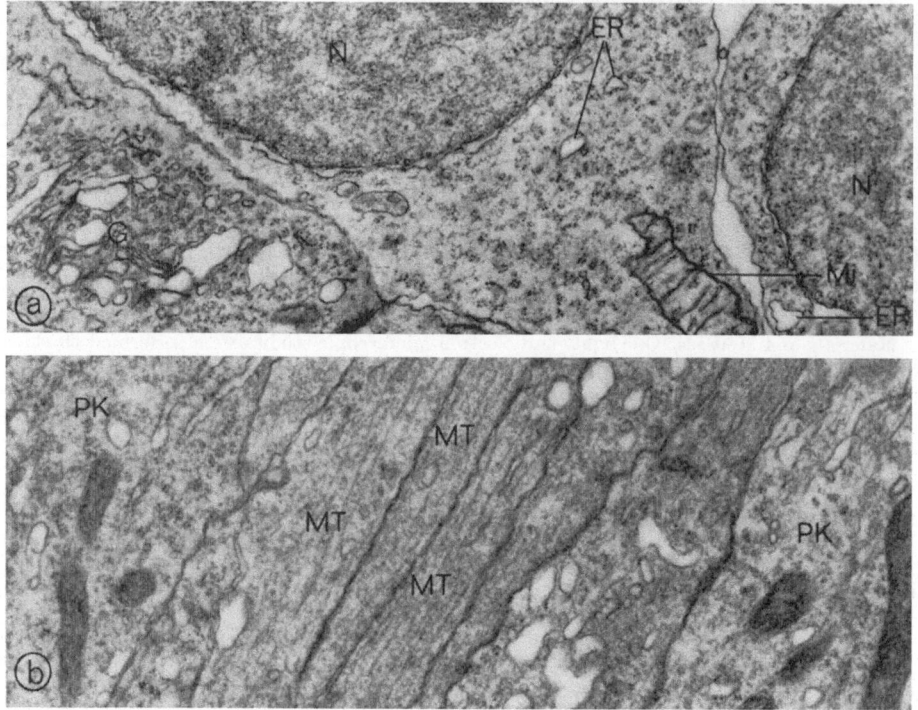

Fig. 2. The fine structure of matrix cells during interphase:
a) The matrix layer of the telencephalon, embryo 6 ½ days old: N – nucleus, ER – endoplasmic
reticulum, MI – mitochondria, G – Golgi-apparatus (No. 1762, magnification 18.000 x)
b) The matrix layer of the tectum opticum, embryo, 8 days old: N.B. different fine structure of
the perikaryon (PK) and the primary processes (PP) with parallel microtubuli (MT) (No. 6731,
magnification 21.000 x)

villi, and terminal bars (Fig. 6 A and B). The outer surface of the marginal layer is composed of undifferentiated processes and covered by a continous basement membrane (WECHS-
LER, 1966 a, b). This observation is important because it indicates a primordial polarisation
of the surfaces of the brain vesicles. From a cytological point of view this is additional
evidence for the *primordial bipolarity of the matrix cells* (WECHSLER, 1966 a, b). An inherent
feature of the polarity of primordial cells is the flexibility to change to different forms.
When matrix cells migrate to the internal surface to undergo mitosis, the characteristic
of polarity is nearly lost. When matrix cells leave the juxta-ventricular position and lose
contact with the ventricular surface, the primordial epithelial characteristics of the internal
processes disappear.

The Differentiation of Nerve Cells

On the 5th and 6th day of incubation the first significant cytoarchitectural changes are observed. Various cellular elements in the substance of the brain proliferate and differentiate resulting
in the establishment of centers and fiber-tracts. Salient concepts are the metamorphosis of the

matrix cells to neuroblasts, the maturation of the nerve cells, and the genesis of nerve fibers. In all species and in all regions of the brain there are patterns of cell migration during cortical evolution. Cells leave the matrix zone (Matrixzone), pass through a migration zone (Schwärmzone), and reach the primary lamellae of the cortex (primärer Cortex). During this migration they undergo differentiation. Studying the fine structure of neuroblasts in the areas of the "Schwärmzone" and the primary cortex of the telencephalon and the tectum lobi optici of chick embryos incubated for 7–9 days (WECHSLER a. MELLER, 1963 a and b) the following observations were made.

The fine structure of *migrating neuroblasts* and the *unipolar neuroblasts* of the primary cortex are similar. The perikaryon of these cells has an oval nucleus containing a nucleolus. There are minor increases in endoplasmic reticulum; sometimes arising from the nuclear envelope (Fig. 3a). The tubuli and vesicles of the GOLGI apparatus are concentrated at either longitudinal pole of the nucleus. There is little difference between the ultrastructural organization of the perikaryon of the matrix cells and the primitive neuroblasts. Studies of *multipolar neuroblasts* in the hemispheres (cortex, corpora striata) and the tectum opticum of chick embryos, incubated for 11, 13 and 16 days, revealed an increase in the volume of the perikaryon. Concurrently there is a differentiation of the cytoplasm, which is characterized by an impressive development of a rough-surfaced endoplasmic reticulum, the initial formation of the ergastoplasm of the NISSL-bodies and the increase in free ribosomes (Fig. 3b). The mitochondria are apparently multiplying and increasing in size. Concomitantly they develop internal cristae (Fig. 3a and b). During the *phase of maturation* of the nerve cells there is a definitive formation of the NISSL-bodies, and an enlargement and distention of the GOLGI apparatus surrounding the nucleus. The so-called secondary structures, osmiophilic bodies or lysosomes, are absent in unipolar, but not in multipolar neuroblasts (Fig. 3b). Neurofilaments and microtubuli are observed in various numbers. Electron-microscopically little information is obtained regarding nuclear metamorphosis. Similar observations with respect to the differentiation of nerve cells have been obtained by FUJITA a. FUJITA (1963) in the brain of chick embryos.

We have stressed that not only the perikarya of developing neurons show fine structural changes but also the axons, the dendrites and the terminal axonal processes (WECHSLER, 1966 b). The fine structure of the minute-primitive *axons* (Fig. 3a) is different from the mature axoplasm of myelinated nerve fibers.

The details of *dendritic differentiation* are not discussed and only general remarks on the *development of the synapses* are made. Some of the structural features of synapses can be seen in the primary cortex of chick embryos incubated for 9 days. The thickening of the pre- and postsynaptic membranes are the first signs. Accumulation of the synaptic vesicles may occur later (Fig. 4a and b). The final steps are the enlargement of the synaptic region, a further increase of synaptic vesicles and a concentration of mitochondria and special structures (Fig. 4c). GLEES a. MELLER (1964) have made similar observations in the spinal cord of chick embryos. VOELLER et al. (1963), PAPPAS a. PURPURA (1964) have reported on the development of the synapses in human and feline immature cortex.

We have come to the following conclusions: (1) Cytodifferentiation of multipolar neurons is a morphological *and* a fine-structural specialization. (2) It is possible to analyze the fine-structural metamorphosis according to quality, time of occurence and localization.

The differentiation and maturation of multipolar nerve cells is characterized by a sequential polymorphous and multifocal cytoplasmic differentiation. This development is paralleled with changes of the nucleus.

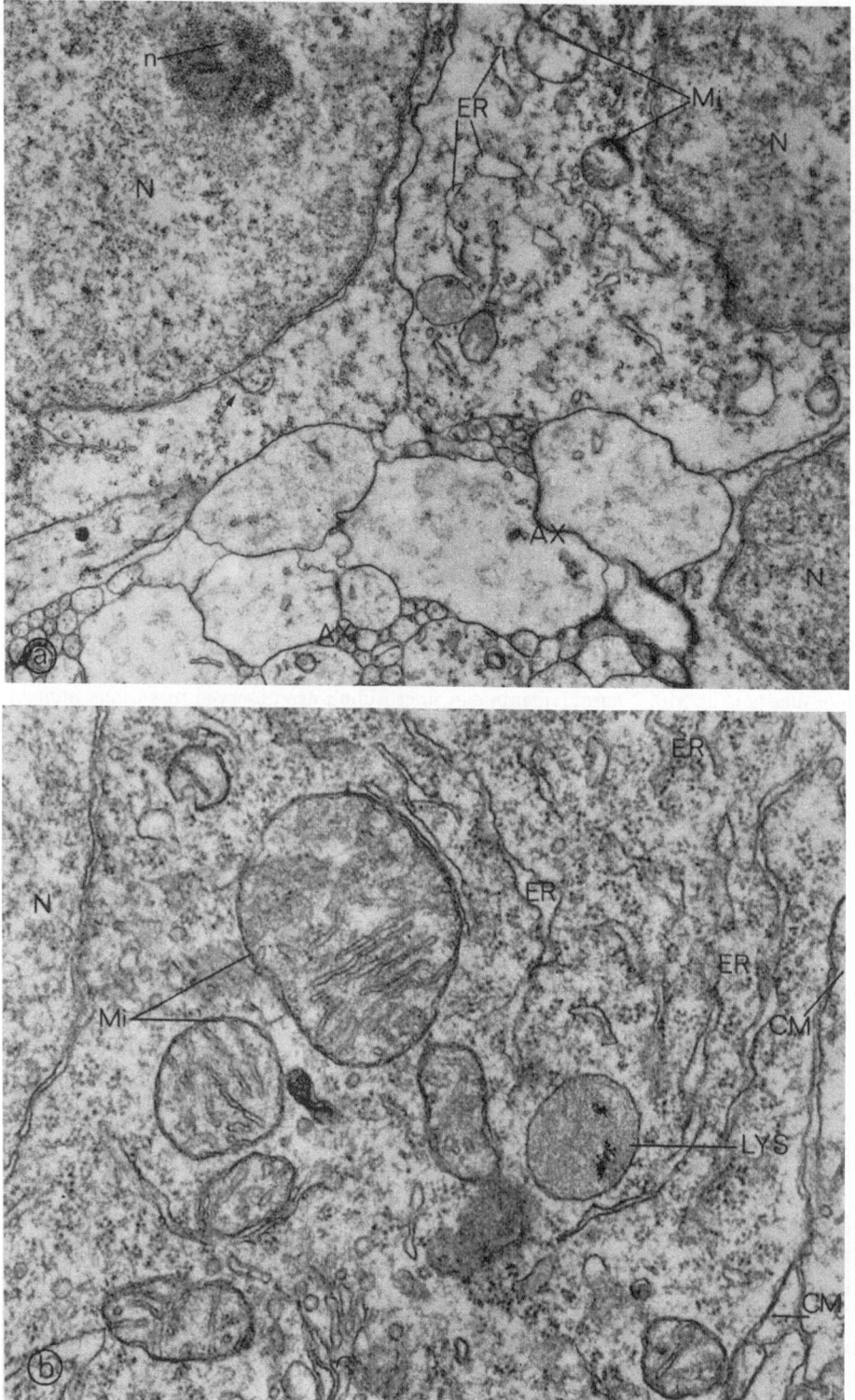

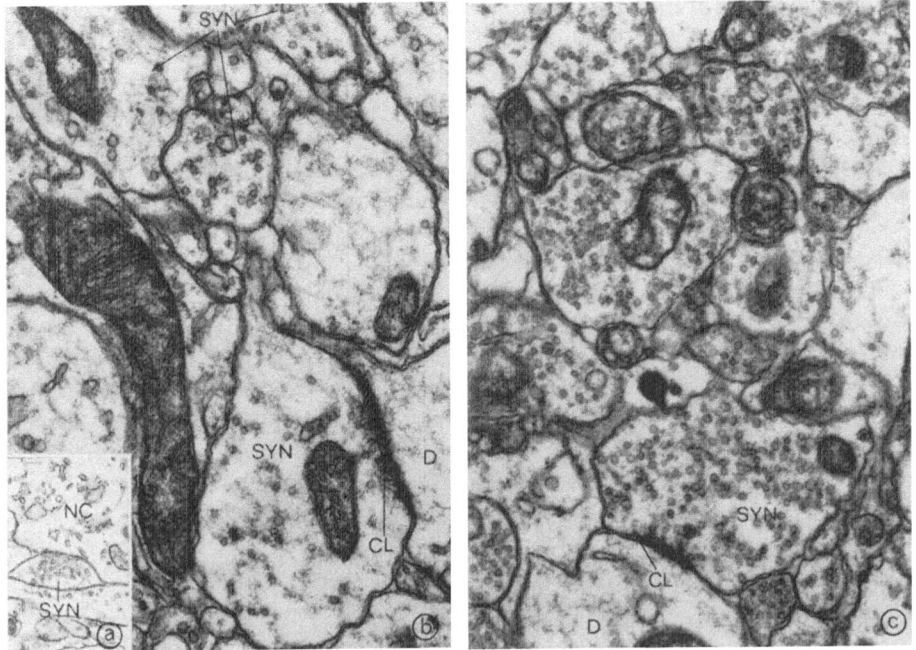

Fig. 4. The general principles of the development of synapses:

a) Striatum, chick embryo, 16 days old: The minute axosomatic synapses (SYN) with accumulation of synaptic vesicles. NC – perikaryon of a nerve cell (No. S 1029, magnification 20.000 x)

b) Molecular layer of the cortex, chicken just upon hatching. Within the neuropil there are synaptic endings (SYN) with a few synaptic vesicles and membrane thickenings. D – dendrite, CL – synaptic cleft (No. 11507, magnification 45.000 x)

c) The molecular layer of the cortex of the hemisphere of a chicken upon hatching with mature synapses (SYN) and a more pronounced accumulation of synaptic vesicles; D – dendrite, CL – synaptic cleft (No. 1491, magnification 40.000 x)

←   Fig. 3. The differentiation of nerve cells in the perikaryon:

a) The "Schwärmzone", tectum opticum of the midbrain of an embryo, 7 days old. The primitive fine structure of migrating neuroblasts and the (axonal) processes (AX). N – nucleus, n – nucleolus, ER – endoplasmic reticulum, ↑ – formation of the endoplasmic reticulum arising from the nuclear envelope, MI – mitochondria (No. 1816, magnification 18.000)

b) The progressive differentiation of a neuron in the striatum of a 16 day-embryo. There is a pronounced rough surfaced endoplasmic reticulum (ER), a typical GOLGI-apparatus (G), a large number of mitochondria (MI) and the first lysosomes (LYS). The amount of free ribosomes is remarkable (No. 4866, magnification 45.000)

The Differentiation of the Neuroglia

The primitive brain vesicles of the telencephalon and mesencephalon consist mainly of the matrix layer. Here, it is not possible to demonstrate primitive and polar glioblasts in EM-micrographs. Whether the cells migrating to the primary cortex should be regarded as neuroblasts, glioblasts or undifferentiated cells is a problem. Because of their fine structure these cells must be regarded as primitive cell types. In the "Schwärmzone" 3 cell types with "embryonic fine structure" can be distinguished; one is regarded as a *primitive glioblast* (WECHSLER a. MELLER, 1963 b). We studied the fine structure of the *polar glioblasts* in the ventral and dorsal position of the central canal of the spinal cord, because of the technical difficulties to study these cells in the brain (WECHSLER 1966 c). The polar glioblasts have an obvious epithelial-glial polarisation. Similar observations have been made by GLEES a. LE VAY (1963). FUJITA a. FUJITA (1964) described the differentiation of ependymal cells and glioblasts in the spinal cord of chick embryos, using H³-thymidine autoradiography and electron-microscopy. Our findings concerning glial development are reported in other communications (WECHSLER 1963, 1966 d).

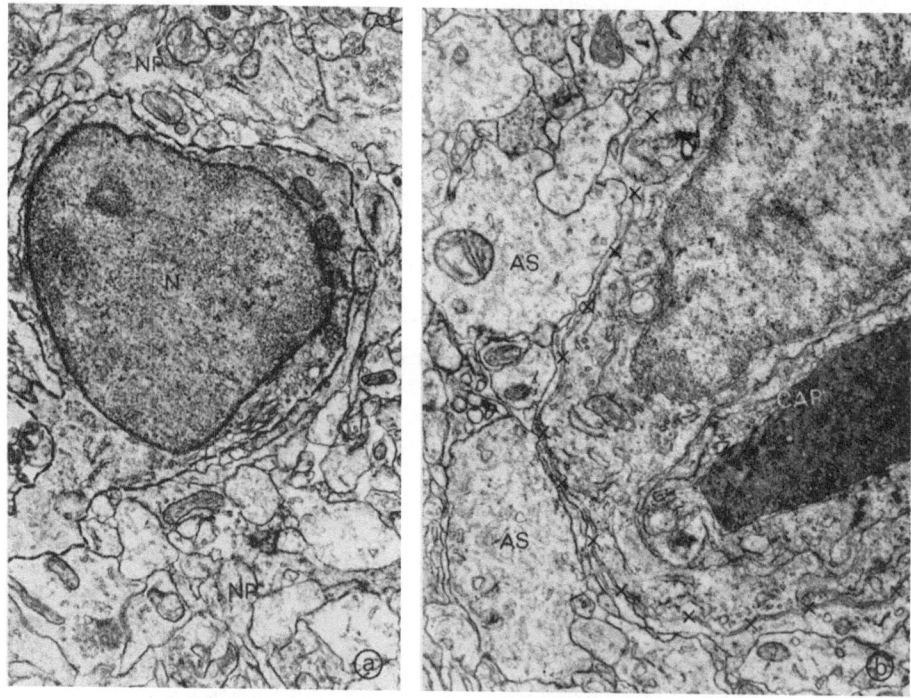

Fig. 5. Differentiation of glial cells:

a) The striatum of a chick embryo, 16 days old; with dotted line showing cell border of a glioblast; N – nucleus, NP – neuropil (No. 4811, magnification 18.000 x)

b) The glial-vascular relationship in the striatum of an embryo incubated for 16 days. This is the phase of the beginning formation of the basement membranes (x). CAP – capillary, AS – processes of astrocytes attached to the vascular basement membrane (No. 2598, magnification 18.000 x)

If we compare the lineage from matrix cells to primitive and polar glioblasts and the differentiating *astro- and oligodendrocytes* (cortex, striatum: Fig. 5a), there is no doubt that glial differentiation is accompanied by changes in the fine structure. When chick embryos are incubated for 13 and 16 days, it is relatively easy to distinguish glial and nerve cells. The glial cell of Fig. 5a is immature. It does not fully resemble a differentiated oligodendrocyte or astrocyte. It is important to realize that the processes of neuroblasts and glioblasts do not form a syncytium. The cellular elements are however extremely close together because of the minimal extracellular space.

The relationship between embryonic capillaries and the elements of the developing nervous tissue is salient. In early stages of vascular proliferation in the CNS the embryonic capillaries have no basement membranes. The neuroectodermal cells are closely attached to the vessel wall. There is no large diffusible perivascular space, and in later stages of development immature clear

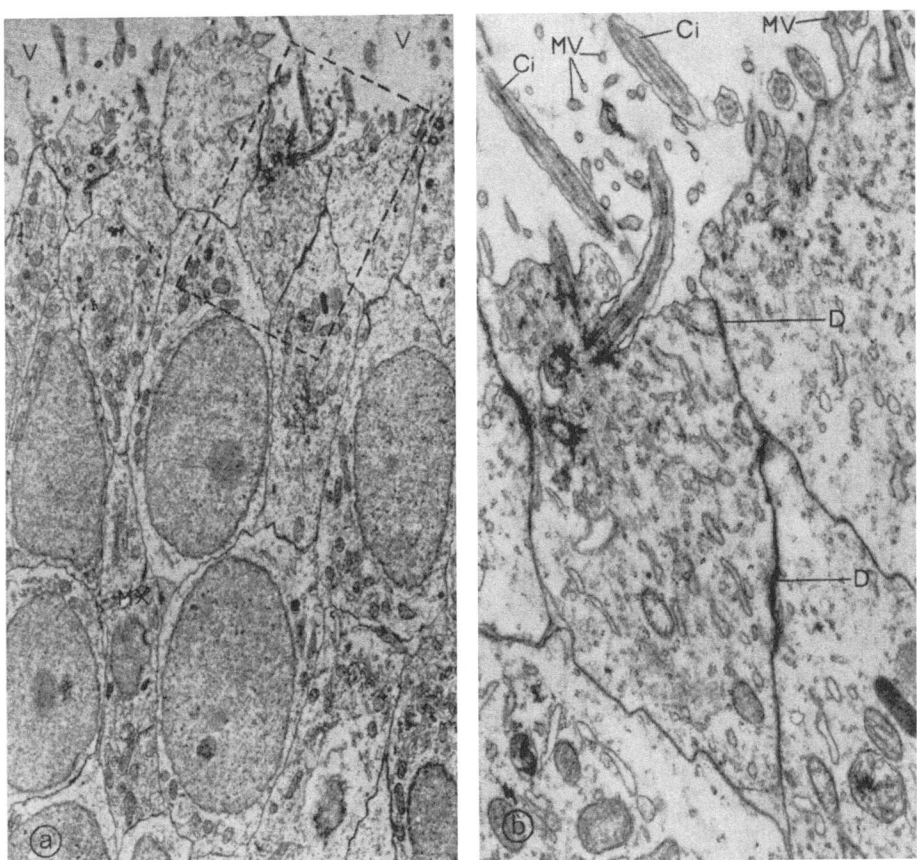

Fig. 6. The phase of the reduction of the matrix layers in the ventricles of the hemispheres of a 13 day-embryo.
a) The matrix zone (MX) is only composed of a few cellular layers and the internal surface has the characteristic feature of apical ependymal cells (No. 4332, magnification 5000 x)
b) A higher magnification of Fig. 6a. Note the typical pecularities of the apical part of the ependymoblasts: CI – cilia, MV – microvilli, D – desmosomes (No. 5343, magnification 20000 x).

processes form a so-called primordial glial sheath surrounding the wall of small vessels. The first time of the formation of vascular basement membranes is visible in the late embryonic period (Fig. 5 b). In both levels of the *blood-brain barrier* structural changes are observed and may be related to the maturation phenomenon of the blood-brain barrier function (WECHSLER, 1965b).

In the adult brain the superficial layer of the cortex is composed of fibrillary astrocytes. Chick embryos with a fully developed marginal layer of the brain have protoplasmic glioblasts and astroblasts, forming a primitive membrane, the *gliae limitans superficialis primitiva*. The external surface is covered by a basement membrane (WECHSLER, 1966a). In sharp contrast the internal or *ventricular surface* is composed of a primitive epithelium and the differentiation of its ependymoblasts occurs in the early phases of the brain development (MELLER a. WECHSLER, 1964). Fig. 6 represents the ventricular surface of a 13-days-old chick embryo. Although the matrix layer is reduced, the differentiated parts of the apex of the ependymal cells are obvious. The basal processes of the ependymoblasts are poorly developed at this stage, but later the fine structures of the basal processes are similar to the processes of protoplasmic astrocytes.

The differentiation of the neuroglial cells reflects a homomorphous (uniform) structural cytoplasmic differentiation, e.g. oligodendrocytes and astrocytes, as well as a *bimorphous (bipolar) structural cytoplasmic differentiation,* e.g. ependyma and plexus epithelium (MELLER a. WECHSLER, 1964, 1965; WECHSLER, 1966a).

Summary and Conclusions

Based on the relationship between fine structure and cytogenesis, we propose the following *types of cytodifferentiation:* (1) Cells with a homomorphous cytoplasmic structural differentiation, e.g. astrocytes and oligodendrocytes; (2) cells with a bimorphous cytoplasmic structural differentiation, e.g. ependyma and plexus epithelium, and (3) cells with a polymorphous cytoplasmic structural differentiation, e.g. multipolar neurons.

Evidence is presented, that there is a primordial polarization of the surfaces of the brain vesicles. In all regions of the developing brain (e.g. the matrix layers, the migration zones, the primary cortex and the marginal layer), the extracellular space is minimal and is a three-dimensional continuous space system. The distances between adjacent cell membranes measure approximately 200 Å. Compared with the adult brain, the volume of the extracellular space of the embryonic brain is minor because of larger but fewer processes. This is important for the development of the neuropil. The increase in extracellular volume during development leads to a more complex labyrinth of the extracellular space. With the differentiation of nerve cells their surfaces increase and synaptic-interrelationships between nerve cells become more abundant.

The distribution pattern of mitochondria and the special organization of ribosomes and endoplasmic reticulum, are discussed with respect to various aspects of cytochemistry in the developing nervous system. The role of electron microscopy in the investigation of functional and experimental neuroembryology is stressed.

Acknowledgements

I thank Dr. K. MELLER (Göttingen), who has participated in our recent studies of the chick brain. More detailed observations will be published in another paper. I also thank PHILIP MACON, M.D., for his help in the preparation of this manuscript.

References

BAILEY, P., H.CUSHING: Die Gewebs-Verschiedenheit der Hirngliome und ihre Bedeutung für die Prognose. Verlag G.Fischer, Jena 1930.

CAJAL, S.R.: Études sur la neurogenèse de quelques vertébrés. Madrid 1929.

CAJAL, S.R.: Histologie du système nerveux de l'homme et des vertébrés. Instituto Ramon y Cajal, Madrid, 1952.

FUJITA, H., S.FUJITA: Electron microscopic studies on neuroblast differentiation in the central nervous system of domestic fowl. Z. Zellforsch. *60* (1963), 463–478.

FUJITA, H., S.FUJITA: Electron microsopic studies on the differentiation of the ependymal cells and the glioblast in the spinal cord of domestic fowl. Z. Zellforsch. *64* (1964), 262–272.

FUJITA, S.: The matrix cell and histogenesis of the central nervous system. Laval Méd. *36* (1965), 125–130.

GLEES, P., S. LE VAY: Some electron microscopical observations on the ependymal cells of the chick embryo spinal cord. J. Hirnforsch. *6* (1964), 355–360.

GLEES, P., K.MELLER: The finer structure of synapses and neurones. Paraplegia *2* (1964), 77–95.

HIS, W.: Die Neuroblasten und deren Entstehung im embryonalen Rückenmarke. Abh. math.-physik. Klasse d. König. Sächs. Ges. Wis. XV (1889), 331–372.

HIS, W.: Histogenese und Zusammenhang der Nervenelemente. Arch. Anat. Physiol., Anat. Abt. Suppl. (1890), 95–117.

KAHLE, W.: Studien über die Matrixphasen und die örtlichen Reifungsunterschiede im embryonalen menschlichen Gehirn. Dtsch. Z. Nervenheilk. *166* (1951), 273–302.

MELLER, K., W.WECHSLER: Elektronenmikroskopische Befunde am Ependym des sich entwickelnden Gehirns von Hühnerembryonen. Acta neuropath. *3* (1964), 309–626.

MELLER, K., W.WECHSLER: Elektronenmikroskopische Untersuchung der Entwicklung der telencephalen Plexus chorioides des Huhnes. Z. Zellforsch. *65* (1965), 420–444.

PAPPAS, G.D., D.P.PURPURA: Electron Microscopy of Immature Human and Feline Neocortex. In: Progr. in Brain Res. *4*, 176–186. Elsevier Publ. Comp., Amsterdam 1964.

PORTER, K.: The ground substance; observation from electron microscopy. In: Brachet, J., A.Mirsky (Eds.), The Cell, Vol. II., 621–676, Acad. Press, New York 1961.

RANSON, ST.W.: The Anatomy of the Nervous System. W.B.Saunders Comp., Philadelphia 1941.

ROMANOFF, A.L.: The Avian Embryo – Structural and Functional Development. The Macmillan Comp., New York 1960.

VOELLER, K., G.D.PAPPAS, D.P.PURPURA: Electron microscopic study of development of cat superficial neocortex. Exp. Neurol. *7* (1963), 107–130.

WECHSLER, W.: Elektronenmikroskopischer Beitrag zur Entwicklung und Differenzierung von Zellen am Beispiel des Nervensystems. Verh. Dtsch. Ges. Path. *47. Tag.,* (1963) 316 bis 322.

WECHSLER, W.: Zur Feinstruktur des peripheren Randschleiers des sich entwickelnden Rückenmarks von Hühnerembryonen. Naturwissenschaften *51* (1964a), 113–114.

WECHSLER, W.: Zur Feinstruktur der Vorderhornregion des sich entwickelnden Rückenmarks von Hühnerembryonen. Naturwissenschaften *51* (1964b), 114.

WECHSLER, W.: Elektronenmikroskopische Untersuchung der Entwicklung von Nervenzellen und Nervenfasern bei Hühnerembryonen. Verh. Anat. Ges. Anat. Anz. Erg. Heft *115* (1965a), 287–302.

WECHSLER, W.: Die Entwicklung der Gefäße und perivasculären Gewebsräume im Zentralnervensystem von Hühnern – Elektronenmikroskopischer Beitrag zur Kenntnis der morphologischen Grundlagen der Bluthirnschranke während der Ontogenese. Z. Anat. Entwickl.-Gesch. *124* (1965b), 367–395.

WECHSLER, W.: Zur Entwicklung der Liquorräume des Gehirns von Gallus domesticus. (Elektronenmikroskopische Untersuchungen) Wiener Zsch.f.Nervenheilk., Suppl.*1* (1966a), 49–69.

WECHSLER, W.: Die Feinstruktur des Neuralrohres und der neuroektodermalen Matrixzellen am Zentralnervensystem von Hühnerembryonen. Z. Zellforsch. *70* (1966b), 240–268.

WECHSLER, W.: Elektronenmikroskopischer Beitrag zur Differenzierung des Ependyms am Rückenmark von Hühnerembryonen. Z. f. Zellforsch. 1966c (in press).

WECHSLER, W.: Elektronenmikroskopischer Beitrag zur Histogenese der weißen Substanz des Rückenmarks von Hühnerembryonen. Z. f. Zellforsch. 1966d (in press).

WECHSLER, W.: Elektronenmikroskopischer Beitrag zur Nervenzelldifferenzierung und Histogenese der grauen Substanz des Rükkenmarks von Hühnerembryonen. Z. f. Zellforsch. 1966e (in press).

WECHSLER, W., K. MELLER: Zur Feinstruktur des primären Cortex von Hühnerembryonen. Naturwissenschaften *50* (1963a), 694–695.

WECHSLER, W., K. MELLER: Zur Feinstruktur der Schwärmzone des Telencephalons von Hühnerembryonen. Naturwissenschaften *50* (1963b), 714–715.

WOHLFARTH-BOTTERMANN, K. E.: Die Kontrastierung tierischer Zellen und Gewebe im Rahmen ihrer elektronenmikroskopischen Untersuchung an ultradünnen Schnitten. Naturwissenschaften *44* (1957), 287–288.

Dr. W. WECHSLER
Abteilung für Allgemeine Neurologie
Max-Planck-Institut für Hirnforschung
Ostmerheimer Straße 200
Köln-Merheim, Deutschland

Structure of Synapses in Evolutionary Aspect
(Optic and Electron-Microscopic Studies)

By S. A. SARKISOV, E. N. POPOVA, and N. N. BOGOLEPOV, Moscow

The problems of the synaptic function mechanism are acquiring now an ever increasing importance.

A voluminous material concerning the structure of synaptic contacts has been accumulated in literature as a result of optic-microscopy studies (CAJAL, 1955; SARKISOV, 1948; POLYAKOV, 1955; SHKOLNIK-YARROS, 1961; ZURABASHVILI, 1958; GRASHCHENKOV, 1948, and others).

The results of electron-microscopy studies have added much new to the submicroscopic morphology of synapses (PALADE, 1954; PALAY, 1956; DE ROBERTIS, 1964; GRAY, 1959). Nevertheless, there is still quite a number of unexplored problems in synaptology, a part of which can be solved by way of evolutionary approach to studying interneuronal contacts. This method was successfully used at our Institute in the course of studying various brain structures, including interneuronal connections. Particularly, as far back as in 1948 SARKISOV has shown, that the transmission of impulses in the lower parts of the CNS is effected mainly via axo-somatic contacts, whereas in higher phylogenetically newer parts of the CNS axo-dendritic transmission is prevailing. There are three types of axo-dendritic interneuronal contacts in the cortex: synapses on large dendrite trunk, synapses on dendrite spines and synapses on terminal branchings. In evolution of particular development achieve the connections on the spines.

The purpose of this communication is to study changes in the post-natal ontogenesis of interneuronal connections, whose formation is closely associated with the cyto-architectonic differentiation and development of the neuronal structure of the cerebral cortex.

Material and Methods. Under study there was the sensomotor cortex of white rats of the following ages: newborn, 2, 7, 14, and 21-day old and adult. The studies were carried out with the use of optic-microscope (by NISSL'S, GOLGI'S and GOLGI-DEINEKA'S methods) and electron-microscope.

Results

The results of the studies have shown, that in newborn rats the synaptic contacts are found very seldom, and the submicroscopic structure of synapses is rather imperfect.

The poor development of interneuronal contacts is in correspondence with the indistinct cyto-architectonic and cytological differentiation of the cortex.

The cortex of newborn rats can cyto-architectonically be subdivided into stratum I, complex of strata II–VI, getting loose in the direction to the depth, and stratum VII (Fig. 1 A). The cells of the superficial strata are represented mainly with "bare" nuclei, but the deeper the stratum, the more frequent are cells with well-developed cytoplasm

at the start of apical dendrites, and among the cells lying still deeper there are some with
a narrow border of cytoplasm around the nucleus. The nuclei of the nerve cells of the
superficial strata are oval and elongated, while in deeper strata they are rounded. The
width of the cortex is 0.59 to 0.63 mm. Studies of the neuronal structure of the sensomotor cortex in newborn rats have revealed great number of bipolar neurons, especially
in the superficial strata. The pyramid neurons are little differentiated: they have apical
dendrites and rudiments of basal dendrites. The ramification of apical dendrites is poorly
developed and the dendrites proper have irregular outlines and varicose swellings (Fig. 1 B).

The spines on the apical dendrites are very scarce. The descending axons have almost
no collaterals.

Electron-microphotographs are in full agreement with the findings obtained by optic
microscopy: they show, that in newborn rats the greatest part of a section is occupied
with nervous cells, while in adult rats the main part of the section shows interlaced
branches of nerve and glial cells, among which the bodies of nerve cells can be found
but only in a very few cases. In newborn rats very frequent are contacts between two or

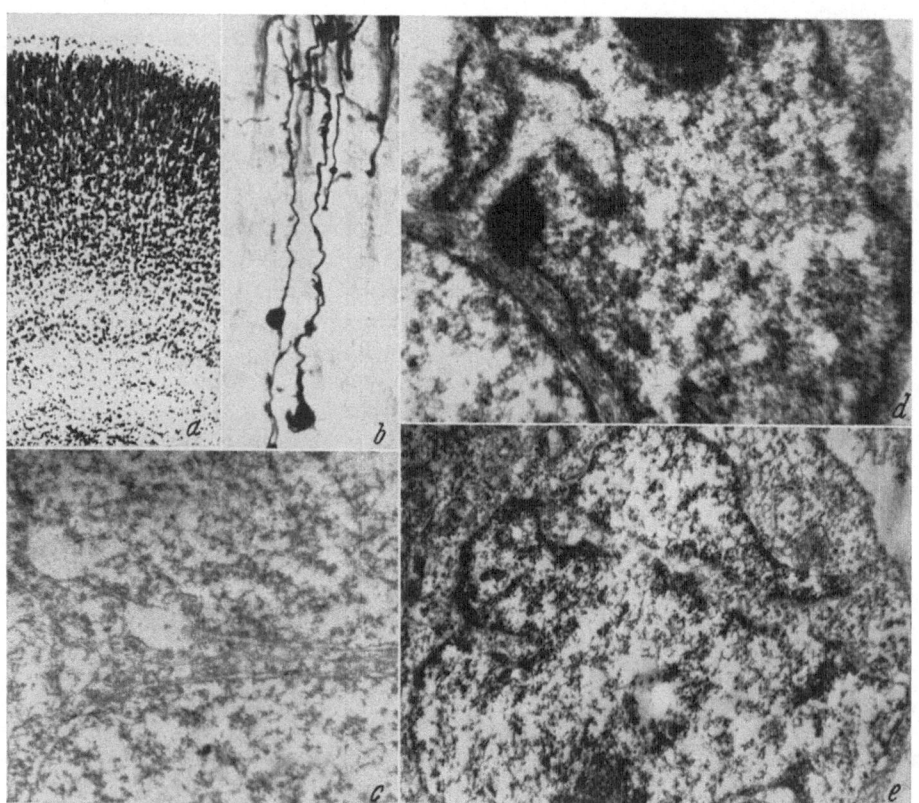

Fig. 1. The sensomotor cortex of the newborn rat.
a) Cytoarchitectonic structure, Nissl stain, 62 x
b) The neuronal structure, Golgi stain, 216 x
c, d, e) some submicroscopical structures of the neurons (close contact, invagination and so on)
20.000 x and 10.000 x (red. to 12.000 x and 6.000 x)

more nerve cells (Fig. 1 C), while in adult rats such contacts are practically never observed. The nerve cells of newborn rats are separated by cellular membranes, which are much less distinct than in adult rats (after fixation in osmium tetroxide). Some cellular membranes are pronouncingly twisted (Fig. 1 D, E). The branches of the nerve and glial cells, which are extremely scarce at this age, are less distinct due to a weaker osmiophilia of cellular membranes.

The nerve cells in newborn rats have large, light nuclei, which contain not one (as in adult rats) but several (3, 4 or more) osmiophilic accumulations of various size. Besides that, if an accumulation of the osmiophilic substance corresponding to the nucleolus is located in adult rats mainly in the centre of the nucleus or slightly eccentrically, the bulk of the osmiophilic material of the nucleus in newborn rats is, as a rule, located near the nuclear membrane.

A characteristic feature of many nuclei in newborn rats is an irregular shape of the nucleus, with protrusions and depressions. In some cases two specific types of increase of the nucleus surface area may be observed. The first type of the increase is due to the development of oval protrusions of the nucleus, which are sometimes rather long and may have wide bases and more or less extended necks. The other type of the nucleus surface-area increase manifests itself in formation of blind channels in the nucleus. These channels may be straight or curved; less frequently they are branched.

As it has already been mentioned, the greatest part of a nerve cell in a newborn rat is occupied with the nucleus surrounded by a narrow border of cytoplasm. The cytoplasm of a newborn rat contains the same principal submicroscopic components as the cytoplasm of an adult rat, i.e. granular and agranular reticulum and mitochondria. The structure of each component, however, has its own peculiarities. The tubules of the granular reticulum are very poorly developed and have an appearance of individual elements not combined into a united system. The granules are often observed to be in indirect contact with the granular reticulum tubules. They often form groups consisting of several granules, and the total number of the granules is insignificant.

The *mitochondria* are rounded or oval, their structure is characterized with poor development of cristae, often observed as small protrusions of the inner membrane. The number of cristae in mitochondria of newborn rats is several times as less as in mitochondria of adult rats.

In 2-day-old rats the number of synapses is also small and their structure is simple.

The sensomotor cortex already shows all cyto-architectonic strata observed in adult animals but in stratum II + III the cells are still in a close contact with each other. Each cell has a narrow border of cytoplasm, which turns to be a distinctly visible apical branch at the top. The nuclei are mainly rounded or oval, and their stratum-by-stratum difference is very small. Some large cells of stratum V clearly show the initial parts of basal dendrites. The width of the cortex increases but insignificantly, to 0.62–0.66 mm.

The general structure of the neurons in 2-day-old rats resembles that of newborns, but the basal dendrites of many neurons are better developed: they are longer than those in newborns and are sometimes covered with rare spines. Short lateral branches can be seen on the apical dendrites. The axons have almost no collaterals. Frequent are cases, when an axon goes in parallel to an apical dendrite and touches it at one or several points. Bipolar neurons are still found rather often.

In a 7-day-old rat the synapses are observed more frequently than in younger animals, and the submicroscopic structure of the synaptic apparatus becomes more perfect. It

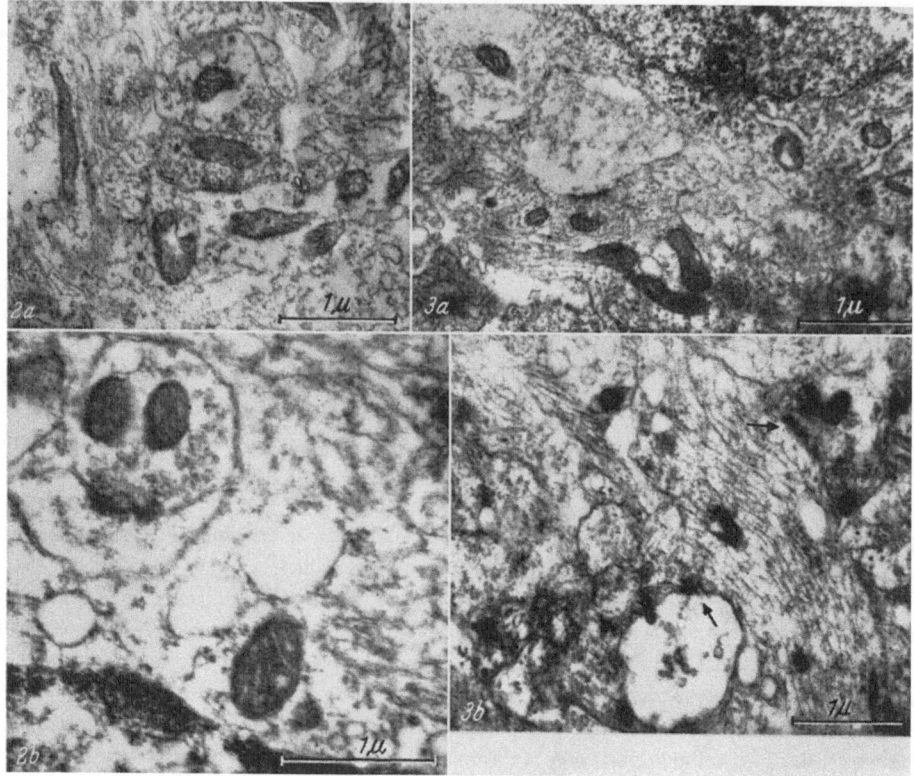

Figs. 2, 3. Submicroscopic structure of synapses in 7 days old rat

begins to approach the synaptic structure of adult animals. The synapses of 7-day-old rats show synaptic vesicles, an increased electronic density, "thickened" pre- and post-synaptic membranes and sub-synaptic network. However, the synapses in 7-day-old rats are still small, their "thickened" parts are very short in most cases, and the intersynaptic fibers and thickened parts of synaptic membranes are less distinct than in adult rats (Figs. 2, 3).

The greatest part of synapses is located on dendrite trunks. As a rule, these are synapses of the 2nd type after GRAY. Synapses on the spines are very scarce, since at this age the spine apparatus is developed poorly.

On the 7th day of the post-natal development the neuronal structure and the cyto-architectonic picture of the cortex become much more complicated (Figs. 4 A, B). The number and size of dendrites increase, and so does the number of spines; the ramification of dendrites gets more complicated. As in adult animals the apical dendrites of stratum V neurons already show three zones differing from each other in density and shape of the spines. In the first zone adjoining immediately the neuron body the spines are rare, wide, short and have a shape of protrusions and bulges (Fig. 4 C). In the second zone the number of spines increases and their shape changes: they become narrower and longer (Fig. 4 D).

Fig. 4. The sensomotor cortex of 7 days old rat.
a) Cytoarchitectonic structure, Nissl stain, 45 x
b) the neuronal structure, Golgi stain, 155 x
c, d, e), the structure of apical dendrite of the neuron from layer Y under high magnification 670 x
c-zone I; d-zone II; e-zone III

On the terminal branches of the apical dendrites the spines are rare, long and narrow (Fig. 4 E). The axons have collaterals. Contacts of axon branches with spines are observed (Fig. 4 D).

The width of the sensomotor cortex in a 7-day-old rat reaches 1.01 to 1.11 mm due to the growth of all cortical cells. The cells show further cytological maturation. But alongside with cells having a more or less developed layer of cytoplasm around the nucleus and clearly visible clumps of Nissl-substance, there are some cells without noticeable cytoplasm and with an intensely stained nucleus. The size of the cells is more variable than that in 2-day-old rats, especially in the deep part of stratum III and in stratum V. The shape of the nuclei is mainly rounded or less frequently oval.

Electron-microscopic findings still reveal the irregularity in the shape of neuronal nuclei in 7-day-old rats, but this irregularity is less pronounced than in newborn or 2-day-old rats.

The lumen of the nuclear membrane is not uniform: the dilatation of the membranes is especially marked in the places of nucleus protrusion. In some cases light rounded vacuoles can be seen in the places of nucleus protrusion or depression: these vacuoles are most often associated with the dilatation of the membrane.

At this age the granular reticulum becomes more distinct and shows an increase in the amount of granular material and in the number of tubules. Another characteristic feature is an improvement of orientation and an increase in the density of granular reticulum tubules arrangement.

The differences in the structure of axons and dendrites become more pronounced: now it is possible to see protoneurofibrils in the axons and tubules in the dendrites.

Two weeks and older rats. The number of synapses considerably increases by the 14th day of the post-natal ontogenesis. By this time the submicroscopic structure of the synapses becomes similar to that of adult animals, while their mutual arrangement becomes much more complicated. If it was mainly the formation of the submicroscopic structure of synapses (up to a stage beginning to resemble the synapses of adult animals), that was observed during the first week of the post-natal development, the 2nd and the 3d weeks are characterized by complication of synapto-architectonics. In parallel to the development of axons and dendrites the number of synaptic contacts and the area of their "active" part increase too. The proportion of various forms of synapses also changes. If the greatest part of the interneuronal synapses in a 7-day-old animal is located on the branches of middle-sized dendrites, synapses in a 14-day-old rat are often found on

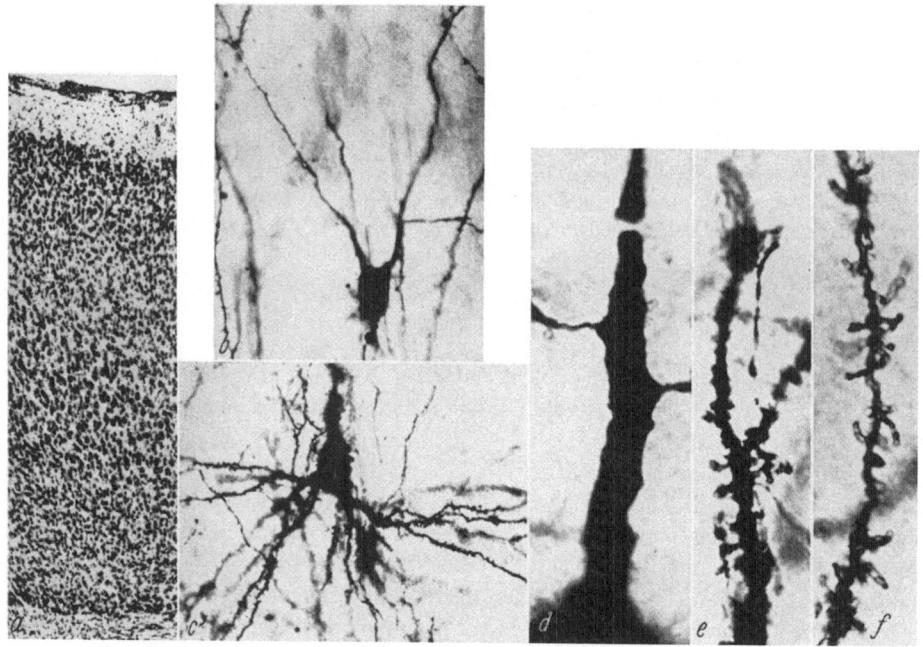

Fig. 5. The sensomotor cortex of 14 days old rat.
a) cytoarchitectonic structure, NISSL stain, 46 x
b) the neuron from layer III with well-developed apical dendrite, GOLGI stain, 160 x
c) basal dendrites of the neuron from layer V, GOLGI stain, 160 x
d-f) the structure of apical dendrite trunk of the neuron from layer V under high magnification.
690 x

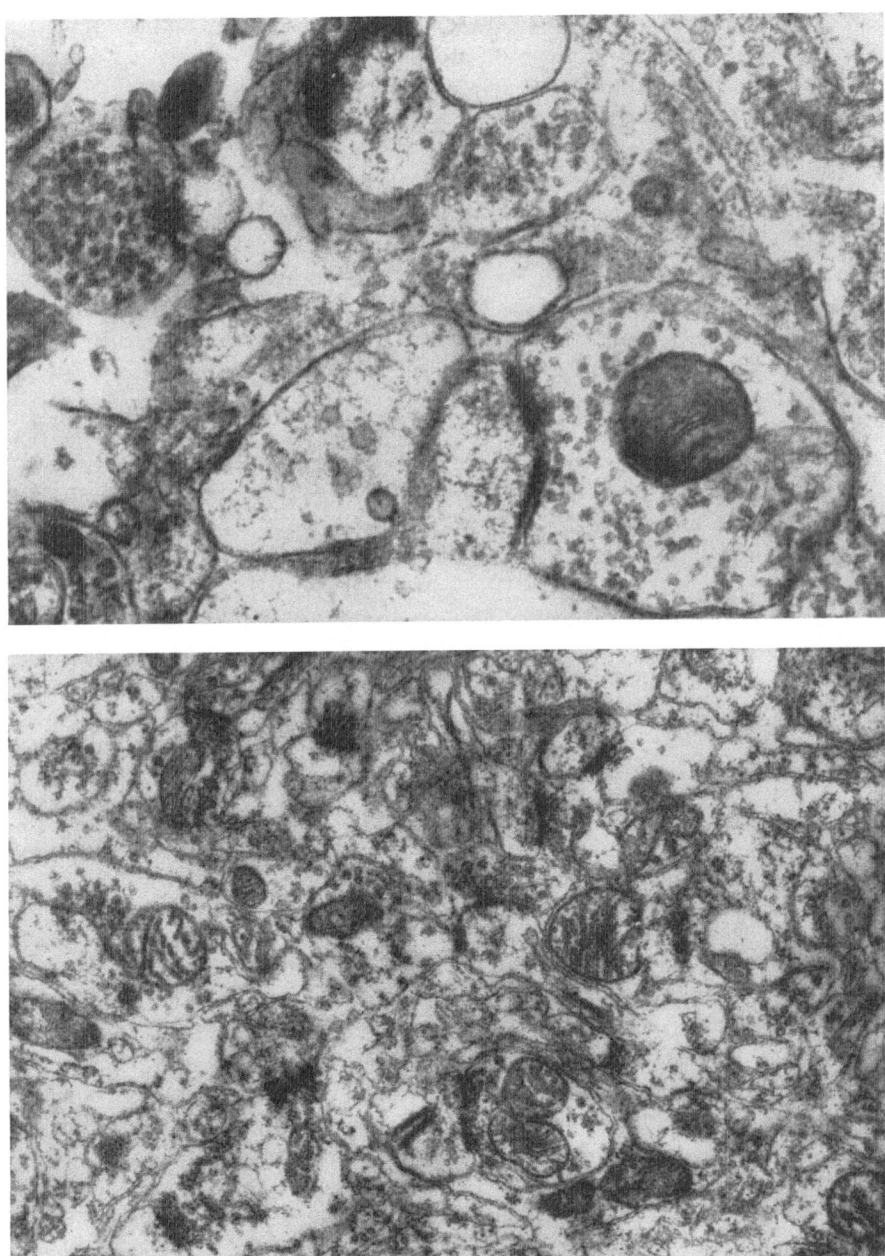

Fig. 6. Submicroscopical structure of synapses of adult rat, 30.000 x (red. to 22.000 x)

spines, whose apparatus already becomes quite distinct. At the age of 14 days the axo-somatic synapses and synapses on large branches of dendrites (structurally similar to the axo-somatic ones) are more frequently found. At the same time the mutual arrangement of the synapses becomes more complicated, and there appear synapses on invaginated branchings and axo-axonal synapses.

The neuronal structure of the cerebral cortex in 14-day-old rats resembles that of adult animals (Figs. 5 B, C). The basal dendrites of most neurons have primary, secondary and tertiary branchings abundantly covered with spines, though in stratum II–III some forms with underdeveloped basal dendrites, but with well-developed apical dendrites can still be found. The apical dendrites of most neurons also show three zones with various density of spine arrangement, but the number of branches and spines on the apical dendrites sharply increases. The spines proper become more variable in shape and size. Many spines have clearly visible heads and stems. One can often observe tangent contacts between branches of axons and spines following each other in line, and in a few cases terminal contacts on one spine can be also found (Fig. 5 D, E, J). The axo-somatic synapses are quite distinct.

Cyto-architectonic findings show further expansion of the cortex up to a width of 1.37 to 1.44 mm, mainly due to the growth of the upper strata (Fig. 5 A). The density of the cell arrangement is diminished, the amount of cytoplasm and the variability of the cells grow up, especially in large cells of stratum V. In all the cells NISSL-substance is clearly visible.

Nevertheless, in the course of further development the synapto-architectonic picture of the cortex gets more and more complicated (Fig. 6). This coincides with the expansion of the cortex up to 1.55–1.59 mm in a 21-day-old rat and up to 1.57–1.62 mm in an adult rat, as well as with the increase of the cell size and of the amount of NISSL-substance.

Discussion

The results of our studies show, that the course of post-natal ontogenesis from neonatal to adult age is characterized by a considerable increase in the number of interneuronal connections, as well as by maturation of the submicroscopic structure of synaptic apparatus and complication of synapto-architectonic picture in general (scheme, Fig. 7).

Especially marked changes in the *number of synapses* (in the direction to its increase) as well as further differentiation of the submicroscopic structures of synapses take place during the second week of the post-natal development. On the age of 2 weeks the structure of neurons and interneuronal connections shows basically all the features characteristic for an adult animal. Further complication of the character of interneuronal connections is observed at subsequent stages, as well.

EAYRS a. GOODHEAD (1959) also point out, that in rats under 6 days of age the cortical neurons have very few dendrites. The greatest part of the dendrites acquires the adult structure by the age of 12 days. Subsequently, the number of dendrites and their ramification increase and by the 18th day the cortical structure has all the features characteristic for adult animals.

The *structure of the cerebral cortex* of rats, just like of the cortex of mice (POPOVA, 1959), cats (ENTIN, 1956; PURPURA et al., 1964), as well as of other animals and man (POLYAKOV,

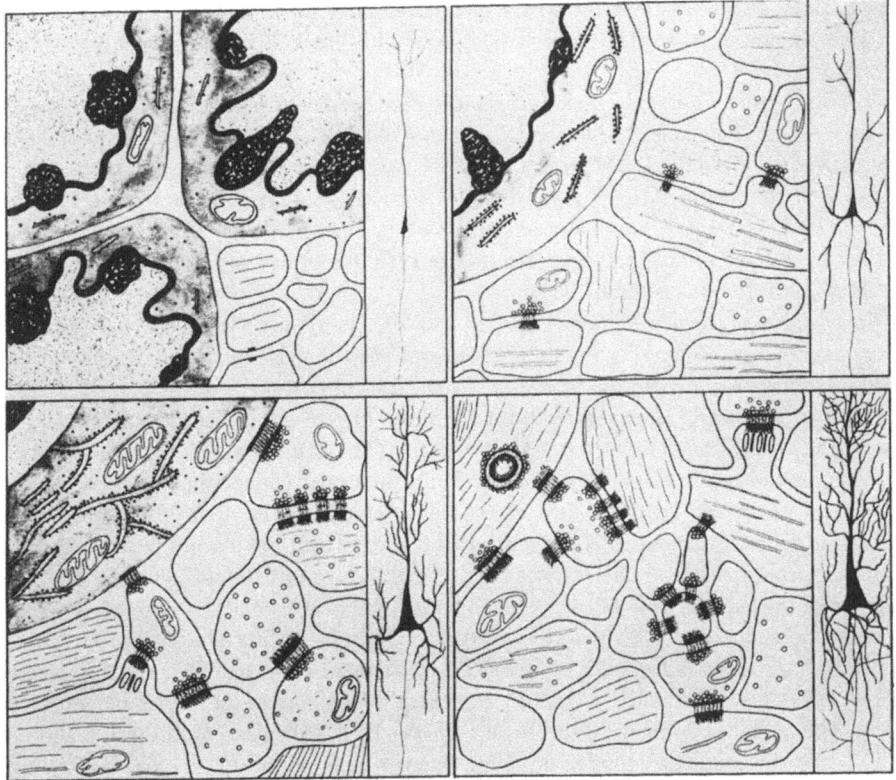

Fig. 7. Scheme illustrating the development of synapses in ontogenesis.

A) Newborn rat, desmosome-like contact, which has no normal structure of synapses

B) 7 days old rat, small, seldom synapses with the short distance of the patch of the synaptic membranes

C) 14 days old rat, the synapses become similar to that of adult rat.

D) Adult rat, the arrangement of the synapses become more complicated.

1946; SARKISOV, 1964) first shows differentiation in apical and later in basal dendrites. This is, probably, an indication to the different phylogenetic age of these structures and to their different functional significance. This is also confirmed by our data on the effects of various neuropharmacological agents, which first of all cause changes in the structure of basal dendrites (POPOVA, 1964; SARKISOV, 1964; POPOVA, BOGOLEPOV, 1965).

Comparing the data obtained by us with the data on functional and biochemical *maturation* of rat cortex published in literature, it is possible to come to a conclusion, that the periods of maturation of interneuronal connections and neurons as a whole to a certain extent coincide with the formation of the functional activity and biochemical maturation of cortical structures.

The immaturity of the functional activity of the cortex corresponds to the period of poor development of interneuronal connections at early stages of post-natal ontogenesis.

CRAIN (1952, 1953) has shown, that the electric activity in newborn rats is irregular and variable. Spontaneous electric activity and strychnine "spikes" appear at the age of 3 to 4 days.

According to our data the period of intense *development of interneuronal connections* at the age of 7 to 14 days is characterized by a marked tendency to an increase of the regularity, rhythm, stability and duration of the electric activity. The direct cortical response develops at the age of 5 to 15 days, and the mature EEG-type characteristic for adult animals is observed in 8- to 14-day-old rats (CRAIN, SCHADÉ, 1959). The electric response of the cortex to sound is first registered on the 14th day of life and becomes regular from the 17th–19th day on (BURES, 1953).

Morphologic-physiological studies carried out by IVANITSKY (1959) have shown, that the appearance of the closing function in ontogenesis coincides in time with the development of spines on the terminal branches of dendrites of cortical neurons.

Physiological studies have revealed, that in the course of ontogenesis the forms and the character of conditioned reflectory activity change (VOLOKHOV, NIKITINA, 1963); and this depends on the maturation of neurons and interneuronal connections in corresponding systems of analysers. In the course of the development of the orientation and defence reflexes to acoustic stimuli in rats aged but a few days and having immature cortex the vegetative components of the response considerably forestall the motor ones. The older the animal the less the difference in the moments of appearance of the vegetative and the motor components of the response (VOLOKHOV, NIKITINA, NOVIKOVA, 1959).

The most detailed *ontogenetic studies of the conditioned reflex activity* in rats have been carried out by PANCHENKOVA (1956). According to her data, the development of the motor-defence reflex becomes possible only on the 10th–15th days of life. From the 17th to 19th day the conditioned reflex in young rats is formed, as a rule, by the same number of combinations as in adult animals. The older the animal the more often the conditioned shaking-off reflex manifests itself as a general motor responce, and less often as off-shaking.

The poor development of interneuronal connections in rats in early ontogenesis coincides with the underdevelopment of excitation and inhibition processes. These processes become the more manifest, the older is the animal (PANCHENKOVA, 1956; SCHLYFER, 1958).

Biochemical studies have shown, that maturation of the enzymatic systems in rats' brain goes on also during the first weeks of their life (POTTER et al., 1945; NACHMANSOHN, 1940; ELKES, TODRICK, 1955; FRIEDE, 1959; GRAVES, HIMWICH, 1959). The levels of biogenic amines and monoamino-oxydase activity in newborn rats are very low, and reach those characteristic for an adult animal at the age of 3 weeks (KARKI, KUNTZMAN, BRODIE, 1962).

The *evolutionary studies* of interneuronal connections enrich our knowledge of the fine submicroscopic structure of synaptic systems. This is very important for correct understanding of the finest mechanisms of the cerebral activity under normal and pathological conditions.

As a whole, the data on the ontogenetic changes of submicroscopic components of nerve cells and on the peculiarities of their structure at various ages supplement and develop the data on maturation of various components of the CNS obtained at Moscow Institute of the Brain before (SARKISOV, POLYAKOV, PREOBRAZHENSKAYA, 1961; KONONOVA, 1961; KUKUYEV, 1961, and others).

Summary

Changes in the neurons and interneuronal connections in sensomotor cortex of white rat during the postnatal ontogenesis were studied. In this investigation the methods of light- and electronmicroscopy were used.

It was shown that during the postnatal ontogenesis from newborn to adult rat the number of the interneuronal connections increased, the maturation of the submicroscopical structure of synapses occurred and the synapto-architecture became more complex. Changes in number of synapses and the differentiation of the submicroscopical synaptic structure were more pronounced during the second week of postnatal life.

The complication of the interneuronal connections is observed in subsequent development.

The evolution of the interneuronal connections is correlated with the evolution of other nerve-cell components and with the development of motor cortex function.

In comparing the ontogenesis of synapses and phylogenesis of synapses the basic tendency towards increase in the number of synapses should be noted, particularyly regarding synapses on the dendrite spines. This is conditioned by some physiological and biochemical features between the organism and the surroundings.

References

BURES, I.: Czechoslov. physiol. *2* (1953), 274.

CAJAL, RAMON S.: Histologie du système nerveux de l'homme et des vertébrés. Madrid 1955.

CRAIN, S.: Proc. Soc. exper. Biol. Med. *81* (1952), 49.

CRAIN, S.: Biol. Abstr. *27* (1953), 1409.

EAYRS, J., B. GOODHEAD: J. Anat. (Lond.) *93* (1959), 385.

ELKES, J., A. TODRICK: Biochem. Development of Nervous System. New York 1955.

ENTIN, T. I.: Problems of Morphology of Nervous System. Leningrad 1956.

FRIEDE, R.: J. Neurochem. *4* (1959), 101.

GRAVES, J., W. HIMVICH: Physiologist *2* (1959), 48.

GRAY, E.: J. Anat. (Lond.) *93* (1959), 420.

GRASHCHENKOV, N. I.: Interneuronal Apparatus of Connection-Synapses and Their Role in Physiology and Pathology. Minsk 1948.

IVANITSKY, A. M.: Bull. exper. Biol. Med. *46* (1958), 27, 118; 87.

KONONOVA, E. P.: Structure and Function of Human's Analyzors in Ontogenesis. Moscow 1961.

KARKI, N., R. KUNTZMAN, B. BRODIE: J. Neurochem. *9* (1962), 53.

KUKUYEV, L. A.: Structure and Function of Human's Analyzors in ontogenesis. Moscow 1961.

NACHMANSOHN, P.: Yale J. biol. Med. *12* (1940), 565.

PALADE, G.: Anat. Rec. *118* (1954), 335.

PALAY, S.: J. Biophys. Biochem. *1* (1956), 68.

PANCHENKOVA, E. PH.: J. high. Nerv. System *6* (1956), 312.

POLYAKOV, G. I.: Cytoarchitectonic of the Human's Cortex. Moscow 1949.

POLYAKOV, G. I.: Arch. Anat. Histol. *32* (1955), 15.

POPOVA, E. N.: In: Psychopharmacology and Treating Nervous and Mental Disorders. Leningrad 1964.

POPOVA, E. N., N. N. BOGOLEPOV: Bull. exper. Biol. Med. *5* (1965), 104.

POTTER, V., W. SCHNEIDER, G. LIEBE: Cancer Res. *5* (1945), 21.

PREOBRAZHENSKAYA, N. S.: Structure and Function of the Human's Analyzors. Moscow 1961.

PURPURA, D., R. SHOFER, E. HOUSEPIAN, C. NOBACK: Progress in Brain Research, Vol. 4: Growth and Maturation of the Brain. Amsterdam–London–New York 1964.

ROBERTIS DE, E.: Histophysiology of Synapses. Oxford–London 1964.

SARKISOV, S. A.: Some Peculiarities of Structure of Neuronal Connections of Brain Cortex. Moscow 1948.

SARKISOV, S. A.: Assays on Structure and Function of Brain. Moscow 1964.

SCHADÉ, I.: J. Neurophysiol. *22* (1959), 245.

SCHLYFER, G. P.: Symposium: Questions of Physiology and Pathology of Nervous System of Animals and Man at Earlier Stages of Ontogenesis. Moscow 1958.

SHKOLNIK-YARROS, E. G.: Neurons of Visual Analyzor. Cortex and Corpus geniculate laterale, Neurons and Interneuronal Connections. Diss., Moscow 1962.

VOLOKHOV, A. A., G. M. NIKITINA: Problems of Physiology and Pathology of the Highest Nervous Activity. Leningrad 1963.

VOLOKHOV, A. A., G. M. NIKITINA, E. G. NOVIKOVA: J. high. Nerv. Activity *9* (1959), 420.

ZURABASHVILI, A. D.: Synapses and Reversible Changes of Nerve Cells. Tbilisi 1958.

Prof. Dr. S. A. SARKISOV
Dr. E. N. POPOVA
Dr. N. N. BOGOLEPOV
Institute of Brain Research
USSR Academy of Medical Sciences
Per. obucha 5
Moscow B-120, USSR

The Primate Globus Pallidus and its Feline and Avian Homologues: A Golgi and Electron Microscopic Study*

By C. A. Fox, D. E. Hillman, K. A. Siegesmund, and L. A. Sether, Detroit, Mich. and Milwaukee, Wis.

The following description of the monkey globus pallidus is based on Golgi impregnations of the macaque monkey *(Macaca mulatta)* and on electron micrographs of the squirrel *(Saimiri sciureus)* and macaque monkeys. It will be seen that the Golgi preparations provide an invaluable guide. Without them interpretation of the electron micrographs would have been difficult, if not impossible.

The description of the cat entopeduncular nucleus and the chicken paleostriatum is based only on electron micrographs. Golgi impregnations were not available. It was assumed that once the pallidal pattern had been established, electron microscopy would be sufficient to determine whether or not a similar pattern exists in homologous regions suggested by comparative studies. From the results obtained, it is now our opinion that electron microscopy can play an important role in comparative neurology and used properly it may resolve some of the controversial issues in this field. Certainly one of its advantages over the Golgi method is: results can always be expected from material prepared for electron microscopy. This is not always true of material prepared by the Golgi method. Its capriciousness is well known; Polyak (1941) likened it to gambling. Perhaps herein lies its fascination – the hope of unexpected good fortune.

Despite its difficulties, the Golgi technique, little used in recent years, is now receiving more and more attention, for investigators, thanks to the development of micro-electrode recording and electron microscopy, are once again interested in the whole neuron with all its processes.

In this account all figures are explained in the text.

The Monkey Globus Pallidus

The low-power electron micrograph. It is not surprising that varying sized myelinated fascicles with varying caliber axons are frequently found in globus pallidus sections prepared for electron microscopy, for the pale cast of the globus pallidus in the fresh condition – hence, its name – is due to its rich myelin content. But our interest is in the gray and not the white matter. The electron micrograph (Fig. 1; *Saimiri sciureus*) shows a band of this interfascicular gray. A pair of arrows, parallel to each other at the left, and a pair of arrows, parallel to each other at the right of Fig. 1, mark the extremities of an array of contiguous synaptic endings, each respectively related to a dendrite. The block arrows point to mosaics of synaptic endings covering dendritic surfaces.

*) Supported by N. I. H. grant NB 05315–02

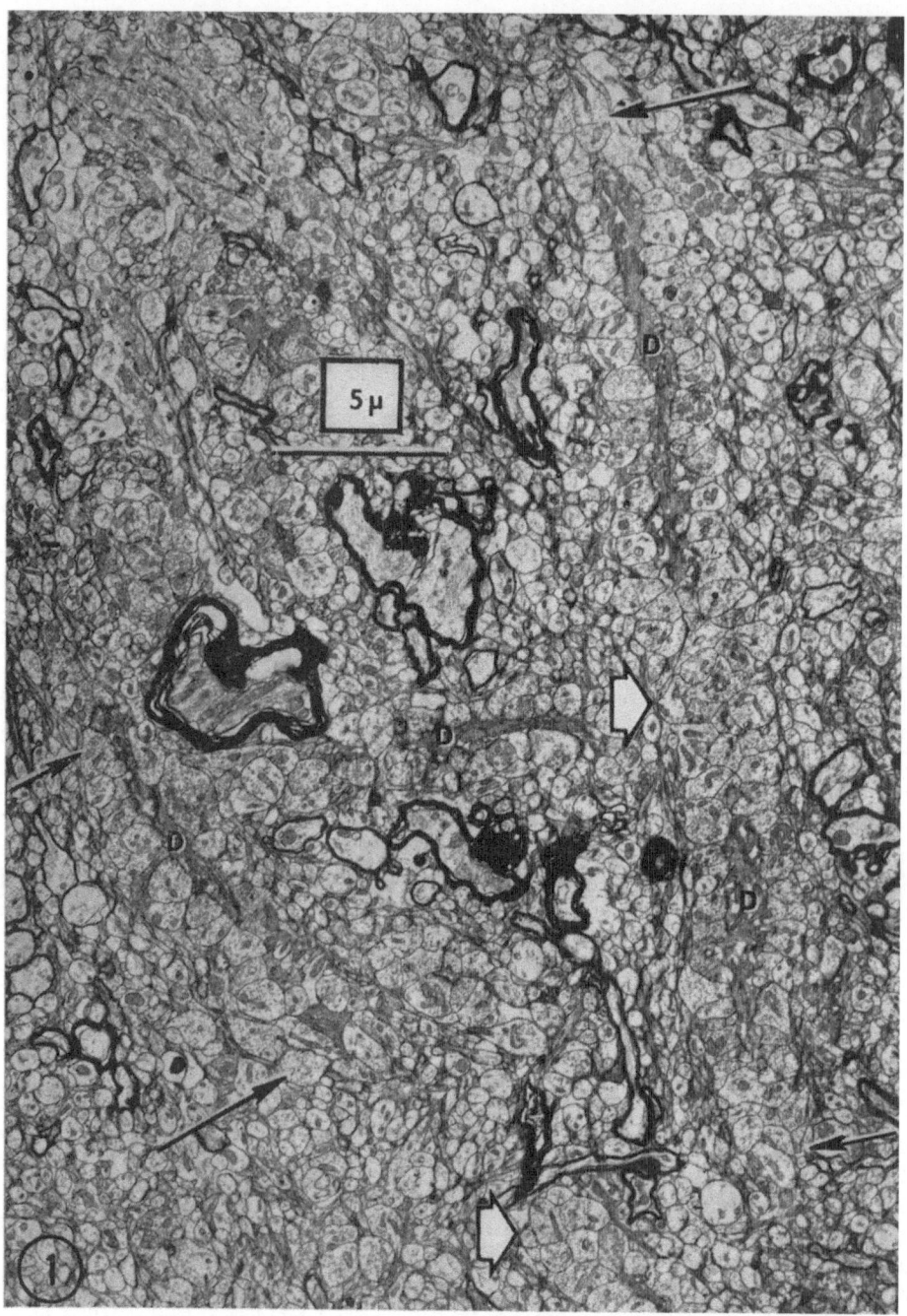

Fig. 1. Explanation see text.

Peripheral to the rows of synaptic endings there is a profusion of thin-caliber, un-myelinated fibers. The significance of this recurring pattern, characteristic of the globus pallidus, will be apparent when the GOLGI material is considered. Cell bodies are rarely encountered during electron microscopic examination; however, this is not unexpected since in the NISSL picture of the lenticular nucleus perikarya are more widely separated in the globus pallidus than in the putamen.

The neurons in Golgi preparations. The globus pallidus has large fusiform nerve cell bodies (Figs. 2, 4 and 5) with long, smooth, radiating, slightly branched dendrites. Branching is usually dichotomous. The triangular-shaped cell bodies, described by FOIX a. NICO-LESCO (1925) and others, are in our opinion the result of sectioning fusiform cells trans-versely in regions where dendrites emerge, as in Fig. 4. The lengths of the dendritic branches of the neuron (Fig. 2), determined by tracing each dendrite under a projector, totaled 4030 micra. The cell-body surface, therefore, is insignificant when compared with its dendritic surface.

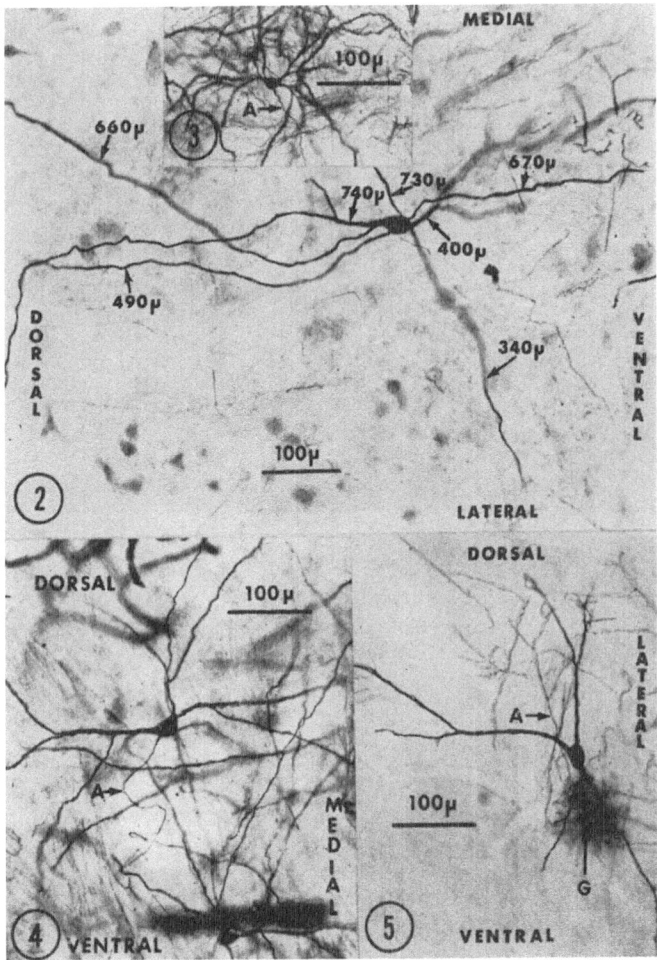

Figs. 2–5. Explanation see text.

The neurons illustrated here are none other than the "radiating cells" KÖLLIKER (1896) found in GOLGI preparations of the human lenticular nucleus. He described the cells as mostly spindle or triangular in form, 36–54 micra, or even 64 micra in length, with three or four or even five principal dendrites, never less than 0.2–0.6 mm. in length and even as long as 0.9–1.00 mm. in length. But erroneously he located these cells in the putamen. This conclusion is not only ours but it is also BIELSCHOWSKY's (1919), who studied the human globus pallidus in material prepared by his well-known technique. To use his words, "the great anatomist of Würzburg was slightly mistaken." BIELSCHOWSKY reasoned that this error is understandable to anyone familiar with the difficulties of orienting exactly the small embryonic objects GOLGI investigators are forced to use.

The neurons (Figs. 2, 4 and 5) are from the medial segment of the globus pallidus, in frontal sections just caudal to the anterior commissure, and their precise positions, dorsal, ventral, medial and lateral are indicated. There can be no doubt as to their location,

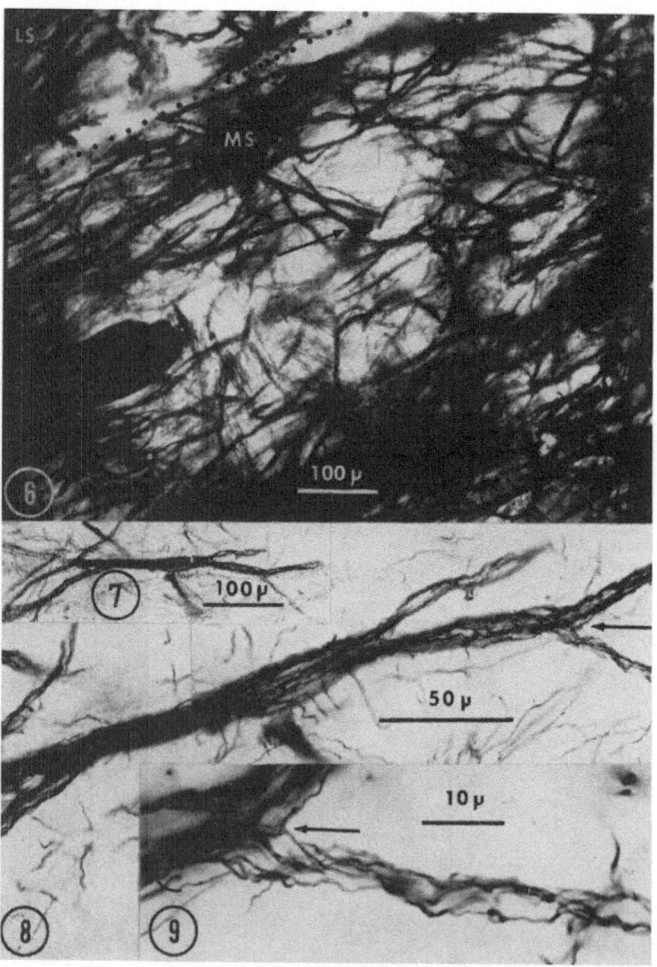

Figs. 6–9. Explanation see text.

for in the method employed (Fox, et al., 1951) whole slices of formalin-fixed adult brains are processed, and the background is sufficient for orientation. The neuron (Fig. 2) lies at the mid-point of the lateral edge of the medial segment of the globus pallidus. The neuron (Fig. 4) is in the medial tip of the medial segment of the globus pallidus and its axon (A), emerging as a conical prolongation of the cell body, descends into the ansa lenticularis. The neuron (Fig. 5) is centrally located in the medial segment of the globus pallidus on the opposite side of the brain, and its axon emerges from a dendrite and soon leaves the plane of section. The portions of the axon (Fig. 5, A) in focus here are collaterals from the main axonic stem. A small glial cell (Fig. 5, G) is just out of the plane of focus.

Putamen. A neuron from the putamen is reproduced (Fig. 3) at the same magnification. Its cell body is smaller and ovoid and its dendrites, which are shorter and radiate in all directions, are initially and for a short distance smooth. Each dendrite then branches three or more times and immediately these branches are covered with the most robust spines found anywhere in the central nervous system. These nerve cells in the putamen are exactly like those in the caudate nucleus. An oil-immersion photomicrograph of a spine-laden dendrite from the caudate nucleus can be seen in the paper by Fox a. BARNARD (1957, Fig. 5).

The afferent plexus in Golgi preparations. The rich afferent plexus of the globus pallidus is shown (Fig. 6) in a horizontally sectioned GOLGI preparation. The diagonally running dotted line at the upper left roughly indicates the boundary between the lateral (LS) and the medial (MS) segments of this nucleus. This field, a fortunate impregnation from a section near the surface of the block – surface sections were too densely impregnated for study – creates the impression that one is viewing a large number of impregnated dendrites and a nerve cell (Fig. 6, arrow) with emerging dendrites. However, closer examination reveals that it is not the neurons but an ensheathment of extremely fine fibers clothing the neurons that is impregnated. In slightly deeper sections where staining is less, this is clearer; for example, as it is in the partial impregnation (Fig. 7) of the fibers investing a nerve cell and its dendrites. The montage (Fig. 8), a photomicrograph taken with a 50x oil-immersion lens and the photomicrograph (Fig. 9) taken with a 100x oil-immersion lens, brings out the finer details of this impregnation. Note at the arrows, pointing to the same region (Figs. 8 and 9), the delicate fibers branching as the dendrite branches. Note also the further branching of these fibers and the swellings which are synaptic endings of the "bouton en passage" variety.

Electron microscopy. Electron micrographs confirm and amplify the GOLGI observations: everywhere the surfaces of the long dendrites are covered with a layer of synaptic endings. Observe the snugly fitting, synaptic endings dove-tailing with each other and with the irregular contour of the dendritic surfaces of the obliquely sectioned dendrite (Fig. 10, D; *Saimiri sciureus*) and of the transversely sectioned dendrite (Fig. 11, D; *Macaca mulatta*). The afferent fibers (Figs. 10 and 11, AF) supplying the synaptic endings are peripherally arranged and oriented in the same direction as the dendrites.

Since GOLGI impregnations (Figs. 14, 16 and 19) and electron micrographs (Figs. 15 and 17; *Saimiri sciureus*) reveal that on any individual fiber the "bouton en passage" endings are separated from each other by some interval and since electron micrographs also reveal that the entire surfaces of dendrites (Figs. 10, 11 and 18) and most of the surface of the perikaryon (Fig. 20; *Saimiri sciureus*) are covered with synaptic endings, a number of afferent fibers must participate in this synaptic arrangement. Packed around the dendrite (D) in the electron micrograph (Fig. 11) are the profiles of 270 fine fibers. It is

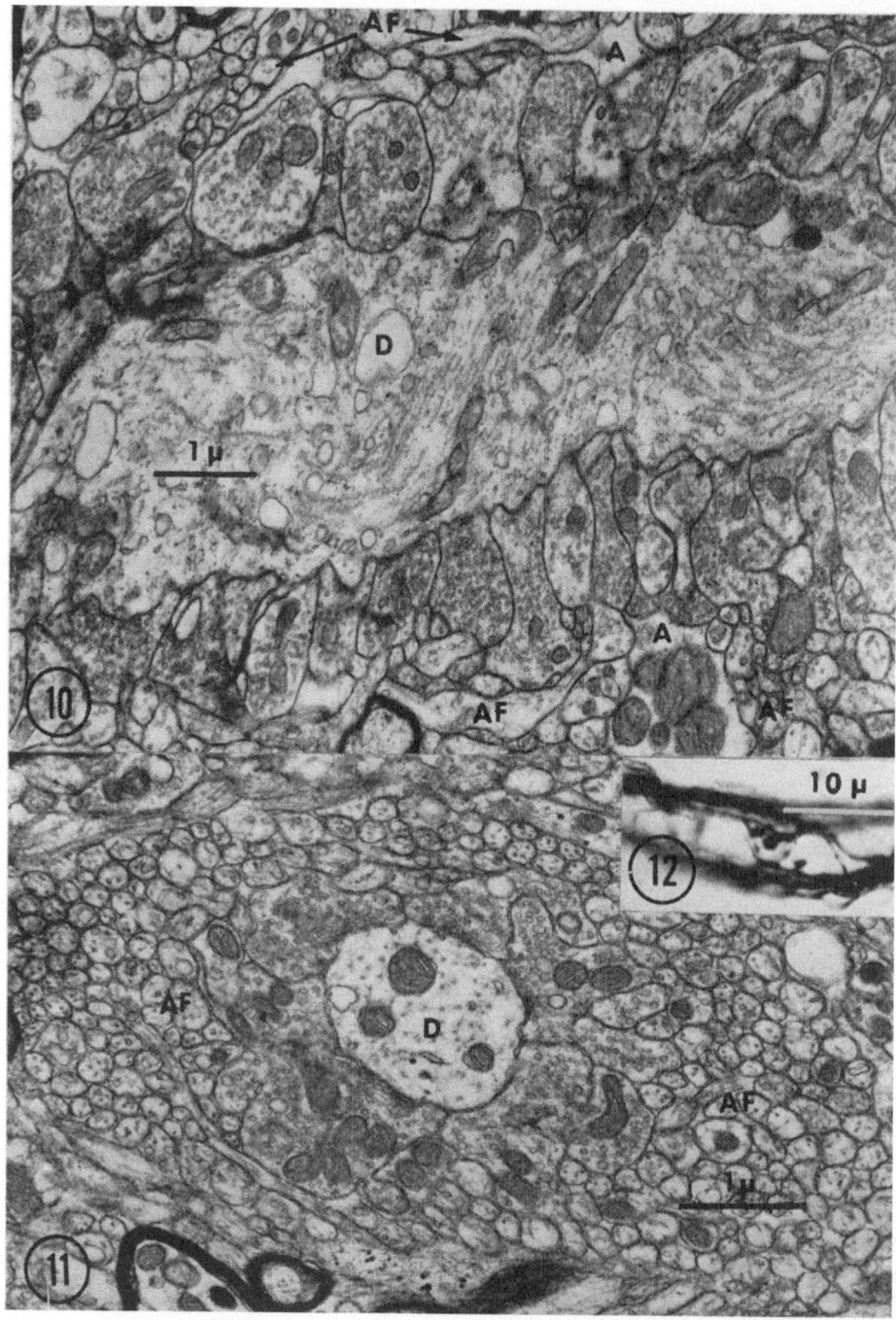

Figs. 10–12. Explanation see text.

their impregnation, either partially or "en masse" that results in the GOLGI pictures seen in Figs. 6, 7, 8, 9, 12, 14 and 16. Most of them are 0.2 of a micron or less in diameter and thus are the same caliber as the fine parallel fibers in the cerebellar cortex. Like the thin parallel fibers (FOX a. BARNARD, 1957) and (FOX, SIEGESMUND a. DUTTA, 1964) they are likely to be displayed only in GOLGI or electron microscopic preparations. BIELSCHOWSKY (1919, 1928) with his own reduced silver technique succeeded in impregnating a few of these fibers.

We are dealing here with the "longitudinal axo-dendritic connection" type of synapse which RAMÓN Y CAJAL (1934, 1954) illustrated with the climbing fiber that winds along the smooth dendritic arborizations of the PURKINJE cell like a vine. An oil-immersion photo-micrograph of a segment of a climbing fiber (Fig. 13) is shown at the same magnification as an afferent fiber of the globus pallidus (Fig. 14). Both are from *Macaca mulatta* and the differences are immediately apparent. The main trunk of the smooth part of the contacting

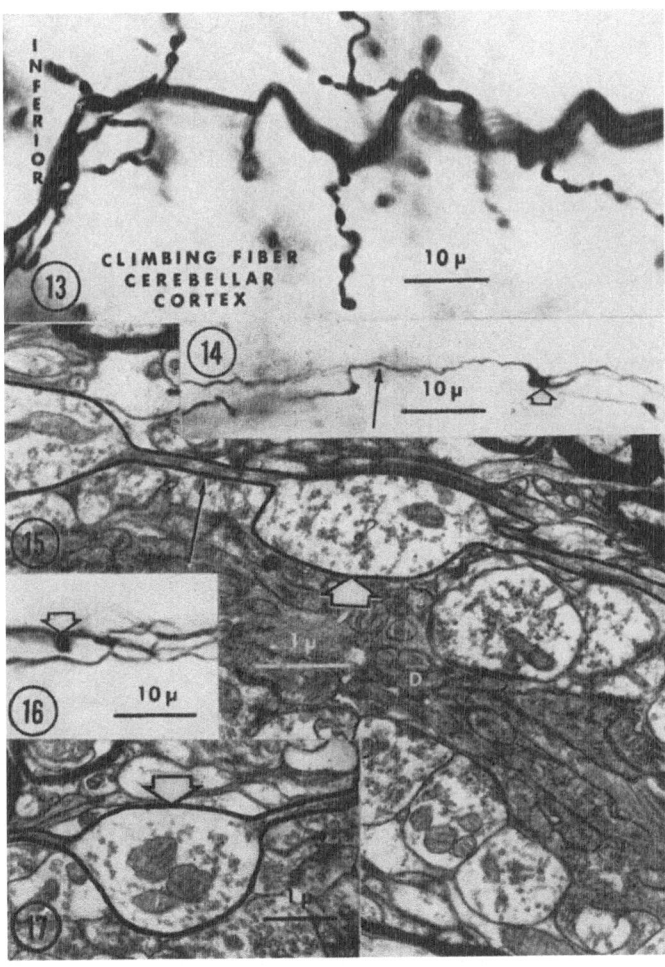

Figs. 13–17. Explanation see text.

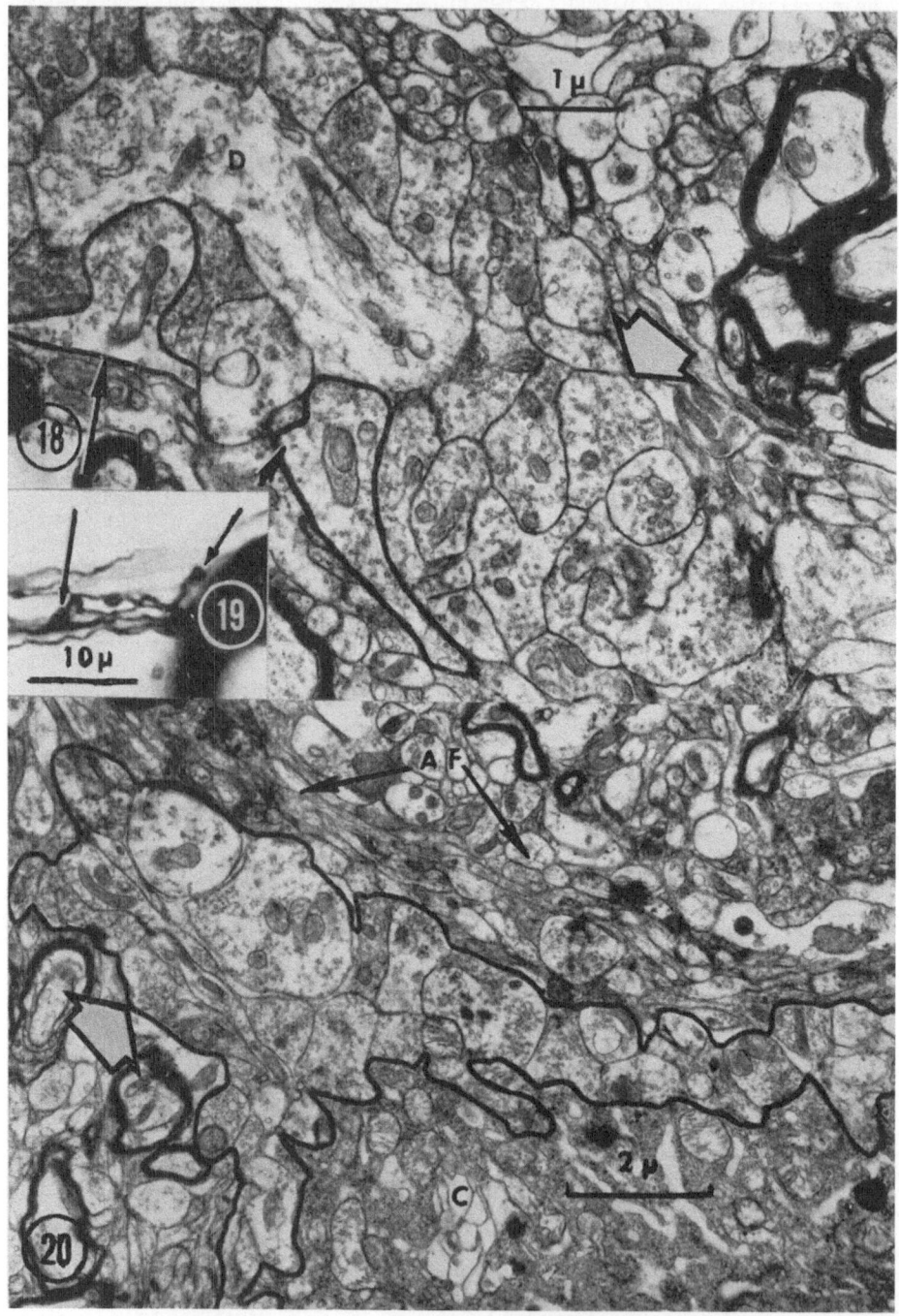

Figs. 18–20. Explanation see text.

portion of the climbing fiber – in places nearly 2 micra – has a considerably heavier caliber; so, too, have its beaded side branches. Thus, within the specific classes of synapses which RAMÓN Y CAJAL (1934, 1954) considered in his now classical classification, there are variations and the conclusion is warranted that, since the globus pallidus has longer dendrites and its afferents are thinner, impulses here must spread more slowly and over longer intervals than they do on the smooth branches of the PURKINJE cell.

It is truly remarkable that despite the supposedly rougher treatment of material prepared for GOLGI preparation and the supposedly more refined treatment of material prepared for electron microscopy, profiles in electron micrographs accord well with GOLGI impregnations. Compare the inked outline of a nerve fiber in the electron micrograph (Fig. 15; *Saimiri sciureus*) with the impregnation (Fig. 14). Arrows point to the constricted conducting portions of the nerve fibers and the block arrows point to synaptic endings with branches emerging from the right side of the endings. Compare also the impregnated fiber and ending (Fig. 16, block arrow) with the inked profile of a fiber and ending in the electron micrograph (Fig. 17; *Saimiri sciureus*). Note, too, the inked endings in the electron micrograph (Fig. 18, arrows; *Saimiri sciureus*) and follow the arrows to the similarly shaped endings in the impregnation (Fig. 19). The dendrite (Fig. 18) is out of the plane of section in the region of the block arrow and here is a surface view of a mosaic covering of synaptic endings.

The electron micrograph (Fig. 20; *Saimiri sciureus*) is one of six micrographs of a montage displaying a complete profile of a perikaryon. Except for a small area of contact with a glial satellite, the entire surface of the cytoplasm (c) is covered with synaptic endings. The array of these endings, inked here (Fig. 20), form a mosaic at the block arrow (left). In our interpretation a dendrite, out of the plane of section, emerges in this region.

Comparative Studies

Entopeduncular nucleus (cat). Comparative neurologists (ARIËNS KAPPERS, HUBER a. CROSBY, 1936) have recognized the close relationship between the globus pallidus and the entopeduncular nucleus. WINKLER a. POTTER (1914) and MORGAN (1927) labeled the entopeduncular nucleus the medial segment of the globus pallidus. OLIVE SMITH (1930) suggests that the large entopeduncular nucleus in *Tamandua* is the homologue of the medial segment of the globus pallidus, and on the basis of certain experimental findings in the cat, Fox a. SCHMITZ (1944) reached the same conclusion. More recently NAUTA a. MEHLER (1961) have demonstrated experimentally that the cat entopeduncular nucleus has the same connections as the medial segment of the monkey globus pallidus. Our present study shows that ultrastructurally the cat entopeduncular and the monkey globus pallidus are similar. This is not illustrated here, but it was demonstrated at the International Symposium on the Phylogenesis and Ontogenesis of the Forebrain.

Paleostriatum primitivum (Gallus domesticus). For a review of the literature on the avian paleostriatum primitivum, ARIËNS KAPPERS, HUBER a. CROSBY should be consulted. ARIËNS KAPPERS (1923) compared the paleostriatum with the human globus pallidus. KARTEN (personal communication), one of NAUTA's collaborators, who has seen samples of our impregnations, has obtained successful GOLGI preparations of the pigeon paleostriatum primitivum and finds that its cells are large and have long, smooth radiating branches with few bifurcations. Thus, the neurons in these homologous regions are similar. Moreover, lesions in the pigeon paleostriatum primitivum (KARTEN; personal communication) result in degeneration to N. spiriformis lateralis via collaterals from the ansa lenticularis which continues on to terminate in N. tegmentosus pedunculo-pontius. According to KARTEN this is the exact pattern of projection NAUTA a. MEHLER (1961) found in the cat from the entopeduncular nucleus, and in the monkey from the medial segment of the globus pallidus.

Ultrastructurally these homologous regions in the primate and avian brains are essentially similar. The low-power electron micrograph (Fig. 21; chicken) shows dendrites (D) sectioned in various planes and covered with synaptic endings. Peripherally there are a few medullated fibers and an enormous number of tightly packed unmedullated afferent fibers (AF). For a striking demonstration of this similarity compare the field of the transversely sectioned dendrite (Fig. 11, D; *Macaca mulatta*) with the field of the transversely sectioned dendrite (Fig. 22, D; *Gallus domesticus*). Finally, in the latter electron micrograph, note the inked contours of afferent fibers with their "bouton en passage" endings.

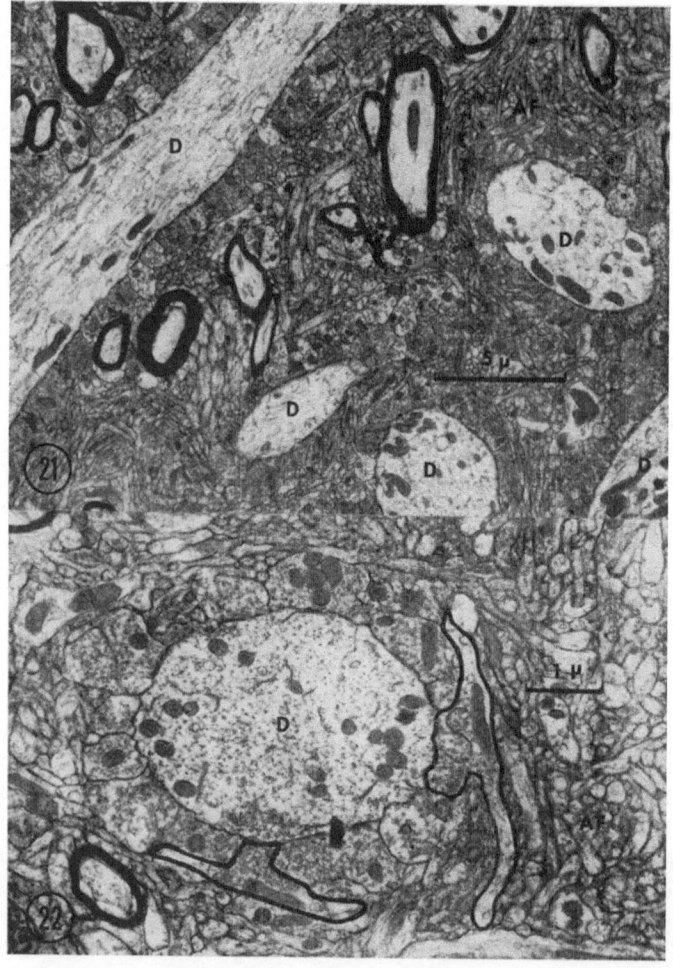

Figs. 21–22. Explanation see text.

Conclusion and Summary

GOLGI impregnations of the basal ganglia of the monkey *(Macaca mulatta)* reveal that the globus pallidus has large fusiform neurons with long, relatively smooth dendrites which are contacted by an afferent plexus of fibers forming longitudinal axodendritic connections. In the present study interpretation of electron micrographs of the globus pallidus of the monkey *(Macaca mulatta)* and of the squirrel monkey *(Saimiri sciureus)* were facilitated by the findings in the GOLGI preparations. The neuropile contains the long dendrites, the afferent plexuses related to these dendrites and the many "bouton en passage" synaptic endings derived from these plexuses. Profiles of the cell bodies and dendrites, sectioned transversely and longitudinally, indicate virtually no receptive surfaces devoid of synaptic endings. Profiles of these fibers and endings match many of the forms exhibited in the GOLGI preparations.

With this ultrastructural information available it was possible to study certain homologies suggested by phylogenetic studies: the cat entopeduncular nucleus, the suggested homologue of the primate medial segment of the globus pallidus and the chicken paleostriatum primitivum, the suggested homologue of the mammalian globus pallidus. Ultrastructurally the striatum primitivum of the chicken and the entopeduncular nucleus of the cat are similar to the monkey globus pallidus.

The present study confirms the observations of BIELSCHOWSKY (1919) who first observed the mantle of fibers, synaptic endings and glia on the globus pallidus neurons. However, the GOLGI technique discloses this in greater detail than the BIELSCHOWSKY method and electron microscopy amplifies these details in a way undreamed of in BIELSCHOWSKY's time. This investigation also suggests the possibility that electron microscopy may be useful in comparative neuroanatomical studies.

Note. The senior author recalls with pleasure the informative discussions he had with R. HASSLER and is grateful to him for calling attention to the important paper of BIELSCHOWSKY (1919) which anticipated some of our results.

References

ARIËNS KAPPERS, C.U.: Le developpement ontogénétique du corps strié des oiseaux en comparaison avec celui des mammifères et de l'homme. Schweiz. Arch. Neurol. Psychiat. *13* (1923), 348–376.

ARIËNS KAPPERS, C.U., G.C. HUBER, E.C. CROSBY: The Comparative Anatomy of the Nervous System of Vertebrates, Including Man. The Macmillan Co., New York 1936.

BIELSCHOWSKY, M.: Einige Bemerkungen zur normalen und pathologischen Histologie des Schweif- und Linsenkerns. J. Psychol. Neurol. *25* (1919), 1–11.

BIELSCHOWSKY, M.: Nervengewebe. In: W. von Möllendorff (Ed.) Handbuch der mikroskopischen Anatomie des Menschen. Vol. 4, p. 1 bis 142. Springer, Berlin 1928.

FOIX, C., J. NICOLESCO: Les Noyaux Gris Centraux et la Région Mésencephalo-sous-optique. Masson et Cie., Paris 1925.

FOX, C.A., J.T. SCHMITZ: The substantia nigra and the entopeduncular nucleus in the cat. J. comp. Neurol. *80* (1944), 323–334.

FOX, C.A., M. UBEDA-PURKISS, H.D. IHRIG, D. BIAGIOLI: Zinc chromate modification of the Golgi technique. Stain Technol. *26* (1951), 109.

FOX, C.A., J.W. BARNARD: A quantitative study of the Purkinje cell dendritic branchlets and their relationship to afferent fibers. J. Anat. (Lond.) *91* (1957), 299–313.

FOX, C.A., K.A. SIEGESMUND, C.R. DUTTA: The Purkinje cell dendritic branchlets and their relation with the parallel fibers: Light and electron microscopic observations. In: Cohen, M.M., R.S.Snider (Eds.) Morphological and Biochemical Correlates of Neural Activity. Vol. 7, p. 112–114. Harper & Row, New York 1964.

KARTEN, H.: Personal Communication.

KÖLLIKER, A.: Handbuch der Gewebelehre des Menschen. Engelmann, Leipzig 1896. Bd. 2, Aufl. 6.

MORGAN, L.O.: The corpus striatum. A study of secondary degeneration following lesions in man and of symptoms and acute degeneration following experimental lesions in cats. Arch. Neurol. Psychiat. *18* (1927), 461–549.

NAUTA, W.J.H., W.R. MEHLER: Some efferent connections of the lentiform nucleus in monkey and cat. Anat. Rec. *139* (1961), 260.

POLYAK, S.L.: The Retina. University of Chicago Press, Chicago 1941.

RAMÓN Y CAJAL, S.: Les preuves objectives de l'unité anatomique des cellules nerveuses. Trab. lab. invest. biol. (Madrid) *29* (1934), 1–137.

RAMÓN Y CAJAL, S.: Neuron Theory or Reticular Theory? Trans. M. Ubeda-Purkiss and C. A. Fox. Consejo Superior de Investigaciones Cientificas, Instituto Ramón y Cajal, Madrid 1954.

SMITH, O.C.: The corpus striatum, amygdala and stria terminalis of Tamandua tetradactyla. J. comp. Neurol. *51* (1930), 65–76.

WINKLER, C., A. POTTER: An Anatomical Guide to Experimental Researches on the Cat's Brain. W. Versluys, Amsterdam 1914.

CLEMENT A. Fox, Ph. D.
D. E. HILLMAN, Ph. D.
Department of Anatomy
Wayne State University
School of Medicine
1400 Chrysler Expressway
Detroit, Michigan, USA

K. A. SIEGESMUND, Ph. D.
L. A. SETHER Ph. D.
Department of Anatomy
Marquette University School of Medicine
561 North Fifteenth Street
Milwaukee, Wisconsin, USA

Embryonal and Postembryonal Development of Neurons of the Human Cerebral Cortex

By G. I. POLIAKOV, Moscow

The most characteristic feature of the cortical organisation, distinct from the nuclear and reticular subcortical formations, is that in its structure a multi-layer and multi-level screen is present, being the most perfect reflective mechanism formed in the course of evolution. The cortical organisation is adapted to effectuating fine analysis and broad synthesis of diverse external and internal stimuli, as well as to differentiating and integrating signals and corresponding responses of various degrees of complexity. The differentiation of the neocortex into cyto- and myeloarchitectonic layers is determined by a number of consecutive transformations of the neuronal structure. It is possible to establish definite features of similarity in the basic laws which govern the development of this process in onto- and phylogenesis.

Corresponding features in onto- and phylogenesis. At the earlier stages of human prenatal ontogenesis, just as in lower mammals (rodents), there are clearly observed *three genetically different complexes of neurons,* out of which corresponding cytoarchitectonic layers are differentiated. These neurono-architectonic complexes are designated by us as the basic zones of the neocortical cross-section: the deep zone – layers VII and VI; the middle zone – layer V, and the superficial zone – layers IV, III and II (POLIAKOV, 1937, 1949, 1965).

We were able to establish that in the embryogenesis of man the above mentioned complexes originate and develop in a definite sequence which is similar to the sequence of progressive differentiation of the neocortical layers in the comparative anatomical chain of mammals. In the ontogenesis of man the superficial zone of the neocortex which becomes separated later than other zones, reaches the highest degree of relative complication of its neuronal organisation. Similarly to this in the evolution of mammals the neuronal structure of the higher level of the cortex (layers IV, III and II) undergoes considerably more pronounced transformations than the neuronal structure of the lower cortical level (layers VII, VI and V) (POLIAKOV, 1958, 1958a; ZHUKOVA, 1953; SHKOLNIK-YARROS, 1961). In particular, highly characteristic of primates in comparison with sub-primates is, along with a considerable absolute and relative growth of the width of layer III, its division into sublayers.

It was also established that in the embryogenesis of man various regions and areas of the neocortex begin to manifest themselves in the peculiarities of the cytoarchitectonics and of the neuronal structure long before the primary convolutions and fissures appear. In particular, the borders of some neocortical formations, which are relatively early differentiated in phylogenesis and early mature in ontogenesis, are distinctly outlined on the still smooth surface of the hemisphere (area striata, areas of the precentral and post-central regions). This may be regarded as a reflection in the ontogenetic development of one of the phylogenetic laws. As it may be concluded from BRODMANN's comparative investigations (1909), a differentiation into cytoarchitectonic formations, which are more or less homologous in the entire chain of mammals, manifests itself among subprimates and primates already in lissencephala, i. e. precedes the complication of the macroscopic relief of the cerebral hemispheres.

Some general structural peculiarities of the cortical formations, which are conditioned by the place occupied by them in the hierarchical system of the entire complex of neocortical regions and areas also begin to manifest themselves at a relative early stage of man's embryogenesis – already during the first half of intrauterine life. As demonstrated by us earlier(POLIAKOV, 1949, 1961 b), one of the essential laws determining the structural and functional organization of the cortex consists in the fact that the delicacy of the neuronal structure gradually increases with the transition from the central projectional zones to the associative zones which are connected with the most complex and higher forms of functional interactions between signals of different modality and different significance. The areas entering the associative zones and participating in realization of the most "delicate" functions are accordingly characterized also by a neuronal structure of highest relative delicacy. As to the projectional zones of the cortex, which become separate and mature earliest, developing in ontogenesis at a highly intensive rate, they are characterized by the greatest "potential strength" of the neuronal organization. Precisely in these divisions of the neocortex, which, according to BRODMANN, belong to the heterotypic formations (gigantopyramidal area, conicortical areas) certain elements of the projectional complex (BETZ a. MEYNERT pyramidal cells, CAJAL star cells) reach gigantic dimensions. The enrichment of the neocortex with elements, more and more delicate as regards their structure and connections, is one of the most important morphological indices of the functional perfection taking place in this part of the central nervous system in the course of progressive evolution.

Results

At relatively early stages of ontogenesis (though in different elements at different periods) there begins to manifest a *structural specialization* of various forms of neurons. This concerns first of all the two basic types of neurons forming the cerebral cortex (Fig. 1), namely the efferent neurons with long axons (pyramidal and fusiform cells) and neurons with short axons which are specialized in accomplishing diverse kinds of intracortical switching (star cells) (POLIAKOV a. SARKISOV, 1949). The structural differentiation of elements of both kinds in ontogenesis (Fig. 2) as well as in phylogenesis of vertebrates proceeds in essentially different ways (POLIAKOV, 1953, 1959, 1964). According to our investigations the star neurons, which are the most delicately and complexly differentiated elements of the entire construction, appear in prenatal ontogenesis only in the fifth month of the intrauterine life, i. e. markedly later than the efferent neurons, of which the neocortex predominantly consists at earlier stages of development. The enrichment of the cortex with these elements grows progressively during the second half of intrauterine life, reaching the highest level in the last months before birth.

In the above mentioned correlations between the time of emergence, rates of development and differentiation of elements of the neuronal organization of the cortex, differing greatly in their functional significance, we can clearly see a strong pronounced parallelism with the complication of the same organization in the course of phylogenesis. A number of investigations carried out in our laboratory (ZHUKOVA, 1953; ZAMBRZHITSKY, 1956, 1959; SHKOLNIK-YARROS, 1959) established that in the comparative-anatomical chain of mammals the elements of intracortical connections become more and more numerous and acquire an ever-increasing diversity of forms; thus, they present one of the most important factors of progressive differentiation of the cortex.

In the whole course of prenatal ontogenesis certain distinctions characteristically manifest themselves in all peculiarities of the development of elements belonging to different neurono-architectonic complexes of the neocortex. A typical example is the *differentiation of the pyramidal neurons* of the higher and lower levels of the cortex. The pyramids of layer III are characterized by a predominant development of ramifications of the main ascending apical dendrite trunk, reaching a high degree of extension. Such correlations of various parts of the dendritic sphere are apparently determined by the specialization of these neurons mainly in coupling the associative (cortico-cortical) connections. On the contrary, characteristic of the pyramids of layer V is a predominant development of ramifications of the basal dendrites and at the same time a relatively weak development of ramifications of the apical dendrite; this indicates that these neurons are mainly specialized in coupling the projectional (cortico-subcortical) connections. Definite peculiarities are observed also in the progressive differentiation of the star cells of various layers.

A systematic study of the formation in ontogenesis of the structural mechanisms of the interneuronal connections caused by the ramifications of the dendrites and axons of

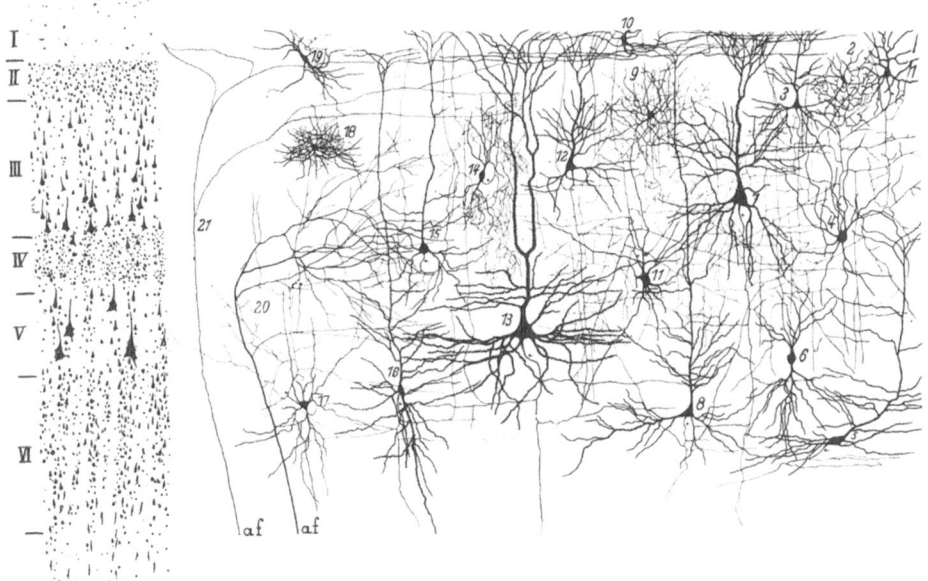

Fig. 1. Chart illustrating the neuronal structure of the neocortex of the human brain. Left: cytoarchitectonic layers of the cortex. af – projectional (20) and associative (21) afferents; 1, 3, 5, 7, 8, 12, 13, 16 – neurons with long axons (pyramidal and fusiform cells); 2, 4, 6, 9, 10, 11, 14, 17, 18, 19 – neurons with short axons (star cells) (After G. I. POLIAKOV, 1953).

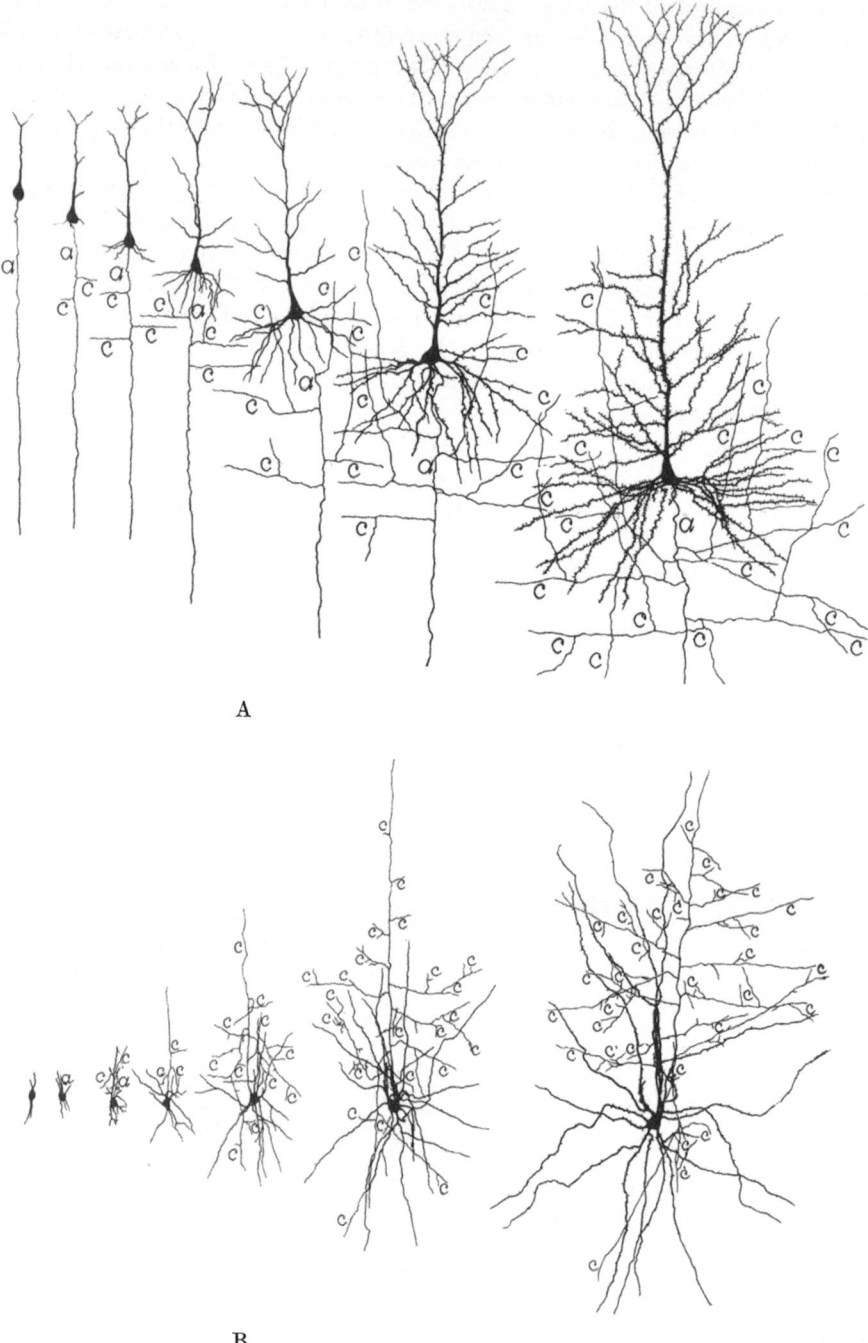

A

B

Fig. 2. Progressive differentiation in man's ontogenesis of a pyramidal neuron with a long axon (A) and a star neuron with short axon (B) in the neocortex. a – axon, c – axon collaterals (After G. I. Poliakov, 1959).

various neurons, as well as the junctions between, is of prime importance for comprehending the *functional organization* of the adult's cerebral cortex. We shall briefly consider here from the ontogenetic point of view only some fragments of this highly delicate and complex organization. We shall use for the purpose of factual illustration only a small part of all the material obtained. First of all, we shall refer here to some data relating to the development of the pericellular plexuses (pericellular nests or baskets according to CAJAL's terminology), formed by the ramifications of short axons of some star cells (POLIAKOV, 1961). In the neocortical formations maturing first the initial phases of organization of

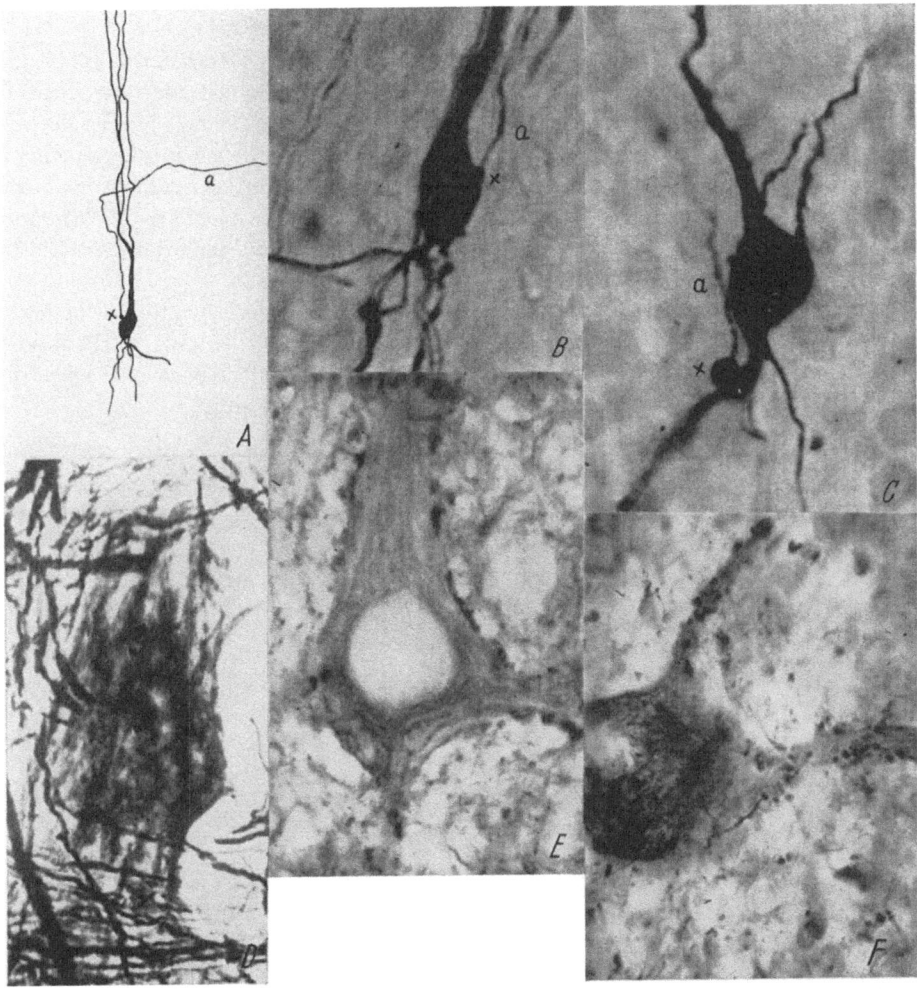

Fig. 3. Formation of axon end contacts (x) in the human cerebral cortex. A, B – contact with the body of a young pyramidal cell (A – drawing, B – photomicrograph). C – contact with the initial division of the dendrite. D – pinshaped end contacts on the Betz gigantic pyramid. E – end buttons on the pyramidal cell. F – end buttons on the star cell. A, B, C – cerebral cortex of a human embryo, 5 ½ months; D, E, F – cerebral cortex of an adult. a – axon (A, B, C, D – POLIAKOV, 1955, 1961; E, F – preparations of BLINOVA).

such structures take place already in the last months of intrauterine life. Thus, there are certain aspects to assume that by the time of birth the given form of interneuronal connections reaches in the corresponding formations of the neocortex a quite distinct level of structural differentiation. However, the above mentioned structures enter the most intensive period of development only after birth. In the cortex of children in the first months of life we found highly complex plexuses formed by numerous mutually overlapping and intertwining ramifications belonging to several different axonal systems (POLIAKOV, 1960).

Development of synapses. Our observations showed that at the early stages of prenatal ontogenesis – already in the first half of intrauterine life – there appear contacts of a relative simple form between the neurons (Fig. 3) resulting from a junction of single-end thickenings of the axon branchings with the bodies and initial divisions of the still insufficiently ramified dendrites of the young pyramidal and fusiform cells. At later stages of prenatal ontogenesis, when the axon and dendrite ramifications show a further growth, there takes place a marked complication of the ways in which the neurons become interconnected. Particularly essential in this respect, along with a further complication of the contact structures formed by the axons themselves, is the development of "counter" contact structures in the shape of appendages ("thorns") on the dendrites described by CAJAL as far back as 1891 (POLIAKOV, 1953).

In the human cerebral cortex such appendages on the dendrites begin to develop in the seventh month of intrauterine life. Before birth they are observed only in the biggest neurons (mainly of layer V); after birth they greatly spread and densely cover the dendrite ramifications of many efferent neurons of the cortex (Fig. 4).

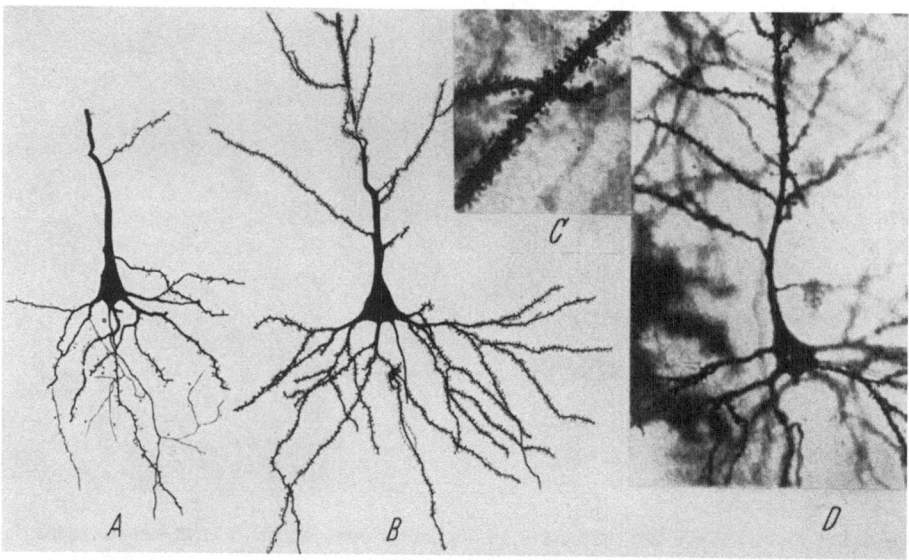

Fig. 4. Development of appendages ("thorns") on the dendrites of a pyramidal cell in the human cerebral cortex. A – brain of a human embryo, 7 months; B, C, D – brain of a child, aged 1, 5 months (After G.I.POLIAKOV, 1953, 1961).

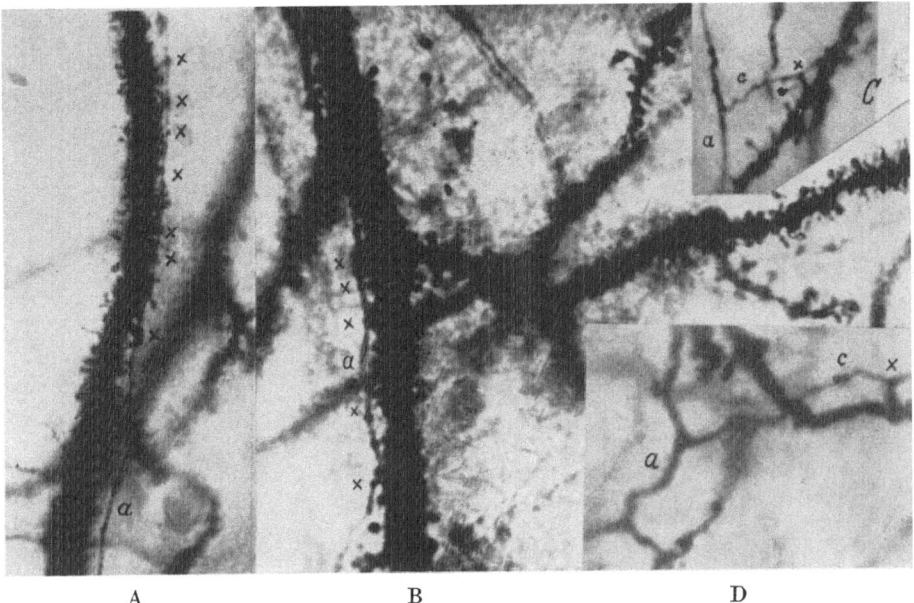

Fig. 5. Formation of tangential contacts (x) between an axon (a) and its ramifications (c) and appendages on the dendrite of a pyramidal (A, B) and star (C, D) cell. Cortex of a child aged 6 months (After G. I. POLIAKOV, 1955, 1961).

Precisely in the first months after birth we were able to discover end-contacts, microscopically in the human cerebral cortex and mainly tangential ones (Fig. 5) between the axon ramifications and the heads of the appendages on the dendrites (POLIAKOV, 1955, 1961 a). Thus, for the first time factual proofs were presented by us corroborating the assumption previously formulated by CAJAL, STEFANOVSKAYA a. LORENTE DE NÓ, according to them the appendages are in their essence contact structures. Such contacts, established through the appendages on the dendrites, were subsequently found also by workers of our laboratory (SHKOLNIK-YARROS, 1963; LEONTOVICH, 1958) in the cortex and subcortical formations of animals. As is known, in the recent years GRAY (1959) proved the existence of such contacts with the help of electron-microscopy.

We consider that this new, previously unknown form of interneuronal connections developing relatively late in ontogenesis supplements the phylogenetically older form of interneuronal contacts which are represented by the axon end structures directly on the bodies and dendrites of the nerve cells (Fig. 6). In the evolution of mammals especially of primates, this form of contacts reaches a particularly high level of development and extension in the cortical formations which are characterized by a high complexity and delicacy of the axodendrite plexuses situated in the corresponding architectonic layers and sublayers of the cortex.

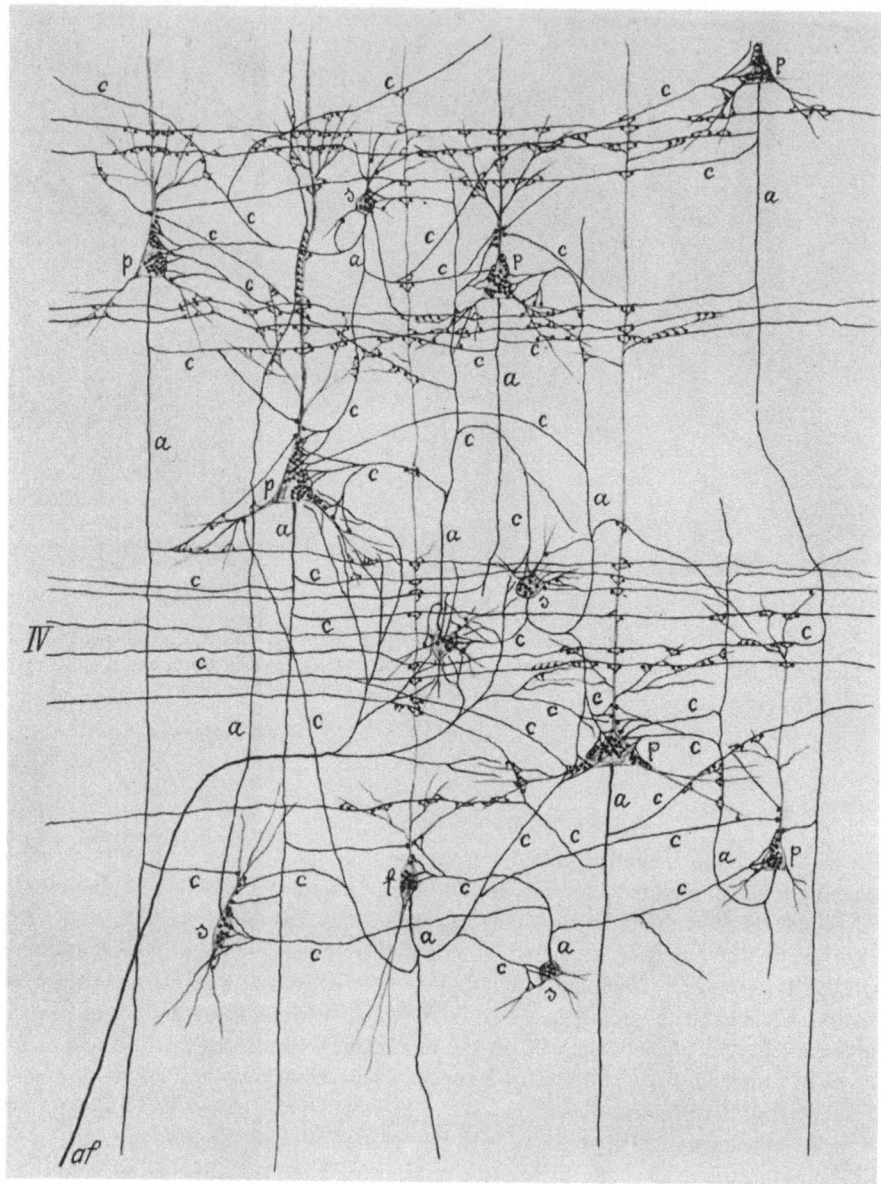

Fig. 6. Scheme illustrating the distribution in various neurons of end contacts and tangential contacts with appendages on the dendrites in the cerebral cortex. af – afferent fibre from the sub-cortex; p – pyramidal neuron; f – fusiform neuron; s – star neuron; IV – level of the cyto-architectonic layer IV. The scheme presents some nests of axon end contacts on the bodies of nerve cells as well as tangential contacts of axons (a) and of their collaterals (c) in points of junction with appendages on the dendrites (After G.I. POLIAKOV, 1961).

Conclusions

From this brief essay presenting the successive development of various forms of interneuronal connections in ontogenesis it follows that in the most differentiated cortical formations of the brain the structural specialization of the neurons, being the morphological basis of the variable analytico-synthetical activity, is best adapted to ensuring the highest functional reactivity and plasticity of the reflex mechanisms of these levels of the central nervous system. It may be assumed that the process of perfection of these mechanisms in the evolution of vertebrates developed precisely in this direction.

Summary

Structural differentiation of the neocortex into cytoarchitectonic layers and areas observed during prenatal and postnatal life is conditioned by a number of consecutive transformations of the neuronal structure.

In early stages of ontogenesis various areas of the neocortex begin to manifest themselves in the peculiarities of cytoarchitectonics and of neuronal structure long before the primary convolutions and fissures appear. Relative early (on the fifth month of the prenatal life) differences in structural specialization of various forms of neurons (pyramidal, fusiform and star cells) are discovered.

Peculiarities typical for corresponding neurons of various layers in various areas of the neocortex are observed clearly during the second part of the prenatal life.

Main elements of the cortical organization, typical for the brain of an adult, reach definite degree of development at the time of birth. This point is valid both for various forms of neurons of which the neocortex is built, and for various forms of interneuronal connections (axosomatic and axodendritic synaptic contacts).

References

BRODMANN, K.: Vergleichende Lokalisationslehre der Großhirnrinde in ihren Prinzipien dargestellt auf Grund des Zellenbaues. Leipzig 1909.

GRAY, E.G.: Axo-somatic and axo-dendritic synapses of the cerebral cortex; an electron microscope study. J. Anat. *93* (1959), 420-433.

LEONTOVICH, T.A.: Peculiarities of interneuronal connections in the subcortical ganglia of mammals. Arch. Anat. *35* (1958), 17–25 (Russisch).

POLIAKOV, G.I.: Early and middle ontogenesis of the human cerebral cortex. State Institute of the Brain, Moscow 1937 (Russisch).

POLIAKOV, G.I.: Structural organization of the human cerebral cortex, according to data concerning its development in ontogenesis. In: Cytoarchitectonics of the Human Cerebral Cortex. Medgiz, Moscow 1949 (Russisch).

POLIAKOV, G.I.: Concerning some delicate peculiarities in the structure of the human cerebral cortex, and the functional interaction of neurons. Arch. Anat. *30* (1953), 48–60 (Russisch).

POLIAKOV, G.I.: Some new data concerning the early embryogenesis of the neurons of the human cerebral cortex. Pavlov J. high. nerv. Activity *4* (1954), 123–13 (Russisch).

POLIAKOV, G.I.: Concerning the structural mechanisms of interneuronal connections in the human cerebral cortex. Arch. Anat. *32* (1955), 15–19 (Russisch).

POLIAKOV, G.I.: Some peculiarities of the complication of neurons in the central nervous system of man, primates and other mammals. Soviet Anthropol. *3* (1958), 35–56 (Russisch).

POLIAKOV, G. I.: Some peculiarities of the complication of the neuronal structure of the cerebral cortex in man, monkeys and other mammals. Soviet Anthropol. *4* (1958a) 69–85 (Russisch).

POLIAKOV, G. I.: Progressive differentiation of neurons of the human cerebral cortex in ontogenesis. In: Development of the Central Nervous System. Medgiz, Moscow 1959 (Russisch).

POLIAKOV, G. I.: The modern state of the neuronal theory. In: Some Theoretical Questions of the Structure and Activity of the Brain. Medgiz, Moscow 1960 (Russisch).

POLIAKOV, G. I.: Some peculiarities of the neuronal structure of the central nervous system revealed by different neurohistological methods. Communication I. Contact mechanisms formed by the axons of the nerve cells. Korsakov J. Neuropath. Psychiat. *61* (1961), 3–10 (Russisch).

POLIAKOV, G. I.: Some peculiarities of the neuronal structure of the central nervous system revealed by different neurohistological methods. Communication II. Contact mechanisms formed by the dendrites of the nerve cells. Korsakov J. Neuropath. Psychiat. *61* (1961a), 271–279 (Russisch).

POLIAKOV, G. I.: Some results of research into the development of the neuronal structure of the cortical ends of the analysers in man. J. comp. Neurol. *117* (1961b), 197–212.

POLIAKOV, G. I.: Development and complication of the cortical part of the coupling mechanism in the evolution of vertebrates. J. Hirnforsch. *7* (1964), 253–273.

POLIAKOV, G. I.: Development of the cerebral neocortex during the first half of intrauterine life. In: Development of the Child's Brain. Meditsina, Leningrad 1965 (Russisch).

POLIAKOV, G. I., S. A. SARKISOV: Neurons and interneuronal connections of the cerebral cortex. In: The Cytoarchitectonics of the Human Cerebral Cortex. Medgiz, Moscow 1949 (Russisch).

SHKOLNIK-YARROS, E. G.: Concerning distinctions between the neurons of the cortical end of the visual analyser in the rabbit and the dog. In: Development of the Central Nervous System. Medgiz. Moscow 1959 (Russisch).

SHKOLNIK-YARROS, E. G.: Peculiarities of the neurons and interneuronal connections of the visual analyser regarded from the comparative-anatomical aspect. In: Proceedings of the 6th All-Union Congress of Anatomists, Histologists and Embryologists. Vol. 1. Kharkov 1961 (Russisch).

SHKOLNIK-YARROS, E. G.: Morphological data on the interneuronal connections in the cerebral cortex. In: The Gagra Conversations. Vol. IV., Academy of Sciences of the Georgian S. S. R. Tbilisi 1963 (Russisch).

ZAMBRZHITSKY, I. A.: The cytoarchitectonics and neuronal structure of the limbic region in some mammals. Arch. Anat. *33* (1956), 41–48 (Russisch).

ZAMBRZHITSKY, I. A.: The cytoarchitectonics and neuronal structure of the limbic region in primates. In: Development of the Central Nervous System. Medgiz, Moscow 1959 (Russisch).

ZHUKOVA, G. P.: Contribution to the question of the development of the cortical end of the motor analyser. Arch. Anat. *30* (1953), 32–38 (Russisch).

Prof. Dr. G. I. POLIAKOV
Laboratory of Neurohistology
Brain Institute
Academy of Med. Sciences
Moscow, USSR

Phylogenesis of the Nerve Cell

By A.ÁBRAHÁM, Szeged

There is not a single question in the biology which could be answered with so great difficulties as the one of phylogenesis. The first appearance, the past and formation of the living world in form and structure of to-day took place for millions and millions of years. The history and shape of the forming work of nature are completely far from the human experience and the possibilities of the real and objective analysis. One thing seems only to be sure for anybody who tries to understand the structure and function of the living matter and intends to follow the courses and ways of the changes which form new worlds and abolish them, that there is the everlasting law, according to which the living world of to-day *developed from primitive forms* innumerable years ago. These forms got due to permanent adaptation, more and more complicated structures and handed them down to their descendants. This part of the question offers no difficulties. It is verified by the comparative anatomy, the comparative physiology, the paleontology, the taxonomy, the serology, the biochemistry and even the molecular biology. But if we try to look for the descendance of the phyla, classes, orders, families and genuses of the animal kingdom, if we want to find the way, the procedure of the evolution of vertebrates from invertebrates, that of the birds and mammals from the reptiles and that of the man from the mammals, we ever encounter almost insurmountable difficulties.

Our task is much easier if we try to follow the shape of evolution of a single organ-system. In such cases we could follow with some certainty for the most part the way of the phylogenesis. We mean here the respiratory organs, the organs of the blood-circulation, the skeleton system, the excretory organs, the sense-organs of vertebrates and others. In the structure and ontogenesis of these organ-systems the course and way of the phylogenesis appear in an acceptable and sometimes excellent form. But of all the organ-systems, the nervous system shows in respect to its comparative anatomy and onto-genesis the long and sometimes complicated course of the phylogenesis in the most excellent form.

The question of the *phylogenesis of the nerve cell* is a much more difficult one, because we do not have any characteristics, the increase or decrease, which could explain the way, whereon the primitive nerve cell underwent up to the highest representants of the animal kingdom. Having not such characteristics I try to give an answer to this question on the basis of our comparative neurohistological investigations considering the form and shape of the nerve cells of the most different animals and man. But I shall give first the genesis of the primitive nerve cell, imaginated after the different hypotheses.

Genesis of the Nerve Cell

The nerve cell as an independent morphological and physiological unity appeared in the epidermis of the primary metazoon at the time when this animal, by means of the so-called independent effectors, began to move forward. The animal moving to one direction encountered with a lot of different stimuli. With this life-form began the transformation of the common epidermis cells into nerve cells.

The transformation of the indifferent epithelial cell into nerve cell started when sensibility of some of the epidermic cells increased, and they began to receive stimuli and to forward excitatory impulses. The peripheral part of the cell became, under such circumstances, a protoplasmatic process of variable length and thickness. In connection with such transformations some of the cells of this nature began to get under the epidermis and obtained a central process. These cells of epidermic origin are the sensory cells or receptor cells. Cells like that moved later into the deeper part of the body and became the bipolar nerve cells. These cells connected synaptically or asynaptically with the independent effectors and the receptor cells of the epidermis rendered possible to start reflexogenic actions. From the bipolar cells the multipolars originated corresponding to the requirement. This theory of the cytogenesis of the nerve cell, originated partly from PARKER (31), does not explain, in PASSANO's opinion, the spontaneous activities which appear very often in the animal kingdom (32). However, to explain the origin of the nerve cell seems to hold good even to-day.

Cells of Diffuse Nervous Systems

Nerve cells of similar appearance as in the supposed primitive metazoon could be seen in the sponges. These cells among which there are uni-, bi- and multipolar forms, according to the data in the literature are in continuity with each other, but their connection with the receptory and effectory elements of the body seems to be synaptic.

The situation is just like that of the coelenterates whose nerve cells are multipolar and the connection with one another and with the receptory and effectory elements pretended to be synaptic (27).

In the higher classes of the Metazoa the nerve cells form centres, ganglia and brain, in which the cells of the most different shape and size come together. I shall deal in the following with the nerve cells in the brain and thereafter with those in the vegetative nervous system, paying attention to the phylogenetical connection.

Nerve Cells of the Brain

Among the cells in the brain of the *Plathelminthes* there are multi-, bi- and unipolar types. So the situation is alike to that in the two first groups of the animal kingdom with the difference that they are forming a centre (1, 22, 28).

In the brain of the *Annelids* the central neuropil is surrounded by the cortex of nerve cells. They possess merely a single process. It is quite thick and its course is roughly straight.

Fig. 1. *Cyprinus carpio:* Midbrain, cross-section. Unipolar nerve cell with synapses. a) cell-body, b) cell-process, c) neurofibrils, d) presynaptic fibre, e) synapses, f) thick nerve fibres, g) thin nerve fibres, h) glial nuclei, i) capillary. BIELSCHOWSKY's method. Magn. 600 x. Photographically reduced to ⅓.

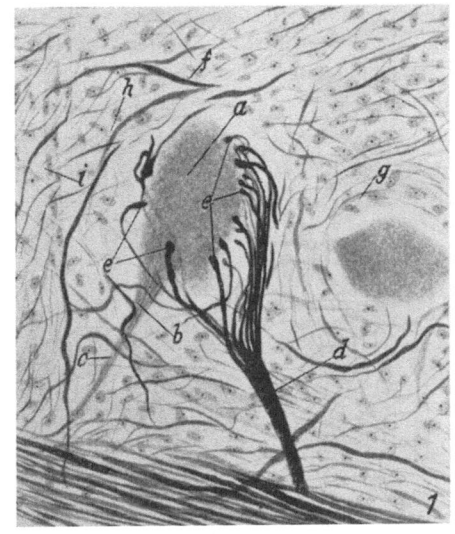

The brain of the *Arthropods* is like that of the Annelids. The outer layer, the cortex, consists of unipolar neurons of different size. They are arranged in several groups, however, well-defined to each other. Among the cells there are special ones producing neurosecret granules (18, 25).

The brain and ganglia of the *Mollusca* consist of unipolar cells of different size. Some are large and their process is very thick. These are the so-called giant cells (24, 26).

According to the foregoing the unipolarity is a basic character of the nerve cells beginning from the Annelids to the Mollusks, but in the classes of the vertebrates there appear again the other basic forms, namely the bipolar and multipolar ones.

In the *Vertebrates* the situation is the following:

The nerve cells in the brain of the *Fishes* are in the greatest part unipolar and bipolar, varying considerably in form and size. Some of the large unipolar cells in the midbrain are very similar to the nerve cells in the brain of the Annelids, Arthropods and Mollusks (Fig. 1). The nerve cells show some characteristics regarding the class, but these are not remarkable enough.

The cells in the brain of the *Amphibians* are for the most part multipolar, but unipolar and bipolar types could be seen in quite great quantity. The multipolar cells are relatively small, the cell-body is like a triangle, the processes are thin and very numerous. The cells belonging to the other two types are larger, the cell-body is longish, the processes are quite thick.

The nerve cells in the brain of the *Reptiles* are peculiar and very different. Some of them are multipolar and form nests of different size below the outer fibrous layer, most of the others are bipolar. The cell-body of the latter is small, the nucleus relatively large and the process thick and straight. Very big cells could be found in the brain of the snakes, very much like that of the Annelids and Mollusks, but they are of multipolar type.

The nerve cells in the brain of the *Birds* form thick cortex on the surface of the brain but no layer at all. The cortex consists of relatively small multipolar cells and nerve fibre bundles, the fibres of the latter are fine and form very straight trabacles vertical to the surface of the brain. Bipolar and unipolar cells occur in all parts of the brain. The multipolar nerve cells in the cortex form baskets very similar to that of the tortoises. The cell-bodies are like triangles, the dendrites are numerous, originating with quite thick basis and ending not far from the cell.

The nerve cells in the brain of the vertebrate classes discussed above show surely some characteristics and some evolutionary connection beginning from the lower to the higher ones in the system, but the transition from one class to the other is not remarkable at all.

The nerve cells in the brain of the *Mammals* show a little different character from the Lissencephala and Gyrencephala. The cells in the cortex of the brain in the Lissencephala do not form any layer. The greatest part of the nerve cells is multipolar and small, however larger pyramidal cells could also be seen in the deeper part of the cortex. The nerve cells in the brain of the Gyrencephala are relatively large and multipolar forming layers which could be easily and well distinguished in most of the cases.

The form and size of the cells are variable in the animals belonging to the different orders, but these differences are not so significant as to demonstrate the way of the

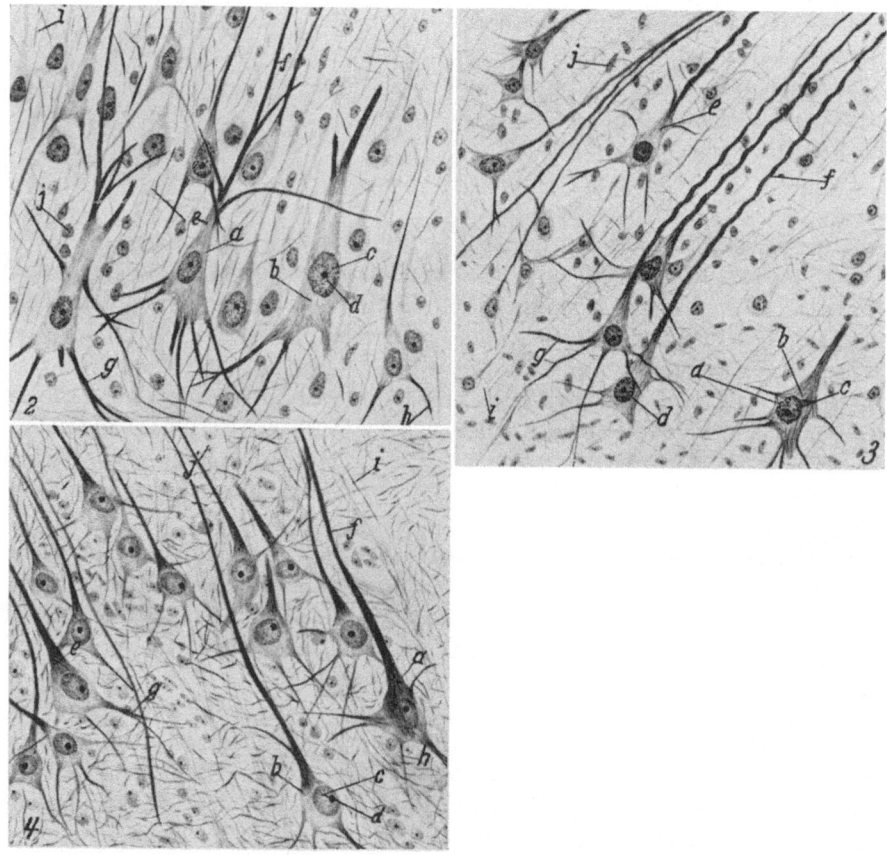

Fig. 2. *Canis familiaris:* Occipital lobe of the forebrain, cross-section. a) pyramidal cell, b) cell-body, c) nucleus, d) nucleolus, e) neurofibrils, f) topdendrit, g) dendrit, h) neurit, i) nerve fibre j) glial nuclei. BIELSCHOWSKY's method. Magn. 600 x. Photographically reduced to ⅓.

Fig. 3. *Sus scrofa domestica:* Occipital lobe of the forebrain, cross-section. a) pyramidal cell, b) cell-body, c) nucleus, d) nucleolus, e) neurofibrils, f) top- dendrit, g) dendrit, h) neurit, i) nerve fibres, j) glial nuclei. BIELSCHOWSKY's method. Magn. 600 x. Photographically reduced to ⅓.

Fig. 4. *Homo sapiens:* Forebrain cortex. Pyramidal cells from the gyrus praecentralis. a) pyramidal cell, b) cell-body, c) nucleus, d) nucleolus, e) neurofibrils, f) topdendrit, g) dendrit, h) neurit, i) nerve fibre, j) glial nucleus. BIELSCHOWSKY's method. Magn. 600 x. Photographically reduced to ⅓.

phylogenesis and the genetical connections between the different groups of mammals. To demonstrate these statements we show the pyramidal cells of the forebrain in three pictures; one originated from the dog, the other from the pig and the last from the man (Fig. 2, 3, 4).

Based on these pictures it can be stated that there is some difference in the cell-form of the three different brains. The differences are the greater, the greater the distance in the taxonomy is. In our opinion it is possible to find so the phylogenetical connection of the

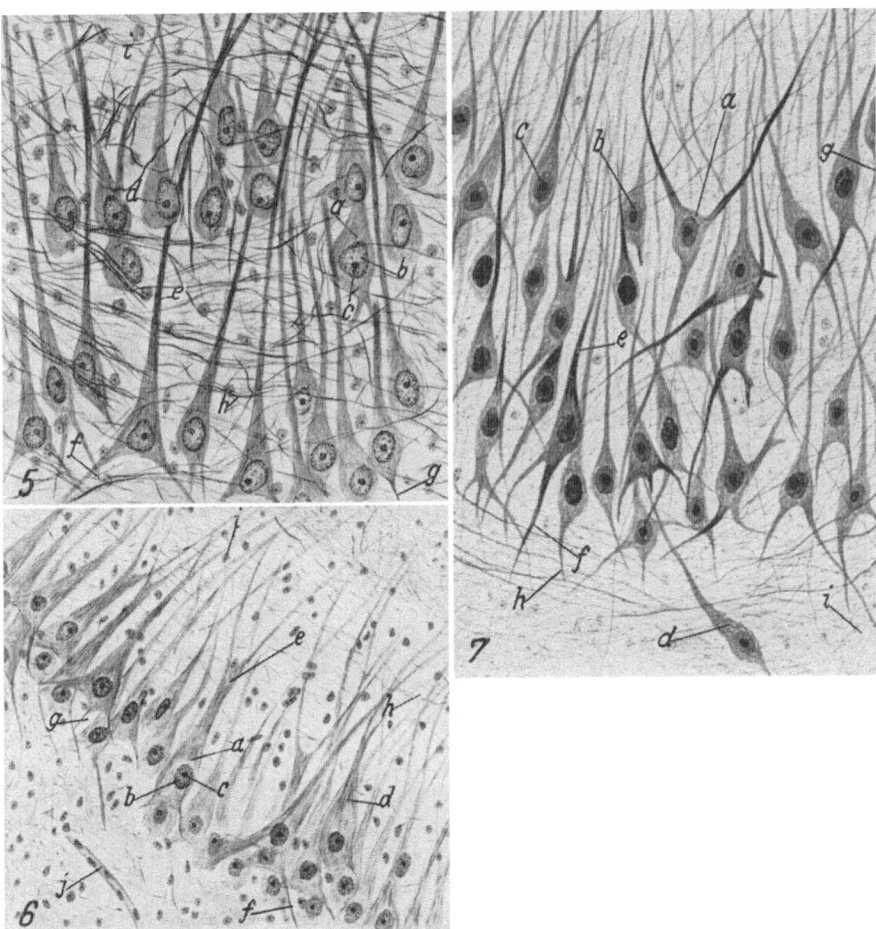

Fig. 5. *Canis familiaris:* Pyramidal cell from the gyrus hippocampi. a) cell-body, b) nucleus, c) nucleolus, d) neurofibrils, e) top-dendrit, f) dendrit, g) neurit, h) nerve fibre, i) glial nuclei. BIELSCHOWSKY's method. Magn. 600 x. Photographically reduced to ⅓.

Fig. 6. *Sus scrofa domestica:* Pyramidal cells from the gyrus hippocampi. a) cell-body, b) nucleus, c) nucleolus, d) neurofibrils, e) top-dendrit, f) dendrit, g) neurit, h) nerve fibre, i) glial nuclei, j) capillary. BIELSCHOWSKY's method. Magn. 400 x. Photographically reduced to ⅓.

Fig. 7. *Homo sapiens:* Pyramidal cells from the gyrus hippocampi. a) cell-body, b) nucleus, c) nucleolus, d) neurofibrils, e) top-dendrit, f) dendrit, g) neurit, h) nerve fibre, i) glial nuclei. BIELSCHOWSKY's method. Magn. 300 x. Photographically reduced to ⅓.

cell in the mammalian brain, but still a great deal is to be done. First of all the careful histological examinations of the brain of the monkeys, their relatives and of many other animals are badly needed.

To find some possibility indicating more clearly the way of the phylogenesis of the nerve cell, I examined also the *gyrus hippocampi* of the dog, the pig, cattle and the man.

According to these examinations the gyrus hippocampi is a place where the phylogenesis of the nerve cell can be followed at least partly. If we look at the cell-forms in the gyrus hippocampi of the dog, the pig, and the cattle we can state the similarity both in form and size of the cells of those animals, which have almost the same type, and the difference between those which are in their system far from each other. Moreover, our material demonstrates that the relations are quite different in the examined animals, and the man (Fig. 5, 6, 7).

The differences found in form and shape of the nerve cells in the gyrus hippocampi indicate, that this part of the brain seems to be a delicate material to investigate the phylogenesis of the nerve cell, the genetical connections between the different forms of mammalians and that between the higher mammalians and the man.

Vegetative Nervous System. In the field of the vegetative nervous system we wish to keep busy with the nerve cells of the digestive tube, ciliary ganglion, of the paravertebral and cardiac ganglia.

Digestive tube. The nerve cells of the digestive tube in *Annelids* are unipolar in the greatest part, but bipolar forms could be found in a considerable quantity (12).

The nerve cells in the wall of the digestive tube of the *Arthropods* belong mainly to the multipolar type, but bipolar forms occur quite frequently.

The nerve cells in the digestive tube of the *Mollusks* are unipolar, but bipolar and multipolar forms do not belong to the most infrequent forms (4).

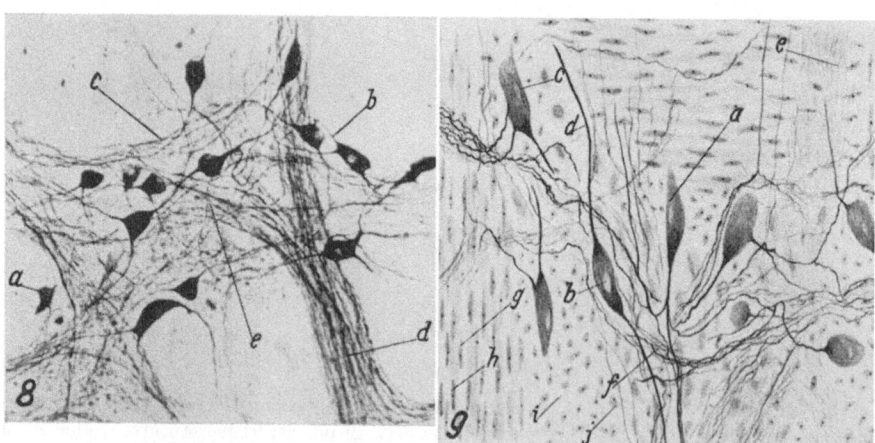

Fig. 8. *Emys orbicularis:* Jejunum. Plexus myentericus. a) cell-body, b) cell-process, c) nerve fibre, d) nerve trunk, e) nerve plexus. BIELSCHOWSKY-ÁBRAHÁM's method. Microphotography.
Fig. 9. *Felis domestica:* Duodenum, plexus myentericus, a) nerve cell, b) nucleus, c) neurofibrils, d) cell process, e) nerve trunk, f) nerve fibres, g) smooth muscle cell, h) nucleus of muscle cell, i) connective tissue, j) nucleus of the connective tissue cell. BIELSCHOWSKY-ÁBRAHÁM's method. Magn. 400 x. Photographically reduced to ⅓.

Among the nerve cells of the plexuses in the digestive tube of the *Fishes* the unipolar, bipolar and multipolar forms are distributed quite equally (2).

The nerve cells of the digestive tube in *Amphibians* belong mostly to the unipolar type, bipolar and multipolar forms occur quite seldom.

In the *Reptiles* multipolar cells play the greatest part in the nerve supply of the digestive tube and most of them belong to the second type of DOGIEL (Fig. 8).

The relations are like in the *Birds* and the *Mammals* (3), but we find digestive tubes where the cell-forms, the processes and the ramifications characterise the species (Fig. 9).

Summarising the situation found in the digestive tube it can be stated that on the nerve cells, starting from the unipolarity to the multipolarity, some steps of the phylogenesis can be seen, but in the higher forms of the Metazoa the cell-forms are so mixed, so no real law can be found.

Ciliary ganglion. The cells in the ciliary ganglion of the *Fishes* are unipolar. The cell-body is rounded, the nucleus is spheric and located in the center of the cell. The cell-process originates with broad basis and becomes gradually thinner. On the surface of the cell there is neither spiral fibre nor pericellular basket.

The cells of the ciliary ganglion in the *Amphibians* are large, rounded and unipolar (35). On the basis of the process on the cell-body there is a spiral fibre which encircles it twice or more.

The cells in the ciliary ganglion of *Reptiles* are large, the nucleus is relatively small. Around the process there is a thick spiral fibre encircling it several times (Fig. 10).

The cells of the ciliary ganglion of the *Birds* are large, rounded, the nucleus is spherical and in the center of the cell-body. The cell has only one process originating with a broad basis from the cytoplasma. On or around the cell-body one can see the ending or end-arborisation of the end-fibres, ending in small end-bulbs on the surface of the cell. The end-fibres form sometimes fibre baskets enclosing the cell-body totally (8).

The cells in the ciliary ganglion of the *Mammals* are relatively large and multipolar (33, 34). The special mammalian character of this organ is the fenestrated cell-form. The latter is the most frequent and the most complicated in the ciliary ganglion of Artiodactyla. The characteristics of this cell-form are the anastomosing processes of the cell except one which independently runs quite far from the cell-body (Fig. 11).

Taking into consideration the results obtained from the comparative histology of the ciliary ganglion we could state that in the development of unipolarity to multipolarity from the spiral to the basket the traces of the phylogenesis of the nerve cell can be noted, but the evolution is interrupted and the continuous transition from one kind of the cell-body to the other could not be stated at all.

Paravertebral ganglia. The nerve cells of the paravertebral ganglia of the *Fishes* are unipolar. The site of the cells is variable. The cell-bodies are surrounded by satellite cells.

The nerve cells in the paravertebral ganglia of the *Amphibians* (29, 30) are relatively large, unipolar, and the basis of the process is encircled by a nerve fibre of considerable thickness (Fig. 12).

The nerve cells in the paravertebral ganglia of the *Reptiles* are arranged in lines between the fibre bundles. The greater part of the cells are unipolar, however bipolar types are seen in considerable quantity. Remarkable in these cells is the relatively small nucleus, chromatin-poor and located quite close to the surface of the cell-body. The speciality of these cells is that the cell-process originates from a well-defined neurofibrillar coil in the

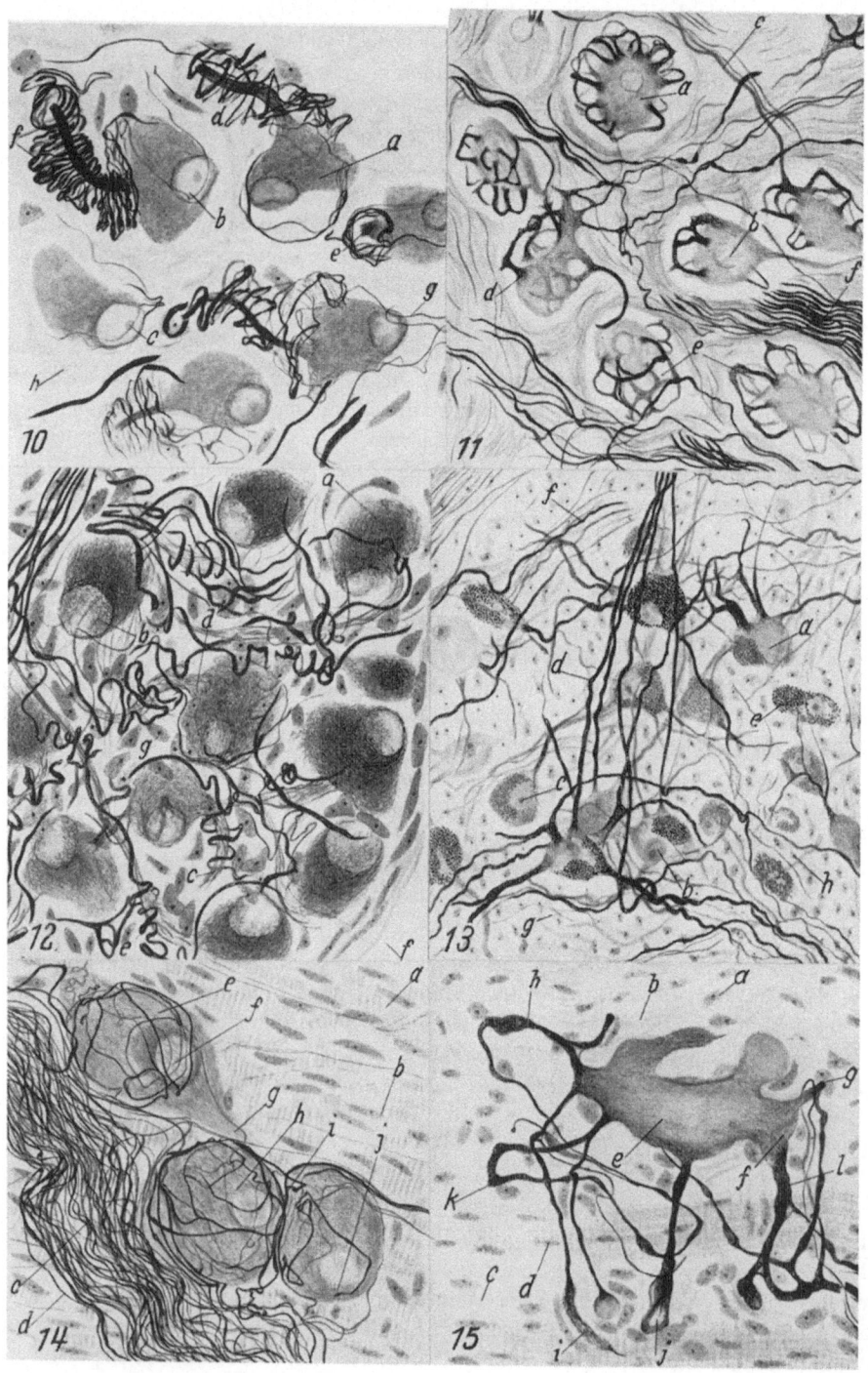

next vicinity of the nucleus. A spiral of nerve fibre occurs neither round the cell nor round the process.

The cells in the paravertebral ganglia of the *Birds* are multipolar, showing a notable variation in form and in size. In some cases between the cells of normal size there are very large cells which show a great affinity to silver nitrate.

The nerve cells in the paravertebral ganglia of the *Mammals* are multipolar, quite large and contain lipofuscin of a considerable quantity in their cytoplasm, in some of the mammalians and in the man. The spheroid granules are dispersed in the cytoplasm or in one of its well-defined part. In other cases they appear accumulated in very different forms. The lipofuscin granules leave sometimes the cell-body and are accumulated around the cells (6) (Fig. 13).

On and around the cells in the human material could be found bulb-like forms. These formations were mainly found in the lumbal ganglia of such people with blood-vessel diseases. Loose nerve-fibre plexuses and pericellular baskets occur round the nerve cells (7, 9, 15).

According to the above findings the structure of the paravertebral ganglia does not show distinctly the way of the phylogenesis of the nerve cell, however there are some signs indicating the degree of the evolution. Such signs are the way from the uni- to the multipolarity and from the spirals to the pericellular baskets.

Cardiac ganglia. To demonstrate the phylogenesis of the nerve cells even the heart ganglia could be used in some respect. They are located in the epicardium of the atria and in the atrial septum except in birds, where the ganglia occur in the epicardium of the ventricles.

The nerve cells of the ganglia in the heart of the *Fishes* are generally unipolar. They are rounded, the nucleus located in the centre of the cell-body. There is no spiral formation

Fig. 10. *Emys orbicularis:* Nerve cells of the ciliary ganglion. a) cell-body, b) nucleus of the nerve cell, c) nucleolus of the nerve cell, d) cell-process, e) neurofibrils, f) nerve fibre spiral, g) nerve fibre, h) connective tissue, i) nucleus of the connective tissue cell. BIELSCHOWSKY-ÁBRAHÁM's method. Magn. 1.600 x. Photographically reduced to ⅓.

Fig. 11. *Bos taurus:* Nerve cells of the ciliary ganglion. a) cell- body, b) cell-nucleus, c) cell-process, d) neurofibrils, e) fenestra, f) nerve fibres. BIELSCHOWSKY-ÁBRAHÁM's method. Magn. 600 x. Photographically reduced to ⅓.

Fig. 12. *Rana ridibunda:* Nerve cells of the paravertebral ganglion. a) cell-body, b) nucleus, c) cell-process, d) neurofibrils, e) nerve fibre spiral, f) connective tissue, g) nucleus of the connective tissue cell. BIELSCHOWSKY-ÁBRAHÁM's method. Magn. 1.200 x. Photographically reduced to ⅓.

Fig. 13. *Homo sapiens:* Nerve cells of the ganglion stellatum. a) cell-body, b) nucleus of the nerve cell, c) nucleolus of the nerve cell, d) cell-process, e) lipofuscin granules, f) nerve fibre, g) connective tissue, h) nucleus of the connective tissue cell. BIELSCHOWSKY-ÁBRAHÁM's method. Magn. 600 x. Photographically reduced to ⅓.

Fig. 14. *Emys orbicularis:* Nerve cells of parasympathetic type in the atrial septum. a) striated muscle fibre, b) nucleus of the muscle fibre, c) nerve trunk, d) nerve fibre, e) nerve cell, f) nucleus of the nerve cell, g) cell-process, h) neurofibrils, i) pericellular apparate, j) synapse. BIELSCHOWSKY-ÁBRAHÁM's method. Magn. 1.000 x. Photographically reduced to ⅓.

Fig. 15. *Homo sapiens:* Nerve cell in the right atrial epicardium. a) nucleus of connective tissue cell, b) collagenic fibres bundle, c) striated muscle fibre, d) nucleus of cardiac muscle fibre, e) nerve cell, f) neurofibrils, g) neurit, h) dendrit, i) dendrit-lamella, j, k, l) special dendrites. BIELSCHOWSKY-ÁBRAHÁM's method. Magn. 1.600 x. Photographically reduced to ⅓.

around the cell-process, however such formation could sometimes be seen on the surface of the cell-body (14).

The cells in the ganglia of the heart in the *Amphibians* are unipolar. They are characterised by fibre-spirals encircling the process especially at their basis. Many nerve cells with spiral formation could be seen in the sinus venosus and in the septum atriorum of the frog (16).

The nerve cells of the heart ganglia of the *Reptiles* are quite large, unipolar cells (17). The pericellular fibre formation on the cells is very frequent, especially in the sinus venosus. Around the nerve cells of the cardiac ganglia in the pond tortoise very rich pericellular plexuses are found (Fig. 14).

The cells in the cardiac ganglia of the *Birds* are multipolar (11, 20). The cell-body is multiangular, the processes originate with broad basis and become thin quite early.

The cardiac ganglionic cells of the *Mammals* are multipolar and of a remarkable size (21, 23). Everywhere, especially in the cardiac ganglia of the pig and of the man are peculiar cell-forms (Fig. 15). The dendrites of such cells are very short, sometimes enormously broad and end just in the close vicinity of the cell-body. Fenestrated cells could also be seen very often in the cardiac ganglia, especially in the wall of the sinus coronarius (5, 7, 10, 13).

The relations found in the cardiac ganglia of Vertebrates show an increase in the number of the cell-processes from the fishes to the mammalians, the pericellular apparates, the fenestrated cells and the peculiar cell-forms appear, however all these formations show no conformity to certain rules.

Summary

According to our foregoing findings, the appearance and the form of the nerve cells are very much alike in the whole animal kingdom from the sponges to the mammals and the man. To find the way of the phylogenesis under such circumstances is the most difficult problem for anyone investigating the laws and history of animal kingdom. However, there are some phenomena and facts, visible in the central nervous and in the vegetative system, which show the signs of the phylogenesis and in certain places quite clearly. These phenomena and facts are to-day generally very few, but still valuable and sometimes useful to elucidate the question, at least in some details. Such a fact is the unipolarity of the nerve cells starting from the Annelids to the Vertebrates, and similarity of the nerve cells in Fishes, Amphibians and Reptiles. However, the similarity between the cells of the Reptiles and the Birds could not be denied, but the same can be said about the Reptiles and Mammals in respect to form and size of the pyramidal cells.

Some signs of the phylogenesis can also be seen in the vegetative nervous system. The development from the unipolarity to the multipolarity in the digestive tube and in the ciliary ganglion, from the spirals to the basket formations, from the cells with free processes to the fenestrated in the ciliary ganglion and in the cardiac ganglia are phenomena which indicate the way of the phylogenesis, at least in some respect. The relations found in the paravertebral ganglia seem to furnish some data concerning the evolution of the vegetative nerve cell. But all these things are insufficient to solve the extremely complicated and important question. Now, what has to be done to give a more favorable answer to the question dealing with the problem ? First of all the structure and the ontogenesis of the nervous system in the whole animal kingdom must be systematically, comparatively and carefully studied, using the lightmicroscope, the electronmicroscope and the different methods of cytochemistry, histochemistry and biochemistry. The very thorough knowledge of the endoplasmatic reticulum, mitochondria, lysosomes, GOLGI bodies, neurofibrils, secretory granules, the chemistry of the different components of the cell-body, especially the structure of the nucleic acids and of the ribosomes will throw light and clearness on the history and evolution of the nerve cell. I believe, when using the methods and instruments needed for this purpose, we may hope to be able to understand the structure, the phylogenesis of the nerve cell and its functions of which is still fairly little known to-day.

References

1. ÁBRAHÁM, A.: Das Nervensystem von Opisthodiscus nigrivasis M. Stud. Zool. *11* (1929), 137–157.

2. ÁBRAHÁM, A.: Über die Innervierung des Verdauungstraktes einiger Knochenfische. Magyar Biol. Kutatóint. I. Oszt. munkáiból. *6* (1933), 1–12.

3. ÁBRAHÁM, A.: Beiträge zur Innervation des Vogeldarmes. Z. Zellforsch. *23* (1936), 737 bis 745.

4. ÁBRAHÁM, A.: Die Innervation des Darmkanals der Gastropoden. Z. Zellforsch. *30* (1940), 273–296.

5. ÁBRAHÁM, A.: Die Sinusgegend des menschlichen Herzens und ihr Nervensystem. Z. Zellforsch. *31* (1940), 146–155.

6. ÁBRAHÁM, A.: The comparative histology of the stellate ganglion. Acta biol. Acad. Sci. hung. *2* (1951), 311–354.

7. ÁBRAHÁM, A.: Blood pressure and peripheral nervous system. Acta biol. Acad. Sci. hung. *4* (1953), 307–365.

8. ÁBRAHÁM, A., A. STAMMER: A madarak szemmozgató izmainak beidegzése tekintettel a ganglion ciliare szerkezetére. Állattani Közl. *44* (1954), 115–134.

9. ÁBRAHÁM, A.: Über die Probleme in der Histologie des vegetativen Nervensystems. Acta biol. Univ. Szeged *2* (1956), 111–135.

10. ÁBRAHÁM, A.: The structure of cardiac ganglia. Proc. First meet. Hungarian Biol. Soc. Budapest. Acta biol. Acad. Sci. hung. Suppl. *8* (1956), 11–12.

11. ÁBRAHÁM, A., A. STAMMER: Die mikroskopische Innervation des Vogelherzens. Acta biol. Univ. Szeged *3* (1957), 247–273.

12. ÁBRAHÁM, A., E. MINKER: Über die Innervation des Darmkanales des medizinischen Blutegels. Hirudo medicinalis. Z. Zellforsch. *47* (1958), 307–311.

13. ÁBRAHÁM, A.: The microscopical innervation of the vertebrate heart. XVth. Intern. Congr. Zool. *33* (1958), 1–3.

14. ÁBRAHÁM, A., I. HORVÁTH: Über die mikroskopische Innervation des Herzens von Süßwasser-Knochenfischen. Z. mikr.-anat. Forsch. *65* (1959), 1–20.

15. ÁBRAHÁM, A.: Zur Frage der interneuronalen Synapsen in den vegetativen Ganglien. Z. mikr.-anat. Forsch. *65* (1959), 574 bis 581.

16. ÁBRAHÁM, A.: Die mikroskopische Innervation des Herzens der Amphibien. Acta biol. Univ. Szeged *7* (1961), 45–64.

17. ÁBRAHÁM, A.: Die mikroskopische Innervation des Herzens der Reptilien. Acta biol. Univ. Szeged. *7* (1961), 95–107.

18. ÁBRAHÁM, A.: Histological, histochemical and cytological examinations on the central nervous system of the swimming beetle. Dytiscus marginalis. Acta biol. Acad. Scient. Hung. Suppl. *4* (1962), 35.

19. ÁBRAHÁM, A.: Die Nervenversorgung der Kranzgefäße des Herzens. Arch. int. Pharmacodyn. *139* (1962), 17–27.

20. ÁBRAHÁM, A.: Die intramurale Innervation des Vogelherzens. Z. mikr.-anat. Forsch. *69* (1962), 195–216.

21. ÁBRAHÁM, A.: Über die Struktur der Ganglien des Säugetierherzens. Acta biol. Univ. Szeged *8* (1963), 121–134.

22. ÁBRAHÁM, A.: Összehasonlító állatszervezettan. Tankönyvkiadó, Budapest 1964.

23. ÁBRAHÁM, A.: Die Innervation des Herzens und der Blutgefäße von Vertebraten. Akad. Kiadó, Budapest 1964.

24. ÁBRAHÁM, A.: The structure of the interneuronal synapses in the visceral ganglion of Aplysia. Acta anat. *54* (1963), 260–263.

25. ÁBRAHÁM, A.: Histochemical investigations on the neurosecretory system of the swimming beetle Dytiscus marginalis. II. Internat. Kongr. für Histo- und Cytochemie. Frankfurt 1964.

26. ÁBRAHÁM, A.: Die Struktur der Synapsen im Ganglion viscerale von Aplysia californica. Z. mikr.-anat. Forsch. 73 (1965), 45–59.

27. BOZLER, E.: Untersuchungen über Nervensystem der Coelenteraten. Z. Zellforsch. 5 (1927), 1–36.

28. HANSTRÖM, B.: Vergleichende Anatomie des Nervensystems der wirbellosen Tiere. Springer, Berlin 1928.

29. HORVÁTH, I.: Histologische Untersuchungen an den paravertebralen Ganglien von Rana ridibunda. Acta biol. Univ. Szeged 8 (1962), 135–143.

30. HORVÁTH, I.: Untersuchungen an den paravertebralen Ganglien von Rana ridibunda. Acta biol. Univ. Szeged. 9 (1963), 117–127.

31. PARKER, G.H.: The Elementary Nervous System. J.B. Lippincott Co., Philadelphia 1919.

32. PASSANO, L.M.: Primitive nervous systems. Proc. Nac. Acad. Sci. 50 (1963), 306–313.

33. STAMMER, A.: Beiträge zur Kenntnis des Ganglion ciliare des Hundes. Acta biol. Univ. Szeged 2 (1957), 219–234.

34. STAMMER, A.: Histological and histochemical examinations on the ciliary ganglion of mammals. Suppl. Biol. Hung. 5 (1965), 93 to 107.

35. STAMMER, A.: Die mikroskopische Struktur des Ganglion ciliare der Frösche. Acta biol. Univ. Szeged 11 (1965), 119–124.

Prof. Dr. A. ÁBRAHÁM
Zoological and Biological Institute
of the József Attila University
Táncsics Mihály-u 2
Szeged, Hungary

Comparative Studies on the Distribution of Monoamine Oxidase and Succinic Dehydrogenase in Vertebrates' Forebrain

By H. Masai, T. Kusunoki, and H. Ishibashi, Yokohama

The purpose of this work is to study the chemoarchitectonics of the forebrain, related to a fundamental plan of the central nervous system of vertebrates. In the architectural organization of the central nervous system of vertebrates the sulcus limitans of His is the most important landmark which runs rostrocaudally to the lateral walls of the neural tube in the embryonal stage.

The visceral areas are derived from the portions along the sulcus limitans, while the somatic areas develop from the parts distant to the sulcus. On the basis of some embryological investigations (1, 3), it is thought that the sulcus limitans ends rostrally in the hypothalamus, though the authors differ in their conclusion about the site of rostral termination of this sulcus within the hypothalamus, where the autonomic centers would be located.

Recently it has been emphasized by many investigators that the limbic system, including the phylogenetically older cortical and subcortical parts of the forebrain, has close connections with the hypothalamus. From many results of electrophysiological experiments on fiber-connections, it was inferred that these parts play a role in the visceral functions. Moreover in the field of neuropharmacology Vogt a. Brodie et al. (2, 6) have found considerable amounts of norepinephrine and serotonin in the hypothalamus which have an important significance as chemical transmitters for the visceral functions in the brain.

Therefore present authors intended to investigate the relationship between the visceral areas of the forebrain and the distribution of monoamine oxidase metabolizing these transmitters, such as catecholamines and serotonin, from phylogenetical standpoints. In addition, the distribution of succinic dehydrogenase participating in aerobic respiration was studied, in order to provide a comparison with monoamine oxidase activity.

Material and Methods. The animals used were the chick embryo, tadpoles of *Xenopus, Rana nigromaculata, Bufo vulgaris japonicus, Uroloncha striata* v. domestica, rat, *Rhinolophus cornutus* and *Carassius carassius auratus*.

For the histochemical procedure, fresh and unfixed brains and spinal cords were cut into serial, frontal and sagittal sections of 30 μ using the cryostat. Glenner's method was utilized for demonstrating monoamine oxidase, and Nachlas' method for succinic dehydrogenase.

Results and Comments

The fundamental distribution of monoamine oxidase activity is shown in chick embryo after 9 days' incubation (5) (Figs. 1, 2). An intensive activity of monoamine oxidase is demonstrated in regions adjacent to the sulcus limitans, and further rostrally, in the

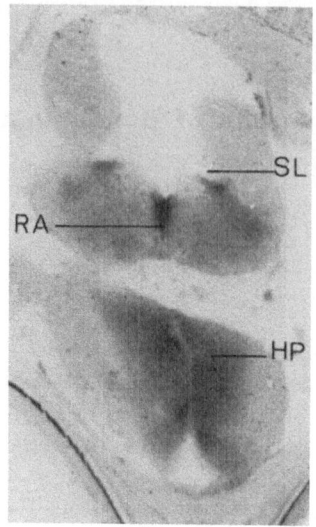

Fig. 1.

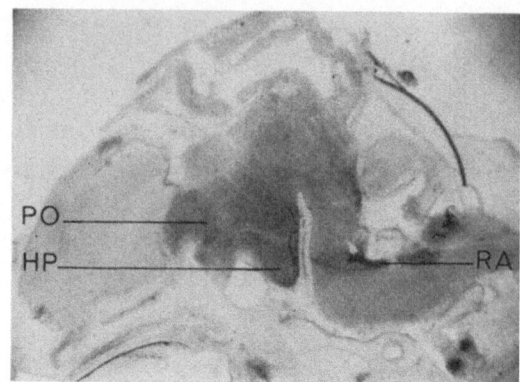

Fig. 2.

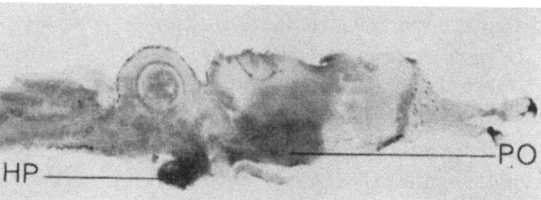

Fig. 3.

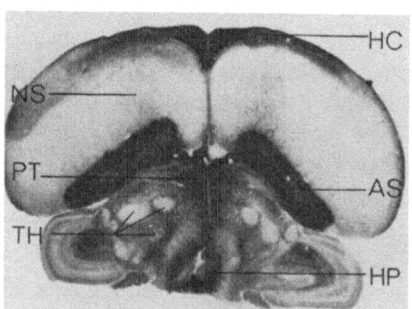

Fig. 4.

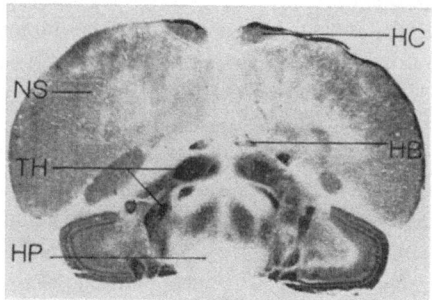

Fig. 5.

Fig. 1. Transverse section of the brain of chick embryo after 9 days' incubation. x 2.5. Monoamine oxidase activity is intense in the areas adjacent to the sulcus limitans (SL), hypothalamus (HP) and raphe (RA) of brain stem.

Fig. 2. Sagittal section of the brain of chick embryo after 9 days' incubation. x 2.5. The hypothalamus (HP), preoptic area (PO) and raphe (RA) show a strong-positive reaction of monoamine oxidase.

Fig. 3. Sagittal section of the brain of *Rana nigromaculata*. x 2.0. Monoamine oxidase activity is strong-positive in the hypothalamus (HP) and preoptic area (PO).

Fig. 4. Transverse section at the caudal level of the forebrain of *Uroloncha striata* v. domestica. x 3.0. Monoamine oxidase activity is intense in the hypothalamus (HP), periventricular parts of the thalamus (PT), archistriatum (AS), and hippocampus (HC), while the neostriatum (NS) and most of the thalamic nuclei (TH) show weak reactions.

Fig. 5. Transverse section at the caudal level of the forebrain of *Uroloncha striata* v. domestica. x 3.0. Succinic dehydrogenase activity is strong-positive in the neostriatum (NS), most of the thalamic nuclei (TH), habenula (HB) and hippocampus (HC), while the hypothalamus (HP) shows a faint reaction.

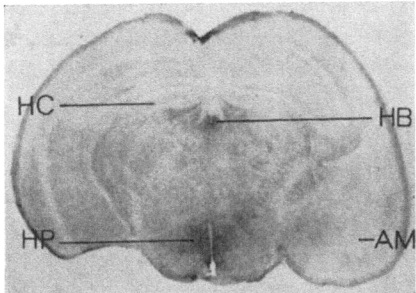

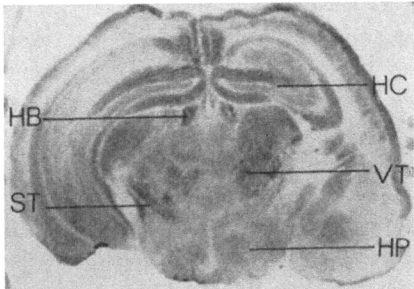

Fig. 6.   Fig. 7.

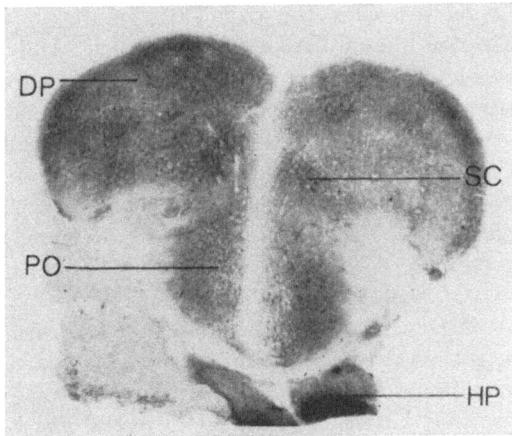

Fig. 8.

Fig. 6. Transverse section at the caudal level of the forebrain of *Rhinolophus cornutus*. x 5.4. Monoamine oxidase activity is relatively strong-positive in the hypothalamus (HP) and habenula (HB), while the hippocampus (HC), amygdaloid body (AM) and most of the thalamic nuclei and subthalamic nucleus show a weak reaction.

Fig. 7. Transverse section at the caudal level of the forebrain of *Rhinolophus cornutus*. x 5.4. Succinic dehydrogenase activity is shown intensely in the hippocampus (HC), habenula (HB), ventral thalamic nuclei (VT), subthalamic nucleus (ST) and neocortex, while the hypothalamus (HP) shows a weak reaction.

Fig. 8. Transverse section at the caudal level of the forebrain of *Carassius carassius auratus*. x 10. Monoamine oxidase activity is strong-positive in the hypothalamus (HP), preoptic area (PO), supracommissural part (SC) and dorsal parts (DP).

hypothalamus and the periventricular part of the thalamus, in the preoptic area and the septal region. Also the raphe showed a strong reaction.

From these results it is thought that monoamine oxidase is distributed in relatively large amounts in the visceral areas, through all parts of the central nervous system. After the incubation has been prolonged to 14 days, a strong reaction of monoamine oxidase is present in the hippocampus, as well as the septal region, the hypothalamus and the preoptic area.

In the amphibia (4) (Fig. 3) the telencephalon, which remains tube-like at the primitive stage of evolution like a matrix, the septal region, preoptic area, hypothalamus and epistriatum represent a strong reaction of monoamine oxidase, as is seen in the chick embryo.

In homoiothermal animals the activity of monoamine oxidase in the bird *(Uroloncha)* (Fig. 4) is relatively strong-positive in the following parts: hippocampus, septal region, preoptic area, hypothalamus, periventricular parts of the thalamus, habenula, paleostriatum and archistriatum. On the other hand, the hyperstriatum, neostriatum and most of the thalamic nuclei, including the nucleus rotundus and nucleus ovoidalis, show a relatively weak reaction of monoamine oxidase. Contrary to the distribution patterns of monoamine oxidase a strong reaction of *succinic dehydrogenase* (Fig. 5) is revealed in the following portions: hyperstriatum, neostriatum, paleostriatum, habenula, hippocampus, and thalamic nuclei, except in the periventricular gray. These gray matters are thought to be related to somatic function, except the habenula. Habenula, hippocampus and paleostriatum show active reactions of both monoamine oxidase and succinic dehydrogenase. However, archistriatum, septal region, preoptic area and hypothalamus show a weak reaction of succinic dehydrogenase. In general, the degree of the enzymic reactions is higher in birds than in mammals.

In relation to mammals many studies have been already carried out and therefore the observation of bat *(Rhinolophus)* is introduced. This species represents one of the most interesting strains, possessing peculiar adaption and specialization, such as hibernation and flight. The activity of monoamine oxidase (Fig. 6) is intensely present in the septal nuclei, preoptic area, habenula, nucleus reuniens of the thalamus and hypothalamus. The hippocampus and amygdaloid body show a faint reaction of monoamine oxidase, differing from rat. The activity of succinic dehydrogenase (Fig. 7) is relatively intense in the neocortex, caudate nucleus, hippocampus, anterior, ventral and lateral nuclei of the thalamus, lateral geniculate nucleus and subthalamic nucleus which are weak-positive in the reaction of monoamine oxidase, except in the habenula.

The telencephalon of *bony fishes* has developed in a special direction of the evolutionary progress and exhibits the eversion of the medial wall towards the dorsolateral side. In another respect fishes show the most different patterns of the brain. In gold fish (Fig. 8) a strong reaction of monoamine oxidase appears in the following areas: olfactory bulb, anterior olfactory nucleus, dorsal area, supracommissural parts of the medial olfactory area, preoptic area and hypothalamus. Contrary to the distribution patterns of monoamine oxidase the reaction of succinic dehydrogenase is strong-positive in the rotundus-complex and lateral geniculate nucleus. The distribution of both enzymes in gold fish is fundamentally similar to that of other classes.

Summary

Monoamine oxidase activity and succinic dehydrogenase activity are studied in verte-brates forebrain by histochemical methods from the phylogenetical point of view.

1) *Monoamine oxidase activity* is recognizable relatively intense and clear in the visceral areas, the hypothalamus, preoptic area, septal area and hippocampus; although the reaction of the hippocampus is variable according to species. The older parts of the striatum show relatively strong reaction of monoamine oxidase. In contrast to the distribution

patterns of monoamine oxidase the activity of succinic dehydrogenase is generally represented relatively intense in most of the somatic areas.

2) The activities of monoamine oxidase and of succinic dehydrogenase is weak in amphibia, which are in the most primitive stage of the metabolic evolution. Birds show the most distinct reactions of both enzymes in all classes observed.

References

1. BERGQUIST, H., B. KÄLLÉN: J. comp. Neurol. *100* (1954), 627.
2. BRODIE, B.B., P.A. SHORE: Ann. N.Y. Acad. Sci. *66* (1957), 631.
3. KINGSBURY, B.F.: J. comp. Neurol. *32* (1920), 113.
4. KUSUNOKI, T., H. ISHIBASHI, H. MASAI: J. Hirnforsch. (in press).
5. MASAI, H., T. KUSUNOKI, H. ISHIBASHI: Experientia *21* (1965), 572.
6. VOGT, M.: Pharmac. Rev. *6* (1954), 31.

Prof. Dr. H. MASAI, Dr. T. KUSUNOKI and Dr. H. ISHIBASHI
Department of Anatomy
Yokohama University School of Medicine
Yokohama, Japan

Some Applications of Enzymo-histochemical Techniques to the Study of the Maturing Allocortex Cerebri

By W. G. M. WITKAM, Nijmegen

Microscopic histochemistry applied to brain sections has afforded during the last decade a lot of information about the repartition of metabolically active substances on a histological level and has given us methods for marking of mitochondria, lysosomes, synaptic membranes etc. by specific staining of the enzymes enclosed in them. (For a review see COLMANT 1961).

At present it seems that interpretation of the results and correlation with neurophysiological and neuroanatomical data is often far from being completed and that with many technics the limit of the power of resolution is reached and further progress must be expected from electron-histochemistry. In the meantime slide-histochemistry of developing animals has been used systematically by only a few investigators and in most instances only with a single technique. Application of several histochemical techniques on a developmental series of brain preparations from a given species can give a better interpretation of the picture seen in adult animals and can be an important contribution to our understanding of brain maturation. Comparative studies on the brains of nidicolous and nidifugous animals ("Nesthocker, Nestflüchter") are very useful in this respect. Certainly the speed of biochemical maturation is quite unlike that of pure morphological development, as has been demonstrated e. g. for the ontogenesis of lactate-dehydrogenase enzymes in rat diencephalon, adult LDH-levels occurring only at 40 days after birth, while morphological differentiation is complete after 2 weeks (BONAVITA, 1964).

Following the general theme of this symposium some preliminary results of our own investigations about the developmental histochemistry of the bulbus olfactorius and the formatio hippocampi will be shown.

Materials and Methods. Fresh, unfixed brains of young Chinese hamsters *(Cricetulus baraben-sis)* and guinea-pigs of several ages, from birth up to about two months, were used for cryostate sectioning. Killing has been done by decapitation. Sections after postfixation or without it were incubated under standard conditions as far as possible.

They were incubated for: *Acetylthiocholinesterase,* AChE (GEREBTZOFF 1959a, ARVY 1962), after 30 sec. postfixation in a glutaraldehyde-cacodylate buffer at ph 7,2.

Incubation 2 h. with or without ethopropazine 10^{-4}m as an inhibitor of nonspecific cholinesterase. After visualisation in dilute $(NH_4)_2S$ they were developed in 1% $AgNO_3$ during 1 min. (MATHISEN a. BLACKSTAD, 1964).

Succinate-dehydrogenase, SDH, without fixation, during 30 sec. acc. to GÖBEL a. PUCHTLER (1955).

Lactate-dehydrogenase, LDH, with or without 15 sec. postfixation in glutaraldehyde-cacodylate buffer. Substrate solution according to ALLEN a. SLATER (1961) as cited by FRIEDE (1963).

Acid phosphatase after 4 hours postfixation in cold ethanol with β-glycerophosphate (FELGENHAUER 1963) or α-naphthylphosphate and hexazonium pararosaniline (ANDERSON 1962).

Thiaminepyrophosphatase, TPP-ase (GOLDFISCHER 1964).

Results

Bulbus olfactorius

The division of AChE in the main olfactory bulb of the adult hamster and guinea-pig shows strong activity of the glomeruli with a periglomerular accentuation; strong activity also of the internal plexiform layer and no activity in the layers of granular and mitral cells. The glomeruli of the accessory bulb show only a very weak AChE activity (Fig. 1). This picture is already complete in the second week and has developed during the first week after birth of the hamster. One day after birth (Fig. 2) we see a more diffuse localisation with accumulation around the glomeruli and in the plexuses external and internal to the mitral cell layer. In the bulb of the newborn guinea-pig, which as a whole contains much more AChE, we find diffuse localisation around the glomeruli, in the somata of the external granular cells, and in external and internal plexiform layers. Negative are the perikarya of tufted and mitral cells – which accounts for the patchy look of the external plexiform layer-, internal granular cells and the glomerular layer of the accessory olfactory bulb (Fig. 3).

The division of SDH and LDH in the main olfactory bulb of the hamster corresponds to the description of ORTMANN (1957). Only a weak dehydrogenase activity is found in the adult accessory bulb. The 7 days old hamster shows already adult SDH and LDH traits. In the neonatal guinea-pig we find strong LDH activity in the layer of the fila olfactoria, and in the glomeruli, whereas SDH activity is found in the glomeruli and in the outer plexiform layer (Figs. 4, 5).

Acid phosphatase in the olfactory bulb is found in the somata of all nerve cells but also in the glomeruli, being weak in the fila olf. and in the glomeruli of the main bulb and very strong in those of the accessory bulb, as can be demonstrated in both the adult hamster and the newborn guinea-pig with the lead salt method (glycerophosphatase) as well as with the azo-coupling technique (naphthylphosphatase) (Fig. 6, 7, 8).

The same relations are found in the 15 days and 11 days old hamsters.

Formatio hippocampi

Concerning the localisation of AChE in the "hippocampus inferior" of the adult hamster our slides show the picture known already from several fine descriptions in the literature (SHUTE a. LEWIS, 1960; GEREBTZOFF, 1959b; PONCELET, 1960; MATHISEN a. BLACKSTAD, 1964). The neuropilum of molecular layers, supra- and infragranular, supra- and infra-pyramidal layers shows strong activity; the stratum oriens moderate, the stratum radiatum weak activity and pyramidal and granular somata are negative with some fine precipitate between the cells which perhaps represents the baskets (MATHISEN a. BLACK-STADT, 1964).

In the hamster the first appearance of AChE is seen on the 9th to 11th day after birth at the level of the supragranular layer, the infrapyramidal layer and the mossy fibre system (Fig. 9); thereafter in the molecular layers on the 14th–15th day (Fig. 10) and at last in the intercellular baskets at the 19th–25th day. In the newborn guinea-pig the hilus fasciae dentatae and the mossy fibre system are still poor in AChE (Fig. 11).

The picture of SDH and LDH in the hippocampal formation of the adult hamster is in accordance with the findings of ORTMANN a. FLEISCHHAUER (1959). Molecular and polymorphic layers are fairly active as well as the mossy fibre system for SDH. When using short postfixation (vide supra) the somata of CA$_4$ pyramidal cells become positive, whereas in unfixed sections they do not show any LDH activity (Fig. 12).

SDH and LDH can be demonstrated already in the first week, for instance LDH in the 5 days old hamster (Fig. 13), although the localisation differs from that in the adult.

When observing our slides of acid phosphatase staining, it is evident that besides the high activity of the pyramidal somata (one of the highest of the brain) there exists activity also at the level of the mossy fibre layer in the adult hamster and in the newborn guinea-pig. When using higher magnifications, one gets the impression that not the mossy fibres themselves, but the apical dendrites of the CA 3 somata with their proximal parts penetrating the mossy fibre layer are stained here. This staining is clearest at the site of the so-called "end bulb" (Fig. 14). Staining of the mossy fibre system can be done with both phosphatase techniques.

In the hamster this staining of the mossy fibre system becomes positive on the 14th–16th day. As TPP-ase could account for the observed activity, we incubated some slides for TPP-ase and found it in the pyramidal and granular somata but not in the mossy fibre system (Fig. 15).

The after-birth development of α-naphthylphosphatase activity in the small somata of the gyrus dentatus neurons in the hamster is very remarkable. At first only the outermost layers of the gyrus dentatus are positive; gradually the inner layers are becoming phosphatase positive in an outside-inside gradient. On the 15th day after birth, when the mossy fibre system is already phosphatase positive, many of the granular somata are still negative (Fig. 16).

Fig. 1. Olfactory bulb of a 47 days old hamster. A Ch E. Horizontal section, 40 x.
    acc.   = accessory olfactory bulb   pl.i.   = internal plexiform layer
    gl.   = glomeruli   tr.olf.l. = lateral olfactory tract

Fig. 2. Olfactory bulb of a one day old hamster. A Ch E. Horizontal section, 40 x.
For abbreviations see Fig. 1

Fig. 3. Olfactory bulb of a newborn guinea-pig. A Ch E. Frontal section, 25 x

Fig. 4. Olfactory bulb of a newborn guinea-pig. SDH. Frontal section, 25 x
fil. = fila olfactoria   pl.e. = external plexiform layer

Fig. 5. Olfactory bulb of a newborn guinea-pig. LDH. Frontal section, 25 x

Fig. 6. Olfactory bulb of a 6 months old hamster. Glycerophosphatase. Horizontal section, 40 x
Positive reaction in the glomeruli of the accessory bulb (↑) and in the mitral cells (↑↑).

Fig. 7. Olfactory bulb of a 15 days old hamster. Naphthylphosphatase. Horizontal section, 40 x
(↑) Glomeruli of the accessory olf. bulb   (↑↑) Mitral cells of main olf. bulb

Fig. 8. Olfactory bulb of a newborn guinea-pig. Naphthylphosphatase. Frontal section, 40 x
(↑) Glomeruli of accessory olf. bulb   (↑↑) Mitral cells of main olf. bulb.

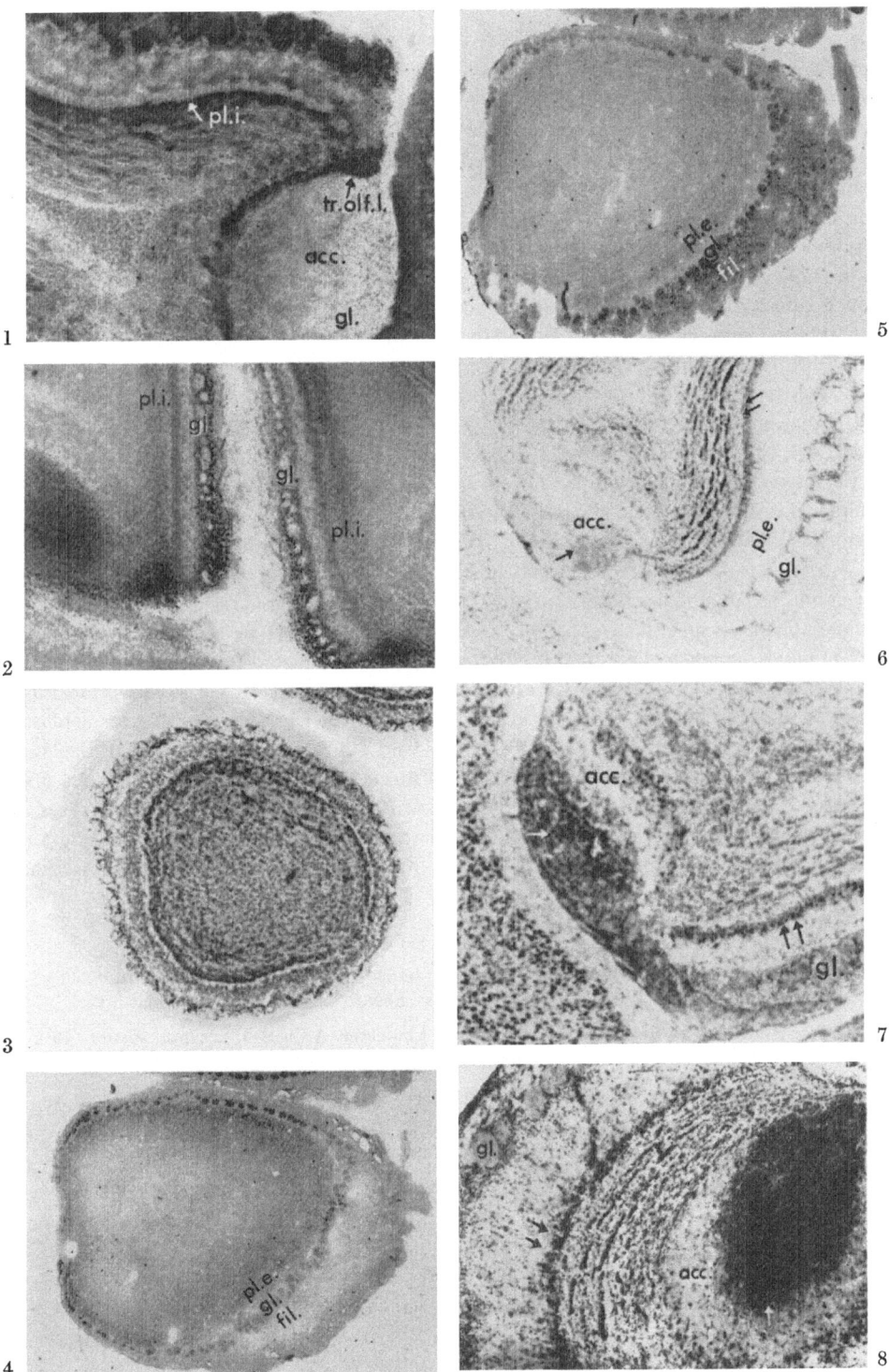

Comments

Hitherto only a few investigations have been devoted to the histochemistry of the bulbus olfactorius: ORTMANN (1957), ZACCHEO (1961), NANDY a. BOURNE (1965). Moreover it seems to us that hitherto in the literature the histochemical differences between main and accessory olfactory bulb have received no attention at all (Tab. 1). The phosphatase staining of the glomeruli with both GOMORI and azocoupling techniques must be a specific one, for it can be suppressed by adding NaF 10^{-2} m as an inhibitor to the substrate solution. AChE activity in newborn animals is rather diffuse (like in the chicken embryo as described by ZACCHEO 1961) but the periglomerular localization in the main bulb of the young hamster and – very clear – in the young guinea-pig indicates the important role of the external granular cells. As from biochemical studies one has concluded that the axo-dendritic glomerular synapses are non-cholinergic (NICCOLINI 1964), we suppose that the external granular cells with their prolongations account for the observed AChE activity around and inside the glomeruli. While only partially the glomeruli can be stained for AChE, they can be stained totally for SDH and LDH immediately after birth.

Further attention deserved the curious fact that in the newborn guinea-pig the axonal components (fila olf.) of the glomeruli can be stained by LDH technique and the dendritic components (outer plexiform layer) by SDH technique (Figs. 4, 5).

When compared with the bulbus olfactorius, enzymatic maturation of the hippocampal formation is much slower; perhaps this fact reflects a fundamental difference between "sinnesabhängige und sinnesunabhängige Hirnsysteme" (STEPHAN 1964).

The rough sequence of e.g. AChE development in the hamster brain seems to be: brain stem – basal ganglia – olf. bulb – cerebellum – hippocampus – isocortex. However, making timetables in this respect is unsatisfactory because of the heterogeneity of gross cerebral areas, the low sensitivity of most techniques and shifts in localisation after the first appearance of a given substance. For SDH (FRIEDE 1959) and LDH (Fig. 13) the first appearance is intracellular. Later on, activity shifts to the neuropilum of the molecular layers. After short fixation the CA 4 somata in the adult hamster are LDH-positive (Fig. 12). The same has been found in *Macaca* brain by FRIEDE (1963). Probably in fixed sections

Fig. 9. Hippocampal formation of a 11 days old hamster. A Ch E. Horizontal section, 40 x
i.p. = infrapyramidal layer   m.f. = mossy fibres   s.gr. = supragranular layer

Fig. 10. Hippocampal formation of a 15 days old hamster. A Ch E. Horizontal section, 40 x
For abbreviations see Fig. 9

Fig. 11. Hippocampal formation of a newborn guinea-pig. A Ch E. Frontal section, 40 x

Fig. 12. Hippocampal formation of a 19 days old hamster. LDH after 15 sec. glutaraldehyde
fixation.   Horizontal section, 40 x   Positive CA_4 somata (↑)

Fig. 13. Hippocampal formation of a 5 days old hamster. LDH. Horizontal section, 40 x

Fig. 14. Hippocampal formation of a newborn guinea-pig. Naphthylphosphatase and celestine
blue. Horizontal section, 40 x
e.b. = end bulb   h. = hilus fasciae dentatae   m.f. = mossy fibres

Fig. 15. Hippocampal formation of a 47 days old hamster. TPP-ase. Horizontal section, 40 x

Fig. 16. Hippocampal formation of a 15 days old hamster. Naphthylphosphatase and celestine
blue. Horizontal section, 40 x

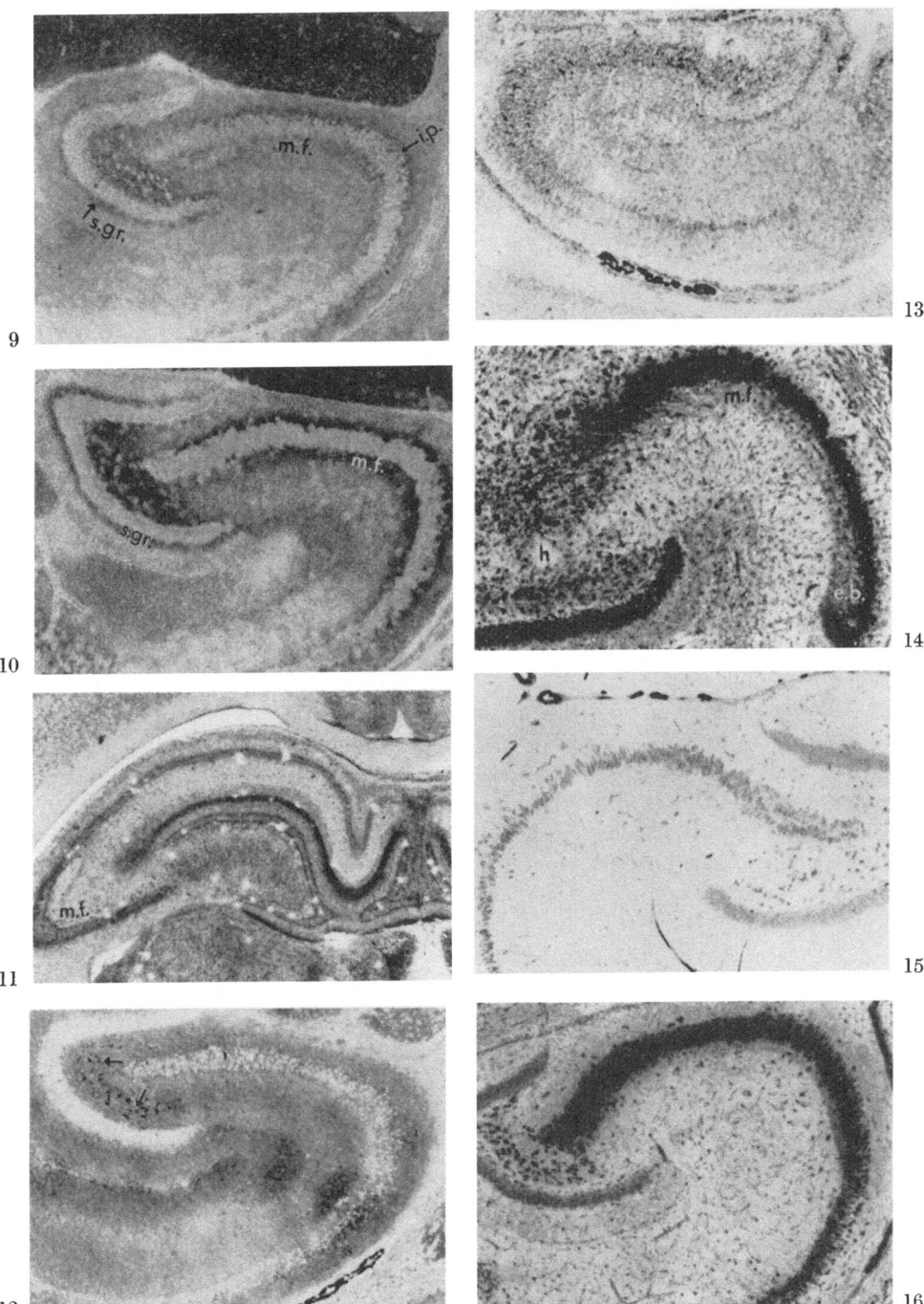

Table 1

	AChE	SDH and LDH	Acid Phosphatase
Glomeruli of bulbus olfact.	+ + +	+ + +	+
Glomeruli of bulbus olfact. accessorius	—	+	+ + +

the cytoplasmic fraction of brain LDH (JOHNSON a. WHITTAKER, 1963) is preserved while in unfixed sections it is washed out. The acid phosphatase activity in the suprapyramidal layer has already been illustrated by FLEISCHHAUER (1959) and can be suppressed by NaF 10^{-2} m. TPP-ase was found in the pyramidal and granular somata, in the GOLGI apparatus (Fig. 15) which according to recent electron-microscopical investigations is well developed in these cells (NIKLOWITZ a. BAK, 1965). Further investigations about TPP-ase development are in progress.

The AChE in the suprapyramidal layer does not coincide with the mossy fibre system, as MATHISEN a. BLACKSTAD (1964) have pointed out and this can be shown more clearly in the newborn guinea-pig (Fig. 11).

AChE differentiation in the molecular layers of the regio inferior hippocampi occurs on the 14th to 15th day, when also acid phosphatase appears in the supra-pyramidal layer and the macroscopical supravital dithizone reaction (TIMM 1958) becomes positive. Perhaps here we have the most critical period of hippocampal chemodifferentiation.

On the 15th day after birth of the hamster, when the mossy fibre layer is already phosphatase-positive, many of the granular somata are still negative: an argument for our statement that not the mossy fibres themselves, but only the apical dendrites of CA 3 possess acid phosphatase, which in these axo-dendritic synapses is located post-synaptically as well as AChE (MATHISEN a. BLACKSTAD) on the proximal parts of the apical dendrites.

The development of acid phosphatase in the gyrus dentatus occurs in the same outside-to-inside gradient as found by ANGEVINE (1964) in his histogenetic study of the area dentata in the mouse by means of tritiated thymidine.

Summary

The brains of young hamsters and guinea-pigs of several ages from birth up to about two months were used for microscopical histochemistry of some dehydrogenases, acetylthiocholinesterase and naphthylphosphatase. Starting from a demonstration of the localisation in sections through the allocortex of adult animals, the appearance of these substances is described in the olfactory bulb, where in the hamster the adult picture is reached already in the first week, as well as the typical differences in enzymatic pattern between the main olfactory bulb and the accessory bulb. In the hippocampal formation of the hamster enzymatic maturation is much slower. Only the dehydrogenases are present in the first days after birth.

Acetylcholinesterase does not appear until the 9–11th day in the infra- and suprapyramidal layers, then in the molecular layers (14–15th day) and last in the interpyramidal baskets (19–25th day).

The differentiation of the naphthylphosphatase-positive granular cells in the gyrus dentatus of the hamster is not completed before the end of the second month and proceeds in a gradient from the outer surface of the gyrus towards the hilus.

The allocortex of the newborn guinea-pig shows the same distribution of these enzymes as in the adult, except for the low level of acetylcholinesterase in the suprapyramidal layer.

References

ANDERSON, P.J.: Acid phosphatase in the nervous system. J. Neuropath. exp. Neurol. *21* (1962), 263–283.

ANGEVINE, J.B.: Autoradiographic study of histogenesis in the area dentata of the cerebral cortex in the mouse. Anat. Rec. *148* (1964), 255.

ARVY, L.: L'A B C de la pratique histoenzymologique. Laval Med. *33* (1962), 708–726.

BONAVITA, V., F.PONTE, G.AMORE: Lactate dehydrogenase isoenzymes in the nervous tissue IV. An ontogenetic study on the rat brain. J. Neurochem. *11* (1964), 39–47.

COLMANT, H.J.: Ergebnisse der Enzymhistochemie am zentralen und peripheren Nervensystem. Fortschr. Neurol. *29* (1961), 61–124.

FELGENHAUER, K.: Der Nachweis von spezifischen und unspezifischen Phosphatasen des Nervensystems. Histochemie *3* (1963), 22 bis 232.

FLEISCHHAUER, K.: Zur Chemoarchitektonik der Ammonsformation. Nervenarzt *30* (1959), 305–309.

FRIEDE, R.L.: Histochemical investigations on SDH in the central nervous system. I. The postnatal development of rat brain. J. Neurochem. *4* (1959), 101–110.

FRIEDE, R.L.: A mapping of the distribution of LDH in the brain of the Rhesus monkey. Amer. J. Anat. *113* (1963), 215–234.

GEREBTZOFF, M.A.: Cholinesterases. Pergamon Press. London 1959.

GEREBTZOFF, M.A.: Apport de l'histochimie à la connaissance de l'écorce cérébrale. Structure and Function of the Cerebral Cortex. Elsevier. Paris 1959.

GÖBEL, A., H. PUCHTLER: Untersuchungen zur Methodik der Darstellung der Succinodehydrase am histologischen Schnitt. Virchows Arch. *326* (1955), 312–331.

GOLDFISCHER, S.: The Golgi apparatus and the endoplasmatic reticulum in neurons of the rabbit. J. Neuropath. exp. Neurol. *23* (1964), 36–46.

JOHNSON, M.K., V.P. WHITTAKER: Lactate dehydrogenase as a cytoplasmic marker in brain. Biochem. J. *88* (1963), 404.

MATHISEN, J.St., TH. W. BLACKSTAD: Cholinesterase in the hippocampal region. Acta anat. *56* (1964), 216–253.

NANDY, K., G.H. BOURNE: Histochemical study on localisation in adenosine-triphosphatase and 5-nucleotidase in the olfactory bulb of the rat. Histochemie *4* (1965), 488 bis 493.

NICCOLINI, P.: cit. in: International Brain Research Organisation Bulletin III (1964), 26–27.

NIKLOWITZ, W., I.J. BAK: Elektronenmikroskopische Untersuchungen am Ammonshorn. I. Die normale Substruktur der Pyramidenzellen. Z. Zellforsch. *66* (1965), 529 bis 547.

ORTMANN, R.: Über Succinodehydrogenase im olfactorischen System. Acta anat. *30* (1957), 542–565.

PONCELET, P.: Convergence de fibres cholinergiques et non cholinergiques sur les cellules pyramidales ammoniques et les grains du fascia dentata. Acta neurol. belg. *60* (1960), 631–637.

SHUTE, C.C.D., P.R. LEWIS: The use of cholinesterase technics combined with operative procedures to follow nervous pathways in the brain in: Histochemistry of Cholinesterase, Bibliotheca Anatomica Fasc. 2, Karger, Basel 1961.

STEPHAN, H.: Die kortikalen Anteile des limbischen Systems. Nervenarzt *35* (1964), 396 bis 401.

TIMM, F.: Zur Histochemie des Ammonshorngebietes. Z. Zellforsch. *48* (1958), 548 bis 555.

ZACCHEO Y E. MONTERO, D.: L'attivita colinesterasica nel corso dello sviluppo del bulbo olfattivo di Gallus Dom, An. Desarrollo (Granada) *9* (1961), 67 bis 71.

Dr. W.G.M.WITKAM
Department of Anatomy and Embryology
(Head: Prof. Dr. H.J.Lammers)
University of Nijmegen
Nijmegen, The Netherlands

Über die Entwicklung des Globus pallidus und des Corpus subthalamicum beim Menschen

Von E. RICHTER, Hamburg

In diesem Vortrag soll zu der immer noch strittigen Frage Stellung genommen werden, ob sich der Globus pallidus in der Ontogenie des menschlichen Gehirns aus dem Diencephalon oder aus dem Telencephalon entwickelt. Anders ausgedrückt: Stammt der Globus pallidus aus der Matrix des 3. Ventrikels oder aus der Matrix des Seitenventrikels?

SPATZ (1921, 1924, 1925) hatte sich früher mit dieser Frage eingehend beschäftigt; er vertrat als erster die Auffassung, daß das Pallidum als ein Derivat des Diencephalon anzusehen sei[1]). KUHLENBECK (1924, 1954), ROSE (1942), SCHNEIDER (1949), KAHLE 1951, 1956, 1958), REINOSO-SUAREZ (1960) und viele andere konnten die Befunde von SPATZ bestätigen, andere dagegen sahen das Pallidum weiterhin als einen Abkömmling des Telencephalon an (ARIËNS KAPPERS 1923, 1947; KODAMA 1927; GRÜNTHAL 1941, 1952 u.a.).

Wir haben zunächst an Hand von Schnittserien menschlicher Embryonengehirne die frühesten, gerade erkennbaren Anlagen der beiden Pallidumglieder sowie des Corpus subthalamicum bei Embryonen vom Ende des 2. Monats aufgesucht und dann deren weiteren Entwicklungsgang verfolgt. Es stellte sich dabei heraus, daß die genannten Kerne erhebliche Lageveränderungen erleiden.

Zweierlei sei vorausgeschickt: 1. Wir meinen, daß unsere Befunde für eine diencephale Genese des Globus pallidus sprechen. 2. Die früheste Anlage des Pallidum internum ist im „Nucleus entopeduncularis" zu suchen, wie es bereits von PAPEZ (1940) u.a. angenommen wurde. (Es sei betont, daß hier unter „Nucleus entopeduncularis" der im embryonalen Säugetier- bzw. Menschenhirn beschriebene Kern gemeint ist, und *nicht* der Nucleus entopeduncularis der niederen Wirbeltiere).

Heterochronie der Entwicklung. Als besonders wichtig erwies sich uns das Prinzip der Heterochronie der Entwicklung. Es besagt, daß sich verschiedene Hirnteile zu verschiedenen Zeiten entwickeln. Dieses Prinzip ist für unsere Betrachtungen in zweierlei Hinsicht besonders interessant. Einmal entwickelt sich das Diencephalon als Ganzes – von gewissen Überschneidungen abgesehen – *früher* als das Telencephalon. Es sei nur daran erinnert, daß sich nach SPATZ die Matrix des Telencephalon über ein halbes Jahr später aufbraucht als die Matrix des Diencephalon. Zweitens gibt es auch innerhalb des Diencephalon eine Heterochronie der Entwicklung. Hier sei besonders auf die Untersuchungen von KAHLE über die längszonale Gliederung des Zwischenhirns hingewiesen. Aus diesen ergab sich, daß die sogenannte „subthalamische Längszone" (im Sinne von KAHLE), aus der Pallidum

---

[1]) Vor SPATZ hatte STRASSER (1920) – ohne nähere Erläuterung – den Globus pallidus als ein Zwischenhirnderivat beschrieben.

und Corpus subthalamicum abstammen, durch besonders frühzeitigen Ablauf der Matrixphasen gekennzeichnet ist.

Vergleicht man die Entwicklung der diencephalen Zentren einerseits mit der Entwicklung des telencephalen Striatum andererseits, so wird die Heterochronie besonders deutlich. Die subthalamischen Zentren eilen nicht nur in bezug auf den Matrixaufbruch und
die Differenzierung, sondern auch bezüglich der Nervenzellreifung und der Myelogenese
weit voraus, während das Striatum in erheblichem Abstand nachfolgt.

Menschliche Embryonen. 24 mm. Wir schildern zunächst die morphologischen Verhältnisse einer frühen Entwicklungsphase. Um die Lage der uns interessierenden Zentren
plastisch darstellen zu können, haben wir Rekonstruktionsmodelle hergestellt. Eines
dieser Modelle ist in Abb. 1 a und b dargestellt; es stammt von einem menschlichen Embryo von 24 mm S.S.L. (einem Alter von etwa 8 Wochen entsprechend).

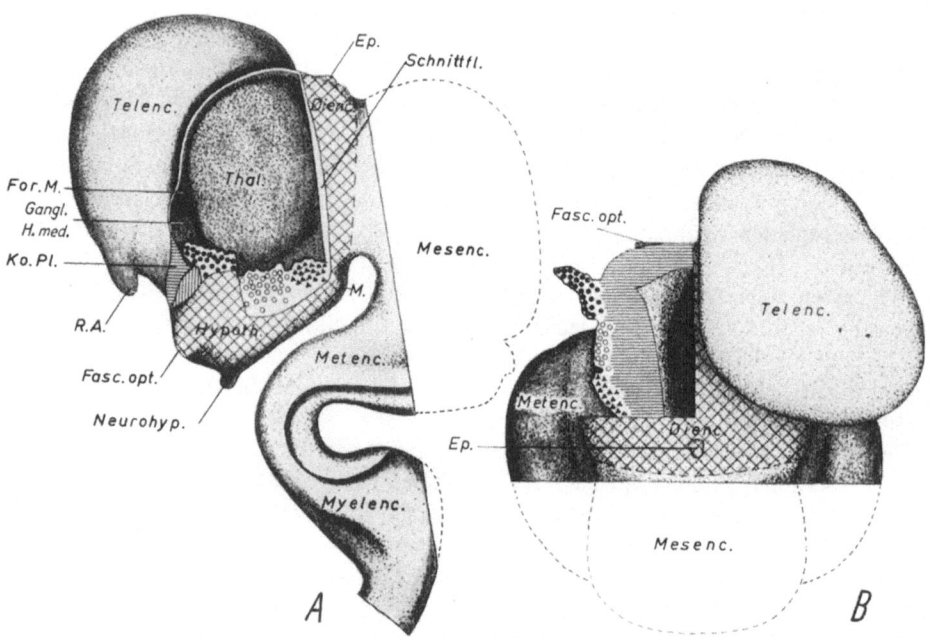

Abb. 1. Hirnmodell von EH II/24 (9 x) (Abb. 1–6 aus RICHTER, 1965).

A) Ansicht von links. Außer der linken Hemisphäre wurden Teile des linken Zwischenhirns
entfernt. Freigelegte subthalamische Zentren hintereinander (von caudal nach oral): Corpus
subthalamicum, Nucleus entopeduncularis, Pallidum externum.

B) Ansicht vom Scheitel her, Einblick in den 3.Ventrikel. Die drei subthalamischen Kerne in
langer Reihe hintereinander. Das Pallidum externum ragt am weitesten nach lateral.
(Dreiecke: Corpus subthalamicum; Kreise: Nucleus entopeduncularis; dicke Punkte: Pallidum
externum; kariert: Oberfläche des Diencephalon; Dienc.: Diencephalon; E.H.: Embryo Homo;
Ep.: Epiphyse; Fasc. opt.: Fasciculus opticus; For. M.: Foramen Monroi; Gangl. H. med.: medi-
aler Ganglienhügel; Ko. Pl.: Kommissurenplatte; M.: Mamillarhöcker; Mesenc.: Mesencephalon;
Myelenc.: Myelencephalon; Neurohyp.: Neurohypophyse; R. A.: Riechhirnausladung; Schnittfl.:
Schnittfläche durch die Zwischenhirnwand; Telenc.: Telencephalon; Thal.: Thalamus der rechten
Seite).

Bei diesem Hirnteilmodell wurde die linke Hemisphäre und außerdem ein großer dorsaler Teil der linken Zwischenhirnwand entfernt, um die drei subthalamischen Zentren freizulegen. Abb. 1a zeigt die Ansicht von links. Im Bereiche des Subthalamus sind drei Kerne in langer Reihe hintereinander angeordnet: am weitesten caudal liegt das durch Dreiecke markierte Corpus subthalamicum, am weitesten oral die Anlage des Pallidum externum (dicke Punkte), und dazwischen die durch Kreise gekennzeichnete Anlage des Pallidum internum, welche dem Nucleus entopeduncularis entspricht. Ventral der Hypothalamus (im engeren Sinne), und dorsal der Thalamus der Gegenseite. Im Bereich des weiten Foramen Monroi taucht der gegenüberliegende Ganglienhügel auf, aus welchem sich das Striatum entfaltet.

Abb. 1b stellt die Aufsicht auf das Hirnmodell vom Scheitel her dar. Auch hier ist die Anordnung der drei subthalamischen Kerne in der Längsrichtung des Diencephalon gut zu erkennen. Der orale Pol des Pallidum externum springt weit nach lateral vor, an dieser Stelle besteht von Anfang an ein Kontakt mit dem telencephalen Putamen.

37 mm. In Abb. 2 und 3 werden einige Frontalschnitte eines Embryo von 37 mm S.S.L. gezeigt, einem Alter von etwa 10 Wochen entsprechend. Wir betrachten die Schnitte in der Reihenfolge von caudal nach oral.

Der Schnitt der Abb. 2a führt durch den caudalen Anteil des Diencephalon, welches eine allseits freie Oberfläche aufweist, die allerdings teilweise von den Hemisphären überlagert wird. Der im ventralen Anteil des Zwischenhirns direkt unter der Oberfläche gelegene linsenförmige Kern stellt die Anlage des Corpus subthalamicum dar.

Gehen wir in der Serie weiter oralwärts (Abb. 2b), so taucht an der Stelle des Corpus subthalamicum der Nucleus entopeduncularis auf, der sich bei näherem Studium als Anlage des Pallidum internum erweist. Der Kern hat dorsolateral einen abgerundeten Pol; nach medial geht er in die übrige Masse des Subthalamus über. Die Matrix des Subthalamus ist hier als ein schmaler, dunkler Streifen gut zu erkennen (zwischen den beiden Pfeilen gelegen), sie ist auf dieser Entwicklungsstufe bereits aufgebraucht. Unterhalb liegt die Matrix des Hypothalamus (im engeren Sinne), oberhalb die des Thalamus ventralis. Noch weiter dorsal fällt die breite, stark proliferierende Matrix des Thalamus dorsalis auf, und ganz oben die des Epithalamus. Auch auf diesem Schnitt sind Diencephalon und Telencephalon noch durch die Fissura telo-diencephalica voneinander getrennt.

Noch weiter oral in der Serie ist der hintere Abschnitt des Hemisphärenstiels getroffen (Abb. 3a). Ventral von seiner kompakten Fasermasse, dem „Stammbündel" von His, liegt eine zackenförmige Zellansammlung. Diese hängt kontinuierlich mit der Masse des Subthalamus zusammen, während sie von der Anlage des Putamen noch deutlich geschieden ist. Spatz hat als erster diese Zacke als Ursprung des Globus pallidus beschrieben. Es hat sich gezeigt, daß es sich hier um den caudalen Abschnitt des Pallidum externum handelt, welcher sich bei fortschreitender Entwicklung nach oro-lateral in Richtung auf die Hemisphäre entfaltet. Auf dieser Entwicklungsstufe ist die Zugehörigkeit zum Diencephalon, genauer zum Subthalamus, deutlich. Auch hier kann man die sehr schmale Matrix des Subthalamus gut erkennen.

Der Schnitt der Abb. 3b liegt am weitesten oral, er führt beiderseits durch das Foramen Monroi. Das eben beschriebene zackenförmige Zellgebiet ist kontinuierlich in ein rundliches und etwas schwerer abgrenzbares Areal übergegangen, welches dem oralen Pol des Pallidum externum entspricht. Hier bestehen von Anfang an enge Beziehungen zu der Anlage des Putamen. Dieser orale Pol des Pallidum externum ragt in den Bereich der

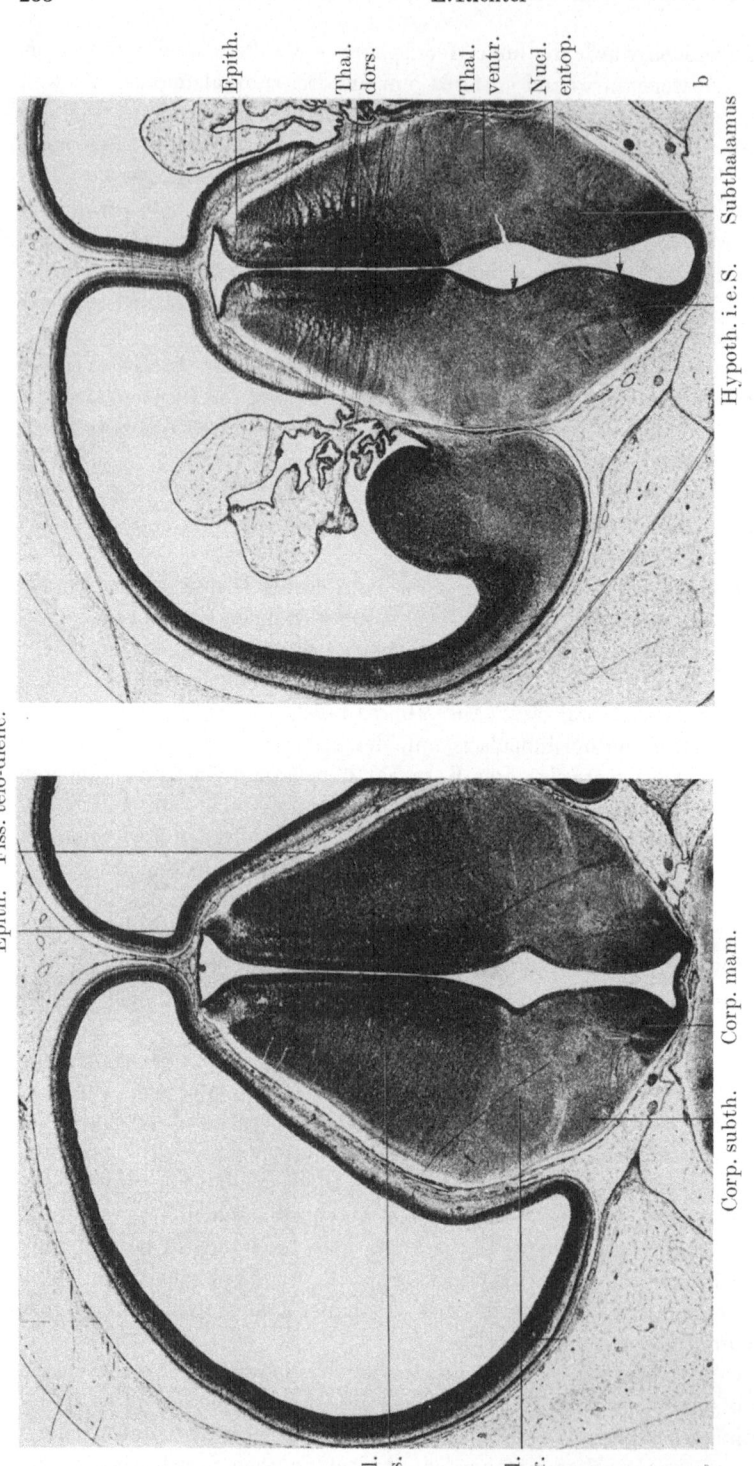

Abb. 2. EH III/37 (frontal; Färbung H.E.; 13 ×; Anordnung von caudal nach oral). a) Schnitt 33/1 durch caudale Anteile des Zwischenhirns mit allseits freier Oberfläche. Im Bereich des Subthalamus liegt das Corpus subthalamicum direkt unter der Oberfläche. b) Schnitt 28/3 dicht hinter dem Hemisphärenstiel. Im Bereich des Subthalamus der Nucleus entopeduncularis. Die Fissura telo-diencephalica trennt das Zwischenhirn völlig vom Endhirn. Matrix des Subthalamus zwischen den beiden Pfeilen. Abkürzungen für Abb. 2–4: Caps.int.: Capsula interna; Corp.mam.: Corpus mamillare; Corp.subth.: Corpus subthalamicum; Epith.: Epithalamus; Fibr. ped.: Fibrae pedunculares; Fiss. telo-dienc.: Fissura telo-diencephalica; For. M.: Foramen Monroi; Gangl. H.: Ganglienhügel; Hypoth. i. e. S.: Hypothalamus im engeren Sinne; Nucl.entop.: Nucleus entopeduncularis; Pall. ext.: Pallidum externum; Pall.int.: Pallidum internum; Pes.ped.: Pes pedunculi; Thal. dors.: Thalamus dorsalis; Thal. ventr.: Thalamus ventralis; Tract. opt.: Tractus opticus.

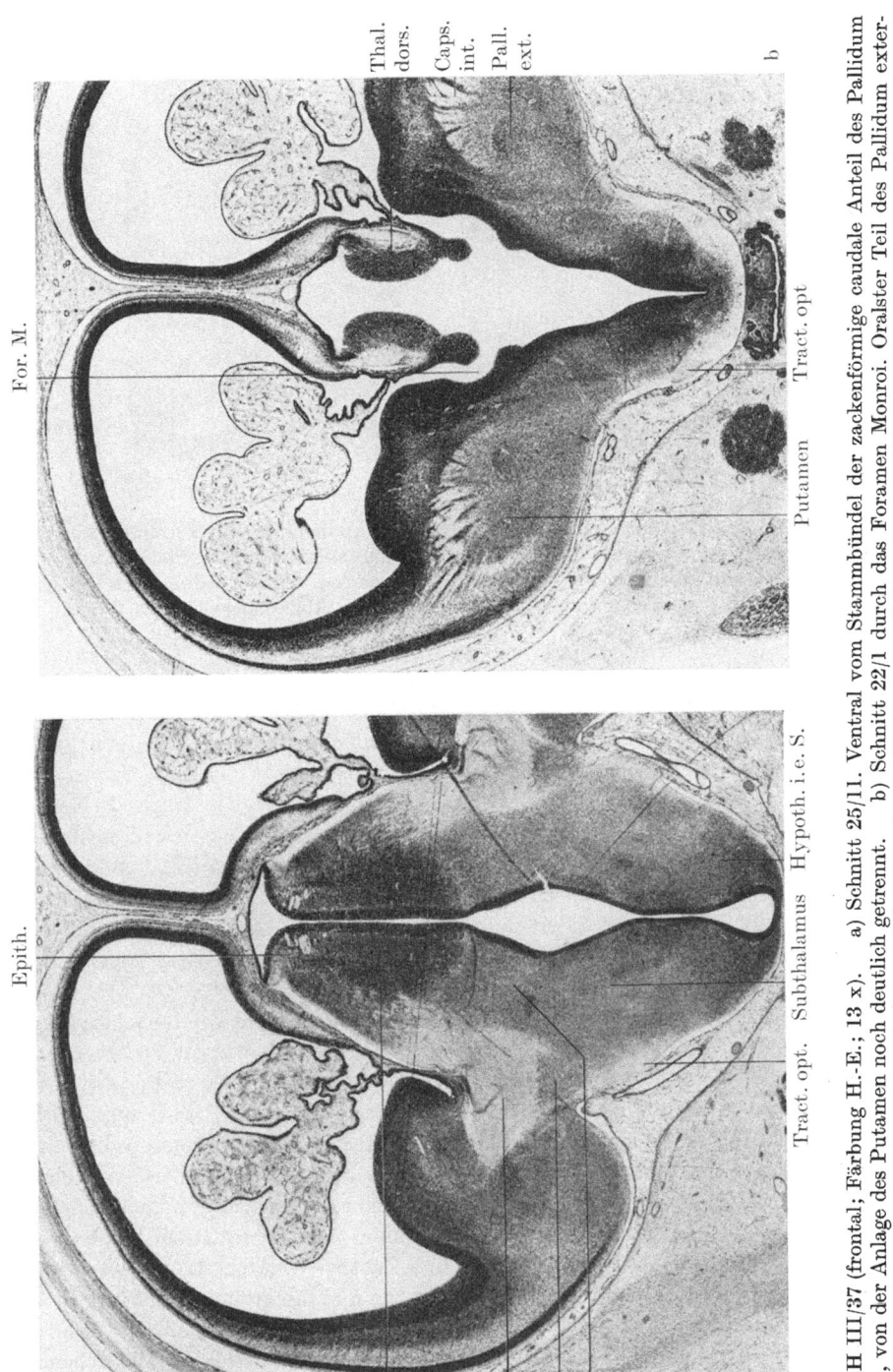

Abb. 3. EH III/37 (frontal; Färbung H.-E.; 13 x).   a) Schnitt 25/11. Ventral vom Stammbündel der zackenförmige caudale Anteil des Pallidum externum, von der Anlage des Putamen noch deutlich getrennt.   b) Schnitt 22/1 durch das Foramen Monroi. Oralster Teil des Pallidum externum in engem Kontakt mit dem Putamen.   Abkürzungen s. Abb. 2.

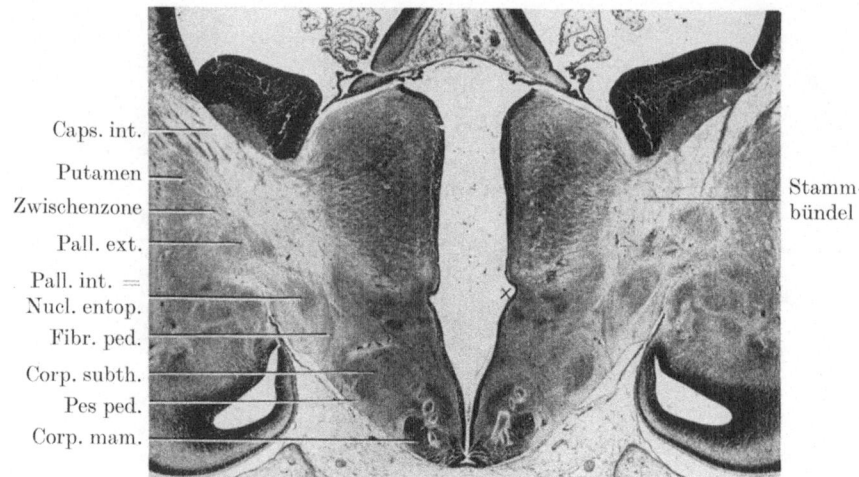

Caps. int.

Putamen
Zwischenzone

Pall. ext.

Pall. int. =
Nucl. entop.

Fibr. ped.

Corp. subth.

Pes ped.

Corp. mam.

Stamm-
bündel

Abb. 4. EH III/58 (schräg-frontal; Färbung H.-E.; 9 x). Schnitt 461: In einer Reihe von
ventro-medial nach dorso-lateral: Corpus mamillare, Corpus subthalamicum, Pallidum internum
(= Nucleus entopeduncularis), Pallidum externum, Putamen. Fibrae pedunculares zwischen
Corpus subthalamicum und Pallidum internum.   Abkürzungen s. Abb. 2.

Hemisphäre vor; der Zusammenhang mit dem Subthalamus ist auf diesem Schnitt nicht
mehr so deutlich wie bei den caudalen Anteilen des Pallidum externum (vgl. Abb. 3a).

58 mm. Abb. 4 zeigt einen schräg-horizontal geführten Schnitt durch das Gehirn eines
Embryo von 58 mm S.S.L. Im Bereiche des Diencephalon sind in langer Reihe von
ventro-medial nach dorso-lateral folgende Kernanlagen getroffen: unten und medial das
Corpus mamillare, dann das Corpus subthalamicum, weiter außen das Pallidum internum
und das Pallidum externum. Noch weiter lateral folgt das telencephale Putamen. Zwischen
Corpus subthalamicum und Pallidum internum treten erstmalig die von der Capsula
interna zum Pes pedunculi absteigenden Fasern hindurch, die das Pallidum internum von
seinem subthalamischen Ursprung abtrennen.

Verlagerung der subthalamischen Anlagen. Bei Abb. 5 handelt es sich um schematische
Darstellungen der 3 subthalamischen Kerne verschiedener embryonaler Phasen bei der
Betrachtung von dorsal, also vom Scheitel her. Die Schemata sind durch Projektion von
Frontalschnitten auf eine Horizontalebene entstanden. Auch hier ist das Corpus subthalamicum durch Dreiecke, das Pallidum internum bzw. der Nucleus entopeduncularis
durch Kreise und das Pallidum externum durch dicke Punkte markiert; das übrige
Zwischenhirn ist kariert eingetragen. Die schwarze Linie, auf welche der Pfeil hindeutet,
veranschaulicht die freie Oberfläche des Diencephalon. – In der frühesten Phase (bei a)
ist wieder die longitudinale Anordnung der drei Zentren deutlich. Der Nucleus entopeduncularis folgt im 3. Monat dem Pallidum externum in oro-lateraler Richtung in den
Hemisphärenstiel hinein und legt sich dabei dem Pallidum externum immer enger an.
Der Nucleus entopeduncularis ist dann zum Pallidum internum geworden. Das Schema g
stellt den weit fortgeschrittenen Zustand eines späten Stadiums aus dem 6. intrauterinen
Monat dar. Die freie Oberfläche des Diencephalon ist jetzt, wie schon HOCHSTETTER (1919)
gezeigt hat, nach caudal gewandt. Die drei subthalamischen Kerne liegen nun nicht mehr

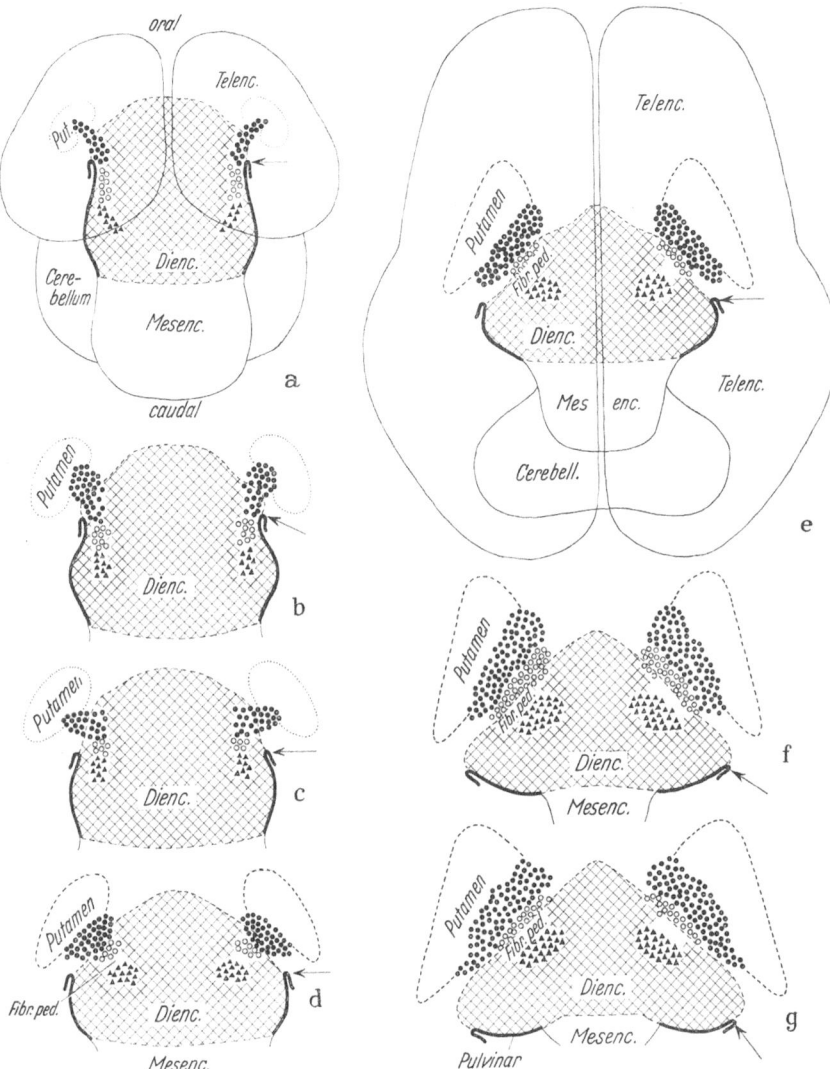

Abb. 5. Schemata von Gehirnen menschlicher Embryonen aus dem II. bis VI. Monat. Das Diencephalon besteht aus dem karierten Gebiet und den drei subthalamischen Kernen. Aufsicht vom Scheitel her; Entstehung der Schemata durch Projektion von Frontalschnittserien auf eine Horizontalebene. Vergrößerungsmaßstab von a bis g abnehmend.

a) EH II/24 Lutz, 3,2fach;   b) EH III/37 Loni, 3fach;   c) EH III/55 Titi, 2,9fach;
d) EH IV/115 Fadö, 1,6fach;   e) EH V/170, 1,2fach;   f) EH VI/190 Eicke F 27, 1,0fach;
g) EH VI Eicke F 8, 0,9fach.

Dicke Linie: Oberfläche des Zwischenhirns, zuerst lateral, später caudal (Pulvinar). Pfeil: Umschlag der Zwischenhirnoberfläche zur medialen Hemisphärenwand im Grund der Fissura telo-diencephalica. Orale gestrichelte Linie: Grund des Sulcus hemisphaericus = äußerer Umfang des Hemisphärenstiels. Caudale gestrichelte Linie: dorsale Grenze Diencephalon gegen Mesencephalon. Schwarze Dreiecke: Corpus subthalamicum Luysi; Kreise: Nucleus entopeduncularis (= Pallidum internum); dicke Punkte: Pallidum externum. Bei c bis g erscheint der Nucleus entopeduncularis relativ zu klein, da zum Teil vom Pallidum externum überdeckt wird. Ausdeh-

hintereinander, sondern sie sind, um eine alte Bezeichnung zu wählen, „konzentriert";
d. h. in diesem Falle ist aus dem lang ausgedehnten Hintereinander mehr ein Nebeneinander geworden. Das Corpus subthalamicum liegt am weitesten medial, das Pallidum
externum am weitesten lateral, dazwischen das Pallidum internum. Zwischen Corpus
subthalamicum und Pallidum internum brechen die von der Capsula interna zum Pes
pedunculi ziehenden Nervenfasern hindurch (hier als Fibrae pedunculares bezeichnet). –
Wenn man die drei subthalamischen Kerne in bezug auf das übrige Diencephalon, welches
hier kariert eingetragen ist, betrachtet, so erkennt man, daß ursprünglich (im Schema a)
das gesamte Pallidum internum und kaudale Anteile des Pallidum externum eindeutig im
Diencephalon liegen. Später sind diese beiden Kerne nach oro-lateral aus dem Diencephalon in den Bereich der Hemisphäre vorgeschoben. Für diese Umlagerung in Richtung
auf die Hemisphäre hat SPATZ die Bezeichnung „*Intussusception*" des Diencephalon in
das Telencephalon geprägt.

Die Intussusception hängt mit dem Verlust der freien Seitenflächen des ursprünglich
langgestreckten Zwischenhirns zusammen. Das Diencephalon bleibt im Längenwachstum
zurück, dabei dehnt sich seine Grenzfläche zum Telencephalon (telo-diencephale Grenzfläche) enorm aus und wird stark nach lateral vorgewölbt.

Heterochrone Reifung der Zellen. Die subthalamischen Zentren (Corpus subthalamicum,
Pallidum internum und externum) unterscheiden sich von Striatum auch durch die
Heterochronie der Entwicklung. Die Matrixreifung und die Abgrenzbarkeit der Zentren
(„Differenzierung") finden, wie u. a. von SPATZ und KAHLE gezeigt wurde, im Dience-
phalon früher statt als im Telencephalon. Innerhalb des Diencephalon eilt die subthala-
mische Längszone voraus. In der gleichen zeitlichen Reihenfolge tritt die Nervenzell-
reifung und die Myelogenese ein: Die Nervenzellen und die Markscheiden reifen im Pallidum und im Corpus subthalamicum wesentlich früher als im Striatum (vgl. hierzu u. a.
FLECHSIG 1920, 1921; C. u. O. VOGT 1909, 1920; RIESE 1924; SPATZ 1924; KÖRNYEY 1926;
KODAMA 1927; ROSE 1942).

Die Abb. 6 von einem Feten aus dem 6. Monat soll die heterochrone Nervenzellreifung
im Corpus subthalamicum und Pallidum einerseits sowie im Striatum andererseits
demonstrieren. Die Nervenzellen des Pallidum (linker Teil der Abb.) sind bereits weitgehend ausgereift, während die Nervenzellen des Putamen (mittlerer Teil der Abb.)
noch vorwiegend einen embryonalen Charakter zeigen. Nur äußerst selten (ca. 2–3 auf
einem Schnitt!) findet man wie hier bei z ein Exemplar einer fortgeschrittenen großen
Nervenzelle (in wesentlich geringerer Anzahl als später im ausgereiften Gehirn die großen
Putamenzellen anzutreffen sind). – Die Nervenzellen im Corpus subthalamicum (rechter
Teil der Abb.) entwickeln sich synchron mit denen im Pallidum.

---

nung des Putamen (nicht des gesamten Ganglienhügels!) bei a bis c punktiert (schwer abgrenzbar),
bei d bis g gestrichelt (gut abgrenzbar). Die drei subthalamischen Zentren ursprünglich (a und b)
in langer Reihe *hintereinander:* am weitesten caudal das sich früh differenzierende Corpus subthalamicum, in der Mitte der Nucleus entopeduncularis und am weitesten oral das Pallidum
externum, dessen vordere Spitze nach lateral „in den Bereich" der Hemisphäre auslädt und von
Anfang an Kontakt mit dem Putamen besitzt. Später Anordnung *zunehmend nebeneinander:* am
weitesten medial das Corpus subthalamicum, in der Mitte – ab IV. Monat durch Fibrae pedunculares (Fibr. ped.) von letzterem getrennt – das aus dem Nucleus entopeduncularis entstandene Pallidum internum und am weitesten lateral das sich stark vergrößernde Pallidum externum, das
jetzt in seiner ganzen Ausdehnung an das Putamen grenzt.

Pallidum	Putamen	Corpus subthalamicum

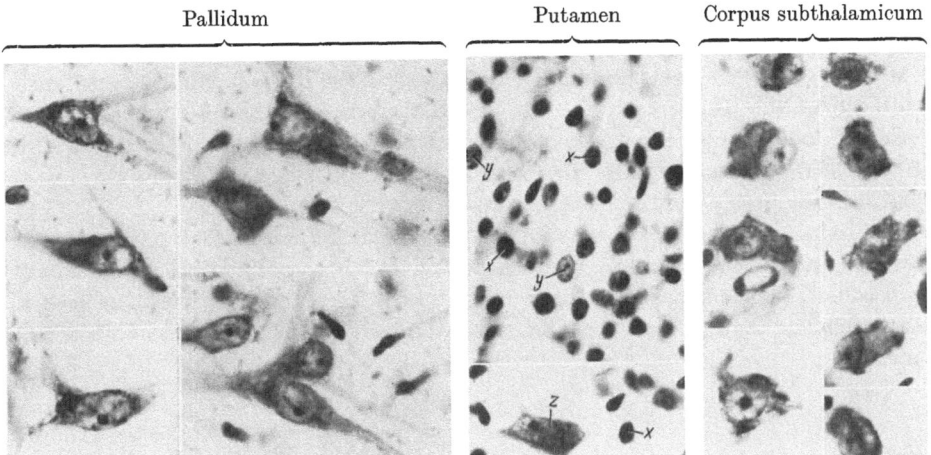

Abb. 6. EH VI/190 Eicke F 27 (Nissl 560 x).
Heterochronie der Zellreifung von Pallidum und Corpus subthalamicum einerseits und Putamen andererseits. Links: Pallidum (internum und externum): In der Reifung fortgeschrittene Nervenzellen mit Nissl-Schollen. Mitte: Putamen: Zellreifung retardiert. Vorwiegend drei Zelltypen (x: kleine dunkle Kerne, sehr zahlreich; y: größere, mehr lockere Elemente; z: *äußerst seltene* reifere Nervenzellen). Rechts: Corpus subthalamicum: Nervenzellreifung weit fortgeschritten, synchron mit Pallidum.

Diskussion

Für die Zugehörigkeit der beiden Glieder des Globus pallidus zum Diencephalon, speziell zur subthalamischen Längszone, sprechen die Schnittserien von verschiedener Richtung und unsere Modelle sowie die Heterochronie der Entwicklung. Die subthalamischen Zentren eilen nicht nur bezüglich des Matrixaufbrauches und der Differenzierung der Zentren, sondern auch hinsichtlich der Nervenzellreifung und der Myelogenese dem telencephalen Striatum weit voraus.

Summary

The ontogenetic development of the two segments of the pallidum and of the corpus subthalamicum was studied by serial histologic sections and reconstructed models of human embryonic brains of different ages.

The "anlagen" of the corpus subthalamicum and of the two pallidal segments arise from the diencephalon, namely from the "*subthalamic longitudinal zone*" (Kahle). The striatum arises from the telencephalon, namely from the colliculus ganglionaris of the lateral ventricle.

The pallidum internum develops from the "*nucleus entopeduncularis*", which is described by some authors in early embryonic brains of mammals and man.

At the onset of development the "anlagen" of corpus subthalamicum, pallidum internum (= nucleus entopeduncularis), and pallidum externum are *arranged in a caudo-oral direction.* In the later course of development this sagittal arrangement becomes almost frontal.

During the early stages the corpus subthalamicum and the pallidum internum still lie within the diencephalon. Only the oral pole of the pallidum externum projects into the region of the hemisphere and is in contact with the "anlage" of the putamen. Later the two pallidal segments advance more and more oro-laterally. This process leads to the "intussusception" (Spatz) of the diencephalon into the telencephalon.

The centers originated from the subthalamic longitudinal zone differ from the striatum by the extreme "heterochrony" of development. In the corpus subthalamicum and the two pallidal segments the exhaustion of the matrix as well as the development of the nerve cells and the myelination of fibers take place very early. The striatum follows much later.

Literatur

FLECHSIG, P.: Anatomie des menschlichen Gehirns und Rückenmarks (auf myelogenetischer Grundlage). Thieme, Leipzig 1920.

FLECHSIG, P.: Die myelogenetische Gliederung der Leitungsbahnen des Linsenkernes beim Menschen. Ber. Verh. sächs. Akad. Wiss. Leipzig, Mathem.-phys. Klasse *73* (1921), 295–302.

GRÜNTHAL, E.: Über die Entstehung des Globus pallidus. Schweiz. med. Wschr. *71* (1941), 498–499.

GRÜNTHAL, E.: Untersuchungen zur Ontogenese und über den Bauplan des Gehirns. In: FEREMUTSCH, K., E. GRÜNTHAL: Beiträge zur Entwicklungsgeschichte und normalen Anatomie des Gehirns. Karger, Basel 1952.

HOCHSTETTER, F.: Beiträge zur Entwicklungsgeschichte des menschlichen Gehirns I. Deuticke, Wien 1919.

KAHLE, W.: Studien über die Matrixphasen und die örtlichen Reifungsunterschiede im embryonalen menschlichen Gehirn. Dtsch. Z. Nervenheilk. *166* (1951), 273–302.

KAHLE, W.: Zur Entwicklung des menschlichen Zwischenhirnes. Dtsch. Z. Nervenheilk. *175* (1956), 259–318.

KAHLE, W.: Über die längszonale Gliederung des menschlichen Zwischenhirns. Pathophysiol. Dienceph. (Wien) 1958, 134–142.

KAPPERS, C. U. ARIËNS: The ontogenetic development of the corpus striatum in birds and a comparison with mammals and man. Koninkl. Akad. Wetensch. Amsterdam, Proc. Sect. Sci. *26* (1923), 135–158.

KAPPERS, C. U. ARIËNS: Anatomie Comparée du Système Nerveux. Masson & Cie, Paris 1947.

KODAMA, S.: Über die Entwicklung des striären Systems beim Menschen. Neurol. psychiat. Abh. (1927), 1–98.

KÖRNYEY, S.: Zur Faseranatomie des Striatum, des Zwischen- und Mittelhirns auf Grund der Markreifung in den ersten drei Lebensmonaten. Z. Anat. Entwickl.-Gesch. *81* (1926), 620–632.

KUHLENBECK, H.: Über den Ursprung der Basalganglien des Großhirns. Anat. Anz. *58* (1924), 49–74.

KUHLENBECK, H.: The Human Diencephalon. Karger, New York 1954.

PAPEZ, J. W.: The embryologic development of the hypothalamic area in mammals. Res. Publ. Ass. nerv. ment. Dis. *20* (1940), 31–51.

REINOSO-SUAREZ, F.: Desarrollo del subtalamo humano. An. Anat. 1960, 5–33.

RICHTER, E.: Die Entwicklung des Globus pallidus und des Corpus subthalamicum. Monographien aus dem Gesamtgebiet der Neurologie und Psychiatrie, H. 108. Springer, Berlin 1965.

RIESE, W.: Beiträge zur Faseranatomie der Stammganglien. J. Psychol. Neurol. (Lpz.) *31* (1924), 81–122.

ROSE, J. E.: The ontogenetic development of the rabbits diencephalon. J. comp. Neurol. *77* (1942), 61–130.

SCHNEIDER, R.: Ein Beitrag zur Ontogenese der Basalganglien des Menschen. Anat. Nachr. *1* (1949), 115–135.

SPATZ, H.: Zur Anatomie der Zentren des Streifenhügels. Münch. med. Wschr. *45* (1921), 1441–1446.

SPATZ, H.: Zur Ontogenese des Striatum und des Pallidum. Dtsch. Z. Nervenheilk. *81* (1924), 185–190.

SPATZ, H.: Über die Entwicklungsgeschichte der basalen Ganglien des menschlichen Großhirns. Erg.-H. Anat. Anz. *60* (1925), 54–58.

SPATZ, H.: Physiologie und Pathologie der Stammganglien. In: Handb. d. norm. u. pathol. Physiologie X, Springer, Berlin 1927.

STRASSER, H.: Anleitung zur Gehirnpräparation. 3. Aufl. Bircher, Bern 1920.

VOGT, C.: La myéloarchitecture du thalamus du cercopithèque. J. Psychol. Neurol. (Lpz.) *12* Erg.-H. (1909), 285–324.

VOGT, C. u. O. VOGT: Zur Lehre der Erkrankungen des striären Systems. J. Psychol. Neurol. (Lpz.) *25,* Erg.-H. 3, (1920), 627–846.

Dr. ERNST RICHTER
Neurologische Universitätsklinik
Martinistraße 52
2 Hamburg 20, Deutschland

Development of the Human Diencephalon

By F. REINOSO-SUAREZ, Pamplona

Among the authors with the most important contributions to this subject are SPATZ (20), KUHLENBECK (12, 13), BERGQUIST (1) and KAHLE (11). They are among those present at this Symposium, and I should like to take this opportunity expressing my profound respect and admiration for the work of all these gentlemen.

I should also like to recall the works of HIS (10), HERRICK (8), FORTUYN (5), MIURA (14), ROSE (19), PAPEZ (15), GILBERT (6), DEKABAN (3) and ESCOLAR (4) and others.

In 1954, I (16) published a study on the earlier stages of the development of the human diencephalon. My main interest was the behavior of the fibre tracts. Later, I studied the development of the subthalamus (18).

In the present study I am dealing with the following questions: the division of the diencephalon in longitudinal zones; the nuclei deriving from each of these zones.

Material. Seventy series of human embryos have been used. They ranged from embryos of two millimetres in length to new-born foetus, cut according to the frontal, horizontal and sagittal planes and stained by hematoxyline-eosine and NISSL methods.

These series are not only from our own collection but also from that of Professor ESCOLAR of Zaragoza and of the Neuroanatomische Abteilung, Max-Planck-Institut für Hirnforschung, where I studied from 1956 to 1958 under the direction of Professor SPATZ. I also studied certain embryos from the collections of Professor STARCK, Frankfurt, and Professor GUIRAO, Granada. I wish to express my gratitude to all above named Professors.

Division of the diencephalon in longitudinal zones. HERRICK (8) divided the diencephalon of the adult amphibian into four longitudinal zones: epithalamus, dorsal thalamus, ventral thalamus and hypothalamus. These zones were separated by three sulci appearing in the lateral walls of the third ventricle: sulcus diencephalicus, dorsalis, medius and ventralis. Since HERRICK made this division, most authors (5, 6, 12, 14) have used it in studies on the embryonic diencephalon in superior vertebrates.

In the human diencephalon at certain stages of its development three longitudinal sulci can be seen. These sulci have been considered as the diencephalic sulci of HERRICK. They cannot be found simultaneously in embryos of all ages, and the nuclei deriving from the longitudinal zones of the matrix situated between them do not correspond exactly to those situated in the epithalamus, dorsal, and ventral thalamus and hypothalamus of HERRICK. As HERRICK (9) affirms, the sulci appearing in the ventricular walls during the embryonic development change. Usually they represent zones of intense growth activity disappearing in some regions, while in others they are displaced, and in the adult they separate cellular groups of different meaning.

Human embryo of 4 mm. Dividing the diencephalon in a human embryo of 4 mm into longitudinal zones, we find a sulcus dividing the walls of the diencephalon into two halves, one dorsal and one ventral. This sulcus ends rostrally in the telencephalic vesicle. The recess and the optic vesicle are in a ventral position to the sulcus. These findings can also be observed in a human embryo of 5 mm.

9 mm. In the human embryo of 9 mm the sulcus can be identified as sulcus diencephalicus medius ending rostrally in the foramen interventriculare. At this stage the matrix layer of diencephalon is compact. In ventral position to this sulcus, some cellular migrations appear, most conspicuously in the caudal region.

10 mm. Fig. 1 A represents a section of an embryo of 10 mm. The diencephalon is divided by the sulcus diencephalicus medius (superior arrow) into two zones, one dorsal to this sulcus with a compact matrix and another ventral to the sulcus with a matrix in the first stage of migration. A small ventral zone with a compact matrix can also be observed. The zone dorsal to the sulcus thalamicus medius corresponds to the thalamus, the zone immediately inferior corresponds to the subthalamus and the most ventral zone to the

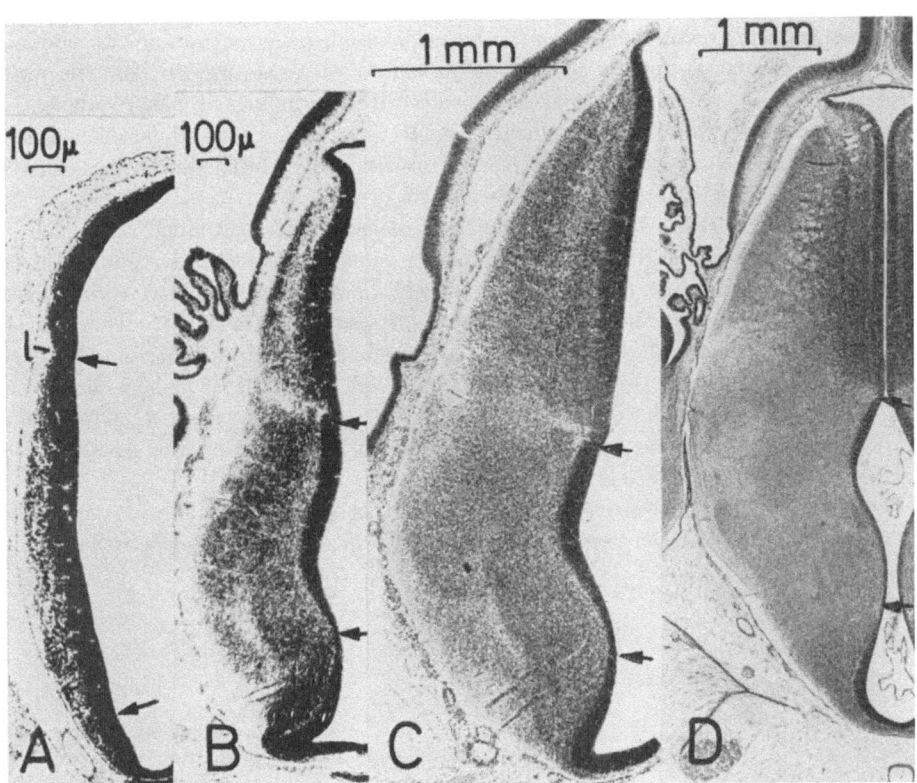

Fig. 1. Frontal sections of the diencephalon of embryos of 10 mm (A) 19,5 mm (B), 24 mm (C) and 35 mm length (D). The separation between the thalamus and subthalamus is determined by the sulcus thalamicus medius (superior arrow), the lamina medullaris externa (l) and the different behavior of their matrices. The dividing line between these two regions is less distinct in the two larger embryos. It is not possible to differentiate the thalamus ventralis in the subthalamic area (the zone between the two arrows). It can be seen that the exhaustion of the matrix in this subthalamic area takes place in a ventral to dorsal direction. The hypothalamus is ventral to the lower arrow.

hypothalamus. The separation between the thalamus and subthalamus can be determined: 1. by the sulcus thalamicus medius; 2. by a marked difference in the form and time of migration from their matrices, and 3. by the existence of a zona limitans (l) which can be observed up to and during the adult stage and which, at this time becomes the lamina medullaris externa of the thalamus (Fig. 6, C, l).

No clear separation can be made in the ventral region. As KAHLE (11) affirmed, only the different state of the matrix permits the differentiation of the hypothalamus from the subthalamus (inferior arrow), neither does there exist a separation between the thalamus and epithalamus.

14 mm. In this embryo the optic recess continues caudally into a short sulcus. In the human embryo of 14 mm length the sulcus originating in the optic recess continues caudally in the sulcus diencephalicus ventralis which is very conspicuous at this stage. In some of our embryos between 14 and 40 mm in length the sulcus diencephalicus ventralis appears to finish rostrally in the foramen interventricularis, while in most embryos it would appear to finish in the optic recess.

The fact that the rostral endings of the sulcus diencephalicus ventralis are to be found in two different places explains the varying opinions of different authors. That the optic recess is the most common ending place coincides with the findings of many workers.

In the ventricular wall the sulcus diencephalicus ventralis is the most conspicuous one; its appearance coincides with a stage of marked growth in the subthalamus which suggests that its formation is a consequence of this growth.

In embryos *more than 14 mm* in length three diencephalic longitudinal zones can be seen: (Figs. 1 and 2) a dorsal zone (unshaded) corresponding to the thalamus and epithalamus and situated dorsally to the sulcus diencephalicus medius, and two other zones ventral to this sulcus. The most dorsal to these is the subthalamus (Fig. 2, close dotting) corresponding to the two labia of the sulcus diencephalicus ventralis. The most ventral zone is the hypothalamus (Fig. 2, spaced dotting). The sulcus diencephalicus medius in these embryos is the only one to separate two different regions. According to COGGESHALL, in the albino rat this sulcus together with the lamina medullaris externa separates the first neuromere from the second.

The caudal origin of the sulcus diencephalicus medius is rostral and dorsal to the sulcus limitans which, in agreement with the findings of KUHLENBECK (13), ends in the mam-

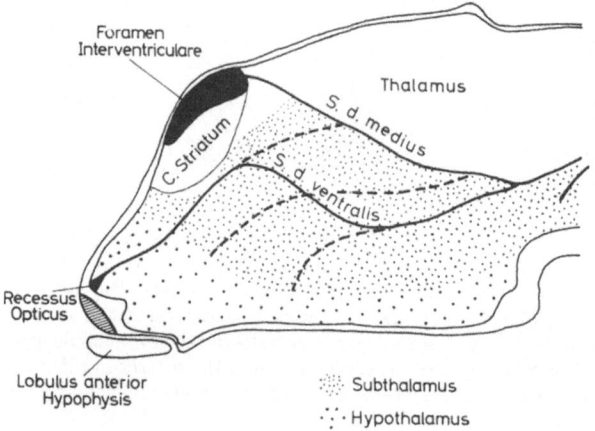

Fig. 2. Scheme of a reconstruction of the ventricular wall in a human embryo of 20 mm length. The sulcus diencephalicus medius originates caudal, rostral and dorsal, to the sulcus limitans and ends rostrally in the foramen interventriculare. The sulcus diencephalicus ventralis ends in the optic recess. Three diencephalic longitudinal zones can be seen.

millary recess. The subthalamus continues caudally in the mesencephalic tegmentum and the thalamus in the tectal region.

The separation made between the subthalamus and hypothalamus coincides with that of KAHLE (11) in all stages of embryonic development. This separation is also the inferior limit of the zone of the diencephalic matrix in which the cellular migration first commences and is the first to reach the stage of exhaustion.

Even when using the method of KAHLE (11), I have not been able to separate the ventral thalamus in the subthalamic zone. In the subthalamic zone the different phases of migration and exhaustion of the matrix take place in a caudal to rostral and ventral to dorsal direction (Fig. 1). In Fig. 2 the broken lines indicate what would be the position of these limits in embryos of different ages. The different limits described by KUHLENBECK (13) and ROSE (19) and others between the ventral thalamus and hypothalamus, and KAHLE (11) between the ventral thalamus and subthalamus, are proof of the difficulty in arriving at this differentiation. A further proof of this difficulty is disagreement of authors (14, 19, 13, 11) regarding the nuclei which derive from each one of these zones.

Differentiation of nuclei and fibre tracts. In agreement with most authors, I find the anlagen of the subthalamic nucleus in *embryos of 18 mm* (already visible in embryos of 14 mm), the globus pallidus (20), the reticular thalamic nucleus and the ventral part of the lateral geniculate body (nucleus praegeniculatum in the human adult), all of which are of subthalamic origin, and also the anlage of the mamillary body of the hypothalamus and the habenula of the epithalamus. A migration in the thalamus has already taken place, which corresponds for some authors (3) to the dorsal part of the lateral geniculate and medial geniculate body.

Many fibre tracts can already be seen in this embryo. To those described by GILBERT (6) I will only add here those, accompanying the hypothalamic tegmental tract and the optic tract, form the more primitive supraoptic decussation.

24 mm. The anlage of the area H_2 of FOREL and the supramamillar commissure can be seen in an embryo of 24 mm length. At this stage the fibre tracts proceeding from the thalamus cross the reticular nucleus. The thalamic matrix is found to be at full migration (Fig. 1 C). The cellular area produced by a first thalamic migration has been displaced caudally. In the hypothalamus the mamillary body can be clearly distinguished. The three diencephalic sulci can be recognized in its ventricular wall.

27 mm. In the human embryo of 27 mm length the two primordia of the globus pallidus are more conspicuous; the most rostral part of the pars lateralis derives from the region dorsal to the sulcus diencephalicus ventralis. I am in agreement with ROSE (19) and ESCOLAR (4) that this is also the origin of the nucleus of the diagonal band. The reticular nucleus occupies a large area of the dorsal and anterior part of the subthalamus. The hypothalamic matrix advances dorsally in its rostral part.

35 mm. In the embryo of 35 mm length (Fig. 1 D), the separation between the epithalamus and thalamus is very conspicuous. The thalamic matrix can be easily distinguished from the subthalamic matrix. Owing to the growth of the thalamus caudally and laterally, its separation from the subthalamus becomes indistinct in the caudal region. The anlage of the areas of FOREL and the zona incerta are clearly visible. The fasciculus H_2 of FOREL can be seen to and rostrally at the level of the primordium of the globus pallidus. In the subthalamic zone the anlagen of the dorsal and dorsocaudal areas of the hypothalamus and of the pars medialis of the globus pallidus can be recognized.

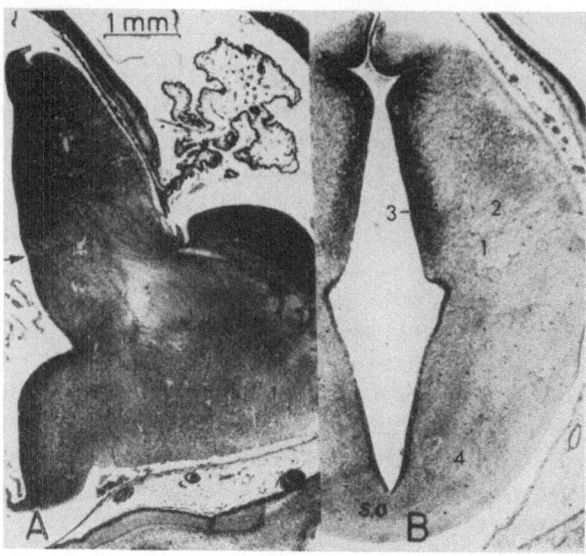

Fig. 3. Oblique sections of the diencephalon of an embryo of 37 mm length, (A) immediately inferior to the interventricular foramen. The separation between the thalamus and subthalamus is determined by the sulcus thalamicus medius (arrow). In the subthalamus the nucleus reticularis and the primordium rostrale of the globus pallidus can be observed. The sulcus diencephalicus dorsalis that separates the thalamus from the epithalamus can be seen in A and B. In B, the subthalamus can be distinguished from the thalamus and hypothalamus by the different behaviour of the matrices. 1 – cells of subthalamic origin which penetrate the thalamus. 2 – thalamic migrations. 3 – the thalamic matrix in the phase of total migration. 4 – nerve fibres heading towards the supraoptic decussation (SO).

37 mm. In the human embryo of 37 mm length the dorsal and medial diencephalic sulci are very conspicuous (Fig. 3). The anlage of the zona incerta, areas H_1 and H_2 of FOREL and other subthalamic formations can be clearly differentiated. I would like to point out the following important facts: 1. the mamillo-thalamic tract is already visible and is about to penetrate the thalamus; 2. at this level the division between the thalamus and the subthalamus is not clearly marked and it appears that owing to the growth of the thalamus, cellular migrations of subthalamic origin are penetrating the ventro-medial part of the thalamus (Fig. 3, 1); 3. the supraoptic decussation is clearly visible (Figs. 3, SO and 4); 4. the great development of the rostral part of the subthalamic zone can be observed (Fig. 3 A).

In this embryo the sulcus diencephalicus dorsalis has been most clearly observed. The sulcus diencephalicus dorsalis separates the epithalamus from the thalamus. According to COGGESHALL (2), this sulcus together with the stria medullaris and tractus habenulo-peduncularis is the limit between the second and third neuromeres. It can now be affirmed that the sulci separating cellular groups of different meanings are acute and they continue laterally with fibre tracts which help to make the separation, as can be observed in the case of the dorsal and medial diencephalic sulci.

55 mm. In embryos of 55 mm the thalamus has increased in size. Caudally, its separation from the subthalamus has become more confused. The same has occurred at the point of

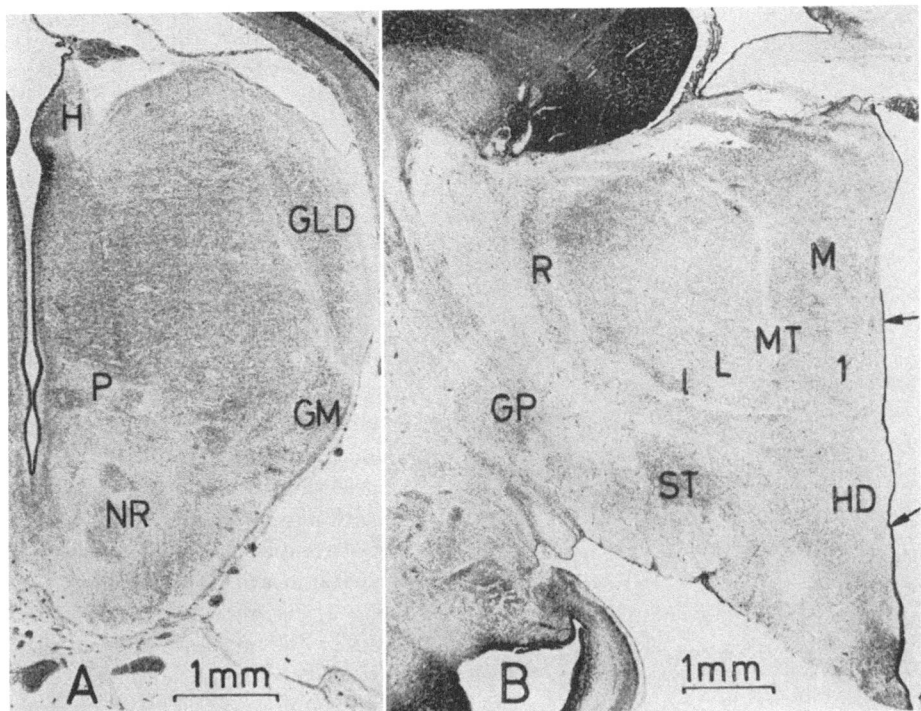

Fig. 4. (A) Frontal section of a human embryo of 64 mm; P – cellular group of subthalamic origin incorporated in the medio-ventral part of the thalamus. (B) Frontal section of a human embryo of 120 mm. A group of cells of subthalamic origin (1) is included in the thalamus. A – anterior groups; CA – commissura anterior; CM – nc. centri mediani; DM – area hypothalamica dorsalis; F – fornix; GLD – corpus geniculatum laterale, pars dorsalis; GM – corpus geniculatum mediale; GP – globus pallidus; GPm – globus pallidus, pars medialis; H – habenula; L – lateral group; l – lamina medullaris externa; M – nc. medialis dorsalis; MD – area hypothalamica dorsalis; MT – fasciculus mamillo-thalamicus; NR – nc. ruber; R – nc. reticularis thalami; ST – nc. sub-thalamicus; VA – nc. ventralis anterior.

penetration of the tractus mamillo-thalamicus in the thalamus. The nuclear differentiations in the subthalamus and hypothalamus and the fibre tracts have become more conspicuous. The boundary between the subthalamus and hypothalamus is clearly defined and advances dorsally in its rostral half.

In the embryos being a little older the migrations from the thalamic matrix cease, and the thalamus acquires an uniform aspect. The habenula is differentiated as are the medial and lateral geniculate bodies (Fig. 4 A). The displacement of the thalamus in caudal, lateral and ventral direction results in subthalamic formations, and left them included in the ventro-medial part of the thalamus. These formations will form the anterior part of the parafascicular centromedian complex (P) of the thalamus.

The telencephalic pedicle grows considerably, the internal and external capsules can be observed (Figs. 4 B and 5). The reticular nucleus of the thalamus has extended laterally and caudally forming a layer of cells (R). This layer decreases in thickness in a rostral to caudal direction and separates the thalamus from the telencephalic pedicle. The growth

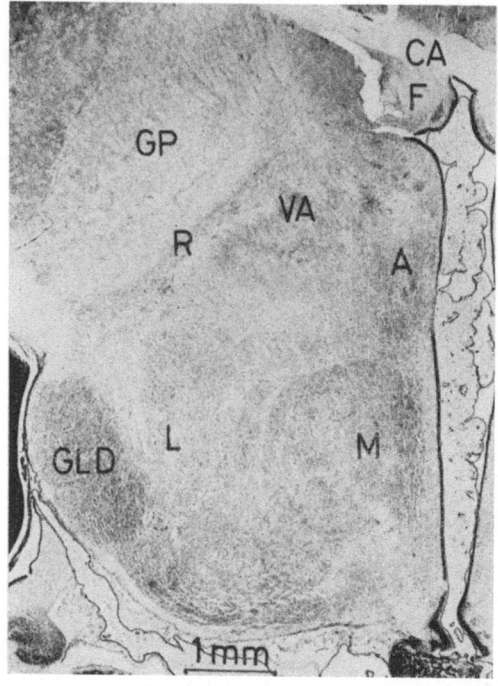

Fig. 5. Horizontal section of a human embryo of 110 mm length. At least part of the nucleus ventralis anterior (VA) has the same place of origin as the rostral part of the nucleus reticularis (R). Abbreviations as in Fig. 4.

of the internal capsule separates the medial part of the globus pallidus from the diencephalon. The latter in an *embryo of 64 mm* has a topographical position, which is very similar to that of the entopeduncular nucleus in carnivora. The mamillo-thalamic tract penetrates the thalamus towards the future position of the anterior thalamic nuclei (Figs. 4, 5 and 6).

110 mm and larger. In the embryo of 110 mm length the first nuclear differentiations in the thalamus appear (Fig. 5). In the embryo of 120 mm length the nuclear differentiations become more conspicuous (Fig. 4 B) and it can be observed that the ventral part of the medial and intralaminar nuclei are formed from the cells, proceeding from the subthalamus, penetrate the thalamus (Fig. 4 B, 1). That the subthalamus is the origin of these nuclei and of the prethalamic nuclei can be demonstrated throughout embryonic development. In larger embryos the greater part of the thalamic and hypothalamic nuclei can be differentiated (Fig. 6). The boundaries between the subthalamus and hypothalamus are fairly distinct. This is not the case in the boundaries between the thalamus and subthalamus since it can be affirmed that the ventral part of the midline and intralaminar nuclei and the parafascicular nucleus are of subthalamic origin. In my opinion the nucleus ventralis anterior and the rostral part of the centro-median originate at least in part from the subthalamic matrix, since their cellular maturation takes place at the same time as that of the near-subthalamic formations. This common origin explains the similarity of the structure, connections and functional meanings of these formations.

Summary

In the diencephalon of the human embryo there are four longitudinal zones: the epithalamus, the thalamus, the subthalamus and the hypothalamus.

The first division made in the embryonic diencephalon is determined by the sulcus diencephalicus medius and the lamina medullaris externa. The matrices dorsal and ventral to this sulcus have a different aspect and evolution.

The zone dorsal to the sulcus diencephalicus medius is divided into two longitudinal zones, the epithalamus and the thalamus. The sulcus diencephalicus dorsalis, the stria medullaris and the tractus habenulo-peduncularis contributed to the division.

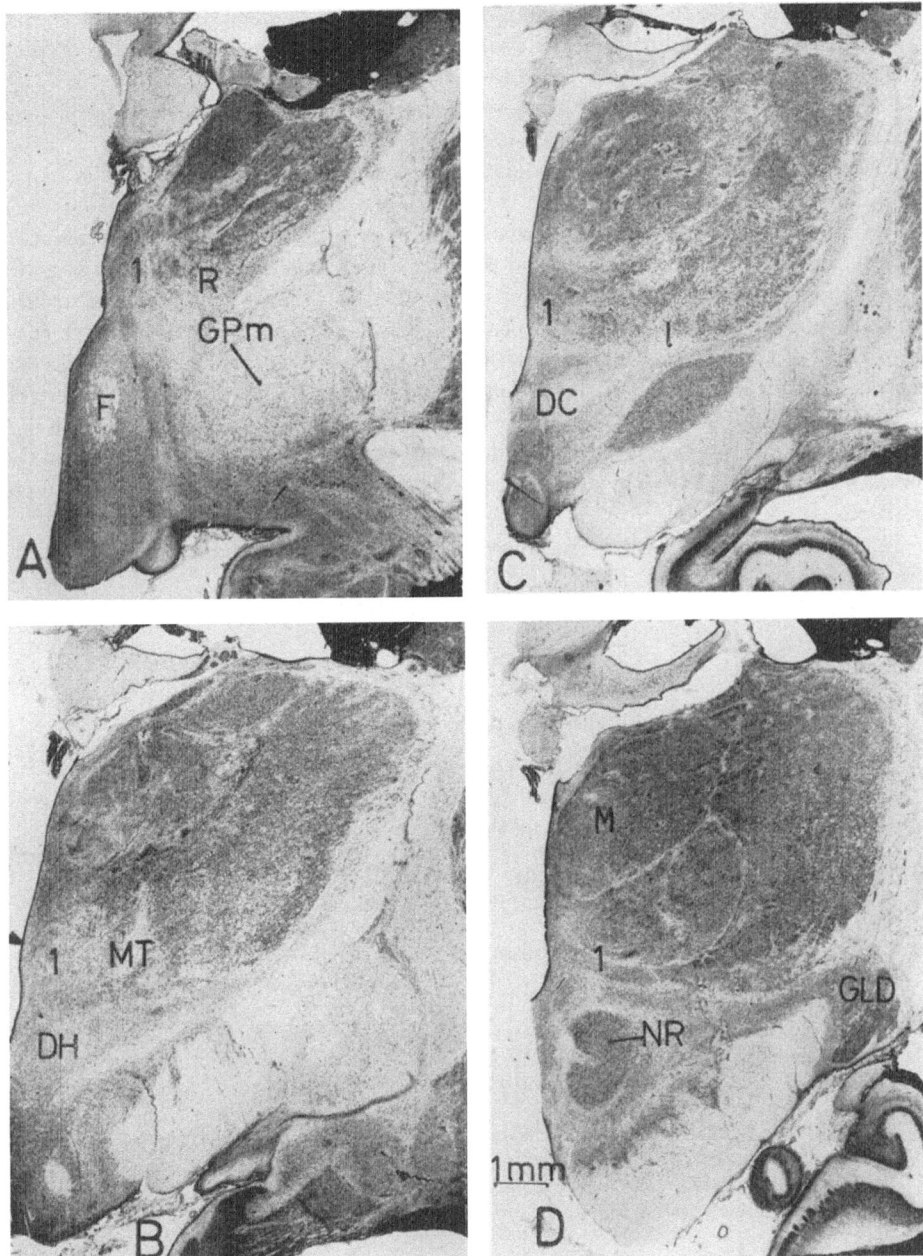

Fig. 6. Frontal sections of a foetus of 6 months. The formations of subthalamic origin (1) which penetrate the thalamus can be distinguished. Abbreviations as in Fig. 4.

The zone ventral to the sulcus diencephalicus medius and lamina medullaris externa is divided into two longitudinal zones, the subthalamus and the hypothalamus.' The division is made on the basis of the behavior of the matrix.

The growth of the subthalamus gives rise to the formation of the sulcus diencephalicus ventralis.

The growth of the thalamus in caudal, ventral and lateral directions results in the inclusion of subthalamic formations in the ventral and medial part of the thalamus.

The following nuclei of the adult thalamus are of subthalamic origin: the reticular nucleus, the pars ventralis of the lateral geniculate body (nucleus praegeniculatus in the adult), the nucleus fasciculosus thalami of HASSLER (7), the nucleus reuniens, nucleus ventralis medialis and the rostral part of the parafascicularis complex. Possibly all, or at least a part of the nucleus ventralis anterior, the pars ventralis of the nuclei medialis and centralis and the rostral part of the nucleus centro-median are also of subthalamic origin.

The following formations in the adult, classically being considered as part of the hypothalamus, are of subthalamic origin: the dorsal part of the lateral hypothalamic area, the dorsal part of the prethalamic nuclei and the dorsal and dorso-caudal areas of the hypothalamus.

The rest of the subthalamic formations consist of the globus pallidus, the zona incerta, the nucleus subthalamicus and the areas of FOREL.

References

1. BERGQUIST, H.: Kgl. Fysiogr. Sällsk. Handl. N.F. 64 (1954), Nr. 20.
2. COGGESHALL, R.E.: J. comp. Neurol., *122* (1964), 241–270.
3. DEKABAN, A.: J. comp. Neurol., *100* (1954), 63–98.
4. ESCOLAR, J.: Ann. Anat. *13* (1964), 5–84.
5. FORTUYN, A.B.D.D.: Arch. Anat. (1912), 303–352.
6. GILBERT, M.S.: J. comp. Neurol. *62* (1935), 81–116.
7. HASSLER, R.: Anatomie des Thalamus. In: Einführung in die stereotaktischen Operationen. Thieme, Stuttgart 1959, 230–290.
8. HERRICK, C.J.: J. comp. Neurol. *20* (1910), 413–547.
9. HERRICK, C.J.: The Brain of the Tiger Salamander, Chicago 1948.
10. HIS, W.: Die Entwicklung des menschlichen Gehirns während der ersten Monaten. Leipzig 1904.
11. KAHLE, W.: Dtsch. Z. Nervenheilk. *175* (1956), 259–318.
12. KUHLENBECK, H.: Anat. Anz. *70* (1930), 122–142.
13. KUHLENBECK, H.: Mil. Surg. *102* (1948), 433–447.
14. MIURA, R.: Anat. Anz. *77* (1933), 1–65.
15. PAPEZ, J.W.: Res. Publ. Ass. nerv. ment. Dis. *20* (1940), 31–51.
16. REINOSO-SUAREZ, F.: An. Anat. *3* (1954), 231–245.
17. REINOSO-SUAREZ, F.: An. Anat. *7* (1958), 239–265.
18. REINOSO-SUAREZ, F.: An. Anat. *9* (1960), 5–33.
19. ROSE, J.E.: J. comp. Neurol. *77* (1942), 61–130.
20. SPATZ, H.: Anat. Anz. *60* (1925), 54–58.

Prof. Dr. FERNANDO REINOSO-SUÁREZ
Departamento de Anatomia
Facultad de Medicina
Universidad de Navarra
Apartado 177
Pamplona, Spain

Zur ontogenetischen Entwicklung der Brodmannschen Rindenfelder

Von W. Kahle, Würzburg

Die praenatale Entwicklung des menschlichen Isocortex wurde bisher vorwiegend unter histogenetischen Gesichtspunkten bearbeitet, wobei die Frage der Schichtenbildung im Vordergrund stand. Brodmann (1909) hatte seine Lehre vom sechsschichtigen Grundtypus aufgestellt, aus dem sich durch Spaltung oder Reduzierung von Schichten der variationsreiche Bau der verschiedenen Areae entwickeln soll. In späteren Arbeiten (Filimonoff 1929, 1947; Beck 1940; Poliakow 1940) wurde die gleiche Frage behandelt. Rabinowicz (1964) verfolgte neben der Histogenese vor allem die Cytogenese in den verschiedenen Schichten. Nur Kononowa (1940) unternahm eine Felderung an fötalen Hemisphären und legte eine Hirnkarte des Frontallappens in verschiedenen Entwicklungsstadien vor.

Die *areale Gliederung des fötalen Isocortex* soll in der vorliegenden Arbeit in erster Linie behandelt werden und die Fragen der Histogenese nur soweit Berücksichtigung finden als sie mit der Entwicklung der Rindenfelder in Zusammenhang stehen. Von besonderem Interesse sind dabei die Veränderungen der arealen Topographie während der Entwicklung. Solche Veränderungen sind zu erwarten, da die successive Reifung der verschiedenen Rindenfelder einen Wechsel ihrer Ausdehnung voraussehen läßt. Allgemein nehmen frühzeitig differenzierte Hirnbezirke anfangs einen größeren Raum ein und werden dann durch die Ausdehnung später reifender Gebiete eingeengt und verlagert. Dieser Prozeß ist bei der Entwicklung des Allocortex, der während der späteren Entfaltung des Isocortex erhebliche Lageveränderungen erleidet, bekannt (Macchi, 1951) und in geringerem Ausmaße auch im Bereiche der homogenetischen Rinde zu erwarten. Die areale Topographie ist außerdem abhängig vom Wachstum der Hemisphärenblase, deren Wand sich nämlich nicht gleichmäßig wie bei einem Ballon ausdehnt, sondern in bestimmten Regionen stärker, in anderen weniger wächst, so daß es unter anderem zu einer Rotationsbewegung der Hemisphärenblase um die Inselregion kommt (Spatz 1949). Derartige Prozesse beeinflussen die Oberflächengliederung und bestimmen wesentlich das endgültige Bild der Hirnkarte.

Material und Methode. Die Untersuchung wurde an 16 Schnittserien von Gehirnen menschlicher Embryonen und Föten in Entwicklungsstadien vom dritten bis zum achten Monat durchgeführt. Die Serien der jüngeren Gehirne waren mit H.E. gefärbt, das Material im Alter von 4 Monaten und älter war durchweg nach Nissl gefärbt. Die Fehlermöglichkeiten, die sich vor allem bei Frontalserien in den Polgegenden der Hemisphären durch Schrägschnitte ergeben, wurden bei der Auswertung berücksichtigt.

Der Isocortex ist in seiner Entwicklung durch das Auftreten einer *Rindenplatte* charakterisiert, die sich als breites zelldichtes Band während der ersten Hälfte des dritten Monats unter dem Randschleier bildet und sich von der Inselregion aus über die ganze

Hemisphäre bis zum Hippocampus ausdehnt. Im Verlaufe der Differenzierung tritt in der Rindenplatte eine helle zellarme Zone auf, die zu einer Spaltung der Rindenplatte in zwei dunkle zelldichte Streifen führt, der späteren äußeren und inneren Körnerschicht. Außerdem wird die Rindenplatte von einer locker gebauten Zellschicht unterlagert, der späteren Lamina multiformis. Damit haben wir bei Einbeziehung des Randschleiers bereits eine sechsschichtige Rinde vor uns, die sich folgendermaßen zusammensetzt: Randschleier (I), äußerer Streifen der Rindenplatte (II), helle zellarme Zone (III), innere Streifen der Rindenplatte (IV), helle zellarme Zone (V), lockere Zellschicht unter der Rindenplatte (VI).

Die sechsschichtige Rinde und ihre Wertung. Eine solche sechsschichtige Rinde ist nun keineswegs erst im sechsten oder siebenten Monat nachweisbar, wie man bisher annahm, sondern bereits gegen Ende des vierten Monats. Sie liegt am frontalen Pol der Hemisphäre und ist auf Frontalserien infolge des schrägen Anschnittes der Rinde in diesem Bereich nicht zu erkennen. Nur bei horizontaler und sagittaler Schnittrichtung kommt die Rinde des Frontalpoles auf einigen Schnitten der Serie klar zur Darstellung. Abb. 1 a zeigt den frontopolaren Cortex eines Föten vom Ende des vierten Monats. Während an der Basis Rindenplatte und anschließende lockere Zellschicht noch keine Gliederung aufweisen, ist im Polgebiet eine Sechsschichtung zu erkennen. Daß wir im Bereiche des Frontalpoles, dessen Rinde als ausgesprochen spät entwickelt gilt, die früheste Schichtenbildung finden, ist ein ungewöhnlicher Befund, auf den wir später noch näher eingehen müssen.

Die Tatsache, daß die Differenzierung des Cortex in sechs Schichten nicht erst im sechsten, sondern schon Ende des vierten Monats beginnt, ist für die Lehre BRODMANNS vom sechsschichtigen Grundtypus bedeutsam. Als nämlich nachgewiesen wurde, daß die Praecentralregion schon im sechsten Monat praktisch agranulär ist, erhoben sich starke Zweifel, ob der sechsschichtige Bau tatsächlich in allen Rindenfeldern als Grundschema des Rindenbaues auftritt (BECK 1940). Beim Nachweis einer sechsschichtigen Rinde im vierten Monat kann man nun ohne weiteres annehmen, daß die Rinde der Areae 4 und 6 bereits vor dem sechsten Monat vorübergehend einen sechsschichtigen Bau besaß.

Wir können jedenfalls aufgrund unserer Befunde die Annahme BRODMANNS bestätigen, daß alle Rindenfelder während ihrer Entwicklung eine Sechsschichtung zeigen. Wir möchten dieser Gliederung aber nicht die Bedeutung eines corticalen Grundschemas beimessen, aus dem sich alle Variationen des Rindenbaues ableiten lassen. Die Gliederung in sechs Schichten ergibt sich zwanglos aus der Spaltung und Unterlagerung der Rindenplatte und ist in vielen Bezirken nur ein flüchtiges Durchgangsstadium, das keineswegs im ganzen Isocortex das gleiche stereotype Bild bietet, sondern von Anfang an lokale Variationen aufweist, wie es von C. u. O. VOGT (1919) schon angenommen wurde.

Differenzierung der Rindenfelder. *Im vierten Monat* ist zwar frontal bereits eine Sechsschichtung zu erkennen, aber Rindenfelder lassen sich trotz lokaler Variationen der Rindenplatte (POLIAKOW 1940) noch nicht identifizieren. Das ist erst *im fünften Monat* möglich. Obwohl auf dieser Entwicklungsstufe noch keine Centralfurche angelegt ist, läßt sich die *Prae-* und *Postcentralregion* und die Grenze zwischen beiden genau bestimmen.

Abb. 1. Rindendifferenzierung im vierten und fünften Monat. a) Frühestes Auftreten einer Sechsschichtung im Bereiche des Frontalpoles bei einem Föten aus der zweiten Hälfte des vierten Monats. Vergrößerung 18fach. b) Die Praecentralregion (linke Bildseite) an der Grenze zum Gyrus cinguli (rechte Bildseite) bei einem Föten aus dem fünften Monat (154 mm Scheitel-Steiß-Länge). Die Grenze ist durch Pfeile markiert. c) Die Postcentralregion (spätere Area 1) beim gleichen Föten. Vergrößerung 39fach.

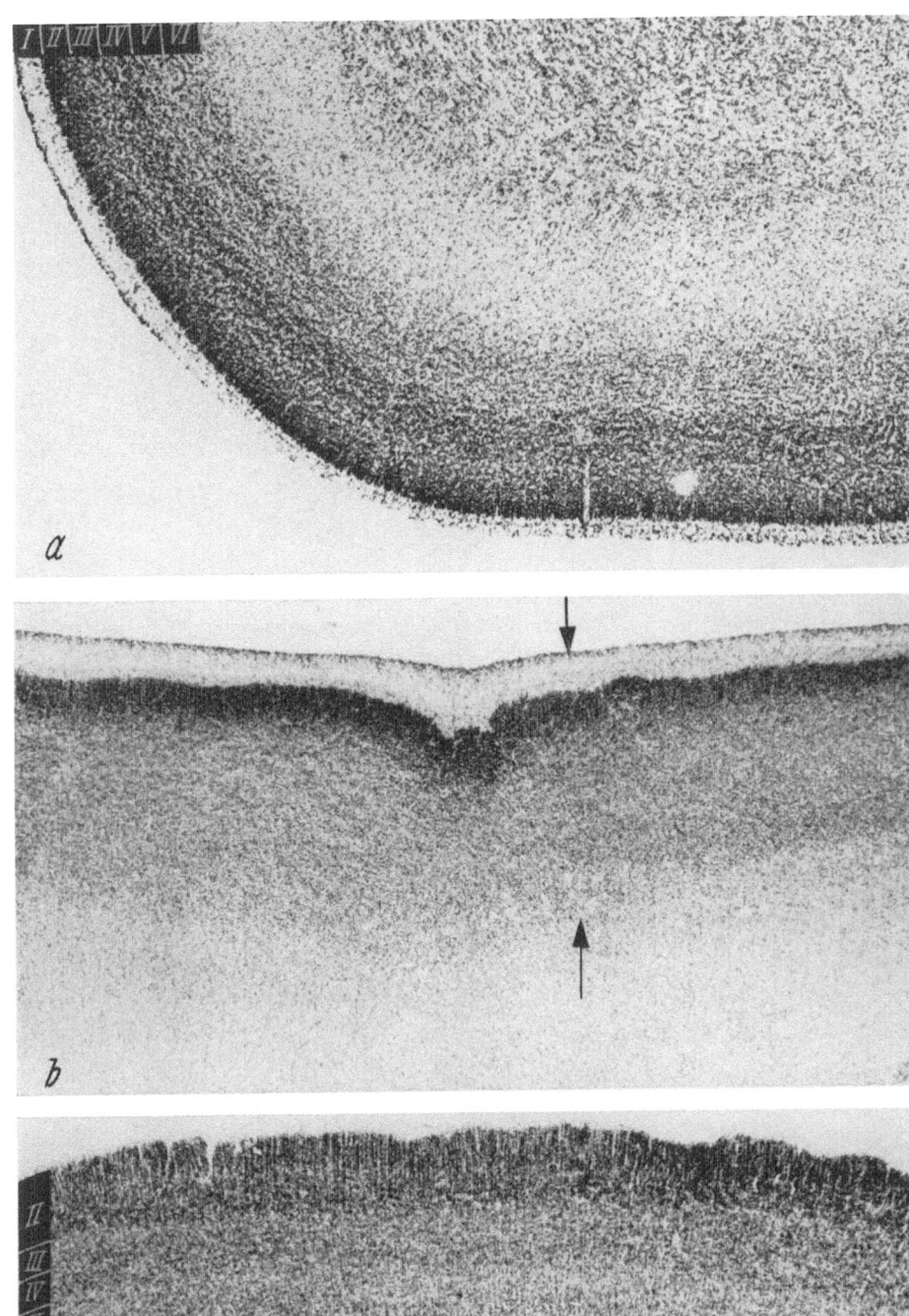

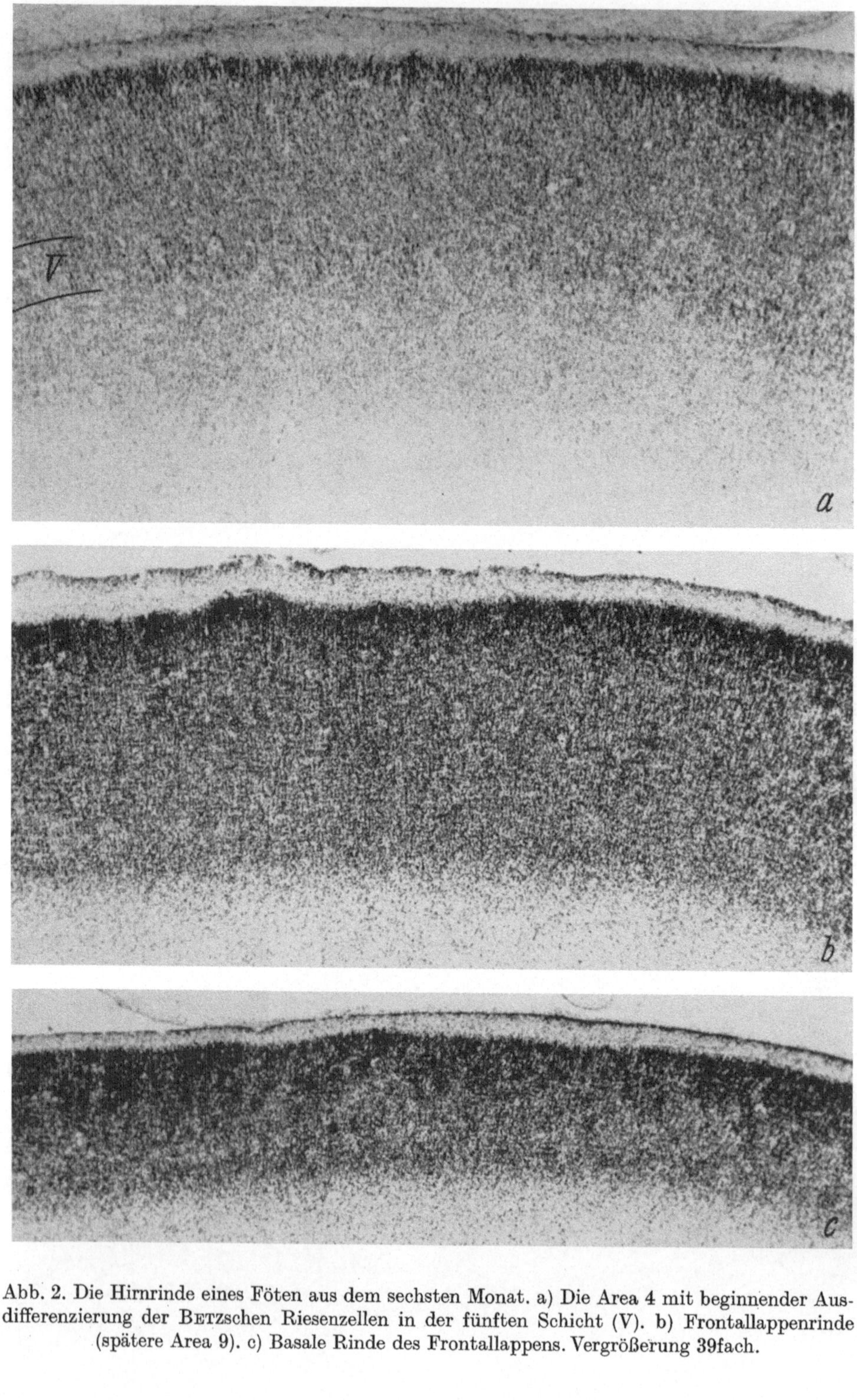

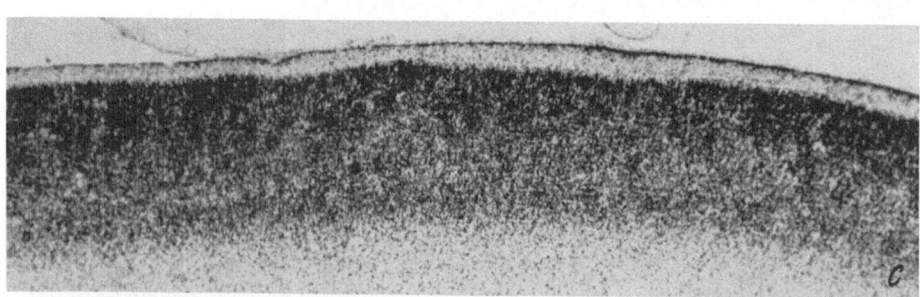

Abb. 2. Die Hirnrinde eines Föten aus dem sechsten Monat. a) Die Area 4 mit beginnender Ausdifferenzierung der BETZschen Riesenzellen in der fünften Schicht (V). b) Frontallappenrinde (spätere Area 9). c) Basale Rinde des Frontallappens. Vergrößerung 39fach.

Die Areae 4 und 6 sind nicht voneinander zu trennen, da sich noch keine Pyramidenzellen entwickelt haben. Aber trotz des Fehlens der Centralfurche und der BETZschen Riesenzellen kann man die praecentralen Felder an ihren sonstigen Merkmalen ohne weiteres erkennen. Auf Abb. 1b ist die Praecentralregion (linke Bildseite) an der Grenze zum Gyrus cinguli (rechte Bildseite) zu sehen. Die Rinde zeigt zwar in Höhe des Randschleiers und der äußeren Körnerschicht keine Unterschiede, aber das anschließende Zellband ist im Abschnitt des Gyrus cinguli zelldicht und schmal, während es auf der linken Bildseite deutlich breiter ist und fließend in das Marklager übergeht; der lockere Bau, die besondere Breite und der fließende Übergang in die weiße Substanz sind charakteristisch für die praecentrale Rinde. Der Cortex der Postcentralregion ist dagegen zelldicht, schmal und klar vom Mark abgesetzt. Die Area 1 (Abb. 1c) besitzt eine Sechsschichtung, die sich durch eine ungewöhnlich breite und kompakte äußere Körnerschicht auszeichnet und dadurch die Rinde schon als künftigen Koniocortex ausweist. Abb. 4a zeigt die Lage der ,,Centralgrenze'' (des späteren Sulcus centralis), die infolge der markanten Strukturunterschiede der Rinde ohne Schwierigkeit festzulegen ist.

Auch im *Parietal-* und *Temporallappen* ist in der Umgebung der *Insel* eine beginnende Schichtenbildung zu beobachten. Um die früh differenzierte Inselrinde liegt halbkreisförmig ein sechsschichtiger Cortex (Abb. 3a), der in eine noch undifferenzierte Rinde (gespaltene Rindenplatte ohne Unterlagerung durch eine weitere Zellschicht) übergeht. Der Rest des Parietal- und Temporallappens wird von einer Rindenplatte bedeckt, die hinsichtlich Breite und Zelldichte lokale Variationen zeigt. Die Verhältnisse an der Konvexität der Hemisphäre gibt Abb. 4a wieder.

Im sechsten Monat erscheinen Pyramidenzellen in der Lamina ganglionaris der Praecentralregion und ermöglichen eine Unterscheidung der Areae 4 und 6. Wie weit die Entwicklung der Praecentralregion im sechsten Monat schon vorangeschritten ist, zeigt ein Vergleich mit anderen frontalen Rindenbezirken (Abb. 2a–c). Am Cortex der Area 4 (Abb. 2a) fällt außer den Pyramidenzellen und der besonderen Rindenbreite die Verschmälerung der äußeren Körnerschicht und der lockere Bau auf. Die Rinde der frontalen Felder, von denen die Area 9 (Abb. 2b) und der basale Cortex (Abb. 2c) als Beispiele gebracht werden, ist wesentlich schmäler, zelldichter und deutlich von der weißen Substanz abgesetzt. In der Area 9 ist eine Sechsschichtung nur schwach angedeutet, Pyramidenzellen sind noch nicht zu erkennen und die basale Rinde bietet praktisch noch das Bild einer Rindenplatte. Im übrigen lassen sich jetzt schon eine Reihe von Feldern abgrenzen und identifizieren; das Resultat der Felderung ist die Hirnkarte auf Abb. 4b.

Im siebenten und achten Monat ist, wie schon BRODMANN betonte, der sechsschichtige Bau der homogenetischen Rinde voll ausgeprägt. Als Beispiel bringen wir einen Rindenabschnitt des Temporallappens (Abb. 3b) mit einer klaren Sechsschichtung und in der Area 22 mit einer schönen, regelmäßigen Radiärstruktur, wie sie in diesem Stadium für den Parietal- und Temporallappen typisch ist. An der Dorsalfläche des Temporallappens ändert sich der Bau der Rinde: Durch die Auflockerung der inneren Körnerschicht (IV) und der sechsten Schicht erscheint die ganze Struktur etwas verwaschen. Weiter medial (rechte Bildseite) wird die Schichtung der Rinde so unscharf, daß die fünfte und sechste Schicht nicht mehr voneinander abzugrenzen sind, sondern ein locker gebautes Zellband bilden, das fließend in das Marklager übergeht. Vor allem aber ist die innere Körnerschicht wolkig aufgelockert und verbreitert. Es handelt sich um die Areae 42 und 41, welche die HESCHLschen Querwindungen bedecken. Ihre Rinde gehört zum Koniocortex, der sich durch das Dominieren der Körnerschichten und die Reduktion der Pyramidenschichten

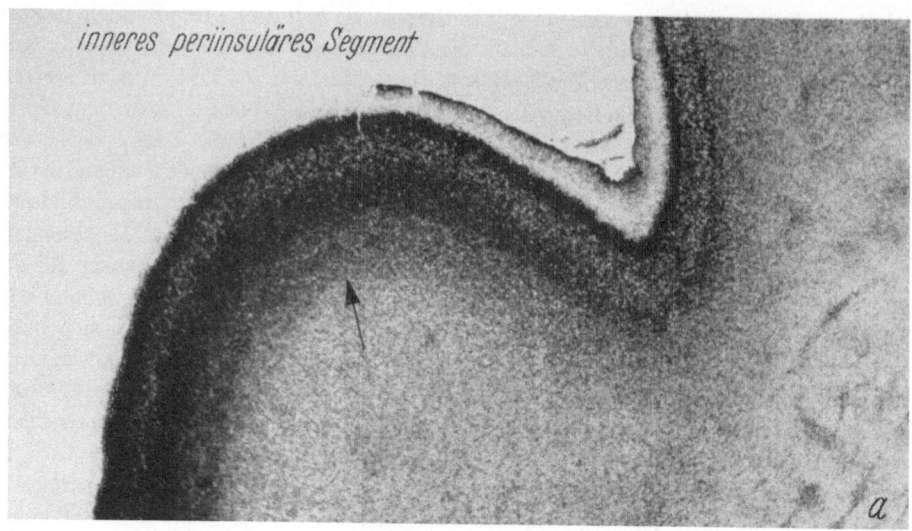

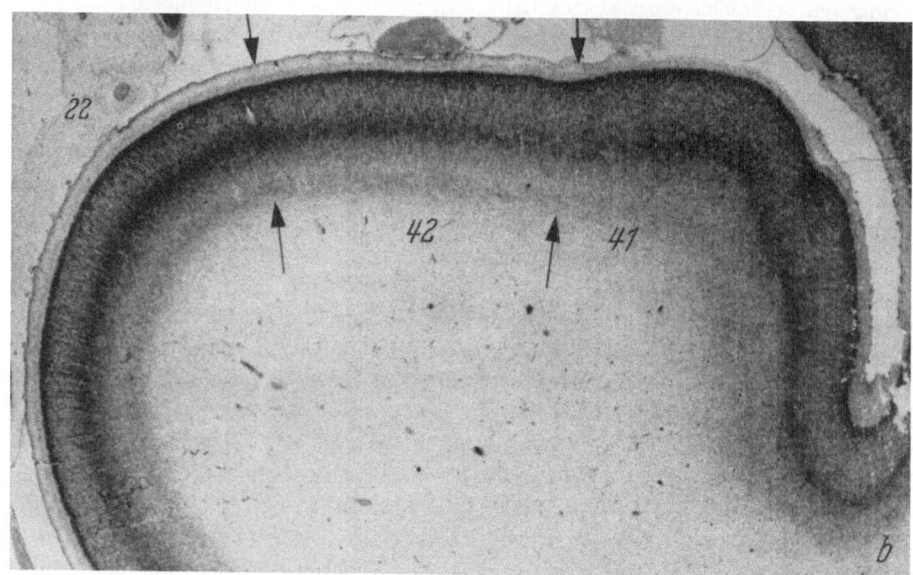

Abb. 3. Temporallappen an der Grenze zur Insel. a) Bei einem Föten aus dem fünften Monat;
der Pfeil gibt die Grenze zwischen Area 42 und 41 an. Vergrößerung 23fach. b) Bei einem Föten
aus dem achten Monat; die Pfeile geben die Feldgrenzen an. Vergrößerung 18fach.

auszeichnet. Im achten Monat sind, wie das Beispiel zeigt, die individuellen Merkmale der
meisten Rindenfelder vorhanden, so daß man fast alle Felder identifizieren und abgrenzen
kann. Entsprechend nähert sich auch die auf Abb. 4c wiedergegebene Hirnkarte aus dem
achten Monat in groben Zügen der des reifen Gehirns.

Reifungsunterschiede. Die auf Abb. 4 wiedergegebenen Hirnkarten von Hemisphären
aus dem fünften, sechsten und achten Monat bringen Rindenfelder und Reifungsunter-

schiede zusammen zur Darstellung. Diese Betrachtungsweise, die sich nur an die tatsächlich vorhandenen Strukturunterschiede hält, macht deutlich, wie aus Gebieten, die
anfangs nur Bezirke gleicher Rindenreife sind, langsam die bleibenden Felder auftauchen.
Es ist dabei aber auch zu erkennen, daß die bleibenden Rindenfelder zum Teil Ausdruck
einer Reifungsdifferenz sind. Am deutlichsten ist das im Frontallappen, bei dem der
Reifungsprozeß von der Area 4 bis zur basalen Rinde stufenweise fortschreitet und dabei
eine sehr erhebliche zeitliche Diskrepanz aufweist. Wie Abb. 2 veranschaulicht, befindet
sich die basale Rinde im sechsten Monat praktisch noch im Stadium der Rindenplatte,
während in der Area 4 bereits die BETZschen Riesenpyramidenzellen erscheinen. Diese
Abstufung der Differenzierung ist sogar noch an der Frontallappenrinde des reifen
Gehirns nachweisbar; unterschiedliche Differenzierungsgrade bleiben hier als architektonische Merkmale erhalten. Eine ähnliche Aufeinanderfolge der Rindenentwicklung findet
sich im Temporallappen (Abb. 3a), in dem der Reifungsprozeß von den Areae 41 und 42
über die Areae 22 und 21 bis zum Feld 20 fortschreitet.

Reifungsgefälle. Wir können daraus schließen, daß das Reifungsgefälle auf der Hemisphärenoberfläche in bestimmten Richtungen verläuft, wobei die Centralregion eine
wichtige Grenze darstellt. Oral von der Centralfurche liegt das am frühzeitigsten differenzierte Gebiet, die Praecentralregion. Von hier aus dehnt sich der Reifungsprozeß in frontobasaler Richtung aus, so daß verschiedene Sektoren von abgestuften Differenzierungs-
graden entstehen. Caudal von der Centralfurche liegen die früh entwickelten Rindenbezirke als Segmente um die Insel herum und der Reifungsprozeß schreitet hier allseitig
von innen nach außen, d. h. in Richtung zur Mantelkante, fort. Die Segmente umschließen
die Insel jedoch nur caudal von der Centralfurche. In den Gebieten, die oral davon
liegen, tritt keine frühzeitige Differenzierung ein, denn hier verläuft die Reifung von der
Praecentralregion aus und erreicht die Rinde des Operculum frontale relativ spät. Der
Frontallappen einerseits und der Parietal- und Temporallappen andererseits zeigen also
in topographischer Hinsicht einen sehr unterschiedlichen Ablauf der Rindendifferenzierung.

Relative Größenänderung einiger Rindenfelder. Bei den Größenveränderungen der
einzelnen Felder während der Entwicklung handelt es sich natürlich nur um relative
Werte, die sich stets auf den Anteil an der Gesamtoberfläche der Hemisphäre beziehen.
Sie sind im Bereiche der homogenetischen Rinde nicht so hochgradig wie wir es vom
Allocortex und von der Inselrinde kennen, die beide einen erheblichen Flächenverlust
erleiden. Das liegt zum Teil daran, daß wir die Rindenfelder in frühen Stadien der corticalen Entwicklung, also im vierten und fünften Monat, nicht sicher identifizieren können
und oft nicht wissen, ob die Grenzen der unterschiedlich entwickelten Rinde oder Rindenplatte mit denen der späteren Felder zusammenfallen. Deutlich ist immerhin der Flächenverlust der Postcentralregion, die im fünften Monat als breites Band über die Konvexität
der Hemisphäre zieht und sich bis zum achten Monat zunehmend verschmälert. Die Area 2
ließ sich zwar nicht eindeutig abgrenzen, aber die Area 1 und 3 sind gut zu bestimmen und
in ihrer Ausdehnung zu verfolgen. Das Auftreten der Centralfurche im sechsten Monat und
ihre zunehmende Vertiefung spielt bei dem Flächenverlust natürlich eine Rolle, denn die
Area 3 verschwindet mehr und mehr in der Tiefe der Furche.

Einen noch stärkeren Rückgang zeigt das innere um die Insel gelegene Rindensegment,
dessen temporaler Abschnitt den späteren Areae 41 und 42 entspricht. Ein Vergleich
dieses Bezirkes auf der Hirnkarte des fünften Monats (auf Abb. 4a kariert) mit der Fläche
der Felder 41 und 42 auf der Karte BRODMANNS demonstriert die Veränderung der Größen-

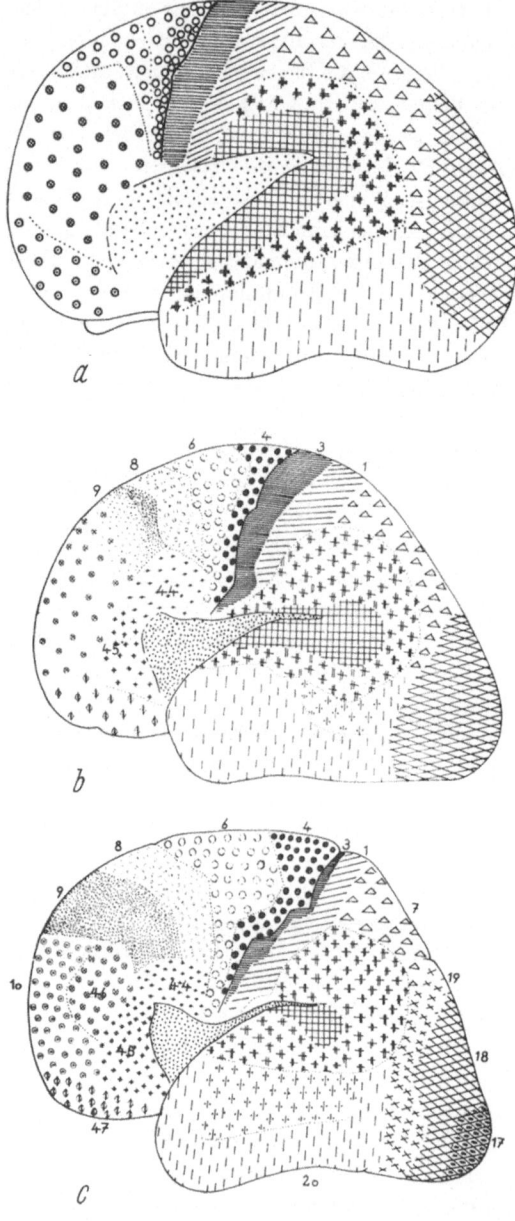

a

b

c

Abb. 4. Hirnkarten fötaler Hemisphären. a) Hemisphäre eines Föten aus dem fünften Monat; das innere periinsuläre Segment ist kariert, das äußere mit Doppelkreuzen versehen. b) Hemisphäre eines Föten aus dem sechsten Monat. c) Hemisphäre eines Föten aus dem achten Monat.

relation. Es handelt sich dabei tatsächlich um den Rückgang des Oberflächenanteils und nicht um das Ergebnis einer Verdrängung des betreffenden Gebietes in die Tiefe der Fossa Sylvii. Eine solche Verlagerung findet zwar statt, aber sie ist auf der BRODMANNschen Karte nicht berücksichtigt; der Bezirk ist bei BRODMANN gewissermaßen aufgeklappt in seiner ganzen Ausdehnung dargestellt. Auf der äußeren Oberfläche erscheinen die Felder 41 und 42 am reifen Gehirn überhaupt nicht, denn sie werden während der Supprimierung der Insel (SPATZ) auf die Innenfläche des temporalen Operculums verlagert. Auf unseren Hirnkarten (Abb. 4a –c), auf denen nur die äußere Oberfläche dargestellt ist, läßt sich das Verschwinden des inneren periinsulären Rindensegmentes in die Tiefe der Fossa verfolgen. Auch auf Abb. 3a und b ist zu sehen, daß die Rinde der künftigen Felder 41 und 42 im fünften Monat nicht nur die Dorsalfläche des Temporallappens einnimmt, sondern sich auch auf die Lateralfläche ausdehnt. Im achten Monat ist sie dann völlig in die Tiefe verlagert.

Lageänderung von Insel und Centralregion. *In der Ontogenese* gehört die Supprimierung der Insel und ihre Überlagerung durch die Opercula zu den großen Massenverschiebungen während der Hemisphärenentwicklung. Die wichtigste von ihnen, die Rotation der Hemisphärenblase um die Insel, läuft im dritten und vierten Monat ab, in einem Zeitraum also, in dem eine Felderung der Hemisphärenoberfläche noch nicht möglich ist. Immerhin gehört die Lageveränderung der Centralregion, die sich zwischen dem fünften und dem achten Monat nachweisen läßt, nach unserer Ansicht zu diesem Prozeß. Auf Abb. 4a–c ist unverkennbar, wie die anfangs relativ weit frontal gelegene Centralregion zunehmend nach caudal abgedrängt wird und im achten Monat schließlich eine Position einnimmt, die den Verhältnissen am reifen Gehirn entspricht.

Diese Wanderung läßt den Schluß zu, daß die Centralregion in früheren Entwicklungsstadien wesentlich weiter rostral gelegen ist. Wir müssen in diesem Zusammenhang auf die früheste Stratifizierung zurückkommen, die wir im vierten Monat am Frontalpol der Hemisphäre gefunden haben. Wir hatten bereits darauf hingewiesen, daß dies ein ungewöhnlicher Befund ist, da der frontopolare und frontobasale Teil des Isocortex als ein ausgesprochen spät entwickeltes Gebiet gilt. Nun kann man aber eigentlich erst vom fünften Monat an von einem wirklichen Frontallappen sprechen, und es ist daher der Schluß naheliegend, daß es sich bei dem geschichteten Cortexbezirk gar nicht um den bleibenden Pol des Frontallappens handelt, sondern um ein Gebiet, das im Verlaufe der Entwicklung verlagert wird. Anders ist es überhaupt nicht zu erklären, daß wir im fünften Monat in der gleichen Position eine ausgesprochen undifferenzierte Rinde antreffen. Bei der frühstratifizierten Rinde kann es sich nur um ein Gebiet handeln, das den übrigen Regionen in der Entwicklung vorauseilt. Das ist im frontalen Bereich allein bei der Praecentralregion der Fall. Daher halten wir diesen Bezirk für identisch mit den Feldern 4, 6 und 8. Zu dieser Annahme berechtigt uns die nachgewiesene Bewegung der Centralregion zwischen dem fünften und achten Monat, die die Fortsetzung und den Abschluß der Wanderung von einer ursprünglich frontalen Position bis in die Parietalgegend darstellt.

Die Phylogenese, für deren Rekapitulation während der Ontogenese auch die Hirnentwicklung zahlreiche Beispiele liefert, zeigt bei der *Entwicklung der Centralregion* einen durchaus vergleichbaren Vorgang. Bei der primitivsten Säugergattung, den Insektivoren, liegt die Praecentralregion direkt über dem Riechhirn und nimmt den frontalen Pol der Hemisphäre ein. Bei *Tupaia* erreicht sie den Pol nicht mehr, und schon bei den Halbaffen hat sich ein relativ großer Frontallappen ausgebildet, der die Praecentralregion occipitalwärts schiebt. Dieser Prozeß ist bei den Affen noch weiter gegangen und hat zu einer Verlagerung der Praecentralregion in die Parietalgegend geführt. Hier begegnen wir also genau der gleichen Lageveränderung, wie wir sie aufgrund unserer Befunde für die Ontogenese des Menschen annehmen.

In die gleiche Richtung weisen auch Reizversuche von HINES (1944) am Cortex von *Macaca*-Föten. Beim frühesten Stadium liegen die Reizerfolge über dem Riechhirn und wandern bei den älteren Föten immer weiter occipitalwärts bis sie vor der Geburt in der endgültigen Position der Praecentralregion liegen. Die anfänglich rostrale Lage der motorischen Reizpunkte führte HINES allerdings zu der Interpretation, daß die vor der Centralregion gelegenen Felder früher reif und funktionstüchtig seien als die Praecentralregion. Diese Annahme steht im eklatanten Widerspruch zu den seit langem bekannten morphologischen Befunden der Myelogenese, der Fibrillogenese und der NISSL-Schollenbildung, nach denen die Centralregion der am frühesten reifende Bezirk des ganzen Isocortex ist (vgl. Abb. 2a). Wir glauben daher daß die Deutung von HINES nicht zutrifft und ihre Experimente für eine Verlagerung der Praecentralregion in unserem Sinne sprechen.

Diskussion

Wir können feststellen, daß bei der Differenzierung der homogenetischen Rinde eine Sechsschichtung in bestimmten Gebieten wesentlich früher auftritt, als man bisher annahm, nämlich bereits Ende des vierten Monats. Die Annahme von BRODMANN, daß die gesamte homogenetische Rinde ein *Sechsschichtenstadium* durchmacht, konnte bestätigt werden.

Allerdings treten dabei schon frühzeitig *lokale Besonderheiten* auf. Die früh entwickelten Felder verlieren später an Ausdehnung und werden zum Teil durch die Gyrifizierung der Hemisphärenoberfläche und durch die Ausbildung der Opercula verlagert. Am bemerkenswertesten ist die *Wanderung der Centralregion* von einer frontalen Position in die Parietalgegend, die in ihrem letzten Abschnitt auf den Hirnkarten fötaler Hemisphären nachweisbar ist und eine Parallele in der Phylogenese hat. Im ganzen betrachtet verläuft der Reifungsprozeß des Isocortex rostral von der Centralfurche von der Praecentralregion aus in frontobasaler Richtung, caudal von der Centralfurche jedoch von der Umgebung der Insel aus in Richtung zur Mantelkante.

Die Untersuchung zeigt also, daß im Bereiche des Isocortex zeitliche Reifungsunterschiede bestehen und daß die architektonische Gliederung während der Entwicklung Veränderungen durchmacht, die zum Teil durch die heterochrone Differenzierung der einzelnen Rindenbezirke, zum Teil durch Massenverschiebungen während des Hemisphärenwachstums verursacht werden.

Summary

The ontogenetic development of the human isocortex has been studied in regard to the topographical changes of the cortical areas. Such changes are caused by the unequal growth of the hemispheres and the heterochronic differentiation of the cortical regions. In the hemispheres of human foetuses from the 5th up to the 8th month cortical regions of different structure representing both areas in different developmental stages or definite areas of the final cortical map have been separated.

Comparing the cortex in hemispheres of different age we are able to observe how regions of certain differentiation grade transform into final areas. This is very clearly seen in the frontal lobe the cortex of which matures step by step beginning in the area 4 and ending in the basal surface. This is a graduation still visible in the final structure of the frontal areas. A similar graduation is seen in the cortex of the temporal lobe which developes gradually from the areas 41 and 42 down to the areas 21 and 20. Thus, we find a graduation of differentiation which progresses in the frontal lobe from the precentral region in frontal and basal direction and caudal of the central fissure from the surroundings of the insula in direction of the border of the sagittal fissure of the hemispheres.

The stratification of the isocortex begins frontally already in the middle of the fourth month. During the process of maturation in all areas an arrangement of six layers can be observed. In this we confirm the findings of BRODMANN however we do not consider the six layer arrangement the basic cortical type but a stage of development. Already from the first appearance the six layer structure of the cortex shows local variations.

During the development of the hemisphere the precentral region moves from a rostral position to its final parietal location. In our findings the precentral region is located during the third month frontally above the rhinencephalon and is displaced during the following months reaching finally the parietal region. This finding is in accordance with the results of other authors in comparative anatomy and in electric stimulation experiments.

Literatur

BECK, E.: Morphogenie der Hirnrinde. Berlin 1940.

BRODMANN, K.: Vergleichende Lokalisationslehre der Großhirnrinde. Leipzig 1909.

FILIMONOFF, J. M.: Zur embryonalen und postembryonalen Entwicklung der Großhirnrinde des Menschen. J. Psychol. Neurol. *39* (1929), 323–389.

HINES, M.: Significance of the precentral motor cortex. In: Bucy, P. C. (ed.): The Precentral Motor Cortex. Urbana, Illinois 1944.

KONONOWA, E. P.: Raswntnje poljej lobnoi oblassti i wariabiljnosstj w strojenii eje kory u tscheloweka. (Die Entwicklung der Stirnhirnfelder und die Variabilität der Rindenstruktur beim Menschen) Nevropat. Psichiat. (Moskwa), *9* (1940), 57–73.

MACCHI, G.: The ontogenetic development of the olfactory telencephalon in man. J. comp. Neurol. *95* (1951), 245–305.

POLIAKOW, G. I.: Rannij ontogenes kory bolschoge mosga tschelowega. (Die Frühentwicklung der menschlichen Hirnrinde) Nevropat. Psichiat. (Moskwa) *9* (1940), 55–56.

RABINOWICZ, TH.: The cerebral cortex of the premature infant of the 8th month. In: Progress in Brain Research *4* (1964).

SPATZ, H.: Über Gegensätzlichkeit und Verknüpfung bei der Entwicklung von Zwischenhirn und basaler Rinde. Z. Psychiat. *125* (1949), 166–177.

VOGT, C. u. O.: Allgemeinere Ergebnisse unserer Hirnforschung. J. Psychol. Neurol. *25*, Erg. H. 1 (1919).

Dr. W. KAHLE
Psychiatrische Univ.-Klinik Burghölzli
Zürich, Schweiz

Phylogenetic and Ontogenetic Aspects of the Lemniscal Systems and the Pyramidal System

By Ch. R. Noback* and J. E. Shriver**, New York

A more comprehensive understanding of encephalization during vertebrate evolution would be uniquely attained if one could compare the actual brains of the ancient common ancestors with those representing a sufficiently complete series of subsequent phyletic stages. Unfortunately, such a panoramic view can never be attained since the former and many of the latter brains are lost in the geological past. Therefore, at present, one can trace the process of the phylogenetic development of the brain only through the evaluation and reconstruction of paleoneurological and comparative neoneurological evidence. The phylogenetic interpretations presented below are made with the full awareness of the pitfalls inherent in evaluating data obtained from living descendents. Often only inferences can be made as to whether apparently similar structures found in modern forms are homologous, or are the products of convergent evolution.

The purpose of this presentation is to outline some aspects related to the evolution and ontogeny of the forebrain of the tetrapods. These include (1) the lemniscal systems, (2) the ontogeny of the pyramidal neuron of the neocortex, and (3) the evolution of the pyramidal system.

The Lemniscal Systems

Many concepts have been advanced in an attempt to gain a more coherent comprehension of the structural and functional organization of the nervous system. Of these, the concept that the major sensory inputs to the forebrain are conveyed by the *lemniscal (specific) systems* and the *reticular (non-specific) system* has been the vehicle for fruitful interpretations. In the following account, the anatomic organization and physiologic characteristics of a lemniscal system are delineated, a lemniscus is redefined, and the evolution of the lemniscal systems is analyzed.

Anatomic organization. The lemniscal pathways in mammals are composed of four orders of neurons. First order neurons (neurons in the spinal and cranial nerve ganglia, and the bipolar neurons of the retina) convey information from the peripheral receptors to the second order neurons (neurons in certain "relay" nuclei of the spinal cord and brain stem, and the ganglionic neurons of the retina). Second order axons transmit impulses to

* Supported in part by Grant B-3473 from the National Institutes of Neurological Diseases and Blindness, Bethesda, Maryland.

** Supported in part by 5T1-NB-5254-05 from the National Institutes of Neurological Diseases and Blindness, Bethesda, Maryland.

the specific "relay" nuclei of the thalamus, i. e., the ventral posterior, dorsal division (neothalamic) of the lateral geniculate and medial geniculate nuclei. The axons from these third order thalamic neurons (specific thalamic radiations) synapse with the fourth order neurons within the primary receptive areas of the neocortex. Specific thalamic neurons can directly excite the cortical pyramidal neurons through axosomatic synaptic connections (BISHOP a. CLARE, 1955). Within each nuclear level of a lemniscal pathway are interneurons with short axons; these neurons are integrated into feedback circuits which exert inhibitory controls (contrast inhibition) on the flow of sensory information (ECCLES, 1962).

Physiologic characteristics. Each lemniscal pathway is organized to convey rapidly a very high degree of place and mode specific information. The data from a relatively restricted peripheral receptive field is transmitted to each nuclear level of the pathway by "point-to-point" ("column-to-column") relays with secure synaptic connections; the rather precise topographical pattern of the periphery is preserved and reduplicated at each nuclear level of the pathway (MOUNTCASTLE, 1961; ALBE-FESSARD a. FESSARD, 1963). In general, when compared with fibers of the reticular system, lemniscal fibers are larger and more heavily myelinated; they have a lower threshold, a shorter latency response and a more rapid conduction rate (BISHOP, 1959). The anatomical specificity of a lemniscal pathway is preserved due to few intercalated synapses in the pathway and few or no collaterals of the lemniscal axons, while the physiologic attributes of temporal facilitation, fringe failure and afferent surround inhibition tune the pathway for discriminatory functions (MOUNTCASTLE, 1961). Lemniscal neurons are quite insensitive to most central anesthetics and are nearly indefatigable to repetitive stimuli (ALBE-FESSARD a. FESSARD, 1963). The lemniscal system has significant interactions with the reticular system at both the cortical and the diencephalic levels of the forebrain.

Definition of lemniscus. The axons of second order lemniscal neurons, many of which decussate to the contralateral side, constitute a lemniscus. A lemniscus is a group of fibers, usually of the second order (there may be a sequence of two neurons of the second order as found in the lateral lemniscus), which arise from neurons located in the terminal nucleus of first order fibers of some specific sensory modality or a group of related submodalities; the fibers of a lemniscal bundle, which may or may not decussate, project directly to and terminate within a specific "relay" nucleus of the thalamus. This definition of a lemniscus is more inclusive than previous ones which characterize a lemniscus as a group of second order fibers which originate from nuclei located in the spinal cord and lower medulla oblongata and ascend to higher levels of the brain (HERRICK a. BISHOP, 1958). The definition proposed here amplifies that of HERRICK a. BISHOP primarily by including all second order fibers of all specific sensory systems. The neurons, whose axons form the lemnisci, represent phylogenetically new additions to the nervous system; in effect, these long lemniscal pathways bypass the persistent primitive connections with the reticular system to establish new connections with more recently evolved terminal nuclei (BISHOP, 1959). According to the definition just proposed, the following pathways are comprised of lemniscal fibers: neospinothalamic tract (MEHLER, FEFERMAN a. NAUTA, 1960), ventral spinothalamic tract – these two tracts are collectively called the spinal lemniscus (protopathic modalities); medial lemniscus, lateral cervical (nucleus) lemniscus (tactile and pressure in cat and raccoon, HA, KITAL a. MORIN, 1965), trigeminal lemnisci, solitario-thalamic fibers (gustatory), lateral lemniscus (auditory), and the retino-geniculate tract (optic lemniscus, BISHOP a. CLARE, 1955).

The ascending lemniscal fibers may be characterized as having few if any collateral branches. This property is established for the medial lemniscus (MATZKE, 1951) and the optic lemniscus. Whether this is characteristic of all lemnisci has not been determined, as illustrated by the following examples. Although many fibers of the spinothalamic tract project collateral and terminal branches to the brain stem reticular formation and the thalamic reticular nuclei, it may be that all of these connections are incorporated into the paleospinothalamic tract (MEHLER, FEFERMAN a. NAUTA, 1960), and that the few fibers of the neospinothalamic tract project directly and exclusively to the ventral postero-lateral thalamic nucleus. Similarly, in the case of the secondary pathways emanating from the spinal trigeminal nucleus, the nucleus caudalis gives origin to bilateral ascending projections which extend to the midbrain and thalamus through the reticular tegmentum as well as in the trigeminal lemnisci (STEWARD a. KING, 1963), while axons of the inter-polaris and oralis nuclei, which are present only in the contralateral trigeminal lemniscus, emit very few collaterals in their ascent to the ventral posteromedial thalamic nucleus (CARPENTER a. HANNA, 1961). Therefore, the caudalis fibers might be considered as the paleotrigeminal pathway while the interpolaris and oralis fibers would constitute the neotrigeminal lemniscus.

Evolution of the Lemniscal Systems

The general thesis is proposed that lemniscal systems, at various stages of evolutionary development, are present in all living tetrapods. The considerable variation in the degree of differentiation of these pathways is related in part to the animal's position in phylogeny and also its habitat. When comparing and correlating the evolutionary levels of the brains of living tetrapods, it is germaine to recall that the phylogenetic tree of the tetrapods is widely branched and in each branch the direction of change is determined by processes of specialization and integration which are distinctive and peculiar to that branch. Therefore, although birds and mammals originated from reptilian ancestors, these three vertebrate classes have evolved along widely divergent evolutionary lines.

One of the problems in formulating the phylogenetic history of the lemniscal system is to identify it in non-mammalian tetrapods. This is crucial because a lemniscal system is generally defined in terms established in mammals. It is proposed that an ascending sensory pathway be recognized as possessing "lemniscal" features, if it exhibits the fundamental characteristic of all lemniscal systems, i.e., a tendency to short circuit the persistent older connections in order to conduct impulses more directly to higher levels of the neuraxis. Although lemniscal systems are present in their entirety only in mammals, a progressive differentiation of several subcortical levels can be recognized in submammalian tetrapods. At the present time it is not possible to determine the degree to which the anatomic and physiologic criteria of mammalian lemniscal pathways apply to those of non-mammalian tetrapods.

Lemnisci in amphibia. Lemniscal pathways are incipient in the most primitive living tetrapods, the amphibia. These pathways are considered to be rudimentary lemnisci because they are comprised of sequences of neurons with long "ascending" axons which bypass some neuropil nuclei before they terminate in a rostrally located nucleus. Several ascending pathways in the amphibia consist of a sequence of neurons of the first order and neurons of the second order, with the latter projecting either to a more rostrally located brain stem nucleus or to a thalamic nucleus. In ambystoma, the dorsal spinal lemniscus and the bulbar lemniscus project many fibers to brain stem nuclei and a few fibers to the thalamus, while the retino-thalamic fibers terminate in the ventral division (paleothalamic) of the lateral geniculate body (HERRICK, 1948). Other lemnisci, which project to brain stem nuclei, include the ventral spinal, lateral, trigeminal, and gustatory lemnisci (HERRICK, 1948). Each of these lemnisci is homologous to, or incorporated within, comparable pathways in mammals.

These "primordial" amphibian lemnisci do not resemble mammalian lemniscal pathways in many respects. The second order neurons do not project to "specific" dorsal thalamic nuclei; these nuclei are absent in amphibia. With the absence of both an identifiable "specific" thalamic nucleus and a neocortex, the lemniscal pathways do not have third and fourth order neurons in the mammalian sense.

Lemnisci in sauropsida. The sauropsida have pathways which exhibit some features of the mammalian lemniscal pathways. A dorsal column-medial lemniscal pathway and a spinothalamic pathway have been described in the reptiles and birds, respectively. The dorsal columns and dorsal nuclei in the alligator are considered to be homologous to the dorsal columns, and nuclei gracilis and cuneatus of mammals (HUBER a. CROSBY, 1926). In addition, these dorsal column nuclei in the alligator exhibit certain "mammalian" lemniscal characteristics, such as some degree of modality specificity (fast-adapting touch and pressure), some qualities of adaptation, and intranuclear topical representation of the body surface (KRUGER a. WITKOVSKY, 1961). In the pigeon, the direct spinothalamic fibers may be considered part of a lemniscal pathway (KARTEN, 1963). The rostral projections of the reptilian medial lemniscus have not been resolved experimentally; however, present evidence, obtained from electrophysiologic (MOORE a. TSCHIRGI, 1960) and degeneration studies, indicates that the somatic afferent projections to the forebrain in the alligator do not exhibit mammalian lemniscal features (KRUGER a. BERKOWITZ, 1960).

Conclusion and remarks. On the basis of the above evidence, it is concluded that *"incomplete lemniscal"* pathways are present in the non-mammalian tetrapods. Pending more analysis with modern neuroanatomical and neurophysiological techniques, only a sketchy outline of the lemniscal system in the sauropsida is indicated at this time. Because the organization of the reptilian and avian brains differs in so many fundamental ways from that of mammals, it would be surprising not to find considerable differences between the sauropsidan and mammalian lemniscal systems.

Ontogeny of a Neuron

The differential maturation of a neuron. A neuron may be visualized as a fundamental morphologic unit which is comprised of major regions or segments, each defined by certain anatomic, ontogenetic, phylogenetic and physiologic criteria. According to BISHOP (1956), a pyramidal neuron possesses a receptive segment (dendrites and possibly cell body), a conductile segment (axon), a trophic segment (cell body) and a transmissive segment (synapse). During the ontogenesis of the mammalian brain, there is a differential rate of maturation of the various segments of a neuron. The development of a large neocortical pyramidal neuron, which contributes its well-myelinated axon to the corticospinal tract, illustrates this concept.

The neocortex of a kitten matures rapidly after birth with the result that the segments of each pyramidal neuron within the neocortex have completed most of their ontogeny by approximately the end of the first postnatal month. The apical dendrites and basilar dendrites, as viewed in Golgi preparations, have developed to a stage where their main branches, collateral branches, terminal branches, and dendritic spines are essentially similar to those observed in the cerebral cortex of the adult cat (NOBACK a. PURPURA, 1961; MARTY, 1962). In addition, the superficial neocortex exhibits ultrastructural characteristics (axodendritic synapses, axosomatic synapses, and fine terminal processes) essentially identical to those observed in the mature mammalian neocortex (VOELLER, PAPPAS a. PURPURA, 1963; PAPPAS a. PURPURA, unpublished). Many electrophysiologic properties of the kitten's neocortex (e. g., activities generated by local stimulation of superficial synaptic pathways, activities generated by corticopetal pathways) undergo rapid changes during the first postnatal month, until these properties resemble those observed in the mature cortex (PURPURA et. al. 1964). The sum total of these observations indicates that the dendritic regions of neocortical pyramidal cells undergo major morphophysiologic maturational changes during the kitten's first postnatal month.

In contrast to the dendritic portion, the heavily myelinated axonal portion of the pyramidal neuron matures later during ontogeny. At progressively later stages of postnatal development, the diameters and conduction velocities (PURPURA, et. al., 1964) of these large axons are as follows: one-week-old 2μ, 1 m/sec.; 2-week-old 2μ, 2–3 m/sec.; 4-week-old 4–6μ, 15–20 m/sec.; 7-week-old 4–6μ, 35–50 m/sec.; 14-week-old 7–10μ, 70–90 m/sec.; adult 10–16μ, 70–90 m/sec. On the basis of this evidence it may be concluded that the structural (increase in axon diameter and myelination) and electrophysiologic (increase in conduction velocity) maturation of the pyramidal axon is completed during the 4th to 5th postnatal month.

Overall conclusion. The dendritic (receptive) segment of a large neocortical pyramidal neuron completes its major anatomic and physiologic maturation before the heavily myelinated axonic (conductile) segment of the same neuron.

Correlation between the differential maturation of a neuron and the phylogenetic age of the physiological characteristics of the neuron. Each region or segment of a pyramidal neuron possesses certain distinguishing electrophysiologic properties. The receptive segment is characterized by local graded, non-propagating responses with no refractoriness, while the myelinated conductile segment is characterized by all-or-none, saltatory conduction at a specific rate with absolute and relative refractoriness (BISHOP a. CLARE, 1955; GRUNDFEST 1959).

The primitive response of excitable tissue is probably of the local graded type, while a later evolutionary adaptation is the all-or-none conduction of the noded myelinated axons of vertebrates (BISHOP, 1955; GRUNDFEST, 1959). These properties indicate that the receptive segment of a large neocortical pyramidal neuron has an older phylogenetic history than the well-myelinated conductile segment.

Overall conclusion. On the basis of the ontogenetic and comparative electrophysiologic evidence presented above, the concept is advanced that the segment of a pyramidal neuron with characteristics of an older phylogeny (dendrites) maturates earlier in ontogeny than that segment with characteristics of a more recent phylogeny (well-myelinated axon).

Evolution of the Corticospinal Tracts

Funicular location. The corticospinal tracts, which are present exclusively in mammals, are not cast on any standard mold. Numerous anatomic and physiologic variations in this system of fibers have been observed among the various orders of mammals (LASSEK, 1954; VERHAART 1962). A constant feature of the corticospinal tract from the phylogenetic point of view is its origin in the neocortex, and its successive course through the internal capsule, the basilar portion of the brain stem, and the medullary pyramids. The level of the pyramidal decussation, the relative number of fibers which cross, and the length of these fibers varies markedly from one mammalian order to the next. Its funicular location within the spinal cord is also variable; the main corticospinal tract (excluding any smaller corticospinal tracts sometimes present) is located within the ventral funiculus in the monotremes, insectivores and elephants, within the lateral funiculus in the ungulates, carnivores, lagomorphs and primates, and within the dorsal funiculus in the marsupials, edentates and rodents. This variability in funicular location is an expression of (1) the concept that phylogenetically new tracts appear to be more subject to individual variations than old ones (NYBERG-HANSEN a. RINVIK, 1963), and (2) the theory that the corticospinal tracts may have evolved independently in each of the different orders of mammals (vide infra).

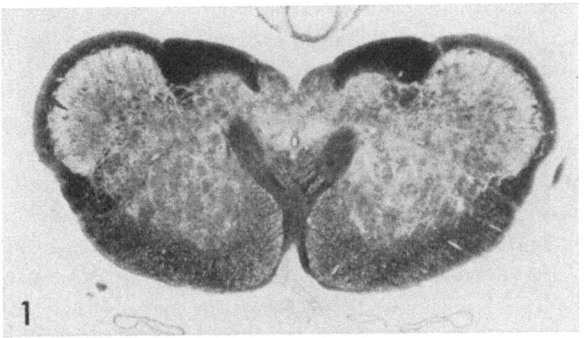

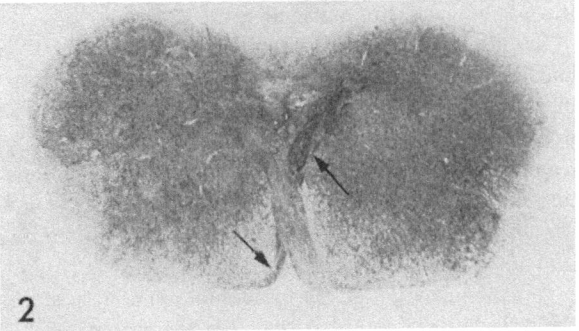

Figs. 1 and 2. Photomicrographs of transverse sections through the lower medulla, at the level of the pyramidal decussation, of *Tupaia glis.*

Fig. 1 *). (HEIDENHAIN-WOELCKE stain) Note the decussating corticospinal fibers coursing dorsally toward the ventrolateral aspect of the dorsal funiculi.

Fig. 2. (NAUTA technique) Note axonal degeneration in diminished left pyramid (arrow), pyramidal decussation (arrow), and in the right dorsal funiculus, following partial ablation of left cerebral cortex.

*) Some observations and Fig. 1 were made from the collection at the Neuroanatomische Abteilung, Max-Planck-Institut für Hirnforschung, Frankfurt/Main, through the courtesy of Professor Dr. R. HASSLER and Dr. H. STEPHAN.

Corticospinal tract as a "lemniscal pathway". In its passage through the brain stem, the corticospinal tract projects collateral branches to the reticular formation. In sub-primates and the lower primates, corticospinal fibers appear to terminate exclusively upon internuncial neurons in the spinal gray, while in higher primates, such indirect connections are paralleled by additional direct pathways (KUYPERS, 1960). Since in phylogeny these monosynaptic projections to the ventral horn neurons become increasingly prominent through the primate series (rhesus monkey, chimpanzee and man), it might be inferred that the termination of the corticospinal tract is related to the evolutionary level of the animal – both to the mammalian order and to the ascending series of evolutionary levels within that order. An analogy exists here between the lemniscal and pyramidal pathways. In both systems, additional fibers are evolved which bypass older internuncial connections in order to convey more directly inputs to and outputs from the neocortex.

Corticospinal tract in Tupaiidae. The examination of myelin-stained serial sections of normal brains and upper cervical spinal cords of *Tupaia glis* (Fig. 1) and *Urogale everetti* indicate that, unlike other insectivores or primates, the corticospinal tracts of these Tupaiidae are located in the dorsal funiculi of the spinal cord. Evidence from degeneration studies (Nauta technique) following neocortical lesions demonstrates that the corticospinal tract of *Tupaia glis* decussates in the lower medulla, passes dorsally to enter the ventro-lateral aspect of the contralateral dorsal funiculus (Fig. 2), and finally extends caudally as far as lower thoracic levels (JANE et al. 1965; SHRIVER a. NOBACK, 1966).

The taxonomic classification of the tree shrews as either insectivores, primates, or in a separate order, has incited considerable debate; the location of the corticospinal tracts in the dorsal funiculi of the spinal cord does not, by itself, suggest that the Tupaiidae are or are not primates.

Comments on the phylogeny of the corticospinal tracts. The following concepts are proposed (1) the common ancestral group (or groups if more than one group existed), which gave rise to the mammals, had not evolved a corticospinal tract, but only possessed the "potential" to form these fibers, (2) the corticospinal tracts evolved and differentiated independently in the various orders of mammals, and (3) the original corticospinal fibers were probably corticoreticular fibers whose terminal branches extended into the spinal cord; the collateral branches of the corticospinal tract terminating in the brain stem reticular formation of living mammals may be the representatives of terminal branches of the original corticoreticular fibers.

The direct evidences to support these concepts are lost in the mammals of the Mesozoic and early Cenozoic eras. In any case, these concepts not only present plausible inter-pretations of many observations, but they offer explanations for the evolution of the pyramidal tracts by utilizing; (1) a pre-existing fiber system – the corticoreticular fibers – from which the pyramidal tracts evolved, (2) a reasonable growth process – formation of terminal (collateral) branches – for the origin of pyramidal fibers, and (3) a common site – the lower brain stem – from which the pyramidal fibers may extend into one or more funiculi within the spinal cord.

Once the location of the pyramidal fibers within a given funiculus was established in the progenitor of an order, that location was probably permanently "determined" for all subsequent species of that order. Stated on the basis of observations made on modern forms, the constant funicular position of the main pyramidal tract in each living species of a particular order suggests that this tract has a similar location both in the progenitor and in the phylogenetically derived successors of that order. In this context, the independent origin of this tract in the Tupaiidae is suggested.

If one assumes that the common ancestral group which gave rise to the mammals did possess a pyramidal tract which occupied a specific funiculus of the spinal cord, then it is difficult to select which funicular location is the most primitive. This selection is complicated because the pyramidal tract's position differs between living primitive mammalian groups (monotremes and insectivores = ventral; marsupials and Tupaiidae = dorsal), and is never in the lateral funiculus. If a primitive location were established, the problem of the shift of pyramidal fibers from one funiculus to another would still remain enigmatic.

The phylogeny of the corticospinal tracts in the mammals seems to exhibit expressions of parallelism, divergence, and convergence. The parallelism is illustrated by the differentiation of corticospinal fibers in all orders of living mammals; the divergence by the different location of the pathways within the spinal cord in different orders; and, the convergence in the fact that the corticospinal tracts primarily have synaptic connections with interneurons in the base of the dorsal horn and intermediate spinal gray.

Summary

Three topics related to the phylogeny and ontogeny of the forebrain are discussed. (1) The organization, characteristics, definition and evolution of the lemniscal systems are outlined. The lemniscal systems, at various stages of evolutionary development, are present in all living tetrapods, from amphibia to mammals. A fundamental characteristic of lemniscal systems in the tetrapods is the tendency of these pathways to bypass the persistent older connections in order to conduct impulses directly to higher levels of the neuraxis, including the forebrain. (2) During the ontogeny of a neuron, such as a neocortical pyramidal neuron, the dendritic (receptor) segment completes its major anatomical and physiological maturation before the heavily myelinated axonic (conductile) segment. The concept is advanced that in this neuron the segment with characteristics of an older phylogeny (dendrites) matures earlier than the segment with characteristics of a more recent phylogeny (well-myelinated axon). (3) Within each mammalian order, the main corticospinal tract is present within one of the three spinal funiculi (for example, dorsal column in Tupaiidae, ventral column in Insectivora, and lateral column in Primates). The phylogeny of these pathways can be plausibly explained as follows: (a) the common ancestral group (or groups) which gave rise to the mammals did not have a corticospinal tract, (b) the corticospinal tracts evolved and differentiated independently in the various orders of mammals and, (c) the original corticospinal tracts were probably corticoreticular fibers whose terminal collateral branches extended into the spinal cord. In this concept, the corticospinal fibers evolve from a pre-existing fiber system (corticoreticular fibers), by a reasonable growth process (formation of terminal collateral branches), and from a common site (lower brain stem). Once the funicular location of this tract was established in the progenitor of an order, that location was permanently "determined" for all subsequent species of that order.

Addendum: After submitting our manuscript, an article discussing the pyramidal tract in Tupaia has been published by VERHAART, W.J.C. (The pyramidal tract of Tupaia, compared to that in other primates. J.comp.Neurol., 126 (1966), 43–50). Data presented by VERHAART is in agreement with those in our paper.

References

ALBE-FESSARD, D., A. FESSARD: Thalamic integrations and their consequences at the telencephalic level. In: Moruzzi, G., A. Fessard, H. H. Jasper (Eds.), Brain Mechanisms. Elsevier, Amsterdam 1963.

BISHOP, G. H.: Natural history of the nerve impulse. Physiol. Rev., *36* (1956), 376 to 399.

BISHOP, G. H.: The relation between nerve fiber size and sensory modality: Phylogenetic implications of the afferent innervation of cortex. J. nerv. ment. Dis., *128* (1959), 89–114.

BISHOP, G. H., M. H. CLARE: Organization and distribution of fibers in the optic tract of the cat. J. comp. Neurol., *103* (1959), 269–304.

CARPENTER, M. B., G. R. HANNA: Fiber projections from the spinal trigeminal nucleus in the cat. J. comp. Neurol., *117* (1961), 117 to 132.

ECCLES, J. C.: Inhibitory controls on the flow of sensory information in the nervous system. In: Gerard, R. W., J. W. Duyff (Eds.), Information Processing in the Nervous System. Excerpta Medica Foundation, Amsterdam 1962.

GRUNDFEST, H.: Evolution of conduction in the nervous system. In: Bass, A. D. (Ed.), Evolution of Nervous Control from Primitive Organisms to Man. Amer. Ass. Advanc. of Science, Washington, D. C., 1959.

HA, H., S. T. KITAI, F. MORIN: The lateral cervical nucleus of the raccoon. Exper. Neurol., *11* (1965), 441–450.

HERRICK, C. J.: The Brain of the Tiger Salamander. University of Chicago Press, Chicago 1948.

HERRICK, C. J., G. H. BISHOP: A comparative survey of the spinal lemniscus systems. In: L. D. Proctor, et al. (Eds.), Reticular Formation of the Brain. Little, Brown and Co., Boston, Mass. 1958.

HUBER, G. C., E. C. CROSBY: On thalamic and tectal nuclei and fiber paths in brain of the American alligator. J. comp. Neurol., *40* (1926), 97–227.

JANE, J. A., C. B. G. CAMPBELL, D. YASHON: Pyramidal tract: A comparison of two prosimian primates. Science, *147* (1965), 153 to 155

KARTEN, H. J.: Ascending pathways from the spinal cord in the pigeon *(Columba livia)*. Proc. XVI Internat. Congr. Zool., Washington 2 (1963), 23.

KRUGER, L., E. C. BERKOWITZ: The main afferent connections of the reptilian telencephalon as determined by degeneration and electrophysiological methods. J. comp. Neurol., *115* (1960), 125–141.

KRUGER, L., P. WITKOVSKY: A functional analysis of neurons in the dorsal column nuclei and spinal nucleus of the trigeminal in the reptile *(Alligator mississippiensis)*. J. comp. Neurol., *,117* (1961), 97–106.

KUYPERS, H. G. J. M.: Central cortical projections to motor, somatosensory and reticular cell groups. In: Tower, D. B., J. P. Schadé (Eds.), Structure and Function of the Cerebral Cortex. Elsevier, Amsterdam 1960.

LASSEK, A. M.: The Pyramidal Tract: Its Status in Medicine. C. C. Thomas, Springfield, Illinois, 1954.

MARTY, R.: Développement post-natal des réponses sensorielles du cortex cérébral chez le chat et le lapin. Arch. Anat. micr. Morphol. *51* (1962), 129–264.

MATZKE, H. A.: The course of the fibers arising from the nucleus gracilis and cuneatus of the cat. J. comp. Neurol., *94* (1951), 439 to 452.

MEHLER, W. R., M. E. FEFERMAN, W. J. H. NAUTA: Ascending axon degeneration following anterolateral cordotomy. An experimental study in the monkey. Brain *83* (1960), 718–750.

MOORE, G. P., R. D. TSCHIRGI: Bilateral organization of the somatic afferent system in lower vertebrates. Physiologist, *3* (1960), 117.

MOUNTCASTLE, V. B.: Some functional properties of the somatic afferent system. In: Rosenblith, W. A. (Ed.), Sensory Communication. The M. I. T. Press, Cambridge, Mass. 1961.

NOBACK, C. R., D. P. PURPURA: Postnatal ontogenesis of neurons in cat neocortex. J. comp. Neurol., *117* (1961), 291–308.

NYBERG-HANSEN, R., E. RINVIK: Some comments on the pyramidal tract, with special reference to its individual variations in man. Acta neurol. scand. *39* (1963), 1–30.

PAPPAS, G. D., D. P. PURPURA: Unpublished.

PURPURA, D. P., R. SHOFER, E. HOUSEPIAN, C. R. NOBACK: Comparative ontogenesis of structure-function relations in cerebral and cerebellar cortex. In: Purpura, D. P., J. P. Schadé (Eds.), Growth and Maturation of the Brain. Elsevier, Amsterdam 1964.

SHRIVER, J.E., C.R.NOBACK: Cortical projections to the brain stem and spinal cord in the tree shrew *(Tupaia glis)*. Anat. Rec. *154* (1966), 423.

STEWART,W.A., R.B. KING: Fiber projections from the nucleus caudalis of the spinal trigeminal nucleus. J. comp. Neurol., *121* (1963), 271–286.

VERHAART, W.J.C.: The pyramidal tract. Its structure and functions in man and animals. World Neurology *3* (1962), 43–53.

VOELLER, K., G.D. PAPPAS, D.P. PURPURA: Electron microscope study of development of cat superficial neocortex. Exper. Neurol., *7* (1963), 107–130.

CHARLES R. NOBACK, Ph. D.

J.E.SHRIVER, Ph.D.

Department of Anatomy

College of Physicians and Surgeons

Columbia University

630 West 168th Street

New York 32, N. Y., USA

Zur vergleichenden Morphologie der spezifischen und retikulären Strukturen des Hirnstamms und ihrer funktionellen Bedeutung

Von W. W. Amunz und N. N. Lyubimov, Moskau

Die Frage nach der strukturell-funktionellen Organisation der Formatio reticularis ist von erheblichem Interesse und bildet den Gegenstand zahlreicher Forschungsarbeiten.

Dieses Interesse stieg besonders nach den bahnbrechenden Untersuchungen von Magoun, Moruzzi, Jasper, Olszewski, Brodal und ihren Mitarbeitern. Sarkisov wies (1959, 1964) auf die methodologischen Schwächen der Konzeption von Magoun und Jasper hin, in der das „unspezifische System" des Hirnstamms den spezifischen afferenten Systemen gegenübergestellt wird.

Im Verlauf weiterer Forschungen richtete sich das Interesse der Neurologen auf die genauere Bestimmung allgemeiner und spezieller Eigenschaften der verschiedenen Strukturen der Formatio reticularis, insbesondere auf das Auffinden spezifischer Eigenschaften im „unspezifischen", d. h. reticulären System.

Die Aufgabe der vorliegenden Mitteilung besteht darin, die *Wechselbeziehungen zwischen* den *spezifischen sensorischen* Systemen *und* dem *reticulären* System des Hirnstamms aufzuzeigen, sowie diese mit den strukturell-funktionellen Besonderheiten des letzteren in Beziehung zu setzen.

Material. Das uns zur Verfügung stehende Material betrifft die Endigungsstätten der verschiedenen Sinnessysteme innerhalb der Formatio reticularis, der Medulla oblongata, der Brücke und des Mittelhirns sowie die Cytoarchitektonik, Angioarchitektonik und Elektrophysiologie dieser reticulären Strukturen, vergleichend bei Katze, Hund und Affen.

Mit neurochirurgischer Technik wurde an Katzen und Hunden der *Tractus opticus* in 6 Fällen, die *laterale Schleife* in 6 Fällen und die Wurzel des *N. intermedius* in 6 Fällen *durchtrennt*. Nach geeigneter Überlebenszeit wurden die Gehirne nach der Marchi- und Nauta-Technik bearbeitet.

Zum Studium der Cytoarchitektonik benutzten wir Serien von mit Kresylviolett gefärbten Präparaten von Katzen-, Hunde- und Affengehirnen; zur Untersuchung der Angioarchitektonik Präparatserien aus Katzen- und Hundegehirnen, die nach der Pfeifferschen Methode in vivo mit Tusche bearbeitet worden waren. Die Dichte des Kapillarnetzes wurde nach Blinkov und Moiseev (1961) gemessen.

Abkürzungen

ctr	Corpus trapezoideum	ncs	Nucleus centralis superior medialis
flp	Fasciculus longitudinalis superior	ncun	Nucleus cuneiformis
gcm	Griseum centrale mesencephali	nD	Nucleus von Deiters
gcp	Griseum centrale pontis	nli	Nucleus intermedius lemnisci lateralis
gVII	Genus nervi facialis	nlv	Nucleus lemnisci lateralis ventralis
lc	Locus coeruleus	npc	Nucleus parvocellularis
lm	Lemniscus medialis	npoc	Nucleus centralis pontis caudalis

npor	Nucleus centralis pontis oralis	pcs	Pedunculus cerebelli superior
npp	Nuclei pontis proprii	pp	Pes pedunculi
nrt	Nucleus reticularis tegmenti von BECH-TEREW oder Nucleus papilioformis	rVd	Radix descendens nervi trigemini
		VI	N. abducens
nVd	Nucleus radicis descendentis nervi trige-mini	VII	N. facialis

Visuelle Afferenzen zum Hirnstamm

Eine *morphologische Untersuchung* der Projektion des Tractus opticus auf die Strukturen der Formatio reticularis des Hirnstamms zeigte folgendes: die degenerierten Fasern lassen sich außer zu ihren bekannten Endigungen im lateralen Kniehöcker, Praetectum, Pulvinar und Colliculus superior, auch zu bestimmten Strukturen der Formatio reticularis verfolgen. Wir fanden solche Degenerationen im *dorso-lateralen Abschnitt der Formatio reticularis* des Mittelhirns, der sich an die vorderen Zweihügel anschließt (Zone des Übergangs der vorderen Zweihügel in die Formatio reticularis), im Gebiet des *Nucleus parabrachialis* des Mittelhirns sowie in anderen Strukturen der Formatio reticularis des Hirnstamms, die in gewisser Entfernung vom vorderen Vierhügel gelegen sind.

In *cyto-* und *angioarchitektonischen* Präparaten läßt sich feststellen, daß die Zell- und Gefäßstrukturen vom Colliculus superior *allmählich* in diejenigen des Nucleus parabrachialis und des anliegenden reticulären Kern des Mittelhirns, des Nucleus cuneiformis, *übergehen*. Das Fehlen einer scharfen Grenze kommt sowohl in der Verteilung und Lageorientierung der Zellen, als auch in der Dichte des Kapillarnetzes und in der Struktur der Neuronen zum Ausdruck. WIKTOROW (1965) beobachtete in GOLGI-Präparaten viele Neurone vom Typus der reticulären Kerne in den unteren Schichten des Colliculus superior.

Auf der Querschnittsebene des Locus coeruleus fehlt die Degeneration der Opticusfasern im dorso-lateralen Abschnitt der Formatio reticularis. Die degenerierten Opticusfasern verlaufen in Richtung auf das ventro-laterale Gebiet des *Nucleus centralis pontis caudalis* und auf den BECHTEREWschen reticulären Kern der Brückenhaube oder *Nucleus papilioformis* weiter (Abb. 1). Die Degeneration verbreitet sich in allen seinen Subnuclei. Cyto- und angioarchitektonisch hat der reticuläre Brückenhaubenkern im Vergleich zu den typischen reticulären Kernen eine kompaktere Lage seiner Zellen und Kapillaren (z.B. die Dichte des Kapillarnetzes bei der Katze beträgt im Nucleus papilioformis 500 mm, im Nucleus centralis pontis caudalis 320 mm) sowie auch schärfere Grenzen. Nach dem Neuronenbau sind in diesem Kern (SHUKOWA 1958) spezifische Nervenzellen vorhanden, die sich von den typischen Reticulärzellen durch reichere Verzweigungen und eine größere Anzahl von dornenförmigen Fortsätzen unterscheiden. Aufgrund der visuellen Afferenz kann man wohl annehmen, daß dieser Kern irgendwelche effektorischen, mit dem Sehen zusammenhängende Funktionen besitzt.

Außer in diesen beiden Endigungszonen sind noch *degenerierte Fasern* aus dem *Tractus opticus* auf der Höhe des Kerns des N. facialis im Gebiet des *Nucleus parvocellularis* lokalisiert. Der Nucleus parvocellularis ist ein typischer reticulärer Kern, der räumlich nicht abgeschlossen ist und im Vergleich zum Nucleus gigantocellularis und centralis pontis caudalis einen Polymorphismus und dichtere Anordnung seiner Zellelemente besitzt, unter denen mittelgroße Zellen überwiegen. Die Dichte des Kapillarnetzes bei der Katze beträgt im Nucleus parvocellularis 420 mm, im Nucleus gigantocellularis 365 mm und im

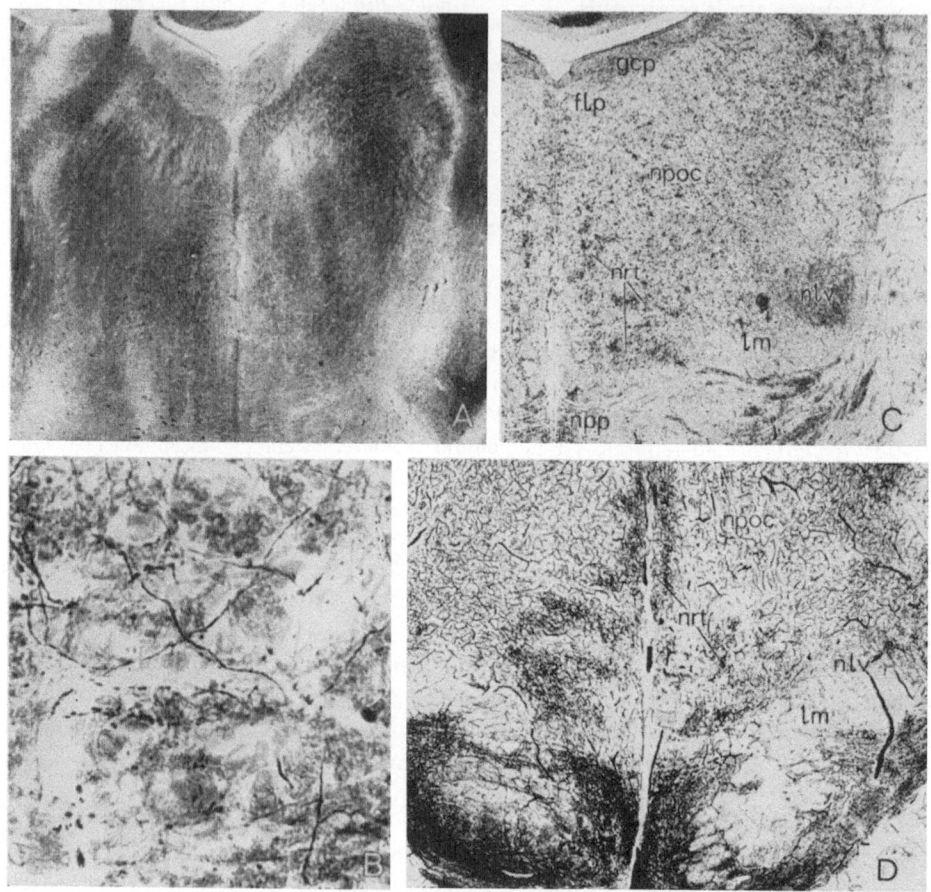

Abb. 1. Schnitte durch den Hirnstamm auf dem Niveau des Nucleus reticularis tegmenti Bechterewi.

A MARCHI-Präparat, Dicke des Schnittes 30 μ, Vergrößerung 30:1 (bei der Reproduktion verkleinert). *Canis familiaris*. Durchtrennung des linken Tractus opticus. Degenerierte Fasern (die schwarzen Punkte stellen mit Osmium imprägnierte Myelinkörner dar) im Gebiet des BECHTEREW-schen Nucl. reticularis tegmenti und im ventro-lateralen Abschnitt des Nucl. centralis pontis caudalis (links).

B NAUTA-Präparat, Schnittdicke 30 μ. Vergrößerung 250:1. *(Felis domestica)*. Durchtrennung des linken Tractus opticus. Reichlich degenerierte Fasern im Gebiet des linken Nucl. reticularis tegmenti Bechterewi.

C Kresylviolett-Färbung. Schnittdicke 20 μ, Vergrößerung 30:1 (bei der Reproduktion verkleinert). *Canis familiaris*.

D Bearbeitung mit Tusche, Schnittdicke 300 μ, Vergrößerung 30:1 (bei der Reproduktion verkleinert). Katze.

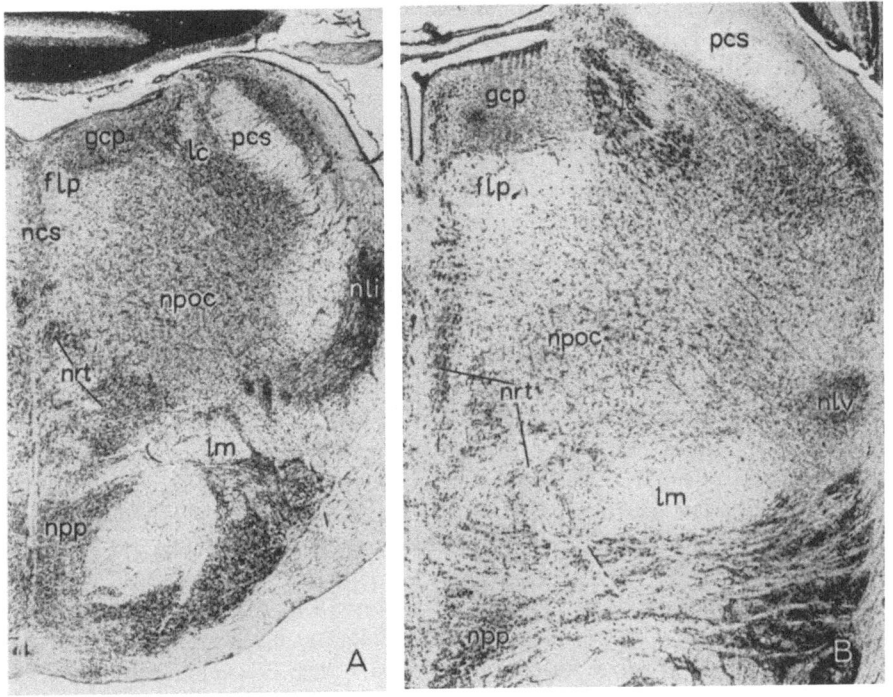

Abb. 2. Schnitte durch den Hirnstamm auf der Höhe des BECHTEREWschen Nucl. reticularis
tegmenti (nrt). Färbung Kresylviolett, Schnittdicke 20 μ, Vergrößerung 30 : 1.
A *Felis domestica*. B Mangabe

Nucleus centralis pontis caudalis 316 mm. Da der Nucleus parvocellularis einen echten
reticulären Bau besitzt, kann man wohl vermuten, daß er die gleichnamigen optischen
Strukturen des Hirnstamms und der Großhirnrinde tonisch aktiviert.

Außer diesen lokalisierten Projektionen des optischen Systems auf die Formatio reticularis des Hirnstamms beobachteten wir auch eine *diffuse Verbreitung degenerierter Fasern*,
die offenbar den zahlreichen Verbindungen des Tractus opticus entspricht, welche zu Zellen der Formatio reticularis ziehen, um wahrscheinlich eine diffuse tonisierende Wirkung
auf die entsprechende Hemisphäre sicherzustellen.

Beim Vergleich der Cyto- und Angioarchitektonik der entsprechenden Gebiete bei
Katzen, Hunden und Affen ergibt sich, daß beim Affen eine deutlichere Differenzierung
sowohl des Zell- als auch des Gefäßbildes des Nucleus parabrachialis, Nucleus papilioformis (Abb. 2) sowie des Nucleus parvocellularis vorhanden ist. Das hängt wahrscheinlich mit der feineren Differenzierung dieser Kerne beim Affen zusammen, die zur Weiter-
gabe und Verarbeitung ankommender Informationen notwendig ist.

Die Ergebnisse unserer morphologischen Untersuchungen, die das Bestehen von Übergangszonen zwischen den spezifischen optischen Strukturen des Hirnstamms und der
Formatio reticularis betreffen, ebenso wie unsere Befunde über die uneinheitlichen Projektionen der sensorischen „Eingänge" (= input) auf die Formatio reticularis entsprechen den Resultaten elektrophysiologischer Untersuchungen.

Visuelle „evoked potentials" im Hirnstamm: Wenn die durch Flackerlicht hervorgerufenen elektrischen Potentiale von verschiedenen Punkten der Formatio reticularis der Katze und des Hundes registriert wurden, sind die „evoked potentials" am ausgesprochensten in den Ableitungen aus den entsprechenden optisch versorgten Strukturen oder aus der Nähe ihrer afferenten Fasern. In den von diesen optischen Projektionen entfernt gelegenen Strukturen der Formatio reticularis beobachtet man entweder eine typische Reaktion einer Blockade der Rhythmen oder undeutliche Veränderungen der elektrischen Aktivität (Abb. 3 A, C).

Im Durchschnitt stellen die „evoked potentials" positive Abweichungen von der Isopotentiallinie dar und entstehen nach einer *Latenz von etwa 1,5–2 msec* nach der elektrischen Reizung des Tractus opticus. (Bei Lichtreizen steigt die Latenzperiode naturgemäß entsprechend stärker an.). Eine zweite Besonderheit dieser Reaktionen in den Zonen der optischen Projektionen auf die Formatio reticularis besteht in ihrem *ipsilateralen* Charakter, der ebenfalls mit unseren hodologischen Experimenten in Einklang steht. Nach der Durchtrennung des linken Tractus opticus verschwinden diese Reaktionen in der linken Formatio reticularis, die sich an die linken vorderen Zweihügel im Gebiet des Nucleus parabrachialis anschließt. In der gegenüberliegenden Seite bleiben sie jedoch in den Strukturen der Formatio reticularis bestehen, welche dem gereizten rechten Tractus opticus entspricht.

Ebensolche Verhältnisse lassen sich bei der Untersuchung der „evoked potentials" beobachten, die in den Strukturen des lateralen Kniehöckers und der ihm anliegenden Abschnitte des Nucleus reticularis thalami zu registrieren sind. Bei chronischen Experimenten an Katzen, Hunden und Affen verschwanden nach Durchtrennung des linken Tractus opticus die „evoked potentials" in den Strukturen des linken lateralen Kniehöckers und der ihm anliegenden Abschnitte des Nucleus reticularis (Abb. 3 D). Besonders hervorgehoben sei folgendes: Die später auftretenden sogenannten „unspezifischen" Reaktionen, die auf Reizung des Tractus opticus in den Strukturen des lateralen Kniehöckers und in des Nucleus reticularis auftreten, unterscheiden sich ihrer Form und Latenz nach von den spezifischen „evoked potentials" sogar bei ein und demselben Versuchstier. Sie besitzen bei Affen zum Unterschied von Katzen und Hunden eine negativ-positive Aufeinanderfolge.

Abb. 3. A und C Ableitungen verschiedener Punkte des Hirnstamms und der Hirnrinde während optischer Reizung (Flacker-Licht – 20mal in der Minute). Bipolare Ableitung. Chronisches Experiment. *(Canis familiaris)*.
B Morphologische Kontrolle der Elektrodenspur. Färbung Kresylviolett. Schnittdicke 20 μ, Vergrößerung 30 : 1. *Canis familiaris*. Die Elektrode lag im ventro-lateralen Abschnitt des Nucl. centralis pontis caudalis.
D "Evoked potentials" in den Kniehöckern und des ihnen anliegenden Nucleus reticularis nach Durchtrennung des linken Tractus opticus, des Balkens sowie der Kommissur der vorderen Zweihügel. *Macaca mulatta*. Chronisches Experiment. CGL Corpus geniculatum laterale; RF Formatio reticularis; L oder (l) links; R oder (r) rechts.

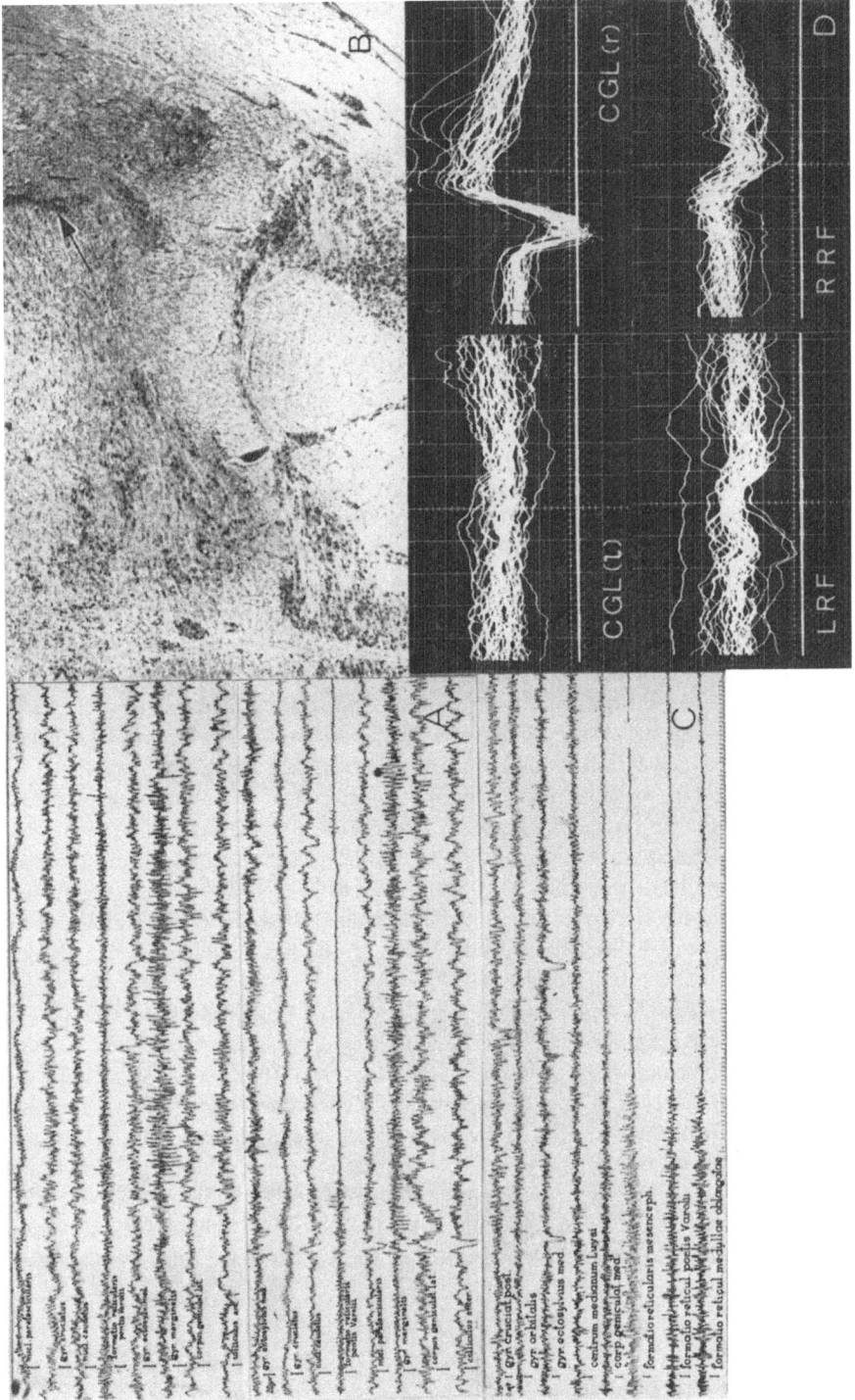

Auditive Afferenzen zur Formatio reticularis

Nach einer *Durchtrennung der lateralen Schleife* waren die degenerierten Fasern in der Formatio reticularis anders lokalisiert als nach Durchtrennung des Tractus opticus. Auf der Höhe der rostralen Abschnitte des Pons Varoli und der hinteren Zweihügel ist eine Degenerationszone sichtbar, die im ventro-lateralen Abschnitt des Nucleus centralis pontis caudalis und des Nucleus centralis pontis oralis (Abb. 4) liegt. Ein Teil dieser Fasern läßt sich in Richtung auf die Pyramidenbahn und die eigenen Fasern der lateralen Schleife hin verfolgen. Wir möchten uns jedoch hier nicht näher mit der Analyse der Endaufsplitterungen dieses Bündels beschäftigen. Zum Unterschied vom übrigen, mehr medial gelegenen Teil der Kerne kann man im Endigungsgebiet gewisse Besonderheiten der Verteilung und Zusammensetzung der Zellelemente bemerken. Die Nuclei centralis pontis caudalis und oralis stellen einen Teil der klassischen Formatio reticularis dar und verwirklichen offenbar lokale tonische Einwirkungen auf die entsprechenden Strukturen des ipsilateralen akustischen Systems.

Mehr rostral sind die degenerierten akustischen Fasern innerhalb der Grenzen des *Nucleus paralemniscalis* von Cajal gruppiert. Ebenso wie der Nucleus parabrachialis liegt dieser Kern nicht im Gebiet der klassischen Formatio reticularis, und unterscheidet sich von den typischen retikulären Kernen durch eine kompaktere Lagerung der Zellen sowie durch ihre einheitlichere Größe und Form. Der Nucleus paralemniscalis von Cajal schließt sich damit eng an die Strukturen mit akustischer Funktion an (Colliculus inferior), was sich auf die Lageorientierung seiner Zellen und die Blutversorgung auswirkt. Das weist auf enge strukturell-funktionelle Beziehungen des Nucleus paralemniscalis zu den ihm anliegenden akustischen Strukturen hin. Man kann annehmen, daß dieser Kern nicht nur die Leitung spezifischer akustischer Einflüsse besorgt, sondern auch tonisierende Wirkungen auf die subcorticalen und corticalen akustischen Strukturen hat.

Außerdem stellten wir unerhebliche diffuse Projektionen der akustischen Fasern des Lemniscus lateralis auf andere Kerne der Formatio reticularis fest. Diese können Kollateralen der langen Hörfasern sein und mit ihrem unspezifischen Einfluß auf die akustischen Rindenstrukturen der entsprechenden Großhirnhemisphäre in Zusammenhang stehen.

Beim Vergleich der Cytoarchitektonik dieses Projektionsgebietes bei Katzen, Hunden und Affen lassen sich einige Unterschiede finden, die wahrscheinlich von den Besonderheiten der Organisation und Funktion dieser Gebiete abhängen.

Die elektrophysiologische Untersuchung der akustischen Projektionen bei Hunden ergibt analoge Verhältnisse wie bei der Untersuchung der optischen Projektionen, abgesehen von der anderen Lokalisation. Auf akustische Reizungen (Anruf) finden sich die in der Formatio reticularis hervorgerufenen Potentiale in Zonen, die den spezifischen akustischen Strukturen anliegen, sowie in den Endigungsgebieten von akustischen Fasern innerhalb des Hirnstamms (Abb. 5 A, B). Entsprechend der wachsenden Entfernung vom Lemniscus lateralis sind die akustischen „evoked potentials" weniger beständig und schwächer ausgeprägt und andere Reaktionen beginnen zu überwiegen: eine Blockade der elektrischen Aktivität oder ihre Regeneration. Die gleichen Verhältnisse beobachteten wir auch bei Katzen und Affen.

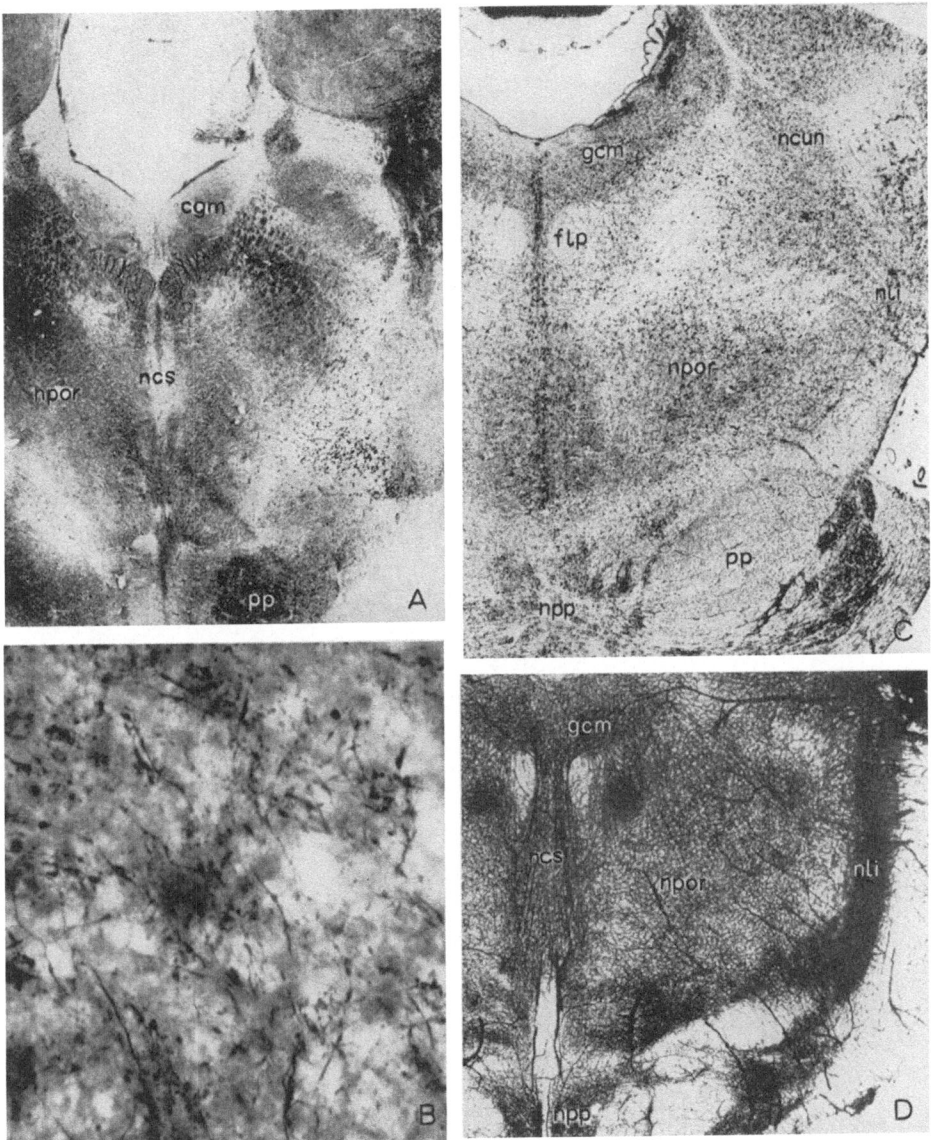

Abb. 4. Schnitte durch den Hirnstamm auf dem Niveau der unteren Zweihügel.
A MARCHI-Präparat, Dicke des Schnittes 30 μ, Vergrößerung 30: 1 (bei der Reproduktion verklei-
nert). Durchtrennung der rechten lateralen Schleife. Degenerierte Fasern im ventro-lateralen
Abschnitt des Nucl. centralis pontis oralis (rechts).
B NAUTA-Präparat, Schnittdicke 30 μ, Vergr. 250: 1. *Felis domestica*. Nach Durchtrennung der
rechten *lateralen Schleife* sind reichlich degenerierte Fasern im ventro-lateralen Abschnitt des
Nucl. centralis pontis oralis (rechts) zu sehen.
C Färbung Kresylviolett, Dicke des Schnittes 20 μ, Vergrößerung 30: 1 (bei der Reproduktion
verkleinert). *Canis familiaris*. Der ventro-laterale Abschnitt des Nucl. centralis pontis oralis ist
ärmer an Zellen als sein ventro-medialer Teil.
D Tusche-Bearbeitung, Dicke des Schnittes 400 μ, Vergrößerung 30: 1 (bei der Reproduktion verkleinert), *Felis domestica*. Der ventro-laterale Abschnitt des Nucl. pontis centralis oralis besitzt
eine geringere Dichte des Kapillarnetzes als sein ventro-medialer Teil.

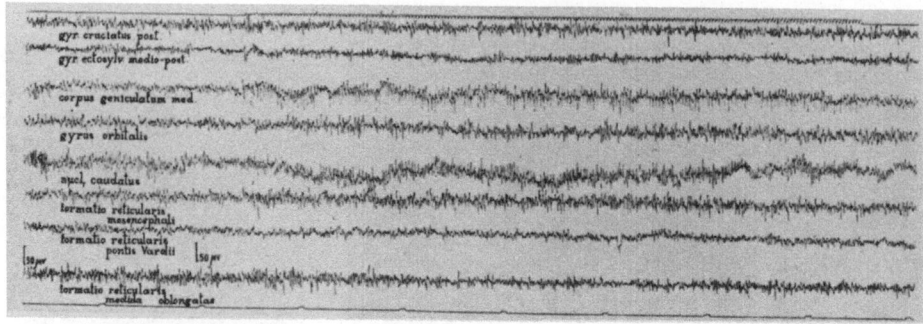

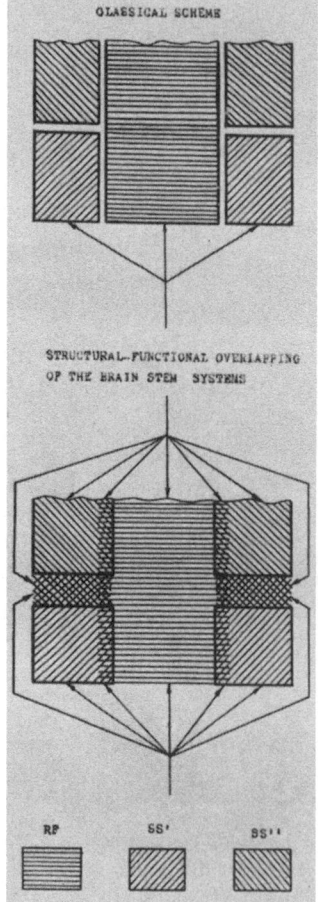

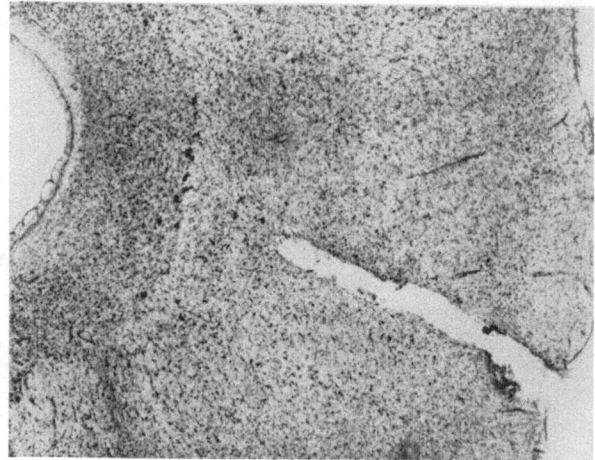

Abb. 5. Potentiale Ableitungen aus verschiedenen Teilen der Formatio reticularis des Hirnstamms bei akustischer Reizung. Bipolare Ableitung. Chronisches Experiment. *Canis familiaris.*
B Morphologische Kontrolle. Färbung Kresylviolett. Schnittdicke 20 μ, Vergrößerung 30 : 1 (bei der Reproduktion verkleinert). *Canis familiaris.* Die Elektrode lag im Nucleus cuneiformis der Formatio reticularis des Mittelhirns.
C Schema. RF – Formatio reticularis; SS[I], SS[II] – spezifische Strukturen des Hirnstamms.

Afferenzen aus der peripheren Geschmacksbahn

Bei der Untersuchung der Projektionen der degenerierten Fasern des eintretenden *Geschmacksnerven* (nach Durchtrennung des N. intermedius) fanden wir eine völlig andere Verteilung. Außer den bekannten Endigungen dieses sensorischen Nerven im entspre-

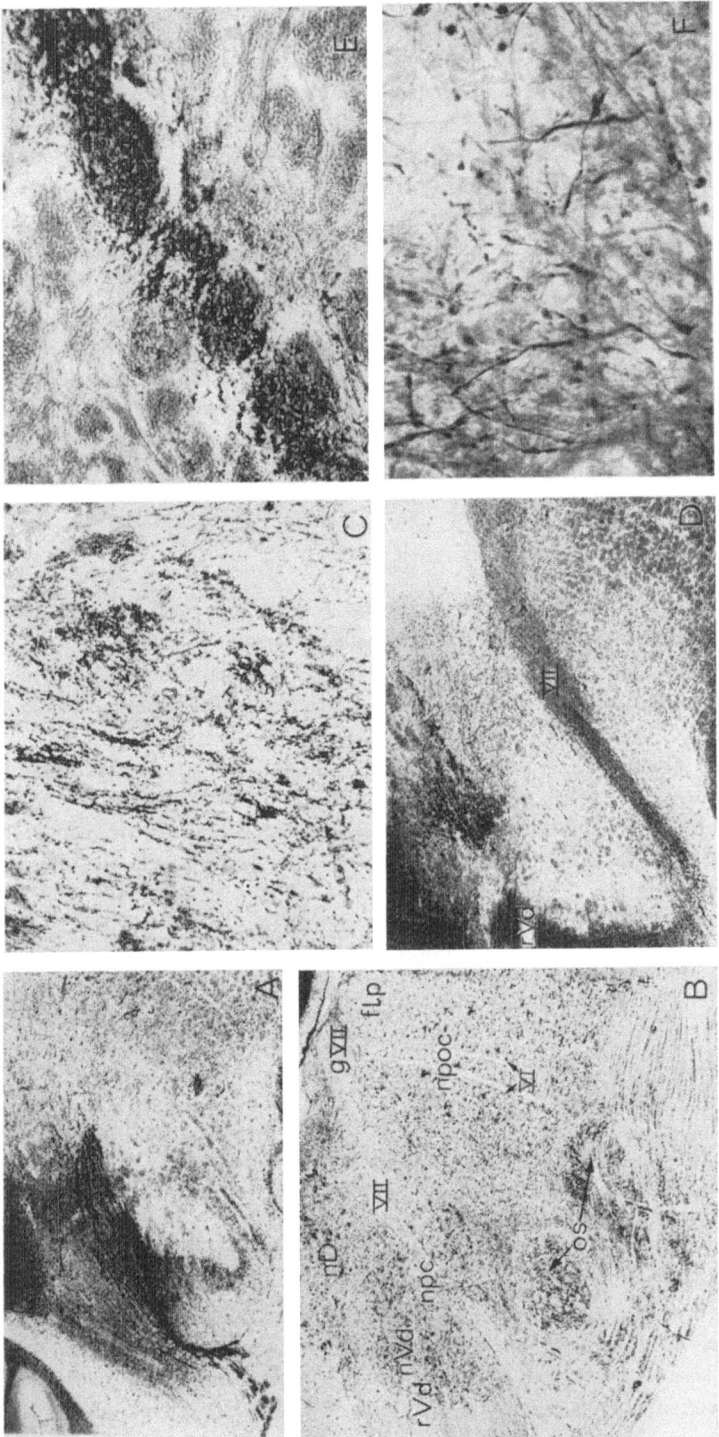

Abb. 6. Schnitte durch den Hirnstamm auf dem Niveau der oberen Olive.   A MARCHI-Präparat, Schnittdicke 30 μ. Vergrößerung 30 : 1 (bei der Reproduktion verkleinert). Nach Durchtrennung der Wurzel des linken N. intermedius sind seine degenerierten Fasern sichtbar. *Canis familiaris*. B Färbung Kresylviolett. Schnittdicke 20 μ, Vergrößerung 3 : 1 (bei der Reproduktion verkleinert). *Canis familiaris*.   C NAUTA-Präparat, Schnittdicke 30 μ, Vergrößerung 40 : 1. Hund. Degenerierte Fasern des Nucleus tractus solitarii.   D MARCHI-Präparat, Schnittdicke 30 μ, Vergrößerung 30 : 1. Hund. Degenerierte Fasern, die in die Abschnitte des Nucl. parvocellularis ziehen, welche dem Kern der absteigenden Trigeminuswurzel und dem DEITERSschen Kern anliegen.   E Präparat nach NAUTA, Dicke des Schnittes 30 μ. Vergrößerung 40 : 1.   F NAUTA-Präparat (Schnittdicke 30 μ, Vergrößerung 250 : 1. *Canis familiaris*). Degenerierte Fasern im Gebiet des Nucl. parvocellularis (Subnucleus oralis).

chenden Tractus solitarius und seinem Kern lassen sich degenerierte Fasern in Abschnitten der Formatio reticularis finden, die diesen spezifischen Geschmacksstrukturen dorso-lateral und ventro-lateral anliegen. Sie ziehen sich über das gesamte Gebiet des Tractus solitarius und seines Kerns hin und werden ihrer Zahl nach in den caudalen Abschnitten des Hirnstamms allmählich geringer. Diese Abschnitte entsprechen dem oralen Teil des kleinzelligen retikulären Kerns *(Subnucleus oralis nuclei parvocellularis)* (Abb. 6). Noch weiter rostral läßt sich eine Zone degenerierter Fasern im kompakten Teil (Subnucleus compactus) Nuclei tegmenti pedunculopontinus feststellen.

Die cyto- und angioarchitektonischen Bilder (1959, 1964) sowie die Neuronenbilder von SHUKOWA (1964) weisen auf enge Wechselbeziehungen zwischen dem Nucleus par-vocellularis und dem sich ihm anschließenden Kern des Tractus solitarius hin. Dies spricht für das Bestehen enger funktioneller Beziehungen im Sinne eines effektorischen Einflusses dieser Zonen auf die Erscheinungen der Salivation. Das stimmt mit den Be-funden von KOHNSTAMM (1902, 1903), YAGITA u. HAYAMA (1909) sowie WANG (1942, 1943) überein, die nach Stimulation dieser Gebiete einen Salivationseffekt beobachteten.

Zusammenfassung

Die Formatio reticularis stellt also ein kompliziertes System dar, daß in seinen ver-schiedenen Abschnitten morphologische und funktionelle Unterschiede aufweist.

Die Verteilung der Sinnessysteme (input) auf die Formatio reticularis ist uneinheitlich. Zwischen den spezifischen und den retikulären Strukturen des Hirnstamms bestehen Übergangszonen (Abb. 5 C).

Wir betrachten die lokalen afferenten Projektionen auf die Formatio reticularis als zusätzliche Kanäle für die Leitung spezifischer und tonischer Einflüsse in Richtung auf die rostralen Strukturen der untersuchten Systeme hin sowie auch für die Kontrolle der effektorischen Systeme des Hirnstamms (insbesondere für die Kontrolle der motori-schen Apparate der Salivation).

Durch Fortsetzung unserer Untersuchungen in dieser Richtung hoffen wir in Zukunft eine rationellere strukturell-funktionelle Klassifikation der Formatio reticularis des Hirn-stamms auszuarbeiten.

Summary

The morphological connections of the optic tract, lateral lemniscal pathway and inter-medial nerve with the reticular formation of the medulla, pons and midbrain have been studied by means of NAUTA's and MARCHI's methods.

Our data show local and general projections of the investigated specific afferent systems to the certain reticular nuclei and transition areas between specific and reticular nuclei.

These correlate with results of our cytoarchitectonic, angioarchitectonic and elektro-physiologic investigations in cats, dogs and monkeys. The combined morphological and physiological results lead us to assume that specific projections of the studied afferent systems are the supplementary channels for conduction of the specific information to the rostral parts of brain.

Literatur

AMUNZ, W. W.: Die Cytoarchitektonik der Formatio reticularis des Hirnstamms bei einigen Säugetieren. In: Struktur und Funktion der Formatio reticularis und ihre Stellung im System der Analysatoren, Moskau, 1959, 27–41 (russisch).

AMUNZ, W. W.: Zur Frage der Wechselbeziehungen zwischen Cyto- und Angioarchitektonik in den Strukturen des Gehirns. In: Die Frage der Gefäßpathologie des Gehirns und der Medulla spinalis, Kischinev, 1964, 31–32 (russisch).

BLINKOV, S. M., G. D. MOISEEV: Die Bestimmung der Dichte des Kapillarnetzes in den Organen und Geweben des Menschen und der Tiere unabhängig von der Dicke des Schnittes. Vorträge Akad. Wiss. USSR, *140,* Nr. 2 (1961), 465–468 (russisch).

KOHNSTAMM, O.: Der Nucleus salivatorius chordae tympani. Anat. Anz. *21* (1902), 362.

KOHNSTAMM, O.: Der Nucleus salivatorius inferior und das cranioviscerale System. Neurol. Cbl. *22* (1903), 699.

SARKISOV, S. A.: In: Struktur und Funktion der Formatio reticularis und ihre Stelle im System der Analysatoren. Moskau 1959. (russisch).

SARKISOV, S. A.: In: Abriß der Funktion und Struktur des Gehirns, Moskau, 1964. (russisch).

SHUKOWA, G. P.: Einige Besonderheiten der Neuronenstruktur der Formatio reticularis des Hirnstamms. In: Struktur und Funktion der Formatio reticularis und ihre Stelle im System der Analysatoren. Moskau 1959. (russisch).

SHUKOWA, G. P., T. A. LEONTOWITSCH: Besonderheiten der Neuronenstruktur und Topographie der Formatio reticularis. (russisch). J. höhere Nerventätigkeit. *14* (1964), 122 bis 147.

WANG, S. C.: Salivatory center in medulla of the cat. J. Neurophysiol. *6* (1943), 195 to 202.

WIKTOROW, J. W.: Der Neuronenbau und die Synapsenarchitektonik der vorderen Zweihügel der Säugetiere. In: Aktuelle Fragen der Neuropathologie. Vortr. I. Konferenz junger Neuropathologen. Moskau 1965. (russisch).

YAGITA, K., S. HAYAMA: Über das Speichelsekretionszentrum. Neurol. Cbl., *28* (1909), 738–753.

Dr. Waleria W. AMUNZ und Dr. N. N. LYUBIMOV
Hirnforschungsinstitut der
Akademie der Medizinischen Wissenschaften
Per. obucha 5
Moscow B–120, USSR

On the Structural Premises of Conditioned Closure

By O. S. ADRIANOV, Moscow

The complicated organization of the excitation at different levels of the brain during conditioned and unconditioned reflex activity has been discovered by series of investigations of the last time. This organization of central interrelationships is changed in a process of settling and stabilizing of conditioned reflexes. Still more scientists in the last years deny one universal mechanism of conditioned closure, which could be realized through strongly fixated apparatus of the cortex and the subcortical structures.

The investigations of our laboratory have shown that the ways and the mechanisms of conditioned closure can be individualized and depend on various factors, such as the complexity of reflex activity, the structural peculiarities of conditioned analysers, the peculiarities of the form of conditioning as well as the level of organization of the central nervous system and so on.

In our short report we only can demonstrate some data about the dependence of reflex conditioned activity (1) on the structural organization of the conditioned signal analyser and (2) on the character of responsive activity (form of conditioning) in dogs and monkeys.

The Pecularities of Conditioned Reflex Depending on the Structural Premises of the Conditioned Signal Analyser

Using the method of conditioned reflexes, most investigators come to the conclusion that the functional power of the *acoustic* analyser in dogs is higher than that of the *visual* analyser. Conditioned reflexes to sound stimuli are established, as a rule, more quickly and have a bigger quantity, in comparision with reflexes to light stimuli.

Our investigations show that such a situation occurs independently from the changes in intensity of visual and sound stimuli (ADRIANOV a. MERING, 1955, 1959; ADRIANOV a. POPOVA, 1963; POPOVA, 1961).

According to our morphological data on the cytoarchitectonics and fine neuronal structure of the cortical and subcortical levels of analysers we have supposed that the dynamic of excited (positive) and inhibited (negative) conditioned reflexes can depend on the same structural signs of visual and acoustic systems.

Some features of construction of visual and acoustic systems are obviously preserved from dogs to monkeys, while other signs are very changed.

The cytoarchitectonical investigations of the cortex showed clear similarities between the central visual area (area 17) in a dog and in a monkey. But such homology between the central acoustic area (area T_3 dog's brain and area 41 monkey's cortex) could not be established. This fact is obviously one of the structural criterions of more pronounced corticalization of visual system in dogs compared with the acoustic one. The new physiological data have confirmed this hypothesis. We want to demonstrate this by a short description of corresponding facts.

MERING (1964) has studied in our laboratory the role of different structures of the acoustic system in the closure of common motor conditioned reactions in dogs (the conditioned running of the animal from the place where it was lying to the food-tray). She came to the conclusion that neither removal of the auditory cortex nor destruction of the medial geniculate body is accompanied by essential disturbances of conditioned reflex activity to single or complicated sound stimuli. On the other side, the lesion of the posterior bigeminal body (which is rather big in dog) has lead to very significant and long-lasting changes of conditioned sound reflexes. As is known, the important role of cortical visual fields and lateral geniculate body in conditioned visual reflexes (especially to object stimuli) in a dog has been shown by many investigators.

ADRIANOV a. POPOVA could conclude that the cortical primary evoked potentials have been very soon diminished and then have at all disappeared after the stabilization of

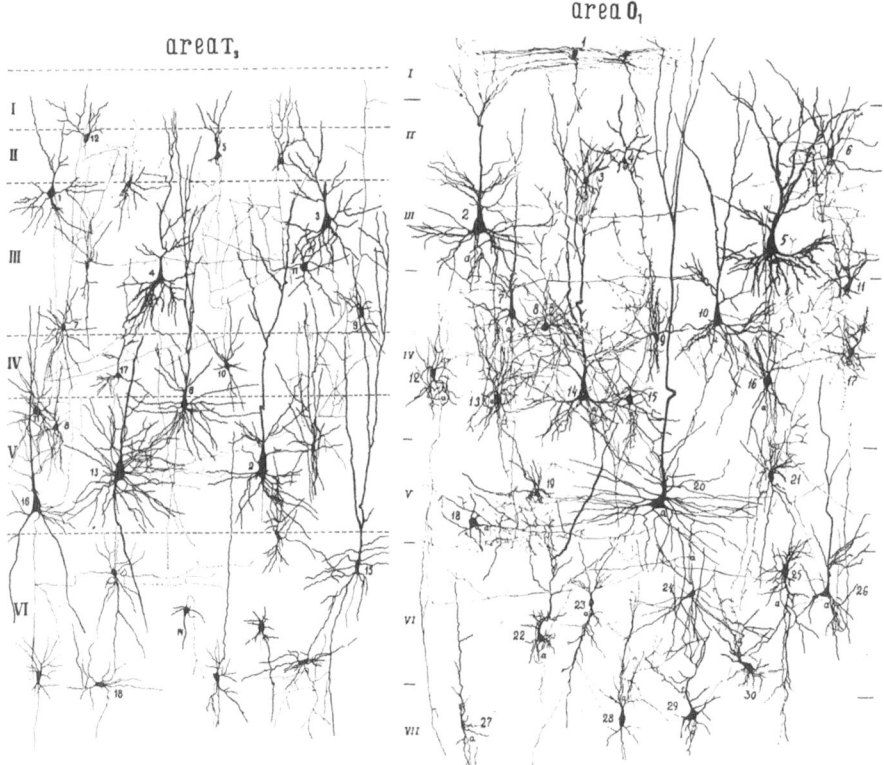

Fig. 1. The scheme of the neuronal structure of the central auditory field – T_3 (POPOVA) and the central visual field – O_I (SKOLNIK-JARROS) of the dog's cortex. GOLGI method.

conditioned defensive reflex to sound (click). But in the case of conditioned defensive reflex to the light (flash) the cortical primary evoked potentials have not disappeared, though the conditioned light stimulus has been given for a long time (more than 500 trials).

We have also seen more protracted preservation of the same nervous process after the application of positive or negative conditioned stimulus in visual analyser (compared to a sound one).

The interinfluence of visual positive and negative stimuli is considerably greater and more continuous than that of sound stimuli. This difference becomes especially distinct if the intervalls between the stimuli are shortened.

In all the dogs the positive sound stimulus could be converted into a negative one and vice versa, whereas such an alteration of the meaning of the visual conditioned stimuli occurred to be impossible in the same dogs.

All these facts may be first of all explained by the peculiarities of the cortical structures of these analysers. So in the central visual area (O_I)*), according to SKOLNIK-JARROS (1954) (Fig. 1, O_I), there are many stellate cells which create the layer IV. Such stellate cells are less numerous in the central acoustic area (POPOVA, Fig. 1, T_3) which has not the isolated layer IV (ADRIANOV a. MERING, 1959).

In the area O_I there are many stellate cells with ramifications of their axons round the cell body. In the area T_3 such elements are much less numerous than in the area O_I.

The idea about the role of the stellate cells as the important receptive and switching elements of cortical analyser is widespread in the investigations of the recent years (SARKISOV, 1948, 1960; POLIAKOV, 1964; BERITASCHWILI, 1961). According to this idea it may be supposed that in the visual cortical structures there are good conditions for a long-lasting circulation of the excitation within a circle of stellate cells and small and middle pyramids in comparison with the acoustic cortical structures.

This hypothesis is confirmed by morphophysiological investigation by POPOVA in the monkey *(Macacus rhesus)*. The acoustic analyser in primates is well corticalized according to the cytoarchitectonical (BLINKOV, 1955) and fine neuronal pictures (typical *koniocortex*). The neurons of area 41 in monkeys have a smaller size and more developing branches. In this area there are many stellate cells with very dense axon ramifications around the cell body (Fig. 2, area 41), whereas in the dog the stellate cells of the central auditory area have sparsely ramificated axons. The character of neuronal organization of the central auditory area in macaca may be obviously compared with a more long-lasting after-effect in the sound-analyser in a monkey compared with a dog. Less distinct differences in character of structural organization of the central visual (SKOLNIK-JARROS, 1955), Fig. 2, area 17) and auditory areas permit to find a distinct similarity in the dynamics of the conditioned reflex processes to visual and to sound stimuli in monkeys compared with dog.

It was also found that the time of stabilization of latency period of food conditioned reflexes in dogs is shorter to the sound-stimuli than to visual ones. The period of stabilization of differentiated stimuli was also shorter to the sound than to light. ZVORYKIN (1961) in the Moscow Brain Institute has obviously shown a clear prevalence of the diencephalic and especially brain-stem structures of the acoustic system in dogs compared with the

---

*) The designation of the cortical fields of the dog's cerebrum has been given according to ADRIANOV a. MERING, 1959 a.

area 17

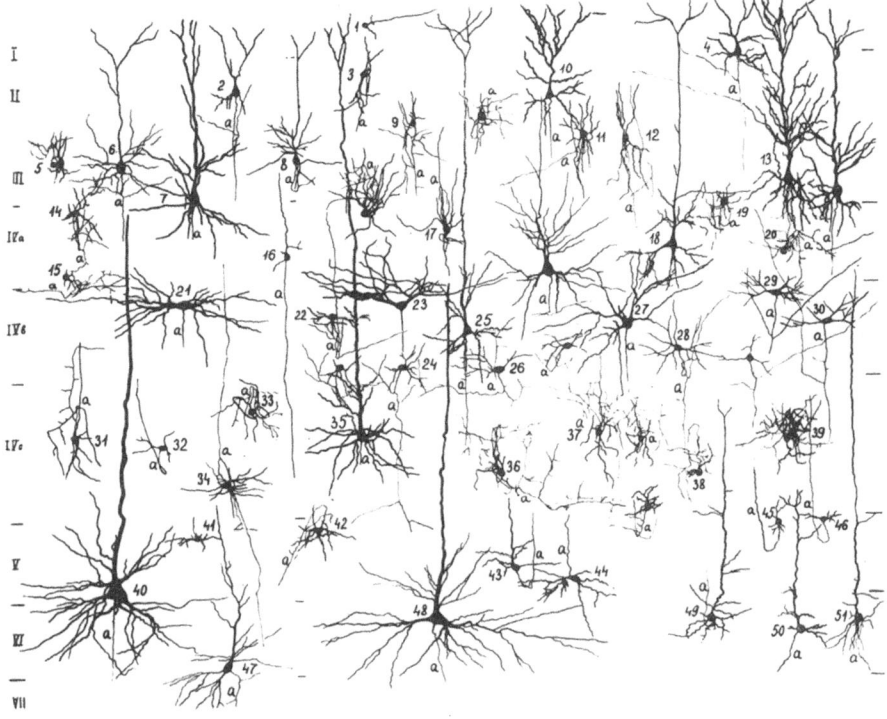

area 41

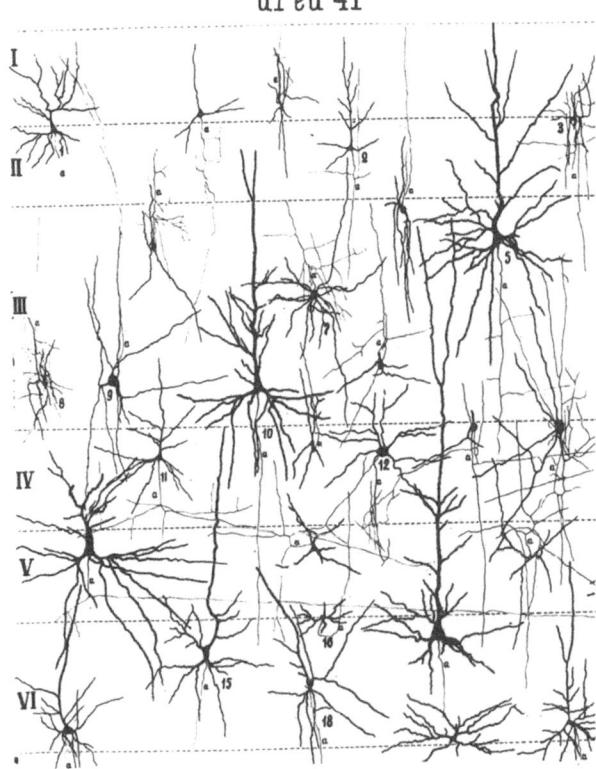

Fig. 2. The scheme of the neuronal structure of the central visual field (SKOLNIK-JARROS) and the central auditory field (POPOVA) of the monkey's cortex. GOLGI method.

visual system. He has found that in dogs and monkeys (Cercopithecidae) with the same brain-weight, the sizes of subcortical structures of acoustic and visual analysers were quite different. For example, the size of corpus geniculatum laterale in dog is 30 mm³, but in monkey 42 mm³, pulvinar correspondingly – 14 mm³ and 101 mm³, corpus geniculatum mediale – 48,3 mm³ and 12 mm³, corpus quadrigeminum posterius – 61,02 mm³ and 10,2 mm³, oliva superior – 8,02 mm³ and 0,63 mm³, nucleus ventralis nervi acustici – 8,19 and 2,58, tuberculum acusticum – 4,01 and 0,53 mm³.

In the monkey there is a relative reduction of lower structure of the acoustic system, on the other side, subcortical levels of the visual system are more developed. Large connections with efferent structures in dogs may create favourable conditions for the transference of the excitation from well-developed levels of acoustic analyser to the motoneurones of the brain-stem and spinal cord.

On the Peculiarities of the Conditioned Reflex Activity Depending on the Character of the Animal's Responsive Activity

We found out that the character and duration of changes of the conditioned reflex activity in dogs and monkeys depended on the specificity of the reflexes, on the basis of which the connection was realized. This conclusion has been, in particular, obtained in the animal in which the operative disconnection of the cortical zones of analysers (total and deep neopalliotomia) was performed. Such an operation was ensuring the separation of all cortical connections between the sensorimotor cortex and visual and acoustic cortex (ADRIANOV, 1960, 1961). The cutting has passed through the white matter of the hemispheres to the lateral ventricle and in some cases even to the thalamus (Fig. 3).

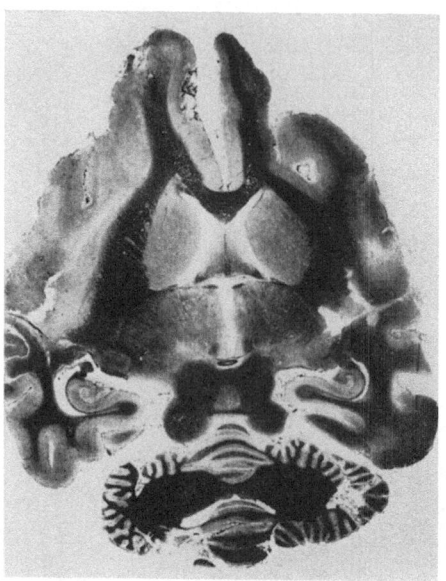

The positive conditioned reflexes on simple or complex stimuli realized on the basis of the simple motor reaction (running) of the animal (dogs and monkeys) to the plate with food, did not, as a rule, disappear after the operation. The operative disconnection was in any cases accompanied by a disturbance of the inhibitory conditioned reflexes.

In contrast to the conditioned motor-nutritive reflexes (running) the motor-defence reflexes in dogs elaborated on the basis of the electric skin stimulation showed more severe and durable changes after such

Fig. 3. Complete neopalliotomy (= interruption of all associative connections) in the dog's cerebral hemispheres. SPIELMEYER's stain preparation.

Fig. 4. Complete neopalliotomy (disconnection) of the monkey's cerebral hemispheres. (NECLUDOVA). SPIELMEYER's stain preparation.

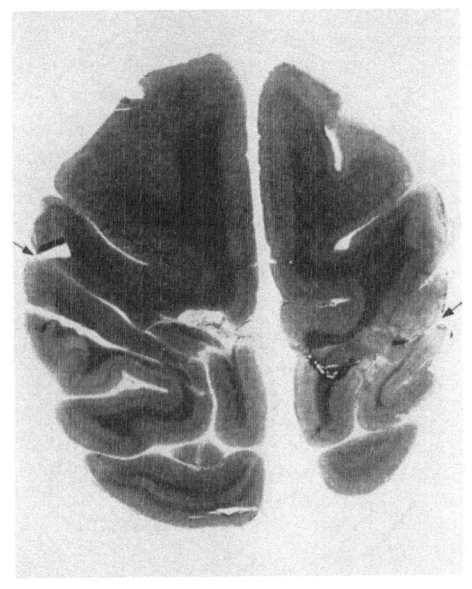

operations. NECLUDOWA (1965) in our laboratory has also shown the dependence of various forms of the conditioned reflexes in a monkey on the basis of the responsive activity. The disconnection of the cortical ends of analysers in monkeys (Fig. 4) as in dogs did not change the conditioned running to the food-tray. But such an operation was followed by a longlasting (3–8 weeks) disappearance of the local feeding-instrumental movements especially in a contralateral forepaw (pressing to the pedal or catching (grasping) for the lever).

The neopalliotomy of the hemispheres in monkeys was also accompanied by disturbances of the visual-motor coordination, by changes of the social behaviour in a group (the loss of ability to be the leader-monkey) and by some emotional disturbances.

It is necessary to say that in dogs after analogous operations there were no changes of behaviour, visual-motor coordination and emotional reactions.

The central mechanisms of interaction of the analysers in monkeys are more complicated than in carnivora. The structural premises of such complication are different. We have a progressive specialization and complication of many associative and projective systems and structures of the brain. Many of these facts are well known today.

The general complication of the hemispheres is particularly followed by a significant development of the associative (intra- and intercortical) systems of connections of the fore-brain. Our morphological investigations in accordance to the data of the literature have shown that in the carnivora the long associative systems are not well developed (Fig. 5). The short systems of connections in dogs have perhaps a relation to a realization of the general (tonic) dynamogenic influences of one cortical region to the other (DUSSER DE BARENNE a. MARSHALL, 1931, BREMER, 1954; ADRIANOV, 1960). But in the monkey, according to investigations of NECLUDOVA and other workers there are well-developed long associative systems of connections especially such as the fasciculus fronto-occipitalis and fasciculus uncinatus (Fig. 6).

The changes of local conditioning after transcortical intersections or circular cutting of the hemispheres (SPERRY, 1947; LAGUTINA a. NORKINA, 1963) and similar clinical observations in man (ARUTIUNOVA a. BLINKOV, 1962; KABELIANSKAJA a. WITING, 1958 and other) permit to point out the complication of functions of the end-brain associative systems in primates.

The specialization and progressive differentiation of the brain is an expression of the continual phylogenetic increase of complexity of the brain which in higher animals insures the most perfect forms of activity.

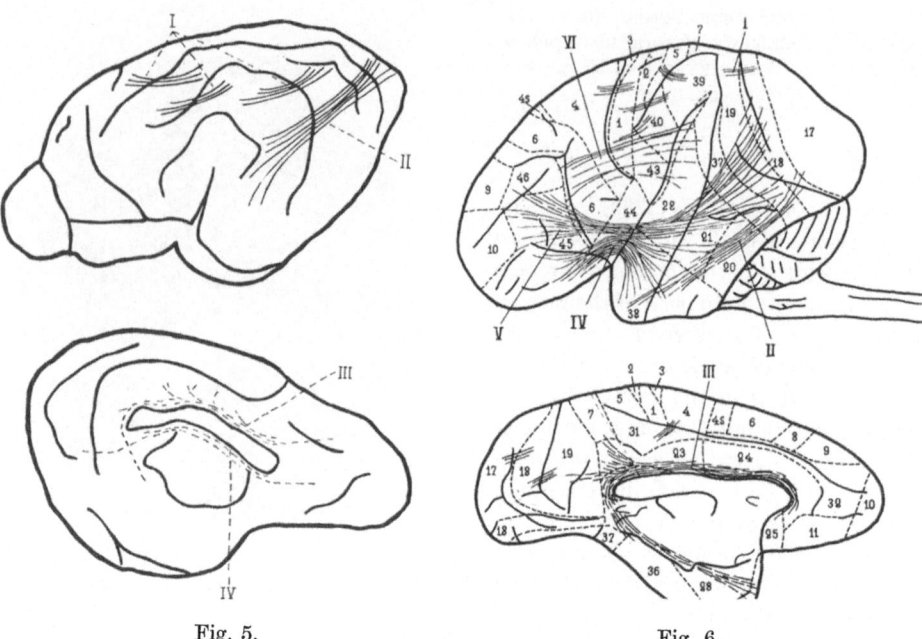

Fig. 5.   Fig. 6.

Fig. 5. Scheme of associative fibres of the dog's cerebral hemispheres. I, II, III – see Fig. 6;
IV stratum subcallosum.
Fig. 6. Scheme of associative fibres of the monkey's cerebral hemispheres. I – fibrae arcuatae;
II – fasciculus longitudinalis inferior; III – cingulum; IV fasciculus uncinatus; V – fasciculus
fronto-occipitalis VI – fasciculus arcuatus.

Summary

The article deals with a comparison of some peculiarities of the formation of conditioned
reflexes to visual and sound stimuli in dogs and monkeys (*Macacus rhesus*) with a general
scheme of the construction (architectonics) and demonstration of the neuronal organisation of the cortical and subcortical levels of visual and acoustic analysers in these animals.

Attention is especially paid to those features of the structural and functional organisation of the analysers which undergo the greatest changes during the evolution.

The article is also concerned with the peculiarities of general conditioned movements
and local forms of food-conditioning in dogs and monkeys after complete bilateral neopalliotomy in the endbrain. Some differences between the results of such an operation in
dogs and monkeys are compared in relation to the specialization and progressive differentiation of the intercortical (associative) and projection connections in primates compared
with those in carnivores.

References

ADRIANOV, O.S.: Motor defensive conditioned
reflexes in dogs after disconnecting cortical
ends of the analysers. J. high. nerv. Activity.
10 (1960), 377–385 (Russ.).

ADRIANOV, O.S.: Morphological and physiological research of the mutual functioning of
analysers in the conditioned reflex activity.
Plzeň lék. Sbor. 3 (1961), 37

ADRIANOV, O.S., T.A. MERING: The materials on the morphology and physiology of the cortical ends of analysers in dog. 8 All-Union Congress of physiologists, biochemistrists, and pharmacologists. Theses, Moscow, 1955.

ADRIANOV, O.S., T.A. MERING: Atlas of the Canine Brain. Moscow, 1959ᵃ (Russ.). English: E.F. DOMINO, ed. Ann Arbor, Michigan, 1964.

ADRIANOV, O.S., T.A. MERING: On morphological peculiarities of the cerebral cortex in dogs. J. high. nerv. Activity, 9 (1959ᵇ), 471 bis 478 (Russ.).

ADRIANOV, O.S., N.S. POPOVA: On the structural premises and the functional peculiarities of the dynamics of the processes in visual, acoustic and somato-sensor analysers. In: Structural and Functional Peculiarities of the Cortical Neurones. Gagra Symposium, Tbilisi, 4 (1963), 13–26 (Russ.).

ARUTUNOVA, A.S., S.M. BLINKOV: Simple reaction time in hemianoptic patients. J. high. nerv. Activity, 12 (1962), 432–436 (Russ.).

BERITOFF, I.S.: The Neuronal Mechanisms of Behavior of Higher Vertebrates. Moscow, 1961. (Russ.).

BLINKOV, S.M.: The Peculiarities of Construction of the Man's Brain. The temporal lobe of the man and monkeys. Moscow, 1955. (Russ.).

BREMER, F.: The neurophysiological problem of sleep. In: Brain mechanisms and consciousness. Oxford, 1954.

DUSSER DE BARENNE, J.G., G. MARSHALL: On a release-phenomenon in electrical stimulation of the "motor" cerebral cortex. Science 73 (1931), 213.

KABELIANSKAIA, L.G., A.I. VITING: Contribution au problème de la dite aphasie sensorielle "transcorticale". J. Neurophysiol. Psychiat. 58 (1958), 39–45. (Russ.).

LAGUTINA, N.I., L.N. NORKINA: To the further analysis of the morphological basis of the motor food conditioned reflexes in monkeys. Theses of XIX. Conference of the Problem of Higher Nervous Activity, Moscow-Leningrad, 1963. (Russ.).

MERING, T.A.: On the role thalamic and brain-stem parts of the acoustic analyser in the closure of motor food conditioned reflexes. X. All-Union congress of physiologists. Theses 2 (1964), 82. (Russ.).

MERING, T.A.: The role posterior colliculi in conditioned activity. J. high. nerv. Activity, 14 (1964), 799–807. (Russ.).

NECLUDOVA, E.S.: The influence of disconnection of cortical analyser zones on the motor nutritive conditioned reflexes in monkeys. In: Structure and Function of Nervous System. The conference of the young scientists of the Moscow Brain Institute. Moscow, 1965. (Russ.).

POLIAKOV, G.I.: Development and complication of the cortical part of the coupling mechanism in the evolution of vertebrates. J. Hirnforsch. 7 (1964), 253–273.

POPOVA, N.S.: Comparative characteristics of the dynamics of nervous processes in dog's acoustic and visual analysers, with regard to the peculiarities of their structure. J. high. nerv. Activity. 2 (1961), 690–696. (Russ.).

SARKISOV, S.A.: The some peculiarities of the construction of the cortical neuronal connections. Moscow, 1948. (Russ.).

SARKISOV, S.A.: The functional interpretation of certain morphological structures of cortex of the brain in the evolutionary aspect. In: D.B. TOWER, (Ed.). Structure and Function of the Cerebral Cortex. Amsterdam 1960.

SKOLNIK-JARROS, E.G.: On the morphology of the visual analyser J. high. nerv. Activity. 4 (1954), 289–304. (Russ.).

SKOLNIK-JARROS, E.G.: On the structure of the cortical end of the visual analyser of the cercopitecs. Probl. physiol. optics. 2 (1955), 162–175. (Russ.).

SPERRY, R.W.: Cerebral regulation of motor coordination in monkeys following multiple transection of sensori-motor cortex. J. Neurophysiol. 10 (1947), 275.

ZVORYKIN, V.P.: Morphological foundations of the unequal role of auditory and optic analysers in the behavior of monkey and dog. Arch. Anat. Histol. Embriol. 41 (1961), 28–37. (Russ.).

Dr. O.S. ADRIANOV
Laboratory of Conditioned Reflexes
Brain Institute AMS
Moscow, USSR

Le développement phylogénétique du thalamus chez les Insectivores

Par R. BAUCHOT, Paris

Les conceptions modernes sur le fonctionnement de l'encéphale font de cet organe le seul témoin réel du niveau évolutif des espèces. La recherche des arguments permettant de fixer pour chaque espèce ce niveau évolutif suppose toutefois qu'a été élucidée la fonction de chaque unité anatomique cérébrale, et qu'on peut juger quantitativement du niveau fonctionnel de ces unités anatomiques. Ces deux conditions ne sont pas encore réalisées, même chez les Mammifères, pourtant si intensément étudiés par les neurophysiologistes. En effet, on ne possède encore que des idées générales sur les fonctions de certains noyaux encéphaliques (noyau = griseum), et les circuits fonctionnels attachés aux principales manifestations physiologiques des organismes ne sont pas tous connus. L'étude quantitative du système nerveux ne fait elle-même que commencer; il est évident que le niveau fonctionnel d'une unité cérébrale est directement lié au nombre de cellules nerveuses et de fibres afférentes ou efférentes de cette unité. Or les méthodes de numération de neurones ou de fibres se font surtout par voie indirecte et laissent planer un doute sur les conclusions qui peuvent être tirées des valeurs ainsi obtenues.

Le volume d'un griseum, ou d'une région corticale, ne traduit donc pas son niveau fonctionnel aussi bien que le ferait le nombre de neurones constituants, de même que les possibilités d'un ordinateur électronique ne dépendent pas tant de son encombrement que du nombre de ses circuits. La densité «cubique» du griseum en cellules intervient donc, et j'ai pu montrer (BAUCHOT, 1963) que cette densité décroît quand le poids somatique augmente (le coefficient de régression moyen des relations densité/poids somatique est : a = — 0,54). La densité décroît encore plus fortement en fonction du poids encéphalique (coefficient de régression moyen des relations densité/poids encéphalique: a = — 0,80; BAUCHOT, 1964). Or ce sont surtout les espèces «évoluées» qui montrent une densification neuronique accentuée, et cela notamment dans les unités structurales elles mêmes évolutives. La méthode utilisée dans cette étude (volume des unités structurales thalamiques et volume néocortical) défavorise donc systématiquement les espèces évoluées; c'est pourtant la seule méthode utilisable pratiquement sur les coupes fines, sur lesquelles le dénombrement direct des neurones ne peut être envisagé.

Si imparfaite soit elle cette méthode a pourtant un grand intérêt; l'anatomie comparée qualitative est en effet incapable de fournir le moindre argument dans une étude de groupe, étude pourtant fort intéressante puisqu'elle utilise les diverses adaptations biologiques des espèces comme autant d'expériences naturelles à analyser.

Nous prenons comme matériel d'étude les Mammifères Insectivores. Le thalamus est divisé en grandes unités structurales, qui n'ont pas toutes une unité fonctionnelle, mais dont les fonctions se situent au même niveau de complexité. L'analyse des corrélations liant chacune de ces unités thalamiques à certains repères montrera celles qui sont stables et celles qui sont «évolutives». Il est entendu que nous qualifierons d'*évolutives* les unités structurales dont l'importance quantitative varie au sein des Insectivores en respectant les données de la phylogenèse.

Matériel et méthodes. Le matériel étudié comporte 14 espèces d'Insectivores et deux Lémuriens lissencéphales. Voici la liste de ces espèces:
Insectivores Tenrecidae: *Tenrec ecaudatus* (Te), *Setifer setosus* (Se) et *Potamogale velox* (Po). Soricidae: *Sorex minutus* (Sm), *S. araneus* (Sa), *Neomys fodiens* (Ne), *Crocidura russula* (Cr) et *C. occidentalis* (Co). Erinaceidae: *Erinaceus europaeus* (Er). Talpidae: *Talpa europaea* (Ta) et *Galemys pyrenaicus* (Gm). Chrysochloridae: *Chlorotalpa stuhlmanni* (Ct). Macroscelididae: *Elephantulus fuscipes* (El) et *Rhynchocyon stuhlmanni* (Rh).
Lemuriens: Tupaiidae: *Tupaia glis* (Tu). Galagidae: *Galago demidovii* (Gg.).
Chaque espèce est définie par un poids somatique moyen adulte (Ps) et un poids encéphalique moyen (Pe). Pour les formations néocorticales comme pour chaque griseum thalamique, le volume est déterminé sur photogramme par planimétrie ou par pesée. Ces méthodes quantitatives ne sont pas à l'abri de toute critique et il est probable en particulier que les valeurs ainsi obtenues ne sont connues qu'avec une certaine marge d'imprécision. Cependant, toutes les espèces ayant subi le même traitement, les conclusions tirées de la *comparaison* de ces valeurs ont une signification.
Le matériel étudié provient en grande partie des collections de l'Institut für Hirnforschung de Francfort, et je remercie les Professeurs SPATZ et HASSLER ainsi que le Docteur STEPHAN qui m'y ont donné libre accès.

Classement évolutif des Insectivores

L'étude de la corrélation liant entre eux poids somatiques moyens (Ps) et poids encéphaliques moyens (Pe) fournit, pour l'ensemble des 16 espèces, un coefficient de corrélation: $r = 0,9145$, et un coefficient de régression Pe/Ps: $a = 0,69$. La corrélation n'est pas très étroite, car les diverses espèces sont écologiquement différentes les unes des autres: on trouve en effet parmi ces 16 espèces les deux Lémuriens *Galago* et *Tupaia* arboricoles; trois espèces d'Insectivores adaptées à la vie semi-aquatique *(Potamogale, Neomys et Galemys)* dont on sait que le degré de céphalisation est supérieur à celui des espèces terrestres apparentées (STEPHAN et BAUCHOT, 1959); deux espèces endogées *(Talpa et Chlorotalpa)* et deux espèces macroptiques *(Elephantulus et Rhynchocyon)*. Il est donc utile de définir, au sein de ces 16 espèces, un groupe d'Insectivores terrestres sans adaptations biologiques très marquées, dont seront éliminées les espèces semi-aquatiques, les Macroscélididés macroptiques et les espèces endogées. On peut donner aux sept espèces restantes le nom d'Insectivores terrestres typiques (BAUCHOT, 1963) ou celui d'Insectivores de base (BAUCHOT et STEPHAN, 1964). Ce sont: *Sorex minutus* et *S. araneus*, *Crocidura russula* et *C. occidentalis*, *Setifer setosus*, *Tenrec ecaudatus* et *Erinaceus europaeus*.
Au sein de ce groupe le coefficient de corrélation entre Ps et Pe devient: $r = 0,9973$ et le coefficient de régression Pe/Ps: $a = 0,64$. L'encéphale a donc, chez les Insectivores, une croissance phylogénétique allométrique négative. Dans la formule générale:

$$Pe = b . Ps^a.$$

a représente la pente de la droite de régression en coordonnées logarithmiques. Le coefficient b de cette formule, dont le logarithme est l'ordonnée à l'origine des droites parallèles à la droite de régression passant par chaque point du diagramme, est égal à Pe si Ps = 1. Le coefficient b représente donc, pour chaque espèce, le poids de l'encéphale lorsque le poids somatique est pris pour unité. Cette façon de procéder permet de ne plus avoir à tenir compte des poids somatiques très divers d'une espèce à l'autre et fournit le *coefficient de céphalisation* de chaque espèce. Ce coefficient est d'un usage plus immédiat s'il est exprimé en pourcentage. En appelant n le coefficient de céphalisation moyen des sept espèces d'Insectivores de base, *l'indice de relation* i = b/n% permet de comparer immédiatement le niveau de chaque espèce au niveau moyen des Insectivores primitifs.

Pour la relation Pe/Ps (a = 0,64), les indices de relation sont les suivants:

Tenrec	82	Talpa	149
S. minutus	89	Potamogale	151
C. russula	90	Chlorotalpa	167
Setifer	104	Elephantulus	233
S. araneus	106	Galemys	238
Erinaceus	107	Rhynchocyon	271
C. occidentalis	122	Tupaia	299
Neomys	133	Galago	477

L'espèce la plus proche du niveau évolutif moyen des Insectivores de base est *Setifer setosus* (i = 104). Les espèces réputées primitives se situent au début de la liste, les Macroscélididés et les Lémuriens, plus évolués, à la fin. Quant aux espèces semi-aquatiques, elles sont toujours plus céphalisées que les espèces terrestres voisines: comparons *Potamogale* (i = 151) avec *Tenrec* (i = 82) ou *Setifer* (i = 104); *Neomys* (i = 133) avec *Sorex minutus* (i = 89) ou *S. araneus* (i = 106); *Galemys* (i = 238) avec *Talpa* (i = 149).

Le classement ainsi obtenu est «évolutif» au sens que nous avons donné à ce terme. Il est d'ailleurs normal qu'il en soit ainsi, puisque l'importance relative du cerveau est certainement liée à la complexité du comportement des espèces. Il est utile de définir en outre *l'amplitude de variation*, égale, pour chaque relation étudiée, au quotient des coefficients b ou des indices de relation extrêmes.

Ainsi, pour la relation Pe/Ps, les espèces extrêmes sont *Tenrec* et *Galago;* l'amplitude de variation est: 477/82 = 5,8. Le cerveau étant la somme arithmétique de ses divers constituants, ce quotient représente la moyenne des amplitudes de variation calculées pour les volumes des diverses parties encéphaliques rapportées à l'unité de poids somatique. Si, pour un griseum donné, et avec des espèces extrêmes identiques, l'amplitude de variation est supérieure à 5,8, ce griseum peut être considéré comme évolutif. Si l'amplitude de variation est comprise entre 5,8 et 1, on obtient tous les stades de la régression «relative», les griseums augmentent bien de taille quand on suit la lignée phylogénétique, mais moins vite que le poids encéphalique global. Dans ce dernier cas, l'utilisation des valeurs relatives, c'est à dire l'étude des quotients: volume du griseum/poids encéphalique conduirait à une interprétation erronée. Enfin, si l'amplitude de variation est inférieure à 1, on a affaire à une structure régressive vraie.

Analyse des unités structurales

Stephan et Andy (1964) ont montré qu'au sein de la lignée Insectivores – Primates, ce sont les formations néocorticales qui manifestent le développement phylogénétique le plus accentué. Les valeurs des volumes du néocortex des 16 espèces étudiées, que m'a communiquées Stephan, permettent de considérer tout d'abord la relation du volume néocortical (NéoC) par rapport au poids somatique (Ps), puis de comparer aux résultats ainsi obtenus ceux que fournissent les relations liant les volumes de chaque unité structurale thalamique au poids somatique, enfin d'étudier les relations liant ces mêmes volumes aux volumes néocorticaux.

Indices de néocorticalisation

Le coefficient de corrélation de la relation NéoC/Ps est: r = 0,8061 pour l'ensemble des 16 espèces, et: r = 0,9792 pour les Insectivores de base. Le coefficient de régression: a = 0,62 est très proche du coefficient de régression de la relation Pe/Ps. Les indices de relation NéoC/Ps, ou indices de néocorticalisation, sont:

Setifer	70	Talpa	209
S. minutus	75	Chlorotalpa	219
Tenrec	78	Potamogale	305
C. russula	90	Elephantulus	324
S. araneus	111	Rhynchocyon	392
C. occidentalis	125	Galemys	468
Erinaceus	150	Tupaia	755
Neomys	177	Galago	1723

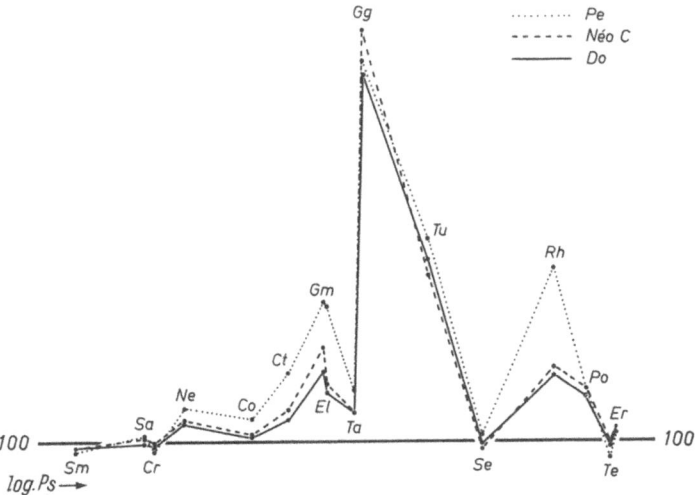

Fig. 1. Valeurs des indices de relation, en fonction du poids somatique, du poids encéphalique (Pe), du volume néocortical (NéoC) et du volume des noyaux thalamiques dorsaux (Do). Les diverses espèces sont distribuées, en abscisse, d'après la valeur logarithmique de leur poids somatique. Les indices sont distribués en ordonnée d'après des échelles telles que l'ajustement des courbes soit maximum.

Comparons ces indices aux indices de céphalisation: les espèces extrêmes sont identiques: *Galago* du côté des valeurs maximales, le groupe *S. minutus – Setifer – Tenrec* du côté des valeurs minimales. La figure N⁰ 1, qui n'est qu'une représentation imagée de ces deux relations, montre que ces indices de relation varient dans le même sens, avec quelques petites fluctuations de détail. Ces fluctuations prouvent que le volume néocortical et le poids encéphalique ne fournissent pas des valeurs interchangeables, et montrent que l'utilisation des poids encéphaliques pour juger du niveau phylogénétique des espèces n'est qu'un pis-aller qui ne saurait dispenser du calcul du volume néocortical, ou d'une structure cérébrale équivalente.

L'amplitude de variation est égale à: 1723/70 = 24,6. Si l'on utilise les mêmes espèces extrêmes que pour les indices de céphalisation *(Galago* et *Tenrec)*, on trouve: 1723/78 = 22,1. Le néocortex est donc, comparé au poids encéphalique global, une structure nettement évolutive, puisque son volume, au sein du même groupe d'espèces, varie quatre fois plus que le poids encéphalique (22,1 : 5,8 = 3,8).

Évolution des unités structurales thalamiques

Le thalamus a été subdivisé dans les grandes unités structurales suivantes: habenula (Hb), noyaux antérieurs (An), noyaux médiaux (Ml), noyaux médians et intralamellaires (Mn), noyaux dorsaux (Do), noyaux ventraux (Ve), ganglion géniculé latéral (Gl) et ganglion géniculé médial (Gm). Le pretectum, qui pose des problèmes d'homologation difficiles entre Insectivores et Lémuriens, n'a pas été étudié.

Les résultats des études des relations liant les volumes de chacune de ces unités structurales au poids somatique sont groupés ci – dessous:

Unité structurale	Coefficient de corrélation		de régression	
	(7 esp.)	(16 esp.)	(7 esp.)	(16 esp.)
Habenula	0,9935	0,9517	0,59	0,62
Noyaux antérieurs	0,8949	0,7393	0,45	0,64
Noyaux médiaux	0,9822	0,8918	0,61	0,67
Noyaux médians	0,9871	0,9272	0,57	0,62
Noyaux dorsaux	0,9885	0,8146	0,64	0,72
Noyaux ventraux	0,9702	0,8706	0,57	0,67
Ganglion géniculé latéral	0,9766	0,8196	0,71	0,80
Ganglion géniculé médial	0,9876	0,8400	0,56	0,67

La valeur moyenne du coefficient de régression Vol/Ps pour tous les noyaux diencéphaliques est: a = 0,58 (BAUCHOT, 1963). Les indices de relation qui en résultent peuvent être comparés aux indices de céphalisation et de néocorticalisation. Ils sont groupés dans le tableau ci-dessous où les espèces sont rangées par indices de néocorticalisation croissants:

Espèces	NéoC	Pe	Hb	An	Ml	Mn	Do	Ve	Gl	Gm
Setifer	70	104	92	37	86	120	83	93	141	72
S. minutus	75	89	79	113	74	91	73	82	53	84
Tenrec	78	82	90	45	74	71	94	55	74	88
C. russula	90	90	114	106	97	87	83	97	96	122
S. araneus	111	106	95	160	97	124	92	110	56	115
C. occidentalis	125	122	118	90	118	94	120	111	106	100
Erinaceus	150	107	111	148	153	113	154	151	173	117
Neomys	177	133	145	148	93	123	170	118	83	277
Talpa	209	149	169	329	186	190	215	298	31	180
Chlorotalpa	219	167	169	104	166	152	183	132	82	267
Potamogale	305	151	132	341	122	124	273	282	87	521
Elephantulus	324	233	158	530	150	166	293	167	414	216
Rhynchocyon	392	271	233	790	291	251	367	268	774	494
Galemys	468	238	175	405	221	159	378	292	68	577
Tupaia	755	299	236	531	344	183	818	266	581	406
Galago	1723	477	180	888	539	353	1646	647	670	464

Des huit relations étudiées, cinq ont pour espèces extrêmes, d'une part un Insectivore de base, d'autre part *Galago*. On peut donc directement comparer ces relations à celles qui lient au poids somatique le poids encéphalique ou le volume néocortical:

- *Noyaux antérieurs*. Amplitude de variation: 888/37 *(Setifer)* = 24,0. Cette amplitude est comparable à l'amplitude de néocorticalisation (24,6).
- *Noyaux médiaux*. Amplitude de variation: 539/74 *(Tenrec)* = 7,3. Cette amplitude est légèrement supérieure à l'amplitude de céphalisation (5,8).
- *Noyaux médians*. Amplitude de variation: 353/71 *(Tenrec)* = 5,0. Cette amplitude est légèrement inférieure à l'amplitude de céphalisation; on peut donc considérer les noyaux médians et intralamellaires comme stables dans la lignée évolutive Insectivores-Primates, ou montrant une légère régression *relative*.
- *Noyaux dorsaux*. Amplitude de variation: 1646/73 *(S. minutus)* = 22,5. Cette amplitude est proche de l'amplitude de néocorticalisation, même si l'on utilise *Setifer* comme valeur inférieure (1646/83 = 19,8). Les noyaux dorsaux représentent donc une unité thalamique très évolutive, que les noyaux antérieurs seuls dépassent.
- *Noyaux ventraux*. Amplitude de variation: 647/55 *(Tenrec)* = 11,8. Cette amplitude est nettement supérieure à l'amplitude de céphalisation (5,8).
  Les trois autres relations doivent être discutées séparément:
- *Habenula*. Amplitude de variation: 236 *(Tupaia)*/79 *(S. minutus)* = 3,0. C'est donc une structure en régression relative nette; si l'on utilise le couple habituel *Galago-Tenrec*, l'amplitude de variation devient: 180/90 = 2,0. L'habenula étant considérée surtout comme un centre à afférentes olfactives, on peut en déduire que la régression olfactive (relative) est nette dans la lignée Insectivores-Primates.

– *Ganglion géniculé latéral.* Amplitude de variation: 774 *(Rhynchocyon)*/31 *(Talpa)* = 25,0. Une telle amplitude est considérable, mais ne peut être comparée à l'amplitude de néocorticalisation. On ne peut en effet considérer le ganglion géniculé latéral comme «évolutif», puisqu'il ne respecte pas les données de la phylogenèse: l'espèce la plus défavorisée n'est pas un Insectivore de base, mais la Taupe endogée, et l'espèce la plus favorisée est un Insectivore macroptique et non un Lémurien. Le couple *Galago-Tenrec* (670/74 = 9,0) fournit une valeur bien plus faible de l'amplitude de variation, encore que supérieure à l'amplitude de céphalisation. Le centre optique primaire a donc bien une signification évolutive des Insectivores aux Lémuriens, mais les variations entre Insectivores endogés et macroptiques sont supérieures aux variations phylogénétiques.
– *Ganglion géniculé médial.* Amplitude de variation: 577 *(Galemys)*/72 *(Setifer)* = 8,0. L'espèce la plus favorisée n'est pas un Lémurien, mais un Insectivore adapté à la vie semi-aquatique. On sait en effet (BAUCHOT, 1963) que le centre auditif thalamique est d'autant plus développé que l'espèce est mieux adaptée au mode de vie semi-aquatique. Le couple *Galago-Tenrec* fournit une amplitude de: 464/88 = 5,3, valeur voisine de l'amplitude de céphalisation. Le centre auditif thalamique est donc une structure phylogénétiquement stable des Insectivores aux Primates, si l'on fait abstraction des espèces adaptées à la vie semi-aquatique.

En conclusion, aucune structure thalamique ne montre de régression réelle au sein des Insectivores; l'habenula montre une régression relative nette; noyaux médians et ganglion géniculé médial sont phylogénétiquement stables, sauf adaptation à la vie semi-aquatique pour le centre auditif. Noyaux médiaux, noyaux ventraux et ganglion géniculé latéral sont évolutifs, mais moins que les structures néocorticales (si l'on ne tient pas compte, pour le centre optique primaire, des espèces endogées aveugles et des Macroscélididés macroptiques). Enfin, noyaux dorsaux et noyaux antérieurs montrent un niveau évolutif comparable à celui du néocortex.

Corrélations thalamo-néocorticales

Il reste à étudier les relations liant au volume néocortical, pris pour variable indépendante, les volumes des diverses unités structurales thalamiques. Les résultats, groupés ci-dessous ne fournissent les coefficients de corrélation et de régression que pour l'ensemble des seize espèces; il est tout à fait inutile de distinguer ici les Insectivores de base, puisque les relations étudiées font abstraction du poids somatique et comparent entre elles des structures encéphaliques.

Unité structurale	Coefficient de corrélation	de régression
Habenula	0,9176	0,66
Noyaux antérieurs	0,9475	0,91
Noyaux médiaux	0,9688	0,80
Noyaux médians	0,9395	0,69
Noyaux dorsaux	0,9961	0,98
Noyaux ventraux	0,9668	0,82
Ganglion géniculé latéral	0,8730	0,94
Ganglion géniculé médial	0,9528	0,84

On peut chercher à classer les unités structurales thalamiques d'après l'étroitesse de la relation qui les lie aux volumes néocorticaux. On trouve alors:

1. les noyaux dorsaux (r = 0,9961), très étroitement associés au néocortex, au point qu'il faut envisager comme probable la projection de ces noyaux sur l'ensemble du néocortex (Figs. 1, 2).

2. les noyaux médiaux (r = 0,9688), ventraux (r = 0,9668), le ganglion géniculé médial (r = 0,9528), les noyaux antérieurs (r = 0,9475) et médians (r = 0,9395) ne sont pas étroitement associés au néocortex, soit parce qu'ils ne se projettent que sur une partie de ce dernier, soit parce qu'ils ont des connexions efférentes extra-corticales.

3. l'habenula (r = 0,9176) a certainement peu de rapports fonctionnels avec le néocortex. Le coefficient de corrélation ne fait que traduire la relation mathématique liant deux centres encéphaliques quelconques, mais évoluant l'un et l'autre phylogénétiquement en fonction du poids somatique.

4. le ganglion géniculé latéral (r = 0,8730) montre la corrélation la plus faible. Il est probable que la raison en est qu'au sein des Insectivores, la projection sur le toit optique est encore prédominante sur la projection néocorticale, et qu'en outre la proportion de fibres allant à l'un ou l'autre centre varie d'une espèce à l'autre.

On peut admettre qu'il y a isométrie entre le développement du néocortex et celui des unités thalamiques (a = 0,98 est voisin de l'isométrie pour les noyaux dorsaux). L'amplitude de variation fournit, ici encore, une méthode d'appréciation de l'étroitesse de la corrélation; l'amplitude est en effet d'autant plus faible que la corrélation est plus étroite. On obtient ainsi:

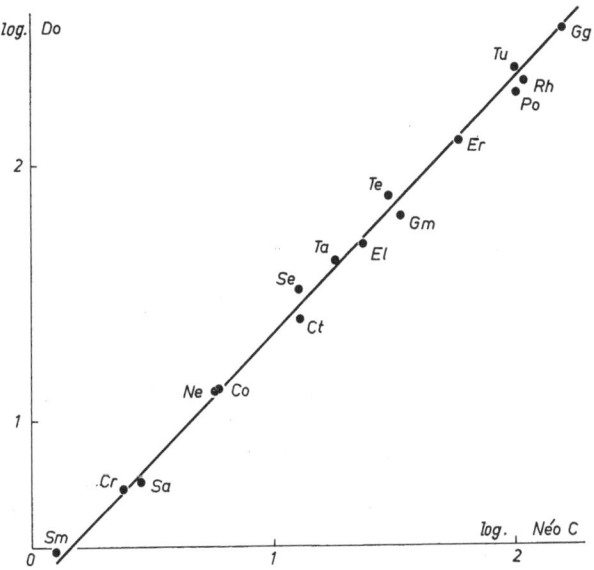

Fig. 2. Coordonnées logarithmiques. Relation «Volume des noyaux thalamiques dorsaux (Do) en fonction du volume néocortical» (NéoC). On notera à la fois l'étroitesse de la corrélation entre ces deux structures et l'isométrie presque parfaite de leur croissance phylogénétique.

Noyaux dorsaux: 2606 *(Talpa)* / 1730 *(Chlorotalpa)* = 1,5. Noyaux médiaux: 3990 *(Setifer)* / 1066 *(Galago)* = 3,7. Noyaux antérieurs: 11.858 *(Rhynchocyon)* / 3097 *(Chlorotalpa)* = 3,8. Noyaux ventraux: 6653 *(Talpa)* / 1589 *(Tupaia)* = 4,2. Ganglion géniculé médial: 3639 *(Neomys)* / 586 *(Galago)* = 6,2. Noyaux médians: 3327 *(Setifer)* / 420 *(Galago)* = 7,9. Habenula: 11. 429 *(C. russula)* / 1146 *(Galago)* = 10,0. Enfin ganglion géniculé latéral: 29.854 *(Setifer)* / 2291 *(Galamys)* = 13,0.

Sont en tête les unités structurales thalamiques à connexions néocorticales prédominantes, en queue celles qui n'ont pas de telles connexions. On constate en outre qu'à aire néocorticale égale, les volumes thalamiques sont plus développés chez les espèces primitives que chez les espèces évoluées (notamment *Galago*); mais comme ces espèces évoluées ont elles-mêmes des griseums thalamiques plus denses, rien ne prouve qu'il n'y ait pas, en fait, correspondance neurone à neurone*) des griseums thalamiques aux zones néocorticales de projection.

Conclusions

Les indices de céphalisation et de néocorticalisation fournissent deux classements des Insectivores légèrement différents mais concordant l'un et l'autre avec la phylogenèse du groupe. Les unités structurales thalamiques peuvent être divisées en quatre catégories:

1. L'habenula est la seule structure thalamique qui progresse moins vite que l'encéphale considéré dans son ensemble; il y a donc régression apparente, ou relative, mais non réelle, de cette structure. Les Insectivores et les deux Lémuriens étudiés restent macrosmatiques, mais avec une prédominance olfactive de plus en plus faible.

2. Noyaux médians et ganglion géniculé médial (espèces semi-aquatiques mises à part) sont des formations phylogénétiquement stables, qui gardent une importance relative constante.

3. Noyaux médiaux, ganglion géniculé latéral (espèces endogées et macroptiques mises à part) et noyaux ventraux sont évolutifs, mais leur croissance phylogénétique est inférieure à celle du néocortex.

4. Seuls les noyaux dorsaux et antérieurs ont une croissance phylogénétique comparable à celle du néocortex. De ces deux unités anatomiques structurales, toutefois, seuls les noyaux dorsaux sont étroitement associés dans leurs variations interspécifiques aux variations néocorticales.

L'étroitesse de la relation liant aux volumes néocorticaux les volumes des noyaux thalamiques dorsaux rend plausible l'hypothèse d'une projection des noyaux dorsaux sur l'ensemble du néocortex, ainsi que celle d'une correspondance neurone à neurone de ces deux formations. Elle autorise également l'utilisation des volumes de ces noyaux dorsaux pour déterminer les indices de néocorticalisation des espèces au sein de la lignée Insectivores-Primates.

---

*) Il faut entendre par là: correspondance de n neurones thalamiques à p neurones néocorticaux, n et p étant des constantes pouvant différer l'une et l'autre de l'unité.

Summary

The quantitative analysis of the thalamic nuclear groups in 14 species of Insectivores and 2 species of lissencephalic Prosimians *(Tupaia* and *Galago)* give information about the evolutionary or regressive position of each group. The thalamic structural units can be divided into four groups:

1. The habenula is the only thalamic structure increasing at a lower rate than the brain as a whole. Thus, its relative size decreases. However, there is still an absolute increase and no reduction.
2. The midline nuclei and medial geniculate body (except in aquatic species of Insectivores) are phylogenetic stable structures, which retain a constant relative importance.
3. The medial nuclei, lateral geniculate body (except in digging and macroptic species of Insectivores) and ventral nuclei are progressive, but their phylogenetic growth is lower than that of the neocortex.
4. Only the dorsal and anterior nuclei have a phylogenetic growth similar to that of the neocortex. However, within these structural units only the volumes of the dorsal nuclei are closely correlated in their interspecific variations with those of the neocortex.

The correlation between the neocortical and the dorsal thalamic volumes is so well established that the latter can also be used as an index of progress in evolution.

Références

BAUCHOT. R.: L'architectonique comparée, qualitative et quantitative, du diencéphale des Insectivores. Mammalia *27*, Suppl. 1 (1963), 1–400.

BAUCHOT, R.: La densité en neurones des noyaux gris diencéphaliques. J. Hirnforsch. *6* (1964), 327–330.

BAUCHOT, R., H. STEPHAN: Le poids encéphalique chez les Insectivores malgaches. Acta zool. *45* (1964), 63–75.

STEPHAN, H.: Vergleichende Untersuchungen über den Feinbau des Hirnes von Wild- und Haustieren. I. Studien am Schwein und Schaf. Zool. Jb. Anat. Abt. *71* (1951), 487 bis 586.

STEPHAN, H., O. J. ANDY: Quantitative comparisons of brain structures from Insectivores to Primates. Amer. Zool. *4* (1964), 59–74.

STEPHAN, H., R. BAUCHOT: Le cerveau de Galemys pyrenaicus, Geoffroy 1811 (Insectivora Talpidae) et ses modifications dans l'adaptation à la vie aquatique. Mammalia *23* (1959), 1–18.

Dr. R. BAUCHOT
Laboratoire de Zoologie
Ecole Normale Supérieure
24, rue Lhomond
Paris 5ème, France

Cytoarchitektonik des Sehhügels (Thalamus dorsalis) der Chiroptera

Von M. Kurepina, Moskau

Eine der allgemeinen Gesetzmäßigkeiten der Evolution ist, daß die Entwicklung von undifferenzierten morphologischen Strukturen mit nicht spezialisierten Funktionsweisen ausgeht und sich allmählich eine weitere Gliederung und Differenzierung in gesonderte Zellgruppen von verschiedenen spezialisierten Funktionen ergibt. Diese *Differenzierung* macht eine Reihe von Etappen durch, und auf jeder Etappe werden die Strukturen immer vollkommener und ihr Zusammenwirken komplizierter. Deswegen ist eines der fruchtbringenden Verfahren bei der Klärung des Zusammenwirkens und der funktionellen Abhängigkeit der morphologischen Strukturen ihre Untersuchung in möglichst niedrigen Evolutionsstufen, wo die alten Bildungen ihre Bedeutung noch nicht eingebüßt haben und die neuen ihrem Entwicklungsursprung noch nahestehen. Auf einem primitiven Evolutionsstadium des Großhirns unter den Mammalia steht das Gehirn der *Chiroptera*.

In dieser Arbeit wird die bisher in der Literatur nicht veröffentlichte Struktur des Thalamus dorsalis der Chiroptera dargelegt. Zwei Vertreter der Fledermäuse wurden untersucht: das Großohr *(Plecotus auritus)* und der Abendsegler *(Nyctalus noctula)*.

Die Fragen der strukturellen und funktionellen Beziehungen innerhalb des Thalamus dorsalis sind noch diskutabel. Das gilt in großem Maße auch für die Beziehungen zwischen seinen sogenannten *spezifischen* und *unspezifischen, reticulären* Strukturen. Die Untersuchung eines primitiven Stadiums seiner Entwicklung, auf dem der Chiropteren-Thalamus steht, kann zur Klärung auch dieses Problems beitragen.

Aufbau und Evolutionsstadium des Thalamus der Chiropteren. Bei den Chiroptera kommt schon der Haupttyp des Aufbaus des Thalamus dorsalis, der für die ganze Klasse der Mammalia charakteristisch ist, zum Vorschein.

Im Frontalschnitt gliedert er sich in *vier Sektoren*. Die zwei basalen Sektoren: die ventrale Formation und der parafasciculäre Kern, von denen bekanntlich somatische und viscerale Impulse aufgenommen werden, erreichen hier ihre maximale Entwicklung. Die dorsalen Sektoren, gebildet von der lateralen Formation und dem Nucleus dorsomedialis, welche die Impulse auf die sich im Laufe der Evolution spät entwickelnde Rinde – parietale und praefrontale – übertragen, sind hier noch sehr schwach ausgebildet.

Der Thalamus dorsalis der Fledermäuse befindet sich in einem Evolutionsstadium, wo einzelne Kerne und ihre Gruppierungen, die wir Formationen nennen, anatomisch noch ganz unscharf voneinander abgehoben sind. Dabei sind die sogenannten spezifischen Strukturen des Thalamus genauso undeutlich auch von den Bildungen abgegrenzt, die von den Elektrophysiologen als unspezifische, reticuläre Strukturen beurteilt werden, d. h. vom centralen Höhlengrau und den intralaminären Strukturen.

Im Chiropteren-Thalamus erreichen diese unspezifischen Strukturen zusammen mit noch einer weiteren Struktur von wahrhaft reticulärem Bau, der Zona reticularis, eine äußerst starke Entwicklung, wogegen die spezifischen Bildungen hauptsächlich bedeutend schwächer ausgebildet sind.

Spezifische Thalamusstrukturen. Der laterale Kern ist die am deutlichsten differenzierte spezifische Struktur des Chiropteren-Thalamus. Besonders groß ist der Kern bei *Nyctalus noctula;* der Größe, Form, Anzahl und Anordnung der Zellen nach kann man in diesem Kern einen vorderen und einen hinteren Teil, und innerhalb von jedem wieder eine laterale und eine mediale Formation unterscheiden.

Die laterale Formation (la), die bei höheren Ordnungen mit der parietalen Rinde verbunden ist, kommt hier ziemlich schwach zur Ausprägung (Abb. 1). Die ventro-anteriore Formation (va), die von einigen Verfassern ohne ausreichende strukturell-genetische Gründe als retikuläre Formation betrachtet wird, und in der laut Literaturangaben der Pedunculus cerebellaris anterior endet, ist bei *Plecotus auritus* klein, bei *Nyctalus noctula* dagegen viel größer, wobei sie in einen medialen (va^i) und einen lateralen (va^e) Teil zerfällt.

Zur ventralen Formation gehört der ventromediale Kern (vm), homolog einem ähnlichen Kern bei Nagetieren und dem submedialen Kern bei Raubtieren (LeGros Clark und Boggon 1933; Adrianov u. Mering 1959), und der von einigen Elektrophysiologen zur retikulären Formation gezählt wird. Er ist sehr groß und steht in unmittelbarer Zellverbindung mit dem centralen Höhlengrau.

Die ventro-posteriore Formation (Abb. 2), homolog dem Hauptrelaiskern des Thalamus in der ganzen Reihe der Mammalia, ist am stärksten entwickelt. Darin werden drei Strukturen ausgesondert (vpl, vpi, vpm), in denen bei höheren Ordnungen in lateromedialer Richtung der Tractus spinothalamicus, Lemniscus medialis und Lemniscus trigeminalis enden und die Impulse auf die prae- und die postcentrale Region der Großhirnrinde projizieren.

Zu spezifischen Thalamuskernen gehört auch der dorsomediale Kern (dm), homolog dem medialen Kern der Primaten, der Impulse vom Hypothalamus empfängt und diese

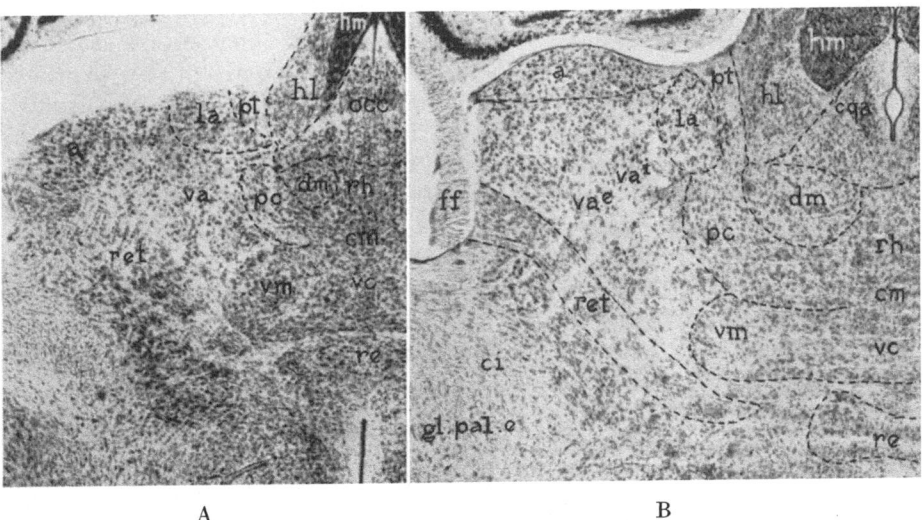

A   B

Abb. 1. Frontalschnitt aus dem oralen Gebiet des Thalamus von *Plecotus auritus* (A) und *Nyctalus noctula* (B). (Microphoto, Vergr. 50×, Nissl).

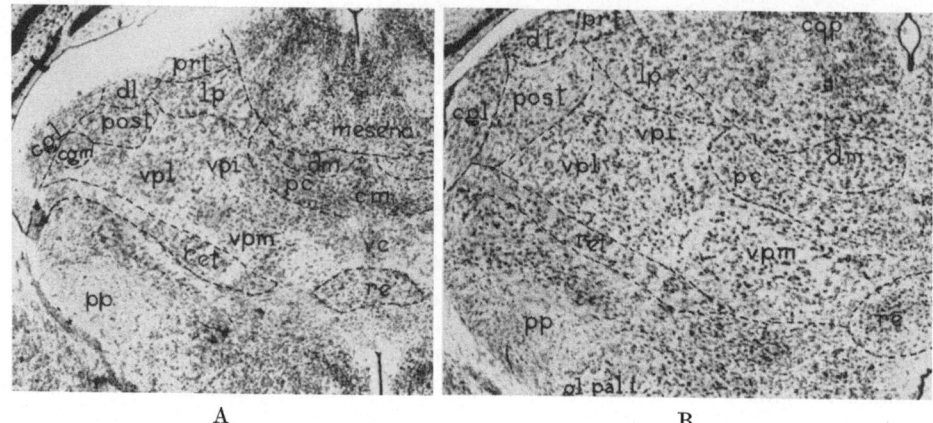

A   B

Abb. 2. Frontalschnitt aus dem caudalen Gebiet des Thalamus von *Plecotus auritus* (A) und
Nyctalus noctula (B). (Microphoto, Vergr. 50×, NISSL.)

auf die praefrontale Rinde überträgt. Genauso wie die praefrontale Rinde bei allen Chiropteren, ist auch dieser Kern bei *Nyctalus noctula* äußerst schwach entwickelt, bei
Plecotus auritus ist er noch im Anfangsstadium seiner Entwicklung. Der Kern ist klein,
von den intralaminären Zellanhäufungen und dem centralen Höhlengrau umgeben und
kann nur wegen der undichten Anordnung seiner etwas größeren Zellen ausgeschieden
werden. Hinten umfassen die intralaminären Zellanhäufungen den Kern in Form einer
Kapsel, vorne geht er in die von mir als „oraler Zellkomplex" genannte Struktur über.

Unspezifische Thalamusstrukturen. Jetzt gehe ich zu den sogenannten unspezifischen
Thalamus-Strukturen über. Das ist vor allem das centrale Höhlengrau (Commissura
mollis).

Das centrale Höhlengrau verbindet die Sehhügel bei *Plecotus auritus* in ihrer ganzen und
bei *Nyctalus noctula* fast in der ganzen Ausdehnung. Auf diese Weise wird der Ventriculus
III nur im Gebiet des „oralen Zellkomplexes" (occ) und Hypothalamus sichtbar. Das stark
entwickelte Höhlengrau wird architektonisch in folgende Kerne eingeteilt (Abb. 1):
Nucleus reuniens (re), commissuralis intraventralis (vc), centralis medialis (cm) und rhom-
boidalis (rh). Diese Zellstrukturen sind aber nicht scharf von der Umgebung abgesetzt,
mit Ausnahme des Nucleus reuniens, der sehr deutlich, besonders in seiner mittleren Ausdehnung umrissen ist. Caudal geht das centrale Höhlengrau des Thalamus unmittelbar
in den reticulären Kern des Mittelhirntegmentum (ret. mes.) über (Abb. 3). Die Fortsetzung des centralen Höhlengraus nach vorne bildet eine Struktur, die von uns als
„oraler Zellkomplex" (occ) des Sehhügels genannt wird. Sie fängt gleich hinter dem
Septum pellucidum (sept) und über der vorderen Kommissur (ca) an. Hinten wird sie
von den Corpora quadrigemina anteriora ersetzt, und geht auf deren Niveau in die reticuläre Formation des Tegmentum über.

Intralaminäre Zellanhäufungen sind durch die paracentrale Zellgruppe (pc), durch den
Nucleus centralis lateralis (cl) und den großzelligen Kern (mcl) dargestellt (Abb. 4). Sie
beginnen schon auf der Querebene des „oralen Zellkomplexes" (occ) und strömen sozu-
sagen aus ihm aus. Beide Kerne (cl, mcl) sind nicht groß, gut umrissen und bleiben im

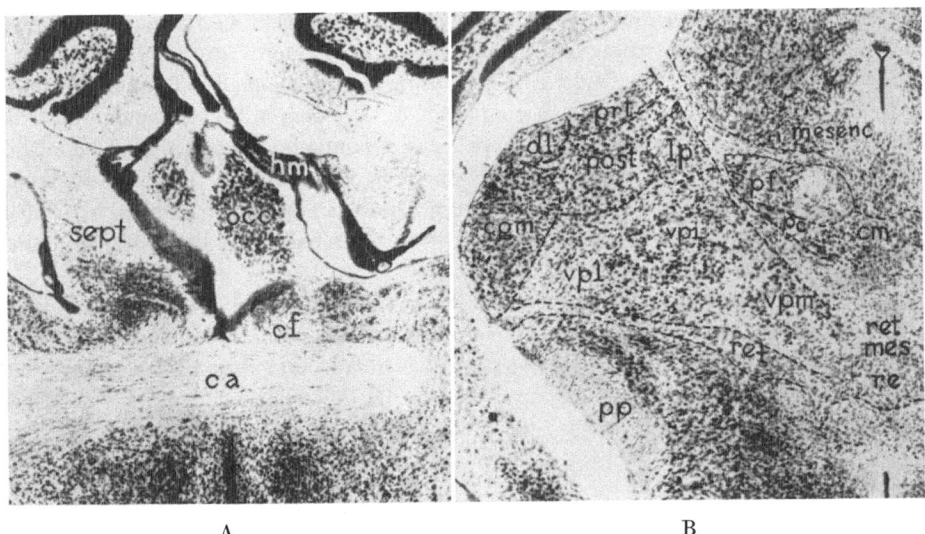

Abb. 3. Frontalschnitt aus oralen (A) und caudalen (B) Pol des Thalamus von *Plecotus auritus*. (Microphoto, Vergr. 50×, NISSL.)

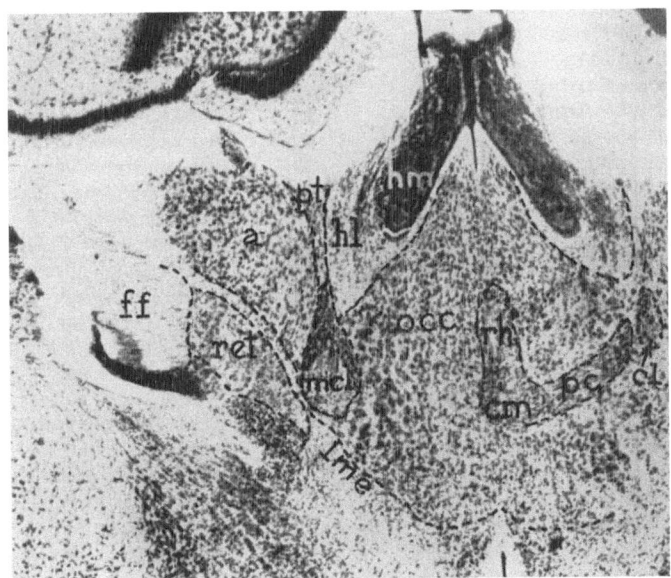

Abb. 4. Frontalschnitt durch den Thalamus von *Plecotus auritus* unweit des oralen Pols. (Micro-photo, Vergr. 50×, NISSL.)

Bereich der vorderen Thalamusteile; die paracentrale Zellanhäufung (pc) aber füllt diffus als ein breiter Streifen die innere medulläre Platte und zieht sich durch die ganze orocau-dale Thalamusausdehnung. Medial vereinigt sie sich mit dem centralen medialen Kern (cm) und an den Grenzen zu den benachbarten spezifischen Strukturen: laterale und ven-

trale Formation, dorsomedialer und parafasciculärer Kern, geht sie fließend in interme-
diäre Übergangsstrukturen über (Abb. 2 u. 3).

Der parafasciculäre Kern (pf), der eine Anhäufung der Zellen um den Tractus habenulo-
peduncularis darstellt, ist nicht allzu groß (Abb. 3). Aus dem Mittelhirntegmentum tritt
der Tractus zusammen mit naheliegenden Zellen in den hintersten Thalamusteil, unmittelbar hinter dem dorsomedialen Kern ein. Er dringt in die Lamina medullaris interna ein;
gerade hier bildet sich der parafasciculäre Kern. Dieses Eindringen der reticulären Zellen
des Mesencephalon in die Lamina medullaris interna des Sehhügels kann bei den Chiropteren sehr deutlich verfolgt werden.

Die reticuläre Zone (ret) begrenzt den Thalamus von der ventrolateralen Seite. Rostral
ist sie besonders breit und zerfällt in einen dorsalen und einen ventralen Teil (Abb. 4).
Die Lamina medullaris externa ist dicht mit Zellen gefüllt, und zwar mit Ausnahme ihres
hintersten Teils, worin die reticulären Zellen des Mittelhirntegmentum eindringen
(Abb. 3).

Abkürzungen

a	Nucleus anterior	mm	Pars medialis nuclei medialis
ca	Commissura anterior	occ	oraler Zellkomplex des Thalamus
cac	Pars accessoria centrum medianum Luysi	pac[1]	Pars accessoria pulvinar prima
		pac[2]	Pars accessoria pulvinar secunda
caud	Nucleus caudatus	pc	Nucleus paracentralis
cf	Columna fornicis	pf	Nucleus parafascicularis
cgl	Corpus geniculatum laterale	post	Nucleus posterior
cgm	Corpus geniculatum mediale	pp	Pes pedunculi
ci	Capsula interna	prt	Nucleus praetectalis
cl	Nucleus centralis lateralis	pt	Nucleus parataenialis
cm	Nucleus centralis medialis	re	Nucleus reuniens
cmL	Centrum medianum Luysi	ret	Zona reticularis
cqa	Corpus quadrigeminum anterior	ret. mes.	Nucl. reticularis mesencephali
cqp	Corpus quadrigeminum posterior	rh	Nucleus rhomboidalis
dl	Nucleus dorsolateralis	sept	Septum pellucidum
dm	Nucleus dorsomedialis	v	ventrale Formation des lateralen Kerns
ff	Fimbria	va	ventro-anteriore Formation
gl. pal. e	Globus pallidus externus	va^e	äußerer Teil der ventro-anterioren Formation
gl. pal. i	Globus pallidus internus		
hl	Ganglion habenulae, pars lateralis	va^i	innerer Teil der ventro-anterioren Formation
hm	Ganglion habenulae, pars medialis		
l	laterale Formation des lateralen Kerns	vc	intraventraler commissuraler Kern
la	latero-anteriore Formation	vm	Nucleus ventromedialis
lme	Lamina medullaris externa	vpi	innerer Teil der ventro-posterioren Formation
lmi	Lamina medullaris interna		
lp	latero-posteriore Formation	vpl	lateraler Teil der ventro-posterioren Formation
mac	accessorischer Teil des medialen Kerns		
mcl	Nucleus magnocellularis	vpm	medialer Teil der ventro-posterioren Formation
mesenc	Mittelhirn		
ml	Pars lateralis nuclei medialis		

Diskussion

Die Untersuchung des Thalamus bei phylogenetisch niedrigstehenden Vertretern der Mammalia, wie den Chiropteren, zeigt eine verhältnismäßig schwache Differenzierung der Kerne mit spezifischer Projektion und eine starke Entwicklung der Strukturen, die von den Elektrophysiologen als ein unspezifisches, retikuläres System des Sehhügels betrachtet werden, nämlich die intralaminären Zellanhäufungen, das centrale Höhlengrau und die retikuläre Zone des Thalamus.

Intralaminäre Zellanhäufungen und das centrale Höhlengrau beginnen rostral mit einer gemeinsamen Zellansammlung, die von uns als „oraler Zellkomplex" des Thalamus bezeichnet wird. Er beginnt auf der Querebene der vorderen Kommissur, liegt unmittelbar vor den Corpora quadrigemina und ist eine direkte Fortsetzung des Mittelhirntegmentum. So haben wir hier einen unmittelbaren Übergang der unspezifischen Strukturen des Thalamus in die retikuläre Formation des Mittelhirntegmentum vor uns.

Der „orale Zellkomplex" fehlt bei den höheren Mammaliern, bei denen der Thalamus nicht auf dem Niveau, sondern caudal von der vorderen Kommissur anfängt. Er ist nur von JASPER (1954) in seinem stereotaxischen Atlas des Katzengehirns unter dem Namen des prothalamischen Kerns erwähnt. Dort hat er eine bedeutend kleinere Größe als bei Chiroptera, und es gibt keine Hinweise auf seine Wechselbeziehungen mit dem Mittelhirn.

Das centrale Höhlengrau endet in caudalen Ebenen des Thalamus. Hier geht es zusammen mit intralaminären Anhäufungen unmittelbar in die retikuläre Formation des Mesencephalon über. Die Zellen des retikulären mesencephalischen Kerns dringen hier auch in die retikuläre Zone des Thalamus ein. Auf diese Weise stehen das centrale Höhlengrau und die intralaminären Anhäufungen der Chiroptera, die vorne zum „oralen Zellkomplex" verschmelzen, sowohl im oralen, als auch im caudalen Pol mit dem Tegmentum des Mesencephalon in direkter Zellverbindung.

Was die *retikuläre Zone* des Thalamus anbetrifft, so kann man beim Vergleich der Angaben für Chiropteren mit den von uns früher durchgeführten Untersuchungen des dorsalen Thalamus der Reptilien (1959) folgern, daß diese Zone einen zweifachen Ursprung hat, nämlich den ventralen Thalamus, eine Übergangsstruktur zwischen dorsalem Thalamus und Hypothalamus und das Mittelhirntegmentum.

Bei Chiropteren und insbesondere bei *Plecotus auritus* sind die *intralaminären Anhäufungen* in ihrer ganzen Ausdehnung innerhalb der Lamina medullaris interna nicht scharf abgegrenzt und gehen fließend in die benachbarten Bildungen über. Infolge dieses Umstands entstehen in der Umgebung der intralaminären Zellanhäufungen Zonen mit Übergangsstruktur. Hier sehen wir einen Entwicklungsvorgang auf tieferem Niveau, der bei Primaten von uns beobachtet wurde, nämlich die Vergrößerung der spezifischen Thalamuskerne auf Kosten der unspezifischen intralaminären Zellanhäufungen. Bei Chiropteren, besonders bei *Plecotus auritus*, kommt es aufgrund dieses Prozesses zur Bildung des spezifischen dorsomedialen Kerns, der sich bei den Mammaliern besonders progressiv entwickelt (Abb. 5). Der Kern (dm) befindet sich hier im ersten Stadium seiner Entwicklung und ist lateral, rostral und caudal von breiten intralaminären Anhäufungen umgeben, die um ihn eine breite Zellkapsel bilden. Sein accessorischer Teil (mac) der den Kern bei Primaten, entsprechend unseren Beobachtungen, stark vergrößert, entsteht auch noch immer innerhalb der Lamina medullaris interna.

Viele Elektrophysiologen betonen den unspezifischen Charakter des *parafasciculären Kerns*. Das wird auch durch morphologische Untersuchungen von LEONTOWICH (1959)

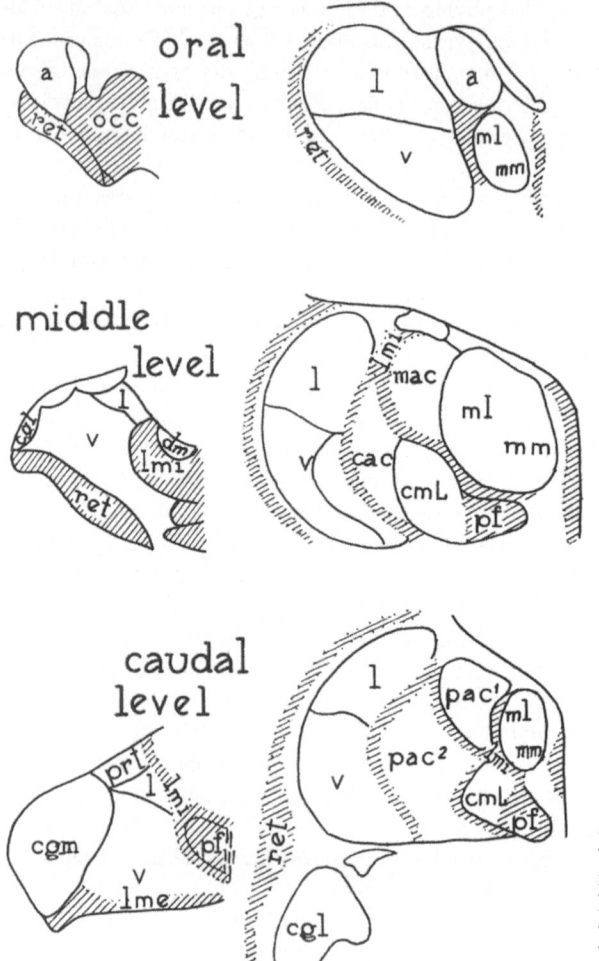

CHIROPTERA   PRIMATES

oral level

middle level

caudal level

Abb. 5. Zwei Stadien der Evolution des Sehhügels. Links: Thalamus der Chiropteren, rechts: der Primaten im oralen, mittleren und caudalen Gebiet. Schema. Unspezifische Strukturen sind gestrichelt.

bestätigt. Sie hat festgestellt, daß der parafasciculäre Kern, wie auch die reticuläre Zone des Thalamus aus besonderen verzweigungsarmen Neuronen vom Charakter der reticulären Hirnstammneurone besteht. Die bei Chiropteren gut beobachtete Einwanderung von Zellen aus dem Tegmentum mesencephali in den Thalamus und die Lamina medullaris interna, dem Tractus habenulo-peduncularis entlang, erklärt gut auch den reticulären Charakter der Neuronen des mit diesem Faserbündel eng verbundenen parafasciculären Kerns.

Der Tractus habenulo-peduncularis, mit den ihm längs anliegenden Zellen, die den parafasciculären Kern (pf) aufbauen, dringt in den Thalamus und die Lamina medullaris interna (lmi) ein. Wie die Literaturangaben und unsere Untersuchungen des Thalamus bei Primaten zeigen, geht die weitere Vergrößerung des parafasciculären Kerns, d.h. die

Entwicklung des Centrum medianum Luysi (cmL) und seines accessorischen Teil (cac) wiederum innerhalb und auf Kosten der Zellen der Lamina medullaris interna. Folglich vervollständigen die Angaben, die die Untersuchungen an den Chiropteren ergaben, das Bild des Evolutionsablaufs des parafasciculären Kerns.

Auf solche Weise steht der *dorsale* Thalamus der Chiropteren im Anfangsstadium der Entwicklung seiner spezifischen Strukturen. Hier fehlt noch das Pulvinar; der dorso-mediale Kern steht seiner Entwicklungsurform nahe, die laterale Formation ist durchaus schwach ausgebildet. Verhältnismäßig gut differenziert ist nur die *ventrale* Formation, die die somatotopische Reception auf die Projektionsfelder der Rinde überträgt. Stark entwickelt sind die unspezifischen Strukturen, deren Neurone schon auf diesem phylogene-tischen Stadium sowie im Laufe der weiteren Mammalier-Evolution den Aufbau der spe-zifischen Thalamuskerne sichern.

Die Thalami von *Plecotus auritus* und *Nyctalus noctula* unterscheiden sich strukturell ziemlich stark. Bei *Plecotus auritus* sind die Neuronen primitiver und liegen viel dichter. Bei *Nyctalus noctula* sind sie mehr differenziert und liegen nicht so dicht, was auf größeren Faserreichtum des Thalamus hinweist. Der dorsomediale Kern und die laterale Formation (Abb. 1 u. 2), die die Impulse auf die sich während der Evolution progressiv entwickelnde praefrontale und parietale Rinde übertragen, sind bei *Nyctalus noctula* besser entwickelt, während sie bei *Plecotus auritus* erst im Anfangsstadium der Entwicklung stehen. Die Primitivität der gesamten Neuronenorganisation des Thalamus bei *Plecotus auritus* sowie der Zentren, die Impulse auf die sogenannten Assoziationsfelder der Rinde übertragen, entspricht seinem niedrigen Stand im Stammbaum der Chiropteren. Die vollkommenere Organisation des Thalamus bei *Nyctalus noctula* sichert die größere morpho-funktionelle Vollkommenheit dieser Art in der Reihe der Fledermäuse.

Aber sowohl *Plecotus auritus* als auch *Nyctalus noctula* haben sich gut an die ihnen eige-nen Existenzbedingungen angepaßt (KUZJAKIN, 1950). Das spiegelt sich in der Struktur des Thalamus wider. So ist bei der schnell fliegenden, virtuos manövrierenden *Nyctalus noctula* die ventrale Formation größer und besser differenziert als beim langsamen, flat-ternd fliegenden *Plecotus auritus*. Gerade in die ventrale Formation kommen die pro-priozeptiven Impulse durch die mediale Schleife und den Pedunculus cerebellaris superior.

In der ventro-anterioren Formation, wo die letzteren enden, macht sich der Entwick-lungsunterschied besonders bemerkbar (Abb. 1), was in unserem Laboratorium von CHATSCHATURJAN festgestellt wurde. Im Gegensatz zu *Plecotus auritus* ist diese Forma-tion bei *Nyctalus noctula* schon in zwei Strukturen differenziert, die den höher organisier-ten Ordnungen eigen sind.

Die Untersuchung des Gehirns der Chiropteren, die ihrer Ökologie nach so eigenartig sind, in so früher Evolutionsperiode und gleichzeitig am Ursprung der Mammalierent-wicklung stehen, kann meiner Meinung nach zur Lösung vieler noch ungelöster Probleme der Entstehung, der Struktur und der Funktion der Hirnmechanismen dienen. Es wäre wünschenswert, der Erforschung des Chiropterengehirns mehr Aufmerksamkeit zu widmen.

Summary

The main features of the structural organisation of the mammalian thalamus are already seen in Chiroptera, but its particular cellular structures are feebly isolated from each other. While the specific structures of the thalamus are only at the roots of their development, the so-called non-specific formations, such as central gray matter, intralaminar structures and zona reticularis show powerful development. They gradually change into intermediate formations at the border with the neighbouring specific structures, creating them so on this phylogenetic stage as in the further evolution of the mammals as well.

The central gray matter, the intralaminar cellular masses and the zona reticularis form a direct cellular continuation of the mesencephalic tegmentum.

The structure of the thalamus of *Plecotus auritus* differs considerably from that of *Nyctalus noctula*. Its primitiveness in *Plecotus auritus* corresponds well to the low position of this bat in the phylogenetic branch of Chiroptera; while its better structural organisation in *Nyctalus noctula* demonstrates the general structural and functional perfection of this species in the order of the bats.

Literatur

ADRIANOV, O.S., A.T.MERING: Atlas of the Canine Brain, Ann Arbor, Michigan, 1964 (übersetzt von russischer Auflage).

CLARK, W.E., LE GROS, H.R.BOGGON: On the connections of the medial cell groups of the thalamus. Brain. *56* (1933).

JASPER, H.H., C.A.AJMONE-MARSAN: Stereotaxic Atlas of the Diencephalon of the Cat. Ottawa 1954.

KUREPINA, M.M.: Reticular formation in the ontophylogenesis of the thalamus. In: Struktur und Funktion der reticulären Formation und ihr Platz im System der Analysatoren. Akad. Med. Nauk, Moskau 1959. (russisch).

KUZJAKIN, A.P.: Die Fledermäuse.Sow.Nauka. Moskau. 1950. (russisch).

LEONTOVICH, T.A.: Reticular formation neurons in the thalamus and the corpus striatum of the dog. In: Struktur und Funktion der retikulären Formation und ihr Platz im System der Analysatoren, Akad. Med. Nauk. Moskau 1959. (russisch).

Prof. Dr. MILIZA KUREPINA
Pädagogisches Lenin-Institut
Malaja Pirogowskaya 1
Moscow, USSR

Les relations thalamo-corticales dans la phylogenèse: Rapport sur les projections thalamo-corticales chez les Mammifères

Par G. MACCHI, Perugia

L'étude anatomo-comparative des relations thalamo-corticales chez les mammifères prend son départ de l'analyse du développement phylogénétique respectif des structures thalamiques et de l'écorce cérébrale.

Mais l'expérimentation animale et la pathologie humaine donnent leur appui aux indications d'ordre anatomo-comparatif en vue d'une systematisation des afférences thalamo-corticales chez les différentes espèces.

Vous savez bien que la méthode expérimentale de choix, dans ce genre d'étude, a été la dégénérescence rétrograde.

Classement des noyaux thalamiques

Un des premiers classements des noyaux thalamiques, rangé sur cette méthode expérimentale, nous vient des expériences d'hémidécortication.

Dans notre opinion, les expériences valables, c'est à dire celles où le neocortex entier a été enlevé, ne sont pas nombreuses. Nous rappelons les travaux de: WALKER 1935, 1938; PAPEZ 1938; LASHLEY 1941; BODIAN 1942; ROSE a. WOOSLEY 1943; METTLER 1943; COMBS 1949; POWELL 1952; PEACOCK a. COMBS 1965.

En suivant les résultats obtenus par les expérimentations d'hémidécortication, nous pouvons classer les noyaux thalamiques en trois catégories distinctes. (v. schéma ci dessous):

GR. ANTERIEUR
Nc. anterior dorsalis (AD)
Nc. anterior ventralis (AV)
Nc. anterior medialis (AM)

GR. POSTERIEUR
Pulvinar (Pul)
Corpus geniculatum mediale (GM)
   magnocellularis (GM mc)
Corpus geniculatum laterale (GL)
   Corpus geniculatum dorsale (GL D)
   Corpus geniculatum ventrale (GL V)
Nc. s u p r a g e n i c u l a t u m (SG)
Nc. limitans (L)

GR. LATERAL
Nc. ventralis anterior (VA)
Nc. ventralis lateralis (VL)
Nc. ventralis intermedius (VI)
Nc. ventralis posterior (ventro-basalis VB)
Nc. lateralis dorsalis (Ld)
Nc. lateralis posterior (LP)

GR. MEDIAL ET INTRALAMINAIRE   Nc. LIGNE MEDIANE

Nc. medialis dorsalis (MD)   Nc. parataenialis (PT)
Nc. submedius (SM)   Nc. paraventricularis (PV)
Nc. medio-ventralis (MV)   Nc. rhomboideus (Rh)
Nc. p a r a c e n t r a l i s (PC)   Nc. centralis medialis (NcM)
Nc. c e n t r a l i s l a t e r a l i s (CL) *Nc. r e u n i e n s* (RE)
Nc. parafascicularis (PF)   *Nc. r e t i c u l a r i s* (R)
Nc. centro-medianus (CM)

Une *première classe* est réprésentée par les noyaux qui *dégénèrent* en voie *rétrograde* après hémidécortication (voire les nc. en caractères italiques).

Une *deuxième classe* comprend les noyaux qui presentent des lésions *rétrogrades partielles* ou douteuses selon les differents auteurs (voire les nc. en caractères espacés).

Une *troisième classe* est formée par des noyaux qui *ne dégénèrent pas* après hémidécortication (voire les nc. en caractères normals).

Les résultats que nous ont donné des ablations plus restreintes de l'écorce cérébrale, ont été utilisés, avant tout, dans l'intention de définir quelques aspects topographiques propres aux connexions efférentes des noyaux thalamiques de notre première classe.

L'étude de la série entière des mammifères a abouti à un schéma topographique des aires corticales auxquelles chaque unité nucleaire thalamique adresse ses projections.

Mais le problème ne siege pas seulement dans l'établissement des rapports topographiques: il faut pousser plus loin nos connaissances sur l'organisation des efférences thalamocorticales par rapport aux différentes aires thalamiques et corticales, et par rapport au niveau phylogénétique atteint.

Ce sont là les points que nous voulons discuter dans la suite de notre rapport, en analysant les différents types de réactions thalamiques à des ablations sélectives de l'écorce cérébrale.

1. Dégénérescence rétrograde focale sévère d'un noyau ou d'une de ses parties, par des lésions limitées à un secteur cortical, (aire cyto-architectonique) anatomiquement ou fonctionnellement unitaire ou à une partie du secteur même (Fig. 1 B).

2. Pas de dégénérescences thalamiques histologiquement appréciables quand la lésion corticale est très discrète, alors que l'on voit paraître la réaction rétrograde par rapport à l'élargissement de l'ablation corticale, toujours entre les limites de l'aire cytoarchitectonique ou de l'unité fonctionnelle rélative (Fig. 1 A).

3. Absence de dégénérescence au niveau du thalamus, à la suite d'ablations isolées d'une aire cyto-architectonique ou d'une unité fonctionnelle corticale, tandis que la dégénérescence rétrograde se revèle lorsque la lésion du cortex atteint simultanément une ou plusieures aires corticales voisines (Fig. 1 C).

4. Absence de dégénérescence rétrograde thalamique en face de n'importe quel type d'ablation corticale étendue.

Dans les lignes d'interpretation que l'on suit habituellement en ce genre d'expérimentation (c'est à dire que la dégénérescence du pyrénophore découle de la lésion axonale et qu'elle peut être évitée par l'integrité de branches axonales collatérales protectrices), la physionomie de nos quatre types réactionnels devrait être interprétée de la façon suivante:

Dans le *premier* cas, la lésion corticale devrait atteindre des axones concentrés au niveau de zones restreintes du cortex: les *collatérales axonales* devraient être également *renfermées dans* l'aire de *la lésion* (Fig. 2 n 1).

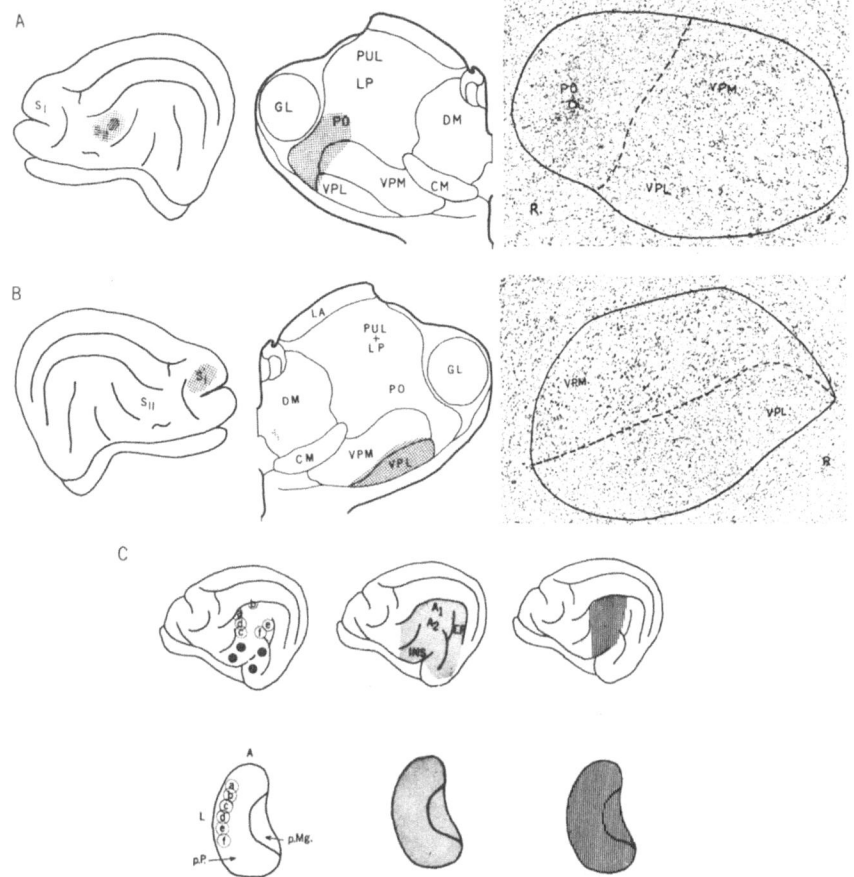

Fig. 1. En A on voit qu'une lésion limitée de l'aire somatique sensitive S II (à gauche) n'est pas suivie de lésions rétrogrades dans le thalamus, pendant qu'un agrandissement de la lésion (marqué plus clair) est suivi par une lésion rétrograde au niveau de PO et des frontières latéro-postérieures de VPM et VPL (schéma au centre, microphot. à droite).

En B on observe une lésion importante rétrograde au niveau de VPL (schéma au centre, microphot. à droite) par une lésion portée dans un secteur de l'écorce sensitive somatique générale S I du chat (à gauche).

En C on voit, à gauche, représentée une série de lésions dans l'aire auditive primaire du chat (le schéma est basé sur les résultats des recherches de ROSE), et les lésions rétrogrades focalisées au niveau des deux tiers antérieurs du corp genouillé médial.

Au centre on voit que la lésion rétrograde atteint aussi la partie caudale du corp genouillé médial et le nouyau magnocellulaire, seulement après une lésion étendue aux aires A 1 et A 2, à l'insula et à la région ectosylvienne postérieure (EP).

A droite on a marqué toute la région corticale où WOLLARD et HARPMAN ont vu terminer les fibres dégénérées après lésion du corp genouillé médial.

A 1	aire auditive I	p. Mg	corpus geniculatum mediale pars magnocellularis
A 2	aire auditive II	p. P.	corpus geniculatum mediale pars principalis
CM	nc. centro-medianus	PUL	pulvinar
DM	nc. dorsalis medialis	R	nc. reticularis
EP	gyrus ectosylvius posterior	S I	aire I de la sensibilité générale
GL	corpus geniculatum laterale	S II	aire II de la sensibilité générale
INS	insula	VPL	nc. ventralis posterior-lateralis
LP	nc. lateralis posterior	VPM	nc. ventralis posterior-medialis
PO	group postérieur		

Il s'agit, selon quelques auteurs, (DIAMOND et UTLEY 1963) d'un aspect absent dans les aires sensorielles chez certains mammifères (opossum), tandis qu'il est propre à une grande partie des aires de projections et associatives corticales des autres mammifères: c'est la *«essential projection»* de ROSE.

Dans le *second* cas, la lésion corticale devrait atteindre une extension suffisante pour entraîner la réaction thalamique: on pourrait y voir l'effet d'une plus grande *diffusion des collatérales protectrices*, ou bien d'une moindre concentration axonale à l'interieur de l'aire en question (Fig. 2 n 2).

C'est ce qu'il arrive, par exemple, à propos de l'aire SII du chat; et, même dans cette condition, il faudrait se réporter aux projections «essentielles» de ROSE.

Dans le *troisième* cas, il faudrait penser à l'existence de *collatérales protectrices à distance*, dans des autres champs corticaux ou à niveau thalamique en jonction synaptique avec des cellules qui se projettent à des autres aires corticales (ROSE 1950); de sorte que seule-

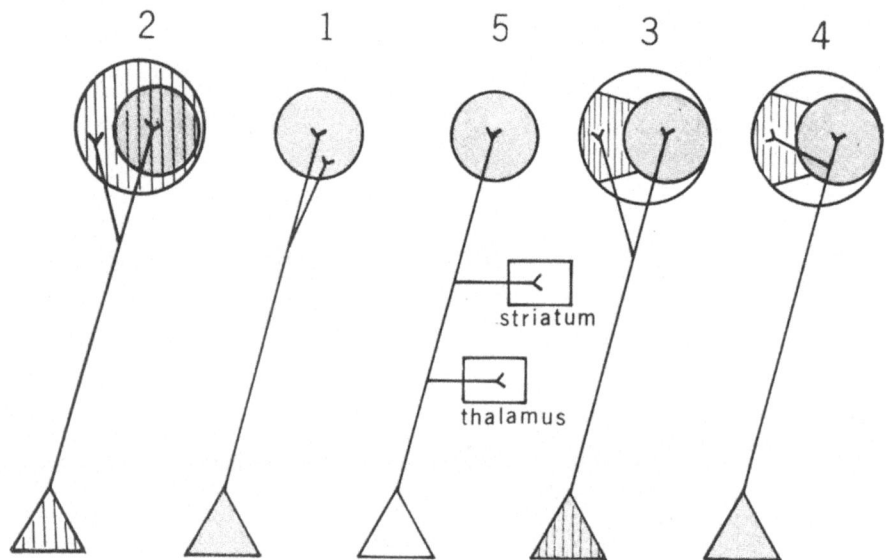

Fig. 2. Dans cette figure on voit schématisée une série des hypothèses des différentes réactions thalamiques, selon l'interprétation courante fondée sur l'action protectrice des collatérales des axones.

En 1 on observe l'exemple d'une *«essential projection»* (ROSE) c'est à dire lésion localisée corticale et réaction rétrograde focale dans le thalamus.

En 2 on observe un deuxième exemple où des collatérales à grande distance protégeraient la cellule qui est atteinte seulement après ablation élargie de l'écorce.

En 3 on observe l'exemple d'une *«sustained projection»* (ROSE) où la lésion simultanée de la collatérale et de l'axone, en deux aires différentes, réalise la réaction rétrograde de la cellule, pendant que la lésion isolée d'une ou de l'autre ne la réalise pas.

En 4 on n'observe pas de dégénérescence rétrograde thalamique aprés lésion d'une aire pourvue seulement de branches collatérales (marquée par des lignes unies verticales).

En 5 on observe le manque de dégénérescence rétrograde pour l'action protectrice de collatérales sous-corticales.

ment l'ablation de ces aires «alliées» permet la réalisation des réactions rétrogrades thalamiques (Fig. 2 n 3).

C'est ce qu'il arrive chez l'*opossum* et, dans de différents noyaux chez les mammifères
supérieurs. On peut le constater, par exemple, dans le tiers caudal du corp genouillé
médial (P. principale), dans le noyau magnocellulaire, dans le noyau ventralis anterior et
dans le PO (groupe postérieur) de ROSE chez le chat. Et même chez les primates LE GROS
CLARK et POWELL (1953) l'auraient demontré par l'aire 2 de l'écorce parietale. On peut
bien voir dans ces aspects, ceux propres aux «*sustained projections*» de ROSE.

Dans le *quatrième* cas, la région corticale lésée serait dépourvue de véritables projections
thalamiques, ou elle serait tout à fait protégée par des *collatérales souscorticales*, ou bien
y parviendraient simplement des collatérales des axones se rendant à des autres régions
du cortex (Figs. 2, n 4, 5).

C'est là la condition de certaines entre les aires temporales, surtout chez les primates,
et rhinencéphaliques (champs dits «*athalamiques*»).

Les limites propres à la méthode expérimentale employée poussent à chercher ailleurs
la confirmation de la classification qu'on a ébouché à propos des projections thalamo-
corticales.

En effet, le premier type réactionnel seulement (en accord avec la définition de «projéctions essentielles» de ROSE) a l'appui d'autres techniques de recherche.

Par conséquent, on peut s'accorder sur la conclusion que, lorsque l'ablation corticale
sélective entraîne une dégénération rétrograde focale sévère au niveau du thalamus, l'aire
corticale doit être le siège d'une *projection spécifique* de la partie du secteur thalamique
en question.

Mais l'interprétation des autres types de comportement réactionnel thalamique est bien
plus équivoque et hypothétique. En fait, nous pourrions émettre des autres interpréta-
tions aussi, en dehors du rôle joué par les collatérales protectrices (Fig. 3).

Le deuxième tableau réactionnel, par exemple, pourrait ressortir du fait que la projection thalamique sur l'aire corticale n'est pas trop concentrée et que les unités cellulaires
thalamiques atteintes sont donc bien peu nombreuses et disseminées pour atteindre
une évidence histologique.

Dans la troisième condition, seul l'atteinte d'une quantité suffisante d'axons thalamiques au niveau cortical s'accorde avec une dégénération histologiquement évidente dans
le thalamus. En d'autres termes, la réaction rétrograde est évidente seulement par ablation parallèle des aires corticales «alliées» qui se partagent la projection d'une unité
nucléaire thalamique.

La projection du complex genouillé (partie caudale et MG), qui se rend à différentes
aires corticales pourrait nous en donner un exemple; les données de WOOLLARD et HARP-
MAN (1939) à ce sujet, au moyen de la dégénération antérograde, nous confirment sur
ce point (Fig. 1 C).

En ce qui concerne la quatrième condition, nous rappelons qu'il y a la possibilité d'une
concentration corticale (au niveau de l'insula et du rhinencéphale, par exemple) de projections venant de différents secteurs thalamiques: dans ce cas, la réaction rétrograde thalamique possible échappe à l'observation au NISSL en raison de sa dispersion.

La méthode de NAUTA montre des projections thalamiques que l'on peut suivre jusqu'à
différentes régions du rhinencéphale (NAUTA et WHITLOCK 1954) tandis que la technique
de dégénération rétrograde ne nous a jamais confirmé l'existence d'un tel genre de connexion (MACCHI, CARRERAS, ANGELERI 1955).

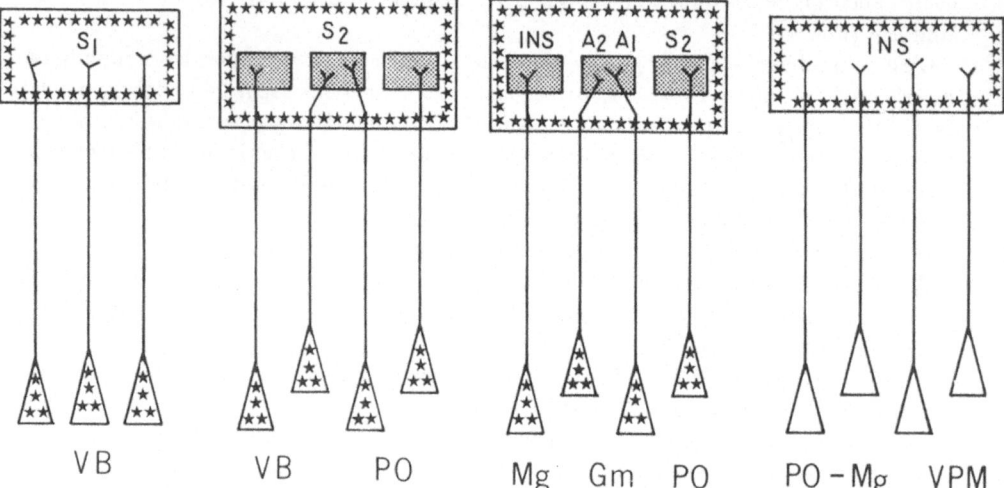

Fig. 3. Dans cette figure on voit schématisées d'autres hypothèses sur les réactions thalamiques, au dehors du rôle joué par l'action protéctrice des collatérales.

Voilà à gauche l'exemple de l'aire S I, où la concentration des projection au niveau de l'écorce et des cellules dégénérées au niveau du thalamus, rend la réaction rétrograde toujours vérifiable.

A la deuxième place, l'exemple de l'aire S II, où une concentration mineure au niveau cortical et plus répandue au niveau du thalamus, entre PO et VB, donnerait une réaction histologiquement positive au Nissl seulement après une lésion étendue.

A la troisième place un exemple plus significatif encore, puisque d'une même région thalamique les projections se dirigeraient à plusieures aires corticales, comme il arriverait pour l'aire de transition qui s'étend entre PO et MG. La dégénérescence rétrograde serait évidente après l'ablation de toutes les aires considerées (A 1, A 2, S 2, INS).

A la quatrième place enfin, les projections d'une même aire corticale arriveraient de plusieures aires thalamiques, et par consequent la réaction rétrograde ne serait pas décélable au niveau du thalamus.

GM: corpus geniculatum mediale: Mg: corpus geniculatum mediale pars magnocellularis.

Fig. 4. Dans cette figure, on voit, en haut, les caractères des unités isolées par Carreras et Andersson au niveau de l'aire S II du chat pendant des pénétrations microélectrodiques.

Les pénétrations ont été définies sur la base des propriétés statiques et dynamiques des neurones rencontrés pendant la pénétration.

A gauche, en bas, la reconstruction d'une pénétration en VPL, avec la représentation des champs periphériques de type lemniscal. A droite la reconstruction d'une pénétration dans la région de PO, avec la représentation des champs periphériques de type antérolatéral.

CL	nc. centralis lateralis	GMpP	Corpus geniculatum mediale
EA	sillon ectosylvien antérieur		pars principalis
EP	sillon ectosylvien postérieur	PS	sillon pseudosylvien
PC	nc. paracentralis	SS	sillon suprasylvien
Ped	pédoncule cérébrale	VL	nc. ventralis-lateralis
GMpMg	Corpus geniculatum mediale	VM	nc. ventralis medialis
	pars magnocellularis		

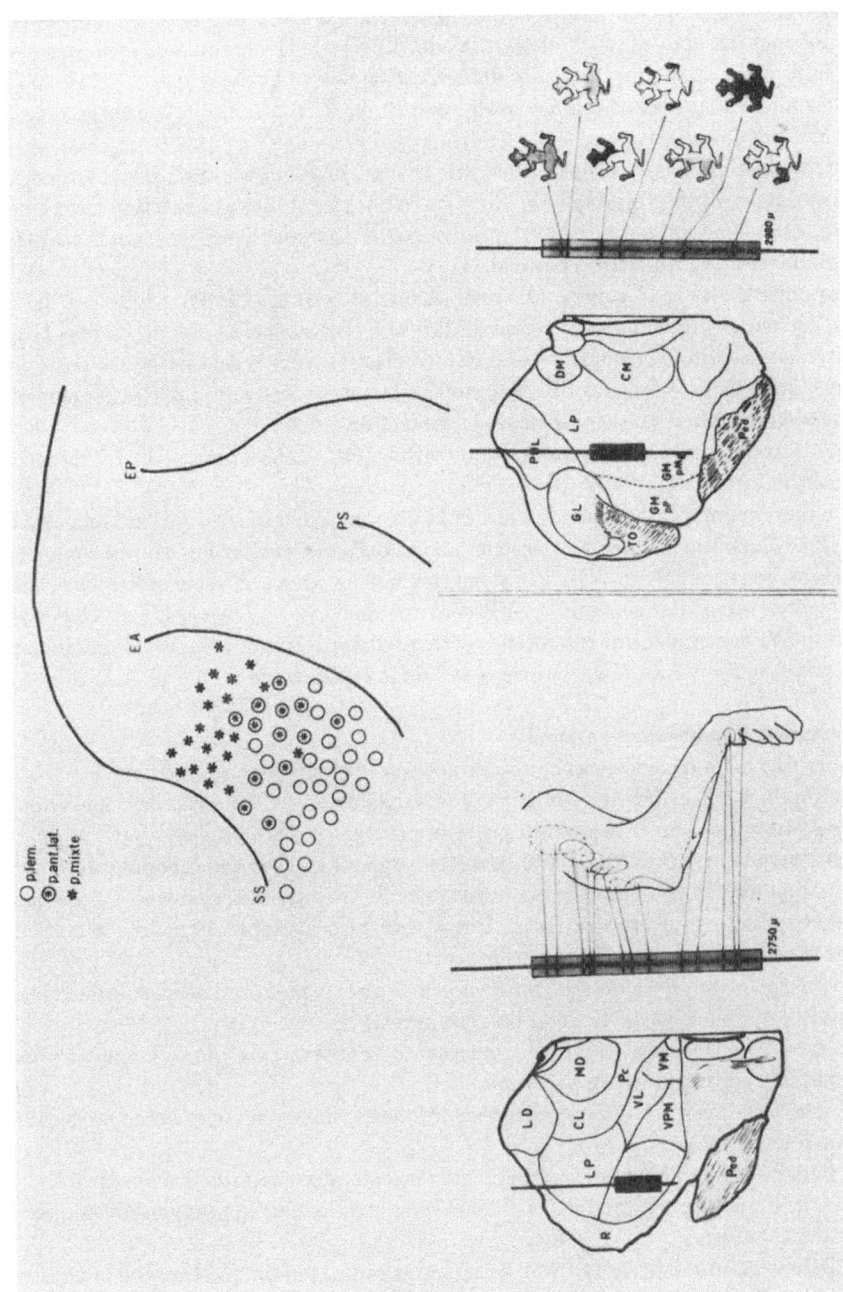

Fig. 4.

Legende auf der gegenüberstehenden Seite.

Tous ces aspects, bien que hypothétiques, confirment la nécessité de nous approcher du problème au moyen de l'intégration de différentes téchniques, parmis lesquelles il faut faire place aussi aux méthodes de dérivation neurophysiologique.

Nous allons maintenant donner quelques exemples des possibilités resortissantes à une telle attitude intégrative.

Un prémier problème concerne les projections thalamiques de l'aire corticale S II. D'après la conception originale de ROSE et WOOLSEY (1958), le caractère «sustained» de la projection thalamique à S II est prouvé par le manque de dégénération en PO après l'ablation sélective du champ cortical.

Nos expériences, par contre, (MACCHI, ANGELERI, GUAZZI 1960) plaidaient en faveur d'une «essential projection» en raison du fait que les lésions élargies de l'aire S II entraînaient une atteinte dégénérative rétrograde certaine dans le secteur latéro-postérieur de VB et antérieur de PO: d'où l'essai d'interprétation de nos résultats, les rapportant à la transection des fibres sous-jacentes de la projection à S I.

Et, de toute façon, même admettant cette thèse, nous étions à la recherche d'une explication différente, à la fin de résoudre le problème.

L'analyse microéléctrodique de l'aire S II du chat accomplie par CARRERAS et ANDERSON (1963) aux Etats-Unis, en décelant une abondante population unitaire pourvue des caractères fonctionnels de VB, PO et mixtes nous a amené à reconnaître la probabilité que quelques-unes des secteurs de VB sont un relais pour l'aire S II; et qu'il s'agit de projections vraisemblement essentielles entre deux aires transitionnelles: l'une thalamique (s'étendant entre VB et MG), l'autre corticale (s'étendant entre S I et A I) (Fig. 4).

Les neurones thalamiques résiduels

dans les foyers de dégénérescence rétrograde sont le deuxième problème.

Bien qu'il y ait différents points de vue sur la nature de ces cellules, une conclusion neuro-anatomique convaincante du problème n'a pas été accomplie.

Nous savons que HASSLER (1958) a même supposé qu'il y a une population neuronale thalamique formée par des cellules de petite taille, de type internuncial, qui ne degenère pas après ablations corticales restreintes et qui réprésenterait la source essentielle des projections thalamo-corticales aspécifiques.

Un même noyau thalamique peut montrer une population neuronale survivante à l'hémidécortication, variable selon l'espèce animale.

La densité en cellules résiduelles augmente, par exemple, dans le noyau ventralis anterior chez les mammifères supérieurs.

Si nous nous tenons à nos expériences sur le chat, nous pouvons classer trois différents clichés réactionnels:

a) Persistence de neurons à morphologie normale, dont la taille encourage à les ranger comme neurones à projection extra-thalamique. C'est ce qui se passe au niveau du noyau ventralis anterior.

b) Persistence limitée de neurons de moindre taille, tandis que les grands neurons ont disparu (ou ils montrent, du moins, des lésions régressives sévères). C'est le cas du noyau ventro-basal.

c) Au niveau de quelques uns entre les noyaux associatifs on ne rencontre pas de survie neuronale.

L'intégration entre de différentes téchniques s'impose pour acheminer le problème à sa solution.

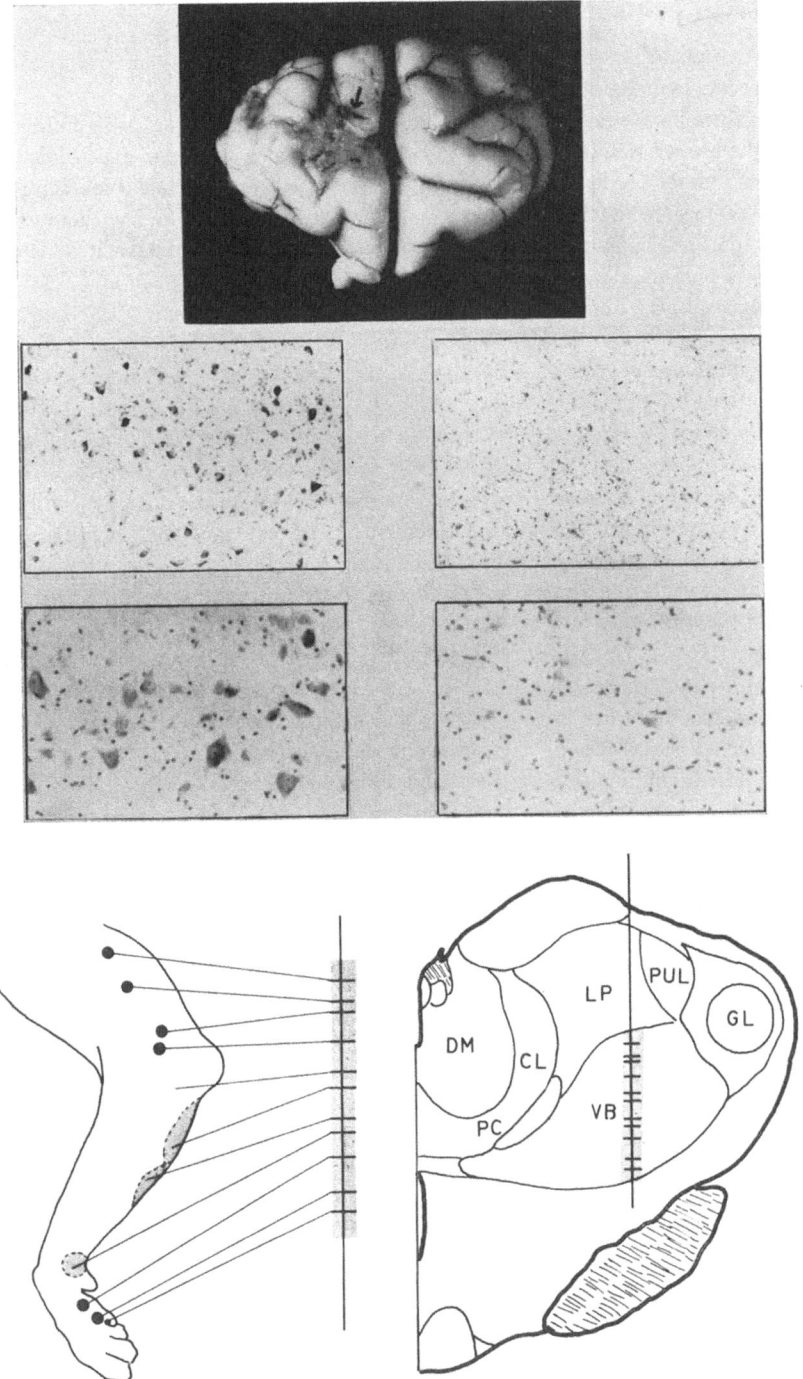

Fig. 5. Dans cette figure on observe, en haut, une lésion de la région sigmoidienne postérieure du chat (flèche). Au centre la dégénérescence rétrograde au niveau du noyau VB (à droite) et le noyau normal (à gauche) en deux différents agrandissements. En bas reconstruction d'une pénétration microélectrodique (CARRERAS) au niveau du nc. VB dégénéré. On observe la localisation des foyers périphériques activateurs du «background» présynaptique et les champs excitatoires périphériques des seules unités post-synaptiques rencontrées pendant la pénétration.

Pour cette raison, la recherche de CARRERAS et al. (1965) dans le nc. VB, porte à la fois sur les aspects hodologiques, quantitatifs et fonctionnels (Fig. 5).

Les conclusions préliminaires, posées dans ces lignes, comportent les points suivants:

1. La population résiduelle du nc. ventro-basal est constituée par des cellules de petite et moyenne taille, tandis qu'on ne peut déceler aucun grand neuron bien conservé.

2. Les propriétés fonctionnelles des neurones résiduels étudiés par l'analyse électro-physiologique unitaire sont en général superposables aux propriétés dites «lemniscales» propres à la plus part des unités du noyau VB indenne.

3. Quelques uns des neurones résiduels résultent de la connexion avec le noyau VB du coté opposé, d'où l'hypothèse de leur caractère commissural.

Un troisième problème, qui concerne les aires corticales dites «athalamiques», implique d'autres occasions de dispute.

Dans l'opinion de ROSE et WOOLSEY (1949), l'extension de ces régions dépourvues de projections thalamiques représente un trait propre aux organisation encéphaliques fort évoluées, tandis que LASHLEY (1949) en admet l'existence même dans l'encephale du rat.

Le manque de réponse définitive à ces questions ressort également de nos propres expériences sur le chat: la réalisation de diverses excisions du cortex rhinencéphalique et differemment étendues, ne nous a donné aucune dégénérescence rétrograde appreciable au niveau du thalamus.

Et, cependant, nous rappelons que NAUTA et WHITLOCK ont suivi jusqu'au rhinencéphale la dégénérescence de fibres axonales à la suite de lésion thalamiques restreintes.

Nous rattachons la disparité de ces conclusions aux limites interprétatives qu'il faut établir pour les techniques de dégénérescence rétrograde.

S'il on admet que les projections thalamiques au cortex rhinencéphalique soient peu nombreuses ou bien dérivées de plus qu'une région du thalamus, une observation histologique insuffisante est justifiée dans le tableau d'une dégénérescence rétrograde.

D'autre part, si l'on admet que les sus-dites projections thalamo-rhinencéphaliques soient des collatérales axonales, on peut paisiblement rapporter la résistance des neurones thalamiques dans les lésions rhinencéphaliques à l'intégrité du tronc axonal principal.

La constatation de l'absence de dégénérescence thalamique en cas de décortication insulaire, et les questions qui en ressortent directement, réprésentent le thème d'une recherche du Laboratoire de Parme.

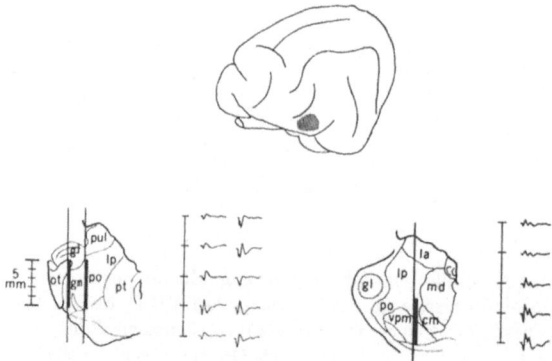

Fig. 6. Dans cette figure on voit des réponses insulaires enregistrées après stimulation du group PO-Mg et VPM. PO-MG group postérieur et corpus geniculatum mediale pars magnocellularis.

Les techniques neurophysiologiques ont décelé la possible pluralité des connexions monosynaptiques entre écorce insulaire et différents régions du thalamus (Fig. 6).

Par consequent la dissémination topographique des neurones thalamiques à projection insulaire, en d'autres termes la faible concentration cellulaire de ces relais thalamiques, pourrait infirmer bien des possibilités pour les techniques de dégénération rétrograde.

En conclusion: notre intention n'était pas de rapporter une série de notions déjà établies, mais de mettre en valeur un point de vue nouveau sur les problèmes posés pour les projéctions thalamo-corticales, et les méthodes propres à resoudre.

Summary

The results obtained by the method of retrograde degeneration are evaluated for the general aspect of the thalamo-cortical projections.

The interruption of "essential thalamo-cortical projections" to a cortical area causes a focal thalamic retrograde degeneration. In respect of the problem of "sustained projections" the surviving axon collaterals are generally considered as essential and decisive. In fact these collaterals, spared from the cortical lesion, are considered to prevent the retrograde degeneration at the thalamic level.

The method of retrograde degeneration, in many cases, may not be able to provide the positive documentation of the thalamo-cortical projections notwithstanding their real existence. It may be that the degenerated thalamic neurons are not numerous enough to give a histological evidence after localized excisions of the cortex; that may happen when a limited thalamic sector has too many projections spread out into one or several cortical areas. Moreover a cortical area may receive projections from several parts of the thalamus.

Therefore it seems to be necessary to apply combined neurophysiological and neuroanatomical methods in order to solve such particular problems corresponding to thalamo-cortical connections.

The second part of this report points out the validity of these associated neuroanatomical and neurophysiological researches.

In the S II area, for exemple, it is evident that some of the thalamo-cortical projections come from VB; moreover in some cortical fields whose ablation do not allow us to recognize the retrograde degeneration at the thalamic level, a direct projection from the thalamus may be suggested.

At last the microelectrodic inquiry allows us to establish some aspects of the surviving thalamic neurons after decortication.

Références

BODIAN, D.: Studies on the diencephalon of the Virginia opossum Part. III: thalamocortical projection. J. comp. Neurol. 77 (1942), 525–571.

CARRERAS et al.: Personal Communication 1965 a, b.

CARRERAS, M., S. A. ANDERSSON: Functional properties of neurons of anterior ectosylvian gyrus of the cat. J. Neurophysiol. 26 (1963), 100–126.

CLARK LE GROS, W. E., T. P. S. POWELL: On the thalamo-cortical connections of the general sensory cortex of Macaca. Proc. Roy. Soc. London *141* (1953), 467–487.

COMBS, C. M.: Fibres and cells degeneration in the albino rat brain after hemidecortication. J. comp. Neurol. *90* (1949), 373–394.

DIAMOND, T. I., A. J. UTLEY: Thalamic retrograde degeneration study of sensory cortex in Opossum. J. comp. Neurol. *130* (1963), 129–160.

HASSLER, R.: Functional Anatomy of the Thalamus. Ann. Vol. Physiol. Exper. Med. Sci. 1957–1958. 56–91.

LASHLEY, K. S.: Thalamo-cortical connections of the rat's brain. J. comp. Neurol. *75* (1941), 67–121.

MACCHI, G., F. ANGELERI, G. GUAZZI: Thalamo-cortical connections of the first and second somatic sensory areas in the cat. J. comp. Neurol. *11* (1959), 387–405.

MACCHI, G., M. CARRERAS, F. ANGELERI: Ricerche sulle connessioni talamo-corticali sulle proiezioni dei nuclei della linea mediana ed intralaminari; il problema delle connessioni talamo-rinencefaliche (studio sperimentale nel gatto). Arch. Anat. *60* (1955), 413–440.

METTLER, F. A.: Extensive unilateral removals in the primate: physiologic effects and resultant degeneration. J. comp. Neurol. *79* (1943), 185–245.

NAUTA, W. J. H., D. G. WHITLOCK: An anatomic analysis of the non-specific thalamic projection system. In: Brain Mechanism and Consciousness pp. 81–116. Blackwell, Oxford 1954.

PAPEZ, J. W.: Thalamic connections in a hemidecorticate dog. J. comp. Neurol. *69* (1938), 103–120.

PEACOCK, J. H., C. M. COMBS: Retrograde cell degeneration in diencephalic and other structures after hemidecortication of rhesus monkeys. Exper. Neurol. 11 (1965), 367–399.

POWELL, T. P. S.: Residual neurons in the human thalamus following hemidecortication. Brain 75 (1952), 571–584.

ROSE, J. E.: Some modern aspects of the functional anatomy of the thalamus. Mschr. Psychiat. Neurol. *120* (1950), 378–387.

ROSE, J. E., C. N. WOOLSEY: The relations of thalamic connections, cellular structure and evocable electrical activity in the auditory region of the cat. J. comp. Neurol. *91* (1949), 441–466.

ROSE, J. E., C. N. WOOLSEY: Cortical connections and functional organization of the thalamic auditory system of the cat. In: Biological and biochemical Bases of Behavior, pp. 127–149. – Univ. Wisconsin Press, Madison 1959.

WALKER, A. E.: The retrograde cell degeneration in the thalamus of Macacus rhesus following hemidecortication. J. comp. Neurol. *62* (1935), 407–419.

WALKER, A. E.: The thalamus of chimpanzee. II Its nuclear structures, normal and following hemidecortication. J. comp. Neurol. *69* (1938), 467–507.

WOOLLARD, H. H., J. A. HARPMAN: The cortical projection of the medial geniculate body. J. Neurol. Neuropsychiat. *2* (1939), 25–44.

Prof. Dr. G. MACCHI
Clinica Neurologica
Via E. del Pozzo, 95
Perugia, Italie

Größenänderungen im olfaktorischen und limbischen System während der phylogenetischen Entwicklung der Primaten

Von H. Stephan, Frankfurt

Aussagen über die phylogenetische Entwicklung einer Ordnung aufgrund vergleichend-anatomischer Untersuchungen an rezenten Arten sind besonders dann gut fundiert, wenn Stadien aus ihrer generellen Entwicklungslinie rezent noch vorhanden sind. Für die Primaten trifft dies voll zu. Sie stammen von Insektivoren ab (Remane 1961, u. a.) und die *evolutive Stufenfolge* – Insektivoren – Halbaffen – Affen – Menschenaffen – Mensch – ist relativ gut gesichert (Starck 1962). Die Insektivoren stellen somit eine gute Bezugsbasis für die Beurteilung der evolutiven Fortschritte der Primaten dar. Besonders gilt dies für die primitiven, wenig spezialisierten, rein terrestrischen Formen, die wir zur Gruppe der „*basalen" Insektivoren* zusammengefaßt haben. Bezüglich des Gehirns zeichnen sie sich durch geringste Cortexgröße bei primitivster Cortexzusammensetzung aus. Hierzu gehören Vertreter der Tenreciden, Soriciden und Erinaceiden. Im Gegensatz dazu zeigen andere Insektivoren, insbesondere die makroptischen Macroscelididen und die semiaquatilen Arten (aus verschiedenen Familien) bereits deutlich Merkmale einer Höherentwicklung. Sie sind deswegen als Basalformen ungeeignet.

In der aufsteigenden Primatenreihe hat der *Neocortex*, der die am höchsten entwickelten Hirnstrukturen enthält, eine ungewöhnlich starke Entfaltung erfahren. Die im Endhirn der basalen Insektivoren noch überwiegenden Gebiete des „Rhinencephalon" (olfaktorisch und limbisch) werden mehr und mehr zurückgedrängt und gelten bei den höheren Primaten recht allgemein als rückgebildet. Daß dies für die limbischen Strukturen nicht zutrifft, konnten wir an einem begrenzten Material bereits zeigen (Stephan u. Andy 1964). Inzwischen haben wir unser Material, insbesondere bei den Primaten, wesentlich erweitern und feinere Strukturen in die Untersuchung einbeziehen können. Die Größenänderungen in der aufsteigenden Primatenreihe werden hier dargestellt für: Bulbus olfactorius, Bulbus accessorius, Nucleus tractus olfactorii lateralis, Hippocampus und Schizocortex (Regio entorhinalis und Regio praesubicularis). Über das Septum telencephali wird zusammen mit Andy eingehender berichtet (s. S. 389).

Material und Methodik

Ermittlung möglichst arttypischer Volumina des Gehirns und seiner Teile

Vermessen wurden insgesamt 67 Gehirne aus 51 Arten, und zwar von 21 Insektivoren, 19 Prosimiern und 11 Simiern.

Basale Insektivoren: Tenrecidae: *Echinops telfairi, Hemicentetes semispinosus, Setifer setosus, Tenrec ecaudatus;* Soricidae: *Sorex minutus, Sorex araneus, Crocidura russula, Crocidura occidentalis, Suncus murinus;* Erinaceidae: *Erinaceus europaeus.*

Progressive Insektivoren: Solenodontidae: *Solenodon paradoxus;* Tenrecidea: *Nesogale talazaci, Limnogale mergulus;* Soricidae: *Neomys fodiens;* Talpidae: *Talpa europaea, Galemys* *pyrenaicus, Desmana moschata;* Chrysochloridae: *Chlorotalpa stuhlmanni;* Macroscelididae: *Elephantulus fuscipes, Rhynchocyon stuhlmanni.*

Prosimier: Tupaiidae: *Tupaia glis, Urogale everetti;* Lemuridae mit Cheirogaleinae: *Microcebus murinus, Cheirogaleus medius, Cheirogaleus major;* Lepilemurini: *Lepilemur ruficaudatus, Hapalemur simus;* Lemurini: *Lemur fulvus, Lemur variegatus;* Indridae: *Avahi laniger, Propithecus verreauxi, Indri indri;* Daubentoniidae: *Daubentonia madagascariensis;* Lorisidae: *Loris gracilis, Perodicticus potto;* Galagidae: *Galago demidovii, Galago senegalensis, Galago crassicaudatus;* Tarsiidae: *Tarsius syrichta.*

Simier: Hapalidae: *Hapale jacchus, Leontocebus oedipus;* Cebidae: *Aotes trivirgatus, Saimiri sciureus, Cebus spec., Ateles ater;* Cercopithecidae: *Colobus badius, Cercopithecus ascanius, Macaca mulatta;* Pongidae: *Pan troglodytes;* Hominidae: *Homo sapiens.*

Das Material wurde auf mehreren Reisen und Expeditionen von uns (STEPHAN, 1958, 1963; ARNOULT u. BAUCHOT, 1963) beziehungsweise von Freunden und Mitarbeitern für uns gesammelt. Fixierung mit BOUINscher Flüssigkeit auf dem Gefäßweg; Einbettung in Paraffin; Serienschnitte 10–20 µ dick; Färbung in gleichmäßigen Abständen wechselweise mit Kresylechtviolett für Zellen (Abb. 1) und mit Eisenalaun-Hämatoxylin nach HEIDENHAIN-WOELCKE für Fasern.

Von jedem Gehirn wurden 60–80 Schnitte (mit gleichem Abstand voneinander) vergrößert auf Fotopapier (extra-hart) projiziert. Die verschiedenen Strukturen wurden auf diesen Negativfotos abgegrenzt, ausgeschnitten und gewogen (= Papiergewicht). Flächengröße der Fotopapiere pro Gewichtseinheit wurden ausreichend getestet (= Papierwert). Die so ermittelten „Schnittserienvolumina"

$$\text{Volumen (mm}^3) = \frac{\text{Papiergewicht g} \cdot \text{Papierwert mm}^2/\text{g} \cdot \text{Schnittabstand mm}}{\text{Quadrat der Vergrößerung}}$$

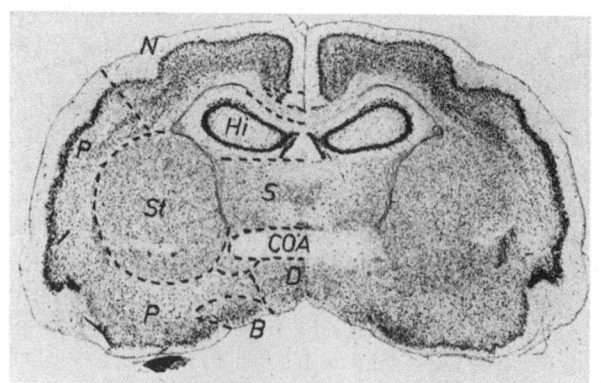

A

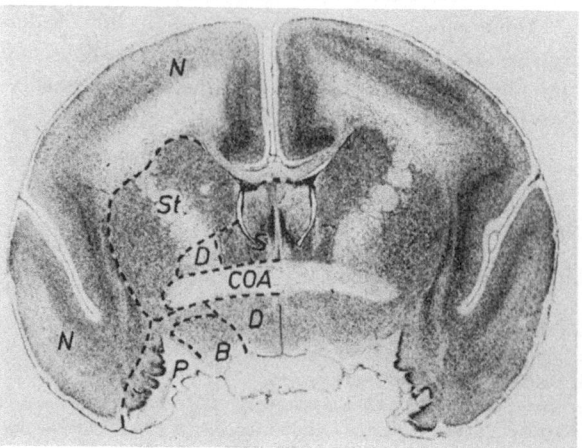

Abb. 1. Gegenüberstellung von Frontalschnitten in Höhe der vorderen Kommissur; Kresylviolett. A. *Hemicentetes semispinosus* M 156, Schnitt 569, 10 µ dick, 8×vergrößert; B. *Leontocebus oedipus* 979, Schnitt B 666, 15 µ dick, 4× vergrößert.

liegen für die Gesamthirne naturgemäß erheblich unter den Frischvolumina (Frischvolumen = Frischgewicht: spezifische Dichte des Gehirns; letztere = 1,036). Der Verlust an Hirnvolumen beträgt bei oben beschriebener Fixierung und Einbettung durchschnittlich 49,1% (42,7–54,6%), also fast die Hälfte. Die gemessenen Werte wurden nun nicht nur auf die Frischgrößen der Gehirne selbst korrigiert, sondern darüber hinaus auch auf die durchschnittliche Hirngröße der Art (Werte bei BAUCHOT und STEPHAN, 1966). Trotz der notwendigen Beschränkung der Zahl der Messungen pro Art können so möglichst arttypische Volumenwerte erhalten werden.

Vergleich der Volumenwerte

Tab. 1. *Vergleich der Hirnzusammensetzung bei niederen Insektivoren und beim Menschen (Gesamthirn = 100%).* Infolge der sehr starken Vergrößerung des Neocortex beim Menschen sinken die Relativwerte aller übrigen Strukturen (abgesehen vom Kleinhirn) z. T. sehr stark ab, obwohl (mit Ausnahme der olfaktorischen Zentren = *) keine dieser Strukturen größenmäßig rückgebildet wird. Dies läßt sich durch den allometrischen Vergleich eindeutig nachweisen.

	Echinops telfairi	Durchschnitt Basale Insektivoren	Mensch
Medulla oblongata	17,83	14,08	0,77
Kleinhirn	9,84	12,08	10,98
Mittelhirn	8,00	6,15	0,65
Zwischenhirn	7,02	7,51	2,66
Endhirn	57,31	60,18	84,94
Bulbus olfactorius *	11,49	10,70	0,01
Bulbus accessorius *	0,05	0,04	0,00
Palaeocortex * + Mandelkern	20,69	18,44	0,73
Nc. tr. olf. lat.*	0,12	0,14	0,00
Septum	1,94	1,79	0,21
Striatum	4,23	4,51	2,28
Schizocortex	1,68	3,17	0,49
Hippocampus	8,71	8,60	0,82
Neocortex	8,52	12,93	80,40

Am einfachsten ist der Vergleich von Relativwerten (prozentualer Anteil eines Teiles am Ganzen). Er ist in Tab. 1 für den primitivsten plazentalen Säuger überhaupt (Echinops telfairi von Madagaskar), für den Durchschnitt der basalen Insektivoren und für den Menschen durchgeführt worden. Wegen der sehr starken Vergrößerung des Neocortex beim Menschen werden auch solche Strukturen relativ kleiner, die absolut ganz erheblich zunehmen, wie etwa das Striatum. Der Vergleich von Relativwerten kann nichts darüber aussagen, ob eine Struktur absolut größer oder ob sie reduziert wird. Solche Aussagen sind nur beim Direktvergleich der absoluten Größen möglich. Bei einem solchen Direktvergleich muß aber die Körpergröße mit berücksichtigt werden; denn es besteht eine deutliche Beziehung zwischen Hirngröße (Hirngewicht bzw. -volumen) und Körpergröße (hier Körpergewicht). SNELL (1892) hat gefunden, daß diese Beziehung im doppelt logarithmischen Koordinatensystem linear ist. Die SNELLsche Methode, die nunmehr als *allometrische Methode* allgemein anerkannt und angewandt wird, ist nach unseren Erfahrungen bei den basalen Insektivoren (oder generell innerhalb verwandter Gruppen von Tieren mit gleicher oder ähnlicher Lebensform), nicht nur für das Gesamthirn, sondern auch für seine Teile und sogar für recht kleine Strukturen anwendbar.

Die Abhängigkeit zwischen Struktur- und Körpergröße wird bei den basalen Insektivoren durch eine *Regressionsgerade**) festgelegt (Abb. 2). An dieser läßt sich für jede beliebige Art ablesen, wie groß die in Frage stehende Struktur bei einem basalen Insektenfresser gleichen Körpergewichts sein würde. Der Vergleich der bei den progressiven Arten wirklich erreichten Größen mit diesen Basalgrößen informiert uns über die jeweilige Größenänderung.

Detaillierte Informationen werden in Form von Tabellen und Skalen gegeben. Die durch die Regressionsgrade gegebene Basis wird = 1 gesetzt. Die Skalenwerte (= Progressionsindices IP) geben an, wievielmal so groß die entsprechenden Strukturen bei den progressiven Arten bzw. systematischen Gruppen sind.

Ergebnisse

Neocortex

Von den ausgemessenen Strukturen wächst der Neocortex in der aufsteigenden Primatenreihe mit Abstand am stärksten. Wir bezeichnen diesen Prozeß als *Neocorticalisation* und sehen in ihm das zur Zeit bestmögliche quantitative Kriterium für die Entwicklungshöhe einer Art. Um beurteilen zu können, inwieweit eventuell vorhandene Trends in den Größenänderungen der untersuchten olfaktorischen und limbischen Strukturen zur Entwicklungshöhe in Beziehung stehen, soll die Rangfolge der untersuchten Gattungen bzw. höheren systematischen Gruppen der Primaten kurz dargestellt werden (Tab. 2).

Tab. 2. *Neocorticalisation der Primaten* (basale Insektivoren = 1)

Prosimier		Simier	
Tupaiidae	8,4	Hapalidae	28
Lepilemurini**)	10	Aotes	34
Cheirogaleinae	12	Colobus	40
Indridae	16	Cercopithecus	55
Galagidae	17	Cebidae	62
Lorisidae	18	Macaca	75
Tarsius	23	Ateles	79
Lemurini**)	25	Pan	84
Daubentonia	33	Homo	214

*) Der Anstieg dieser Regressionsgeraden kann für verschiedene Strukturen verschieden sein. Generell scheint der Zuwachs sinnes*ab*hängiger Zentren bei steigender Körpergröße stärker zu sein als der sinnes*unab*hängiger Zentren (s. a. STEPHAN 1961).

Bei einem von WIRZ (1950) mit der allometrischen Methode durchgeführten Vergleich wurden unterschiedliche Anstiegswerte dadurch umgangen, daß die Regressionsgerade nur für einen „Stammrest" ermittelt wurde und alle Strukturen auf die Stammrestwerte bezogen wurden. Die Resultate eines solchen Vergleichs zeigen dann, wie groß die Zuwachsraten bei den diversen Strukturen sein würden, wenn sich das Gehirn eines mittelgroßen Insektenfressers proportional, und zwar im gleichen Ausmaß wie der Stammrest vergrößern würde. Wir sind dem nicht gefolgt, weil wir der Überzeugung sind, daß die unterschiedliche Vergrößerung der verschiedenen Hirnstrukturen physiologische Ursachen hat und nicht vernachlässigt werden sollte.

Die methodischen Schwierigkeiten bringen es mit sich, daß an die Genauigkeit solcher Vergleiche keine unerfüllbaren Forderungen gestellt werden dürfen. Wir glauben jedoch, daß die von uns ermittelten Werte brauchbare Anhaltspunkte für das Ausmaß der Größenänderungen geben können, sowohl was die verschiedenen Hirnstrukturen, als auch was die verschiedenen Arten und systematischen Gruppen betrifft.

**) Bezüglich der Untergliederung der Unterfamilie Lemurinae in die beiden Tribus Lemurini und Lepilemurini s. STEPHAN u. BAUCHOT (1965).

Abb. 2. Volumina von Hirnstrukturen bezogen auf das Körpergewicht im doppelt logarithmischen Koordinatensystem. A. Bulbus olfactorius; B. Hippocampus.

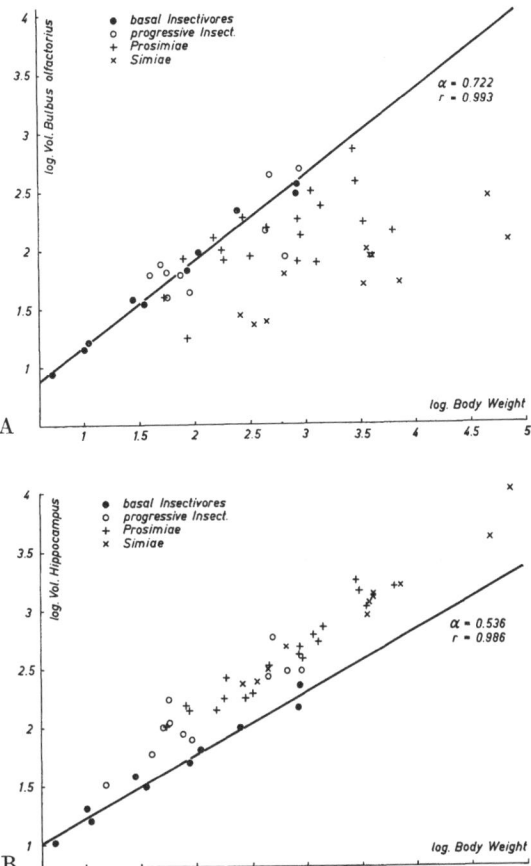

Überraschend ist der sehr hohe Wert des Prosimiers *Daubentonia*, durch den die niedrigsten Simiae, die Hapaliden, deutlich übertroffen werden. Bei den Untersuchungen über das Hirngewicht (STEPHAN u. BAUCHOT 1965; BAUCHOT u. STEPHAN 1966) deutete sich dies schon an. Innerhalb der Cebiden und Cercopitheciden bestehen sehr große Unterschiede in der Neocortexgröße. *Aotes* und *Colobus* haben sehr niedrige Werte, *Macaca* und *Ateles* hingegen sehr hohe, die fast an jene des Schimpansen heranreichen. Die Zuwachsrate vom Schimpansen zum Menschen ist sehr groß und größer als der Gesamtzuwachs von den basalen Insektivoren zum Schimpansen.

Bulbus olfactorius

(Abb. 2 A, 3). – Der Bulbus olfactorius erfährt in der aufsteigenden Primatenreihe eine deutliche Reduktion. Diese setzt in der Prosimierphase ein, doch gibt es noch Formen, wie die Tupaiiden und *Galago demidovii*, die gegenüber dem Durchschnitt der basalen Insektivoren sogar einen leichten Größenanstieg erkennen lassen. Die stärkste Reduktion unter den Halbaffen findet sich bei den Indriden. Sie ist stärker als bei dem Simier *Aotes*. Generell sind die Bulbi bei allen Simiern sehr stark reduziert, am stärksten beim Menschen, bei dem der Bulbus nur noch etwa $1/80$ so groß ist, wie bei einem basalen Insektenfresser.

Eine klare Abhängigkeit von der Entwicklungshöhe in dem Sinne, daß mit zunehmender Neocorticalisation der Bulbus kleiner wird, besteht nicht. Von den primitivsten Halbaffen haben die Tupaiiden sehr große, Lepilemurini und Indridae sehr kleine Bulbi. Der am höchsten neocorticalisierte Halbaffe *Daubentonia* hat einen sehr gut entwickelten Bulbus olfactorius, der nahezu die Größe bei basalen Insektivoren erreicht.

Bei keiner der untersuchten Formen fehlte der Bulbus olfactorius ganz. Bemerkenswert ist, daß bereits bei den Insektivoren eine Anpassung an semiaquatile Lebensweise in jedem Falle eine Reduktion des Bulbus olfactorius nach sich zieht, die beträchtliche Ausmaße erreichen kann *(Potamogale)*. Den vergleichsweise größten Bulbus hat *Rhynchocyon*.

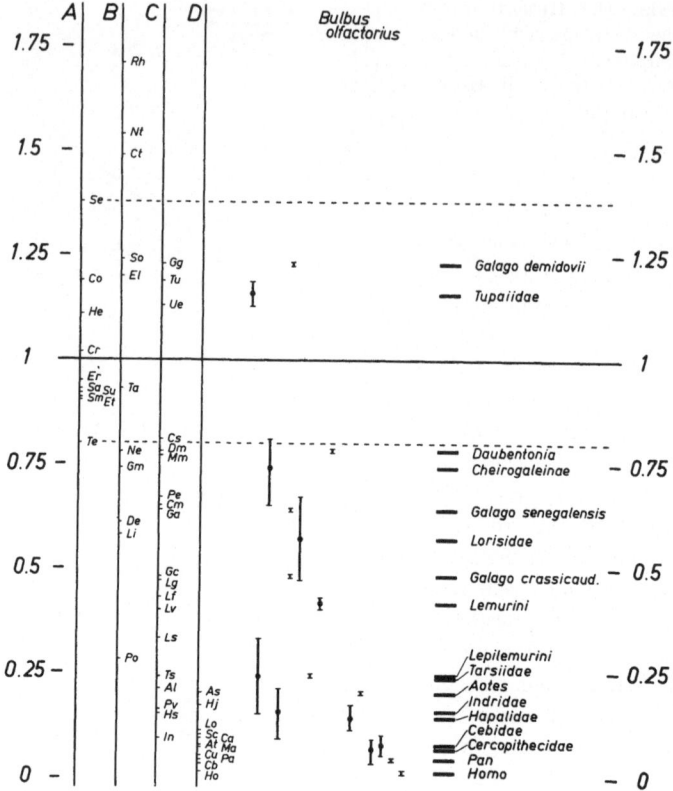

Abb. 3. Größenvergleich des Bulbus olfactorius bei verschiedenen Arten und systematischen Gruppen; links Einzelarten (A basale Insektivoren, B progressive Insektivoren, C Prosimiae, D Simiae); Mitte systematische Gruppen der Primaten in ihrer Variationsbreite, rechts Durchschnittswerte dieser systematischen Gruppen. Durchschnittliche Größe bei den basalen Insektivoren = 1.

Al	Avahi laniger	Gc	Galago crassicaudatus
As	Aotes trivirgatus	Gg	Galago demidovii
At	Ateles ater	Gm	Galemys pyrenaicus
Ca	Cercopithecus ascanius	He	Hemicentetes semispin.
Cb	Colobus badius	Hj	Hapale jacchus
Cm	Cheirogaleus major	Ho	Homo sapiens
Co	Crocidura occidentalis	Hs	Hapalemur simus
Cr	Crocidura russula	In	Indri indri
Cs	Cheirogaleus medius	Lf	Lemur fulvus
Ct	Chlorotalpa stuhlmanni	Lg	Loris gracilis
Cu	Cebus spec.	Li	Limnogale mergulus
De	Desmana moschata	Lo	Leontocebus oedipus
Dm	Daubentonia madagascar.	Ls	Lepilemur ruficaudatus
El	Elephantulus fuscipes	Lv	Lemur variegatus
Er	Erinaceus europaeus	Ma	Macaca mulatta
Et	Echinops telfairi	Mm	Microcebus murinus
Ga	Galago senegalensis	Ne	Neomys fodiens

Nt	Nesogale talazaci
Pa	Pan troglodytes
Pe	Perodicticus potto
Po	Potamogale velox
Pv	Propithecus verreauxi
Rh	Rhynchocyon stuhlmanni
Sa	Sorex araneus
Sc	Saimiri sciureus
Se	Setifer setosus
Sm	Sorex minutus
So	Solenodon paradoxus
Su	Suncus murinus
Ta	Talpa europaea
Te	Tenrec ecaudatus
Ts	Tarsius syrichta
Tu	Tupaia glis
Ue	Urogale everetti

Histologie: Die starke Reduktion bei den höheren Simiern geht einher mit einer Verwischung der Schichten und einer deutlichen Abnahme der inneren Körner. Die innere plexiforme Schicht wird hingegen breiter.

Bulbus accessorius

(Abb. 4 B). – Die Größe des Bulbus accessorius (Nebenbulbus) ist sehr variabel. Bei den progressiven Insektivoren gibt es solche mit sehr gut entwickeltem Nebenbulbus *(Elephantulus)* neben solchen mit weitgehend oder ganz fehlendem *(Solenodon)* *), und selbst bei enger verwandten Formen kann er sehr unterschiedlich ausgeprägt sein *(Elephantulus-Rhynchocyon; Galago-Arten).* Dies deutet an, daß das kleine olfaktorische Nebensystem in hohem Maße von Anpassungen und Spezialisation abhängig ist. Die Anpassung an semiaquatile Lebensweise hat jedoch keinen offensichtlichen Einfluß auf seine Größe.

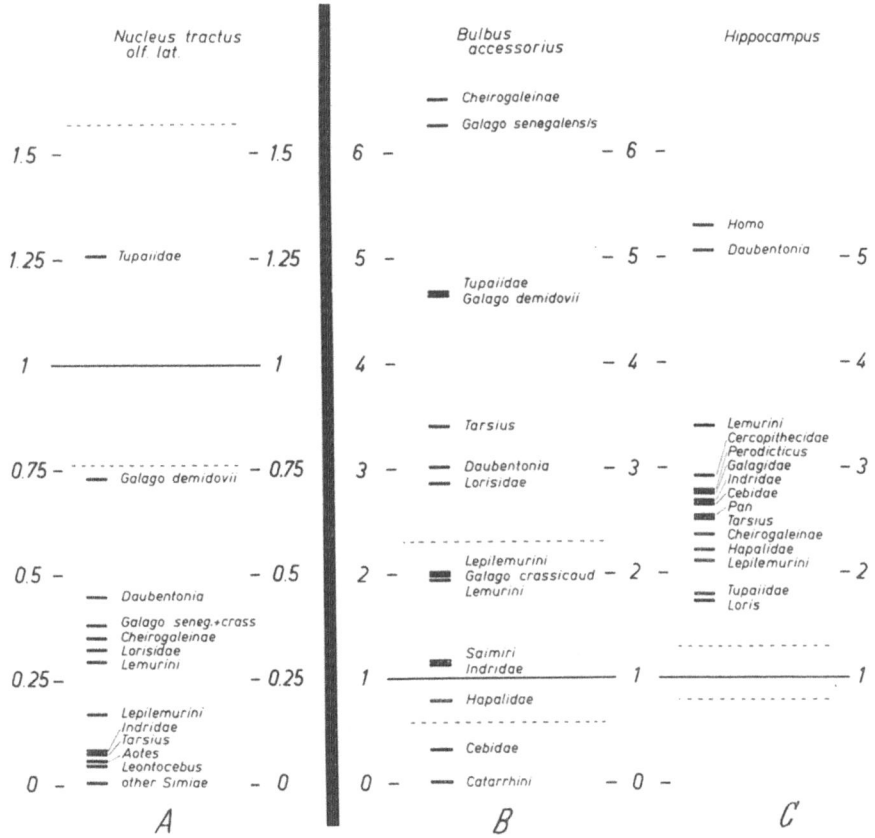

Abb. 4. Vergleich der durchschnittlichen Größenänderung von Hirnstrukturen (Nucleus tractus olfactorii lateralis, Bulbus olfactorius accessorius, Hippocampus) in den verschiedenen systematischen Gruppen der Primaten; durchschnittliche Größe bei den basalen Insektivoren = 1.

*) Wir konnten bei *Solenodon* mit unseren normalen cyto- und myeloarchitektonischen Methoden einen Nebenbulbus nicht mit Sicherheit feststellen. Möglicherweise helfen hier die von WITKAM (s. S. 282) aufgezeigten selektiven histochemischen Methoden weiter.

Bei allen Halbaffen (mit Ausnahme des Indri) ist der Nebenbulbus ebenso groß oder größer als bei den basalen Insektivoren, und er erreicht bei *Microcebus* (ca. 8 × Basalgröße) einen Höchstwert. Innerhalb der Simier fehlt er bei den Altweltaffen*); bei den Neuweltaffen ist er gering bei *Aotes* und *Cebus*, gut entwickelt hingegen bei *Ateles, Saimiri* und den Hapaliden. Es besteht also bezüglich der Größenänderung des Nebenbulbus in der aufsteigenden Primatenreihe keine einheitliche Tendenz von den basalen Insektivoren über die Prosimier zu den Simiern. Eine solche muß auch in der echten Phylogenese nicht bestanden haben. Auch hier kann einer progressiven Prosimierphase eine regressive Phase gefolgt sein, die schließlich bei den rezenten Simiern zu ganz unterschiedlicher Ausbildung der Nebenbulbi geführt hat.

Die nicht gleichsinnig gerichteten Größenänderungen in der aufsteigenden Primatenreihe dürfen also *nicht* in dem Sinne gedeutet werden, daß den rezenten Halbaffen ähnliche Formen in der Phylogenese der Simier keine Rolle gespielt haben können, weil für die echten halbäffischen Vorfahren kleinere Nebenbulbi gefordert werden müssen, damit sie sich einer einheitlichen Änderungstendenz von den Insektivoren bis zu den Simiern einpassen. Weiterhin ist es möglich, daß der Nebenbulbus bei den echten insektivoren Vorfahren der Primaten eine größere Rolle gespielt hat als bei den rezenten Insektivoren. Bei vielen Reptilien ist er ungewöhnlich stark ausgeprägt (vgl. CROSBY u. HUMPHREY 1939a, b).

Histologie: Wie beim Bulbus olfactorius geht eine starke Größenreduktion *(Desmana, Aotes)* mit einer Verwischung der Schichten und einer deutlichen Abnahme der inneren Körner einher.

Funktion: Unmittelbare Aussagen über die Funktion lassen sich unseren Messungen nicht entnehmen, wohl aber Hinweise, wo weitere Untersuchungen ansetzen können: z.B. bei einem Vergleich der olfaktorischen bzw. oralen Verhaltensweisen von Neuweltaffen (mit Nebenbulbus) und Altweltaffen (ohne Nebenbulbus), oder von enger verwandten Tieren mit sehr unterschiedlichem Nebenbulbus, wie z.B. *Elephantulus* und *Rhynchocyon.*

Größenbeziehungen zwischen Haupt- und Nebenbulbus: Solche existieren nicht. Es gibt Formen mit sehr gut entwickeltem Hauptsystem und schwachem oder fehlendem Nebensystem *(Rhynchocyon, Solenodon)* und solche mit schwach entwickeltem Hauptsystem bei gut entwickeltem Nebensystem (viele Halbaffen, *Potamogale*). Diese Größenunabhängigkeit erlaubt es uns zu prüfen, zu welchem der beiden Systeme der Nucleus tractus olfactorii lateralis engere Beziehungen hat.

Nucleus tractus olfactorii lateralis

(Abb. 4A) – Abgesehen von den Tupaiidae ist dieser Kern bei allen Halbaffen zumeist sehr deutlich reduziert. Eine Größenbeziehung zum Bulbus accessorius ist nicht vorhanden, hingegen eine deutliche zum Bulbus olfactorius (vgl. Abb. 4A mit 4B und 3). Aber bereits bei den Halbaffen zeigt dieser Kern deutlich stärkere Reduktionserscheinungen als der Hauptbulbus, was schließlich dazu führt, daß er bei den Simiern von einer gewissen Reduktionsstufe des Hauptbulbus ab nicht mehr zu erkennen ist. Möglicherweise ist er noch in Resten vorhanden. Diese sind jedoch mit unseren normalen cyto- und myeloarchitektonischen Methoden nicht sicher nachweisbar.

---

*) Wir haben ihn bei keinem Altweltaffen mit Sicherheit feststellen können. CROSBY u. HUMPHREY (1939b) und LAUER (1945) haben bei *Macaca* je ein sicheres Rudiment gefunden.

ALLISON (1953) hatte vermutet, daß der Nucleus tractus olfactorii lateralis seine Projektionen vom Nebenbulbus bekommt. Unsere vergleichenden Messungen machen dies
unwahrscheinlich; denn dieser Kern kann bei gut entwickeltem Nebenbulbus sehr gering
sein bzw. fehlen *(Saimiri, Ateles, Callithrix)* und bei fehlendem Nebenbulbus gut ausgebildet sein *(Solenodon)*. Er gehört sicherlich zum olfactorischen Hauptsystem, wird aber
in der aufsteigenden Primatenreihe stärker reduziert als dessen Primärzentrum.

Hippocampus

(Abb. 2 B und 4 C). – Gegenüber den basalen Insektivoren ist der Hippocampus bei allen
progressiven Arten größer geworden. Innerhalb der Halbaffen zeigt er eine klare Anstiegs-
tendenz mit zunehmender Neocorticalisation. Nur *Loris* und *Tarsius* haben vergleichsweise zu geringe Werte. Bei *Daubentonia* ist der Hippocampus besonders groß. Bei den
Simiern sind die Verhältnisse sehr ähnlich wie bei den Prosimiern, sowohl was die Zuwachsraten, als auch was die Tendenz zur Vergrößerung bei den höher entwickelten
Formen anbelangt (abgesehen vom Schimpansen). Eine weitere Größensteigerung gegenüber den Halbaffen besteht jedoch nicht. Ähnlich wie *Daubentonia* bei den Halbaffen
übertrifft der Mensch bei den Simiern die übrigen Formen bei weitem. Mit 10,5 cm³
Volumen (für beide Hippocampi zusammen) hat der Mensch mit Abstand den größten
Wert bei allen hier untersuchten Arten.

Bezüglich der Größenentwicklung des Hippocampus bilden die rezenten Simier keine
kontinuierliche Fortsetzung der rezenten Prosimier. Es kann wohl angenommen werden,
daß die echten halbäffischen Vorfahren der Simier einen schwächer entwickelten Hippocampus hatten, und daß sowohl die rezenten Prosimier als auch die rezenten Simier in

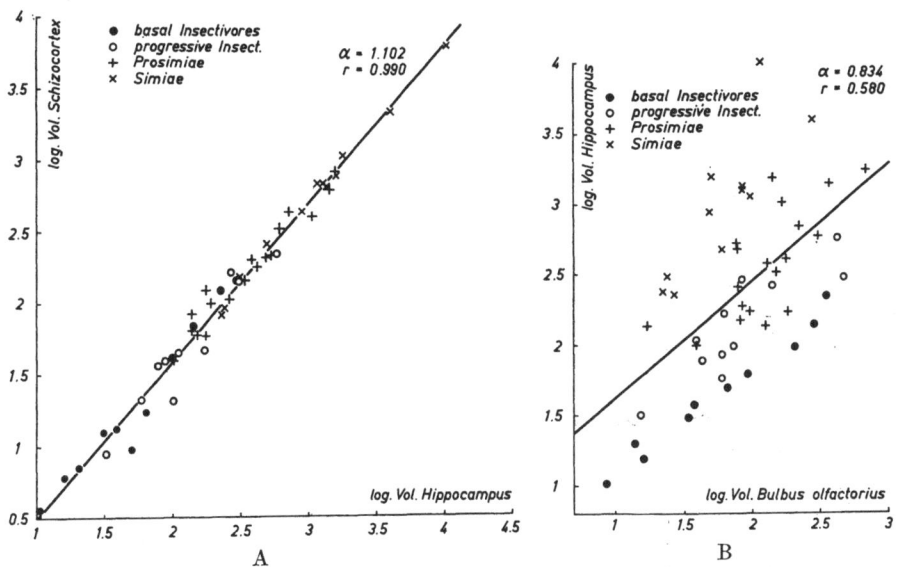

Abb. 5. Größenkorrelation zwischen Schizocortex und Hippocampus (A) und zwischen Hippocampus und Bulbus olfactorius (B). Bei A sehr geringe, bei B sehr starke Streuung um die Regressionsgerade.

ihrer Hippocampusentwicklung progressiv sind. Die weitgehend übereinstimmenden Zuwachsraten bei Prosimiern und Simiern müssen jedoch noch eingehender analysiert werden.

Nach ersten Messungen ergibt sich, daß die Zusammensetzung des Hippocampus in beiden Gruppen deutlich verschieden ist: In der aufsteigenden Primatenreihe nehmen vor allem die breiten, locker gelagerten Zellgebiete des Subiculum und des Feldes h_1 zu, während die schmalen, dichtzelligen Gebiete der Felder h_2 und h_3 eine deutliche relative Abnahme zeigen. Erstere steigen von *Setifer* mit 10% über *Crocidura* und *Galemys* (13–14%) und *Galago* (15%) bei *Cercopithecus* auf 44% an. Das dichtzellige schmale Band fällt umgekehrt von 55 auf 19% ab, wobei auch hier der Hauptabfall zwischen Prosimiern und Simiern liegt, nicht zwischen Insektivoren und Prosimiern. Der Anteil der Fascia dentata zeigt im Verhältnis zum gesamten Hippocampus leicht fallende Tendenz (22 auf 15%), im Verhältnis zum dichtzelligen schmalen Band hingegen eine relative Vergrößerung. Die bisher durchgeführten Messungen reichen für den direkten Vergleich der absoluten Werte noch nicht aus. Da der Hippocampus im Vergleich zu den basalen Insektivoren bei den Primaten größer wird, bedeutet eine relative Zunahme eines Hirnteiles eine noch stärkere Vergrößerung, eine relative Abnahme hingegen nicht zwangsläufig auch eine absolute Verkleinerung.

Schizocortex

Der Schizocortex zeigt ein dem Hippocampus ganz entsprechendes Verhalten. Die Größenkorrelationen zwischen diesen beiden Strukturen (Abb. 5A), ebenso wie zwischen dem Septum und dem Hippocampus (ANDY und STEPHAN, s. S. 397), sind sehr gut. Da bekannt ist, daß zwischen all diesen Strukturen sehr enge funktionelle Beziehungen bestehen, ist dieser Befund zugleich ein Hinweis darauf, daß aus dem Vorhandensein oder Fehlen von Größenkorrelationen Rückschlüsse auf funktionelle Beziehungen möglich sind. Dies ist von Bedeutung für den folgenden Größenvergleich zwischen olfaktorischen und limbischen Strukturen.

Vergleich der Größenentwicklung olfaktorischer und limbischer Strukturen

Vergleichen wir Bulbus olfactoricus und Hippocampus, also die beiden Hauptzentren des olfaktorischen und limbischen Systems miteinander, so ergibt sich keinerlei Übereinstimmung bezüglich ihrer Größenentwicklung in der aufsteigenden Primatenreihe. Die beiden Größenskalen sind ganz verschieden (vgl. Abb. 3 mit 4C) und im doppelt logarithmischen Koordinatensystem (Abb. 5B) zeigen die Werte eine weite Streuung um die Regressionsgrade; der Korrelationskoeffizient ist sehr niedrig (0,580).

Während das olfaktorische System von den Halbaffen ab eine stets vorhandene und sehr deutliche Rückbildung erfährt (beim Bulbus olfactorius des Menschen bis auf $1/_{80}$ der Basalgröße), wird der *Hippocampus deutlich größer* (beim Menschen auf etwa das 5fache). Wenn wir die Volumina von Bulbus olfactorius und Hippocampus bei den basalen Insektivoren mit 1:1 ansetzen, dann beträgt dieses Verhältnis beim Menschen etwa 1:400. Daraus läßt sich mit ziemlicher Sicherheit schließen, daß die limbischen Strukturen kein Teil des olfaktorischen Systems sind. Sicherlich wird das limbische System auch olfaktorisch beeinflußt, aber dieser Einfluß scheint selbst bei den primitivsten makrosmatischen Insektivoren nicht eben groß zu sein, obwohl bei diesen direkte Verbindungen über die noch relativ gut ausgebildete Pars praecommissuralis des Hippocampus am ehesten möglich sind. Bei den höheren Primaten wird die Pars praecommissuralis nahezu völlig reduziert.

Summary

The volumes of bulbus olfactorius, bulbus accessorius, nucleus tractus olfactorii lateralis, hippocampus and schizocortex (entorhinal and praesubicular cortex) were measured in 51 species of insectivores and primates and compared by the allometric method. Reference base was the size of the corresponding structures in the primitive "basal" insectivores.

The size of the neocortex (neocorticalization) gives an indication of the degree of development of the various species. It was determined whether the changes in size of the olfactory and limbic structures went parallel with the development of the neocortex.

The olfactory bulb undergoes a distinct reduction in the ascending primate scale, which is especially evident in man. However, there exists no clear reciprocal relation to the neocorticalization. This is especially clear within the prosimians, where the forms with a high neocorticalization may have distinctly larger olfactory bulbs than those with low neocorticalization *(Daubentonia* in comparison with Lepilemurini and Indridae).

The accessory bulb is well developed in all prosimians, varies greatly in the new-world monkeys and is absent or extremely small in the old-world monkeys. Its size is independent from the development of the main olfactory system.

The size of the *nucleus of the lateral olfactory tract* is related to the size of the main bulb, but not to that of the accessory bulb. However, its reduction is proportionally greater than that of the main bulb itself and therefore in many of the higher primates it can not be demonstrated by the usual histological methods.

The hippocampus is greater in all primates than in the basal insectivores. It shows in both prosimians and simians an increase in size with an increase in neocorticalization. However, the simian level does not represent a steady continuation of the prosimian level, but begins anew with relatively low values. The composition of the hippocampus changes very distinctly from the prosimians to the simians. The detailed analysis of these changes has not yet been completed.

The size of the *schizocortex* correlates very well with that of the hippocampus.

The size of the *limbic structures* is completely independent of that of the primary olfactory center (olfactory bulb). If there exists an olfactory influence on the limbic system it is probably very low in the higher mammals.

Literatur

ALLISON, A.C.: The morphology of the olfactory system in the vertebrates. Biol. Rev., *28* (1953), 196–244.

ANDY, O.J., H.STEPHAN: Phylogeny of the primate septum telencephali. In: HASSLER-STEPHAN (ed.): Evolution of the Forebrain. Thieme, Stuttgart 1966, 389–399.

ARNOULT, J., R.BAUCHOT: Compte rendu de mission a Madagascar (Octobre 1962 – Janvier 1963). Bull. Mus. Nat. Hist., Série 2, *35* (1963), 219–227.

BAUCHOT, R., H.STEPHAN: Données nouvelles sur l'encéphalisation des Insectivores et des Prosimiens. Mammalia, *30* (1966), 160–196.

CROSBY, E. C., T. HUMPHREY: A comparison of the olfactory and the accessory olfactory bulbs in certain representative vertebrates. Papers Mich. Acad. Sci. Arts Let., *24* (1939a), 95–104.

CROSBY, E. C., T. HUMPHREY: Studies of the vertebrate telencephalon. I. The nuclear configuration of the olfactory and accessory olfactory formations and of the nucleus olfactorius anterior of certain reptiles, birds and mammals. J. comp. Neurol., *71* (1939b), 121–213.

LAUER, E. W.: The nuclear pattern and fiber connections of certain basal telencephalon centers in the macaque. J. comp. Neurol. *82* (1945), 215–254.

REMANE, A.: Probleme der Systematik der Primaten. Z. wiss. Zool. (Lpzg) *165* (1961), 1–34.

SNELL, O.: Die Abhängigkeit des Hirngewichts von dem Körpergewicht und den geistigen Fähigkeiten. Arch. Psychiat. (Bln) *23* (1892), 436–446.

STARCK, D.: Die Evolution des Säugetiergehirns. S.-B. wiss. Ges. J. W. Goethe-Univ., Frankfurt, *1* (1962), 23–60.

STEPHAN, H.: Bericht über eine Belgisch-Deutsche Gemeinschaftsexpedition in Belgisch-Kongo. Mitt. Max-Planck-Ges. (1958), 96 bis 106.

STEPHAN, H.: Vergleichend-anatomische Untersuchungen an Insektivorengehirnen. V. Die quantitative Zusammensetzung der Oberflächen des Allocortex. Acta anat. *44* (1961), 12–59.

STEPHAN, H.: Forschungsreise nach Lemuria (Madagaskar). Mitt. Max-Planck-Ges. (1963), 318–333.

STEPHAN, H., O. J. ANDY: Quantitative comparisons of brain structures from Insectivore to Primates. Amer. Zool., *4* (1964), 59–74.

STEPHAN, H., R. BAUCHOT: Hirn-Körpergewichtsbeziehungen bei den Halbaffen (Prosimii). Acta Zool. *46* (1965), 209–231.

WIRZ, K.: Zur quantitativen Bestimmung der Rangordnung bei Säugetieren. Acta anat. *9* (1950), 134–196.

WITKAM, W. G. M.: Some applications of enzymo-histological techniques to the study of the maturing allocortex cerebri. In: HASSLER-STEPHAN (ed.): Evolution of the Forebrain. Thieme. Stuttgart 1966. 276–284.

Dr. HEINZ STEPHAN
Max Planck-Institut für Hirnforschung
Neuroanatomische Abteilung
Frankfurt/M.-Niederrad
Deutschordenstr. 46

Phylogeny of the Primate Septum Telencephali

By O. J. ANDY*, Jackson, Miss. and H. STEPHAN, Frankfurt

Information concerning the phylogenetic evolution of the brain can be obtained indirectly from comparative neuroanatomical studies on living species (see also contribution STEPHAN, in this volume). Primates originate from insectivores and consequently, for our quantitative studies, insectivores were utilized to provide a base line of reference.

It has been erroneously thought, the septum undergoes a marked reduction in higher primates and especially in the human in whom it should have undergone an almost total reduction and nearly have become functionless (ROSE, 1927 b). That impression has predominantly resulted from comparing the development of the septum to that of other brain structures or the brain as a whole. Such a comparison is obviously misleading especially if made in relation to the neocortex, a structure which undergoes tremendous evolutionary growth and is primarily responsible for the increased size of the brain as a whole (STEPHAN a. ANDY, 1962, 1964 a). Since some investigators associated the septal area predominantly with olfactory function, it was believed to have undergone a simultaneous reduction with the decreased need for olfaction (SMITH, 1895). In view of this commonly accepted impression and the obvious importance of the septum to the anatomical and functional integrity of the limbic system, a study of its cytoarchitectonics and phylogenetic growth was made in insectivores and primates (ANDY a. STEPHAN, 1959, 1961, 1965; STEPHAN a. ANDY, 1962, 1964a, 1964b). In the present study the material previously utilized is expanded by including chimpanzee, human and many more prosimians.

Material and Technique. Comparative quantitative studies of the total septum were done in 21 species of insectivores, 19 species of prosimians and 11 species of simians, including man (for species used and methods see STEPHAN, in this volume).

The cytoarchitectonics of the septum and volume measurements of its various components were established in the following species: *Sorex araneus* and *Crocidura russula*, belonging to the family Soricidae, one of the most primitive within the insectivores; *Galago demidovii*, a primitive prosimian; *Colobus badius* and *Cercopithecus ascanius*, representatives of typical monkeys; *Pan troglodytes* (Chimpanzee), a representative of the anthropomorphs; and *Homo sapiens* (Man), representing the highest level of primate.

Results

Boundry and configuration of the septum

The septum arises embryologically from the commissural plate and extends anteriorly a short distance into the medial wall of the hemisphere. However, it must be emphasized that this septal extension does not possess a typical cortical organization. It belongs to our nucleus septalis medialis and to the diagonal band of BROCA.

*) Supported in part by NIH Grant (0451–01).

The septum is bounded laterally by the lateral ventricle, dorsally by the corpus callosum, rostrally by the precommissural hippocampus, caudally by the anterior commissure and the third ventricle and ventrally by the nucleus accumbens and the olfactory tubercle (Figs. 1, 2). These boundaries are not all sharp and rigid. Some structures extend beyond them. For example, the diagonal band of BROCA extends ventro-caudo-laterally behind the olfactory tubercle. The bed nucleus of the stria terminalis extends into the preoptic area and anterior hypothalamic region. The nucleus of the anterior commissure is located at the rostro-ventral border of the anterior commissure.

The nucleus preopticus medianus is not included in the septum because the bulk of this nucleus is in the preoptic area and only a small projection extends dorsally along the anterior margin of the anterior commissure to the ventral portion of the septum. The subfornical organ which has been considered as a paraventricular organ and the nucleus accumbens which has at times been called the nucleus accumbens septi, also have not been included in the septum. According to ARIENS KAPPERS (1921) and BROCKHAUS (1942) the nucleus accumbens belongs to the striatum and was identified as fundus striati by BROCKHAUS.

In the lower insectivores, the septum is relatively flat and wide (Fig. 2). It progressively becomes narrower in the ascending primate scale and is especially slender and elongated in the sagittal plane in the highest primates. It appears, as if this alteration is consequent to the enormous development of the corpus callosum and its marked expansion anteriorly. The narrow dorsal component which is predominantly composed of glial cells and fiber

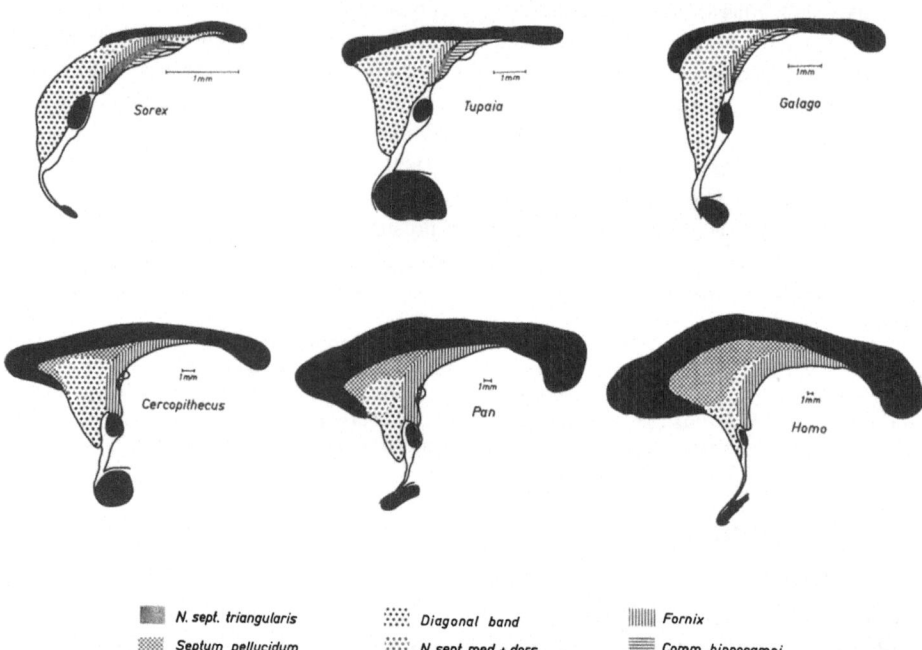

Fig. 1. Diagramatic sagittal views of the septum representing various stages in primate evolution
Note the progressively enlarging corpus callosum with its relatively marked forward expansion
and accompanying formation of the septum pellucidum in primates.

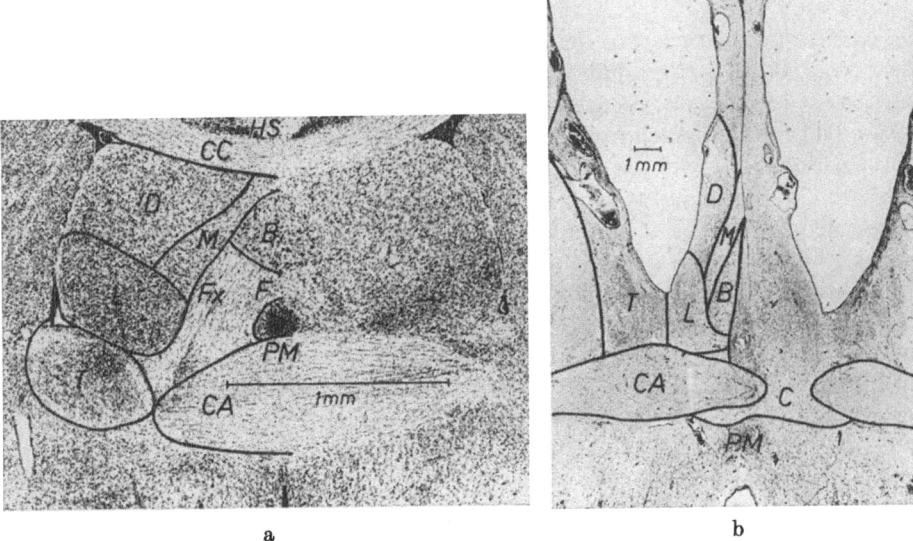

a   b

Fig. 2. Frontal view of the septum in (a) *Sorex araneus* and (b) human. Note the relatively wide septum in *Sorex* and the very narrow septum with a septum pellucidum (immediately above nucleus septalis dorsalis) in the human; cresylviolet. *Sorex araneus,* brain 461, slide 152; enlargement 28.3×. *Homo sapiens* "Allocortex", slide 1935; enlargement 3.3×.

B	Diagonal band of BROCA	Fx	Fornix
C	Bed nucleus of the anterior commissure	HS	Hippocampus, Pars supracommissuralis
CA	Commissura anterior	L	Nucleus septalis lateralis
CC	Corpus callosum	M	Nucleus septalis medialis
D	Nucleus septalis dorsalis	PM	Nucleus praeopticus medianus
F	Nucleus septalis fimbrialis	T	Bed nucleus of the stria terminalis

bundles is here identified as *septum pellucidum* (Figs. 1, 2). This structure is extensive in the higher primates, especially in man. That part of the septum which primarily contains neurons and fiber tracts is termed *"septum verum"* (true septum). Thus one may consider the septum pellucidum as providing the septum verum with a glial and fiber attachment to the corpus callosum. It is being proposed that the term *"septum telencephali"* be used to include both the septum verum and pellucidum. It is hoped this terminology will clarify the confusion that exists with the use of the terms septum and septum pellucidum. The term septum was used by BROCKHAUS (1942) to denote that part of the septum which we call septum verum, whereas his use of the term septum pellucidum corresponds to that of GASTAUT a. LAMMERS (1961) and ours. In contrast, ROSE (1927a, b) utilizes the term septum pellucidum to include both the nerve-cell containing basal component and the nerve-cell free dorsal part.

The spatial organization of the septal nuclei in higher forms is in general, similar to that in lower forms. However, as a consequence of the elongation of the total septum, there has been an accompanying change of septal nuclei from rather plump and round formations in lower forms to narrow and elongated shapes in higher forms.

Cytoarchitectonics of the septum

The various *nuclei of the septum* contain cells which are different in size, shape and density. Thus, there is in general no difficulty to differentiate and homologize the different nuclei between species. However, associated with the progressively increasing size of animals in the ascending scale, the cell sizes become larger. These tendencies are demonstrated best by comparing the cells of the diagonal band of BROCA in the various species (Fig. 3). In addition, the cell shapes tend to change from round contours to triangular and other multangular forms. This alteration in shape may in part be due to differences of time intervals between sacrificing and perfusion of the brains. Large animals, in general, necessitate more time for handling and transport before perfusion can be performed, thus permitting a longer period of postmortem cell change to take place. However, within the same brain, the cell differences existing in various nuclei appear to be influenced very little, if any, by fixation. The cells of the dorsal and lateral septal nuclei, for example, demonstrate similar criteria for their differentiation throughout the whole ascending scale (Fig. 3). The cells of the lateral nucleus are smaller, paler and rounder than those in the dorsal nucleus.

Medial group of nuclei. The septum contains a fiber rich medial region and a fiber poor lateral region. The medial group of septal nuclei, which are in the fiber-rich region, are divided into nucleus septalis medialis anterior and posterior and the nucleus of the diagonal band of BROCA. The greater part of the diagonal band extends ventrally and

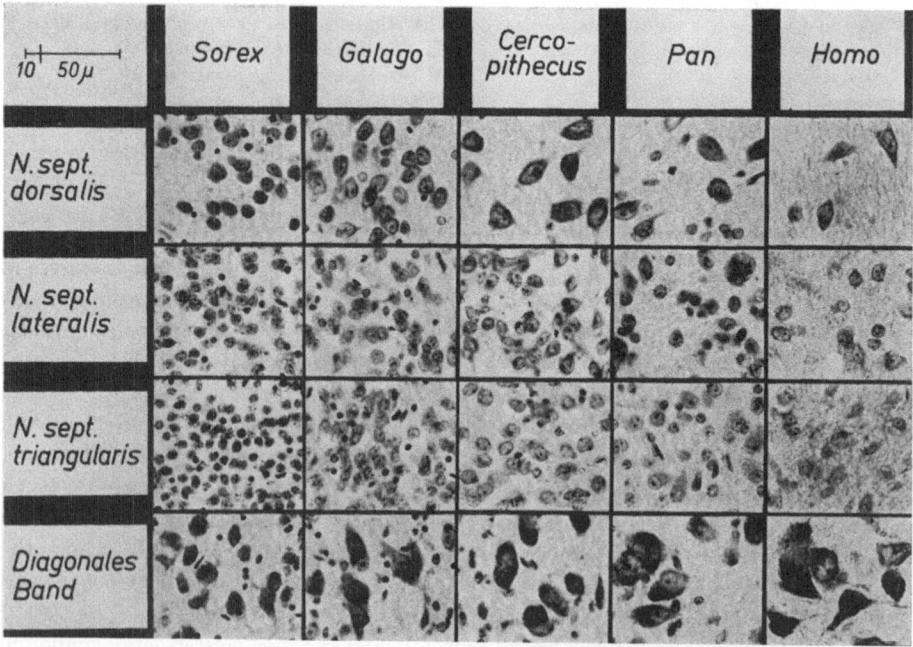

Fig. 3. Microscopic views of characteristic cells in 4 different nuclei among species representing various stages of evolution. Note that cells retain their characteristics although they progressively become larger. Enlargement 200×, cresyl-violet stain.

caudally behind the olfactory tubercle. Pars posterior of nucleus septalis medialis is sparsely populated with small spindle cells and easily differentiated from the diagonal band which contains large dark cells. Pars anterior, which is more densely populated with somewhat larger cells than pars posterior, lies rostrally between the precommissural hippocampus and the islands of CALLEJA (of the tuberculum olfactorium). In this position, the nucleus septalis medialis may provide a connection to olfactory centers.

In insectivores and lower primates the dorsal component of the diagonal band predominantly extends caudally above the level of the anterior commissure along the anterior margin of the main body of the fornix. In these lower forms the anterior border of the fornix is poorly delineated, presumably due to a relatively large amount of precommissural fibers streaming out of it. In contrast, in the human the main body of the fornix is much better delineated and precommissural fibers appear to be scant. In the human septum there are many fibers which reach the septum verum by way of the septum pellucidum. A large amount of these fibers come down from the ventral border of the anterior extension of the corpus callosum and are thought to constitute part of the fornix longus. These fibers appear to influence the orientation of the dorsal component of the diagonal band of BROCA, which is predominantly directed rostrally in contrast to lower forms, as described above.

Lateral group of nuclei. The nuclei in the lateral region have been grouped together by most other investigators. In our studies they were differentiated into a distinct ventral nucleus of thickly populated small cells and a dorsal nucleus of heterogeneous larger and, in some instances, darker cells. The cells of the lateral nucleus are smaller, paler and rounder than those in the dorsal (Fig. 3).

Caudal group of nuclei. The nuclei of this group are closely related to fiber tracts that run through or close to the posterior borders of the septum. These are the nucleus septalis fimbrialis which consists of cells distributed among the loose fibers of the fornix that enter the septum, the nucleus septalis triangularis, which contains small round cells in close relation to the ventral hippocampal commissure, the nucleus of the anterior commissure which contains sagittally oriented spindle cells and the bed nucleus of the stria terminalis. In the bed nucleus of the stria terminalis it is easy to differentiate the large cells of pars externa from the smaller cells of pars interna. The last two nuclei, which are related to the fiber tracts identified by their respective names, may be considered as not strictly belonging to the septum telencephali. However, they appear to be more closely related to this structure than to any other.

Size of the total septum in relation to brain volume

In comparing the development of the septum verum with that of the corpus callosum (Fig. 1), it appears as if the size of the septum becomes reduced during phylogeny. This also appears to be true when volumetric comparisons are made with the total brain or neocortex. In the highest primates the septum undergoes a reduction to nearly $^1/_{10}$ of that in the lowest insectivores (1.79 to 0.20%, Tab. 1). This result provides the background for accepting the common impression that the septum in man has undergone clear regression. However, this relative reduction is due to the extreme enlargement of the neocortex and related structures and such a comparison obviously provides an erroneous impression of its true or actual size change in primate evolution. This can be demonstrated by the following allometric evaluation.

Table 1. *The relative size of the septum in relation to the volume of the brain in various groups of insectivores and primates*

Basal Insectivores	1.79	(1.55–2.05)
(*Sorex,Crocidura,Suncus,Echinops,Hemicentetes,Setifer,Tenrec,Erinaceus*)		
Progressive Insectivores	1.57	(1.18–2.15)
(*Solenodon, Nesogale, Limnogale, Potamogale, Neomys, Talpa, Galemys, Desmana, Chlorotalpa, Elephantulus, Rhynchocyon*)		
Tupaiidae	1.07	(1.05–1.08)
(*Tupaia, Urogale*)		
Prosimians	0.76	(0.55–1.07)
(*Microcebus, Cheirogaleus, Lepilemur, Hapalemur, Lemur, Avahi, Propithecus, Indri, Daubentonia, Loris, Perodicticus, Galago, Tarsius*)		
Simians	0.41	(0.20–0.71)
Hapalidae	0.70	(0.69–0.71)
(*Hapale, Leontocebus*)		
Aotes	0.61	
Cebidae and Cercopithecidae	0.35	(0.31–0.40)
(*Saimiri, Cebus, Ateles, Colobus, Macaca, Cercopithecus*)		
Pan	0.22	
Homo	0.20	

Size of the total septum in relation to body-weight

Within the basal insectivores the size of the septum is closely related to the body size. This relationship can be expressed through a regression line as demonstrated in Fig. 4a (black circles). The size alterations in the various progressive species in relation to the basal insectivores are determined by the distance of the various points from the line. They are given in the form of scales in Fig. 4b. In prosimians the increase in size is up to 2.9 × that found in the basal insectivores. Higher primates have enlargement volumes from 1.6 to 2.8 × the basal forms and in man the increase is about 4 × (4.5 × with septum pellucidum, 3.8 × without SP). It is evident that the septum undergoes a definite increase in size in phylogeny. However, there is no close correlation between the development of the septum and the evolutionary level of the species, which is based on the enlargement values of the neocortex. (For further information about neocortex and limbic structures see STEPHAN, in this volume.)

Septal components in relation to the size of the total septum

We have stated that during development from insectivores through primates the septum undergoes a definite enlargement. It now is of interest to determine which parts of the septum undergo the greatest alteration in size.

Adjacent nuclei occasionally were difficult to outline because their cells intermingled at the borders. In these narrow border zones it was thus necessary to arbitrarily delineate the nuclei in order to make possible the quantitative analyses.

The relative composition of the septum telencephali undergoes clear changes (Tab. 2). Since the septum pellucidum in man contains a large amount of fiber tracts which are thought to be associated with the fimbria/fornix complex, the values of both have been combined. Thus the fimbria/fornix plus the septum pellucidum reveals a tremendous

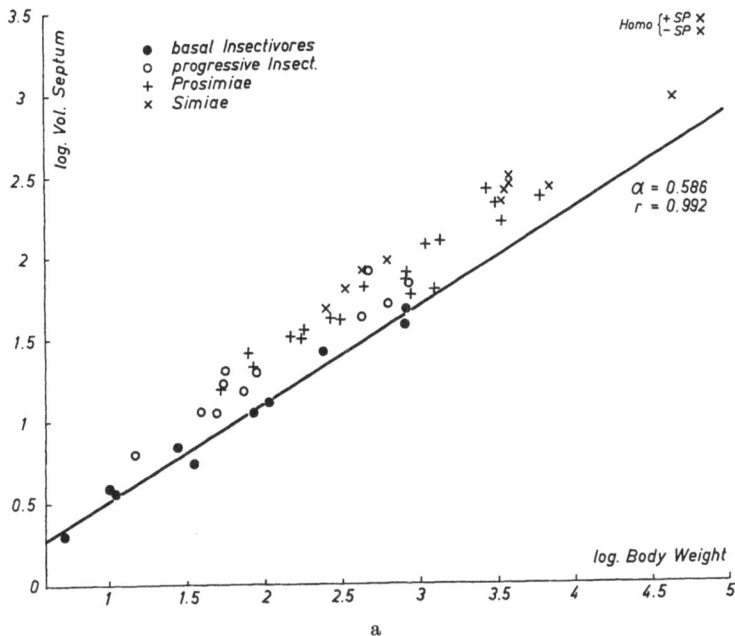

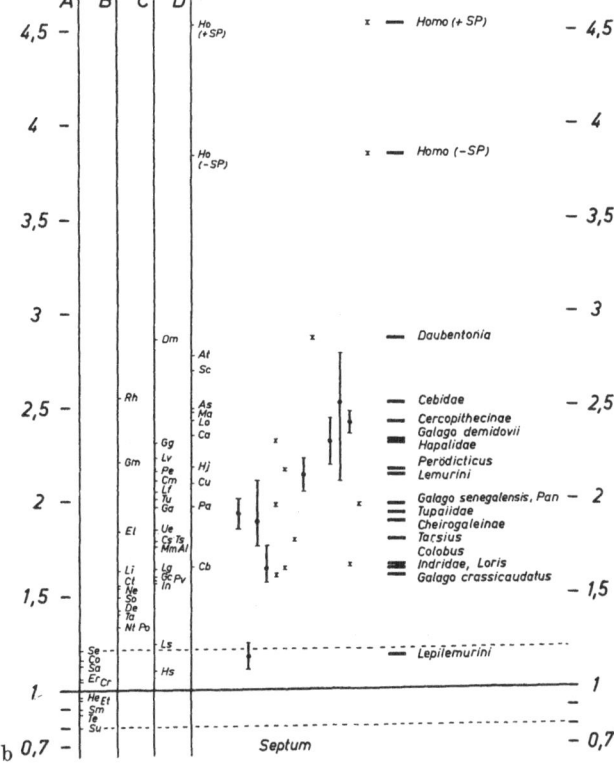

Fig. 4. a. Septum plotted against body-weight in a double logarithmic scale. The regression lines in this and Fig. 5 are constructed from the values obtained from the basal insectivores (solid circles). In human, values for the septum verum (— SP) and septum telencephali (+ SP, i.g. septum verum + septum pellucidum) are given (right upper corners of the figures). b. Volume of the septum in the various species in relation to the average of basal insectivores (= 1). On the left side the different species: A basal insectivores, B progressive insectivores, C prosimians, D simians; in the middle systematic groups of primates; on the right side averages within these groups. SP Septum pellucidum. Abbreviations of species see page 382.

enlargement (8.5% in insectivores to 30.5% in human). Even if the septum pellucidum is excluded, there is still a very definite relative enlargement (from 8.5% to 18.3%). The percentage value for the bed nucleus of the stria terminalis also becomes larger (13.2 to 19.4%). The value of the diagonal band of BROCA remains relatively the same from the lower to the higher forms. Thus the absolute enlargement of this component was equivalent to that of the total septum. The following structures becomes relatively smaller: Nc. dorsalis, Nc. lateralis, Nc. fimbrialis, Nc. triangularis and Nc. anterior commissure. However, since the total septum increases in absolute size, a relative reduction of one of its components does not necessarily signify an absolute reduction. This fact frequently is overlooked.

Table 2. *Percentage composition of the septum telencephali (total septum = 100%)*

	Nc. dorsalis	Nc. lateralis	Nc. medialis	Diagnonal band	Nc. fimbrialis	Nc. triangularis	Nc. anterior commissure	Nc. stria terminalis	Fimbria-Fornix
Basal Insectivores	20.4	8.3	10.8	29.2	3.7	3.0	2.9	13.2	8.5
Galago	17.0	13.6	8.8	31.8	3.8	1.8	2.5	12.1	8.6
Cercopithecidae	17.8	9.9	7.8	29.4	3.1	0.9	1.7	16.4	13.1
Chimpanzee	12.0	4.3	8.0	34.0	2.2	0.2	1.4	20.6	17.3
Human (without SP)	10.5	3.4	7.2	35.6	1.4	< 0.1	0.8	22.8	18.3
Human (with SP)	8.9	2.9	6.1	30.3	1.2	< 0.1	0.7	19.4	30.5*)

Septal components in relation to body-weight

Because of the inadequacies of relating septal components to that of the total septum we have determined the absolute size changes in relation to body weight by again employing the allometric method. This type of comparison demonstrates that only the nucleus septalis triangularis displayed a reduction, which was marked and well defined. The

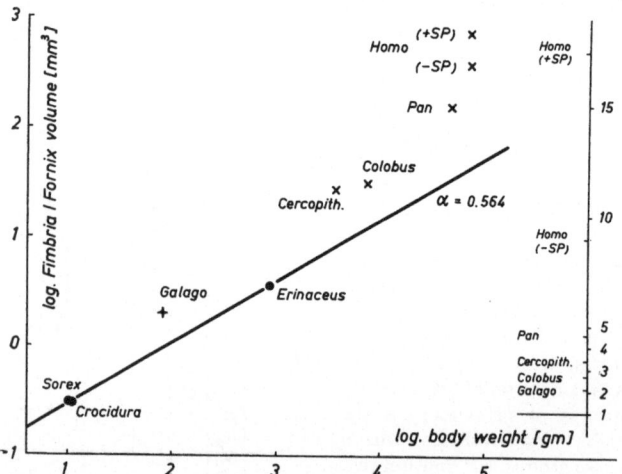

Fig. 5. Fimbria/fornix complex plotted against body-weight in a double logarithmic scale; on the right side distances of various species from the regression line in terms of considering the base line = 1.

*) In this value are included 15.6% for Fimbria/Fornix and 14.9% for Septum pellucidum.

nucleus becomes so small in the human brain, that it was difficult to identify. It is located adjacent to the ventral hippocampal commissure, a structure which also underwent a marked reduction. These combined factors tend to imply the existence of an interdependance between them.

Only a slight reduction or no change occurred in the absolute size of the nucleus septalis fimbrialis, nucleus septalis lateralis and bed nucleus of the anterior commissure. The nucleus septalis dorsalis and medialis showed a slight increase. A definite increase was present, as already mentioned, in the nucleus of the diagonal band of BROCA, the bed nucleus of the stria terminalis and the fimbria/fornix complex, which demonstrated the greatest enlargement rate up to 9 × in human (Fig. 5). The decreased size of the nucleus triangularis and ventral hippocampal commissure (psalterium ventrale) and the marked increased size of the fornix are well demonstrated in Fig. 1.

Diskussion

Based on the differentiation of the septum BROCKHAUS (1942) and others suggested, that it belongs to the phylogenetically older parts of the forebrain. This is in accord with our observations, which revealed the septum to be already very well differentiated in the lowest insectivores. It is also pertinent to note that the septum is well developed in many submammalian forms (ARIENS KAPPERS et al., 1936), and a clear nuclear differentiation has recently been described in amphibians, reptiles and marsupials (HOFFMAN, 1966; CAREY, 1966; HAMEL 1966, and CROSBY et al., 1966). Within the very early differentiation

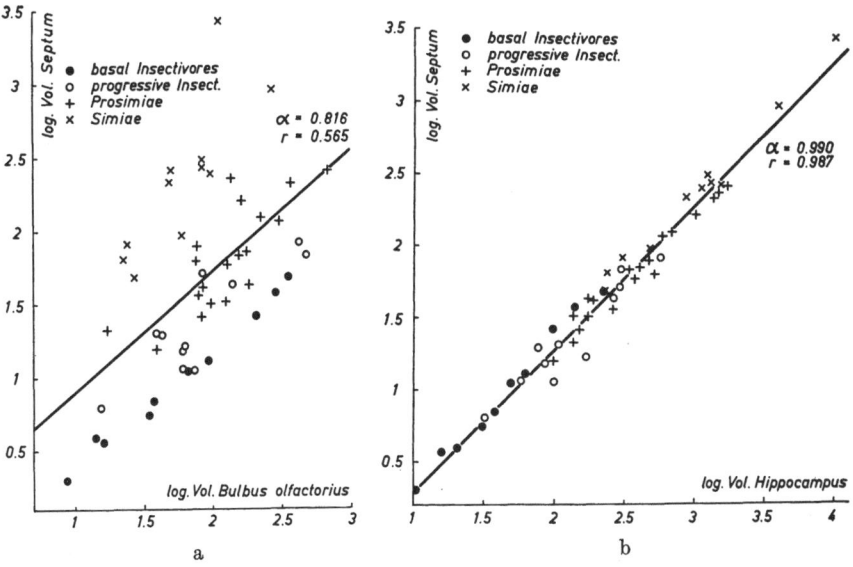

Fig. 6a. Septum plotted against olfactory bulb in a double logarithmic scale. Note the lack of size correlation (r = 0.565) between these two structures.
b. Septum plotted against hippocampus. Note the very good size correlation (r = 0.987) between these two structures.

of the septum, which remains essentially the same throughout the whole ascending primate scale up to man, there obviously existed a fundamental physiologic function, which is still not well understood. However, we can now say that its function is not predominantly concerned with olfaction, although the septum is contiguous with olfactory structures.

In primates the olfactory structures undergo a marked reduction whereas the septum undergoes a slight but distinct enlargement. The reduction of the olfactory system reveals no sign of having influenced the development of the septum. The lack of correlation between the primary olfactory center (olfactory bulb) and the septum has been well demonstrated in Fig. 6a. Thus it is obvious that the septum (at least in higher mammals) is predominantly concerned with functions other than those of olfaction. In contrast there is a very good correlation between the septum and the hippocampus (Fig. 6b), the central substrate of the so-called limbic system, to which many diverse vegetative and autonomic functions have been attributed. Furthermore, the septum and hippocampus have been demonstrated to be very closely associated physiologically and anatomically. Thus we may assume that the septum is integrated more with the limbic than with the olfactory system.

Summary

Comparative quantitative studies of the septum were done in 51 species representing the evolutionary scale from insectivores through primates, including man.

In insectivores the septum is extremely broad and entirely consists of nerve cell elements *(septum verum)*. In the ascending primate scale it becomes progressively narrower and elongated in the ventro-dorsal direction. A *septum pellucidum* begins to appear dorso-rostrally in monkeys and is largest in the human. It only contains glial cells and fibers, and serves as an attachment between the septum verum and the corpus callosum.

The term *septum telencephali* has been proposed to include both the septum verum and the septum pellucidum.

The *septal nuclei* retain characteristic features (size, shape and density of cells and fibers) which identify them throughout the ascending scale from insectivores to man. Their shapes and spatial relationships undergo slight changes. They are plump in insectivores, start to become narrow in lower primates and attain maximum narrowing in man (consequent to the altered shape of the total septum).

In comparison to basal insectivores the septum undergoes an absolute and *progressive enlargement in primate* evolution. In the human its size becomes 4 × greater than in basal insectivores.

The various components of the septum possess different growth rates. Growth rates greater than that of the septum as a whole were displayed by the fimbria/fornix complex and the bed nucleus of the stria terminalis. A marked reduction was present in the nucleus septalis triangularis and the adjacent ventral hippocampal commissure.

There is *no size correlation* between the *septum* and the *olfactory bulb*. In contrast to the septum, the olfactory bulb underwent a marked reduction. Thus it is suggested that the septum is predominantly concerned with functions other than olfaction.

However, there is a very impressive size correlation between the *septum* and the *hippocampus*, which implies their having a close functional relationship within the limbic system.

References

ANDY, O.J., H.STEPHAN: The nuclear configuration of the septum of Galago demidovii. J. comp. Neurol., *111* (1959), 503–545.

ANDY, O.J., H.STEPHAN: Septal nuclei in the Soricidae (Insectivors); Cyto-architectonic study. J. comp. Neurol., *117* (1961), 251–274.

ANDY, O.J., H.STEPHAN: Septal nuclei in primate phylogeny (a quantitative investigation). J. comp. Neurol. *126* (1966), 157–170.

ARIENS KAPPERS, C.U.: Die vergleichende Anatomie des Nervensystems der Wirbeltiere und des Menschen. II. Band. De Erven F. Bohn, Haarlem 1921.

ARIENS KAPPERS, C.U., G.C.HUBER, E.C. CROSBY: The Comparative Anatomy of the Nervous System of Vertebrates, including Man. The MacMillan Co., New York 1936.

BROCKHAUS, H.: Zur feineren Anatomie des Septum und des Striatum. J. Psychol. Neurol. *51* (1942), 1–56.

CAREY, J.H.: The nuclear pattern of the telencephalon of the black snake (Coluber constrictor constrictor). In: HASSLER-STEPHAN: The Evolution of the Forebrain. Thieme, Stuttgart 1966. 73–80.

CROSBY, E.C., B.R.DEJONGE, R.C.SCHNEIDER: Evidence for some of the trends in the phylogenetic development of the vertebrate telencephalon. In: HASSLER-STEPHAN: The Evolution of the Forebrain. Thieme, Stuttgart 1966. 117–135.

GASTAUT, H., J.H.LAMMERS: Anatomie du rhinencéphale. In: ALAJOUANINE: Les Grandes Activités du Rhinencéphale. Masson et Cie, Paris 1961.

HAMEL, E.G.: A study of the hippocampal and septal areas in the opossum. In: HASSLER-STEPHAN: The Evolution of the Forebrain. Thieme, Stuttgart 1966. 81–91.

HOFFMAN, H.H.: The hippocampal and septal formations in some anurans. In: HASSLER-STEPHAN: The Evolution of the Forebrain. Thieme, Stuttgart 1966. 61–72.

ROSE, M.: Der Allocortex bei Tier und Mensch. I. Teil. J. Psychol. Neurol. *34* (1927a), 1–111.

ROSE, M.: Die sog. Riechrinde beim Menschen und beim Affen. J. Psychol. Neurol. *34* (1927b), 261–401.

SMITH, G. ELLIOT: Morphology of the true limbic lobe, corpus callosum, septum pellucidum and fornix. J. Anat., *30* (1895), 157 bis 167; 185–205.

STEPHAN, H.: Größenänderungen im olfaktorischen und limbischen System während der phylogenetischen Entwicklung der Primaten. In: HASSLER-STEPHAN: Evolution of the Forebrain. Thieme, Stuttgart 1966. 377–388.

STEPHAN, H., O.J.ANDY: The septum (a comparative study of its size in Insectivores and Primates). J. Hirnforsch. *5* (1962), 229–244.

STEPHAN, H., O.J.ANDY: Quantitative comparisons of brain structures from Insectivores to Primates. Amer. Zool. *4* (1964a), 59–74.

STEPHAN, H., O.J.ANDY: Cytoarchitectonics of the septal nuclei in old world monkeys (Cercopithecus and Colobus). J. Hirnforsch. *7* (1964b), 1–23.

ORLANDO J. ANDY, M. D.
Department of Neurosurgery
University of Mississippi
Medical Center
Jackson 6, Miss., USA

Dr. HEINZ STEPHAN
Max-Planck-Institut für Hirnforschung
Neuroanatomische Abteilung
Deutschordenstr. 46
6 Frankfurt-Niederrad
Deutschland

A Supplementary Motor Pattern in the Precommissural Septotubercular Area of Macaca mulatta

By J. A. TAREN, Ann Arbor, Mich.

A long-held view that the brain was composed of centers that rigidly subserved one function has gradually been eroded by the demonstration of the numerous supplementary motor patterns. Therefore, it was not with any great surprise that we evoked motor responses from the region of the precommissural septum and tuberculum olfactorium of *Macaca mulatta*. During the course of further experimentation, it became apparent that there was a pattern of motor behavior elicitable by stimulation of the septotubercular area (Fig. 1), and that these responses were contralateral and appeared to be associated with disagreeable behavior.

Material and Method. Seven specimens of *Macaca mulatta* were used. In all cases the area to be stimulated was directly visualized through an appropriate surgical exposure under general anesthesia. A right parasagittal frontal craniectomy with resection of the medial portion of the right frontal lobe, including the right septotubercular area and the right tuberculum olfactorium, permitted visualization of the left septotubercular area and the left tuberculum olfactorium.

Using a GRASS stimulator (S 4 A), monopolar stimulation with a rectangular pulse of 44 pps., stimulus duration of 10 milliseconds, and amplitude of 4 volts was then applied to these areas. The indifferent electrode was rectally positioned in the anesthetized animals or inserted into the left frontal bone by a trephine for future stimulation in the awake animal. In those animals used for chronic stimulation, a fine wire electrode ($\#$ 36 insulated silver wire with an exposed tip of 2 mm.) was implanted and brought out through skin.

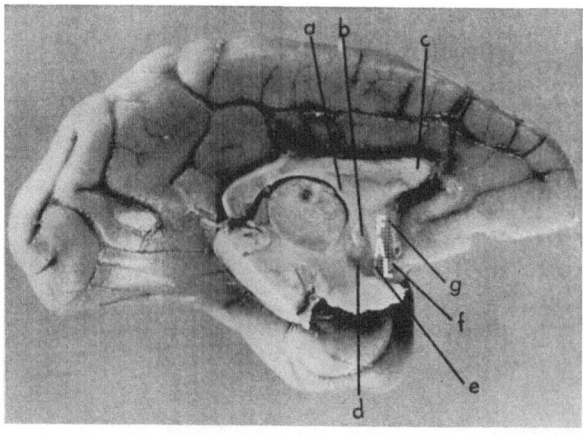

Fig. 1. A sagittal section of a normal brain *(Macaca mulatta)* to illustrate the motor pattern of response obtained by stimulating the region of precommissural septum (g) and the tuberculum olfactorium (f). The strippling indicates the region of stimulation from which face (F), arm (A), and leg (L) movements were elicited. a, fornix; b, anterior commissure; c, genu of the corpus callosum; d, third ventricle; e, lamina terminalis.

Results

Contralateral movements of the face, the arm, and the leg were obtained in all cases (Fig. 2). The responses were obtained with greater ease in the unanesthetized or in the lightly anesthetized animal. The sacrifice of the animals within three weeks of the procedure permitted histologic verification of the electrode positions (Fig. 3).

Fig. 2 A1.   Fig. 2 A2.

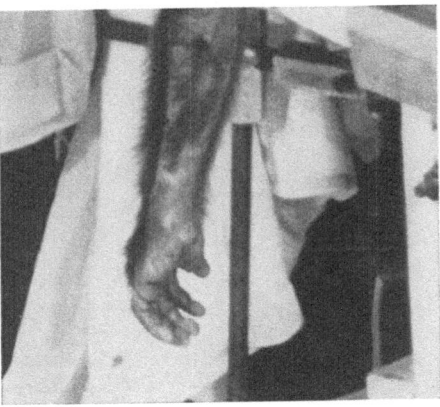 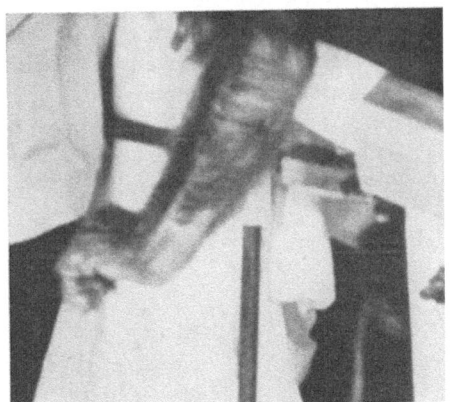

Fig. 2 B1.   Fig. 2 B2.

Fig. 2. Photographs made during stimulation. A. an unanesthetized *Macaca mulatta* with an electrode in the superior precommissural septal region on the left side. 1. prestimulation. 2. during stimulation. Note the closing of the right eye and the turning of the head to the left. B. Photographs from a film recording stimulation of the anesthetized animal. The electrode is in the precommissural septotubercular region in the "arm" area. 1. prestimulation state. 2. Note the movement of the upper right extremity during stimulation.

Fig. 3. Cross section of a MARCHI preparation of a Macaque brain at the level of the genu of the corpus callosum. The arrow marks the position of penetration of the electrode, the tip of which extended slightly deeper into the left septotubercular area. The MARCHI degeneration in the right side of the genu of the corpus callosum and in the right precommissural septal region and right tuberculum olfactorium is the result of resection of the right frontal lobe to gain exposure.

Pathways for motor response. Once it was ascertained that there was a reproducible pattern for motor response, we attempted to delineate the pathways involved. We found that fornicotomy, section of the medial forebrain bundle, and cingulotomy would not ablate the evoked response despite doing these procedures unilaterally, bilaterally, singly, or in various combinations. Left temporal lobectomy also failed to eliminate the response but, when the right temporal cortex was removed as well as the left temporal lobe cortex, a response could not be obtained. It should be noted that the amygdala on the right was spared, which is important since it is possible to evoke facial movement from amygdala. Thus, it would appear that amygdala was not involved in this pattern; this is also suggested by the presence of lower extremity movements which suggest temporal lobe cortex.

Therefore, *both temporal lobe cortices* must be involved in this pathway and presumably are connected to the septotubercular area through the anterior commissure. Fig. 4 depicts a well known supplementary motor pattern on the temporal lobe of the monkey. The descending pathways from temporal lobe cortex to facial nuclei are equally well known. There is a direct corticotegmental (temporotegmental) pathway (the evidence for this is largely physiologic, WALL a. DAVIS), as well as a pathway from temporal lobe to basal ganglia and to the tegmentum of the midbrain by the lenticular fasciculus. Relay from the tegmentum in the midbrain to the appropriate motor nuclei can occur over the medial longitudinal fasciculus and the tegmentobulbar pathway.

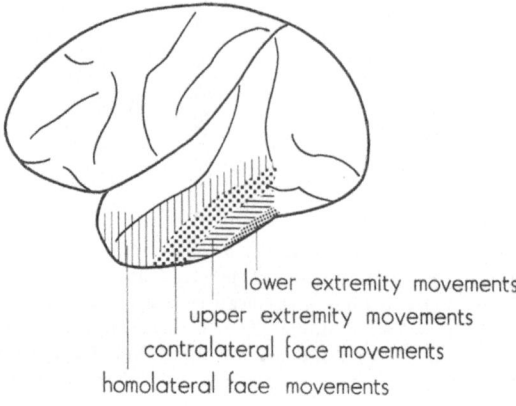

lower extremity movements
upper extremity movements
contralateral face movements
homolateral face movements

Fig. 4. A diagram taken from CROSBY et al., based partly on experimental results of SCHNEIDER, CROSBY and DE JONGE (unpublished), illustrating supplementary motor pattern on the temporal lobe of monkey. Note that the region from which homolateral face movements can be evoked is more superior to the region from which contralateral face movements can be evoked.

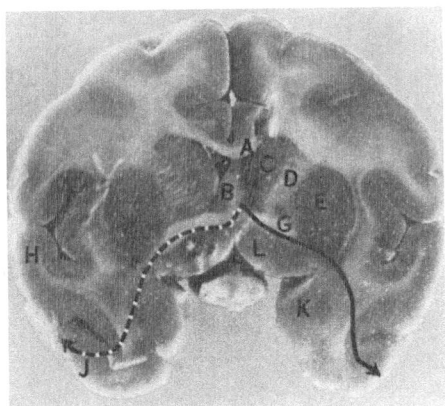

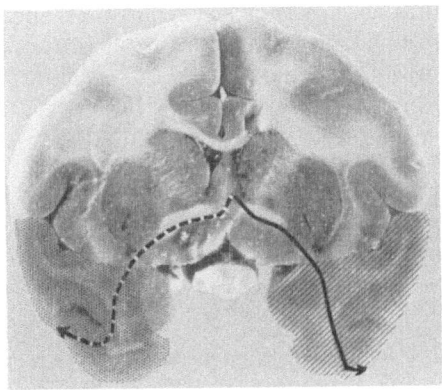

Fig. 5.   Fig. 6.

Fig. 5. A diagram of a cross-section of a Macaque brain to illustrate the suggested extralimbic connections from the septotubercular region to the temporal lobe cortices. A. corpus callosum. B. precommissural septum. C. head of the caudate nucleus. D. anterior limb in the internal capsula. E. region of the basal ganglia. G. anterior commissure. H. superior temporal gyrus. J. middle temporal gyrus. K. amygdala. L. diagonal band of BROCA. The solid black arrow outlines the pathway from the precommissural septal region on the left to the inferior portion of the middle temporal gyrus on the left, the region where contralateral face movements can be evoked. The dotted arrow represents the pathway from the left precommissural septal region through the anterior commissure crossing to the superior portion of the right middle temporal gyrus, the region from which homolateral face movements have been obtained. Stimulation of the left precommissural septal region would then be expected to produce right-sided face movements mediated by impulses travelling over both of these pathways.

Fig. 6. A similar cross-section of a Macaque brain at the level of the anterior commissure to illustrate how bilateral temporal lobectomy prevents right facial movements after stimulation of the left precommissural septal region. The black arrow is the pathway from the left precommissural septal region to the inferior portion of the left middle temporal gyrus. The diagonally shaded area represents the extent of removal of left temporal lobe. The broken arrow is the pathway from the left precommissural septal region over the anterior commissure to the superior portion of the right middle temporal gyrus. The stippled area of shading represents extent of removal of the right temporal lobe, sparing amygdala.

Fig. 5 postulates the extralimbic connections from the septotubercular region to the temporal lobe cortex. Fig. 6 illustrates how bitemporal lobectomies interrupt the pathway, thus preventing the motor response.

Chronic stimulation of this region in a patient who had electrodes implanted during medically-indicated frontal lobotomies corroborates this work and, more importantly, indicates that these evoked responses are unconscious or on the verge of consciousness and appear to be associated with disagreeable behavior. This work will be published elsewhere.

The purpose of supplementary motor pathways, of which this must be one, is to supplement the highly voluntary movements and provide the accompanying movements that are on the verge of consciousness. The appearance of *highly disagreeable behavior* in our human cases suggests that the hypothalamus may be set off as well with the appearance of parasympathetic and sympathetic effects such as erection of the hair (SPIEGEL et al.). It would appear that the motor-pattern described may be a supplementary system for displaying the emotional accompaniments of disagreeable behavior. The face has a greater area of representation than the extremities in the human precommissural septotubercular area. This is in accord with the readily observable fact that human emotions tend to be displayed by facial movements rather than the more primitive and violent movements of arms and legs.

Summary

A definite reproducible motor pattern of response was evoked from the septotubercular area of *Macaca mulatta*. Contralateral face, arm, and leg movements were observed. The responses tended to be those motor movements that are often associated with disagreeable behavior. The pathway for the response appears to be from septotubercular area to both temporal lobe cortices through the anterior commissure and then over well known routes to the appropriate lower motor neurons.

Acknowledgements

This work would not have been possible without Dr. ELIZABETH C. CROSBY. I also wish to acknowledge the assistance of Dr. B. R. DEJONGE. The work was supported in part by a contribution from The Kresge Foundation to the Edgar A. Kahn Neurosurgical Fund, # 33280.

References

CROSBY, E.C., T. HUMPHREY, E.W. LAUER: Correlative anatomy of the nervous system. The Macmillan Co., New York 1962.

SCHNEIDER, R.C., E.C. CROSBY: Stimulation of "second" motor areas in the macaque temporal lobe. Neurology *4* (1962), 612–622.

SPIEGEL, E.A., H.R. MILLER, M.J. OPPENHEIMER: Forebrain and rage reactions. J. Neurophysiol. *3* (1940), 538–548.

WALL, P., G.P. DAVIS: Three cerebral cortical systems affecting autonomic functions. J. Neurophysiol. *14* (1951), 507–517.

JAMES A. TAREN, M. D.
Section of Neurosurgery
Department of Surgery
University of Michigan Medical Center
Ann Arbor, Michigan, USA

Maturation post-natale de l'aire visuelle du cortex cérébral chez le Chat

Par R. Marty et R. Pujol, Montpellier*

Les recherches consacrées à la maturation des systèmes afférents de l'écorce cérébrale, tant sur le plan anatomique que sur le plan fonctionnel, offrent un intérêt qui déborde largement le cadre strict du développement du système nerveux. En effet, dans un grand nombre de cas, elles permettent de mieux comprendre l'organisation et le fonctionnement des systèmes sensoriels de l'animal adulte. La recherche systématique de corrélations anatomo-fonctionnelles au cours du développement post-natal du système nerveux des Mammifères a ainsi abouti à quelques publications récentes (Purpura et al., 1960, 1964; Marty, 1962; Marty et Scherrer, 1964; Scheibel et Scheibel, 1964). Mais ces recherches corrélatives conduisent elles-mêmes à approfondir les investigations, en particulier dans le domaine structural (Voeller et al., 1963). L'étude résumée ci-dessous se propose de préciser certains aspects de la maturation post-natale de l'aire visuelle du cortex cérébral chez le Chat et de compléter ainsi les informations obtenues dans une série de recherches antérieures sur le système visuel (Marty, 1962). Basée principalement sur la méthode de Golgi-Cox, elle est à rapprocher de celle consacrée par Noback et Purpura (1961) au développement post-natal du néocortex cérébral du Chat considéré dans son ensemble.

Matériel et Méthodes. a) Ce travail repose sur l'emploi de méthodes d'imprégnation métallique du tissu nerveux et, accessoirement, de la méthode de Nissl. Les observations ont porté exclusivement sur les cellules nerveuses de l'écorce cérébrale et de quelques structures sous-jacentes. Elles ont été effectuées chez 46 chats répartis de la façon suivante:

Méthode de Golgi-Cox: 15 chats âgés de quelques heures à deux semaines, et 6 chats âgés de 1 mois ½ à 4 mois, que l'on peut considérer comme «adultes» avec certaines restrictions qui seront exposées plus loin.

Variantes des procédés classiques de Golgi ou de Cajal: 12 chats.

Méthode de Nissl (thionine ou crésyl violet): 13 chats.

Dans les deux derniers groupes, l'âge a été choisi en fonction des problèmes posés par les résultats obtenus par la méthode de Golgi-Cox.

b) L'aire visuelle de l'écorce cérébrale du Chat faisant l'objet de cette étude a été délimitée selon les critères de Winkler et Potter (1914) et d'Otsuka et Hassler (1962). Ainsi définie, elle correspond dans l'ensemble à l'aire visuelle «primaire» des neurophysiologistes classiques.

Résultats

Dans un travail précédent (Marty 1962), l'étude de la maturation corticale avait été centrée sur le développement post-natal des cellules pyramidales de l'aire visuelle du Lapin et du Chat. Le problème est considéré ici d'un point de vue stratigraphique par

*) Travail effectué avec le concours du C.N.R.S. (R.C.P. n⁰ 75), en collaboration avec le Centre de Recherches Neurophysiologiques de l'INSERM à Paris et avec l'aide technique de P. Sibleyras

l'analyse de la maturation neuronique aux différents niveaux de l'écorce visuelle du Chat,
en conservant dans la mesure du possible les repères classiques des «couches».

Nous envisagerons d'abord la maturation post-natale de l'écorce d'un point de vue
stratigraphique avant de confronter son développement avec celui d'autres structures,
corticales et sous-corticales.

Etude stratigraphique de la maturation post-natale de l'écorce visuelle

Etude d'ensemble. L'absence de différence appréciable dans la maturation des trois
subdivisions classiques de l'aire visuelle (aires 17, 18 et 19) nous a incité à faire une description globale du développement post-natal dé cette aire.

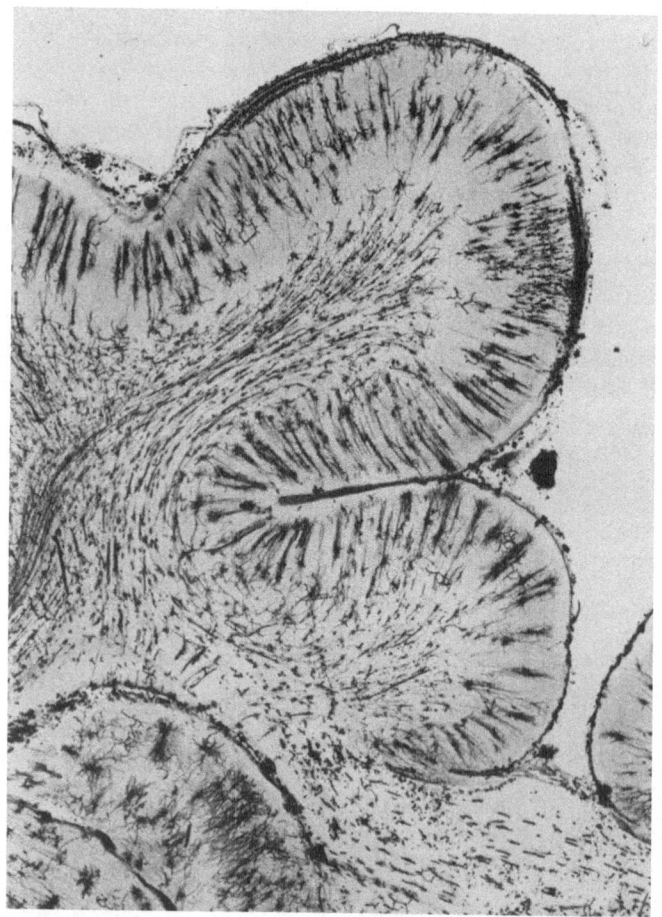

Fig. 1. Aire visuelle de l'écorce cérébrale du Chat dans la période néo-natale.
Chat âgé de 2 jours. L'aire visuelle est limitée en dehors par le sillon latéral, en dedans et en bas
par le sillon splénial. Le sillon supra-splénial n'est pas encore apparent à cet âge. La corne d'Ammon
et la fascia dentata sont visibles en bas et à gauche. Méthode de Golgi-Cox. Gr.: ×9.

Période néo-natale (chats âgés de quelques heures à 3 jours): Sur la plupart des préparations obtenues à cet âge, les couches supérieures de l'écorce apparaissent mieux imprégnées que les couches profondes, phénomène qui tient peut être à la plus grande densité cellulaire à ces niveaux (Fig. 1). De nombreuses cellules présentent les signes de maturation corticale devenus classiques depuis les travaux de RAMÓN Y CAJAL (1911) et de LORENTE DE Nó (1933), en particulier le développement précoce de la dendrite apicale qui atteint la couche moléculaire avant même que l'arborisation basilaire ne se soit nettement individualisée. Toutefois, si l'on compare la maturation dans les différentes couches mis à part la couche moléculaire, on s'aperçoit que les neurones les plus développés sont les cellules pyramidales profondes qui présentent un soma bien différencié et une arborisation basilaire déjà bien développée (Fig. 2, e et f). En dessous de ces grandes cellules, principalement au niveau des sillons, on remarque des neurones moins évolués et orientés en tous sens.

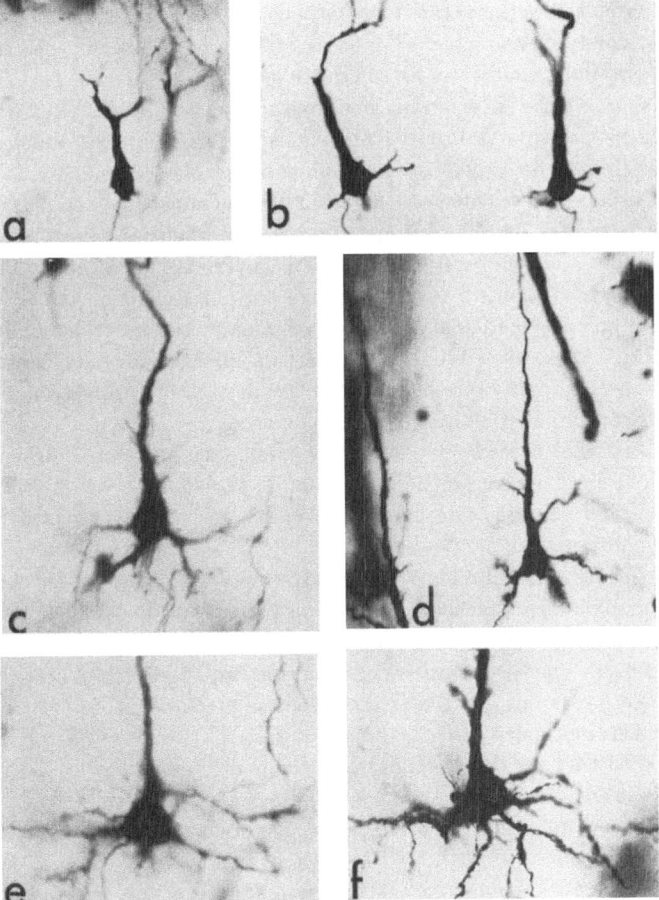

Fig. 2. Aire visuelle de l'écorce cérébrale du Chat dans la période néo-natale.
(a) et (b): chats âgés de 48 heures. Neurones situés au voisinage immédiat de la couche moléculaire.
(c) et (d): chats âgés de 10 heures. Neurones de la partie moyenne de l'écorce.
(e) et (f): chats âgés de 48 heures. Cellules pyramidales profondes. Méthode de GOLGI-COX. Gr.: × 140.

Dans les couches supérieures, où la densité cellulaire est plus forte, les neurones pyramidaux ont un soma encore fusiforme avec des dendrites basilaires plus courtes et peu ramifiées (Fig. 2, c et d). Dans la zone corticale immédiatement sous-jacente à la couche moléculaire, on distingue des neurones très particuliers caractérisés par leur soma arrondi, leur dendrite apicale souvent bifurquée après un court trajet vertical et l'absence de ramifications basilaires (Fig. 2, a et b). Les cellules stellaires, à cet âge, sont très peu ou pas imprégnées.

Fin de la première semaine (chats âgés de 6 à 8 jours): Dans les différentes couches, les cellules subissent quelques changements. Le soma se rapproche de sa forme adulte, les ramifications des dendrites apicales se multiplient surtout dans la zone sous-jacente à la couche moléculaire, enfin le plexus dendritique basilaire se développe à tous les niveaux. Il en résulte une diminution de la densité cellulaire dans les couches supérieures, et une distribution cellulaire plus homogène sur toute la hauteur de l'écorce.

Deuxième semaine néo-natale (chats âgés de 9 à 14 jours): Le processus que nous venons de décrire se poursuit. A la limite inférieure de la couche moléculaire, les dendrites apicales s'épanouissent, donnant des images en pinceau («brush»). Ce phénomène est bien visible dès le 10ème jour, en particulier au niveau des sillons. Simultanément les grandes cellules pyramidales de la couche V achèvent leur maturation somatique (Fig. 3a). Dès cette période, leur soma présente sa forme définitive et parfois même ses dimensions adultes, soit 30 microns environ de largeur basale et de hauteur. Mais, à ce stade, la prolifération des dendrites basilaires de ces mêmes cellules n'a pas encore atteint son maximum. Quand aux cellules pyramidales des couches moyennes et superficielles, elles présentent encore un retard global dans leur maturation. A la partie supérieure de la couche II, on note des neurones de forme triangulaire à base supérieure portant à chaque extrémité une épaisse dendrite apicale qui pénètre dans la couche moléculaire par un trajet oblique. Au pôle inférieur du corps cellulaire, l'axone se détache en direction des couches sous-jacentes (Fig. 3, b et c). Ces neurones représentent peut être des cellules primitivement incluses en totalité dans la couche moléculaire (Fig. 6 c).

Dans les deux étapes précédentes, les cellules stellaires mises en évidence étaient à la fois peu nombreuses et mal imprégnées. A la fin de la deuxième semaine néo-natale, on en trouve un certain nombre, déjà bien développées, en particulier à la partie moyenne de l'écorce (Fig. 3 d).

Age «adulte» (chats âgés de 1 mois ½ à 4 mois): Nous avons considéré les chats de cet âge comme des animaux adultes, bien que la croissance pondérale du cerveau et probablement aussi, dans une certaine mesure, la croissance en hauteur de l'écorce se poursuivent au-delà du 4ème mois (le chat est considéré comme adulte à l'âge de 12 mois). Cette attitude nous a été dictée par le fait que toutes les descriptions classiques de l'écorce visuelle du Chat basées sur les méthodes de GOLGI (RAMÓN Y CAJAL, 1922; O'LEARY, 1941; SHOLL, 1955) ont été effectuées sur des chats âgés de moins de 1 mois. Fait à souligner d'autre part, ces descriptions ont été généralement limitées à la partie inter-hémisphérique de l'aire visuelle, dans l'intention délibérée de restreindre les observations à l'area striata (aire 17).

L'examen de nos préparations dans cette région aboutit à une description comparable d'un point de vue analytique à celle des auteurs précités. On retrouve ainsi, à la partie supérieure de l'écorce, la couche moléculaire dans laquelle les dendrites apicales des neurones pyramidaux de tous niveaux pénètrent plus ou moins profondément, mais où se trouvent encore des cellules nerveuses, de type étoilé pour la plupart. Dans les couches sous-jacentes, les cellules pyramidales de toutes tailles et les cellules étoilées apparaissent avec netteté. Dans la couche polymorphe, de très nombreux neurones pyramidaux se trouvent en position oblique, horizontale ou même renversée (Fig. 4).

Une analyse détaillée de cet ordre, étendue aux autres territoires de l'aire visuelle (c'est-à-dire avant tout à la partie externe du gyrus latéral), ne fait pas apparaître de distinction bien systématisable par rapport à la description précédente, si ce n'est une différence dans l'épaisseur totale de l'écorce (l'épaisseur moyenne de l'écorce visuelle, à

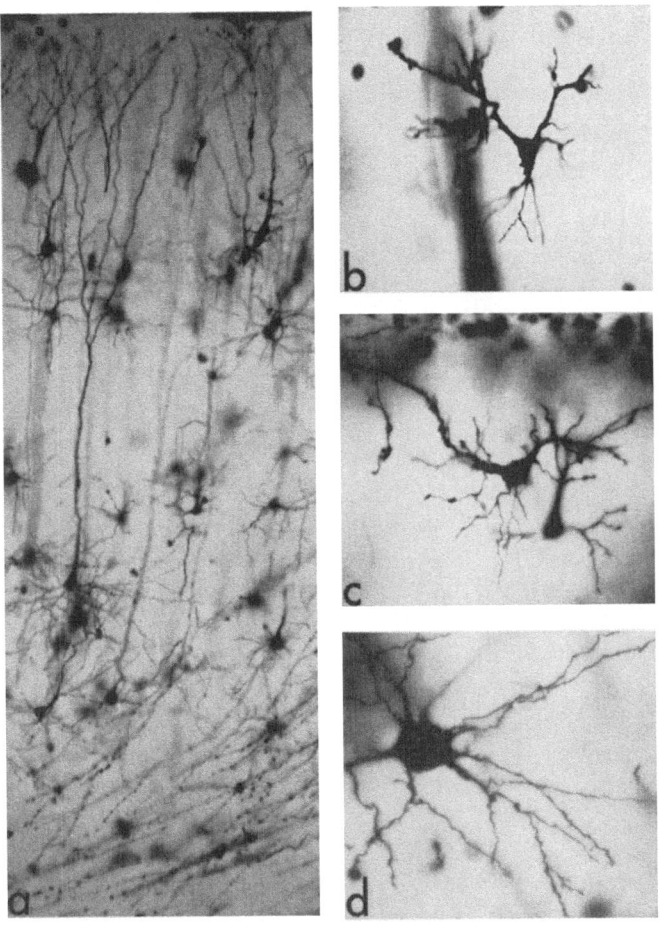

Fig. 3. Maturation de l'aire visuelle corticale au cours de la 2ème semaine néo-natale.
(a): chats âgés de 10 jours. Epaisseur totale de l'écorce au niveau du gyrus splénial. La maturation est déjà très avancée dans toutes les couches de l'écorce. Noter en particulier l'arborisation basilaire de la grande cellule pyramidale profonde et l'épanouissement («brush») des dendrites apicales au niveau de la couche moléculaire. Gr.: ×50
(b), (c), (d): chats âgés de 14 jours. En (b) et (c), cellules triangulaires situées à la jonction des couches I et II (comparer avec la fig. 6c). En (c), on aperçoit également un neurone typique de la couche II. En (d), cellule étoilée de la partie moyenne de l'écorce. Gr.: ×115(b) et (c) et×180(d).
Méthode de GOLGI-COX.

l'état frais ou sur les préparations de GOLGI-COX, est de 1,9 mm à la partie interhémisphérique et de 2,15 mm à la partie externe du gyrus latéral). Cette constatation nous a incité à effectuer l'étude synoptique de l'aire visuelle dans son ensemble, non seulement sur des préparations bien imprégnées, mais aussi sur leurs agrandissements photographiques. Compte tenu du fait que la méthode de GOLGI-COX ne permet pas une étude détaillée des plexus afférents, il nous apparaît que la structure cyto-architectonique de l'aire visuelle «primaire» est relativement homogène dans son ensemble (Fig. 5).

Résultats complémentaires. Nous envisagerons dans cette rubrique un certain nombre de résultats complémentaires concernant les axones, les épines dendritiques, les fibres afférentes et la couche moléculaire.

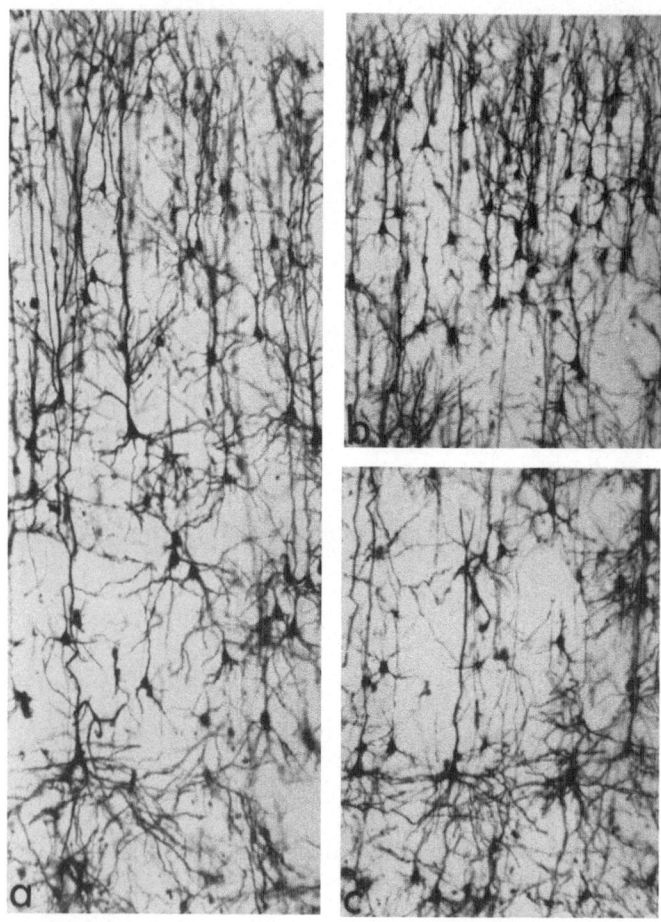

Fig. 4. Architectonie de l'aire visuelle corticale. Chat âgé de 41 jours.
(a): épaisseur totale de l'écorce visuelle à la partie haute du gyrus splénial. Gr. ×40.
(b) et (c): sommet du gyrus latéral. Couches supérieures (b) et profondes (c) de l'écorce visuelle.
Gr.: ×35. Méthode de GOLGI-COX.

Les axones: Contrairement à une opinion couramment répandue, l'imprégnation des axones sur les préparations de GOLGI-COX est assez bonne (Figs. 2, 7). Cependant une étude systématique de ces éléments n'a pas été entreprise ici.

Les épines dendritiques: Les épines dendritiques et les gemmules, qui en constituent une variété morphologique, sont considérées par GRAY (cf. 1964) comme témoignant systématiquement d'un contact synaptique. Elles s'observent dès les premiers stades sur les dendrites des cellules de tous niveaux, sauf peut-être dans la couche moléculaire où leur identification est malaisée.

Les fibres afférentes: La méthode de GOLGI-COX ne permet qu'une appréciation indirecte des fibres afférentes. Aussi, malgré quelques imprégnations complémentaires réussies avec la méthode de CAJAL et sur lesquelles la strie de GENNARI, entre autres, apparaît avec netteté, nous n'avons pas abordé plus longuement ce problème qui fait l'objet d'un travail en cours.

La couche moléculaire: Encore appelée plexiforme, cette couche a retenu particulièrement notre attention. Chez le jeune chat (âgé de quelques heures à 3 jours), elle est assez

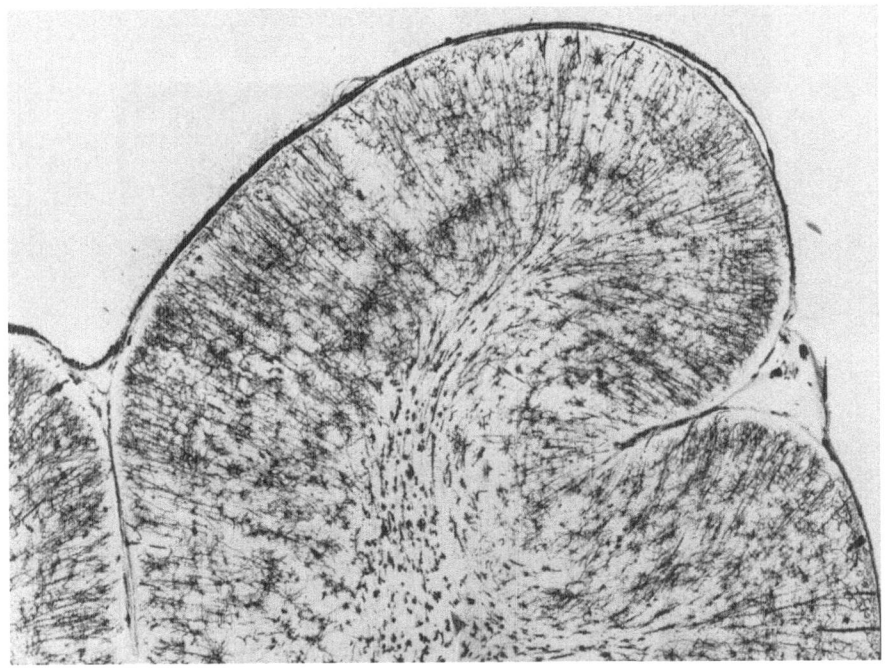

Fig. 5. Panorama de l'aire visuelle du cortex cérébral.
Chat âgé de 41 jours. On aperçoit à gauche le sillon latéral, à droite le sillon supra-splénial. Ce panorama a été obtenu par la juxtaposition de 9 microphotographies. Méthode de GOLGI-COX.
Gr.: ×8.

riche en cellules nerveuses. Ce fait est encore plus net sur les imprégnations argentiques.
Sur les préparations de GOLGI-COX, on reconnaît la plupart des types cellulaires décrits
par RAMÓN Y CAJAL (1929), en particulier les cellules horizontales typiques (Fig. 6a), les
cellules dont le soma se trouve accolé à la surface piale (Fig. 6b) et les neurones de forme
étoilée identifiables aux cellules à cylindre-axe court de CAJAL. On y voit en outre des
cellules de morphologie différente occupant toute la hauteur de la couche (Fig. 6, d et e)
et des cellules triangulaires à base supérieure (Fig. 6c). Au cours de l'évolution post-
natale, les éléments cellulaires de la couche moléculaire se raréfient, sans cependant
disparaître totalement. On trouve ainsi quelques cellules, surtout de type étoilé, chez des
chats de 1 et 2 mois (Fig. 6f).

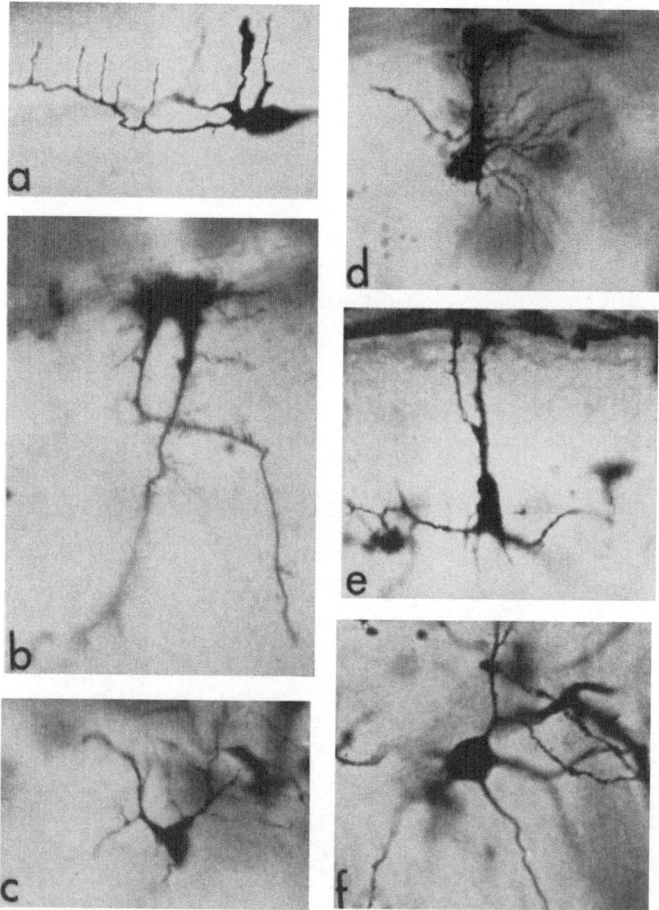

Fig. 6. Couche moléculaire de l'aire visuelle corticale.
(a), (b), (c), (d), (e): chat âgé de 2 jours. On aperçoit divers types cellulaires dont certains ont une
morphologie classique (a) et (b).
(f): chat âgé de 28 jours. Cellule de type étoilé. Méthode de GOLGI-COX. Gr.: ×225.

Comparaison de la maturation de l'aire visuelle avec celle de structures voisines, corticales et sous-corticales

Au niveau de l'écorce. Il existe à la naissance des différences évolutives entre le degré de maturation de l'aire visuelle et celui d'autres territoires néo-corticaux, en particulier le cortex sensitivo-moteur (Marty, 1962). Une différence apparaît également lorsqu'on compare à cette époque le développement de l'aire visuelle avec celui d'un territoire archicortical, l'aire cingulaire, qui s'étend chez le Chat tout le long du bord inférieur du sillon splénial. Le développement dendritique y est en effet notablement plus avancé (Fig. 7 a).

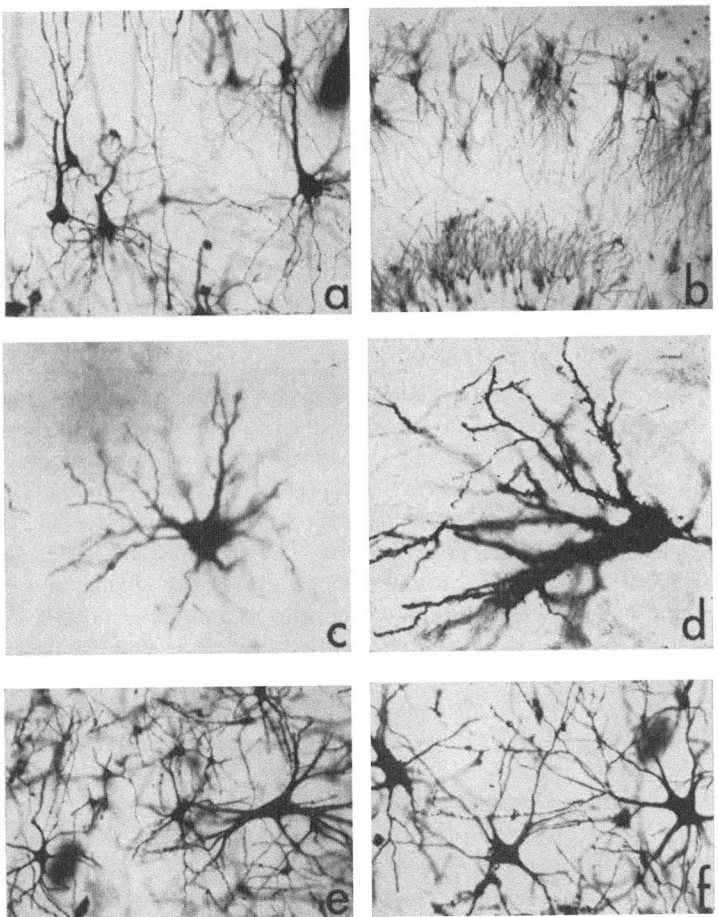

Fig. 7. Archicortex et grands relais visuels sous-corticaux dans la période néo-natale.
(a) et (b): chats âgés de 3 jours. (a): grandes cellules profondes du gyrus cingulaire. Gr.: × 40.
(b): corne d'Ammon et fascia dentata. Gr.: ×20.
(c) et (d): neurones du noyau dorsal du C.G.L., à 3 jours (c) et 7 jours (d). Gr. :×135.
(e) et (f): neurones du stratum intermediale du colliculus superior, à 3 jours (e) et 41 jours (f).
Gr.: ×30.
Méthode de Golgi-Cox pour toutes les figures, sauf pour la figure (e) (méthode de Golgi-Hortega).

Grands relais visuels sous-corticaux. Nous avons limité nos recherches au C. G. L. et au colliculus superior, tout en notant que cette dernière structure, au sens strict du terme, n'est pas une formation prosencéphalique.

1. Noyau dorsal du C. G. L.: Son imprégnation par la méthode de GOLGI-COX étant souvent excessive, nous avons complété nos informations par quelques imprégnations argentiques. Dès les premiers jours, le champ dendritique et la densité des épines ne laissent aucun doute sur l'antériorité de la maturation des cellules du noyau dorsal du C. G. L. par rapport à celles de l'écorce visuelle (Fig. 7, c et d).

2. Colliculus superior: L'imprégnation métallique de cette structure s'est montrée excellente dans la plupart des cas. Dès la naissance, le colliculus superior présente des cellules dont la disposition et la structure sont très proches de celles de l'adulte (Fig. 7, e et f). C'est ainsi que les grandes cellules de la couche intermédiaire, dont le plus grand diamètre chez l'adulte est de l'ordre de 50 microns, atteignent déjà à cet âge une quarantaine de microns.

Discussion

Cinétique de la maturation dans l'aire visuelle corticale du Chat

D'une manière générale, dans la première semaine post-natale, l'imprégnation métallique des neurones est plus riche dans les zones superficielles que dans les parties basses du cortex cérébral. L'étude de l'écorce sur toute sa hauteur révèle cependant que, mis à part les neurones de la couche moléculaire, ce sont les cellules pyramidales profondes qui présentent le degré le plus avancé de développement. L'antériorité de la maturation des cellules pyramidales profondes sur celle des autres neurones corticaux est une notion apparemment bien établie (RAMÓN Y CAJAL, 1911; LORENTE DE NÓ, 1933; CONEL, cf. 1952; RABINOWICZ, 1964) qui concorde dans une certaine mesure avec les résultats des explorations autoradiographiques de l'écorce dans la période périnatale, effectuées récemment chez la Souris (ANGEVINE et SIDMAN, 1961) et chez le Rat (BERRY et al., 1964). Ces travaux viennent en effet de montrer qu'au cours des migrations cellulaires constitutives du cortex cérébral, les couches inférieures étaient mises en place avant les couches moyennes et superficielles, contrairement à la description classique de TILNEY (TILNEY et KUBIE, 1931). Il peut donc paraître logique que la différenciation cellulaire débute à la partie basse de l'écorce pour s'étendre ensuite en hauteur, en direction de la surface.

En fait, cette appréciation globale de la cinétique de la maturation nécessite d'importants correctifs intéressant au moins trois cas particuliers.

La couche moléculaire. – A la naissance, cette couche est riche en éléments cellulaires variés et en fibres nerveuses. Des contacts synaptiques s'y organisent précocement (VOELLER et al., 1963). Au cours des semaines suivantes, les neurones de la couche moléculaire subissent apparemment peu de modifications morphologiques, mais on assiste à leur raréfaction progressive. L'évolution de cette couche est de ce fait considérée comme régressive. Le problème est probablement plus complexe. En effet, si certains types cellulaires disparaissent au cours de l'évolution post-natale, d'autres variétés de neurones persistent à l'âge adulte.

La couche inférieure de l'écorce. – Dans la période néo-natale, la partie basse de l'écorce est remarquable par sa faible densité cellulaire. Ce fait est encore plus évident sur les colorations de Nissl que sur les imprégnations de GOLGI-COX. Cette partie basse de l'écorce peut elle-même être subdivisée en 2 étages. L'étage supérieur est caractérisé par la présence des grands neurones pyramidaux dont la précocité de maturation explique en partie la faible densité cellulaire observable à ce niveau. L'étage inférieur représente la future couche polymorphe. Les caractéristiques de la population cellulaire qu'on y observe à la naissance sont difficiles à systématiser, et la faible densité neuronique de cette couche dépend sans doute de facteurs complexes. En effet, les neurones y sont moins évolués que les grandes cellules pyramidales sus-jacentes et la plupart d'entre eux ont une orientation atypique. Des recherches complémentaires sont nécessaires pour préciser l'évolution de cet étage au cours des dernières étapes de la croissance en hauteur de l'écorce.

Les cellules étoilées ou stellaires. – Ces neurones ne peuvent être mis en évidence avec certitude qu'au cours de la deuxième semaine néo-natale. Leur imprégnation par la méthode de GOLGI-COX est peut-être moins sélective que celle des cellules pyramidales, mais leur différenciation est probablement plus tardive. Utilisant également la méthode de GOLGI-COX, MITRA (1955) a ainsi observé, chez le Lapin, le Chat et le Singe, que le pourcentage des neurones stellaires par rapport à l'ensemble de la population neuronique de l'écorce visuelle n'atteignait sa valeur définitive que progressivement. C'est ainsi que, chez le Lapin, le nombre des cellules stellaires représente à l'âge adulte 31 p. 100 des neurones corticaux; dans la période néo-natale, ce pourcentage est notablement inférieur, soit 13 p. 100 à l'âge de 7 jours et 23 p. 100 à 17 jours. De toutes façons, l'identification des cellules stellaires est parfois délicate et leur discrimination avec des cellules pyramidales coupées tangentiellement peut être malaisée; l'absence ou la rareté des épines dendritiques orientent la distinction en faveur des cellules stellaires (RAMÓN-MOLINER, 1961), mais ce critère distinctif peut semble-t-il prêter à discussion.

Chronologie de la maturation de l'aire visuelle corticale

L'inégalité de développement de l'écorce cérébrale à la naissance dans les diverses espèces animales a déjà été soulignée (RIESE, 1944; NOBACK et PURPURA, 1961; MARTY, 1962). Ce phénomène est en rapport direct avec la durée propre de la gestation dans chaque espèce, alors que la vitesse de la maturation post-natale du cortex dépend probablement en partie de la durée totale de la vie. Ce phénomène, observable à la fois sur le plan structural et le plan fonctionnel, apparaît nettement lorsque l'on compare par exemple l'évolution post-natale du fonctionnement cortical chez le Chat et chez le Lapin (MARTY, 1962).

L'inégalité de la maturation dans les diverses aires corticales à la naissance est un autre aspect chronologique du développement cérébral. CONEL (cf. 1952) puis RABINOWICZ (1964) y ont insisté chez l'Homme. Nous avons pour notre part mis ce décalage en évidence chez le Lapin et chez le Chat en établissant une hiérarchie chronologique de l'organisation des systèmes afférents du néocortex (MARTY, 1962). Dans cette hiérarchie, la somesthésie occupe la première place et la vision la dernière. Nous ne reviendrons pas sur cette discrimination, pour souligner maintenant le décalage de la maturation de l'aire visuelle par rapport à d'autres territoires.

Comparaison de la maturation dans l'aire visuelle et l'aire cingulaire. L'aire cingulaire fait partie du cortex limbique, défini par ses connexions avec les noyaux thalamiques antérieurs et latéral dorsal (LOCKE et al., 1964). Au point de vue embryologique, elle appartient à l'archicortex «au sens large» (FEREMUTSCH, 1952). A la naissance, on l'a vu, les neurones de l'aire cingulaire présentent un degré de maturation plus avancé que ceux de l'aire visuelle. Ce fait est à rapprocher du haut degré de développement observé dans l'archicortex proprement dit (corne d'Ammon et fascia dentata) à la naissance (MARTY, 1962 et Fig. 7b). Il apparaît ainsi que deux structures phylogénétiquement plus anciennes que le néocortex présentent une plus grande rapidité d'évolution que celui-ci. Le fait est net pour l'archicortex proprement dit; il est plus nuancé pour l'aire limbique. Le caractère transitionnel de cette structure est ainsi mis en évidence. Des observations analogues ont été faites chez le Lapin (CHEVREAU et MARTY, 1962).

Comparaison de la maturation dans l'aire visuelle corticale et dans les grands relais sous-corticaux du système visuel. *Noyau dorsal du corps genouillé latéral.* – Malgré une imprégnation souvent excessive, les neurones du noyau dorsal du C. G. L., à la naissance, apparaissent plus développés que ceux de l'écorce visuelle. Dans cette structure diencéphalique, les champs dendritiques sont très denses et les épines dendritiques nombreuses. Précisons que les neurones susceptibles d'appartenir au noyau réticulaire thalamique, dont la différenciation est antérieure à celle des noyaux du thalamus dorsal (ROSE, 1942), on été dans la mesure du possible écartés de nos observations. Nos informations ne sont cependant pas suffisantes pour décrire l'évolution cellulaire dans cette structure. Récemment, TUNG et LU (1964) chez le Lapin ont signalé que le C. G. L. et l'écorce visuelle atteignaient leur maturité dans des délais comparables. Il est intéressant de rappeler enfin que l'analyse structurale classique du C. G. L. réalisée par O'LEARY (1940) chez le Chat a été effectuée chez des animaux âgés de 8 à 18 jours.

Colliculus superior. – Dans cette structure, la précocité du développement par rapport à celui de l'aire visuelle corticale est encore plus manifeste. Une observation identique a été faite chez le Lapin nouveau-né (SIOU, 1963). Il apparaît ainsi que la maturation cellulaire des secteurs centraux du système visuel se fait de façon ascendante, du mésencéphale à l'écorce cérébrale. Les études de maturation basées sur la cinétique du processus de myélinisation avaient déjà abouti à une systématisation analogue où l'on décrivait un stade mésencéphalique de la vision caractérisé par la myélinisation précoce des fibres rétino-colliculaires et l'apparition du réflexe photo-moteur, précédant l'étape thalamo-corticale (TILNEY et CASAMAJOR, 1924).

Ces investigations comparées à trois étages successifs de la voie visuelle ne nous permettent pas encore de proposer un substratum anatomique défini pour le «système visuel primitif» dont nous avons antérieurement postulé l'existence chez le Chat sur la base d'arguments électrophysiologiques (1962). Des investigations complémentaires basées sur l'exclusion de certaines structures sous-corticales chez l'animal nouveau-né sont nécessaires à ce propos.

Corrélations anatomo-fonctionnelles dans l'aire visuelle corticale du Chat au cours de sa maturation

Nos résultats font apparaître sans équivoque l'immaturité de l'aire visuelle du Chat à la naissance. L'expérimentation neurophysiologique en démontre pourtant avec précision les propriétés fonctionnelles. C'est ainsi que la stimulation du nerf optique à cette époque provoque l'apparition systématique de réponses électro-corticales de caractères bien définis (MARTY, 1962).

Les étapes de la maturation de l'écorce visuelle se traduisent par l'apparition progressive des caractères définitifs des réponses électrocorticales. L'acquisition d'un signe électrique positif constitue la manifestation fondamentale de la maturité de ces réponses. On a pu démontrer, chez le Chat et chez le Lapin, que ce phénomène survient lorsque la surface du champ dendritique basilaire des cellules pyramidales a atteint une valeur critique évaluée à environ 50 p. 100 de la surface totale du neurone (MARTY et SCHERRER, 1964). Il n'est pas inutile de rappeler à ce propos qu'à l'âge adulte, la surface dendritique basilaire représente 70 p. 100 de l'aire réceptrice de la cellule pyramidale (SCHADÉ et BAXTER, 1960).

Summary

The post-natal maturation of the visual area of the cerebral cortex has been studied in the cat with the method of GOLGI-COX and, secondarily, with several other histological technics.

The visual area of the cat's cerebral cortex is entirely immature at birth. The stratigraphic study of its maturation yields various precise points concerning the kinetics of its development in the diverse cortical strata. The evolution of the molecular stratum and the maturation of the lower part of the cortex raise particular problems.

The comparison of the development of the cortical visual area at birth with that of neighbouring cortical territories of a different embryologic significance (archicortex), and with that of the great sub-cortical relays of the visual system (C. G. L. and colliculus superior) brings out the proportionate slowness of the maturation in the neocortex.

The observations collected on the histological plan are confronted with the results of the physiological experiments carried out during the same period.

Références

ANGEVINE, J.B., R.L.SIDMAN: Autoradiographic study of cell migration during histogenesis of cerebral cortex in the mouse. Nature *192* (1961), 766–768.

BERRY, M., A.W.ROGERS, J.T.EAYRS: Pattern of cell migration during cortical histogenesis. Nature *203* (1964), 591–593.

CHEVREAU, J., R.MARTY: Développement postnatal de la région limbique du cerveau du Lapin. C. R. Acad. Sci. *255* (1962), 1316–1318.

CONEL, J. LE ROY: Histologic development of the cerebral cortex. In "The biology of mental health and disease". Hoeber, New York 1952.

FEREMUTSCH, K.: Die Morphogenese des Palaeocortex und des Archicortex. In: FEREMUTSCH, K., E.GRÜNTHAL. Beiträge zur Entwicklungsgeschichte und normalen Anatomie des Gehirns. Karger, Basel 1952.

GRAY, E.G.: Tissue of the central nervous system. In: KURTZ, S.M. (Ed.), Electron Microscopic Anatomy. Academic Press, New York 1964.

LOCKE, S., J.B.ANGEVINE, P.I.YAKOVLEV: Limbic nuclei of thalamus and connections of limbic cortex. VI Thalamocortical projection of lateral dorsal nucleus in Cat and Monkey. Arch. Neurol. *11* (1964), 1–12.

LORENTE DE NÓ, R.: Studies on the structure of the cerebral cortex I. The area entorhinalis. J. Psychol. Neurol. *45* (1933), 381–438.

MARTY, R.: Développement post-natal des réponses sensorielles du cortex cérébral chez le Chat et le Lapin. Arch. Anat. *51* (1962), 129–264.

MARTY, R., J.SCHERRER: Critères de maturation des systèmes afférents corticaux. In: PURPURA, D.P., J.P.SCHADÉ, (Ed.), Progress in Brain Research, Vol. 4. Elsevier, Amsterdam 1964, p. 222–234.

MITRA, N.L.: Quantitative analysis of cell types in mammalian neocortex J. Anat. (Lond.), *89* (1955), 467–483.

NOBACK, C.R., D.P.PURPURA: Postnatal ontogenesis of neurons in cat neocortex. J. comp. Neurol. *117* (1961), 291–307.

O'LEARY, J.L.: A structural analysis of the lateral geniculate nucleus of the cat. J. comp. Neurol. *73* (1940), 405–430.

O'LEARY, J.L.: Structure of the area striata of the cat. J. comp. Neurol. *75* (1941), 131–164.

OTSUKA, R., R. HASSLER: Über Aufbau und Gliederung der corticalen Sehsphäre bei der Katze. Arch. Psychiat. Neurol. *203* (1962), 212–234.

PURPURA, D.P., M.W. CARMICHAEL, E.M. HOUSEPIAN: Physiological and anatomical studies of development of superficial axodendritic synaptic pathways in neocortex. Exper. Neurol. *2* (1966), 324–347.

PURPURA, D.P., R.J. SHOFER, E.M. HOUSEPIAN, C.R. NOBACK: Comparative ontogenesis of structure-function relations in cerebral and cerebellar cortex. In: PURPURA, D.P., J.P. SCHADÉ, (Ed.), Progress in Brain Research. Vol. 4, p. 187–221. Elsevier, Amsterdam 1964.

RABINOWICZ, TH.: The cerebral cortex of the premature infant of the 8th month. In: PURPURA, D.P., J.P. SCHADÉ, (Ed.), Progress in Brain Research Vol. 4, p. 39–86. Elsevier, Amsterdam 1964.

RAMÓN Y CAJAL, S.: Histologie du système nerveux de l'Homme et des Vertébrés, tome II. C.S.I.C. Madrid 1911, 1955.

RAMÓN Y CAJAL, S.: Studien über die Sehrinde der Katze. J. Psychol. Neurol. *29* (1922), 161–181.

RAMÓN Y CAJAL, S.: Studies on vertebrate neurogenesis (Transl. L. Guth). Thomas, Springfield 1929, 1960.

RAMÓN-MOLINER, E.: The histology of the postcruciate gyrus in the cat. III. Further observations. J. comp. Neurol. *117* (1961), 229–249.

RIESE, W.: Structure and function of the mammalian cerebral cortex at the time of birth. Virginia med. Mthly *71* (1944). 134–139.

ROSE, J.E.: The ontogenic development of the rabbit's diencephalon. J. comp. Neurol. *77* (1942), 61–130.

SCHADÉ, J.P., C.F. BAXTER: Changes during growth in the volume and surface area of cortical neurons in the rabbit. Exper. Neurol. *2* (1960), 158–178.

SCHEIBEL, M., A. SCHEIBEL: Some structural and functional substrates of development in young cats. In: HIMWICH, W.A., H.E. HIMWICH, (Ed.), Progress in Brain Research, Vol. 9, p. 6–25, Elsevier, Amsterdam 1964.

SHOLL, D.A.: The organization of the visual cortex in the cat. J. Anat. (Lond.), *89* (1955), 33–46.

SIOU, G.: Changements morphologiques des neurones colliculaires consécutifs à l'énucléation oculaire chez le Lapin nouveau-né. Arch. Anat. *52* (1963), 53–74.

TILNEY, F., L. CASAMAJOR: Myelinogeny as applied to the study of behavior. Arch. Neurol. Psychiat. *12* (1924), 1–66.

TILNEY, F., L.S. KUBIE: Behavior in its relation to the development of brain. Part I. Ann. Neurol. Inst. N. Y. *1* (1931), 229–313.

TUNG, H.W., T.W. LU: The postnatal development of the lateral geniculate body of the rabbit (en chinois, résumé anglais). Acta anat. sinica *7* (1964), 87–94.

VOELLER, K., G.D. PAPPAS, D.P. PURPURA: Electron microscope study of development of cat superficial neocortex. Exper. Neurol. *7* (1963), 107–130.

WINKLER, C., A. POTTER: An Anatomical Guide to Experimental Researches on the Cat's Brain. Versluys. Amsterdam 1914.

Dr. R. MARTY et Dr. R. PUJOL
Laboratoire de Neurophysiologie
Faculté des Sciences
Chemin des Brusses
Montpellier 34, France

Comparative Anatomy of the Central Visual Systems in Day- and Night-active Primates

By R. HASSLER, Frankfurt

The sense systems for distant objects provide the orientation in the environment. The development of such sense organs is therefore often used as criterion of the bearer's grade of evolution.

The *visual system* is supposed to characterize the primates, as it does the olfactory system for the insectivores. The visual system, however, is composed of *many functional apparatus:* One for vision in twilight or skotopic vision, another for colour vision, and another apparatus for the stereoscopic vision. The performances of the visual system comprise, however, also the vision of forms, of movements, the recognition of symbols and images and of physiognomies as well as the constancy of visual space and the co-ordination of retinal pictures and signals from the gaze movements. It is evident that this diversity of performances corresponds to a similar multiplicity of central brain structures responsible for each of them. Rather the central relay stations of the visual system than the extension of visual receptors account for efficiency of vision in its broadest sense.

Just the visual sense with its multiple central structures is suited for the study of the *ways of brain evolution,* all the more as the central visual system was the first which developed subcortical *laminated* structures. On the other hand, from the state of development of the central visual systems conclusions could be drawn about the functional meaning of special visual structures. Especially the localization of the central apparatus of the black-white vision in twilight and night in contrast to the colour vision in the day should be detected. During this study it resulted, that some features have rather a taxonomic than a functional meaning.

The observations concern the first, the second and the third cerebral relay stations of the visual system in the basal ganglia as well as in the cortex.

Material and Methods. The cerebral visual centers were investigated in very different forms of primates as man, *Papio, Colobus, Cercopithecus, Ateles, Cebus, Saimiri, Callithrix, Lemur fulvus, Propithecus, Urogale* and *Tupaia* as day-active primates and *Aotes, Tarsius, Loris, Microcebus, Cheirogaleus, Lepilemur, Avahi* and *Daubentonia* as night-active forms. For reason of comparison two macroptic insectivores *(Rhynchocyon* and *Elephantulus)* and a microptic one *Solenodon* also were studied.

For *visual centers* were considered all those brain structures which receive primary or secondary optic fibre terminals. Merely the superior colliculus is not regarded. Visual centers in this sense are in the diencephalon the lateral geniculate, the pregeniculate body, the intergeniculate nucleus of the pulvinar and one of its lateral nuclei together with some parts of the reticulate nucleus of the pulvinar; and in the cortex the striate area, the area 18 and 19.

The comparison is based on the *cytoarchitectonic* and *myeloarchitectonic structure,* the relative position of the special parts and their fibre connections. – In the lateral geniculate body moreover the volume and the numbers of nerve cells were determined for each single layer by H.-D. SCHULZ. The operations were performed by WAGNER and most of the serial sections of brains of different unoperated primates are from STEPHAN's collection.

Results

The day- and night-active species of Prosimiae, including *Tupaia*, and of Simiae differ significantly in the size of the eyes. To the large and sometimes *huge eyes of the night-active* primates correspond *optic nerves* which are markedly *smaller* than the optic nerves of comparable *day-active* species. That means, to the relative small retinae of the day-active primates is (in relation to the number of receptors) a much larger number of optic nerve fibres at their disposal for central conduction than to the large retinae of the night-active primates. The crepuscular (twilight) receptors, the rods, have, as it is known, a greater convergence to one cell of the ganglion cell layer than the colour receptors, the cones. I know that these statements contain some not proved generalizations in regard to the morphology of visual receptors but not unproved in regard to their functional properties.

Accordingly also the subcortical relay station of optic nerve fibres, the *lateral geniculate* body, has a *smaller* size *in the night-active* primates with their huge eyes than in the day-active with smaller eyes. The optic nerve fibres with the stronger convergence of crepuscular receptors were established to be thicker fibres with faster conduction speed (DOTY, KIMURA and MORGENSON) than the fibres with more restricted convergence of the impulses from the colour receptors. The physiological consequence is that the impulses of the very sensitive black and white receptors, which are accumulated in the periphery of the retina, arrive earlier in the subcortical visual centers than the impulses from the colour receptors crowded in the centre of the retina.

The lateral geniculate bodies of the primates including the subprimates *Tupaia* and *Urogale* are clearly *laminated* and composed of four to six layers. Merely in a few places the layers are connected by bridges of cells which pierce the separating fibre strata. From all the insectivores *Elephantulus* is the only one in which the stratification of the lateral geniculate is foreshadowed, not in *Rhynchocyon* and less then ever in the microptic *Solenodon*.

Tupaia and all the *old world primates* investigated – with the single exception of *Tarsius* but including all the Lorisids and all the Lemurids – have *six layers* in the lateral geniculate body. The same is true for many of the new world monkeys but not for the *Hapalids*, *Saimiri* (Fig. 1b) and not for *Aotes* (Fig. 1a). Thus both the peculiar forms of night-active primates *Tarsius* and *Aotes* as well as the *small new world monkeys* have *four layers* in the lateral geniculate.

Ipsi- and contralateral layers. – Basing on experiments with transneuronal degeneration after unilateral interruption of the optic nerve or after exstirpation of one eye or section through parts of the retina, it can be confirmed for all the primates as well as for man, that always one *half of the layers* is supplied by the *contralateral* eye, the *other half* by the *ipsilateral* eye. That is also true for the four-laminated geniculate bodies (Fig. 1a+b). If the layers are numbered, as usually, from the surface to the depth, the most superficial and the deepest layer are supplied from the contralateral eye (Fig. 1a + b), and the second and third layer from the ipsilateral eye. In contrast to all simians and human the pattern of *contralateral* supply is in the *Lemuriformes* and *Lorisiformes* (s. Fig. 1c from night-active *Cheirogaleus*) not 1, 4, 6 but *1, 5, 6* and of *ipsilateral* supply *2, 3, 4*, instead of 2, 3, 5. *Tupaia*, however, has yet another pattern than all primates with six-laminated geniculates. This fact has probably a taxonomic value.

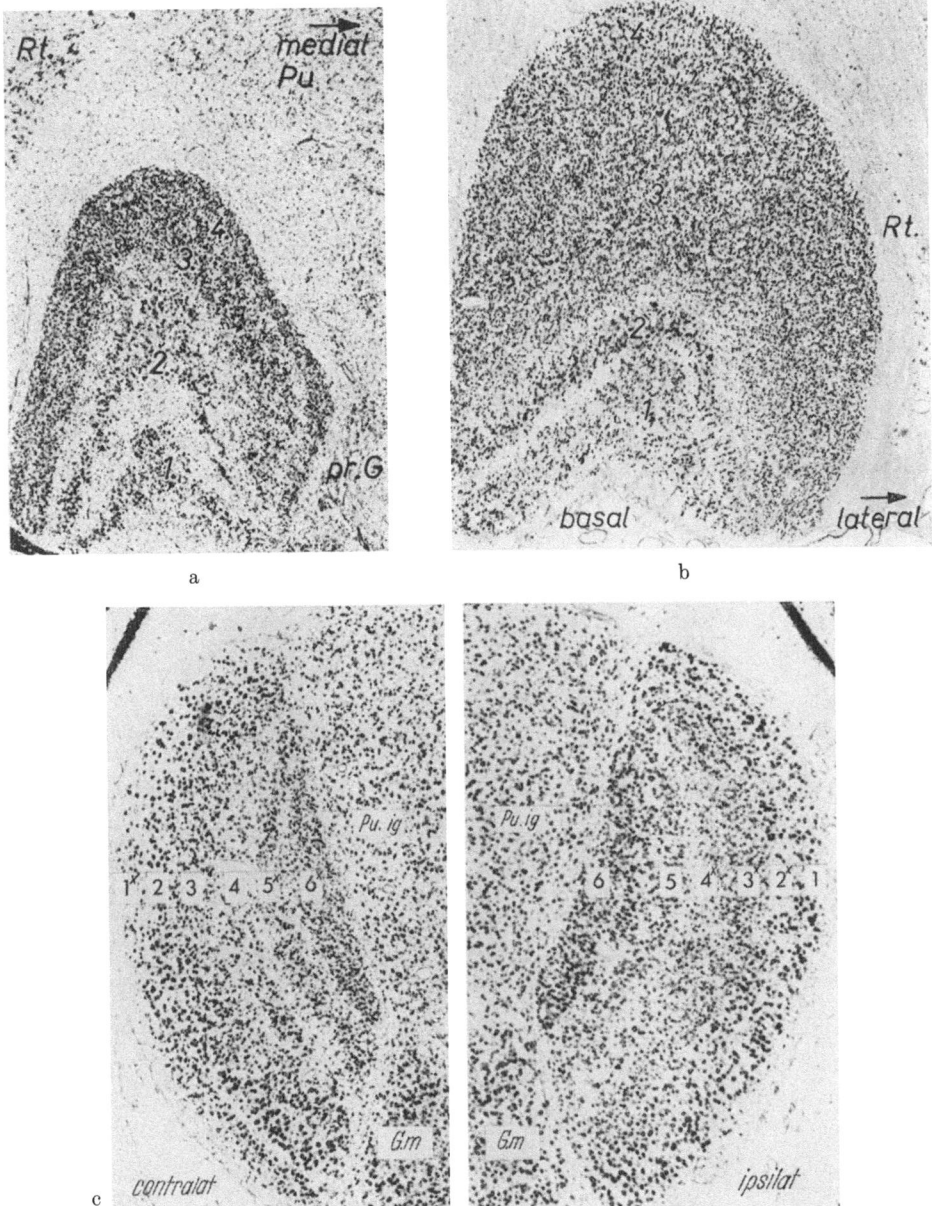

Fig. 1. a) The left lateral geniculate body of *Aotes trivirgatus* in cross section. The pair of magno-cellular layers (1.2.) is very extensive in comparison with the pair of the 3rd and 4th layers. Medial in connection with the peripeduncular nucleus the fibrous part of *pregeniculate nucleus* (pr. G). In consequence of the destruction of the upper retinal quadrants in the contralateral eye the nerve cells of the medial part of the 1st and 4th layer are atrophied. Magn. 18 : 1.

b) Right lateral geniculate body of *Saimiri sciureus*. According to the day-activity the parvo-cellular layers 3 and 4 prevail above the magno-cellular layers (1 and 2). Magn. 18 : 1.

c) Frontal section of both lateral geniculate bodies of *Cheirogaleus*, a night-active Prosimian from Madagascar. Following the interruption of the right optic nerve transneuronal degeneration in the ipsilateral (right side) geniculate body, occurring in the 2., 3. and 4. layer. On the contralateral side (left) the degeneration concerns the 1., 5., 6. layer. Pu.ig.: intergeniculate nucleus of the pulvinar, G.m: medial geniculate body. Magn.: 30 : 1.

Cell structure of different layers. – The layers of the lateral geniculate body are not uniform in their cell structure. The two superficial layers (1 and 2) contain larger nerve cells than all the inner layers (3 to 6). This difference of a *magnocellular* and a *parvocellular* or principal system exists in all primates with the single exception of *Tupaia* in which the cell-size of the superficial layers does not exceed clearly and constantly that of the deeper layers (Fig. 6c).

The fourth and fifth layers of the lateral geniculate body are the new layers added to the four originally. These two layers are less dense and have less dark nerve cells than the others. This is most distinctly marked in *Daubentonia*, where these layers consist merely of a few rather large nerve cells, less than in *Cheirogaleus* (Fig. 1c).

The input to the magnocellular layers has *thicker fibres* with a higher conduction-speed than the input to small cell layers (DOTY). And even in the output the same difference in fibre size occurs as projections to the cortex. These differences in the spectrum of the afferent and efferent fibres and in the sizes of the neurones suggest some functional differences between the magnocellular and parvocellular systems in the lateral geniculate body. Since the *macula-representation spares,* as it is known, almost completely the layers of *larger cells,* it could be supposed that the magnocellular layers are related in the first line to the skotopic receptors and to the periphery of the retina.

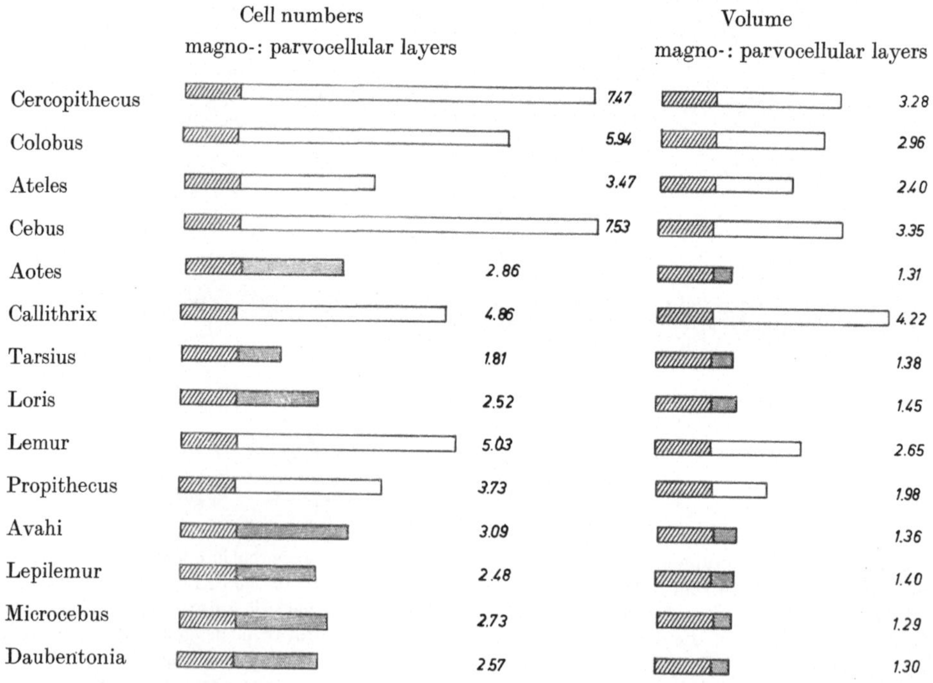

	Cell numbers magno-: parvocellular layers		Volume magno-: parvocellular layers	
Cercopithecus		7.47		3.28
Colobus		5.94		2.96
Ateles		3.47		2.40
Cebus		7.53		3.35
Aotes		2.86		1.31
Callithrix		4.86		4.22
Tarsius		1.81		1.38
Loris		2.52		1.45
Lemur		5.03		2.65
Propithecus		3.73		1.98
Avahi		3.09		1.36
Lepilemur		2.48		1.40
Microcebus		2.73		1.29
Daubentonia		2.57		1.30

Fig. 2. Diagram of the ratios of the principal layers of the lateral geniculate body to its magnocellular layers, set as 1. The left column demonstrates the ratio of cell numbers, the right column that of layers volume. In the night-active species (dark) the principal layers prevail only slightly, in the day-active species (white) considerably the magno-cellular layers. The quotient is given for each species.

Representation of skotopic vision. – If the volumes of the large cell layers are compared to that of small cells, a clear-cut correlation results between the large cell layers and night-activity of different primate species. In all day-active monkeys the volume of small cell layers does exceed the volume of the large cells by the factors 2.0–5.0; in contrast to the night-active primates having a factor of less than 1.5 (Fig. 2). Comparing the number of nerve cells the strong correlation between these two neuronal apparatus and the skotopic or the colour vision is confirmed.

Integrative subcortical visual centres. In all primates there are some other subcortical visual centres. The first is a nucleus of far less cell-density always immediately adjacent to the lateral geniculate body and often not separated by a fibre capsule. That is the *intergeniculate nucleus of the pulvinar* (Pu. ig in Fig. 3, 6b+c) in higher primates. This nucleus is constantly present in all primates, in all the prosimians and even in *Tupaia* and *Urogale* (Fig. 3c + d). In all lower forms this intergeniculate nucleus adjoins to the optic tract

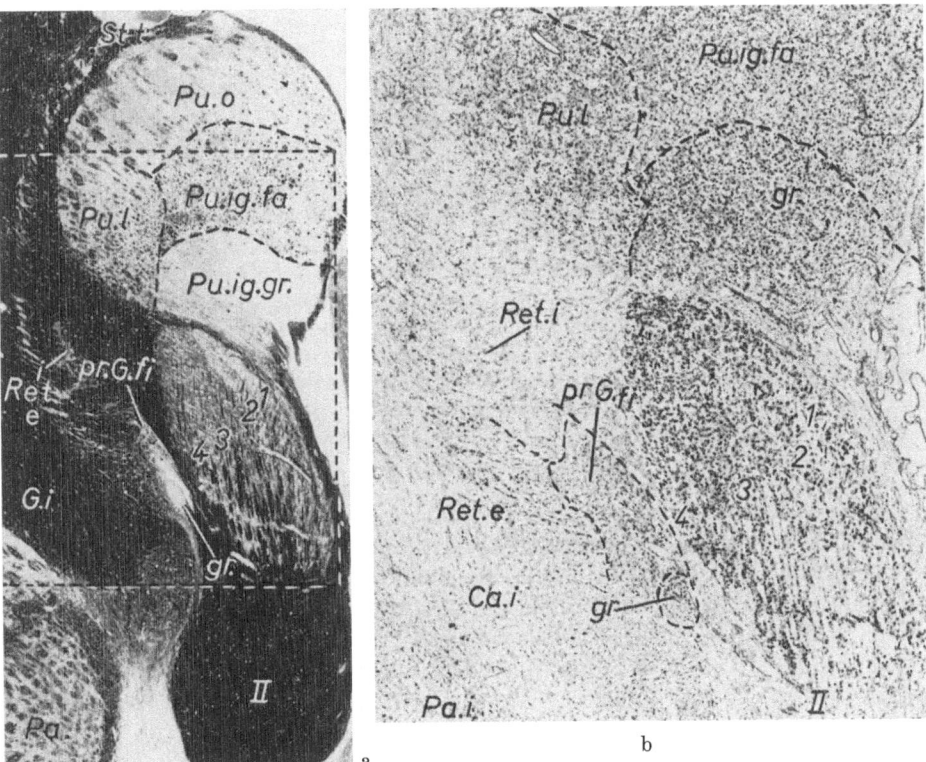

Fig. 3. a + b) Sagittal section of the thalamus of *Saimiri sciureus* in fibre-staining (left; magn. 13:1) and in NISSL staining (right; magn.: 20:1). The myeloarchitecture of the 1. and 2. layer of the lateral geniculate is less dense than in the 3. and 4. layer. In front of the lateral geniculate the pregeniculate nucleus (pr. G) with its 2 parts: fasciculosus (fi) and griseus (gr.). The pr. G. fi is contiguous with the inner layer of the reticulate zone (Ret.i.). In the pulvinar, the two parts of the intergeniculate nucleus (Pu.ig.fa and Pu.ig.gr) and the pulvinaris lateralis (Pu.l) are demonstrated. Above these the nucleus pulvinaris oralis (Pu.o). II = optic tract; Ca.i. (in 3a erroneously G.i) = internal capsule.; Pa. i = Pallidum internum.

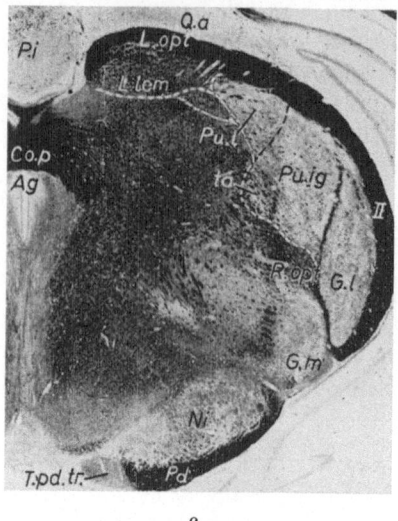
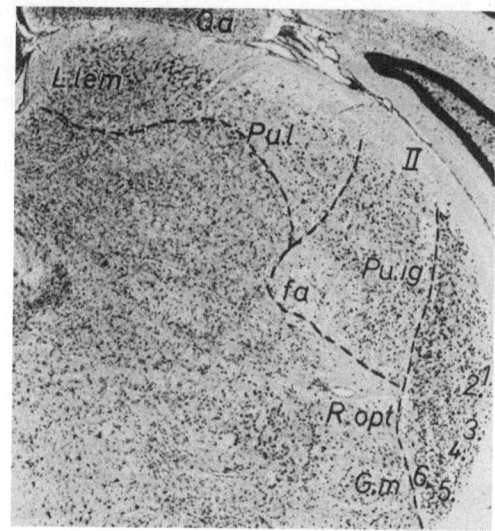

c   d

Fig. 3. c) Cross-section of the diencephalon of *Urogale* (magn.: 10: 1). The optic tract (II) surrounds the dorsal diencephalon. Below the lateral geniculate body (G.l.), the medial geniculate
body (G.m.) is situated. Dorsally to the lateral geniculate and contiguous with it there lies the
intergeniculate nucleus (Pu.ig) with a very small fibrous part (fa).
The medially adjoining nucleus is distinguished by less single fibres: Pu.l. The dark nucleus more
medially belongs to the lemniscal layer (L.lem) of the superior colliculus (Q.a,). The cerebral
peduncle (Pd) is continued medialwards by the nucleus tractus peduncularis transversus (T.pd.tr.).
R.opt = Radiatio optica.

d) Same section in cresyl-violet staining; (magn.: 16: 1). The lateral geniculate nucleus is composed of 6 layers (1–6). Dorso-medially adjoins the intergeniculate nucleus (Pu.ig.), separated by
a sharp borderline, with a small fasciculosus part (fa), continued dorso-medially by a lateral
pulvinar (Pu.l) nucleus with less cell density. More medial follows the lemniscal layer (L.lem) of
the superior colliculus (Q.a).

too. Even in the macroptic insectivores this intergeniculate pulvinar nucleus can be
identified: In most forms – exclusively *Tupaia* – there exist one intergeniculate nucleus
with bundled fibres *(fasciculosus)* and the other without fibre bundles *(griseus)*. Whether
this intergeniculate nucleus receives direct optic fibres or not in all forms should be decided after further experiments. Collaterals are often described (O'LEARY).

A subcortical *visual* center of *higher integration* is represented by the *lateral superior
pulvinar nucleus* (Pu. l.s.) of man. Its partial *degeneration* in two cases of human *anophthalmia* suggests strongly its visual – integrative – function (HASSLER). This integrative
visual nucleus exists in all simians (Fig. 3a + b) and Lorisiformes and Lemuriformes but is
small in *Tupaia* and *Urogale* (Fig. 3c + d) and doubtful or completely absent in macroptic
insectivores. Regarding the fibre connections this lateral pulvinar nucleus receives collaterals of the optic radiation fibres to the occipital lobe (HASSLER). It has a separate cortical
projection outside the area striata.

Cortical visual fields. This *triad* of subcortical visual nuclei is so important, because
they correspond with three visual centres in the cortex, which were first stated by
BRODMANN: Area striata = 17, area occipitalis = 18, area praeoccipitalis = 19.

Each of these visual fields receives its own subcortical projection. As the striate area is supplied by a two-fold projection from the lateral geniculate, one thick-fibrous and one thin-fibrous, the *visual projection to the cortex is quadruple* (HASSLER u. WAGNER 1965).

The primary visual field, the striate area, distinguished by the characteristic strip of GENNARI is present in all primates including the subprimate *Tupaia* (Fig. 5b). The typical accumulation of parallel fibres in the IV. (granular) layer and the inner half of IIIc sublayer stands out mostly in the day-active animals. This GENNARI-VICQ D'AZYR strip is

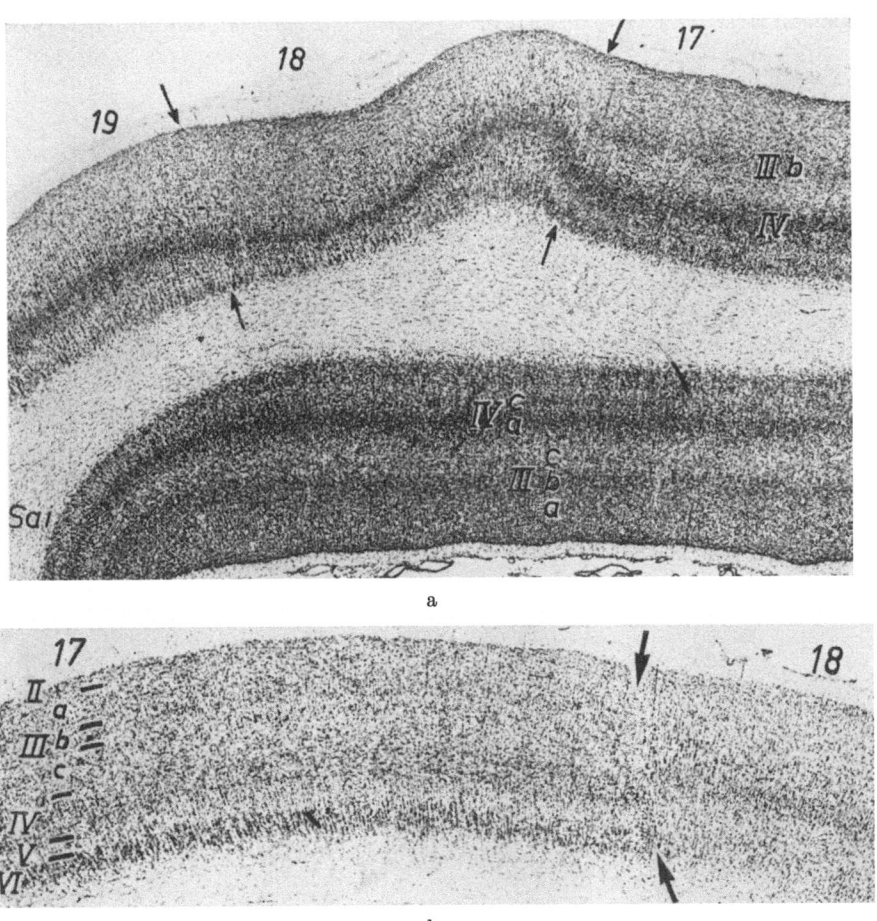

a

b

Fig. 4. a) Sagittal section of the occipital lobe of *Saimiri*. The calcarine fissure is bordered by the very dense striate area (17). Note that the IIIrd layer is divided in 3 sublayers with a very dense middle part (III b) and a light broad inner part (c). The IVth layer is divided only on some places in 3 parts by splitting in the midst of the layer (in the lower part of the figure). Cresyl-violet staining. Magn.: 18 : 1.

b) Cross-section through the occipital lobe of *Aotes*, the night-active monkey. The outer layers of the area 17 are less distinguished, although the dense III b can be seen. The IVth layer is continued in area 18 and not splitted in sublayers. On the borderline to area 18 the density of the VIth layer is strongly diminished and in the IIIc many large pyramidal cells appear. Magn.: 23,5 : 1.

less developed and less dense in most of the *night-active* primates, so that for its demonstration myelin-impregnation methods are needed.

In all the day-active monkeys and prosimians the organization of the striate area is clear and distinct also in the cell preparation (Fig. 4a). But the splitting off of the inner granular layer in three or four sublayers does not occur in the usual sense. This is mostly a misinterpretation of an enlargement of the third (III) (pyramidal) layer, which shifts the granular layer more to the white matter (Fig. 4a). The broad but very sparse IIIc-sublayer contains a great part of the afferent terminations and a half of the GENNARI strip (Fig. 5a + b). Thus the *striate area* is conspicuous by an *enlargement of the III* (pyramidal)-layer with three distinct sublayers, on both sides of a slender but very dense

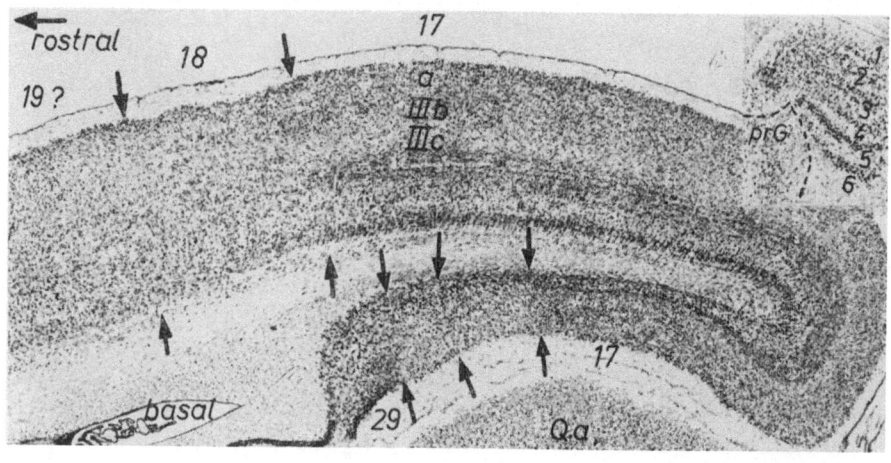

a

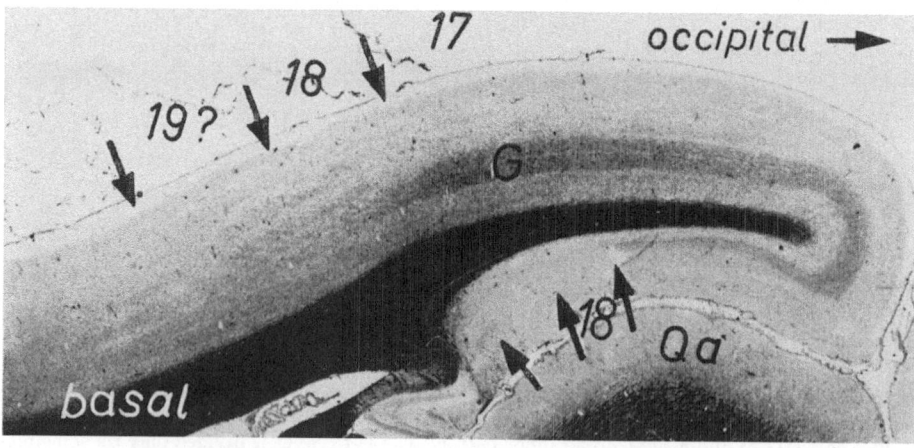

b

Fig. 5. a + b) Sagittal section of the occipital lobe of *Tupaia* in cresyl-violet Magn. 17: 1 (a) – and in HEIDENHAIN-WOELCKE (b) staining (Magn. 12: 1). The strip of GENNARI (G) which is splitted longitudinally corresponds to the IVth layer and to the lower light part of IIIc. Area 17 is surrounded by area 18. The identification of area 19 is doubtful. The insert demonstrates the 6 layers of the lateral geniculate body; in front of it the pregeniculate nucleus (pr. G).

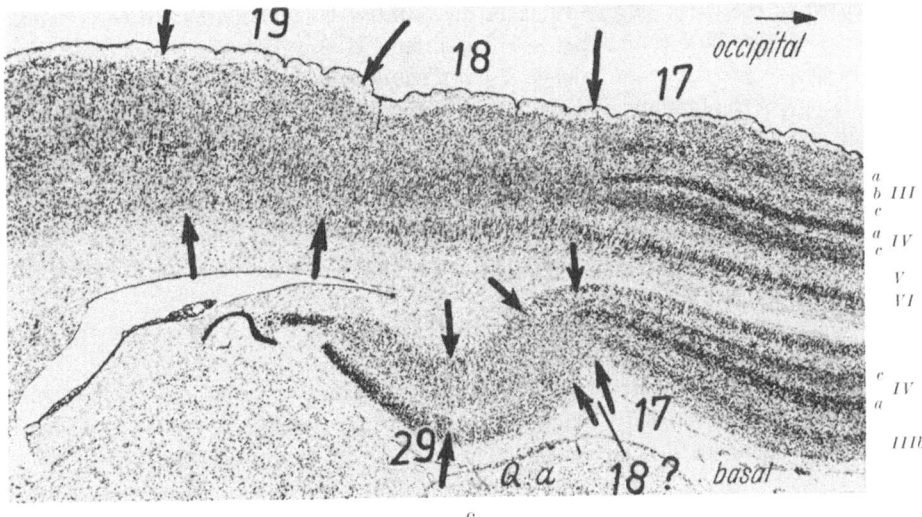

c

Fig. 5. c) Sagittal section to the occipital lobe of *Tarsius*. The stratification of area 17 is much more marked. The IVth layer is devided in a very dense superficial (IVa), a light middle (IVb) and a broad dense deep sublayer (IVc). Note the accumulation of small nerve cells in IIIb-sublayer. Area 18 and 19 are clearly differentiated. Area 29 is a part of the retrosplenial region. Magn.: 19:1.

middle sublayer (III b) composed by minute pyramids (Fig. 4a, 5a), and by a very dense inner granular layer (IV) shifted to the white matter and only sometimes longitudinally splitted by a narrow zone of clearing (Fig. 4a basal, Fig. 5c). Another characteristic of the striate area for comparative studies is the very distinct and dense sixth (VI) layer (Figs. 4, 5).

Comparing the striate areas of day- and night-active primates, or prosimians, respectively, the striate structure is as a general rule in day-active animals much more distinct and the outer layers prevail the inner (Fig. 4a, 5c). In night-active primates, simians *(Aotes)* or lemurs and loris, the inner layers IV, V, VI. are prevailing and the outer layers appear blurred (Fig. 4b). There is only one exception in the night-active primates: *Tarsius*, the extreme example of a visual animal. It has the most strongly marked articulated visual area. The outer as well as the inner layers are distinct and separated in sublayers (Fig. 5c). The probable reason is that *Tarsius* lives exclusively arboreal and gains its food by jumping. *Tarsius* needs therefore a very exact stereoscopic vision for distance estimation. In this respect *Tarsius* is unique between all night-active primates.

A quantitative study of the surface of the striate area comparing night- and day-active primates has still to be done.

The occipital (18) and preoccipital (19) areas can be differentiated in all primates from *Tupaia* (Fig. 5a + b) to the chimpanzee and man. The border of the striate area is occupied in perhaps all regions by a narrow strip of vertical structure, the marginal block, first described by BENDRAT (Fig. 4a, + b; 5a + c).

Area 19 is always differentiated from area 18 by the more loose arrangement of nerve cells and by the larger cells in IIIc and V-layer (Fig. 4. 5). The inner granular layer is always less dense than in area 18.

In regard to the areas 18 and 19 exact quantitative comparison has to be done and exact map of distribution over the cerebral surface. However the fact that these three visual areas are *present in all primates* from *Tupaia to man* demonstrates impressively the constancy of the fundamental configurational pattern of the central visual system in primates.

The question, whether the areas 17, 18 and 19 are homologous in primates, can be answered affirmative, as I believe. The reasons are the very similar architectonic structure independently from species and especially the constant positional relations between these three areas according to the topological principle and also the same fibre relations with the same afferent projections.

With these cortical projections of the visual system the central visual structures are not exhausted.

Pregeniculate nucleus: In the diencephalon there is another visual nucleus belonging to the lateral geniculate body which, however, does not project to the cerebral cortex. That is the pregeniculate or lateral geniculate nucleus pars ventralis. But the position to

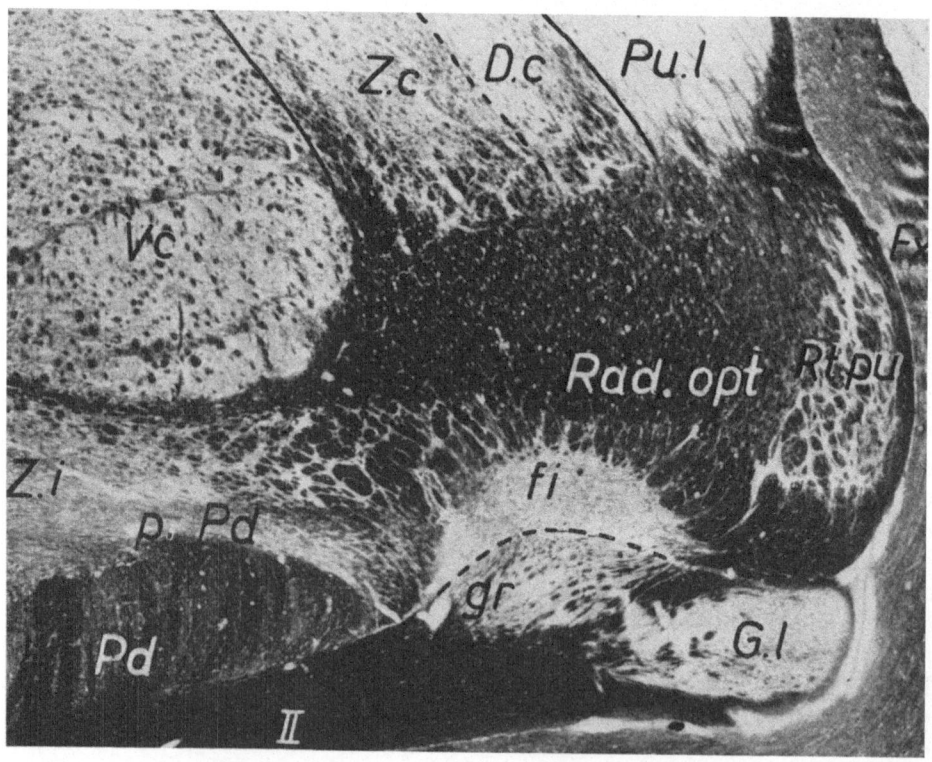

a

Fig. 6. a) Cross-section of the thalamus of *Tarsius* in fibre staining. Above the rostral pole of the lateral geniculate (G.l) the 2 parts of the pregeniculate nucleus: the fibrous part (fi) with strands of single fibres is contiguous to the inner part of the reticulate zone. The grey part of pregeniculate nucleus (gr) is poor in single fibres. Lateral the pulvinar part of the reticulate nucleus can be seen (Rt.pu). Above the optic radiation (Rad. opt) the lateral pulvinar nucleus (Pu.l) is situated. II: optic tract. p. Pd = Nucleus peripeduncularis. Vc = Ventrocaudal nucleus of the thalamus. Z.i = Zona incerta. Magn.: 32 : 1.

the main part of the lateral geniculate can also be reversed from ventral to dorsal as in the human and all higher apes. The configurational pattern of the pregeniculate is constant between the optic tract and the lateral geniculate nucleus, therefore in my opinion pregeniculate is the best name. The *pregeniculate* nucleus *decreases* in size *in the ascending scale* of primates. Thus the prosimians have larger pregeniculates than the simians, the largest occuring in *Tupaia* (Fig. 6 b +c) and *Urogale*, the subprimates. In all primates the pregeniculate is divided in a smaller part, rich in cells (gr), and a larger fibrous (fi) part (Fig. 3 a + b, 6 a + c).

The two parts of the pregeniculate: The *fibrous* part, provided with a dense strand of single fibres and larger scattered nerve cells, passes and fuses into the occipital and inner layer of the reticulate nucleus of the thalamus (Fig. 6 b + c), which has the same struc-

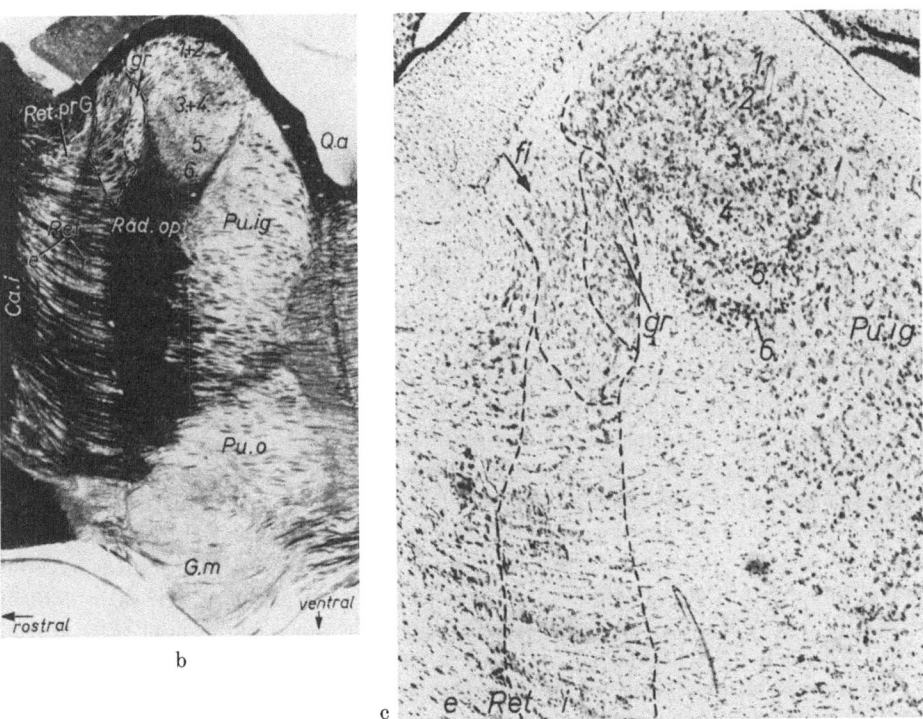

Fig. 6.b) Fibre preparation of a sagittal section through the thalamus of *Tupaia glis.* Magn.: 16 : 1. The two layers of the *reticulate zone* (Ret) in front of the thalamus are clearly differentiated by the strand of single fibres in the inner layer (i). The external layer of the reticulate zone (Ret.e) borders to the internal capsule (Ca.i). Dorsally it is narrower and crossed by less fibre bundles (Ret.pr.G). The inner layer (Ret.i) fuses dorsally to the fibrous part of the pregeniculate nucleus (pr.G.fi). Between the pr.G.fi and the lateral geniculate nucleus the grey part of the pregeniculate nucleus (gr). Caudal to the optic radiation (Rad.opt) lies the intergeniculate nucleus of the pulvinar (Pu.ig); on the basis of the thalamus the medial geniculate body (G.m) and another nucleus of the pulvinar (Pu.o).

c) The same situation in NISSL staining but in higher magnification (35 : 1). The 2 parts e and i of the reticulate zone (Ret) differ by the larger nerve cells in the outer (e) layer. The 6 layers of the lateral geniculate are clearly distinguished. The fibrous part (fi) of pregeniculate nucleus is continued by the inner layer of reticulate zone.

tural equipment with slender light nerve cells and especially with the same tape (strand) of single fibres. It is continued to the midbrain by the peripeduncular nucleus (Fig. 6a). The *fibreless* light grey part, provided with small light nerve cells, packed closely together (Fig. 3b, 6c), is protruded from the fibrous part in direction to the lateral geniculate body. In *Tupaia* and *Urogale* it fuses lateral and rostral into the *outer layer* of the thalamic *reticulate nucleus,* just as poor in single fibres. This outer reticulate layer with an arrangement of coarse and dark nerve cells forms the third cover-layer frontally and laterally to the geniculate (Fig. 6b + c). These positional relations of the two parts of pregeniculate nucleus to two parts of thalamic reticulate nucleus are really surprising. – Relation of the pregeniculate nucleus or of one of its parts to the day- or night-activity of a species could not be detected.

Afferents to this pregeniculate nucleus are not definitely established. In spite of the constant positional relation to the optic tract, terminals of optic fibres could not be demonstrated (CHACKO) with the usual methods, including MARCHI. Even using the NAUTA-GYGAX-method, TIGGES saw merely fibres of passage from the degenerated optic tract in the very extensive pregeniculate nucleus of *Tupaia*. With the GOLGI-method, however, O'LEARY was able to demonstrate fine terminal collaterals of optic tract fibres in the pregeniculate nucleus of the cat. According to the constant bordering on the optic tract it should be expected that the pregeniculate undergoes transneuronal degeneration in the experiments. But the reverse is true in two cases of (human) *anophthalmia;* the *pregeniculate* presents a marked *hyperplasia* while the lateral geniculate is aplastic (HASSLER, 1965). After cortical exstirpation retrograde degeneration does not occur in the pregeniculate. Its efferent pathway seems to go to the reticular formation of the midbrain or to the pallidum.

The presence of the pregeniculate with its two parts always in the same relative position between the entering optic fibres and the main laminated part of the lateral geniculate through the entire ascending scale of primates is very suggestive for the constancy of the fundamental configurational pattern of the central visual systems: the optic tract, the reticulate subnucleus, the fibrous part of the pregeniculate, the dense part continued by the four- or six-layered lateral geniculate in topographic neighborhood of the intergeniculate nucleus (often divided in two parts) and the highest integration level: the superior lateral pulvinar nucleus. This repeated sequence of five to seven nuclei, each of them with very similar histological characteristics, makes the homologization of these central visual structures quite probable.

The nucleus of the tractus peduncularis transversus should be the last described subcortical structure supplied by optic tract fibres. The structure of this optic by-path to

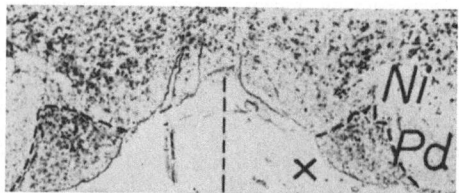

Fig. 6d) Cross section of the midbrain of *Tupaia* after interruption of all left optic fibres. Degeneration of the contralateral nucleus tractus peduncularis transversi with shrinkage and reduction of the cell size and cell number. Ni = Nigra. Pd = Pedunculus cerebri. Magn.: 26 : 1.

the midbrain and of its nucleus differs scarcely in different primates. But its size and the size of its nucleus vary considerably, even in closely related forms. There is no relation of the size of this nucleus to the night- or day-adaptations of the primate species: *Tupaia*, *Tarsius*, *Saimiri* and *Galago* have extensive nuclei of tractus peduncularis transversus. In the most lower forms the nucleus has its position in the medial border of the substantia nigra (Figs. 3c, 6d), but in *Saimiri* and *Aotes* the nucleus lies on the superior border of the cerebral peduncle below and behind the peripeduncular nucleus. In *Galago crassicaudatus* the same nucleus is attached laterally to the cerebral peduncle. – In *Tupaia* the degeneration of the nucleus with shrinkage and diminution of the nerve cells occurs exclusively contralateral to the optic nerve interruption (Fig. 6d), confirming TIGGES' findings.

Discussion

The visual system, which seems to be characteristic for primates, has eight subcortical and three cortical relay stations. The primary relays are the optic tectum (not dealt with), the lateral geniculate body, the nucleus of tractus peduncularis transversus and the pregeniculate nucleus. For its subnucleus griseus and the special pregeniculate subnucleus of the reticulate zone (Ret. pr. G) of the thalamus (Fig. 6b) the supply with optic tract fibres or collaterals is not yet established. The intergeniculate nucleus is the *first level of visual integration* as it receives collaterals of optic tract fibres. In the other integrative visual nucleus (Pu. l) the supply with optic fibre collaterals is lacking.

In all primates the *lateral geniculate nucleus* is a *laminated* structure. This is barely foreshadowed in *Elephantulus* but not in other insectivores, not even in *Rhynchocyon*, although this lamination is present in other strains of mammals. Because these forms are also derived from the insectivores, e. g. carnivores and lagomorpha, the lamination of the geniculate body in different orders of mammals is not homologous, but merely a convergence or interesting parallel development. In primates the laminated geniculate bodies must be considered as homologous. All species of primates investigated have at least four layers, two of which are composed of large nerve cells. These magnocellular layers are without exception the superficial layers. In *Tupaia*, however, the cells of the two superficial layers do not clearly differ in size from those of the inner four layers.

The number of layers in the lateral geniculate does not correlate with the position of the species in the ascending series of primates. *Four layers* may occur in the new world monkeys, in *Callithrix*, *Aotes* and *Saimiri*, but only in one prosimian i. e. *Tarsius*. All other prosimians have *six layers*, as all old world monkeys, except *Tarsius*. The number of layers does not correlate to the adaptation to night- or day-activity. Among the night-active primates the two highest developed, *Aotes* and *Tarsius*, have four layers; all others have six layers, even the most primitive. The *night-activity* of primates is clearly correlated to the *magnocellular layers*, if their volume and their number of nerve cells are compared with the same values of the small cell layers (Fig. 1, 2).

Each layer of the lateral geniculate receives its optic fibre-input exclusively *from one eye*. The 4th and 5th layer are inserted between the third and the deepest layer in the lateral geniculate bodies with six layers. They have always very similar histological characteristics. Together they form the third pair in addition to the first pair of magnocellular layers (1st and 2nd) and the second pair of 3rd, and the deepest (4th or 6th) layer. The pattern of layers with contralateral optic fibre supply is 1, 4, 6 and ipsilateral 2, 3, 5, as known since MINKOWSKI. But this is true only for simians and man, not for prosimians, Lemurs and Loris. If they have six layers, the contralateral pattern is 1, 5, 6 and the ipsilateral 1, 3, 4; *Tupaia* and *Urogale* are exceptions.

The first level of visual integration in the diencephalon is formed by the *intergeniculate* nucleus, which is a part of the pulvinar in higher forms. The intergeniculate is present without exception in all primates, even in macroptic insectivores. Its subnucleus, poor in fibre bundles, is always immediately adjoining to the lateral geniculate, sometimes it fuses with the first layer of the latter.

The second subcortical level of visual integration is formed by another pulvinar nucleus, in the human called *"pulvinaris lateralis superior."* It is a constant nucleus in all primates, including *Tupaia* and *Urogale*, even if it is poorly developed and small in the subprimates. This nucleus is missed in the diencephalon of insectivores, even of macroptic ones. In contrast to *Tupaia*, all prosimians and simians are endowed with posterior outgrowth of the diencephalon: a real *pulvinar*. Hodologically the pulvinaris lateralis is connected with the visual system via collaterals of the optic radiation fibres.

The pregeniculate nucleus is a constant diencephalic manifestation of the visual system in primates. Contrary to the lateral geniculate it is cortex-independent. It decreases in size in the ascending series of primates and is most developed in *Tupaia* and *Urogale*. It is always composed of two parts (first described by BALADO and FRANKE): one rich in single fibres and one fibreless. The *fibrous part fuses* in all primates *with the reticulate zone* of the thalamus, especially with its inner layer, having the same strand of single fibres. The reticulate zone forms regularly another cover layer, poor in single fibres in front of lateral geniculate body against the internal capsule.

The tractus peduncularis transversus is an offshoot of the optic tract running to the midbrain. Its size differs considerably in different primates; in lower primates it is more conspicuous. In some new world monkeys, e.g. *Aotes* and *Saimiri*, the tractus peduncularis terminates lateral and dorsal to the cerebral peduncle, fusing with the peripeduncular nucleus. In *Tarsius*, *Urogale* and *Tupaia* its terminal nucleus is located in the interpeduncular fossa as the medial, most fibrous part of the substantia nigra. After interruption of one optic nerve the nucleus tractus peduncularis transversi degenerates ecxlusively on the *contralateral* side, as does the nucleus isthmo-opticus of birds, starting the efferents to the retina.

Three cortical visual areas (17, 18, 19) correspond to the three diencephalic visual centres with separate cortical projections. The lateral geniculate nucleus projects exclusively to the striate area (17), the intergeniculate nucleus to the area 18 (occipitalis) and the nucleus pulvinaris lateralis corresponds to the preoccipital area 19.

The clear *differentiation of the striate area* is not correlated to the grade of the ascending series of primates: The clearest arrangement and differentiation of cell layers is manifested in *Tarsius*, the extreme example of a visual animal, the brainweight of which exceeds scarcely the weight of one of its eyes. It is the only night-active primate with a strong differentiation of the outer layers of the striate cortex. Apart from *Tarsius*, all night-active primates, from which seven species were investigated, show a less distinct differentiation of the outer cortical layers and well-differentiated inner layers (V and VI).

The occipital area (18 BRODMANN) can be identified in all primates, including *Tupaia*. The border between area 18 and striate area is marked by a marginal block of larger cells and coarse radii, which blurred the stratification of the cortex in all primates and in the macroptic insectivores. This marginal block is missed on the border near the retrosplenial cortex.

In most primates, especially in *Tarsius*, in front of the area 18 another cortical field with a less dense population of granular and other nerve cells can be separated. It corresponds to the preoccipital area (19) of higher apes and man. But its homologization becomes doubtful in descending the scale of primates, so that its existence cannot be ascertained in subprimates.

Summary

The subcortical and cortical visual centres of 20 different species of primates, including *Tupaia* and *Urogale* (12 day-active and 8 night-active animals), were compared. Among all subcortical visual centres only the *magnocellular pair* of layers of the lateral *geniculate* body has a positive correlation to night-activity or skotopic vision (regarding number of cells and volume of layers). In the cortical visual fields the night-active primates differ from the day-active in that the outer layers in the former are less differentiated and developed and the strip of GENNARI is more diffuse.

The three cortical visual fields (17, 18, 19 BRODMANN) can be identified by their structure, their configurational pattern and their fibre connections in all species of primates investigated. The area 17 and 18 are present even in macroptic insectivores. The identification of a preoccipital area (19) is uncertain in *Tupaia*.

All *subcortical visual centres*, as lateral geniculate nucleus, magnocellular and principal part, the pregeniculate nucleus with its fibrous and dense grey parts, the pregeniculate subnucleus of the reticulate zone of the thalamus, the intergeniculate nucleus with the subnuclei fasciculosus and griseus, the lateral superior pulvinar nucleus, and the nucleus tractus peduncularis transversi can be differentiated in all species of primates investigated. Considering their fundamental configurational pattern and architectonic structure and fibre connections they seem to be homologous in all primates.

The *layer pattern* of the lateral geniculate body for ipsilateral and contralateral supply by optic fibres differs in prosimians from that in simians; *Tupaia* has yet another pattern.

The *intergeniculate nucleus* which borders the lateral geniculate is composed of two parts in all primates, except *Tupaia* and *Urogale*, in which it is uniform. – The *pulvinar* as a caudo-dorsal outgrowth of the diencephalon is missed in *Tupaia* and *Urogale*, although they have a lateral pulvinar nucleus. The pulvinar is a new acquisition in prosimians above subprimates and a typical diencephalic structure of all primates.

References

BALADO, M., E. FRANKE: Das Corpus geniculatum externum. Berlin 1937.

BENDRAT, M.: Über die Angioarchitektonik einer parastriatalen Randzone im Okzipitalhirn. Inaug.-Diss., Leipzig 1935.

BRODMANN, K.: Vergleichende Lokalisationslehre der Großhirnrinde in ihren Prinzipien dargestellt auf Grund des Zellenbaues. 1. Aufl. Barth, Leipzig 1909.

CHACKO, L.W.: The laminar pattern of the lateral geniculate body in the primates. J. Neurol. *11* (1948), 211–224

DOTY, R.W.: Functional significance of the topographical aspects of the retino-cortical projection. In: JUNG, KORNHUBER (Hrsg.), Neurophysiologie und Psychophysik des visuellen Systems. Springer, Berlin 1961.

DOTY, R.W., D.S. KIMURA, G.J. MORGENSON: Photically and electrically elicited responses in the central visual system of the squirrel monkey. Exp. Neurol. *10* (1964), 19–51.

GIOLLI, R.A.: An experimental study of the accessory optic system in the cynomolgus monkey. J. comp. Neurol. *121* (1963), 89–108.

HASSLER, R.: Über die bei der optischen Wahrnehmung beteiligten Hirnsysteme. Zbl. Neurol. *112* (1951), 152

HASSLER, R.: Functional anatomy of the thalamus. Ann. Physiol. exp. Med. Sci. India *1* (1958), 56–91.

HASSLER, R.: Anatomy of the thalamus. In: SCHALTENBRAND-BAILEY (Ed.), Introduction to Stereotaxis with an Atlas of the Humain Brain. Thieme, Stuttgart 1959.

HASSLER, R.: Die zentralen Systeme des Sehens. Ber. ophthal. Ges. *66* (1965), 229–251.

HASSLER, R., A.WAGNER: Experimentelle und morphologische Befunde über die vierfache corticale Projektion des visuellen Systems. Proc. 8. Internat. Congr. Neurol., Nr. 3: Disturbances of the Occipital Lobe. pp. 77 bis 96 Wien 1965.

HENSCHEN, S. E.: Lichtsinn- und Farbensinnzellen im Gehirn. In: Pathologie des Gehirns, VIII. Teil. Stockholm 1930.

LORENTO DE NÓ, R.: Cerebral cortex: architecture and intracortical connections. In: FULTON, J. F.: Physiology of the Nervous System 3. ed. pp. 288–330, Oxford University Press, New York 1949.

MINKOWSKI, M.: Über den Verlauf, die Endigung und die zentrale Repräsentation von gekreuzten und ungekreuzten Sehnervenfasern bei einigen Säugetieren u. beim Menschen. Schweiz. Arch. Neurol. Psychiat. *6* (1920), 201–252.

O'LEARY, J. L.: A structural analysis of the lateral geniculate nucleus of the cat. J. comp. Neurol. *73* (1940), 405–430.

O'LEARY, J. L.: Structure of the area striata of the cat. J. comp. Neurol. *75* (1941), 131–164.

O'LEARY, J. L., G. H. BISHOP: Electrical activity of the lateral geniculate nucleus of the cat. Amer. J. Physiol. *126* (1939), 594–595.

OTSUKA, R., R. HASSLER: Über Aufbau und Gliederung der corticalen Sehsphäre bei der Katze. Arch. Psychiat. Nervenkr. *203* (1962), 212–234.

POLYAK, ST.: The vertebral visual system. University of Chicago Press, Chicago 1957.

PREOBRASHENSKAJA, N.: Störungen und Wiederherstellung der optischen Funktionen bei Verletzung der Occipitallappen des Großhirns. Proc. 8. Internat. Congr. Neurol., Nr. 3: Disturbances of the Occipital Lobe. Wien 1965.

STEPHAN, H., H. SPATZ: Vergleichend-anatomische Untersuchungen an Insektivorengehirnen. IV. Gehirne afrikanischer Insektivoren. Versuch einer Zuordnung von Hirnbau und Lebensweise. Morphol. Jb. *103* (1961), 108 bis 174.

TIGGES, J.: Ein experimenteller Beitrag zum subcorticalen optischen System von Tupaia glis. Folia primat. *4* (1966), 103–123.

WALLS, G. L.: The lateral geniculate nucleus and visual histophysiology. Univ. Califor. Publ. Physiol. *9* (1953). 1–100.

Prof. Dr. R. HASSLER
Max-Planck-Institut für Hirnforschung
Neuroanatomische Abteilung
Deutschordenstr. 46
Frankfurt/Main-Niederrad

Entwicklungsprinzipien der menschlichen Sehrinde

Von H. GRÄFIN VITZTHUM und F. SANIDES, Frankfurt

Die phylogenetisch bis zu den Prosimiern zurückgehende Fissura calcarina ist auch die ontogenetisch am frühesten angelegte Fissur des menschlichen Großhirns und zugleich, einer allgemeinen Regel entsprechend, eine der tiefsten. Darüber hinaus umkleidet die in die Tiefe der Fissura calcarina supprimierte (SPATZ 1959) Area striata das komplizierteste versenkte Windungsgebilde, das einer genaueren architektonischen Erforschung stets besondere Schwierigkeiten entgegengesetzt hat, ebenso wie der genaueren elektrophysiologischen Erforschung der Area striata bei den höheren Primaten dadurch besondere Schwierigkeiten begegneten. So wissen wir physiologisch sehr viel genauer über die Sehrinde der Katze Bescheid als über die der Primaten.

Der singulostriäre Typus der Sehrinde. Eine echte Area striata, d. h. eine durch den *Gennarischen Streif* ausgezeichnete Sehrinde besitzen jedoch nur die Primaten. Dieser faserdichteste Markfaserstreifen der menschlichen Hirnrinde stellt zugleich die älteste

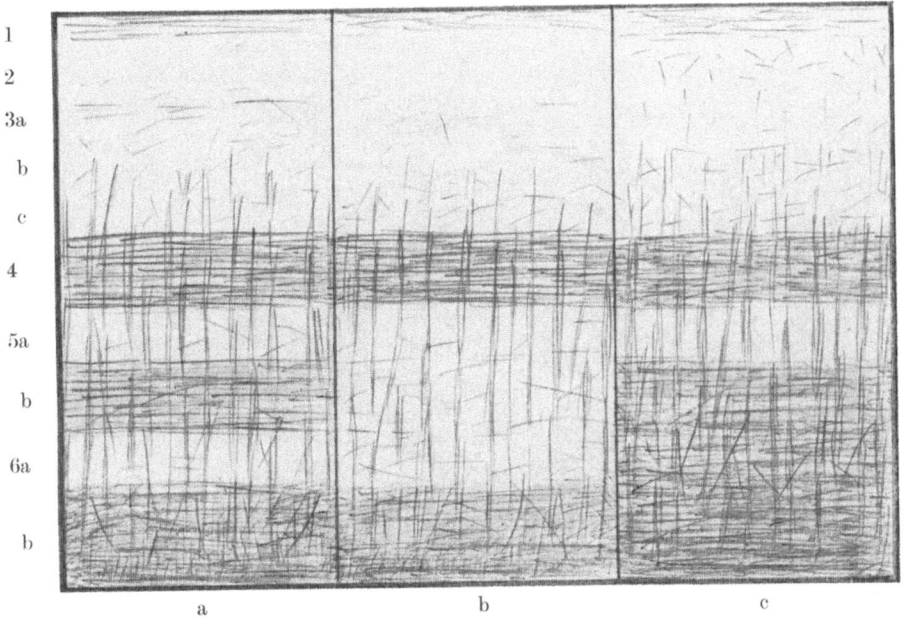

Abb. 1. Myeloarchitektonisches Schema des Isocortex cerebri. a) bistriärer Durchschnittstyp b) singulostriärer Typ (Grundtyp der Area striata) c) unistriärer Typ.

rindenarchitektonische Entdeckung in der Geschichte der Hirnforschung dar, weswegen auch an seinem Namen festgehalten wird, obwohl in den übrigen Teilen des Isocortex die beiden bekannten horizontalen Markfaser-Striae nach BAILLARGER benannt sind. Der GENNARISche Streif entspricht dem äußeren der beiden und stellt wie dieser den mit der IV. Zellschicht koinzidenten afferenten Plexus dar. Myeloarchitektonisch hat die Area striata bei dem Fehlen eines inneren BAILLARGERschen Streifens den sogenannten *singulostriären* Charakter (Abb. 1) und damit eine Sonderstellung unter den sensorischen Primärgebieten. Aber auch im übrigen Isocortex ist dieser Typus in reiner Ausprägung nur einmal gefunden worden, und zwar von HOPF als *Feldergürtel* an der Basis des Temporallappens, der, am Temporalpol beginnend, die *entorhinale* Region von oral, ventral und caudal umfaßt.

Beziehungen zu limbischen Rindenabschnitten und ihre Deutung. Bei unseren myeloarchitektonischen Untersuchungen konnten wir nachweisen, daß dieser singulostriäre periarchicorticale Feldergürtel sich bis zum Isthmus lobi limbici fortsetzt und in der Tiefe des Truncus communis fissurae calcarinae et parieto-occipitalis direkten Kontakt mit der hier auf der unteren Furchenwand beginnenden Area striata hat (Abb. 2). Diese architektonische Tatsache ist in mehrfacher Hinsicht wesentlich: sie durchbricht die von der BRODMANNschen Felderkarte ausgehende Faszination, daß wir es mit einem geschlossenen doppelten ringförmigen Feldergürtel um die Area striata zu tun haben (Felder 18 und 19 BRODMANNS). Dabei hatte sich übrigens BRODMANN in seinem Hauptwerk sehr zurückhaltend über diesen oralen Schluß des doppelten Felderringes geäußert. Als erster war es dann PFEIFER gewesen (1940), der bei der angioarchitektonischen Bearbeitung des *Macacus rhesus* einen direkten oralen Kontakt der Striata mit Feldern limbischen Typs nachweisen konnte.

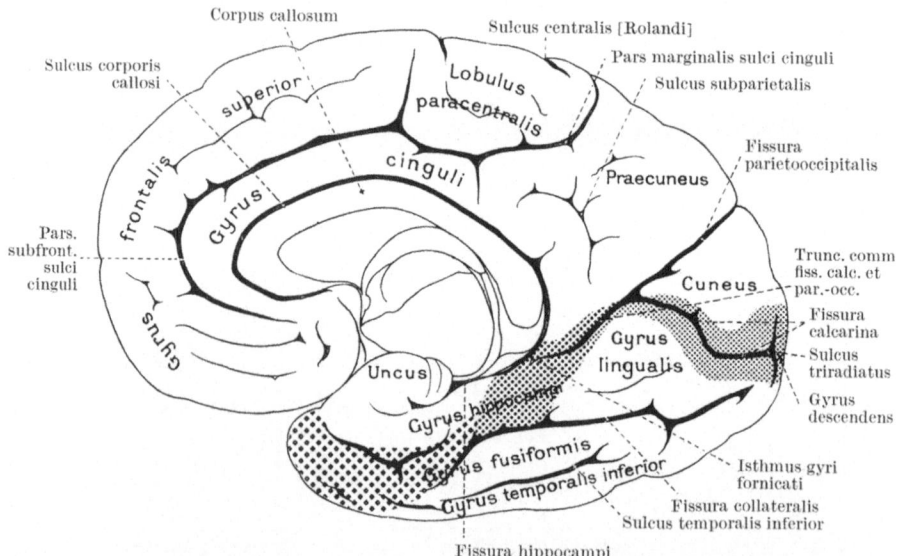

Abb. 2. Schema der Topographie des periarchicorticalen singulostriären Feldergürtels in Verbindung mit der Area striata am menschlichen Großhirn (Medianfläche). Dünne Punkte: Area striata; mittlere Punkte: sog. *Prostriata;* dicke Punkte: weitere singulostriäre Felder des Schläfenlappens.

Diese direkte Verbindung der oralen Striata mit einem besonderen Feldertyp des limbischen Ringes in der Tiefe des Truncus communis erfährt im Rahmen der von einem von uns zuerst am menschlichen Stirnhirn dargestellten *Entwicklungsprinzipien der Großhirnrinde* eine besondere Bedeutung: die zyto- und myeloarchitektonische Gliederung des menschlichen Stirnhirns läßt eine Folge von architektonischen Differenzierungsstufen erkennen, die vom vorderen limbischen Cortex einerseits und von der vorderen Insel andererseits ausgehen und die Architektonik des Frontallappens bestimmen*). Die Einzelfelder konnten auf diese Weise zu Zonen, das bedeutet Gürtel verwandter Felder, zusammengefaßt werden, die von den beiden Differenzierungsrichtungen, sogenannten „Gradationen" nach C. u. O. Vogt (1919), determiniert sind. Wie die in den Frontalhirnkarten als Pfeile symbolisierten Differenzierungsrichtungen die Architektonik der Feldergürtel (Zonen) bestimmen, sei an einem Einzelbeispiel dargelegt: So hat die mediale Differenzierungsrichtung in Form des Proisocortex des Gyrus cinguli, d. h. eines unausgereiften Isocortex, ein ausgesprochen markarmes und kleinzelliges Ursprungsgebiet mit einer sehr schwachen III. Schicht, die über die paralimbische Zone (PlZ), die frontomotorische Zone (FmZ) und die paramotorische Zone (PmZ) stufenweise gesteigert wird bei gleichzeitiger Steigerung des Markfasergehaltes und der Durchschnittszellgröße. Der Einfluß der von der dorsalen Inselrinde, ebenfalls einem markarmen und kleinzelligen Proisocortex, ausgehenden Differenzierungsrichtung bestimmt zunächst den Charakter von Feldern der Innenseite des Operculum und sodann den Charakter der frontopercularen Zone (FoZ) auf der 3. Frontalwindung. Die limbische Rinde ebenso wie die Inselrinde, die aus grundsätzlichen phylo- und ontogenetischen Erwägungen von Yakovlev (1959) zum limbischen Cortex gezählt wird, sind aber ältere Rinden als der Isocortex der Hirnlappen. Wir haben sie daher als *Urgebiete* von den sensorischen und motorischen *Primärgebieten* und den *Sekundärgebieten* oder sogenannten Assoziationsrinden, abgetrennt. Da die beiden von den limbischen bzw. insulären Rinden ausgehenden Differenzierungsrichtungen damit in *älteren* Rinden des Neocortex, seinen Urgebieten beginnen, und die architektonische Differenzierung von *jüngeren* Gebieten des Neocortex bestimmen, können wir sie als evolutionäre Differenzierungsrichtungen auffassen, die einen zweifachen Ursprung des Neocortex von den palaeo- und archicorticalen Urrinden widerspiegeln, wie er erstmalig von Abbie für die Monotremen (1940) und Marsupialier (1942) dargestellt worden ist.

Belege aus der Neurophysiologie. Die Organisation der sensorischen und motorischen Rinden läßt sich mit Hilfe dieser Entwicklungsprinzipien jedoch noch weiter erhellen. Insbesondere sind es die von Adrian, Woolsey und Penfield erschlossenen zusätzlichen sensorischen und motorischen Repräsentationen, die hierdurch ihre Einordnung erfahren: Die sogenannte sekundär-somatosensorische Repräsentation in parinsulärer Lage und die supplementär-motorische in paralimbischer Lage erscheinen, im Strome der beiden Urgradationen gelegen, als den klassischen Repräsentationen der Centralregion vorgeschaltet und damit als phylogenetisch älter als diese. In bezug auf die sekundärsensorische Repräsentation wurde die Vorstellung, daß wir es hier mit einer phylogenetisch älteren zu tun haben, auch von neurophysiologischer Seite (Woolsey 1946) entwickelt, indem auf die geringere Ausarbeitung und Differenzierung und den hohen ipsilateralen

---

*) Zum Verständnis der folgenden Darstellung sind die Abb. 55–58 und 59–60 aus F. Sanides „Die Architektonik des menschlichen Stirnhirns" Monograph. Gesamtgebiet Neurol. Psychiat. 98, Springer, Berlin-Göttingen-Heidelberg 1962 oder Abb. 1 aus „Structure and function of the human frontal lobe" Neurophysiologia *2* (1964), 209–219 heranzuziehen.

Anteil dieser Repräsentationen neben ihrer parinsulären Lage hingewiesen wurde. Auch CROSBY vertritt, besonders aufgrund klinischer Erfahrung, den Standpunkt, daß die zusätzlichen sensomotorischen Repräsentationen phylogenetisch älter sind als die klassischen (persönliche Mitteilung). Wir konnten mittlerweile im Labor von WOOLSEY in Zusammenarbeit mit WELKER und KRISHNAMURTI diese Stellung der sogenannten sekundär-sensorischen Repräsentationen bei einem Prosimier, dem Plumplori, architektonisch und funktionell bestätigen. Parinsulär und caudal an die sekundär-sensorische Repräsentation (Sm II) anschließend, liegt aber auch die von ROSE und WOOLSEY besonders bei der Katze erforschte sekundär-auditorische Repräsentation (A II), die als Ausdruck einer geringeren Differenzierungshöhe teilweise funktionelle Überlappung mit der ersteren zeigt (BERMANN 1961). Zugleich haben wir es hier wie in dem sekundär-sensorischen Feld mit einer architektonischen Zwischenstellung zwischen der dorsalen Inselrinde und dem Koniocortex, hier der Hörrinde (A I) zu tun. Aufgrund dieser architektonischen Zwischenstellung der parinsulären Felder haben wir bei der Bearbeitung des Plumplori den *architektonischen* Begriff eines *Prokoniocortex* für die sensorischen Felder Sm II und A II eingeführt.

Die Prostriata. Die singulostriäre Rinde, die der Area striata limbisch vorgelagert ist, trägt ebenfalls Züge einer Zwischenstellung, hier zwischen dem entorhinalen und parasplenialen Allocortex einerseits und der Striata andererseits, indem sie sowohl limbische Zeichen wie Akzentuierung der II. Schicht, einzelne limbische Zellformen, wie überschlanke und Gabelzellen, in der V. Schicht und einen sehr geringen Markfasergehalt*), als auch Striata-Ähnlichkeit besitzt. Letztere bezieht sich nicht nur auf den singulostriären Charakter, sondern auch auf einen gewissen Grad der Verkörnelung und eine für den Koniocortex typische Aufhellung der V. Schicht. So haben wir es auch hier mit einer Form des *Prokoniocortex* zu tun, die wir *Prostriata* benannt haben und deren Differenzierung auf den Archicortex zurückweist.

Belege aus der Ontogenese. Für ontogenetische Entwicklungszusammenhänge zwischen *Archicortex* und *Area striata* sprechen Ergebnisse von KAHLE, die seiner Monographie „Zur pränatalen Entwicklung der menschlichen Großhirnhemisphäre"**) zu entnehmen sind. KAHLE stellt im 4. Fötalmonat eine räumlich enge Beziehung der entorhinalen Region und der Area striata fest, wobei letztere der ersteren noch direkt anliegt. Erst mit der Bildung des Occipitallappens und des Hinterhorns werde die Area striata nach caudal ausgezogen und verliere ihre engen Beziehungen zum Archipallium weitgehend. Eine auffallende Parallele zwischen Archicortex und Striata betrifft nach KAHLE außerdem noch die geringe Intensität der Migration und die im Verhältnis zum frühen Matrixaufbrauch relativ spät beginnende Differenzierung dieser Rinde.

Aufgrund cytoarchitektonischer Kontrollen am Menschen- und Schimpansengehirn begrenzen wir die Prostriata *oral* mit der *caudalen* Grenze der entorhinalen Region, da die weiter oral befindlichen Teile des *singulostriären*, die entorhinale Region umfassenden

---

*) Die Bewertung des Markfasergehaltes der Rinde an sich im phylogenetischen Sinne erfährt durch die Ergebnisse von BISHOP eine weitere Bestätigung. Seine jahrzehntelangen systematischen Untersuchungen der Spektra der Nervenfaserdurchmesser von peripheren Nerven und zentralen Faserbahnen und ihren neurophysiologischen Eigenschaften führten ihn zu dem Schluß des grundsätzlich größeren phylogenetischen Alters der feineren Markfasern gegenüber den dickeren. Auf diesem Wege kam auch er schließlich zu der Auffassung, daß die sog. sekundärsensorischen Rindengebiete phylogenetisch älter sind als die Primärgebiete (G. BISHOP: My life among the axons. Annual Rev. Physiol. 27 (1965), 1–18).

**) Die Arbeit kommt in den Monographien aus dem Gesamtgebiet der Neurologie und Psychiatrie heraus.

Feldergürtels im Gegensatz zu den caudalen eine stärkere V. Schicht mit zum Teil bandartigem Charakter besitzen.

Prostriata als Urgebiet des Sehens. Die von uns aufgrund dieser Befunde und ihrer Einordnung in die oben dargestellten Entwicklungszusammenhänge gemachte Annahme, daß wir es bei der Prostriata mit einem Urgebiet des Sehens zu tun haben, steht aber in guter Übereinstimmung mit Befunden, wie sie von MacLean u. Mitarb.[**]) elektrophysiologisch beim Totenkopfäffchen *(Saimiri)* erhoben worden sind. Es kommen ähnliche Befunde von Monnier u. Mitarb. beim Kaninchen (Vision Research 1964) und von Montero (in Vorbereitung) bei der Ratte hinzu. Beide Male wurden medial von der klassischen Sehrinde, d. h. *paralimbisch*, photisch ausgelöste Potentiale in diesem Teilabschnitt der retrosplenialen Region, die hier noch bis auf die Konvexität reicht, gefunden.

Die architektonische Gliederung der Area striata selbst ordnet sich der Abgrenzung einer *Prostriata* oral von ihrem schmalen Beginn in der Tiefe des Truncus communis sinnvoll ein. Primär mit der myeloarchitektonischen Methode arbeitend, die allein das Überblicken mehrerer Windungen bei schwacher Vergrößerung erlaubt, konnten wir eine Gradation, d. h. eine Differenzierungsrichtung *innerhalb* der Striata nachweisen, die von der sehr myelinarmen oralen Truncus-Striata ausgeht und in Richtung auf den Occipitalpol und die mediane Oberfläche verläuft, wobei es zu einem *praepolaren* Fasermaximum kommt (Abb. 3). Diese Gradation besitzt eine gute Korrelation zu der RetinaStriata-Topik, indem die faserarme orale Truncus-Striata die Repräsentation des temporalen Halbmondes des Gesichtsfeldes entspricht, der den phylogenetisch alten Rest monokulären Sehens darstellt, während die phylogenetisch junge Macula-Repräsentation den faserdunklen praepolaren und Polgebieten entspricht.

Es ist noch hervorzuheben, daß die Abstufung des Markfasergehaltes der Striata eine durchgehende Parallele im Verhalten des parastriatalen Feldes 18 (Brodmann) besitzt (Abb. 3). Auch dieses Feld Brodmanns bedarf, da es a- bis unistriär oder aber unitostriär ist, einer weiteren Untergliederung. Es stellt jedoch stets ein Maximum des Fasergehaltes auf allen okzipitalen Schnittflächen dar, wobei sein Gesamtfasergehalt sich parallel zur Area striata ändert. Der myeloarchitektonische Aspekt mit sehr starken Radiärfasern ist der eines efferenten Feldes, was mit Erfahrungen der Strychnin-Neuronographie übereinstimmt (McCulloch 1949). Da die für die Funktion des Auges ausschlaggebenden feinen Augenbewegungen schon in der Peripherie der Macula selbst ausgelöst werden sollen, könnte das präpolare Maximum des Fasergehaltes der Striata, das in etwa der Peripherie der Macula-Repräsentation entspricht, zusammen mit dem korrelierten Fasermaximum der Parastriata (Area 18) für die Optomotorik im Dienste der Macula verantwortlich sein (s. Sanides u. Vitzthum 1965 b). Cytoarchitektonische Kontrollen an mehreren Schnittserien haben gezeigt, daß die myeloarchitektonisch dargestellte Differenzierungsrichtung der Striata auch eine Entsprechung im Nissl-Bild besitzt, indem die nicht rein granulären Schichten, wie die III, IVb und insbesondere die VI. Schicht eine Korrelation in der Zellgröße zeigen mit Minimum oral und Zunahme nach polar und medial[**]). Schließlich fand auch der Befund des größeren Faserreichtums der Striata dorsal

---

[*]) Cuénod, M., K. L. Casey, P. D. MacLean, "Unit analysis of visual input to posterior limbic cortex. I. Photic stimulation" J. Neurophysiol. *28* (1965), 1101–1117 und Casey, K. L., M. Cuénod, P. D. MacLean, "Unit analysis of visual input to posterior limbic cortex. II. Intracerebral stimuli." J. Neurophysiol. *28* (1965), 1118–1131.

[**]) Ein entsprechendes Zellgrößengefälle der VI. Schicht der Striata zwischen oralen und Polabschnitten konnte bei Nachuntersuchungen auch beim Schimpansen gefunden werden.

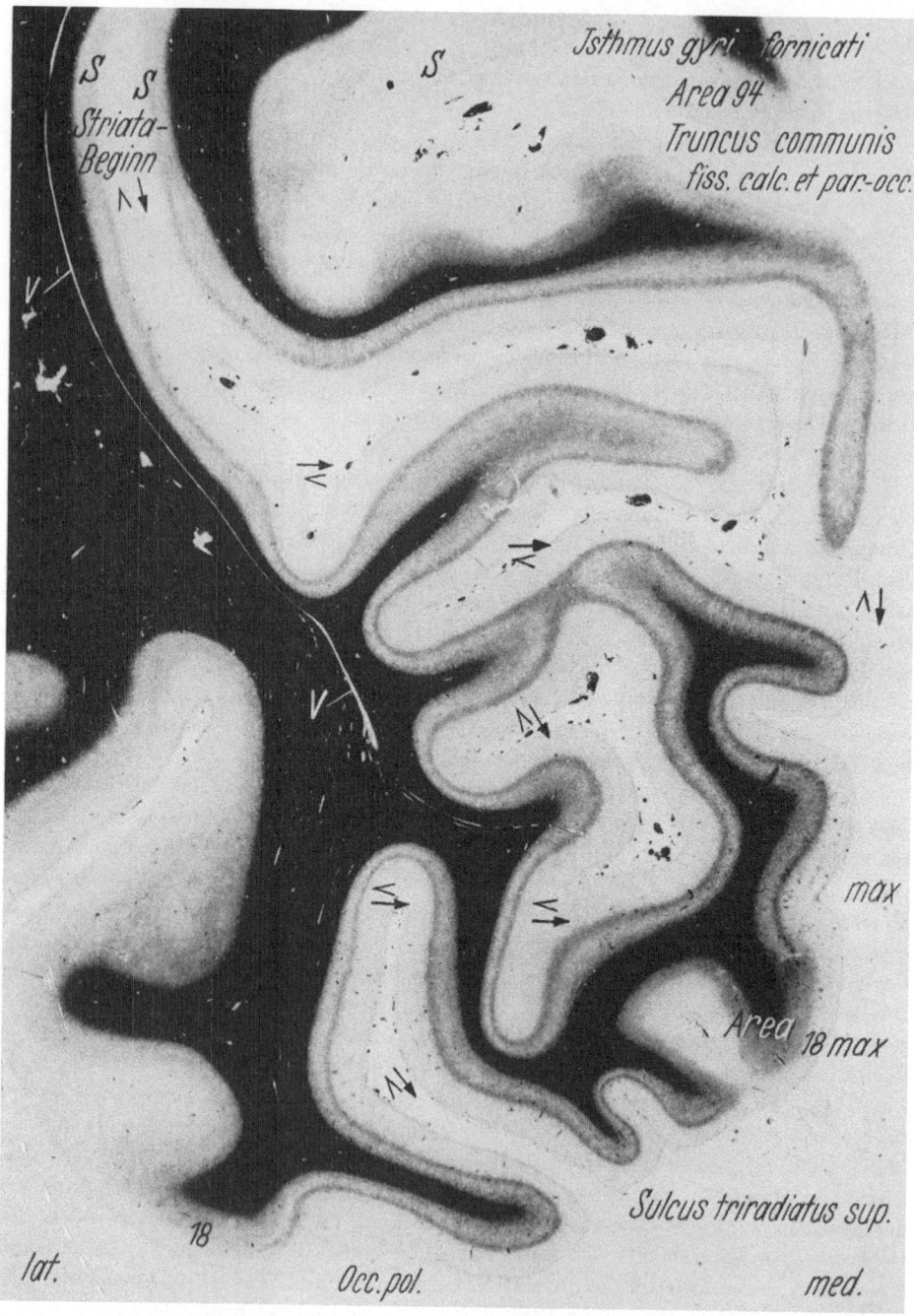

Abb. 3. Subsplenialer Horizontalschnitt durch die Fissura calcarina (Area striata) der linken Hemisphäre von oben. Übersichtsbild mit oralem Area striata-Beginn aus heller singulostriärer Rinde (S) des Truncus communis fiss. calc. et par.-occip. Innerhalb der Area striata Zunahme des Markgehaltes (siehe GENNARIscher Streif!) nach medial und polar mit Maximum (max) präpolar. Die Pfeile geben die Gradationsrichtung an, die Vergrößerungszeichen die stufenweise Markfaserzunahme ($\nearrow$). S singulostriär, entspricht der „Prostriata". V = Hinterhorn des Seitenventrikels, spaltförmig bzw. verklebt. Das parastriatale Feld, BRODMANNS Area 18, ist entsprechend dem Striatamaximum präpolar ebenfalls extrem dunkel (Area 18 max).

vom Fundus calcarinae, dem bevorzugten unteren Gesichtsfeld entsprechend, gegenüber der Striata ventral vom Fundus seine Parallele in der Durchschnittszellgröße der nicht granulären Schichten.

Summary

Myeloarchitectonic and cytoarchitectonic studies of the occipital lobe in man and chimpanzee have shown that the area striata is not surrounded orally by BRODMANN's area 18 and 19. On the contrary, the striate area is orally contiguous with a region showing both limbic and area striata-characteristics. Here we deal with a *proisocortex*, extending from the isthmus gyri fornicati to the entorhinal region. In myeloarchitecture for such a proisocortex faint myelinisation is typical. The striate area presents in its range from its paralimbic beginning orally to the occipital pole caudally several grades of myelinisation with a maximum in the prepolar site. These grades are paralleled by a corresponding increase of cell size in the nongranular layers. The polar and prepolar area striata with their highest myelin fibrillarity correspond to central vision, which is phylogenetically most recently acquired; on the other hand, the faintest myelinized oral part of the area striata corresponds to the temporal crescent, which forms in the primates the remainder of ancient monocular vision. These relations can be seen on the background of the principles of cortical development presented by SANIDES for the human frontal lobe. The grades of myelinisation within the area striata, forming a poleward direction of differentiation can be interpreted as a phylogenetic differentiation. The proisocortex with limbic features and striata similarities located orally to the striate area was termed *"prostriata"*, since we believe it represents a phylogenetic more ancient limbic-visual cortex. This hypothesis is supported by neurophysiological results of MACLEAN and co-workers in the squirrel monkey.

Literatur

ABBIE, A. A.: Cortical lamination in the Monotremata. J. comp. Neurol. *72* (1940), 429–467.

ABBIE, A. A.: Cortical lamination in a polyprodont Marsupial, Perameles natusa. J. comp. Neurol. *76* (1942), 509–536.

BERMAN, A. L.: Overlap of somatic and auditory cortical response fields in anterior ectosylvian gyrus of cat. J. Neurophysiol. *24* (1961), 595–607.

BERMAN, A. L.: Interaction of cortical responses to somatic and auditory stimuli in anterior ectosylvian gyrus of cat. J. Neurophysiol. *24* (1961), 608–620.

BRODMANN, K.: Vergleichende Lokalisationslehre der Großhirnrinde. Leipzig 1909.

McCULLOCH, W. S.: Cortico-cortical connections. In: BUCY, P. C. The precentral Motor Cortex. The University of Illinois Press, Illinois 1949.

PFEIFER, R. A: Die angioarchitektonische areale Gliederung der Großhirnrinde. Thieme, Leipzig 1940.

ROSE, J. E., C. N. WOOLSEY: Cortical connections and functional organization of the thalamic auditory system of the cat. In: HARLOW, H. F., C. N. WOOLSEY (eds.). Biological and Biochemical Bases of Behavior. University of Wisconsin Press, Madison 1958.

SANIDES, F.: Die Architektonik des menschlichen Stirnhirns. Monograph. Gesamtgebiet Neurol. Psychiatr. *98*, Springer, Berlin-Göttingen-Heidelberg 1962.

SANIDES, F.: The cyto-myeloarchitecture of the human frontal lobe and its relation to phylogenetic differentiation of the cerebral cortex. J. Hirnforsch. *6* (1964), 269 bis 282.

SANIDES, F., A. KRISHNAMURTI: Cytoarchitectonic subdivisions of the sensorimotor and prefrontal regions and of the bordering insular and limbic fields in the slow loris (Nycticebus coucang coucang). J. Hirnforsch. (im Druck.)

SANIDES, F., H. Gräfin VITZTHUM: Zur Architektonik der menschlichen Sehrinde und den Prinzipien ihrer Entwicklung. Dtsch. Z. Nervenheilk. *187* (1965a), 680–707.

SANIDES, F., H. Gräfin VITZTHUM: Die Grenzerscheinungen am Rande der menschlichen Sehrinde. Dtsch. Z. Nervenheilk. *187* (1965b), 708–719.

SPATZ, H.: Die vergleichende Morphologie des Gehirns vor und nach L. Edinger. Schr. d.

wiss. Ges. J. W. Goethe Universität Frankfurt a. M., Naturwiss. Reihe. Steiner, Wiesbaden 1959.

VOGT, C., O. VOGT: Allgemeinere Ergebnisse unserer Hirnforschung. J. Psychol. Neurol. (Lpz.) *25* (1919), 279–462.

WOOLSEY, C. N., D. FAIRMAN: Contralateral, ipsilateral, and bilateral representation of cutaneous receptors in somatic areas I and II of the cerebral cortex of pig, sheep, and other mammals. Surgery, *19* (1946), 684 bis 702.

YAKOVLEV, P. I.: Pathoarchitectonic studies of cerebral malformations, III. Arhinencephalies (Holotelencephalies). J. Neuropath. exp. Neurol. *18* (1959), 22–25.

Dr. H. Gräfin VITZTHUM
Neuropathologische Abteilung
Max-Planck-Institut für Hirnforschung
Deutschordenstr. 46
6 Frankfurt-Niederrad, Deutschland

Dr. F. SANIDES
Neuroanatomische Abteilung
Max-Planck-Institut für Hirnforschung
Deutschordenstr. 46
6 Frankfurt-Niederrad, Deutschland
z. Zt.: Laboratory of Neurophysiology
University of Wisconsin
Madison Wis. U.S.A.

The Limbic and Visual Cortex in Phylogeny: Further Insights from Anatomic and Microelectrode Studies

By P. D. MacLean, Bethesda, Md.

In this presentation, I shall summarize the results of our further attempts to learn whether or not the *limbic cortex* receives significant *connections from visual structures*. An answer to this question is of basic importance, particularly from the standpoint of primate behavior, because of research which has shown that the limbic system, in addition to olfactory functions, influences hypothalamic and other brainstem mechanisms involved in neurovegetative functions and the expression of emotion. Furthermore, an affirmative answer to this question would help to shed light on the continuing enigma as to how the visual system itself establishes a working relationship with the hypothalamus and thereby influences circadian and other rhythms through variations in the total luminous flux.

Phylogenetically, the question is also of special significance because of the progressively important role of vision in guiding behavior. In the evolution of primates, this is anatomically reflected in the great development of the geniculostriate system of the brain which culminates in man. On the basis of classical anatomy, one would presume that in this evolutionary process visual cortical projections bypassed the phylogenetically old limbic cortex and established connections only with that highly specialized part of neocortex recognized as the striate area. The microelectrode and neuroanatomic findings, however, which I shall describe provide evidence contrary to such an assumption.

The traditional emphasis on the close topographical relationship of the hippocampal formation to the olfactory cortex and rhinal fissure appears to have diverted attention from what may be an equally, and perhaps in primates even more, important relationship to the visual cortex and calcarine sulcus. In his paper of 1902 on the cerebral sulci, ELLIOT SMITH emphasized that, phylogenetically, the calcarine ranks with the hippocampal and rhinal fissures as being one of three most constant cerebral markings. In the first figure of that paper, it is shown as corresponding to the retrosplenial stem of the splenial sulcus. KAPPERS, HUBER a. CROSBY (1936) accept this interpretation. CAMPBELL confirmed SMITH's observation that this sulcus defines the limit between the visual and the limbic cortex. SMITH showed that essentially the same relationship holds in subhuman primates and man, as in a number of lower mammals. Despite the great expansion of the parieto-occipital and occipito-temporal cortex, the line between the striate and limbic cortex is held close to this primitive furrow. Fig. 1 illustrates in the squirrel monkey *(Saimiri sciureus)* the characteristic abrupt transition from limbic to striate cortex. In the experiments to be described, we have used this small New World primate because it has a well-developed visual system, and its brain is of very desirable size for neuroanatomic and neurophysiologic studies. The lower photograph in the figure also illustrates that the limbostriate transition occurs near the junction of the anterior and posterior calcarine sulci.

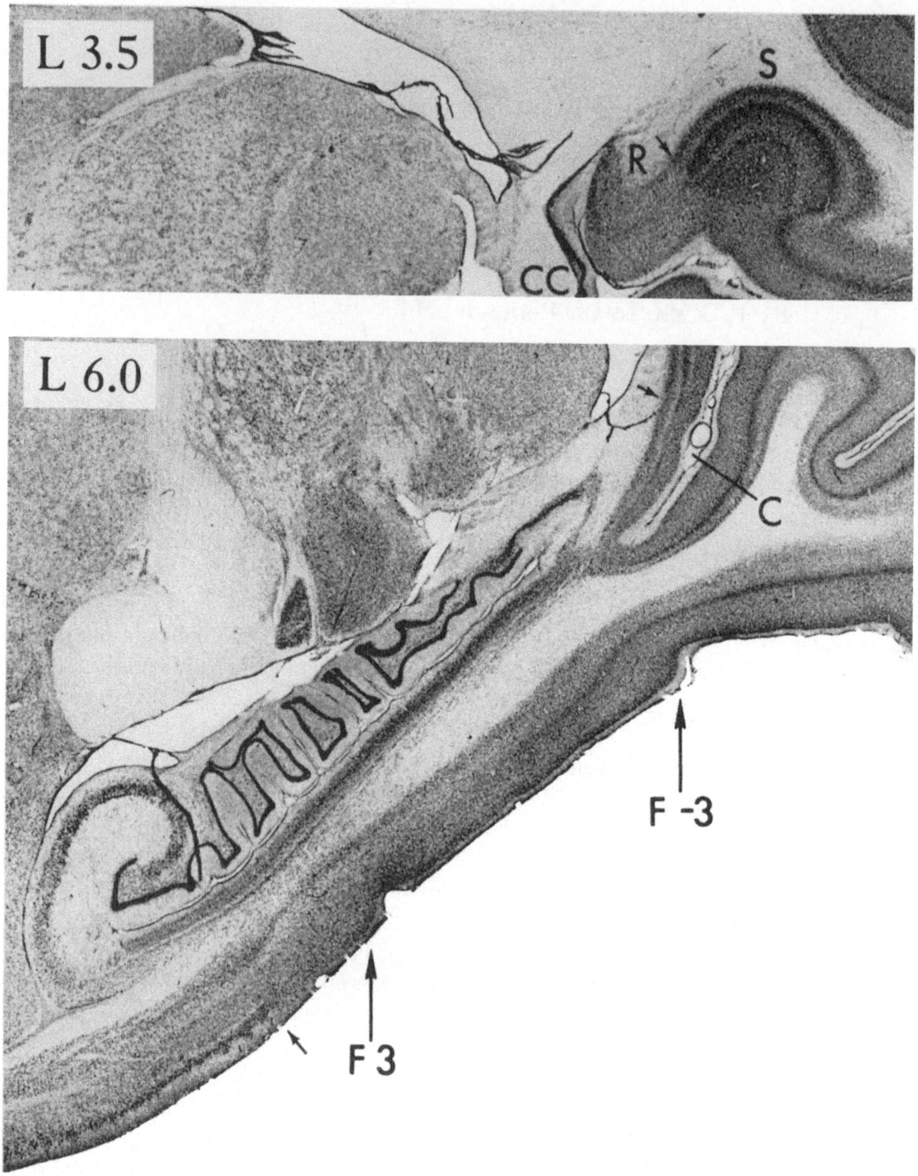

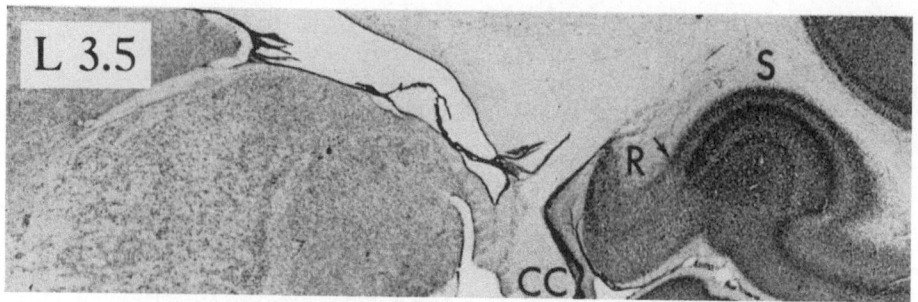

Fig. 1. Sagittal sections from brain of squirrel monkey, showing representative cortical regions explored with microelectrodes and illustrating anatomical relationship of the limbic and striate cortex above and below level of the splenium of corpus callosum (CC). Small arrow in upper photograph (L 3.5 indicates mm lateral to midline) points to the characteristic abrupt transition between retrosplenial (R) and striate cortex (S). In lower photograph (L 6.0) the label "C" is placed near the junction of the anterior and posterior calcarine sulci. As indicated by small arrow above, the transition between the striate and limbic cortex takes place near the junction of the sulci. Exploration with microelectrodes was carried out between frontal planes F 3 and F—3 indicated by large arrows. At present, exploration is being extended into the typical entorhinal cortex of BRODMANN (area 28) which, as indicated by the small arrow to left of F 3, begins abruptly near frontal plane 4.5 of the brain atlas. See this atlas also for fiber stained sections adjacent to those shown above.

As in the macaque and in man, the anterior calcarine sulcus in the squirrel monkey terminates rostrally as a shallow groove between the lingual gyrus and the presubiculum of the hippocampal formation, giving to the latter the appearance of a microgyrus. In microscopic sections, the anterior extremity of this groove is a helpful landmark in identifying the beginning of the lingual gyrus. Cross-sectional drawings which compensate for differences in brain size of the squirrel monkey, macaque, and man reveal the striking expansion of the convolutions in this region that has taken place in evolution. This would appear to be significant in connection with the development of the fusiform gyrus lying lateral to these convolutions and which, as KAPPERS, HUBER and CROSBY (1936) have pointed out, represents a new structure in the brain of primates.

Electrophysiologic Studies

Background. Although classical neuroanatomy provides no basis for visual projections to the limbic cortex, it leaves open the possibility that visual impulses might be relayed to it from the surrounding visual cortex (MACLEAN, 1949). Support for this inference was gained from our strychnine neuronographic studies in 1952 (PRIBRAM a. MACLEAN). It was found in the macaque that the cortex above and below the calcarine fissure "fired" into the posterior hippocampal gyrus. The hippocampal gyrus was found in turn to "fire" into the hippocampus. In the light of this potential visual pathway to the limbic cortex, we were disappointed to find subsequently that photic stimulation failed to evoke consistent electrical responses in the hippocampus of either the cat (MACLEAN) or monkey (GERGEN a. MACLEAN, 1961). Although some workers reported photically evoked slow potentials, it was possible that they were attributable to volume-conducted responses in neighboring visual structures. Consequently, we decided to re-examine this question with the use of microelectrodes which by recording activity of single neurons at the site of the electrode tip dispel doubt about the origin of evoked potentials.

Methods. For this work we have adapted the stereotaxic technique used in our laboratory for brain stimulation and recording with macroelectrodes in chronically prepared waking animals (MACLEAN a. PLOOG 1962; CUÉNOD et al., 1965). It eliminates the need for open surgery at the time of an experiment and affords an essentially closed system. A small platform with guides for electrodes is fixed *above the scalp* on four screws cemented in the skull. The holes for the guides are in rows 1 mm apart and are stereotaxically oriented to conform with planes of the brain atlas (GERGEN a. MACLEAN, 1962). After selecting a hole at desired co-ordinates, a drill of exact size is introduced, and an opening is made in the skull for the electrode guide. The guide provides a protective sheath for the microelectrode when it is inserted into the brain and maintains it in the correct orientation as it is lowered by a micromanipulator.

For showing a one-to-one relationship between stimulus and response and obtaining accurate latency measurements, we have routinely used a brief, intense flash of light from a stroboscope as the visual stimulus. As a check on specificity of unit responses to a flash of light, a comparison has been made with the effect of auditory and somatic stimulation. In continuing these studies, we are also testing the effects of steady illumination of various duration and of moving patterns.

Experiments have been performed on more than 50 squirrel monkeys. Most of the early observations were made in animals anesthetized with α-chloralose. In a further attempt to delineate the responsive areas, we are recording unit activity in waking animals given a tranquilizing dose of nembutal and sitting in a special type of chair.

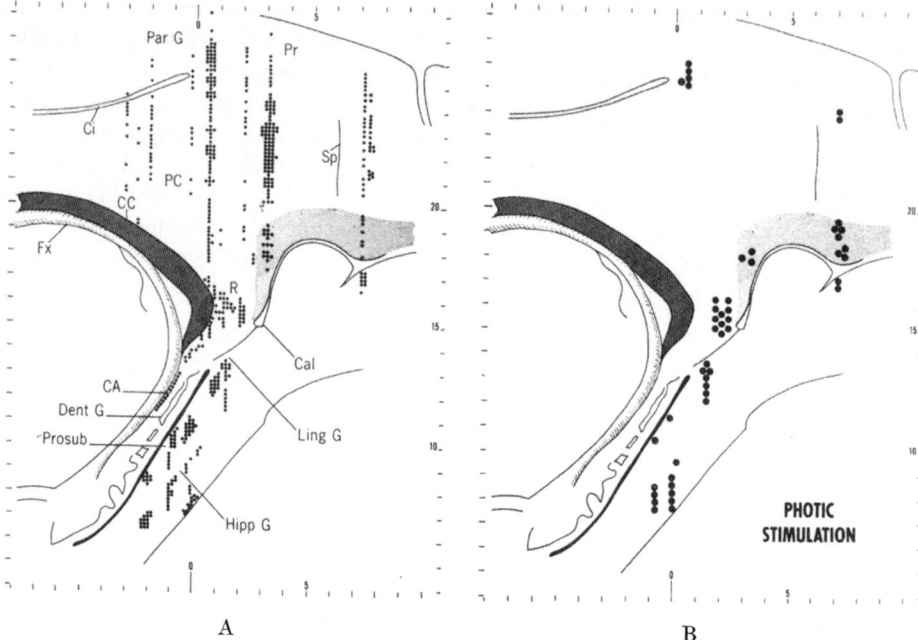

A   B

Fig. 2. A Cortical distribution of units (small dots) isolated and tested in microelectrode study. Lightly shaded area indicates portion of striate cortex in superior bank of posterior calcarine sulcus. Abbreviations: CA, pyramidal layer of hippocampus; Cal, posterior calcarine sulcus; CC, corpus callosum; Ci, cingulate (callosomarginal) sulcus; Dent G, dentate gyrus; Fx, fornix; Hipp G, hippocampal gyrus; Ling G, lingual gyrus; Par G, paracentral gyrus; PC, posterior cingulate gyrus; Pr, precuneus; Prosub, prosubiculum; R, retrosplenial cortex; Sp, subparietal sulcus. Marginal scales in mm correspond to coordinates of stereotaxic atlas. (From CUÉNOD et al., 1965).

B Location of units (large dots) activated by photic stimulation (from CUÉNOD et al., 1965).

Results. In the first study with GERGEN (1964) attention focused on the hippocampus itself. Under chloralose anesthesia, photic stimulation evoked in the majority of animals a large slow potential which reached maximum amplitude in the *stratum radiatum*. This depolarization, however, was not associated with unit discharge. *Evoked unit activity* was observed only when the microelectrode was lowered into the *hippocampal gyrus*.

Accordingly, in experiments with CUÉNOD a. CASEY (1965), we proceeded to explore the posterior hippocampal gyrus, as well as the posterior cingulate gyrus and retrosplenial cortex lying above the calcarine fissure. For comparative purposes we also sampled units in the striate and peristriate cortex. Fig. 2A shows the distribution of 518 units that were examined and Fig. 2B the location of the *photically responsive cells*.

Of 249 units tested while exploring the medial parietal and posterior cingulate cortex, only 5 (2%) responded to photic stimulation. All were located near the cingulate sulcus. Two of these also responded to somatic stimulation.

The most significant observations pertained to the retrosplenial region and to the posterior hippocampal gyrus and adjoining lingual cortex (see Fig. 2B). Of 51 units tested in the *retrosplenial region*, 11 (22%) responded to a flash with a mean latency of

53 msec, with a range of 42–67 msec. In the *hippocampal gyrus* and *adjoining lingual cortex,* 17 (26%) of 66 units responded at a mean latency of 80 msec, with a range of 39 to 120 msec. All of these photically activated units appeared to be modality specific insofar as they were unaffected by auditory or somatic stimulation. Fig. 3 shows responses of representative units.

The percentage of responding units in these supra- and infracalcarine limbic areas was not significantly different from that of a sample population of units in the striate cortex. The range of latencies was also comparable and was compatible with the possibility of both a direct and indirect relay, respectively, from subcortical and cortical visual structures. In an attempt to gain information about conducting pathways, we tested the responsiveness of the same group of units shown in Fig. 2 A to electrical stimuli applied to the optic chiasm, lateral geniculate bodies, and striate cortex, with bilaterally implanted electrodes (CASEY et al., 1965). The most significant outcome was the finding in the waking animal that stimulation of the ipsilateral lateral geniculate body evoked short latency unit responses in the retrosplenial cortex and in the cortex near the junction of the hippocampal and lingual gyri. Fig. 2 of CASEY, CUÉNOD a. MacLEAN, 1965 shows the locus and oscillographic record of a unit near the junction of the hippocampal and lingual gyri that responded at a 2 msec latency with stimulation by an electrode well within the lateral geniculate body of the same side. The inability of the unit to respond to stimuli exceeding 77/sec was presumptive evidence of a direct orthodromic conduction.

PHOTIC STIMULATION

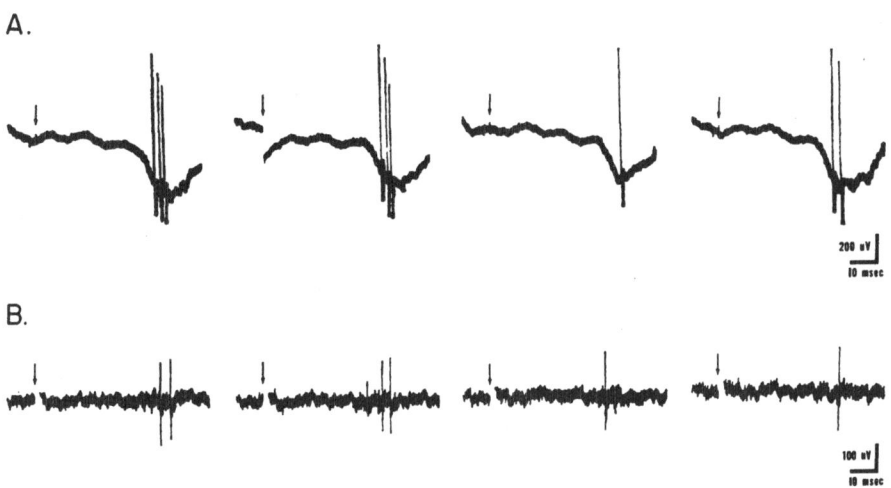

Fig. 3. Four successive tracings showing unit and slow wave responses to a flash of light recorded with microelectrode in retrosplenial cortex. Photic stimulation at 1/sec. B: Same, but with slow wave filter, for a unit in lingual cortex bordering the presubiculum. Negativity upwards. Stimulus artifacts replaced by arrows.

Neuroanatomic Findings

There is of course no neuroanatomical basis for the temporo-occipital pathway suggested by the electrophysiologic findings. In an attempt to resolve the possibility of main or collateral fibers from the optic radiation, CRESWELL and I are carrying out degeneration studies with JACOBSON's (1963) and NAUTA's own modification of the NAUTA-GYGAX stain for demonstrating fine cortical fibers. Thus far we have placed lesions with a WYSS coagulator in or near the geniculate body in ten squirrel monkeys.

The lesion shown in Fig. 4 was well within the lateral geniculate body. It illustrates that in sections cut normal to the hippocampal gyrus, one finds a continuous band of degenerated fibers leading from the geniculate body into the core of the posterior hippocampal gyrus. Using either of the mentioned modifications of the NAUTA-GYGAX stain, one can follow degeneration into the cortex of hippocampal gyrus and the adjoining part of the lingual gyrus. Fig. 5 shows such degeneration in another animal with a lesion of the geniculate body. In keeping with the electro-physiologic findings, there is no significant degeneration in the hippocampus itself. Traced caudally, the part of the radiation projecting to the postlimbic striate cortex appears to interweave with the lateral fibers of the cingulum. At this level some fine degenerating fibers can be followed into the retrosplenial cortex.

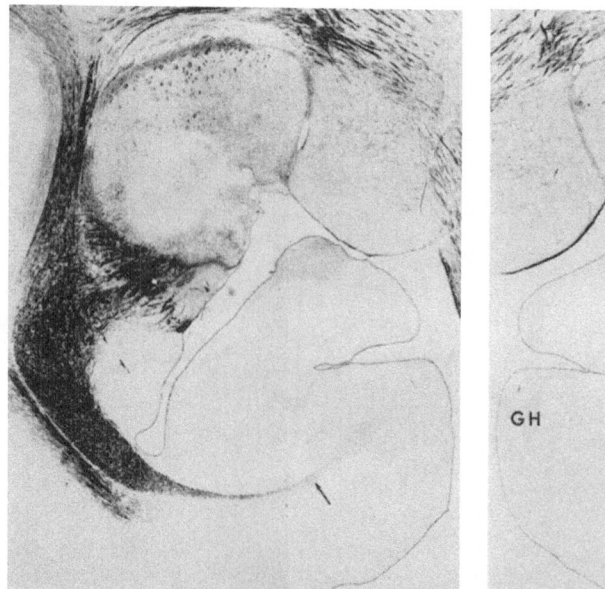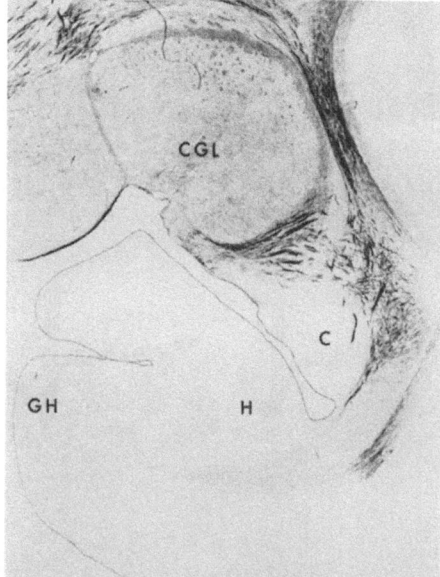

Fig. 4. Section on left shows temporal band of degenerating fibers typically seen following lesions placed within the lateral geniculate body. Unoperated side (from same histological section) is shown on the right with key structures labeled. CGL, *corpus geniculatum laterale*, is placed in a position corresponding to the site of the lesion seen on the left. GH, gyrus hippocampi; H, hippocampus; C, caudate. The temporal band of degeneration would appear to correspond to that part of the optic radiation called the "temporal loop" in man. The lower margin of the loop, in POLYAK's words, "slips medially underneath the ventricle and becomes the core of the hippocampal gyrus". This description would apply to the fascicle of degeneration indicated by the arrow. NAUTA's modification of NAUTA-GYGAX stain for showing fine cortical fibers.

Fig. 5. Section from another animal with lesion of lateral geniculate body, illustrating degenerating fibers that appear to leave the main degenerating band depicted in Fig. 4 and terminate in the cortex of the posterior hippocampal gyrus and adjoining parts of the fusiform and lingual gyri. Same stain as in section shown in Fig. 4.

Followed in a rostral direction, the degenerating temporal band of fibers from the latera geniculate body appears to stop short of the so-called entorhinal cortex.

In cross-section the hippocampo-fusiform gyrus resembles a shoe, with the toe lying underneath the hippocampus. In addition to the degeneration found in the parts of the hippocampal gyrus corresponding to the sole and toe, one also traces degenerating fibers into the angle of the fusiform resembling the heel.

In a control animal in which a needle was introduced as far as the lateral geniculate body by the usual approach, no degeneration was seen in the areas in question.

Discussion

The inability to introduce an electrode into the lateral geniculate body without injuring non-visual structures makes it difficult to control the possibility that the degenerating fibers seen in the cortex of the hippocampal and lingual gyri are not of some other origin than this nucleus. The finding, however, that they appear to stem from the main degenerating band of the optic radiation argues in favor of their true visual origin. The electro-physiologic findings also support this inference.

The temporal band of degenerating fibers that has been described would appear to correspond to that part of the optic radiation in man which has been variously referred to as FLECHSIG's "temporal knee" or MEYER's *"temporal loop."* POLYAK's (1957) descrip-tion of the course of the loop in man could be used to apply to the degenerated band shown in Fig. 4. "The bent-in lower edge of [the loop]", he states, "thins out into a narrow fiber sheet, which slips medially underneath the ventricle and becomes the core of the hippocampal gyrus." He goes on to say, "It is very probable that the most inferior and medial bundles of this fiber lamina of the temporal lobe are in no way concerned with the visual function."

In the light of the present findings it becomes of added comparative interest that the "temporal knee," as PUTNAM (1926) has shown in his reconstructions of the optic radiation, is found not only in various primates, but also in carnivores. A further consideration of comparative aspects requires clarification of terms used in different species to refer to cortical areas of the medial temporo-occipital cortex.

CAJAL (1955) described a special cortical area which he called spheno-occipital ganglion and which in macrosmatic animals occupies the posterior hippocampal gyrus, forming a triangular wedge between the piriform and occipital cortex. He pointed out that this cortex has certain similarities to the visual cortex and emphasized that it is the major source of afferents to the hippocampus. LORENTE DE NÓ (1933, 1934) claimed that this area is coextensive with BRODMANN's entorhinal and perirhinal areas which are numeri-cally identified, respectively, as 28 and 35. He used the term "gyrus entorhinalis" to refer to the caudal part of the hippocampal gyrus. This and other statements about homologies in the monkey has perhaps influenced some neuroanatomists to use the term entorhinal in a broad sense to refer to the posterior hippocampal gyrus in primates, as well as in macrosmatic animals. It is evident, however, from Fig. 1 that if the term ento-rhinal is used in the strict cytoarchitectural sense, it would fail to include in primates a large segment of the hippocampal gyrus adjoining the hippocampus.

The abrupt ending of the entorhinal area indicated by the small arrow to the left in Fig. 1 is seen in transverse sections to occur close to the same level as the caudal termina-tion of the rhinal fissure. Behind this level the cortex of the hippocampal gyrus is of a transitional type, showing gradational changes between the archicortex medially and the fully developed temporal neocortex laterally. It resembles the areas which BECK (1934) labeled "thi" in the macaque and which he presumed to correspond to what BRODMANN (1909) alternately referred to as "ectorhinal" or "retrosubicular" and numbered "36" in man and "48" in the marmoset. Caudal to "thi", and appearing approximately at frontal plane AP 1 of the squirrel monkey atlas (GERGEN a. MACLEAN, 1962), the cortex of the hippocampal gyrus again shows transitional changes, assuming features of occipital cortex. In this region the cortex at the crown of the hippocampal gyrus and the adjoining lingual gyrus next to the presubiculum shows a thin, relatively superficial granular layer and poorly developed supragranular layers. It resembles the type which BRODMANN (1904–05) identified as Area 26 in his original study on the macaque, and it would appear to corre-spond to the transitional cortex that BECK (1934) included under the designation "sub-regio praestriatalis ventralis".

The transitional prestriate cortex and the areas which BECK labeled "thi" would fall within a cortical area which MAUSS (1908) differentiated on the basis of its myeloarchitecture and numbered "29" in the brain of *Cercopithecus*. He grouped area 29, in turn, with areas comprising what he called "Regio hippocampica." As we have just heard in the preceding paper on their myeloarchitecture studies of human material, SANIDES a. VITZTHUM (1966) refer to the prestriate limbic cortex as "singulostriate proisocortex," an expression calling attention to both its primitive features and similarities to visual cortex. Remarking that it may represent a primitive form of visual cortex, they suggest that "prostriata" would be an appropriate designation for it. In his study of 1934, BECK speculated that the prestriate and visual cortex were on the one side functionally related to the entorhinal area by "thi," and on the other, to the hippocampus by the retrosplenial and presubicular cortex.

It would be a reasonable inference that the prestriate retrosplenial cortex found to be photically responsive in the squirrel monkey is homologus to the postsplenial cortex of macrosmatic animals (CAMPBELL, 1905). But it is not at all clear what the homologies would be in regard to the photically responsive cortex of the posterior hippocampal gyrus and adjoining part of the lingual gyrus. It would be my hope that this question will receive comment in the general discussion. As already noted in the introduction, the convolutions in this part of the brain undergo great expansion in evolution, and, according to KAPPERS, HUBER a. CROSBY (1936), the laterally lying fusiform gyrus represents a new structure in the brain of primates.

In our continuing electrophysiologic survey of the hippocampal gyrus, we are carrying exploration forward into the typical entorhinal cortex. In the light of CAJAL's (1955) comments about the similarities of this area to the visual cortex, the results thus far have been disappointing.

If eventually it can be anatomically confirmed that the so-called "temporal loop" gives off terminals to the cortex of the posterior hippocampal gyrus and adjoining part of the lingual, it will greatly add to the significance of the electrophysiologic findings by showing that they are not dependent on some indirect and possibly minor pathway. It would also provide a solid basis for reconsidering many clinical and physiologic time-worn questions about the role of the temporal lobe in visually related functions.

But regardless of the anatomical outcome, the electrophysiologic findings make evident at least one functional pathway by which visual information at the cortical level can influence the hypothalamus. Such a pathway, of course, presupposes that visual impulses reaching the cortex of the retrosplenial region and hippocampal gyrus are relayed through the hippocampus. Evidence for this was presented and discussed in conjunction with a model in a preceding study (GERGEN a. MACLEAN, 1964). In continuing our microelectrode investigation in waking animals, it has been of particular interest to identify a class of cells in the posterior hippocampal gyrus that responds similarly to slowly adapting "on"

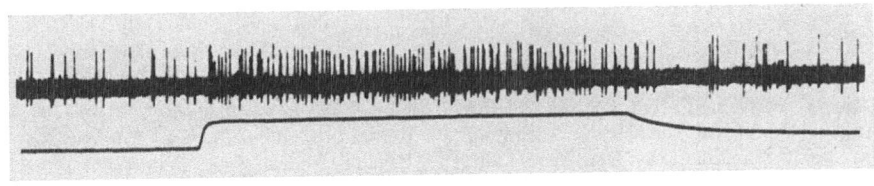

Fig. 6. Record of a unit in posterior hippocampal gyrus (approximate coordinates AP 2.5, L 8, H 6 [cf. GERGEN a. MACLEAN, 1962]) responding throughout a 2.5 sec. period of illumination signaled in lower tracing by a photocell. Recording from an awake, sitting, dark-adapted monkey.

units of the retina. Fig. 6 shows the record of a responding unit of this type. It is conceivable that such cells are implicated in affecting hypothalamic mechanisms responding to changes in the total luminous flux.

Summary

With the traditional emphasis on the close topographical relationship of the hippocampal formation to the olfactory apparatus and rhinal fissure, it would seem that insufficient attention had been paid to what may be an equally, and perhaps in primates even more, important relationship to the calcarine sulcus and visual cortex. This is a matter of basic importance from the standpoint of *visual-hypothalamic mechanisms*. As recognized since the original observations of ELLIOT SMITH (1902), the calcarine is phylogenetically one of the most constant of the cerebral sulci. The anterior calcarine sulcus of primates corresponds to the retrosplenial stem of the splenial sulcus of macrosmatic forms. This primitive furrow, as SMITH also noted (1903–04), forms the boundary between the striate and limbic cortex.

Using microelectrode techniques, we have found in experiments on squirrel monkeys that *photic stimulation* activates a relatively large proportion of units in the *prestriate retrosplenial cortex* and in the cortex of the *posterior hippocampal gyrus* and the adjoining part of the *lingual gyrus* next to the anterior calcarine sulcus (CUÉNOD et al., 1965). The units appear to be modality specific insofar as they are unaffected by somatic or auditory stimulation. The proportion of activated units for these areas, as well as the range of latencies, is not significantly different from a sample population in the visual cortex.

Results of electrical stimulation of the lateral geniculate body suggest that some limbic units may be activated by a *direct orthodromic pathway*. Accordingly, we have employed a modification of the NAUTA-GYGAX stain to investigate the possibility of direct connections from the lateral geniculate body to the posterior limbic cortex (MACLEAN a. CRESWELL, 1966). Thus far lesions have been placed in or near this nucleus in ten squirrel monkeys. In sections cut normal to the hippocampal gyrus, one sees a continuous band of degenerated fibers leading from the lateral geniculate body into the core of the posterior hippocampal gyrus. Degenerating fibers are seen in the cortex of this portion of the gyrus, as well as in adjoining parts of the fusiform and lingual gyri. Some degeneration is also found in the prestriate retrosplenial cortex.

The temporal band of degenerating fibers would appear to correspond to that part of the optic radiation that has been variously referred to in man as FLECHSIG's "temporal knee" or MEYER's "temporal loop." Comparative and other implications of the present findings are discussed.

References

ARIENS KAPPERS, C.U., G.C.HUBER, E.C. CROSBY: The Comparative Anatomy of the Nervous System of Vertebrates, Including Man. Vol. 2, MacMillan Co., New York 1936.
BECK, E.: Der Occipitallappen des Affen (Macacus Rhesus) und des Menschen in seiner cytoarchitektonischen Struktur. J. Psychol. Neurol. (Lpz.) *46* (1934), 193–323.

BRODMANN, K.: Beiträge zur histologischen Lokalisation der Großhirnrinde. Dritte Mitteilung: Die Rindenfelder der niederen Affen. J. Psychol. Neurol. (Lpz.) *4* (1904–05), 177–226.
BRODMANN, K.: Vergleichende Lokalisationslehre der Großhirnrinde in ihren Prinzipien dargestellt aufgrund des Zellenbaues. Barth, Leipzig 1909.

CAMPBELL, A.W.: Histological Studies on the Localization of Cerebral Functions. University Press, Cambridge 1905.

CASEY, K.L., M.CUÉNOD, P.D.MACLEAN: Unit analysis of visual input to posterior limbic cortex. II. Intracerebral stimuli. J. Neurophysiol. 28 (1965), 1118–1131.

CUÉNOD, M., K.L.CASEY, P.D.MACLEAN: Unit analysis of visual input to posterior limbic cortex. I. Photic stimulation. J. Neurophysiol., 28 (1965), 1101–1117.

GERGEN, J.A., P.D.MACLEAN: Hippocampal seizures in squirrel monkeys. Electroenceph. clin. Neurophysiol. 13 (1961), 316–317.

GERGEN, J.A., P.D.MACLEAN: A Stereotaxic Atlas of the Squirrel Monkey's Brain (Saimiri sciureus). U.S. Government Printing Office. Washington, D.C. 1962.

GERGEN, J.A., P.D.MACLEAN: The limbic system: Photic activation of limbic cortical areas in the squirrel monkey. Ann. N.Y. Acad. Sci., 117 (1964), 69–87.

JACOBSON, S.: Adaptation of the Nauta method for use on degenerating fibres in rat cortex and thalamus. Stain Technology, 38 (1963), 275–279.

LORENTE DE NÓ, R.: Studies on the structure of the cerebral cortex. I. The area entorhinalis. J. Psychol. Neurol. (Lpz.), 45 (1933), 381–438.

LORENTE DE NÓ, R.: Studies on the structure of the cerebral cortex. II. Continuation of the study of the ammonic system. J. Psychol. Neurol. (Lpz.) 46 (1934), 113–177.

MACLEAN, P.D.: Psychosomatic disease and the "visceral brain." Recent developments bearing on the Papez theory of emotion. Psychosom. Med., 11 (1949), 338–353.

MACLEAN, P.D.: Unpublished observations.

MACLEAN, P.D., G.F.CRESWELL: In preparation.

MACLEAN, P.D., D.W.PLOOG: Cerebral representation of penile erection. J. Neurophysiol. 25 (1962), 29–55.

MACLEAN, P.D., T.YOKOTA, M.KINNARD: Unpublished observations.

MAUSS, T.: Die faserarchitektonische Gliederung der Großhirnrinde bei den niederen Affen. J. Psychol. Neurol. (Lpz.) 13 (1908), 263–325.

NAUTA, W.J.H.: Personal communication.

POLYAK, S.: The Vertebrate Visual System. University of Chicago Press, Chicago 1957.

PRIBRAM, K.M., P.D.MACLEAN: Neuronographic analysis of medial and basal cerebral cortex. II. Monkey. J. Neurophysiol. 16 (1953), 324–340.

PUTMAN, T.J.: Studies on the central visual system. II. A comparative study of the form of the geniculostriate visual system of mammals. Arch. Neurol. Psychiat. 16 (1926), 285–300.

RAMÓN Y CAJAL, S.: Studies on the cerebral cortex (limbic structures). (Trans. by L. M. Kraft), Lloyd-Luke Ltd, Year Book, London, and Chicago 1955.

SANIDES, F., H.G.VITZTHUM: Zur Architektonik der menschlichen Sehrinde und den Prinzipien ihrer Entwicklung. Dtsch. Z. Nervenheilk. (im Druck).

SMITH, G.E.: On the homologies of the cerebral sulci. J. Anat. 36 (1902), 309–319.

SMITH, G.E.: The morphology of the occipital region of the cerebral hemisphere in man and ape. Anat. Anz., 24 (1903–04), 436–451.

SMITH, G.E.: The morphology of the retrocalcarine region of the cortex cerebri. Proc. Roy. Soc. Lond. 73 (1904), 59-65.

PAUL D. MACLEAN, M.D.
Section on Limbic Integration and Behavior
Laboratory of Neurophysiology
National Institute of Mental Health
National Institutes of Health
Bethesda, Maryland, USA

Index